THE ELECTRICAL ENGINEERING
AND APPLIED SIGNAL PROCESSING SERIES

Handbook of
Multisensor
Data Fusion

Theory and Practice

Second Edition

THE ELECTRICAL ENGINEERING
AND APPLIED SIGNAL PROCESSING SERIES
Edited by Alexander Poularikas

MIMO System Technology for Wireless Communications
George Tsoulos

Signals and Systems Primer with MATLAB®
Alexander Poularikas

Adaptation in Wireless Communications - 2 volume set
Mohamed Ibnkahla

Handbook of Multisensor Data Fusion, Second Edition: Theory and Practice
Martin E. Liggins, David L. Hall, and James Llinas

THE ELECTRICAL ENGINEERING
AND APPLIED SIGNAL PROCESSING SERIES

Handbook of
Multisensor Data Fusion

Theory and Practice

Second Edition

Edited by

Martin E. Liggins
David L. Hall
James Llinas

CRC Press
Taylor & Francis Group
Boca Raton   London   New York

CRC Press is an imprint of the
Taylor & Francis Group, an **informa** business

CRC Press
Taylor & Francis Group
6000 Broken Sound Parkway NW, Suite 300
Boca Raton, FL 33487-2742

© 2009 by Taylor & Francis Group, LLC
CRC Press is an imprint of Taylor & Francis Group, an Informa business

No claim to original U.S. Government works

ISBN 13: 978-1-4200-5308-1 (hbk)

Library of Congress Cataloging-in-Publication Data

Handbook of multisensor data fusion : theory and practice / editors, Martin E. Liggins, David L. Hall,
  James Llinas. -- 2nd ed.
    p. cm. -- (Electrical engineering and applied signal processing series ; no. 22)
    Includes bibliographical references and index.
    ISBN 978-1-4200-5308-1 (hbk. : alk. paper)
    1. Multisensor data fusion--Handbooks, manuals, etc. I. Hall, David L. (David Lee), 1946- II.
Liggins, Martin E. III. Hall, David L. IV. Llinas, James. V. Title. VI. Series.

TK5102.9.M86 2008
681'.2--dc22
                                        2008025446

Visit the Taylor & Francis Web site at
http://www.taylorandfrancis.com

and the CRC Press Web site at
http://www.crcpress.com

Dedicated by James and Sara Llinas to the memory of their beloved son

James Craig Llinas, too soon gone but living in their hearts.

Dedicated by David L. Hall to the memory of his late wife, Mary Jane Hall,

whose spirit and courage provided a gift of faith to all who knew her.

Dedicated by Martin Liggins to his wife, Shigeko, a wonderful

role model, without whom I would not be where I am today.

Contents

Preface

Multisensor data fusion seeks to combine information from multiple sources (including sensors, human reports, and data from the Internet) to achieve inferences that cannot be obtained from a single sensor or source, or whose quality exceeds that of an inference drawn from any single source. Historically, the U.S. Department of Defense (DoD) and other defense establishments around the world invested enormous amounts of funds for data fusion systems in applications such as automatic target recognition, target tracking, automated situation assessment, identification-friend-foe-neutral (IFFN) systems, and *smart* weapons. This investment has resulted in an extensive legacy of data fusion technological capability that combines process models, algorithms, evolving tool kits, and system engineering methodologies (e.g., system design and algorithm selection). Of particular note is the U.S. Joint Directors of Laboratories (JDL) Data Fusion Working Group process model, which has been the foundation for characterizing hierarchical levels of fusion processing, categorized fusion functions, and candidate algorithm approaches. The JDL model continues to evolve and expand to meet the demands of DoD and non-DoD applications.

During the 1990s, the data fusion research and development community commonly termed multisensor fusion as an *emerging technology*, meaning that there were numerous outstanding issues related to algorithm design and selection, fusion system architectures, test, and evaluation, establishing the utility of fusion systems, and other issues. Indeed, the creation of the JDL model in the early 1990s was in part motivated by a need to establish common definitions and understandings of this emerging technology to guide researchers and system designers. Although all issues have not yet been resolved, multisensor fusion can no longer be considered as an emerging technology. Fusion development has had to keep pace with a proliferation of micro- and nanoscale sensors, increasing speed and availability of wireless communications, and increased computing speeds, enabling the assembly of information from sensors, models, and human input. Fusion research and applications have adapted to service-oriented architectures, pushed the boundaries of situational modeling of human behavior, and expanded into fields such as chemical and biological sensing, medical diagnostics, environmental engineering, crisis management, monitoring and control of manufacturing processes, and intelligent buildings. In the defense and national security arenas, requirements and capabilities have moved from traditional force-on-force, nation-state-centric problems to a broad array of new and yet more challenging problems of asymmetric warfare, regional conflict, counterterrorism, and a host of specialized problems. The information spaces of these application domains are extremely large, and their demands for timely and *sufficiently accurate* inferences have required extraordinary creativity and agility from the data fusion community to respond with a multitude of inventive solutions.

This second edition to the original *Handbook of Multisensor Data Fusion* captures the latest data fusion concepts and techniques drawn from a broad array of disciplines including statistical estimation; signal and image processing; artificial intelligence; and biological, social, and information sciences. Leading experts in the fusion community have contributed to this new edition. This new edition has been expanded to 31 chapters and adds a wide range of new topics that represent the new technology directions. Many of the chapters brought into this edition from the first edition have been rewritten and updated. New material involves extensions into service-oriented networks and data mining to expand applicability to the dynamic growing field of information technology; automated detection fusion that adapts decision thresholds based on the variable performance of multiple sensors; an introduction of particle filtering, which provides a look into the background theories that gave rise to this important direction in target tracking; random set theory, which has been completely rewritten to provide an extensive development of this generalization of Bayesian and non-Bayesian approaches; situation and impact assessment theory and concepts, which have been significantly extended to build on research directions that have been previously glossed over; new techniques in visualization that have been developed; and finally, a new chapter in commercial off-the-shelf (COTS) software tools, which provides the reader with a wealth of fusion research tools and techniques. Additionally, areas of application have expanded to include electromagnetic systems, chemical and biological sensors, and army command and combat identification techniques. Finally, an appendix provides links to current web sites related to multisensor fusion.

This book follows the same theme of the first edition, dividing the material into four topical themes: Introduction to Multisensor Data Fusion provides the latest fundamental concepts and techniques; Advanced Tracking and Association Methods builds on the fundamentals to provide important research directions of fusion processing; Automated Reasoning and Visualization for Situation and Threat Refinement pushes the state-of-the-art concepts of fusion to situational behavior modeling, cognitive methods for human–machine interaction, and software sources for the reader to build on current technologies; and finally, Sample Applications demonstrate that the field of fusion research is no longer an emerging field, but a mature, strong contributor to information processing and understanding.

Acknowledgment

The editors are grateful for the extensive help provided by Tracy Ray. Her dedication, attention to detail, and efforts to work with the chapter authors greatly assisted in the development of this second edition.

Editors

Martin E. Liggins II is an engineer with The MITRE Corporation working with the Office of Naval Research. He has more than 20 years of research and development experience in industry and with the Department of Defense. Mr. Liggins has performed fusion research in a number of areas including sensor and data fusion, multisensor and multitarget tracking, radar, high-performance computing, and program management. He is the author of more than 30 technical and research papers. Mr. Liggins has served as the chairman of the National Symposium of Sensor and Data Fusion (1995, 2002, and 2003) and has been an active senior committee member, since 1990. He has also been active in the SPIE Aerosense Conference on Signal Processing, Sensor Fusion, and Target Recognition since 1992. He was awarded the Veridian Medal Paper Award in fusion research (2002) and the first Rome Air Development Center Major General John J. Toomay Award for advances in multispectral fusion technology (1989).

Dr. David L. Hall is a professor in the College of Information Sciences and Technology at the Pennsylvania State University, where he also leads the Center for Network Centric Cognition and Information Fusion. He has more than 30 years of experience in research, research management, and systems development in both industrial and academic environments. Dr. Hall has performed research in a wide variety of areas including celestial mechanics, digital signal processing, software engineering, automated reasoning, and multisensor data fusion. During the past 15 years, his research has focused on multisensor data fusion. He is the author of more than 200 technical papers, reports, book chapters, and books. Dr. Hall is a member of the Joint Directors of Laboratories (JDL) Data Fusion Working Group. He serves on the Advisory Board of the Data Fusion Center based at the State University of New York at Buffalo. In addition, he has served on the National Aeronautics and Space Administration (NASA) Aeronautics and Space Transportation Technology Advisory Committee. In 2001, Dr. Hall was awarded the Joe Mignona award to honor his contributions as a national leader in the data fusion community. The Data Fusion Group instituted the award in 1994 to honor the memory of Joseph Mignona. Dr. Hall was named as an IEEE Fellow in 2003 for his research in data fusion.

Dr. James Llinas has more than 30 years of experience in multisource information processing and data fusion technology extending over research, teaching, and business development activities. He is an internationally recognized expert in sensor, data, and information fusion; coauthored the first integrated book on multisensor data fusion; and has lectured internationally for over 20 years on this topic. He was coeditor of the first edition of the *Handbook of Multisensor Data Fusion* also Dr. Llinas received the definitive U.S. defense community award from the data fusion community, the Joe Mignona Award, in 1999. In addition, reflecting his international interests and stature, Dr. Llinas was voted as the first president of the International Society for Information Fusion in 1998, and maintains a position on the Executive Board. He has frequently provided high-level assessments of the state-of-the-art in data fusion, most recently to the U.S. Army for their Future Combat Systems program, and previously to the U.S. Air Force Scientific Advisory Board. Dr. Llinas has provided similar high-level guidance to international clients including an invited personal review of the data fusion programs of the Australian Defense Science and Technology Organization (DSTO), and consultations to the Swedish Defense Agency FOI, and the Canadian Defense R&D Canada organization. Dr. Llinas created the concept for and is the executive director for the Center for Multisource Information Fusion located at the State University of New York at Buffalo. This first-of-its-kind, university-based research center has received sponsorship from a broad base of defense and industrial R&D organizations, and is conducting basic research in distributed situational estimation, distributed learning, and correlation science, among a wide variety of other programs.

Contributors

Richard Antony
SAIC
Arlington, Virginia

Paul Applegate
University of Buffalo
Buffalo, New York

Stan Aungst
The Pennsylvania State University
University Park, Pennsylvania

Todd Bacastow
The Pennsylvania State University
University Park, Pennsylvania

Yaakov Bar-Shalom
Electrical Computation Department
University of Connecticut
Stanford, Connecticut

David Beyerle
The Pennsylvania State University
University Park, Pennsylvania

Erik Blasch
Sensors Directorate
Air Force Research Lab (AFRL/RY)
Wright-Patterson Air Force Base
Dayton, Ohio

Christopher L. Bowman
Data Fusion & Neural Networks
Broomfield, Colorado

Richard R. Brooks
Holcombe Department of Electrical and
    Computer Engineering
Clemson University
Clemson, South Carolina

Carl S. Byington
Impact Technologies, LLC
State College, Pennsylvania

Mark Campbell
The Pennsylvania State University
University Park, Pennsylvania

Kuo-Chu Chang
SEOR Department
George Mason University
Fairfax, Virginia

Amulya K. Garga
Lockheed Martin IS&S
Philadelphia, Pennsylvania

Lynne Grewe
California State University, East Bay
Hayward, California

Cristin M. Hall
Department of Educational Psychology,
    School Psychology, and Special
    Education, College of Education
The Pennsylvania State University
University Park, Pennsylvania

David L. Hall
Center for Network-Centric
    Cognition and
    Information Fusion
The Pennsylvania State University
University Park, Pennsylvania

J. Bockett Hunter
Lockheed Martin Maritime
    Systems & Sensors
Moorestown, New Jersey

Simon Julier
University College London
London, England

Otto Kessler
The MITRE Corporation
McLean, Virginia

T. Kirubarajan
McMaster University
Hamilton, Ontario, Canada

Jason Knox
United States Army
Fort Indiantown Gap, Pennsylvania

Jeff Kuhns
The Pennsylvania State University
University Park, Pennsylvania

Martin E. Liggins
The MITRE Corporation
McLean, Virginia

James Llinas
State University of New York
Buffalo, New York

Suihua Lu
Synopsys
Sunnyvale, California

Ronald Mahler
Lockheed Martin
Eagan, Minnesota

Capt. Lori McConnell
USAF/Space Warfare Center
Schriever, Colorado

Sonya A. Hall McMullen
Tech Reach, Inc.
State College, Pennsylvania

Shikha Miglani
The Pennsylvania State University
University Park, Pennsylvania

Mary L. Nichols
The Aerospace Corporation
El Segundo, California

Aubrey B. Poore
Numerica Corporation
Loveland, Colorado

Tod M. Schuck
Lockheed Martin Maritime
    Systems & Sensors
Moorestown, New Jersey

Richard R. Sherry
The Pennsylvania
    State University
Port Matilda, Pennsylvania

Alan N. Steinberg
Independent Consultant
Palm Coast, Florida

Lawrence D. Stone
Metron Inc.
Reston, Virginia

Brian J. Suchomel
Northrop Grumman Corporation,
    Mission Systems
Aurora, Colorado

David C. Swanson
Applied Research Laboratory
The Pennsylvania
    State University
University Park, Pennsylvania

Jeffrey K. Uhlmann
University of Missouri
Columbia, Missouri

Ed Waltz
Intelligence Innovation
    Division
BAE Systems Advanced
    Information Technologies
Arlington, Virginia

Tim Waltz
Space and Intelligence Operations
Applied Systems Research, Inc.
Fairfax, Virginia

Frank White
The MITRE Corporation
San Diego, California

Daniel D. Wilson
Economic & Decision Analysis
    Center (EDAC)
The MITRE Corporation
Bedford, Massachusetts

1

Multisensor Data Fusion

David L. Hall and James Llinas

CONTENTS

1.1 Introduction

Over the past two decades, significant attention has been focused on multisensor data fusion for both military and nonmilitary applications. Data fusion techniques combine data from multiple sensors and related information to achieve more specific inferences than could be achieved by using a single, independent sensor. Data fusion refers to the combination of data from multiple sensors (either of the same or different types), whereas information fusion refers to the combination of data and information from sensors, human reports, databases, etc.

The concept of multisensor data fusion is hardly new. As humans and animals evolved, they developed the ability to use multiple senses to help them survive. For example, assessing the quality of an edible substance may not be possible using only the sense of vision; the combination of sight, touch, smell, and taste is far more effective. Similarly, when vision is limited by structures and vegetation, the sense of hearing can provide advanced warning of impending dangers. Thus, multisensory data fusion is naturally performed by animals and humans to assess more accurately the surrounding environment and to identify threats, thereby improving their chances of survival. Interestingly, recent applications of data fusion[1] have combined data from an artificial nose and an artificial tongue using neural networks and fuzzy logic.

Although the concept of data fusion is not new, the emergence of new sensors, advanced processing techniques, improved processing hardware, and wideband communications has made real-time fusion of data increasingly viable. Just as the advent of symbolic processing computers (e.g., the Symbolics computer and the Lambda machine) in the early

1970s provided an impetus to artificial intelligence, the recent advances in computing and sensing have provided the capability to emulate, in hardware and software, the natural data fusion capabilities of humans and animals. Currently, data fusion systems are used extensively for target tracking, automated identification of targets, and limited automated reasoning applications. Data fusion technology has rapidly advanced from a loose collection of related techniques to an emerging true engineering discipline with standardized terminology, collection of robust mathematical techniques, and established system design principles. Indeed, the remaining chapters of this handbook provide an overview of these techniques, design principles, and example applications.

Applications for multisensor data fusion are widespread. Military applications include automated target recognition (e.g., for smart weapons), guidance for autonomous vehicles, remote sensing, battlefield surveillance, and automated threat recognition (e.g., identification-friend-foe-neutral [IFFN] systems). Military applications have also extended to condition monitoring of weapons and machinery, to monitoring of the health status of individual soldiers, and to assistance in logistics. Nonmilitary applications include monitoring of manufacturing processes, condition-based maintenance of complex machinery, environmental monitoring, robotics, and medical applications.

Techniques to combine or fuse data are drawn from a diverse set of more traditional disciplines, including digital signal processing, statistical estimation, control theory, artificial intelligence, and classic numerical methods. Historically, data fusion methods were developed primarily for military applications. However, in recent years, these methods have been applied to civilian applications and a bidirectional transfer of technology has begun.

1.2   Multisensor Advantages

Fused data from multiple sensors provide several advantages over data from a single sensor. First, if several identical sensors are used (e.g., identical radars tracking a moving object), combining the observations would result in an improved estimate of the target position and velocity. A statistical advantage is gained by adding the N independent observations (e.g., the estimate of the target location or velocity is improved by a factor proportional to $N^{1/2}$), assuming the data are combined in an optimal manner. The same result could also be obtained by combining N observations from an individual sensor.

The second advantage is that using the relative placement or motion of multiple sensors the observation process can be improved. For example, two sensors that measure angular directions to an object can be coordinated to determine the position of the object by triangulation. This technique is used in surveying and for commercial navigation (e.g., VHF omni-directional range [VOR]). Similarly, sensors, one moving in a known way with respect to another, can be used to measure instantaneously an object's position and velocity with respect to the observing sensors.

The third advantage gained using multiple sensors is improved observability. Broadening the baseline of physical observables can result in significant improvements. Figure 1.1 provides a simple example of a moving object, such as an aircraft, that is observed by both a pulsed radar and a forward-looking infrared (FLIR) imaging sensor. The radar can accurately determine the aircraft's range but has a limited ability to determine the angular direction of the aircraft. By contrast, the infrared imaging sensor can accurately determine the aircraft's angular direction but cannot measure the range. If these two observations are correctly associated (as shown in Figure 1.1), the combination of the two sensors provides a

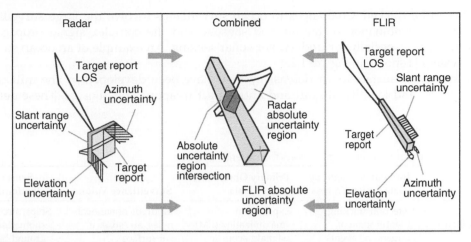

FIGURE 1.1
A moving object observed by both a pulsed radar and an infrared imaging sensor.

better determination of location than could be obtained by either of the two independent sensors. This results in a reduced error region, as shown in the fused or combined location estimate. A similar effect may be obtained in determining the identity of an object on the basis of the observations of an object's attributes. For example, there is evidence that bats identify their prey by a combination of factors, including size, texture (based on acoustic signature), and kinematic behavior. Interestingly, just as humans may use spoofing techniques to confuse sensor systems, some moths confuse bats by emitting sounds similar to those emitted by the bat closing in on prey (see http://www.desertmuseum.org/books/nhsd_moths.html—downloaded on October 4, 2007).

1.3   Military Applications

The Department of Defense (DoD) community focuses on problems involving the location, characterization, and identification of dynamic entities such as emitters, platforms, weapons, and military units. These dynamic data are often termed as order-of-battle database or order-of-battle display (if superimposed on a map display). Beyond achieving an order-of-battle database, DoD users seek higher-level inferences about the enemy situation (e.g., the relationships among entities and their relationships with the environment and higher-level enemy organizations). Examples of DoD-related applications include ocean surveillance, air-to-air defense, battlefield intelligence, surveillance and target acquisition, and strategic warning and defense. Each of these military applications involves a particular focus, a sensor suite, a desired set of inferences, and a unique set of challenges, as shown in Table 1.1.

Ocean surveillance systems are designed to detect, track, and identify ocean-based targets and events. Examples include antisubmarine warfare systems to support navy tactical fleet operations and automated systems to guide autonomous vehicles. Sensor suites can include radar, sonar, electronic intelligence (ELINT), observation of communications traffic, infrared, and synthetic aperture radar (SAR) observations. The surveillance volume for ocean surveillance may encompass hundreds of nautical miles and focus on air, surface, and subsurface targets. Multiple surveillance platforms can be involved and numerous targets can be tracked. Challenges to ocean surveillance involve the large surveillance volume, the combination of targets and sensors, and the complex signal propagation environment—especially for underwater sonar sensing. An example of an ocean surveillance system is shown in Figure 1.2.

Air-to-air and surface-to-air defense systems have been developed by the military to detect, track, and identify aircraft and antiaircraft weapons and sensors. These defense

TABLE 1.1

Representative Data Fusion Applications for Defense Systems

Specific Applications	Inferences Sought by Data Fusion Process	Primary Observable Data	Surveillance Volume	Sensor Platforms
Ocean surveillance	Detection, tracking, identification of targets and events	Expectation maximization (EM) signals, acoustic signals, nuclear-related, derived observations	Hundreds of nautical miles, air/surface/ subsurface	Ships, aircraft, submarines, ground-based, ocean-based
Air-to-air and surface-to-air defense	Detection, tracking, identification of aircraft	EM radiation	Hundreds of miles (strategic), miles (tactical)	Ground-based, aircraft
Battlefield intelligence, surveillance, and target acquisition	Detection and identification of potential ground targets	EM radiation	Tens of hundreds of miles about a battlefield	Ground-based, aircraft
Strategic warning and defense	Detection of indications of impending strategic actions, detection and tracking of ballistic missiles and warheads	EM radiation, nuclear-related	Global	Satellites, aircraft

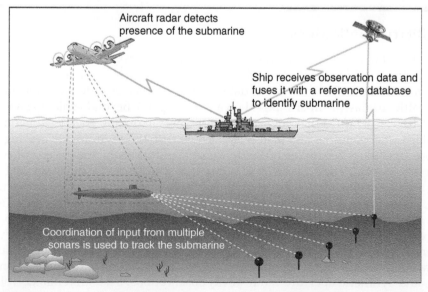

FIGURE 1.2
An example of an ocean surveillance system.

systems use sensors such as radar, passive electronic support measures (ESM), infrared identification-friend-foe (IFF) sensors, electrooptic image sensors, and visual (human) sightings. These systems support counterair, order-of-battle aggregation, assignment of aircraft to raids, target prioritization, route planning, and other activities. Challenges to these data fusion systems include enemy countermeasures, the need for rapid decision making, and potentially large combinations of target-sensor pairings. A special challenge for IFF systems is the need to confidently and noncooperatively identify enemy aircraft. The proliferation of weapon systems throughout the world has resulted in little correlation between the national origin of a weapon and the combatants who use the weapon.

Finally, battlefield intelligence, surveillance, and target acquisition systems attempt to detect and identify potential ground targets. Examples include the location of land mines and automatic target recognition. Sensors include airborne surveillance via SAR, passive ESM, photo-reconnaissance, ground-based acoustic sensors, remotely piloted vehicles, electrooptic sensors, and infrared sensors. Key inferences sought are information to support battlefield situation assessment and threat assessment.

1.4 Nonmilitary Applications

Other groups addressing data fusion problems are the academic, commercial, and industrial communities. They involve nonmilitary applications such as the implementation of robotics, automated control of industrial manufacturing systems, development of smart buildings, and medical applications. As with military applications, each of these applications has a particular set of challenges and sensor suites, and a specific implementation environment (see Table 1.2).

TABLE 1.2

Representative Nondefense Data Fusion Applications

Specific Applications	Inferences Sought by Data Fusion Process	Primary Observable Data	Surveillance Volume	Sensor Platforms
Condition-based maintenance	Detection, characterization of system faults, recommendations for maintenance/corrections	EM signals, acoustic signals, magnetic, temperatures, x-rays, lubricant debris, vibration	Microscopic to hundreds of feet	Ships, aircraft, ground-based (e.g., factories)
Robotics	Object location/recognition, guide the locomotion of robot (e.g., "hands" and "feet")	Television, acoustic signals, EM signals, x-rays	Microscopic to tens of feet about the robot	Robot body
Medical diagnoses	Location/identification of tumors, abnormalities, and disease	X-rays, nuclear magnetic resonance (NMR), temperature, infrared, visual inspection, chemical and biological data, self-reports of symptoms by humans	Human body volume	Laboratory
Environmental monitoring	Identification/location of natural phenomena (e.g., earthquakes, weather)	Synthetic aperture radar (SAR), seismic, EM radiation, core samples, chemical and biological data	Hundreds of miles, miles (site monitoring)	Satellites, aircraft, ground-based, underground samples

Remote sensing systems have been developed to identify and locate entities and objects. Examples include systems to monitor agricultural resources (e.g., to monitor the productivity and health of crops), locate natural resources, and monitor weather and natural disasters. These systems rely primarily on image systems using multispectral sensors. Such processing systems are dominated by automatic image processing. Multispectral imagery—such as the Landsat satellite system (http://www.bsrsi.msu.edu/) and the SPOT system—is used (see http://www.spotimage.fr/web/en/167-satellite-image-spot-formosat-2-kompsat-2-radar. php). A technique frequently used for multisensor image fusion involves adaptive neural networks. Multiimage data are processed on a pixel-by-pixel basis and input to a neural network to classify automatically the contents of the image. False colors are usually associated with types of crops, vegetation, or classes of objects. Human analysts can readily interpret the resulting false color synthetic image.

A key challenge in multiimage data fusion is coregistration. This problem requires the alignment of two or more photos so that the images are overlaid in such a way that corresponding picture elements (pixels) on each picture represent the same location on earth (i.e., each pixel represents the same direction from an observer's point of view). This coregistration problem is exacerbated by the fact that image sensors are nonlinear and they perform a complex transformation between the observed three-dimensional space and a two-dimensional image.

A second application area, which spans both military and nonmilitary users, is the monitoring of complex mechanical equipment such as turbo machinery, helicopter gear trains, or industrial manufacturing equipment. For a drivetrain application, for example, sensor data can be obtained from accelerometers, temperature gauges, oil debris monitors, acoustic sensors, and infrared measurements. An online condition-monitoring system would seek to combine these observations to identify precursors to failure such as abnormal gear wear, shaft misalignment, or bearing failure. The use of such condition-based monitoring is expected to reduce maintenance costs and improve safety and reliability. Such systems are beginning to be developed for helicopters and other platforms (see Figure 1.3).

FIGURE 1.3
Mechanical diagnostic test-bed used by The Pennsylvania State University to perform condition-based maintenance research.

1.5 Three Processing Architectures

Three basic alternatives can be used for multisensor data: (1) direct fusion of sensor data; (2) representation of sensor data via *feature vectors*, with subsequent fusion of the feature vectors; or (3) processing of each sensor to achieve high-level inferences or decisions, which are subsequently combined. Each of these approaches utilizes different fusion techniques as described and shown in Figures 1.4a through 1.4c.

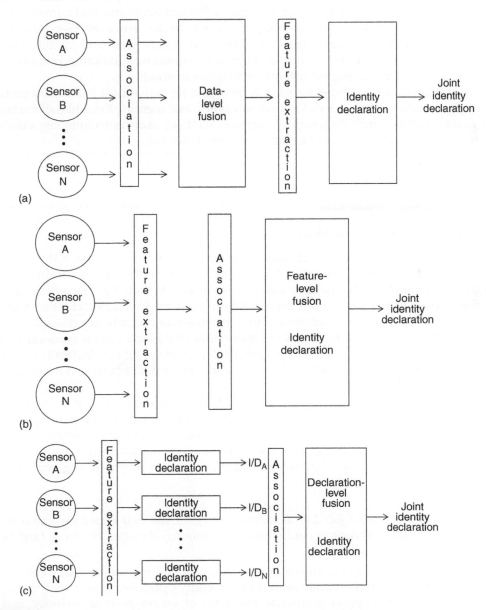

FIGURE 1.4
(a) Direct fusion of sensor data. (b) Representation of sensor data via feature vectors, with subsequent fusion of the feature vectors. (c) Processing of each sensor to achieve high-level inferences or decisions, which are subsequently combined.

If the multisensor data are commensurate (i.e., if the sensors are measuring the same physical phenomena such as two visual image sensors or two acoustic sensors) then the raw sensor data can be directly combined. Techniques for raw data fusion typically involve classic estimation methods such as Kalman filtering.[2] Conversely, if the sensor data are noncommensurate then the data must be fused at the feature/state vector level or decision level.

Feature-level fusion involves the extraction of representative features from sensor data. An example of feature extraction is the cartoonist's use of key facial characteristics to represent the human face. This technique—which is popular among political satirists—uses key features to evoke recognition of famous figures. Evidence confirms that humans utilize a feature-based cognitive function to recognize objects. In the case of multisensor feature-level fusion, features are extracted from multiple sensor observations and combined into a single concatenated feature vector that is an input to pattern recognition techniques such as neural networks, clustering algorithms, or template methods.

Decision-level fusion combines sensor information after each sensor has made a preliminary determination of an entity's location, attributes, and identity. Examples of decision-level fusion methods include weighted decision methods (voting techniques), classical inference, Bayesian inference, and Dempster–Shafer's method.

1.6 Data Fusion Process Model

One of the historical barriers to technology transfer in data fusion has been the lack of a unifying terminology that crosses application-specific boundaries. Even within military applications, related but distinct applications—such as IFF, battlefield surveillance, and automatic target recognition—used different definitions for fundamental terms such as correlation and data fusion. To improve communications among military researchers and system developers, the Joint Directors of Laboratories (JDL) Data Fusion Working Group (established in 1986) began an effort to codify the terminology related to data fusion. The result of that effort was the creation of a process model for data fusion and a data fusion lexicon (shown in Figure 1.5).

The JDL process model, which is intended to be very general and useful across multiple application areas, identifies the processes, functions, categories of techniques, and specific techniques applicable to data fusion. The model is a two-layer hierarchy. At the top level, shown in Figure 1.5, the data fusion process is conceptualized by sensor inputs, human–computer interaction, database management, source preprocessing, and six key subprocesses:

Level 0 processing (subobject data association and estimation) is aimed at combining pixel or signal level data to obtain initial information about an observed target's characteristics.

Level 1 processing (object refinement) is aimed at combining sensor data to obtain the most reliable and accurate estimate of an entity's position, velocity, attributes, and identity (to support prediction estimates of future position, velocity, and attributes).

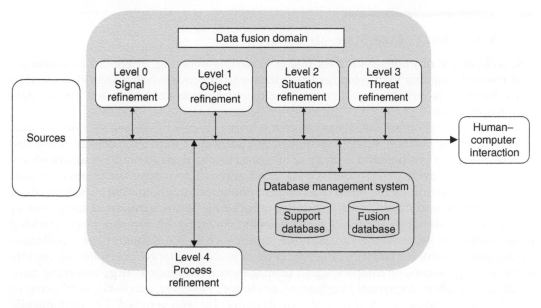

FIGURE 1.5
Joint Directors of Laboratories process model for data fusion.

Level 2 processing (situation refinement) dynamically attempts to develop a description of current relationships among entities and events in the context of their environment. This entails object clustering and relational analysis such as force structure and cross-force relations, communications, physical context, etc.

Level 3 processing (significance estimation) projects the current situation into the future to draw inferences about enemy threats, friend and foe vulnerabilities, and opportunities for operations (and also consequence prediction, susceptibility, and vulnerability assessments).

Level 4 processing (process refinement) is a meta-process that monitors the overall data fusion process to assess and improve real-time system performance. This is an element of resource management.

Level 5 processing (cognitive refinement) seeks to improve the interaction between a fusion system and one or more user/analysts. Functions performed include aids for visualization, cognitive assistance, bias remediation, collaboration, team-based decision making, course of action analysis, etc.

For each of these subprocesses, the hierarchical JDL model identifies specific functions and categories of techniques (in the model's second layer) and specific techniques (in the model's lowest layer). Implementation of data fusion systems integrates and interleaves these functions into an overall processing flow.

The data fusion process model is augmented by a hierarchical taxonomy that identifies categories of techniques and algorithms for performing the identified functions. An associated lexicon has been developed to provide a consistent definition of data fusion terminology. The JDL model is described in more detail in Chapters 2 and 3, and by Hall and McMullen.[3]

1.7 Assessment of the State-of-the-Art

The technology of multisensor data fusion is rapidly evolving. There is much concurrent research ongoing to develop new algorithms, to improve existing algorithms, and to assemble these techniques into an overall architecture capable of addressing diverse data fusion applications.

The most mature area of data fusion process is level 1 processing—using multisensor data to determine the position, velocity, attributes, and identity of individual objects or entities. Determining the position and velocity of an object on the basis of multiple sensor observations is a relatively old problem. Gauss and Legendre developed the method of least squares for determining the orbits of asteroids.[2] Numerous mathematical techniques exist for performing coordinate transformations, associating observations to observations or to tracks, and estimating the position and velocity of a target. Multisensor target tracking is dominated by sequential estimation techniques such as the Kalman filter. Challenges in this area involve circumstances in which there is a dense target environment, rapidly maneuvering targets, or complex signal propagation environments (e.g., involving multipath propagation, cochannel interference, or clutter). However, single-target tracking in excellent signal-to-noise environments for dynamically well-behaved (i.e., dynamically predictable) targets is a straightforward, easily resolved problem.

Current research focuses on solving the assignment and maneuvering target problem. Techniques such as multiple-hypothesis tracking (MHT) and its extensions, probabilistic data association methods, random set theory, and multiple criteria optimization theory are being used to resolve these issues. Recent studies have also focused on relaxing the assumptions of the Kalman filter using techniques such as particle filters and other methods. Some researchers are utilizing multiple techniques simultaneously, guided by a knowledge-based system capable of selecting the appropriate solution on the basis of algorithm performance.

A special problem in level 1 processing involves the automatic identification of targets on the basis of observed characteristics or attributes. To date, object recognition has been dominated by feature-based methods in which a feature vector (i.e., a representation of the sensor data) is mapped into feature space with the hope of identifying the target on the basis of the location of the feature vector relative to *a priori* determined decision boundaries. Popular pattern recognition techniques include neural networks, statistical classifiers, and vector machine approaches. Although numerous techniques are available, the ultimate success of these methods relies on the selection of good features. (Good features provide excellent class separability in feature space, whereas bad features result in greatly overlapping feature space areas for several classes of target.) More research is needed in this area to guide the selection of features and to incorporate explicit knowledge about target classes. For example, syntactic methods provide additional information about the makeup of a target. In addition, some limited research is proceeding to incorporate contextual information—such as target mobility with respect to terrain—to assist in target identification.

Level 2 and level 3 fusions (situation refinement and threat refinement) are currently dominated by knowledge-based methods such as rule-based blackboard systems, intelligent agents, Bayesian belief network formulations, etc. These areas are relatively immature and have numerous prototypes, but few robust, operational systems. The main challenge in this area is to establish a viable knowledge base of rules, frames, scripts, or other methods to represent knowledge about situation assessment or threat assessment. Unfortunately, only primitive cognitive models exist to replicate the human performance of these functions. Much research is needed before reliable and large-scale knowledge-based systems can be developed for automated situation assessment and threat assessment. New approaches that offer promise are the use of fuzzy logic and hybrid architectures, which extend the concept of blackboard systems to hierarchical and multi–time scale orientations. Also, recent work by Yen and his associates[4] on team-based intelligent agents appears promising. These agents emulate the way human teams collaborate, proactively exchanging information and anticipating information needs.

Level 4 processing, which assesses and improves the performance and operation of an ongoing data fusion process, has a mixed maturity. For single-sensor operations, techniques from operations research and control theory have been applied to develop effective systems, even for complex single sensors such as phased array radars. By contrast, situations that involve multiple sensors, external mission constraints, dynamic observing environments, and multiple targets are more challenging. To date, considerable difficulty has been encountered in attempting to model and incorporate mission objectives and constraints to balance optimized performance with limited resources, such as computing power and communication bandwidth (e.g., between sensors and processors), and other effects. Methods from utility theory are being applied to develop measures of system performance and effectiveness. Knowledge-based systems are being developed for context-based approximate reasoning. Significant improvements would result from the advent of smart, self-calibrating sensors, which can accurately and dynamically assess their own performance. The advent of distributed network-centric environments, in which sensing resources, communications capabilities, and information requests are very dynamic, creates serious challenges for level 4 fusion. It is difficult (or possibly impossible) to optimize resource utilization in such an environment. In a recent study, Mullen et al.[5] have applied concepts of market-based auctions to dynamically allocate resources, treating sensors and communication systems as suppliers of services, and users and algorithms as consumers, to rapidly assess how to allocate system resources to satisfy the consumers of information.

Data fusion has suffered from a lack of rigor with regard to the test and evaluation of algorithms and the means of transitioning research findings from theory to application. The data fusion community must insist on high standards for algorithm development, test, and evaluation; creation of standard test cases; and systematic evolution of the technology to meet realistic applications. On a positive note, the introduction of the JDL process model and the emerging nonmilitary applications are expected to result in increased cross-discipline communication and research. The nonmilitary research in robotics, condition-based maintenance, industrial process control, transportation, and intelligent buildings would produce innovations that will cross-fertilize the entire field of data fusion technology. The challenges and opportunities related to data fusion establish it as an exciting research field with numerous applications.

1.8 Dirty Secrets in Data Fusion

In the first edition of this handbook, a chapter entitled "Dirty Secrets in Data Fusion" was included. It was based on a article written by Hall and Steinberg.[6] This original article had identified the following seven challenges or issues in data fusion:

1. There is no substitute for a good sensor.
2. Downstream processing cannot absolve the sins of upstream processing.

3. The fused answer may be worse than the best sensor.
4. There are no magic algorithms.
5. There will never be enough training data.
6. It is difficult to quantify the value of data fusion.
7. Fusion is not a static process.

Subsequently, these "dirty secrets" were revised as follows:

- There is *still* no substitute for a good sensor (and a good human to interpret the results)—This means that if something cannot be actually observed or inferred from effects, then no amount of data fusion from multiple sensors would overcome this problem. This problem becomes even more challenging as threats change. The transition from the search for well-known physical targets (e.g., weapon systems, emitters, etc.) to targets based on human networks causes obvious issues with determining what can and should be observed. In particular, trying to determine intent is tantamount to mind reading, and is an elusive problem.

- Downstream processing *still* cannot absolve upstream sins (or lack of attention to the data)—It is clear that we must do the best processing possible at every step of the fusion/inference process. For example, it is necessary to perform appropriate image and signal processing at the data stage, followed by appropriate transformations to extract feature vectors, etc., for feature-based identity processing. Failure to perform the appropriate data processing or failure to select and refine effective feature vectors cannot be overcome by choosing complex pattern recognition techniques. We simply must pay attention at every stage of the information chain, from energy detection to knowledge creation.

- Not only may the fused result be worse than the best sensor, but failure to address pedigree, information overload, and uncertainty may really fowl up things—The rapid introduction of new sensors and use of humans as "soft sensors (reporters)" in network operations places special challenges on determining how to weight the incoming data. Failure to accurately assess the accuracy of the sensor/input data would lead to biases and errors in the fused results. The advent of networked operations and service-oriented architectures (SOA) can exacerbate this problem by rapidly disseminating data and information without understanding the sources or pedigree (who did what to the data).

- There are still no magic algorithms—This book provides an overview of numerous algorithms and techniques for all levels of fusion. Although there are increasingly sophisticated algorithms, it is always a challenge to match the algorithm with the actual state of knowledge of the data, system, and inferences to be made. No single algorithm is ideal under all circumstances.

- There will never be enough training data—However, hybrid methods that combine implicit and explicit information can help. It is well-known that pattern recognition methods, such as neural networks, require training data to establish the key weights. When seeking to map an n-dimensional feature vector to one of m classes or categories, we need in general $n \times m \times (10\text{--}30)$ training examples under a variety of observing conditions. This can be very challenging to obtain, especially with dynamically changing threats. Hence, in general, there will never be enough training data available to satisfy the mathematical conditions for pattern recognition techniques. However, new hybrid methods that use a combination of sample data, model-based data, and human subject explicit information can assist in this area.

- We have started at "the wrong end" (viz., at the sensor side vs. at the human side of fusion)—Finally, we note that extensive research has been conducted to develop methods for level 0 and level 1 fusions. In essence, we have "started at the data side or sensor inputs" to progress toward the human side. More research needs to be conducted in which we begin at the human side (viz., at the formation of hypotheses or semantic interpretation of events) and proceed toward the sensing side of fusion. Indeed, the introduction of the level 5 process was recognition of this need.

The original issues identified (viz., that fusion is not a static process, and that the benefits of fusion processing are difficult to quantify) still hold true.

Overall, this is an exciting time for the field of data fusion. The rapid advances and proliferation of sensors, the global spread of wireless communications, and the rapid improvements in computer processing and data storage enable new applications and methods to be developed.

1.9   Additional Information

Additional information about multisensor data fusion may be found in the following references:

- D.L. Hall, *Mathematical Techniques in Multisensor Data Fusion*, Artech House, Inc. (1992)—Provides details on the mathematical and heuristic techniques for data fusion.
- E. Waltz and J. Llinas, *Multisensor Data Fusion*, Artech House, Inc. (1990)—Presents an excellent overview of data fusion especially for military applications.
- L.A. Klein, *Sensor and Data Fusion Concepts and Applications*, SPIE Optical Engineering Press, Vol. TT 14 (1993)—Presents an abbreviated introduction to data fusion.
- R. Antony, *Principles of Data Fusion Automation*, Artech House, Inc. (1995)—Provides a discussion of data fusion processes with special focus on database issues to achieve computational efficiency.
- A multimedia computer-based training package, *Introduction to Data Fusion, A Multimedia Computer-Based Training Package*, available from Artech House, Inc., Boston, MA, 1995.

References

1. T. Sundic, S. Marco, J. Samitier, and P. Wide, Electronic tongue and electronic nose data fusion in classification with neural networks and fuzzy logic based models, *IEEE*, 3, 1474–1480, 2000.
2. H.W. Sorenson, Least-squares estimation: From Gauss to Kalman, *IEEE Spectrum*, 7, 63–68, July 1970.

3. D. Hall and S.A.H. McMullen, *Mathematical Techniques in Multisensor Data Fusion*, Artech House Inc., Boston, MA, 2004.
4. G. Airy, P.-C. Chen, X. Fan, J. Yen, D. Hall, M. Brogan, and T. Huynh, Collaborative RPD agents assisting decision making in active decision spaces, in *Proceedings of the 2006, IAT'06, IEEE/ WIC/ACM International Conference on Intelligent Agent Technology*, December 2006.
5. T. Mullen, V. Avasarala, and D.L. Hall, Customer-driven sensor management, *IEEE Intelligent Systems*, Special Issue on Self-Management through Self-Organization in Information Systems, March/April 2006, 41–49.
6. D.L. Hall and A. Steinberg, Dirty secrets in multisensor data fusion, *Proceedings of the National Symposium on Sensor Data Fusion (NSSDF)*, San Antonio, TX, June 2000.

2

Data Fusion Perspectives and Its Role in Information Processing

Otto Kessler and Frank White*

CONTENTS

---

* The authors' affiliation with the MITRE Corporation are provided only for identification purposes and is not intended to convey or imply MITRE's concurrence with, or support for, the positions, opinions, or viewpoints expressed by the authors.

2.1   Operational Perspective of Fusion

In the past decade, the word *fusion* has become familiar in most households. There is "Fusion," the car; "Fusion," the cuisine; bottled "Fusion" drinks, cold "Fusion," and a multitude of other uses of the word, including its being occasionally associated with data and information.* It is this latter domain where fusion may be the most important. The immense changes in the nature of the information environment, driven by rapid evolution in communications technology and the Internet, have made raw data and processed information available to individual humans at a rate and in volumes that are unprecedented. But why is fusion important? Why do we fuse? We fuse because fusion is a means to deal with this glut of readily available data and information that we might organize and present information in ways which are accessible and capable of supporting decisions, even decisions as simple as where to go for dinner.

2.1.1   Introduction: Fusion in Command and Control and Decision Processes

In the context of the information domain, fusion is not a thing or a technology but a way of thinking about the world and the environment that focuses on data and information content relevant to a human and the decisions that must be made. In traditional military usage, it is the means by which data from one or a multiplicity of sensors, along with data or information from a variety of nonsensor sources, can be combined and presented to satisfy a broad range of operational goals. At the simplest level, a sensor system may detect and report aspects of a target or the environment, which when correlated over time may be sufficient to support a decision. For example, a radar system detecting an approaching aircraft can trigger a decision to fire on the aircraft when it is within range. However, even this simple example requires the fusion of a time-series analysis of observations that can then be associated unambiguously with a single object and by some other data (visual observation, lack of an identification-friend or foe [IFF] transponder, additional related and fused data) can be identified as an enemy.

The essence of command and control (C2)[1] is humans making timely decisions in the face of uncertainty, and acting on those decisions. This essence has changed little over history, though the information domain and the possibilities for mission success and failure have changed dramatically. Today, with greater dependence on technology, the military goal of detecting, identifying, and tracking targets in support of a decision process involves taking some type of action, which may not be achievable with only a single sensor system. Most tactical decision processes and virtually all operational and strategic C2 decisions require a wide range of sensors and sources. Reaching a decision or an execution objective is unachievable without the benefit of a disciplined fusion process. The role of fusion extends to many diverse nonmilitary domains as well. Data fusion has the highest priority in many homeland security domains (such as maritime domain awareness and border security among others). Medical equipment is reaching a degree of sophistication that diagnosis (traditionally a human fusion process) is based on standard fusion processes to provide quality levels of automation. The growth of fusion awareness across the civil sector is a boon and a challenge to the data fusion community.

---

* The terms *data and information* will appear throughout this handbook and often will appear to be used interchangeably. Their definitions and distinctions are discussed at greater length in Section 2.1.3.

2.1.2 History of Fusion in Operations

The modern concept of fusion as a key enabler of C2 decision-making dates to the period between World War I and II. It received recognition and emphasis just before the outbreak of World War II. The Royal Navy, the Admiralty of the United Kingdom, had a worldwide collection process consisting of agents, ship sightings, merchant ship crew debriefing, and other means to attain information,[2] all of which had served them well over the span of the empire. However, in the run-up to World War II, the British Admiralty recognized that "no single type of information could, without integration of data from many other sources, provide a sufficiently authoritative picture of what to expect from their enemies"[2] and subsequently set up an all-source fusion center, the Operational Intelligence Center (OIC). Many other similar intelligence centers dedicated to multisource fusion followed, including the U.S. Tenth Fleet and R.V. Jones's Scientific Intelligence apparatus.[3] Historically, fusion has been associated with the intelligence cycle and the production of "actionable" intelligence. In fact fusion has often been synonymous with intelligence production. Intelligence "fusion centers" with access to data at multiple levels of security have been and continue to be producers of all-source or, more accurately, multisource intelligence products. This association has been so strong that a cultural conflict has developed over the ownership of the fusion process. Characterizing fusion as the exclusive domain of the intelligence process is too narrow and the continuing political battles are not useful, particularly in today's environment. The reality is that fusion is fundamental to the way human beings deal with their environment and essential to both intelligence and C2 processes. This essential quality was recognized early in World War II by the British who moved the Convoy Routing section, a C2 component, into the OIC (Intel) to further integrate and fuse intelligence with operational information in support of strategic and tactical planning. This move produced such positive outcomes in the Battle of the Atlantic that even today it is held up as an example of the right way to do fusion. The need to fuse many sources of data in the right way is even more imperative in our modern multilateral environment with the primacy of the global war on terror and homeland security concerns extant not just in the United States but in all the nations of the civilized world. Former special assistant to the director of Central Intelligence Agency (CIA) and vice chairman of the CIA's National Intelligence Council, Herbert E. Meyer,[4] describes intelligence as *organized information* in his book *Real World Intelligence*. His definition is very useful in a world where C2 and intelligence distinctions are blurring and the demand for organized information, while at an all time high, continues to increase. It is critical that individuals, government, and military organizations not allow dated patterns of thinking or personal and cultural biases to get in the way of managing information to support fusion and decision processes.

The OIC, the Battle of the Atlantic, and the Battle of Britain are success stories that attest to the value of multisensor/multisource data and information, as is the decades-long struggle against Soviet and Bloc submarine forces during the cold war. Using the OIC as a model, the U.S. and Allied navies established a network of fusion centers that successfully fused traditional intelligence data with operational data, effectively countering a dangerous threat for many years.

Along with the successes, however, military history is replete with examples of the failure of intelligence to provide commanders with the information needed to conduct military operations effectively. In many cases, such as the Japanese attack on Hawaii's Pearl Harbor and the World War II Battle of the Bulge, the failures began with a breakdown of data and information distribution as well as access. The data and information needed to anticipate the attacks were arguably available from a wide variety of sources but were dispersed over organizational hierarchies, times, and locations, and thus the information was not fused and made available to the total intelligence community and the responsible commanders. In the case of the Battle of the Bulge, for example, ULTRA intercepts and information from other highly sensitive sources—in particular, human intelligence (HUMINT)—were provided to senior echelons of intelligence organizations but not passed to the lower echelons, principally for security reasons (protection of sources and means). At the same time, information obtained by patrols behind enemy lines, through interrogation of prisoners of war, from line-crossers, and via direct observation from frontline U.S. and Allied forces often remained at the local level or was distorted by a lower echelon intelligence officer's personal bias or judgment. On many of the occasions when locally collected data and information were reported up the chain, the multiple analysts from different commands participating in the chain of fusion and analysis introduced many opportunities for bias, distortion, or simple omission in the intelligence summaries and situation reports delivered to higher headquarters. When these reports were consolidated and forwarded to upper echelon intelligence organizations and command headquarters to be fused with the very sensitive, decoded messages from ULTRA, there was no way to identify the potential errors, biases, or completeness of the reports. In some cases, individuals along the chain simply did not believe the information or intelligence. The process also required too much time to support the evolving tactical situation and resulted in devastating surprises for our forces. These examples illustrate some of the many ways fusion processes can break down and thereby fail to adequately support the decision process. The consequences of these failures are well known to history.

Admiral William O. Studeman, former Deputy Director of Central Intelligence and a strong yet discerning voice for fusion, once told a meeting of the Joint Directors of Laboratories/Data Fusion Subpanel (JDL/DFS) that "the purpose of this fusion capability is to allow some predictive work on what the enemy is doing. The job of the intelligence officer is not to write great history, it is to be *as predictive as possible* in support of decision making for command."[5] To be predictive requires urgency and speed, and unfortunately much of the fusion production in the past has been forensic or constituted historical documentation. The reasons are complex, as illustrated in the example. Often key data is not accessible and much of what is accessible is not in a form that is readily understood. Sometimes the human analysts simply lack the time or the ability to sort and identify relevant data from a wide variety of sources and fuse it into a useful intelligence product that will be predictive as early in the cycle as possible. While fusion is difficult, making predictions is also risky, which leads to a propensity to "write great history" among risk averse analysts and intelligence officers. Very often failures of both command and intelligence are failures of fusion, and are not confined to the distant past, as reports from the 9/11 Commission[6] and the Commission on Weapons of Mass Destruction[7] make abundantly clear.

2.1.3 Automation of Fusion Processes in Operation

It is important to remember that fusion is a fundamental human process, as anyone crossing a busy street will quickly become aware. In stepping from the curb, a person must integrate and fuse the input from two visual sensors (roughly 170° of coverage), omnidirectional acoustic sensors, olfactory sensors, and tactile as well as vibration sensors. Humans do this very well and most street crossings have happy outcomes because a lot of fusion is occurring. The human mind remains the premier fusion processor; from the emergence of modern humans as a species up to the past 35 years, all fusion was performed exclusively by the human mind, aided occasionally by simple tools such as calculators and overlays. However, in the modern era, a period Daniel Boorstin has characterized as the age of the *mechanized observer*, raw data is available totally unprocessed and not yet interpreted by a human: "Where before the age of the mechanized observer, there was a tendency for meaning to outrun data, the modern tendency is quite the contrary as we see data outrun meaning."[8] This is a recent phenomenon and, with the amount of data from mechanized observers increasing exponentially, the sheer volume of data and information is overwhelming the human capacity to even sort the incoming flow, much less to organize and fuse it.

The terms *data and information* are often used interchangeably. This is not surprising particularly since the terms are used this way in operational practice, still it can be disconcerting. It is wise to avoid becoming embroiled in circular semantic arguments on this matter, for very often, one person's information is another's data and there is no absolute definition of either term. *Knowledge* is another loaded word in military and fusion parlance. Several definitions of data and information have been used over the years, such as *information is data with context* or *information is data that is relevant over some definable period of time*. Both of these provide useful distinctions (relevance, time, and context). In addition, the business community has developed a set of definitions that are practical:

Data is a set of discrete, objective facts—often about events.

Information is a message that is intended to inform someone. Information is built from data through a value-adding transformation. This may be, for example, categorizing or placing in context.

Knowledge is a fluid mix of framed experience, values, contextual information, and expert insight that provides a framework for evaluating and incorporating new experiences and information. It originates and is applied in the knowers' minds. It is built from information through a value-adding transformation. This may be, for example, comparison with other situations, consideration of consequences and connections, or discussion with others.[9]

The automation of fusion functions is imperative but at the same time is very new and proving to be difficult, even though the concept of fusion is a very ancient capability. Conversely the automation of sensors has been relatively easy and there is no indication this trend will be slowing with the introduction of micro- and nanotechnologies and swarms of sensors. The automation of fusion processes is falling further behind.

The process of automating some fusion functions began in late 1960s with the operational introduction of automated radar processing and track formation algorithms, and the introduction of localization algorithms into sound surveillance system (SOSUS) and the high frequency direction finding (HFDF) systems. The first tracker/correlator (a Kalman filter) was introduced into operational use in 1970. These first introductions were followed with rapid improvement and automation of basic correlation and tracking functions in many systems; their impact has been widespread and substantial. Automation has greatly speeded up the ingestion, processing, and production of all types of time-series sensor system data and provided timely reporting of what have become vast volumes of sensor reports.

2.1.3.1 Automation of Fusion in Operations, the SOSUS Experience

As every problem solved creates a new problem, the introduction of automated functions has added new dimensions of difficulty to the fusion equation. Analysts may no longer have to spend as much time sorting raw data but they now receive automated reports from trackers where the underlying fusion processes are not explicit and the analysts and operators are not trained to understand how the processes work. Thus, they may not be able to recognize and correct clear errors nor are they able to apply variable confidence metrics to the output products. For example, the SOSUS multisensor localizer (MSL) algorithms provided a statistically significant localization and confidence ellipse with the use of concurrent bearing and time-measurement reports from multiple sensors. This was demonstrably faster and better than the manual methods, creating a statistically significant confidence ellipse. The localization ellipse was an improvement over the manual process that generated an awkward polygon with a variable number of sides that prosecuting forces had to search. The automated product was simpler to understand and reduced the size of the search area the forces had to deal with. However, the humans generating the report still associated measurement reports with a new or existing submarine contact, basing the association on the contact's previous location and acoustic signature. The automated function made it much more difficult to identify mis-associations and thus the SOSUS products contained more association errors. Still, the benefits of automation outweighed the problems by a wide margin.

Upon its introduction into SOSUS, the multisensor tracker (MST) greatly improved the use of the confidence values from individual sensor reports of bearing and time than did the occasional individual localizations, even though they were generated from the identical series of reports. In addition, the many nonconcurrent measurements could be used to maintain and even improve a continuous track. The error ellipses of the same statistical significance were often orders of magnitude smaller than those of the individual localizations, a great boon to the prosecuting forces. On the downside, mis-associations had a disastrous effect on the tracker that was difficult to identify initially. Further, when a submarine or surface contact maneuvered, the tracker's motion model assumptions drove the tracker for some period with increasing residual error until it began to catch up with the maneuver. Compounding the difficulty, a submarine on patrol, maneuvering randomly, deviated so far from the underlying motion model that the MST was rendered useless.

Nevertheless, the value of automation in this case was so profound that developers and operators worked together diligently to resolve the problems. The developers modified the software to expose the residual errors for every measurement association that greatly facilitated finding and correcting mis-associations. The developers also introduced alternate motion models and built in the ability to interrupt the tracker and run multiple recursive iterations when required. This ability became an additional analysis tool for predicting submarine behavior. The analysts and operators, for their part, took courses to learn the underlying theory and mathematics of the combinatoric algorithms and the Kalman filter tracker. This training armed them to recognize problems with the automation and further allowed them to recognize anomalies that were not problems with automated tools but interesting behaviors of submarines and other target vessels.

This automation good news story did not last, however. The next developer (new contract) did not expose residual errors and took the algorithm "tuning" and iteration controls away from the operator/analysts. At the same time the operators and analysts stopped receiving any training on the automated tools beyond simple buttonology. In short order, the automation became a fully trusted but not understood "black box" with the tracker output generated and disseminated with little or no oversight from the human. Informal studies and observation of operational practice revealed that obvious errors of mis-association and model violation were unrecognized by the operators and that error-full reports were routinely generated and promulgated. Despite obstacles and set-backs in the incorporation of automation, when used under the oversight of trained humans, automation produces much better results at an exponentially faster and more accurate rate than human analysis alone. Better integration of the automated processes with the human is a continual requirement for improving overall fusion performance.

2.1.3.2 Operational Fusion Perspectives

The SOSUS experience illustrates the potential advantages and disadvantages of introducing automated fusion processes into operational use. The operators and analysts—in fact, all users—must be fully cognizant of what automation can do for and, perhaps most importantly, to an individual or project. Captain W. Walls, the first Commander of the PACOM Joint Intelligence Center and a fusion-savvy individual, once stated that to develop a successful fusion process (automated or manual) all of the following are required:[10]

1. Knowledge and understanding of the physics of the sensor/collector and the phenomena being observed
2. Knowledge and understanding of data fusion processes in general and the approaches being applied to this problem
3. Knowledge and understanding of the warfare mission area
4. Knowledge and understanding of the customer/user, his/her information needs, decision processes, etc.

Although it is impossible for any individual to possess all the requisite knowledge and understanding, it is important to have it available to the process. In the past, this was accomplished by colocating as much of the expertise and resources as possible, the OIC being an excellent example. Modern information technology offers promising ways to accomplish colocation within a networked enterprise, and the potential impact of new information concepts on fusion will be a consistent theme throughout this chapter.

Historical examples of fusion automation are instructive, many of them illustrating great success or great failure, some even wonderfully amusing. The overwhelming message is that while automating multisensor fusion processes is essential, it is difficult and must be undertaken with care. It is important to remember Captain Wall's four elements. They provide a good high-level checklist for any fusion endeavor, particularly one that intends to automate a functional component of a fusion process. Approaches that are purely technical or exclusively engineering will likely fail.

2.1.4 Fusion as an Element of Information Process

An additional lesson to be taken from the operational experience of fusion and attempts to automate its processes is that the notion that somehow a fully automated system will "solve" the data fusion problem must be abandoned. Data fusion is an integral part of all our systems, from the simplest sensor and weapons systems to the most complicated, large-scale information-processing systems. No single fusion algorithm, process capability, system or system of systems will "solve" the problem. Operational experience, experimentation, and testing all confirm that there is no "golden algorithm"—it is an expensive myth. Further, we should desist from building fusion systems and rather focus on building fusion capabilities into systems or, better still, developing fusion capabilities as services in a networked enterprise that presents new challenges and opportunities.

The examples presented and the lessons of operational fusion experience, while confirming the importance and central role of fusion processes, also make clear that fusion is only part of an overall information process that includes data strategy issues, and data mining and resource management. Fusion is first and foremost dependent on the existence of data and information of known quality from sensors and multiple sources. Fusion processes must then have the ability to access the data and information whether from data streams themselves, from various databases and repositories, or through network subscriptions. The precise mechanisms for access are less important than the fact that mechanisms exist and are known or can be discovered. Means of gaining access to and defining *known* quality of data and information are important and complex issues, and they constitute the subject of many current research efforts. Although many sensors and sources are potential contributors to fusion, their ability to contribute depends on how they exploit the phenomena, what they are able to measure, what and how they report, and how capable they are of operating in a collaborative network environment.

Also, data fusion is essentially a deductive process dependent on *a priori* models (even if only in the analyst's mind) that form the basis for deducing the presence of an object or relationship from the data. In the World War II examples, analysts had to rapidly develop models of U-boat behavior and communications patterns; in cold war operations, an immense set of models evolved through collection and observation of our principal opponents over many years. However, in the post–cold war era, very few models exist on opponent behavior or even basic characteristics and performance of opponent's equipment; because such behavior is dynamic and unpredictable, pattern discovery and rapid model development, and validation (a process called data mining) are essential. The fusion process provides a measure of what we know and, by extension, what we do not know. Therefore, a properly structured information environment would couple fusion with resource management to provide specific collection guidance to sensing assets and information sources, with the aim of reducing levels of ignorance and uncertainty. All these processes—data and information access, fusion, mining and resource management—are mutually dependent. Fusion is no more or less important than any other, although it can be argued that access to data is the *sine qua non* for all higher-level information processes. The remaining sections of this chapter explore these processes, their functionality, their relationships, and the emerging information issues.

2.2 Data Fusion in the Information-Processing Cycle

Information, by any definition, represents the set of collated facts about a topic, object, or event. For military activity, as for business and routine human interactions, information is an essential commodity to support decision making at every level of command and every aspect of activity planning, monitoring, and execution. Data fusion, as we shall soon see, has been described as a process that continuously transforms data from multiple sources into information concerning individual objects and events; current and potential future situations; and vulnerabilities and opportunities to friendly, enemy, and neutral forces.[11] The JDL defines data fusion as follows:

> *Data fusion.* A process dealing with the association, correlation, and combination of data and information from single and multiple sources to achieve refined position and identity estimates, and complete and timely assessments of situations and

threats, and their significance. The process is characterized by continuous refinements of estimates, assessments and the evaluation for the need of additional sources, or modification of the process itself, to achieve improved results.

Definitions continue to evolve and, although the original has been legitimately criticized as too long and not just a definition but a definition and a modifier, it still captures the essence of fusion as a purposeful, iterative estimation process. Indeed, data fusion is an essential, enabling process to organize, combine, and interpret data and information from various sensors and sources (e.g., databases, reports) that may contain a number of objects and events, conflicting reports, cluttered backgrounds, degrees of error, deception, and ambiguities about events or behaviors. The significance of data fusion as a process to minimize uncertainty in information content can be measured in department of defense (DoD) circles by the large number of research and acquisition programs devoted to building fusion capability for specific platform and sensor combinations to achieve specified mission goals. A survey conducted in 2000 identified over 60 data fusion algorithms in operational usage alone.[12]

Despite the attention, and investment, devoted to the data fusion process, it is important to keep in mind that fusion is only one element of a set of enabling processes all of which are necessary to produce information relevant to a particular problem or decision. Fusion, along with the other elements—resource management, data mining, and an enabling information environment—are all essential elements in an overall cycle we call the information-processing cycle (IPC). How these elements have been identified and characterized and the importance of ensuring their integration will constitute the bulk of this chapter.

2.2.1 Functional Model of Data Fusion

The thinking that led to the recognition of the IPC and its elements grew out of a progression of functional models, beginning with the JDL data fusion model. Functional models have proven useful in systems engineering by providing visualization of a framework for partitioning and relating functions and serving as a checklist for functions a system should provide. In systems engineering, they are only a single step in a many step process. In this progression, a functional model lays out functions and relationships in a generic context whereas a process or process flow model performs a selection and mapping of functions, including inputs and outputs in a specific way to meet a particular goal. A design model goes deeper, utilizing the process model then taking the additional step of specifying what methods and techniques will be used. It is important to be cautious when using these models and to understand the limitations of each model type; in particular, using a functional model as a design specification is inappropriate. Process flows, system design models, and specifications are developed for application to a specific problem or set of problems.

2.2.1.1 *Joint Directors of Laboratories*

This chapter introduces fusion from an operational perspective and addresses its place as a central function in a broader information-processing structure, supporting decision making at all levels. As also discussed, data and information fusion are immature and complex disciplines. The level of immaturity did not become widely recognized until the early and mid-1980s when many acquisition programs and research projects throughout DoD and the broader intelligence community were established to perform sensor, fusion, data fusion, correlation, association, etc. The JDL model and surveys contributed to this awareness and led to a unification of the fusion community through the creation of the JDL.*

2.2.1.2   JDL Data Fusion Subpanel and the JDL Model

Upon the establishment of the JDL/DFS by the JDL in 1984, the growing recognition of the central role of fusion in intelligence and C2 processes had led to a broad but unstructured collection of individuals, programs, and projects with interest in fusion matters but no mechanisms for finding each other or establishing common ground. The Data Fusion Subpanel, later renamed the Data Fusion Group (DFG), recognized the immaturity of the fusion discipline and suggested that the creation of a community with common interests would help speed the maturation process. To begin, the DFS established a common frame of reference by developing a taxonomy and lexicon to bring discipline to the definition of terms and usage. The definition presented in Section 2.2 was from the original lexicon and captures the essence of fusion as a purposeful, iterative estimation process. The DFS also created a visual representation of the fusion process, a functional model of the data, and an information fusion domain to develop a framework for a common understanding of fusion processes. The original JDL model, shown in Figure 2.1, provided a common understanding of fusion that facilitated the creation of a fusion community providing guidance to theoreticians, field users, and developers. The model also made possible the partitioning of functions for organizing surveys of community development practices, research programs, and system development. Surveys based on the original JDL model, for example, exposed the lack of fielded level 2/3 fusion capability and a corresponding paucity of research in the domain leading to a redirection of critical research and development (R&D) resources. The model has been accepted nationally and internationally and is still in widespread use. Evolved versions and extensions of the model are having continuing utility as fusion, resource management and data mining functionality advances, and the implications of modern information services, architectures, and network-enabled environments are being appreciated.

The JDL/DFG has organized fusion into five levels as follows:

Level 0: *Source preprocessing/subobject refinement.* Preconditioning data to correct biases, perform spatial and temporal alignment, and standardize inputs.

---

* The original need for the JDL arose from a 1982 Defense Science Board study that found that

• The three services' research and technology bases were not coordinated at the headquarters level
• The various service laboratory investigators had no direct path to each other without bureaucratic permission
• There was no coherent and coordinated program of C2 research anywhere in DoD

As a result of these deficiencies, the Assistant Secretary of Defense for Command, Control Communications and Intelligence (ASDC3I) directed the Services to work together and develop a solution. In response to this direction, the navy, the army, and the air force created a new organization the JDL so that they could work as a team and respond specifically to these and other deficiencies that might arise. ASDC3I accepted this proposal and chartered the JDL to address the above deficiencies. JDL did this by conducting a survey of managerial and structural weaknesses and then creating processes to remedy them. They established a basic research group (BRG), a technology applications and demonstration group (TAD), and four subpanels corresponding to what were considered to be the driving initiatives in C3. They were

• Data fusion subpanel
• Networks and distributed processing subpanel
• Decision aids subpanel
• Dynamic spectrum management subpanel

Although the JDL as a specific organization has been replaced, the infrastructure and collaborative processes they created remain active under the Deputy Secretary of Defense for Research and Engineering (DDR&E).

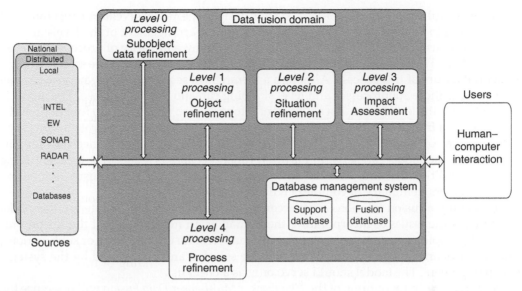

FIGURE 2.1
Joint Directors of Laboratories data fusion model.

Level 1: *Object refinement.* Association of data (including products of prior fusion) to estimate an object or entity's position, kinematics, or attributes (including identity).

Level 2: *Situation refinement.* Aggregation of objects/events to perform relational analysis and estimation of their relationships in the context of the operational environment (e.g., force structure, network participation, and dependencies).

Level 3: *Impact assessment.* Projection of the current situation to perform event prediction, threat intent estimation, own force vulnerability, and consequence analysis.
Routinely used as the basis for actionable information.

Level 4: *Process refinement.* Evaluation of the ongoing fusion process to provide user
advisories and adaptive fusion control or to request additional sensor/source data.

2.2.1.3 Use of JDL Model

The JDL model as it is currently represented focuses exclusively on the fusion domain and
thus has contributed to the misperception that fusion could be an independent process.
Practically speaking, in operations, fusion never stands alone nor does it theoretically make
sense from an information-processing perspective. Another shortcoming of the JDL model is
that the human role in the process is not represented except as an interface. These two aspects
have had unintended consequences, including inter alia army and navy attempts to build
centralized fusion processes as monolithic, stand-alone systems, attempts which have been
largely, and not surprisingly, unsuccessful. With regard to the human role, The Technical
Cooperation Program's (TTCP) technical panel for information fusion, in particular, Lambert,
Wark, Bosse, and Roy have pointed out that the JDL model provides no way of explicitly
tying automated data fusion processes to the perceptual and cognitive needs of the decision-
makers, the individuals the automation is expected to support. This is a serious deficiency in
the model, one that needs to be addressed. More valuable discussion on this topic is available
in Bosse et al.'s 2007 book *Concepts, Models, and Tools for Information Fusion.*[13]

As the model is applied, useful modifications, iterations, and implications continue to be
debated in numerous fusion community forums. One area of discussion concerns whether

the model usefully represents a potential system design. The model's creators never intended for the model to be used in this way. The diagrammatic arrangement of the levels and the nomenclature used were not intended to imply a hierarchy, a linearity of processes, or even a process flow. The term *level* may have been a poor choice because it can be seen to imply a sequential nature of this process. Although a hierarchical sequence often emerges, parallel processing of these subprocesses and level skipping can and does take place. The model is still useful, for the levels serve as a convenient categorization of data fusion functions and it does provide a framework for understanding technological capability, the role fusion plays, and the stages of support for human decision making. In information or decision support applications, the model can serve to partition functions and as a checklist for identifying fusion capabilities that are essential to the implementation of mechanisms for aiding decision makers in military or complex civil environments.

Automation of fusion solutions must account for complexity not illustrated in the model and address the issues of incorporating the human as an integral part of the fusion process. In developing a system design for the automation of fusion functions or subfunctions, process flows and design specifications must be selected and engineered by the systems design engineers. The model should serve only as a checklist.

Many chapters in this edition of the *Handbook of Multisensor Data Fusion* will reference the JDL model, taxonomy, and lexicon, including the model's utility or deficiencies in system design. Much of the evolution of the model, taxonomy, and lexicon will be discussed and the healthy controversies that have accompanied the evolutionary process will be exposed in the succeeding chapters.

2.2.2   Information-Processing Cycle

To reiterate from Section 2.1, fusion is a fundamental human process and, as an automated or manual process, occurs in some form at all echelons from rudimentary sensor processing to the highest levels of decision making. Fusion is such a foundational process that it is linked or integrated to myriad other processes involved in decision making. Fusion by itself makes no sense; it is a tool, an enabler that is useful only as an element of a larger information process, an IPC.

In its most basic form, an IPC must transform a decision-maker's information needs through the processes and mechanisms that will create information to satisfy those needs. Elements of the IPC include (a) resource management to translate decision-maker information needs to the real-world devices (sensors, databases, etc.) that will produce data relevant to the stated needs and (b) data fusion to aggregate that data in a way that satisfies the decision maker. These relationships are illustrated in Figure 2.2. The benefits of coupling these elements and the related issues will be discussed in Section 2.4.

Since the raw material for fusion processes is data from sensors and information from multiple sources, we must recognize that orchestrating the collection of required data with appropriate sensing modalities, or data and information from appropriate (possibly archival) sources, is an essential element of the larger information process. Such orchestration of sensors and identification of sources to produce relevant input for a given fusion process is often referred to as *resource management*. The resources are the technical means (e.g., sensors, platforms, raw signal processing) as well as archived data and human reporting employed to gather essential data. The function of resource management is fundamentally one of control, and should be distinguished from the essential role of data fusion, which is to perform estimation. In control theory, the mathematical relationship between control and estimation is understood and expressed mathematically as a duality. In more general information processes that are the means to develop relevant information of significance

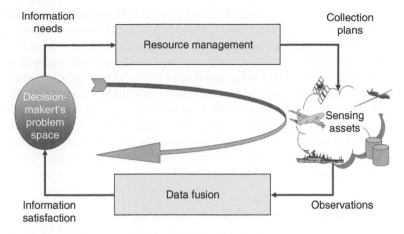

FIGURE 2.2
Basic information-processing cycle.

for human decision making, a comparable complementary relationship of data fusion and resource management can be recognized in achieving the goal of decision quality information. This relationship is neither well understood nor does it have the mathematical underpinnings of a duality. The resource-management process, like fusion, has been a manual process traditionally broken into collection requirements management (coupled to decision-maker's needs) and a collection-management process that plans and executes sensor and source collection. In practice, the processes are often only loosely coupled and too often completely independent of the fusion process. Like fusion, the exponential increase in the number of sensors, sensor data, and information sources requires some level of automation in the resource-management process.

For environments in which behavior of the sensing resources, their interaction with the real world, and the interpretation of observations is well understood, these elements are sufficient. Such understanding provides the basis for modeling the underlying resource-management and fusion processes with sufficient fidelity to produce satisfying results. Under conditions where real-world behaviors, or the related observations, are not understood, an additional information-processing element is required. That element is a data mining process to discover previously unknown models of activity or behavior. Data mining complements resource-management and data fusion processes by monitoring available data to identify new models of activity. Those newly derived models are used to update fusion and resource-management processes and to maintain their validity. Such a process of discovery, model creation, update, and validation is necessary for processes and algorithms that operate under conditions of incomplete understanding or problem space dynamics.

The following sections provide an overview of the roles of the three major information-processing elements: data fusion, resource management, and data mining. The overviews discuss the elements in relation to each other and in the context of the driving concept underlying the transformation of military and commercial information technology—network-centric operations.

A consideration of emerging notions of net-centricity and enterprise concepts is important because, independent of the particulars of enterprise implementation, future fusion algorithms will have access to increasing volumes and diversity of data, and in turn, increasing numbers and types of users will have access to algorithmic products. Net-centric enterprises must become sensitive to both appropriate use and aversion to misuse of data and products. Arbitrary access to sensor and source data, and to algorithmic products, requires a discipline for characterizing data and products in a manner that is transferable among machines and usable by humans. The characterization of sensors and algorithms provides the context necessary to support proper interpretation by downstream processes and users. Such context was naturally built into traditional platform-based systems as a result of systems engineering and integration activity, which typically entailed years of effort, and resulted in highly constrained and tightly coupled (albeit effective) solutions. In net-centric enterprise environments the necessary discipline will likely be captured in metadata (sensor/source characterization) and pedigree (process characterization and traceability) as an integral part of the IPC. The discipline for executing that role has not yet been established but is being addressed within a number of developmental efforts; some of these issues are further discussed in Section 2.3.

2.2.2.1 Data Fusion in the Information-Processing Cycle

As noted in the preceding text, the role of data fusion in information processing is to aggregate raw data in a way that satisfies the decision-maker's information needs, that is, to minimize the user's uncertainty of information. Data fusion is a composition process that leverages multiple sources of data to derive optimal estimates of objects, situations, and threats. The process involves continual refinement of hypotheses or inferences about real-world events. The sources of data may be quite diverse, including sensor observations, data regarding capability (e.g., forces, individuals, equipment), physical environment (e.g., topography, weather, trafficability, roads, structures), cultural environment (e.g., familial, social, political, religious connections), informational environment (e.g., financial, communications, economic networks), and doctrine as well as policy. Data fusion provides the means to reason about such data with the objective of deriving information about objects, events, behaviors, and intentions that are relevant to each decision-maker's operational context. Users or decision-makers' need can be characterized in many ways; this chapter often addresses operational and tactical decision makers. The fusion process applies at all echelons, however, and information needs can be long-standing requirements established by strategic, intelligence, or National Command Authority (NCA) components. In the modern environment, multiple decision makers must be satisfied in near-simultaneous, asynchronous fashion. Fusion processes that perform these functions are described in detail throughout this handbook.

Section 2.2.1.1 introduced a functional model of fusion that provides a framework for partitioning more detailed fusion roles and technology solutions. The current practice in algorithm development is to incorporate models of phenomenology and behavior as the basis for accumulating evidence (e.g., observational data) to confirm or discount objective hypotheses. Representation of the model basis for each fusion algorithm (along with key design assumptions that are not currently exposed) will need to be a part of the algorithm characterization. Capturing such meta-information in the form of metadata or pedigree will provide the basis for net-centric enterprise management of information processing that will likely engage combinations of similar and diverse algorithms. Section 2.3 describes related net-centric implications that will drive the need for stronger, more robust fusion algorithm development. It is apparent that the net-centric environment will enhance sharing of common contextual information (e.g., models, metrics, assumptions), given the interdependence of data fusion and resource-management processes in the IPC.

2.2.2.2 Resource Management in the Information-Processing Cycle

Automated resource management has been addressed in military and commercial environments with considerable effectiveness. Credit for success in the field is more a consequence of well-structured manual procedures than of automation technology.[14] The scale of investment (judged qualitatively) has not matched that underpinning data fusion development. As a result, representation of resource-management processes and functionality is less mature than equivalent data fusion representations. The following section provides a summary discussion of current perceptions and practices within the discipline.

The role of resource management is to transform decision-maker's information needs to affect real-world actions in a manner that produces data and information that will satisfy those needs—in short, to maximize the value of data and information that is gathered. Resource management is a decomposition process that breaks down information needs (or requests) into constituent parts that include method selection, task definition, and plan development. These stages correspond to the levels of fusion as identified in the JDL data fusion model (Section 2.2.1) and are shown in an extension of the functional model in Figure 2.3. This extension illustrates fusion in a much broader and more realistic information framework, a representation that is instructive because it reveals mutual dependencies. The extended model can also serve as a checklist for identifying and partitioning functions required in specifying process flows and design parameters in engineering a specific application. Also evident in the figure is the central importance of context in capturing attributes of the processing cycle that must be shared among all process elements to achieve consistency across the IPC.

Attributes of the IPC that provide context for maintaining performance include the models that underpin algorithm design, the metadata required to characterize sensors and sources, and the pedigree needed to explain (and couple) data collection purposes to the exploitation processes invoked to transform that data into usable information. Other attributes, such as metrics, may also be identified as enablers of processing cycle consistency. Notions of models include representations of the platforms, the sensors, and

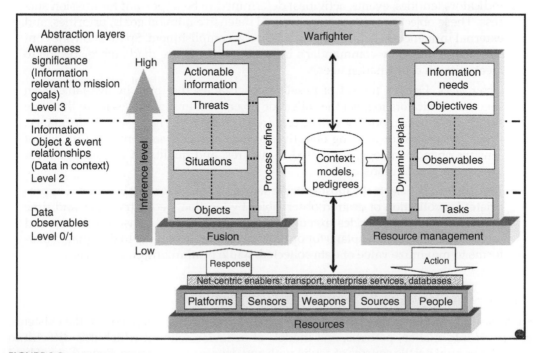

FIGURE 2.3
An information framework.

the real-world phenomenology they observe. Algorithms designed for all stages of decomposition (resource management) and composition (data fusion) functions are based on such models. It is clear that shared model can be used and understanding is required to "close the loop" in the IPC and maintain focus on meeting decision-maker's needs. Similarly the use of pedigree and metadata provide mechanisms for relating resource-management goals to sensor characteristics and, in turn, the exploitation required of data fusion processes. Proper use of pedigree and metadata can inform downstream processes of their role, provide traceability or accountability of system performance, and support explanation of automated results as needed by decision makers.

This model also highlights the issues of information access, the need for a unifying information infrastructure discipline, the parallelism of the composition and decomposition processes, the information content management, and the retention of context throughout the process.

Section 2.4 provides a more detailed discussion of the parallels between data fusion and resource-management functionality. The following provides a brief discussion of the functional role and significance of the resource-management stages. As with the fusion model, the discussion describes functional relationships and is not intended to constrain architectural or design options.

- *Information needs.* Reflects the commander's intent in the context of mission goals and priorities, through a statement of the situation and conditions that are expected to provide key indicators of adversary actions or own force progress. Such needs are generally characterized by coarse specification of space and time constraints on objects, events, and activities deemed to be of particular importance. Indicators are generally expressed in operational terms, independent of phenomenology.

- *Collection objectives.* Defines focused goals for collection of data relevant to the indicators (entities, events, activities) determined to be important for mission success. These objectives reflect offensive and defensive mission goals, priorities, and external influences that may impact mission accomplishment. Space and time constraints identified by commander's information needs are made more specific to meet mission coordination needs.

- *Observables.* Derives the set of possible observables that may be available from specified indicators to meet the collection objectives. In some cases more than one sensor, or sensing modality, will be required to achieve the necessary performance. Provides options for meeting objectives among available sensors as a function of sensing phenomenology, associated signatures, and quality of measurement for the expected conditions.

- *Tasks and plans.* Selects the most suitable sensor or source (or combination if appropriate) for collection of desired observables by location, geometry, time, and time interval. This also provides executable tasks for specified sensors/sources, and includes development of plans for optimized allocation of tasks to designated platforms to maximize value of data collected, while minimizing cost and risk.

2.2.2.3   Data Mining in the Information-Processing Cycle

Successful fusion processing, and particularly automated fusion, depends on the existence of *a priori* information, patterns, and models of behavior, without which it is difficult to predict potential hostile courses of action with any degree of confidence. Regardless of how well or poorly we have performed data fusion and resource management in the past, it is clear that we no longer have the long-term observation and analysis of our opponents on which to base the models underlying fusion and collection processes. For example, current homeland security operations lack patterns for counterterror and counternarcotics activities and the troops in Iraq and Afghanistan face an insurgency for which there are no preexisting models. Our level of success is correspondingly low in these situations. This will remain a problem, for in these domains complete knowledge of a comprehensive set of threatening entities, behaviors, and activities, which are always changing, will elude us. One of the central challenges associated with future operations of all types, from combat operations to humanitarian operations, is to discover previously unknown behaviors based on signatures, and indicators buried in immense volumes of data. A new set of functions will be required that enable behavior and pattern discovery as well as rapid model construction and validation. Data mining is becoming the focus of research on many fronts.

As an automated process, data mining is relatively immature, although the practice, particularly in the hands of analysts, has a long history. Data mining has been described as *Discovery of previously unrecognized patterns in data (new knowledge about characteristics of an unknown pattern class) by searching for patterns (relationships in data) that are in some sense interesting.*[15]

As we have noted earlier, current data fusion and resource-management practice are model-based. Data mining provides the means for maintaining the currency of those models in dynamic or poorly understood environments. Its role therefore is to identify new models of activity or relationships, or significant variation in existing models. These updates serve to alert users, to provide resource management with new objective opportunity for data collection, and to update the models employed in data fusion processes. This extension of the basic IPC is represented in Figure 2.4.

From the perspective of the evolving functional model it is important to distinguish between the terms *data fusion* and *data mining*. Unfortunately, the terms are often used interchangeably or to describe similar processes. In reality, they represent two distinct, complementary processes. Data fusion focuses on the detection of entities and relationships by deductive reasoning using previously known models. Data mining, however, focuses on the discovery of new models that represent meaningful entities and relationships; it does this by using abductive and inductive reasoning to discover previously unknown patterns. The process then uses inductive generalization from particular cases observed in the data to create new models. Since data mining is a complementary function, to fusion it is also part of the broader IPC and can be included in the functional model

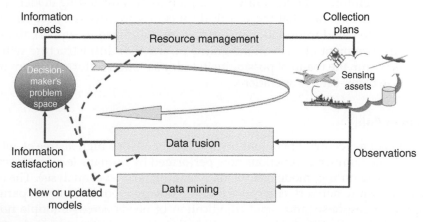

FIGURE 2.4
Extended information-processing cycle.

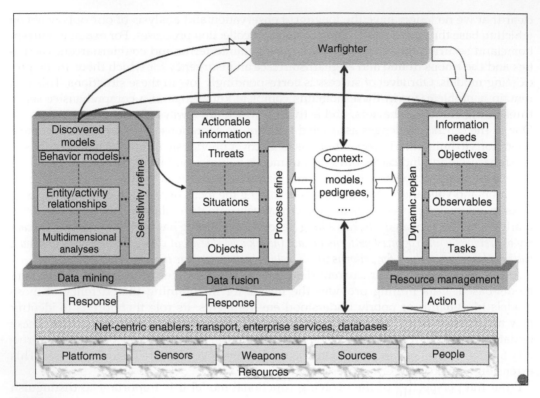

FIGURE 2.5
Extended information environment.

as shown in Figure 2.5. Some stages of data mining appear to parallel the levels identified in data fusion and resource management: multidimensional analyses, entity relationships, and behavior models.

The addition of the data mining functionality provides additional insight into the overall information process, reinforcing the dependencies of resource-management and fusion processes, and calling attention to the importance of the integration of data fusion and data mining for future applications since they are mutually dependent. This representation of data mining also reinforces the centrality of common context (including models, pedigree, etc.) and emphasizes the fact that these critical functions are necessary for effective performance of an integrated information infrastructure optimized to support all aspects of the IPC. The development of such an integrated information infrastructure will be very complex but is within the realm of possibility in today's modern information environment and the promise of net-centric operations.

2.2.3 Human Role

The implication of net-centric operations involves the need for a high degree of automation to perform a large number of functions now performed by humans (e.g., platform/sensor routing, sensor scheduling, mode control, data tagging, low-level analysis). The complexity envisioned for any network enterprise and the obvious requirement for parallel processing, speed of response, and rapid adjudication of needs across multiple nodes and users suggest that processing by humans will be inappropriate for many roles in a fully evolved net-centric environment. Humans will have to interact with automated capabilities to provide supervision and confirmation of critical information products. The difficulty of assessing the appropriate degree of automation and the appropriate mechanisms to interface automated products with humans (i.e., to achieve human–machine collaboration), creates a challenge that calls for implementation of a coevolutionary process (interactive development of technology with tactics, techniques, and procedures) with an emphasis on experimentation to properly introduce and integrate technology that enhances rather than inhibits the human user and creates an advanced and integrated capability.

2.3 Challenges of Net-Centricity

Net-centric environments hold great potential to provide an enabling infrastructure for implementing the information process flow and related algorithms, which are needed to support decision makers at distributed nodes of the military or commercial enterprises. The movement to net-centric operations, built on an underlying service-oriented architecture (SOA), is mandated for the military by DoD[16] and could provide distinct advantages for fielding an improved information-processing capability. This section addresses some of the implications of net-centricity for fusion (and other) algorithm design and implementation.

It is recognized by analysts in the field that the state of net-centric development and fielding, particularly for military environments, is very much in the embryonic state. Much of the effort to date has addressed net-centric infrastructure issues, including the difficult challenges of security and transport in the wireless environments that dominate military situations. Infrastructure development has focused on addressing enterprise services (see for instance net-centric enterprise services (NCES) description of required services[17]), service architectures and types, computing, and enterprise service bus issues. These efforts are concerned with *how* the volume of data will be moved timely across the network, and with the general functions necessary to manage that data flow. Less attention has been paid to content and management of information to support specific mission goals, independent of time and location. The details of *what* data and information will be moved and processed lie above the infrastructure layers and grouped into what is often referred to as application services. The issue for fusion algorithm designers is to anticipate the unique demands that net-centricity will place on design practice and to take advantage of the opportunities created by a new and different system engineering environment, so that fusion algorithms are properly configured to operate in net-centric environments. Given the immature state of net-centric development it is important to anticipate the implications for algorithm design practice and to participate in the evolution of net-centric system engineering practices.

2.3.1 Motivation

In the DoD, the issuance of a mandate for net-centricity signals a recognition of the importance of information as the foundation for successful military activities and missions. The mandate is driven by a desire to achieve multiple key benefits, which include increased affordability, reduced manpower, improved interoperability, and ease of integration, adaptability, and composability. The goals stated for enabling these gains are to make data and information visible, accessible, and understandable. Clearly, to attain these goals will require significant effort beyond the scope of fusion algorithm design but will nevertheless benefit from also considering the impact that current fusion algorithms provide in support of this endeavor. For any existing fusion algorithm we might ask, "Does it satisfy the goal of making its data and information *visible, accessible,* and *understandable?*" Taking the perspective of an average user, and without quibbling over the definition of terms, we might attempt to answer the following sample questions: Is the algorithm product understandable without knowledge of the context for which it was specifically designed? Are the constraints or assumptions under which the algorithm performs visible or accessible? Are the sensor data types, for which the algorithm was designed, made obvious (visible, accessible, or understandable)? Is the error representation of the information product either accessible or understandable? Is the error model of the data input (whether provided by the source or assumed) visible, accessible, or understandable? Given appropriate connectivity, could a remote user (with no prior involvement) invoke it and understand its capability, limitations, and performance? Can the algorithm's information product (for specific areas or entities) be compared with a competitive algorithm's product without ambiguity? Does the algorithm provide an indication (visible or understandable) when it is broken? For the preponderance of extant fusion algorithms, the answers to these questions will be negative. One or more negative answers provide a clear indication that algorithm design must evolve to satisfy net-centric goals.

2.3.2 Net-Centric Environment

The implementation of algorithms in net-centric environments includes benefits of modularity, and common information-processing discipline. Modularity is a natural consequence of the vision of net-centric operations. Net-centricity incorporates a SOA framework, with open standards and common data strategy (e.g., data model, data store control, information flow) accompanying a defined enterprise infrastructure and enterprise services (e.g., security, discovery, messaging) that provide the high-level functional management of a layered enterprise structure. The technical view of this layered enterprise infrastructure provides a plug-and-play environment for various applications, as shown in Figure 2.6. The figure suggests that the infrastructure, in combination with selected applications, will support desired

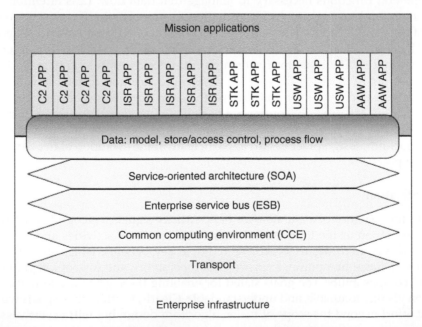

FIGURE 2.6
Technical view of net-centric enterprise structure.

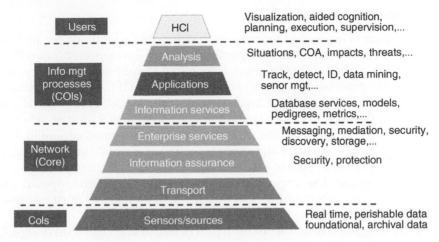

FIGURE 2.7
Layered technical view of functionality within a networked enterprise.

mission capability; it also suggests that a common infrastructure will enable selection of a set of applications to support execution of multiple missions.

A functional view of the overall enterprise is provided in Figure 2.7. The layers correspond to significant functional partitions such as sensors, transport, security, enterprise services, applications, and others. Within the application layer, fusion algorithms of various types, as well as other elements of the IPC (e.g., resource-management and data mining algorithms), may be provided as building blocks or modules. Analysis functions that currently remain largely as manual functions are shown as a separate layer within the application area. The figure also includes the notion of information services; these services are intended to capture those functions believed to be common to a broad set of algorithms or applications and which are needed to manage detailed information content. These information services are necessary to maintain an IPC, as discussed in Section 2.2. Design and development of such services are a challenge that should, logically, be met by the algorithm-development community. It is likely that as the design and performance of such information services becomes better understood, their implementation will migrate to become part of the infrastructure. The algorithm-development community could be quite influential in this role. This modularity supports both developmental growth and operational adaptability—attractive features for enabling spiral development of overall enterprise capability.

Developmental growth is enabled with the implementation of a robust SOA that provides common services, data layer (e.g., access, storage), and visualization support for plug-and-play of disciplined applications. Because of the modular approach, operational capability can be achieved in incremental stages, through an efficient coevolutionary process that integrates engineering and operational test and evaluation while reducing development (nonrecurring) cost and time. Operational adaptability results from modularity because the building blocks may be logically connected in a variety of ways to achieve changing mission goals. An additional benefit comes from the fact that the approach does not require long periods of engineering integration (typically 2 years or more) resulting in lower cost solutions. As a result there are very few negative consequences of discarding older applications when they become outmoded by advancing technology or by changing requirements.

2.3.3　Implications for Fusion

The implementation of automated techniques in net-centric operations requires a degree of implementation commonality, or disciplined algorithm development that will enhance both user (decision maker) ability to select among like algorithms, and enterprise ability to couple necessary processing stages to create a goal-driven IPC. This requirement that a common discipline be developed and invoked across the network enterprise (all nodes, all users) arises from the need to ensure consistent use and interpretation of data and information among machines and humans. Algorithmic functions must adhere to a common set of standards (e.g., data strategy) and expose significant internal functions (e.g., models, constraints, assumptions) to ensure consistent data and information processing (fusion, etc.) across the enterprise. This imposition on algorithms in net-centric operations is consistent with the dictum underpinning SOA design—that all services must be self-describing and usable without any prior background or knowledge.

The discipline suggested here has not been part of traditional algorithm design. Current algorithms (fusion and others) have always performed in systems that are engineered to provide selected information functions; these algorithms operate within the confines of specific systems, with known underlying but generally hidden assumptions. In net-centric environments, where machine-to-machine data passing is needed to meet data volume and speed requirements, such hidden assumptions must be exposed to ensure that all processes are mathematically and logically consistent with source sensing and context characteristics, and intermediate processing updates. Furthermore, in current practice, information requirements (and information flow) have been designed to support specific (generally local) requirements—in net-centric operations algorithmic products will be called on to support multiple distributed nodes, for multiple distributed users with differing roles and perspectives. Such environments will require standards and discipline for implementing algorithms, and for sharing sufficient descriptive information, so that the information products are usable by machines and humans across the enterprise.

The implementation of algorithms for net-centric operations is likely to require new practices in exposing previously internal functions, and in aggregating functions that are common to large classes of algorithms. Such implications for net-centric algorithm design are not well-understood. Research to identify best practices for either decomposing algorithmic functions or exposing needed internal data is being conducted by organizations such as the Office of Naval Research.[18] The expectation is that, in addition to exposing algorithmic functions needed for efficient net-centric operation, a number of enabling functions that are common to a range of algorithms will be identified as items for enterprise control and monitoring. Candidate functions include the following:

- Models that capture and share knowledge of phenomenology, platforms, sensors, and processes—insofar as these form the basis for algorithmic computation and inform algorithm usage and human interpretation, it may be advisable to maintain and distribute such models from a common (and authoritative) enterprise source. Issue: Can algorithms be decoupled from the model they employ?

- Metadata and pedigrees to identify sensor or source characteristics and intermediate processing actions, including time—needed to avoid redundant usage (especially data contamination), to enable validation and provide explanation of information products, and to support error correction when necessary. Issue: Can a standard for metadata and pedigree representation be developed and promulgated for all enterprise participants including platforms, sensors, data sources, and processors?

- Metrics to qualify data and processing stages on some normative scale for consistent and (mathematically) proper usage—provides enterprise awareness of algorithm health and readiness, and to support product usage as a function of enterprise operating conditions and operational goals. Issue: Can metrics that reflect operational effectiveness (as opposed to engineering measures) be developed and accepted for broad classes of algorithms?

Although the modern information environment and the notions of network-centric or network-enabled operations offer great promise for fusion and the related IPC functionality, the potential remains to be demonstrated. Information technology advances at an almost unimaginable pace and predicting when or if some breakthrough will occur that may render all things possible is only an exciting exercise. The difficulties with realizing the potential of net-centricity for operational use lie not with the technology but rather with the adaptation of cultural practice in intelligence, operations, research, and, most importantly, in the acquisition process. Moving into a new networked information era will mean dramatic change in the way we conduct operations, acquire capabilities, and allocate resources, particularly our funds. Whether all these nontechnical barriers can be socialized and overcome presents the greatest challenge to advancing network centric operations.

2.4   Control Paradigm: TRIP Model Implications for Resource Management and Data Fusion

The IPC, as described in Section 2.3, consists of two basic elements: resource management and data fusion. If the benefit of coupling these elements is significant, as asserted, one might ask why it is not done routinely. In fact there are a number of examples of colocated or platform-based systems that include both collection of essential data and processing in a timely, tightly coupled manner. Fire control systems provide one such example.* In retrospect, it is apparent that data fusion and resource-management business and technology practices have developed independently. The result, as we move toward greater degrees of automation (as demanded by net-centricity), is that we need to identify and capture the common context and dependencies that enable a functional understanding (and implementation) of coupled decomposition and composition processes and the potential

---

* Many of these systems are designed for operation in tactical environments where sensors are applied to search areas of suspected hostile activity and to respond when objects are detected in a manner consistent with mission goals that could vary from signature collection to weapon deployment. The issue of note for extant examples of such closed systems is that, in general, they operate within a larger framework that formulates the "big picture" in a manner that exploits independent (largely uncoupled) system products. For example, the decision to deploy a particular weapon-bearing platform to a specific area in a given time period, to find and engage threat vehicles, is the product of an independent information process. How was selection made of platform type, weapon type, area, and time period? This section contends that such decision making can be improved through closer coupling of the supporting IPC and, further, that the movement toward net-centricity can be leveraged positively to aid that improvement. Further, virtually all of these systems were the product of separate, highly structured, system engineering processes that tuned each system to perform well against a limited set of mission goals. Although those systems perform well in the roles for which they were designed, they have little adaptivity to changing missions, and have great difficulty being interoperable with new partners—without costly (time and dollars) re-engineering. This provides another argument for net-centricity.

benefit that might be gained. This section will provide a high-level description of a model developed under defense advanced research projects agency (DARPA) advanced ISR management program (concluded in 2002) to explore the functional relationships between resource management and data fusion and to identify the common functions necessary to enable coupling in an IPC. This discussion will outline a functional model of a coupled process and identify the drivers and constraints for linking the control and estimation elements.

2.4.1 Resource-Management Model

Resource management (or collection management, in military parlance) is a transformation process that starts with understanding mission objectives and mapping the information needed by decision-makers to clarify desired objectives, the observables that are expected to satisfy those objectives, and the tasks required to gather those observables. This control process includes the articulation of discrete tasks in a form that can be understood by real-world actors, and the execution of those tasks by designated platforms, sensors, and agents of various types. This resource-management process serves to decompose information needed to satisfy mission objectives into one or more tasks that, in sum, are expected to answer the decision-maker's queries. The tasks are articulated in terms that can be executed by a designated sensor type or agent; that is, it contains specific, appropriate requirements for the observations (e.g., time, location, geometry, accuracy, signatures) expected to satisfy the information need. A functional description of the stages of this process appears in Figure 2.8. The relevance of resource management for data fusion purposes is apparent—fusion processes are employed to take the observations (or data) that result from the resource-management process, and compose them into information abstractions that satisfy the stated need. This is seldom a static process. Most environments are inherently dynamic, requiring ongoing adjustments in the collection-management process—and therefore a continuing process of data fusion refinement.

2.4.1.1 TRIP Model

Transformations of "requirements for information" process (TRIP) model was developed under a DARPA program addressing collection-management technology. The model shown in Figure 2.8 portrays a framework that relates the process stages for decomposing mission objectives into information needs, and for selecting the appropriate sensors and task assignments to meet those information expectations. The TRIP model[19] describes the

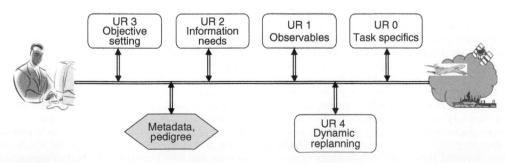

FIGURE 2.8
Resource-management functions.

information transformation needed to identify the observation set and associated tasking, necessary to satisfy the user's needs. The transformation refers to what literally is a translation of information needs (generally expressed in operational and linguistic terms) to the specific direction (e.g., location, time, geometry, conditions) required for control of technical means or database searches. The model captures the significant stages of the control paradigm including the information elements, the operational influences, and the underlying assumptions (or dependencies) used in effecting control. The reference provides a very detailed discussion of TRIP and its relation to fusion processes in satisfying information needs. A functional representation of the resource-management process, which provides and abstracts the decomposition stages captured by TRIP, is illustrated in Figure 2.8.

Operational decision makers are not focused on the details of which resources are needed to gather information, much less the process for decomposing their needs into executable tasks. Decision makers are interested in understanding some aspect of the environment or situation, for which they believe sensor observations (or archived data) can provide explanatory evidence. This desire for evidence is expressed as a user requirement (UR). The decomposition of URs from a high-level statement of information needs is represented in multiple functional levels including articulation of specific objectives, related observables, and the tasks and plans to gather the needed data.

UR Level 3: *Information needs.* The need for information about environments or situations is generally expressed as a request for information about the location, behavior, or condition of key indicators (e.g., objects, events, or entities) at a given space and time. Such requests are agnostic to the means of gathering the data, but may include implications for accuracy and timeliness. UR Level 3 functions provide articulation of the situation indicators believed necessary to confirm or discount hypotheses that will guide courses of action, and the constraints under which the indicators will be relevant or useful.

UR Level 2: *Collection objectives.* URs are driven by the understanding of overall mission goals and the contribution of specific objectives for attaining those goals. The objectives may be partitioned into separable mission tasks, but the dependencies must be understood and captured as context for the stated objectives or tasks, in the form of relationship, priority, and value. Such context is necessary to adjudicate conflicting demands for resources, which is generally the case for environments that are characterized by some combination of cost, risk, and supply factors. Briefly, all resources are consumable and so there is a cost (including opportunity cost) associated with allocating a resource to perform a given task; the use of resources entails risk that it will fail or become unavailable; and, in most cases, the supply-to-demand ratio is a fractional number. These factors combine to dictate that planned usage of assets be managed in a way that maximizes the expected value of collected information. The UR Level 2 function provides motivation, in the form of high-level goals, and necessary context to drive the resource-management process.

UR Level 1: *Observables.* For each key indicator, one or more sensing modalities, or archival sources, may be adequate to provide the set of measurements necessary to support estimation of the requested information. Assessing which sources are adequate for the desired measurement requires knowledge of sensing phenomenology and parametrics, along with expected background conditions, to determine the sensor type (or combination), parametrics, and conditions that will enable observation of a given indicator with the required quality. The UR Level 1 function provides options specifying alternative mechanisms for observation, which include sensing modality, relevant parameters, and constraining physical conditions, required to observe (and ultimately estimate) the desired measurements or signatures.

UR Level 0: *Tasks and plans.* Once the observability options and associated constraints have been specified it is possible to apply specific knowledge of the performance and behavior of each platform, sensor, or archival source to determine which combination is capable of providing the required measurement set. The UR Level 0 function specifies one or more collection tasks to a selected platform/sensor/source combination and the constraints (including time, location, geometry, or other variables) under which the task is to be accomplished.

And to provide the basis for replanning in response to situation dynamics:

UR Level 4: *Dynamic replanning*. Given the nature of planning in even minimally dynamic environments, it is imperative that a response to contingencies be available. Causes for contingencies may range from natural causes (e.g., weather effects, equipment failure), to adverse action (e.g., cyber attack, unforeseen threat), or to meet the need for additional data from a user (or algorithm) to resolve an ambiguity or refine an estimate. UR Level 4 functions provide control and updates for UR Levels 3 through 0, appropriate to the priority and constraints associated with the contingency. The response must include consideration for the nature of the immediate response and for longer-term ripple effects that may result.

Although in general we seek to automate the resource-management process, the identification of these functional elements is not intended to imply that full or even partial automation is feasible for each of them. The state-of-the-art for the higher-level URs is often a manual capability depending on the complexity of the domain. It should also be noted that this functional description is not intended to imply a process or design hierarchy. Depending on domain or information need particulars, a variety of process flows or design configurations might be envisioned. The functional breakout is intended to identify relationships and to partition activities as an aid to understanding applicable techniques (manual or automated) and to recognizing the utility of new technology as a solution option.

Throughout this discussion there have been references to implicit information (e.g., "knowledge of ...") and to background influences (e.g., context). The former are intended as references to the models, or metadata, which underpin both human- and machine-based interpretive and inferential activity. The functional capabilities discussed here rely on embedded knowledge (or models) of the entities from which they intend to interact or control (e.g., platforms, sensors, phenomenology, data archives). To the extent that such knowledge is absent or in error, the results of the control or interaction will fail to achieve the desired goal.

The reference to influences or context is intended to emphasize the point that, to the extent the functions discussed here are aligned to form a process (as will almost always be the case), the assumptions made, and dependencies identified, in earlier process stages must be carried forward to inform successive stages. These assumptions and dependencies have explanatory power for motivation, purpose, focus of attention, and semantic relationships. They are, in fact, the pedigree for the decision process that the functional elements will support. Pedigree captures the context under which decisions in the process chain are made. Pedigree must be linked and accessible in a manner that informs all stages of the process so that decisions across the processing chain can be made in a consistent manner. Humans fulfill this role (often in unstructured fashion) through explanation, discussion, or training sessions. In automated environments the discipline for generating and maintaining pedigree will become essential to provide machines with the appropriate constraints for algorithms of virtually any type to perform as intended. Pedigree also becomes the mechanism for machine-based processes to adapt to changing conditions, since the constraints under which they operate (and therefore the process goals) can be changed in a uniform and consistent manner. The TRIP model paper[19] provides considerable discussion on the significance of pedigree and the errors that can occur if it is absent or ignored.

The inclusion of metadata and pedigree in the functional breakout of Figure 2.8 is meant as a reminder that these functions have an essential enabling role in the implementation of any resource-management process. Their inclusion here is not meant to imply a unique resource-management role. Indeed the utility of metadata and pedigree extends to every part of the IPC.

2.4.2 Coupling Resource Management with Data Fusion

The duality between resource management and data fusion is discussed in Section 2.2. The complementary relationship can be noted with the recognition that resource management seeks to maximize the total value of collected information, whereas data fusion seeks to minimize the uncertainty of its estimates. The essence of resource management, then, is large-scale optimization and the essence of data fusion is uncertainty management. The complementary nature of the two can be illustrated by juxtaposing the well-known levels of fusion with the UR levels. Table 2.1 provides a summary representation of these levels.

TABLE 2.1

Juxtaposition of Data Fusion and Resource-Management Functionality

Level (Lvl)	Estimation Data Fusion	User Requirement (UR)	Control Resource Management
0	Source preprocessing/subobject refinement: preconditioning of data to correct biases, perform spatial and temporal alignment, standardize inputs	0	Task specification: selection of sensor/source for collection of desired observables by location, geometry, time, and interval. Provides executable tasks for specified sensors/sources
1	Object refinement: association of data (including products of prior fusion) to estimate an object or entity's position, kinematics, or attributes (including identity)	1	Observability context: derivation of observables available from specified indicators as a function of quality, vulnerability, availability, or feasibility phenomenologies and associated signatures
2	Situation refinement: aggregation of objects/events to perform relational analysis and estimation of their relationships in the context of the operational environment (e.g., force structure, network participation, and dependencies)	2	Information needs: specification of space and time constrained objects, events, and conditions, which are expected to provide key indicators for own or adversary courses of actions (COAs): Indicators are independent of phenomenology
3	Impact assessment: projection of the current situation to perform event prediction, threat intent estimation, own force vulnerability, and consequence analysis. Routinely used as the basis for actionable information	3	Objective setting: focused goals and objectives, including offensive and defensive mission tasks, priorities, and external influences that may impact accomplishment. Provides mission context for planning, including cost/risk factors
4	Process refinement: evaluation of the ongoing fusion process to provide user advisories, adaptive fusion control, or to request additional sensor/source data	4	Dynamic replanning: modification of the collection plan in response to contingencies due to various causes: (a) natural (e.g., weather, equipment failure), (b) man-made (e.g., cyber attack, threat), (c) dynamic information needs (e.g., fusion request to resolve ambiguity). Includes recovery for both short- and long-term needs

As a follow-up to the discussion of metadata and pedigree it is also apparent that these functions enable consistency across the IPC. The data fusion community understands the use of metadata and pedigree, although it has not always been employed in a disciplined fashion. As with resource management, metadata provides pointers to the models, or aspects thereof, that characterize the data sources and conditions under which the data were collected. Pedigree provides the traceability across the processing chain and the explanation of what processing was performed, and when. In a coupled environment, as discussed in Section 2.2, resource-management pedigree can inform data fusion processes about information objectives, effectively setting goals for the estimation process and enabling adaptivity (within limits) by impacting processing goals and architecture. Similarly, data fusion pedigree can inform resource-management processes with near real-time feedback on the adequacy of collection: requesting more or different data when needed to meet information objectives, and indicating when specific collection can be terminated, releasing scarce assets to perform another task.

Metadata and pedigree, along with models of the underlying sensors, platforms, phenomenology, and the processes themselves, provide the context that describes the IPC. Clearly this context needs to be shared to assure consistency across the total IPC. The separate functional breakouts of resource management and data fusion, and the shared use of context are illustrated in Figure 2.9.

The data mining functions discussed in Section 2.2 and illustrated in Figure 2.4 could also be shown to relate to the shared context in a similar fashion.

2.4.3 Motivational Note

It is worth observing that the emergence of net-centric enterprises, and the potential for coupling information processes with shared context over distributed sensing and processing

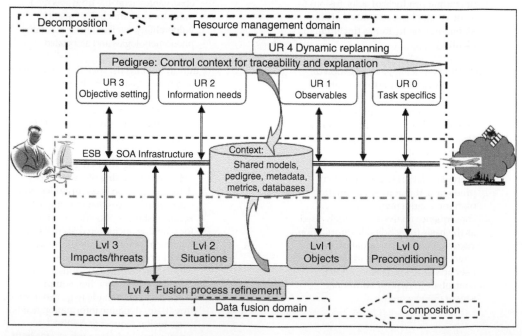

FIGURE 2.9
Functional relationships in the information-processing cycle.

nodes, gives rise to significant new possibilities. In the past, algorithms were designed to fit into static environments and produce results that were independent of decision-makers' needs. For example, trackers were designed to ingest data from the sensor it was engineered to work with, and to keep processing target tracks independent of situation dynamics or changing information needs. It performed a single function with no significant operational adaptivity. The future vision, enabled by a net-centric composability environment, is that different algorithms or algorithmic functions can be invoked as needed, in response to changing information needs that are communicated via common access to pedigree. In this vision, pedigree is employed both to explain decision choices and to inform or coordinate dependent processing stages (control and estimation) of changing mission objectives as well as information needs to manage the details of what data to collect, and what fusion processes to invoke to best satisfy mission needs.

The end result is common awareness of information needs across all processes, better use of available resources and the data they generate, and improved satisfaction for users. The aim is an IPC that is responsive to dynamic changes in mission goals and the environment.

References

1. JCS-Joint Pub 1-02. *Department of Defense Dictionary of Military and Associated Terms-12*, April 2001.
2. Ford, C.A. and Rosenberg, D.A. *The Admirals Advantage*, Naval Institute Press, Annapolis, MD, 2005.
3. Jones, R.V. *Most Secret War*, Harnish, Hamilton Ltd, London, 1978.
4. Meyer, H.E. *Real World Intelligence*, Weidenfield and Nicolson, New York, 1987.
5. Studeman, W.O. RADM. Keynote Address to Data Fusion Symposium DFS 88, November, 1988.
6. The 9/11 Commission Report, U.S. Government Printing Office, 2004.
7. Report of the Commission on the Intelligence Capabilities of the United States Regarding Weapons of Mass Destruction. Report to the President of the United States, March 31, 2005.
8. Boorstin, D. *Cleopatra's Nose: Essays on the Unexpected*, Vintage Books, New York, NY, 1994.
9. Davenport, T.H. and Prusek, L. *Working Knowledge: How Organizations Manage What They Know*, Harvard Business School Press, Boston, MA, 2000.
10. Captain William Walls, personal communication.
11. Kessler, O. *Functional Description of the Data Fusion Process*, Office of Naval Technology, Naval Air Development Center, Warminster, PA, 1991.
12. Nichols, M. *A Survey of the Current State of Data Fusion Systems*, Joint C4ISR Decision Support Center, OSD, 30 September 2000.
13. Bosse, E., Roy, J., and Wark, S. *Concepts, Models, and Tools for Information Fusion*, Artech House Inc., 2007.
14. JCS-Joint Publication 2-01. *Joint and National Intelligence Support to Military Operations*, Washington, DC, October 7, 2004.
15. Ed Waltz. *Information Warfare Principles and Operations*, Artech House, Boston, MA, 1998.
16. Department of Defense Directive. *Data Sharing in a Net-Centric Department of Defense*, 8320.2, Washington, DC, Dated December 2, 2004.
17. DISA. *Net-Centric Enterprise Services (NCES) User Guide*, Version 1.1/ECB 1.2, April 20, 2007.
18. BAA 05-015. *Combat ID: Information Management of Coordinated Electronic Surveillance*, Office of Naval Research, May 2004.
19. Fabian, B.E. and Eveleigh, T.J. *Estimation and ISR Process Integration*, National Symposium on Sensor and Data Fusion, Proceedings, June 2001.

3

Revisions to the JDL Data Fusion Model

Alan N. Steinberg and Christopher L. Bowman

CONTENTS

3.1 Objective

This chapter presents refinements to the well-known Joint Directors of Laboratories (JDL) data fusion model. Specifically, the goals of the chapter are to

1. Reiterate the motivation for this model and its earlier versions, viz., to facilitate an understanding of the types of problems for which data fusion is applicable. Such understanding should aid in recognizing commonality among problems and determining the applicability of candidate solutions.

2. Refine the definition of the data fusion *levels* to establish a clearer basis for partitioning among such problems and concepts.

3.1.1   Background

The data fusion model, developed in 1985 by the U.S. Joint Directors of Laboratories (JDL) Data Fusion Group, with subsequent revisions, is the most widely used system for categorizing data fusion-related functions. The goal of the JDL data fusion model is to facilitate understanding and communication among acquisition managers, theoreticians, designers, evaluators, and users of data fusion technology to permit cost-effect system design, development, and operation.[1,2]

The stated purpose for that model and its subsequent revisions have been to

- Categorize different types of fusion processes
- Provide a technical architecture to facilitate reuse and affordability of data fusion and resource management (DF&RM) system development
- Provide a common frame of reference for fusion discussions
- Facilitate understanding of the types of problems for which data fusion is applicable
- Codify the commonality among problems
- Aid in the extension of previous solutions
- Provide a framework for investment in automation.

It should be emphasized that the JDL model was conceived as a functional model and not as a process model or as an architectural paradigm.[3] An example of the latter is the dual node network (DNN) technical architecture developed by Bowman[4] and described in Chapter 22.

3.1.2   Role of Data Fusion

The problem of deriving inferences from multiple items of data pervades all biological cognitive activity and virtually every automated approach to the use of information. Unfortunately, the universality of data fusion has engendered a profusion of overlapping research and development in many applications. A jumble of confusing terminology (illustrated in Figure 3.1) and *ad hoc* methods in a variety of scientific, engineering,

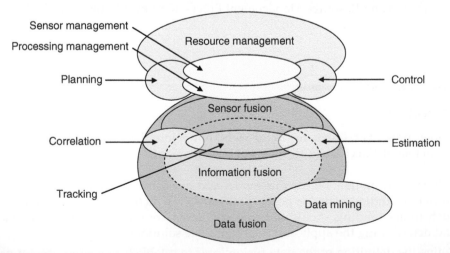

FIGURE 3.1
(Con)fusion of terminology. (From Steinberg, A.N. and Bowman, C.L., *Handbook of Multisensor Data Fusion*, CRC Press, London, 2001.)

management, and educational disciplines obscures the fact that the same ground has been covered repeatedly.

The role of data fusion has often been unduly restricted and its relevance limited to particular state estimation problems. For example, in military applications such as targeting or tactical intelligence, the focus is on estimating and predicting the state of specific types of entities in the external environment (e.g., targets, threats, or military formations). In this context, the applicable sensors or sources that the system designer considers are often restricted to sensors that directly collect observations of targets of interest.

Ultimately, however, such problems are inseparable from other aspects of the system's assessment of the world. In a tactical military system, these will involve estimation of the state of one's own assets in relation to the relevant external entities: friends, foes, neutrals, and background. Estimation of the state of targets and threats cannot be separated from the problems of estimating one's own navigation state, of calibrating one's sensor performance and alignment, and of validating one's library of target models. The data fusion problem, then, becomes that of achieving a consistent, comprehensive estimate and prediction of some relevant portion of the world state. From such a view, data fusion involves exploiting all available sources of data to solve all relevant state estimation or prediction problems; where relevance is determined by utility in forming plans of action.

The data fusion problem therefore encompasses a number of interrelated problems: estimation and prediction of states of entities, both external and internal (organic) to the acting system, and the interrelations among such entities. Evaluating the system's models of the characteristics and behavior of all of these external and organic entities is, likewise, a component of the overall problem of estimating the actual world state.

The complexity of the data fusion system engineering process is characterized by difficulties in

- Representing the uncertainty in observations and in models of the phenomena that generate observations
- Combining noncommensurate information (e.g., the distinctive attributes in imagery, text, and signals)
- Maintaining and manipulating the enormous number of alternative ways of associating and interpreting large numbers of observations of multiple entities.

Deriving general principles for developing and evaluating data fusion processes—whether automatic or manual—can benefit from a recognition of the similarity in the underlying problems of associating and combining data that pervade system engineering and analysis, as well as human cognition. Furthermore, recognizing the common elements of diverse data fusion problems can provide extensive opportunities for synergistic development. Such synergy—enabling the development of information systems that are cost-effective and trustworthy—requires (a) common performance evaluation measures, (b) common system engineering methodologies, (c) common architecture paradigms, and (d) multispectral models of targets and of information sources.

3.1.3 1998 Revision

In 1998, Steinberg et al.[5] published a article formally addressing various extensions to the model. That article began by revisiting the basic definition(s) of data fusion both

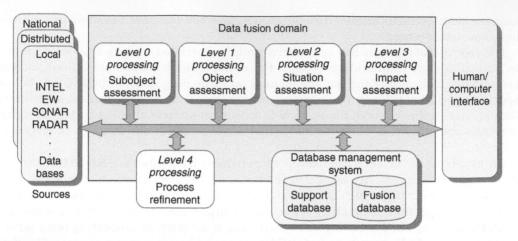

FIGURE 3.2
The 1998 revised Joint Directors of Laboratories data fusion model (with process refinement and database management system partially outside the data fusion domain). (From Steinberg, A.N., Bowman, C.L., and White, F.E., *Proceedings of the SPIE*, 3719, 1999.)

conceptually and in terms of the *levels* that are characterized in the original JDL model. This article was republished in 2000 with minor revisions.

Figure 3.2 depicts the 1998 version of the JDL data fusion model. The goals of the 1998 revision—as in the present proposed revision—were to

- Clarify some of the concepts that guided the original model
- Refine the definition of the JDL data fusion *levels* to better reflect different classes of data fusion problems—that is, systematic differences in types of inputs, outputs and techniques
- Broaden the definitions of fusion concepts and functions to apply across as wide a range of problems as possible, beyond the initial focus on military and intelligence problems
- Maintain a degree of continuity by deviating as little as possible from the usage of concepts and terminology prevailing in the data fusion community

The 1998 revision included the following actions:

1. The definitions of data fusion levels 1–3 were broadened to accommodate fusion problems beyond the military and intelligence problems that had been the focus of earlier versions of the JDL model.

2. A level 0 was introduced to address problems of detecting and characterizing signals, whether in a one-dimension time-series or transform spaces or in multiple spatial dimensions (as in imagery or video feature extraction).

3. The 1998 revision article emphasized that process refinement, as originally conceived, involves not only data fusion functions but planning and control functions as well. Given the formal duality between estimation and control and the similar duality between data association and planning, as discussed in Section 3.6, the present recommended revision clarifies these distinctions: (a) it retains within

level 4 fusion the process assessment functions such as performance and consistency assessment and (b) it treats the process refinement management functions as a part of the resource management functional partitioning.

4. The database management system (DBMS) is treated as a support application accomplished within a layer of the common operating environment below the applications program interface.

5. The notion of estimating informational and psychological states in addition to the familiar physical states was introduced, citing the work of Waltz.[6]

6. An approach to standardization of an engineering design methodology for data fusion processes was introduced, citing the earlier works of Bowman,[7] Steinberg and Bowman,[8] and Llinas et al.[9] in which engineering guidelines for data fusion processes were elaborated.

3.1.4   Definition of Data Fusion

The initial JDL data fusion lexicon (1985) defined *data fusion* as

A process dealing with the association, correlation, and combination of data and information from single and multiple sources to achieve refined position and identity estimates, and complete and timely assessments of situations and threats, and their significance. The process is characterized by continuous refinements of its estimates and assessments, and the evaluation of the need for additional sources, or modification of the process itself, to achieve improved results.[1]

As theory and applications have evolved over the years, it has become clear that this initial definition is rather too restrictive. A definition was needed that could capture the fact that similar underlying problems of data association and combination occur in a very wide range of engineering, analysis, and cognitive situations. Accordingly, the initial definition requires a number of modifications:

1. Although the concept *combination of data* encompasses a broad range of problems of interest, *correlation* does not. Statistical correlation is merely one method for generating and evaluating hypothesized associations among data.

2. Association is not an essential ingredient in combining multiple pieces of data. Work in random set models of data fusion provides generalizations that allow state estimation of multiple targets without explicit report-to-target association.[6-8]

3. 'Single or multiple sources' is comprehensive; therefore, it is superfluous in a definition.

4. The reference to position and identity estimates should be broadened to cover all varieties of state estimation.

5. Complete assessments are not required in all applications; 'timely', being application-relative, is superfluous.

6. 'Threat assessment' limits the application to situations where threat is a factor. This description must also be broadened to include any assessment of the impact (i.e., the cost or utility implications) of estimated situations. In general, data fusion involves refining and predicting the states of entities and aggregates of entities and their relation to one's own mission plans and goals. Cost assessments can include variables such as the probability of surviving an estimated threat situation.

7. Not every process of combining information involves collection management or process refinement. Thus, the definition's second sentence is best construed as illustrative, not definitional.

Pruning these extraneous qualifications, the model revision proposes the following concise definition for data fusion:[5]

The process of combining data or information to estimate or predict entity states.

Data fusion involves combining data—in the broadest sense of 'data'—to estimate or predict the state of some aspect of the universe. Often the objective is to estimate or predict the physical state of entities: their identity, attributes, activity, location, and motion over some past, current, or future time period.

However, estimation problems can also concern nonphysical aspects. If the job at hand is to estimate the state of people (or any other sentient beings), it may be important to estimate or predict the informational and perceptual states of individuals or groups and the interrelation of these with physical states.

Arguments about whether 'data fusion' or another label—say, some other label shown in Figure 3.1—best describes this very broad concept are pointless. Some people have adopted terms like 'information integration' in an attempt to generalize earlier, narrower definitions of 'data fusion' (and, perhaps, to dissociate themselves from old data fusion approaches and programs).

Nonetheless, relevant research should not be neglected simply because of shifting terminological fashion. Although no body of common and accepted usage currently exists, this broad concept is an important topic for a unified theoretical approach and, therefore, deserves its own label.

3.1.5 Motivation for Present Revision

Like 'data fusion' itself, the JDL data fusion *levels* are in need of definitions that are at once broadly applicable and precise. The suggested revised partitioning of data fusion functions is designed to capture the significant differences in the types of input data, models, outputs, and inferencing appropriate to broad classes of data fusion problems. In general, the recommended partitioning is based on different aspects of a situation for which the characterization is of interest to a system user. In particular, a given entity—whether a signal, physical object, aggregate, or structure—can often be viewed either (a) as an individual whose attributes, characteristics, and behaviors are of interest or (b) as an assemblage of components whose interrelations are of interest.* For effective fusion system implementation, the fusion system designer needs to determine what would be the types of basic entities of interest from which the relationships of interest can be defined.

Sections 3.2 and 3.3 respectively present and discuss the recommended data fusion level definitions. Section 3.4 discusses the system integration of data fusion processes across the levels. Section 3.5 examines some prominent alternatives and extensions to the JDL model. Sections 3.6 and 3.7 extend the revised JDL data fusion levels to corresponding dual levels in resource management, which together are used to decompose DF&RM problems and processes for effective system design. The revised level 4 of data fusion retains the "process assessment" fusion functions in the original JDL fusion level 4 and includes process

---

* This distinction in which the ontology derives from users' interests, originates in the work of Quine[18] (whose answer to *What exists?* was *Everything*) and Strawson.[19]

and sensor management as part of the corresponding dual resource management levels. It is proposed that these extended DF&RM levels will serve much the same purpose for resource management problems as the original JDL levels do for data fusion problems: facilitating understanding, design and integration.[10]

3.2 Recommended Refined Definitions of Data Fusion Levels

The original JDL model suffered from an unclear partitioning scheme. Researchers have sometimes taken the model as distinguishing the fusion levels on the basis of process (identification, tracking, aggregation, prediction, etc.); sometimes on the basis of topics (objects, situations, threats); products (object, situation, cost estimates) levels. In some ways, this ambiguity is a virtue: correlating significant differences in engineering problems and uses.

Nonetheless, the conflation of processes, topics and products has tended to blur important factors in data fusion problems and solutions. Numerous examples can be cited in which problems of one level find solution in techniques of a different level. For example, entity recognition—a paradigmatically level 1 problem—is often addressed by techniques that assess relationships between an entity and its surroundings, these being paradigmatically level 2 processes. Also relations among components of an object can be used in level 0 fusion to improve object recognition products.

It is the goal of the recommended revised definitions of the data fusion functional levels to provide a clear and useful partitioning while adhering as much as possible to current usage across the data fusion community. With this goal in mind, the recommended revised definitions partition data fusion functions basis entities of interest to information users. The products of processes at each level are estimates of some existing or predicted aspects of reality.

The recommended revised model diagram is shown in Figure 3.3, with fusion levels defined as follows:

- Level 0: *Signal/feature assessment*. Estimation of signal or feature states. Signals and features may be defined as patterns that are inferred from observations or measurements. These may be static or dynamic and may have locatable or causal origins (e.g., an emitter, a weather front, etc.).
- Level 1: *Entity assessment*. Estimation of entity parametric and attributive states (i.e., of entities considered as individuals).
- Level 2: *Situation assessment*. Estimation of the structures of parts of reality (i.e., of sets of relationships among entities and their implications for the states of the related entities).
- Level 3: *Impact assessment*. Estimation of the utility/cost of signal, entity, or situation states, including predicted utility/cost given a system's alternative courses of action.
- Level 4: *Process assessment*. A system's self-estimation of its performance as compared to desired states and measures of effectiveness (MOEs).

These concepts are compared and contrasted in Table 3.1.

In general, the benefit of this partitioning scheme is that these levels distinguish problems that characteristically involve significant differences in types of input data, models, outputs, and inferencing. It should be noted that the levels are not necessarily processed

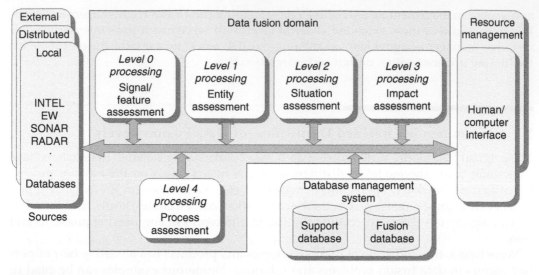

FIGURE 3.3
Recommended revised data fusion model, Bowman '04. (From Bowman, C.L., *AIAA Intelligent Systems Conference*, Chicago, September 20–22, 2004.)

TABLE 3.1

Characteristics of the Recommended Data Fusion Levels

Data Fusion Levels	Association Process	Estimation Process	Product
L.0—signal/feature assessment	Observation-to-signal or feature*[a]	Feature extraction	Estimated signal/feature states and confidences
L.1—entity assessment	Signal/feature-to-entity or entity state-to-entity*	Attributive entity state estimation	Estimated entity states and confidences
L.2—situation assessment	Entity*-to-entity*, entity*-to-relationship,* or relationship*-to-relationship*	Relational state estimation	Estimated relationships, situation (set of relationships), and confidences
L.3—impact assessment	System courses of action to situation and utility*	Cost/benefit analysis	Estimated/predicted entity and situation utilities and confidences
L.4—process assessment	System states* to goals and truth estimates	Performance analysis	Estimated measures of performance (MOPs) and measures of effectiveness (MOEs) and confidences

[a] More precisely, where indicated by asterisk, association is with features, signals, entities, relations, and situation that are inferred by the system.

in order and any one can be processed on its own given the corresponding inputs. In addition, a system design may involve addressing more than one fusion level within a single processing node.

3.3 Discussion of Data Fusion Levels

3.3.1 Level 0: Signal/Feature Assessment

The 1998 revision[5] recommended defining a level 0 fusion to encompass various uses of multiple measurements in signal and feature processing. These include processes of feature extraction from imagery and analogous signal detection and parameter estimation in electro-magnetic, acoustic, or other data. In the most general sense, these problems concern the discovery of information (in Shannon's sense) in some region of space and time.

Level 0 processes include inferences that do not require assumptions about the presence or characteristics of entities possessing such observable features. They concern the structure of measurement sets (their syntax), not their cause (i.e., their semantics). In cases where signal characteristics are conditioned on the presence and characteristics of one or more presumed entities, the latter are treated as a context, or situation, in which the signals or features are inferable, for example, through a likelihood function $\lambda(Z|x)$.

These functions include DAI/FEO (data in/features out) and FEI/FEO (features in/features out) functions as defined in Dasarathy's model,[11] as well as DEI/FEO (decisions in/features out) in our expanded version of that model.*[12]

Level 0 processes are often performed by individual sensors, or with the product of individual sensors. That certainly is the case with most tactical systems. However, when communications bandwidth and processing time allow, much can be gained by feature extraction or parameter measurement at the multisensor level. Image fusion characteristically involves extracting features across multiple images, often from multiple sources. Multisensor feature extraction is also practiced in diverse systems that locate targets using multistatic or other multisensor techniques.

3.3.2 Level 1: Entity Assessment

The notion of level 1 data fusion—called variously *object assessment* or *entity assessment*—was originally conceived as encompassing the most prominent and most highly-developed applications of data fusion: detection, identification, location, and tracking of individual physical objects (aircraft, ships, land vehicles, etc). Most techniques involve combining observations of a target of interest to estimate the states of interest. Thus, it was convenient to conflate process, target and product in distinguishing *level 1 fusion*. However, states of physical objects can be estimated using indirect, non-attributive information. Also, non-physical objects are amenable to attributive state estimation. Accordingly, we prefer to construe level 1 data fusion as the estimation of states of entities considered as individuals. JDL level 1, as in earlier versions, is concerned with the estimation of the identity, classification (in some given taxonomic scheme), other discrete attributes, kinematics, other continuous parameters, activities, and potential states of the basic entities of interest for the fusion problem being solved.

Individuation of entities implies an inclusion, or boundary, function such as is found in such paradigmatic entities as

- Biological organisms
- Purpose-made entities; for example, a hammer, an automobile, or a bird's nest
- Other discriminated spatially contiguous uniformities that persist over some practical time interval, for example, a mountain, coral reef, cloud, raindrop, or swarm of gnats
- Societally discriminated entities, for example, a country, tract of land, family, or other social, legal or political entity.

---

* Dasarathy's original categories represent constructive, that is data-driven, processes, in which organized information is extracted from relatively unorganized data. In the earlier edition of this handbook[12] we defined additional processes that are analytic, or model-driven, such that organized information (a model) is analyzed to estimate lower-level data (features or measurements) as they relate to the model. Examples include predetection tracking (an FEI/DAO process), model-based feature extraction (DEI/FEO), and model-based classification (DEI/DAO). Dasarathy's model is discussed further in Section 3.5.

A level 1 estimation process can be thought of as the application of one-place relations $R^{(1)}(a) = x$, where x may be a discrete or a continuous random variable, or a vector of such variables, so that $R^{(1)}(a)$ may be the temperature, location, functional class, an attribute, or activity state of an entity a.

3.3.3   Level 2: Situation Assessment

Level 2 data fusion concerns inferences about *situations*. Devlin[13] provides a useful definition of a situation as *any structured part of reality*. We may parse this in terms of a standard dictionary definition of a structure as *a set of entities, their attributes, and relationships* (Merriam-Webster Dictionary, 3rd edition). Once again, data fusion is *the estimation of the state of some aspects of the world on the basis of multiple data*.

Abstract situations can be represented as sets of relations; actual situations can be represented as sets of instantiated relations (or, relationships).* Thus, situation assessment involves inferencing from the estimated state(s) of one entity in a situation to another and from the estimated attributes and relationships of entities to situations. Attributes of individuals can be treated as one-place relationships.

Methods for representing relationships and for inferring entity states on the basis of relationships include graphical methods (e.g., Bayesian and other belief networks).

Situation assessment, then, involves the following functions:

1. Inferring relationships among (and within) entities
2. Recognizing/classifying situations basis estimates of constituents (entities, attributes, and relationships)
3. Using inferred relationships to infer entity attributes, to include attributes (a) of entities that are elements of a situation (e.g., refining the estimate of the existence and state variables of an entity on the basis of its inferred relationships) and (b) of the entities that are themselves situations (e.g., in determining a situation to be of a certain type)

These functions implicitly include inferring and exploiting relationships among situations, for example, predicting future situation basis current and historical situations.

As noted above, an entity—whether a signal, physical object, aggregate, or structure—can often be viewed either (a) as an individual whose attributes, characteristics, and behaviors are of interest or (b) as an assemblage of components whose interrelations are of interest. From the former point of view, discrete physical objects (the paradigm targets of level 1 data fusion) are components of a situation. From the latter point of view, the targets are themselves situations, that is, contexts for the analysis of components and their relationships.

A more detailed discussion on situation assessment and its relationship to entity and impact assessment is given in Chapter 18.

---

* It is useful to distinguish between relations in the abstract (e.g., *trust*) and instantiated relations (e.g., *Othello's trust in Iago*), which we will call *relationships*. There is no obvious term in English to mark the analogous distinction between abstract and instantiated *attributes*. In any case, it is convenient in some applications to consider these, respectively, as single-place relations or relationships.

3.3.4 Level 3: Impact Assessment

It must be admitted that level 3 data fusion has been the most troublesome to define clearly and consistently with common usage in the field. The initial name of "threat assessment" is useful and relates to a specific, cohesive set of operational functions, but only within the domain of tactical military or security applications. The 1998 version attempted to broaden the concept to nonthreat domains; with a change of names to *impact assessment*. This was defined as *the estimation and prediction of effects on situations of planned or estimated/predicted actions by the participants (e.g., assessing susceptibilities and vulnerabilities to estimated/predicted threat actions, given one's own planned actions).*

We may simplify this definition by taking impact assessment to be conditional situation estimation. Questions to be answered include those of the following form:

"If entities $x_i \in X$ follow courses of action α_i, what will be the outcome?"

Many problems of interest involve a given agent x operating in a reactive environment (i.e., one that responds differentially to x's actions). Furthermore, such reaction is often the result of one or more responsive agents, often assumed to be capable of intentional activity. As such, impact assessment problems can be amenable to game theoretic treatment, when the objective functions of the interacting agents can be specified.

In military threat assessment, we are usually concerned with the state of a set of such agents that constitute *our* forces as affected by intentional actions and reactions of hostile agents. In other words, we are concerned with inferring the concept:

"If we follow course of action α, what will be the outcome (and cost)?"

Impact assessment so construed can also include counterfactual event or situation prediction:

"If x had followed this course of action, what would have been the outcome?"

In general then, level 3 fusion involves combining multiple sources of information to estimate conditional or counterfactual outcome and cost. It is amenable to cost analysis, whether it is Bayesian-based or otherwise.

Because the utility of a fused state in supporting a user generally needs to be predicted based on the estimated current situation, known plans, and predicted reactions, level 3 fusion processing typically has different inputs and different outputs (viz. utility predictions) than the other fusion levels.

An example of a level 3 type of association problem is that of determining the expected consequence of system plans or courses of action, given the current estimated signal, entity, and situational state. An example of a level 3 type of estimation problem is the prediction of the impact of the estimated current situational state on the mission utility.

Level 3 issues and methods are discussed further in Chapter 18.

3.3.5 Level 4: Process Assessment

In early versions of the JDL model, level 4 was designated as *process refinement*. This was meant to encompass both assessment and adaptive control of the fusion process and—in many interpretations—in the data collection process as well. We have argued (a) that a more fundamental partitioning of functionality involves a distinction between data fusion functions (involving data association and state estimation) and resource management functions (involving planning and control), and (b) that functions that determine how the system resources should respond are clearly resource management functions.[5,8,12]

There is, however, a need for a system-level category of level 4 fusion processes, one that is involved in associating system states and actions to desired system states and estimating or predicting the performance of the system. We therefore propose a data fusion level 4, concerning this process performance. Level 4 data fusion functions combine information to estimate a system's measures of performance (MOPs) and MOEs based on a desired set of system states and responses. In the case of a data fusion system, MOEs can involve correlation of system state estimates to truth.

In level 4 data fusion, the primary association problem to be solved is that of determining which system outputs correspond to which of the desired goal states.

In the case of a data fusion system, the level 4 association problem may be that of determining a correspondence between system tracks and real-world entities.

In level 4 fusion, estimation may include such MOPs as sensor calibration and alignment errors, track purity and fragmentation, etc. MOEs could include utility metrics on target classification and location accuracies. Such estimates may be made on the basis of the system's fusion multisensor product, or by an external test-bed system on the basis of additional fiducial information.

For an integrated DF&RM system, there can be several levels of planning and control (discussed in Sections 3.6 and 3.7).

RM association problems may involve determining a correspondence between resource actions and desired responses. A hierarchy of MOPs and MOEs can be estimated based on these associations.

3.4 Information Flow Within and Across the "Levels"

Processing at each of these data fusion levels involves the batching of available data for fusion within a network of fusion nodes where, paradigmatically, each fusion node accomplishes

- Data preparation (data mediation, common formatting, spatiotemporal alignment, and confidence normalization)
- Data association (generation, evaluation, and selection of association hypotheses, i.e., of hypotheses as to the applicability of specific data to particular aspects of the state estimation problem)
- State estimation and prediction (estimating the presence, attributes, interrelationships, utility, and performance or effectiveness of entities of interest, as appropriate to the data fusion node)*

A comparison of the data association problems that need to be solved at each fusion level is given in Figure 3.4. In all the fusion levels, the accuracy of the fused state estimates

---

* We use the word "paradigmatically" because some systems may not require data preparation in every fusion node and data association may not be separable from estimation in some implementations (e.g., segmentation leads to less complexity but usually gives up some performance). Furthermore, not every multitarget state estimation processes requires explicit data association at the target level; rather, data can be determined to be relevant at the situational (e.g., multitarget) level.

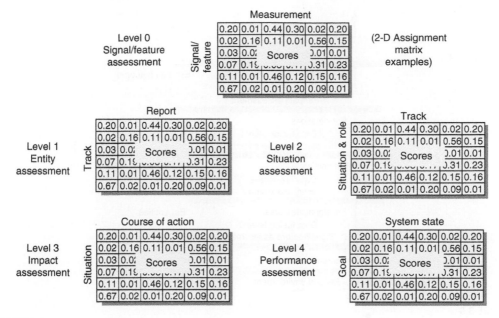

FIGURE 3.4
Data association problems occur at each fusion level.

tends to increase as larger batches of data (e.g., over more data types, sources or times) are fused; however, the cost and complexity of the fusion process also increases. Thus, a knee-of-the-curve of performance versus cost fusion node network is sought in the system design and operation.

As noted earlier, the data fusion levels are not necessarily processed in order and any one can be processed on its own or in combination given the corresponding inputs. Therefore, the term *level* can be misleading in this context. The early data fusion model descriptions rightfully avoided the sense of hierarchy by the use of functional diagrams such as Figures 3.2 and 3.3, in which various data fusion functions are represented in a common bus architecture.

Nonetheless, there is often a natural progression from raw measurements to entity states, to situation relationships, to utility prediction, and to system performance assessment, in which data are successively combined as suggested by the level ordering. Composition of estimated signals (or features), entities, and aggregates per levels 0–2 is quite natural, with a corresponding reverse flow of contexts as illustrated in Figure 3.5.* Utility is generally assessed as a function of an estimated or predicted situational state, rather than of the state of a lone entity within a situation. That is the reason that level 3 fits naturally above level 2, but level 2 outputs are not always required for level 3 processing. Similarly, system performance can also be assessed on the basis of estimated or predicted outcomes of situations and executed or planned system actions, consistent with our ordering of levels.

---

* In many or most data fusion implementations to date, the predominant flow is upward; any contextual conditioning in a data fusion node being provided by corresponding resource management nodes (e.g., during process refinement). Steinberg and Rogova[20] provide a discussion of contextual exploitation in data fusion.

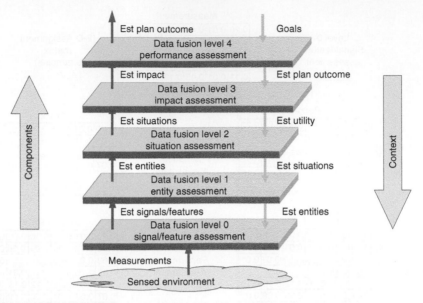

FIGURE 3.5
Characteristic functional flow across the data fusion levels.

3.5 Model Extensions and Variants

3.5.1 Level 5: User Refinement

A process identified as "user refinement" has been proposed variously by Hall et al.[14] and by Blasch and Plano.[15] The latter give the following definition:

Level 5: Visualize. *This process connects the user to the rest of the fusion process so that the user can visualize the fusion products and generate feedback or control to enhance/improve these products.*[15]

Hall and McMullen[16] define level 5 data fusion as *cognitive refinement and human–computer interaction.* Cognitive refinement includes the transformation from sensor data to a graphics display, but encompasses all methods of presenting information to human users. User refinement processes can involve adaptive determination of which users have needs for information and which have access to information. User refinement can also include knowledge management to adaptively determine the data that is most needed to be retrieved and displayed to support cognitive decision making and actions.

While there is no denial of the essential character of such functions in many systems involving both human and automated processes, it can be argued that these are not specifically data fusion functions. In fact they usually need to be managed by higher resource management processes such as objective management and resource relationship management. As a functional model, the JDL data fusion model is partitioned according to the functions performed, not by the performance and response management mechanisms, whether automatic, human, or other mechanisms.

TABLE 3.2

Interpretation of Dasarathy's Data Fusion Input/Output Model

	Data	Output features	Objects
Data	Signal detection DAI/DAO	Feature extraction DAI/FEO	Gestalt-based object characterization DAI/DEO
Features	Model-based detection feature extraction FEI/DAO	Feature refinement FEI/FEO	(Feature-based object characterization) FEI/DEO
Objects	Model-based detection estimation DEI/DAO	Model-based feature extract DEI/FEO	Object refinement DEI/FEO

Input (left vertical label)

Level 0 (spanning Data and Output features) — Level 1 (Objects)

The level 0–4 data fusion processes as we have defined them—involving the combination of information to estimate the state of various aspects of the world—can and have been performed by people, by automated processes, and by combinations thereof.*

We therefore argue against the inclusion of such a level 5 in the data fusion model. Like data archiving, communications, and resource management, such functions as data access, data retrieval, and data presentation are ancillary functions that often occur in systems that also do data fusion. As these descriptions indicate, some of these functions are resource management functions that are formal duals of certain data fusion functions.

3.5.2 Dasarathy's Input/Output Model

The partitioning basis process products (as distinct, e.g., from process types) allows a direct mapping from Dasarathy's input/output (I/O) model,[11] as extended in Chapter 2 of the previous edition of this handbook.[12] Accordingly, the latter can be seen as a refinement of the JDL model.

Dasarathy[11] categorized data fusion functions in terms of the types of data/information that are processed and the types that result from the process. Table 3.2 illustrates the types of I/O considered. Processes corresponding to the cells in the highlighted matrix region are described by Dasarathy, using the abbreviations DAI/DAO, DAI/FEO, FEI/FEO, FEI/DEO, and DEI/DEO.† A striking benefit of this categorization is the natural manner in which technique types can be mapped into it.

---

* As stated by Frank White, the first JDL panel chairman, "much of [the JDL model's] value derives from the fact that identified fusion functions have been recognizable to human beings as a 'model' of functions they were performing in their own minds when organizing and fusing data and information. It is important to keep this 'human centric' sense of fusion functionality since it allows the model to bridge between the operational fusion community, the theoreticians and the system developers" (personal communication, quoted in Ref. 4).

† See Section 3.3.1 for the definition of these acronyms.

In the previous edition of this handbook,[12] this categorization was extended by

- Adding labels to these cells, relating I/O types to process types
- Filling in the unoccupied cells in the original matrix

Note that Dasarathy's original categories represent constructive, or data-driven, processes, in which organized information is extracted from less organized data. Additional processes—FEI/DAO, DEI/DAO, and DEI/FEO—can be defined that are analytic, or model-driven, such that organized information (a model) is analyzed to estimate lower-level data (features or measurements) as they relate to the model. Examples include predetection tracking (an FEI/DAO process), model-based feature extraction (DEI/FEO), and model-based classification (DEI/DAO). The remaining cell in Table 3.2—DAI/DAO—has not been addressed in a significant way (to the authors' knowledge), but could involve the direct estimation of entity states without the intermediate step of feature extraction.

In Ref. 12, we extended Dasarathy's categorization to encompass level 2, 3, and 4 processes, as shown in Table 3.3. Here rows and columns have been added to correspond to the object types listed in Figure 3.5.

With this augmentation, Dasarathy's categorization is something of a refinement of the JDL levels. Not only can each of the levels (0–4) be subdivided basis of input data types, but level 0 can also be subdivided into detection processes and feature-extraction processes.

Of course, much of Table 3.3 remains virgin territory; researchers have only seriously explored its northwest quadrant, with tentative forays southeast. Most likely, little utility will be found in either the Northeast or the Southwest. However, there may be gold buried somewhere in those remote stretches.

TABLE 3.3

Expansion of Dasarathy's Model to Data Fusion Levels 0–4

		Output					
		Data	Features	Objects	Relations	Impacts	Responses
Input	Data	Signal detection DAI/DAO	Feature extraction DAI/FEO	Gestalt-based object extract DAI/DEO	Gestalt-based situation assessment DAI/RLO	Gestalt-based impact assessment DAI/IMO	Reflexive response DAI/RSO
	Features	Model-based detection/ feature extraction FEI/DAO	Feature refinement FEI/FEO	Object characterization FEI/DEO	Feature-based situation assessment FEI/RLO	Feature-based impact assessment FEI/IMO	Feature-based response FEI/RSO
	Objects	Model-based detection/ estimation DEI/DAO	Model-based feature extraction DEI/FEO	Object refinement DEI/DEO	Entity-relations situation assessment DEI/RLO	Entity-based impact assessment DEL/IMO	Entity-relation based response DEI/RSO
	Relations	Context-sensitive detection/est RLI/DAO	Context-sensitive feature extraction RLI/FEO	Context-sensitive object refinement RLI/DEO	Micro-macro situation assessment RLI/RLO	Context-sensitive impact assessment RLI/IMO	Context-sensitive response RLI/RSO
	Impacts	Cost-sensitive detection/est IMI/DAO	Cost-sensitive feature extraction IMI/FEO	Cost-sensitive object refinement IMI/DEO	Cost-sensitive situation assessment IMI/RLO	Cost-sensitive impact assessment IMI/IMO	Cost-sensitive response IMI/RSO
	Responses	Reaction-sensitive detection/est DEI/DAO	Reaction-sensitive feature extraction RSI/FEO	Reaction-sensitive object refinement RSI/DEO	Reaction-sensitive sit assessment RSI/RLO	Reaction-sensitive impact assessment RSI/RLO	Reaction-sensitive response RSI/RSO
		Level 0		*Level 1*	*Level 2*	*Level 3*	*Level 4*

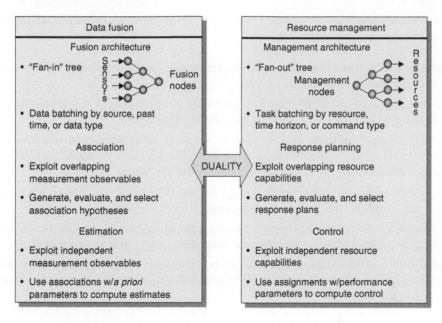

FIGURE 3.6
The data fusion and resource management duality: key concepts.

3.5.3 Other Data Fusion Models

Bedworth and O'Brien[21] provide a synthesis of data fusion models. Their resultant *Omnibus* model maps the early JDL and Dasarathy's models into an OODA (observe, orient, decide, act) model.

It should be noted that the omnibus model—like the OODA loop—is a process model, whereas the JDL model and Dasarathy's I/O model are functional models.* Among the most influential variants and alternatives to the JDL model are those presented by Endsley,[22] Salerno,[23] and Lambert.[24] These are discussed in some detail in Chapter 18.

3.6 Data Fusion and Resource Management Levels†

Just as a formal duality exists between estimation and control, there is a more encompassing duality between data fusion and resource management (DF&RM).[10] The DF&RM duality incorporates the association/planning duality as well as the estimation/control duality, as summarized in Figure 3.6. The planning and control aspects of process refinement—level 4 in early versions of the JDL model—are inherently resource management functions.

A functional model for resource management was proposed in Ref. 17 in terms of functional levels that are, accordingly, the duals of the corresponding data fusion levels. This model has been refined in Ref. 4.

---

* See Ref. 3, pp. 331ff for definitions of various types of engineering models.
† See Chapter 22 for an in-depth discussion.

A dual scheme for modeling DF&RM functions has the following beneficial implications for system design and development:

- Integrated DF&RM systems can be implemented using a network of interacting fusion and management nodes.
- The duality of DF and RM concepts provides insights useful for developing and evaluating techniques for implementing both DF and RM.
- The technology maturity of data fusion can be used to bootstrap resource management technology, which is lagging fusion development by more than 10 years, much as estimation did for control theory in the 1960s.
- The partitioning into *levels* reflects significant differences in the types of data, resources, models, and inferencing necessary for each level.
- All fusion levels can be implemented using a fan-in network of fusion nodes where each node performs: data preparation, data association, and state estimation.
- Analogously, all management levels can be implemented using a fan-out network of management nodes where each node performs: task preparation, task planning, and resource state control.

Given the dual nature of DF&RM, we would expect a corresponding duality in functional levels. As with data fusion levels, the dual resource management processing levels are based on the user's resources of interest. The proposed dual DF&RM levels are presented in Table 3.4, and the duality concepts that were used in defining these resource management levels are summarized in Figure 3.7.

As with the corresponding partitioning of data fusion functions into levels, the utility of these management levels results from the significant differences in the types of resources, models, and inferencing used in each level. Resource management includes applications-layer functions such as sensor management, target management, weapons management, countermeasure management, flight management, process management, communications management, and so on. As with the data fusion levels, the resource management levels are not necessarily processed in order and any one can be processed on its own or in combination given the corresponding inputs.

All the management levels can be implemented and integrated using a fan-out network of management nodes where each node performs the functions of task preparation, task planning, and resource state control. DF&RM systems can be implemented using a

TABLE 3.4

Dual Data Fusion and Resource Management Processing Levels

Level	Data Fusion Level	Resource Management Level
0	Signal/feature assessment: estimation of signal/feature states	Signal management: management of resource emissions/observables
1	Entity assessment: estimation of entity attributive states	Resource response management: management of individual resources
2	Situation assessment: estimation of entity relational/situational states	Resource relationship management: management of resource relationships
3	Impact assessment: estimation of the impact of fused states on mission objectives	Mission objective management: management of mission objectives
4	Process assessment: estimation of MOP/MOE states	Design management: management of system engineering and operational configuration

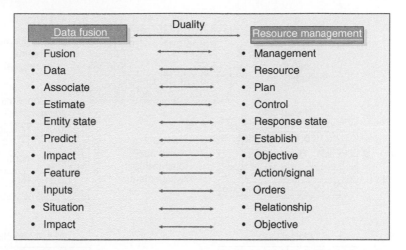

FIGURE 3.7
Duality between DF and RM elements.

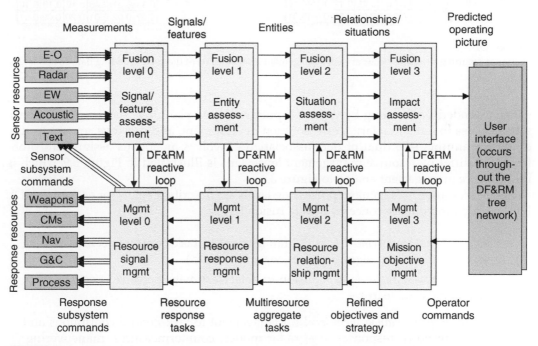

FIGURE 3.8
Illustrative multilevel data fusion and resource management system network with sequential level interactions.

network of interacting fusion and management nodes. These node interactions can occur across any of the levels. However, for illustrative purposes, Figure 3.8 shows a sequential processing flow across the DF&RM levels 0–3.*

---

* Level 4 fusion and management processes have typically been performed off-line, during system design and evaluation. Their use in on-line self-assessment and reconfiguration has potential for improved efficiency in information exploitation and response.

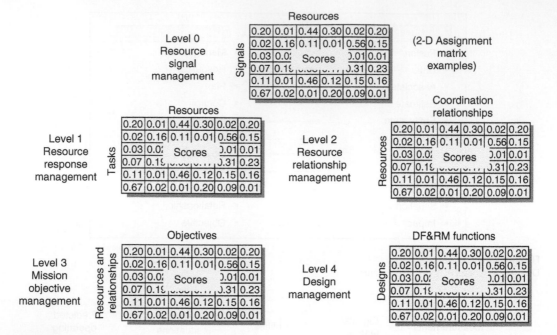

FIGURE 3.9
Response planning problems occur at each management level as duals of the corresponding data fusion levels.

The extended partitioning by type of inputs, response planning, and output response states strives to enhance clarity, ease solution development, and preserve respect for the duality. Planning—the dual of data association—involves analogous assignment problems that differ by resource management level. This is illustrated in Figure 3.9, which is the resource management analog to Figure 3.4.

The five resource management processing levels are described below for each of the JDL levels comparing duality concepts to data fusion levels:

- *Resource signal management level 0.* Management to task/control specific resource response actions (e.g., signals, pulses, waveforms, etc.)

- *Signal/feature assessment level 0.* Fusion to detect/estimate/perceive specific source entity signals and features

- *Resource response management level 1.* Management to task/control continuous and discrete resource responses (e.g., radar modes, countermeasures, maneuvering, communications)

- *Entity assessment level 1.* Fusion to detect/estimate/perceive continuous parametric (e.g., kinematics, signature) and discrete attributes (e.g., entity type, identity, activity, allegiance) of entity states

- *Resource relationship management level 2.* Management to task/control relationships (e.g., aggregation, coordination, conflict) among resource responses

- *Situation assessment level 2.* Fusion to detect/estimate/comprehend relationships (e.g., aggregation, causal, command/control, coordination, adversarial relationships) among entity states

- *Mission objective management level 3.* Management to establish/modify the objective of level 0, 1, 2 action, response, or relationship states
- *Impact assessment level 3.* Fusion to predict/estimate the impact of level 0, 1, or 2 signal, entity, or relationship states
- *Design management level 4.* Management to task/control the system engineering (e.g., problem-to-solution space algorithm/model design mapping, model discovery, and generalization)
- *Process assessment level 4.* Fusion to estimate the system's MOPs and MOEs.

3.7 Data Fusion and Resource Management Processing Level Issues

The user's entities of interest can be the basis of all five levels of fusion processing. The features of an entity can be estimated basis attributes inferred from one or more entity signal observations (e.g., through a level 0 data preparation/association/estimation process). For example, signal-level association and estimation problems occur in Electronic Intelligence (ELINT) pulse train deinterleaving or feature extraction of an entity in imagery. This involves inferring the existence and characteristics of the features of an entity by attributive or relational state estimation from observations and measurements specific to each sensor/source modality (e.g., radar pulses, hyperspectral pixel intensities).

The identity, location, track, and activity state of an entity of interest (whether it be a man, a molecule, or a military formation) can be estimated basis attributes inferred from one or more signal or entity observations (e.g., through one or a network of data preparation/association/estimation fusion nodes). The same entity's compositional or relational state (e.g., its role within a larger structure and its relations with other elements of that structure) can be inferred through level 2 processes. The behavior of the same entity can also be projected to assess the impact of the utility of an estimated or predicted situation relative to the user's objective.

The fused states can then be compared to desired states at all fusion levels to determine the performance of the fusion system and similarly for response states and the resource management system performance assessment.

The declaration of features, entity states, their relationships, their conditional interactions and impacts, or their correspondence to truth is a data association (i.e., hypothesis generation, evaluation, and selection) function within the particular fusion level.

Fused state projection at the impact assessment level (data fusion level 3) needs to determine the impact of alternative projected states. This requires additional information concerning the conditional actions and reactions of the entities in the relevant situation (e.g., in the battlespace). Furthermore, utility assessment requires additional information on the mission success criteria, which also is unnecessary for level 0–2 fusion.*

Utility assessment is the estimation portion of level 3 processing. Predicting the fused state beyond the current time is of interest for impact assessment whose results are of interest to mission resource management.

The objective of the lower levels of resource management in using impact assessments is to plan responses to improve the confidence in mission success, whereas the objective of

---

* Generally speaking, Level 0–2 processes need only to project the fused state forward in time sufficient for the next data to be fused.

level 4 design management in using the performance evaluation (PE) outputs is to improve confidence in system performance. PE nodes tend to have significant interaction with their duals in design management, in that they provide the performance estimates for a DF&RM solution that are used to propose a better DF&RM solution. This resource management function for optimizing the mapping from problem to solution space is usually referred to as system engineering (i.e., equated to design management here). The design management architecture provides a representation of the system engineering problem that partitions its solution into the resource management node processes; that is, of

- Problem alignment (resolving design needs, conflicts, mediation)
- Design planning (design generation, evaluation, and selection)
- Design implementation and test (output-specific resource control commands)

The dual management levels are intended to improve understanding of the management alternatives and facilitate exploitation of the significant differences in the resource modes, capabilities, and types as well as mission objectives.

Process refinement—the old data fusion level 4—has been subsumed as an element of each level of resource management that includes adaptive data acquisition and processing to support mission objectives (e.g., sensor management and information systems dissemination).

User refinement (which, as noted above, has been proposed as a data fusion level 5)[15] has been subsumed as an element of knowledge management within resource management.

References

1. Llinas, J., C.L. Bowman, G. Rogova, A.N. Steinberg, E. Waltz, and F. White, Revisiting the JDL data fusion model II, *Proceedings, Seventh International Conference on Information Fusion*, Stockholm, 2004.
2. White, F.E., Jr., A model for data fusion, *Proceedings of the First National Symposium on Sensor Fusion*, 1988.
3. Buede, D.M., *The Engineering Design of Systems*, Wiley, New York, 2000.
4. Bowman, C.L., The dual node network (DNN) data fusion & resource management (DF&RM) architecture, *AIAA Intelligent Systems Conference*, Chicago, September 20–22, 2004.
5. Steinberg, A.N., C.L. Bowman, and F.E. White, Revisions to the JDL Model, *Joint NATO/IRIS Conference Proceedings, Quebec, October, 1998 and in Sensor Fusion: Architectures, Algorithms, and Applications*, Proceedings of the SPIE, Vol. 3719, 1999.
6. Waltz, E., *Information Warfare: Principles and Operations*, Artech House, Boston, MA, 1998.
7. Bowman, C.L., The data fusion tree paradigm and its dual, *Proceedings of the Seventh National Symposium on Sensor Fusion*, 1994.
8. Steinberg, A.N. and C.L. Bowman, Development and application of data fusion engineering guidelines, *Proceedings of the Tenth National Symposium on Sensor and Data Fusion*, 1997.
9. Llinas, J. et al. Data fusion system engineering guidelines, *Technical Report 96-11/4*, USAF Space Warfare Center (SWC) Talon-Command project report, Vol. 2, 1997.
10. Bowman, C.L., Affordable information fusion via an open, layered, paradigm-based architecture, *Proceedings of Ninth National Symposium on Sensor Fusion*, Monterey, CA, March 1996.
11. Dasarathy, B.V., Sensor fusion potential exploitation-innovative architectures and illustrative applications, *IEEE Proceedings*, Vol. 85, No. 1, 1997.
12. Steinberg, A.N. and C.L. Bowman, Revisions to the JDL data fusion model, Chapter 2 of *Handbook of Multisensor Data Fusion*, D.L. Hall and J. Llinas (Eds.), CRC Press, London, 2001.

13. Devlin, K., *Logic and Information*, Press Syndicate of the University of Cambridge, Cambridge, 1991.
14. Hall, M.J., S.A. Hall, and T. Tate, Removing the HCI bottleneck: How the human computer interface (HCI) affects the performance of data fusion systems, *Proceedings of the MSS National Symposium on Sensor Fusion*, 2000.
15. Blasch, E.P. and S. Plano, Level 5: User refinement to aid the fusion process, in *Multisensor, Multisource Information Fusion: Architectures, Algorithms, and Applications 2003*, B.V. Dasarathy (Ed.), Proceedings of the SPIE, Vol. 5099, 2003.
16. Hall, D.L. and McMullen S.A.H., *Mathematical Techniques in Multisensor Data Fusion*, Second Edition, Artech House, Boston, 2004.
17. Steinberg, A.N. and C.L. Bowman, Rethinking the JDL data fusion levels, *Proceedings of MSS National Symposium on Sensor and Data Fusion*, Laurel, Maryland, 2004.
18. Quine, W.V.O., *Word and Object*, MIT Press, Cambridge, MA, 1960.
19. Strawson, P.F., *Individuals: An Essay in Descriptive Metaphysics*, Methuen Press, London, 1959.
20. Steinberg, A.N. and G. Rogova, Situation and context in data fusion and natural language understanding, *Proceedings of Eleventh International Conference on Information Fusion*, Cologne, 2008.
21. Bedworth, M. and J.O'Brien, The Omnibus model: a new model of data fusion?, *Proceedings of the Second International Conference on Information Fusion*, Sunnyvale, CA, 1999.
22. Endsley, M.R., Toward a theory of situation awareness in dynamic systems, *Human Factors*, Vol. 37, No. 1, 1995.
23. Salerno, J.J., Where's level 2/3 fusion—a look back over the past 10 years, *Proceedings of the Tenth International Conference on Information Fusion*, Quebec, Canada, 2007.
24. Lambert, D.A., Tradeoffs in the design of higher-level fusion systems, *Proceedings of the Tenth International Conference on Information Fusion*, Quebec, Canada, 2007.

4

Introduction to the Algorithmics of Data Association in Multiple-Target Tracking

Jeffrey K. Uhlmann

CONTENTS

4.1 Introduction

When a major-league outfielder runs down a long fly ball, the tracking of a moving object looks easy. Over a distance of a few hundred feet, the fielder calculates the ball's trajectory to within an inch or two and times its fall to within milliseconds. But what if an outfielder was asked to track 100 fly balls at once? Even 100 fielders trying to track 100 balls simultaneously would likely find the task an impossible challenge.

Problems of this kind do not arise in baseball, but they have considerable practical importance in other realms. The impetus for the studies described in this chapter was the strategic defense initiative (SDI), the plan conceived in the early 1980s for defending the United States against a large-scale nuclear attack. According to the terms of the original proposal, an SDI system would be required to track tens or even hundreds of thousands of objects—including missiles, warheads, decoys, and debris—all moving at speeds of up to 8 km/s. Another application of multiple-target tracking is air-traffic control, which attempts to maintain safe separations among hundreds of aircraft operating near busy airports. In particle physics, multiple-target tracking is needed to make sense of the hundreds or thousands of particle tracks emanating from the site of a high-energy collision. Molecular dynamics has similar requirements.

The task of following a large number of targets is surprisingly difficult. If tracking a single baseball, warhead, or aircraft requires a certain measurable level of effort, then it might seem that tracking 10 similar objects would require at most 10 times as much effort. Actually, for the most obvious methods of solving the problem, the difficulty is proportional to the square of the number of objects; thus, 10 objects demand 100 times the effort, and 10,000 objects increase the difficulty by a factor of 100 million. This combinatorial explosion is the first hurdle to solving the multiple-target tracking problem. In fact, exploiting all information to solve the problem optimally requires exponentially scaling effort. This chapter, however, considers computational issues that arise for any proposed multiple-target tracking system.*

Consider how the motion of a single object might be tracked, on the basis of a series of position reports from a sensor such as a radar system. To reconstruct the object's trajectory, plot the successive positions in sequence and then draw a line through them (as shown on the left-hand side of Figure 4.1). Extending this line yields a prediction of the object's future position. Now, suppose you are tracking 10 targets simultaneously. At regular time intervals 10 new position reports are received, but the reports do not have labels indicating the targets to which they correspond. When the 10 new positions are plotted, each report could, in principle, be associated with any of the 10 existing trajectories (as illustrated on the right-hand side of Figure 4.1). This need to consider every possible combination of reports and tracks makes the difficulty of all n-target problem proportional to—or on the order of—n^2, which is denoted as $O(n^2)$.

Over the years, many attempts have been made to devise an algorithm for multiple-target tracking with better than $O(n^2)$ performance. Some of the proposals offered significant improvements in special circumstances or for certain instances of the multiple-target tracking problem, but they retained their $O(n^2)$ worst-case behavior. However, recent results in the theory of spatial data structures have made possible a new class of algorithms for associating reports with tracks—algorithms that scale better than quadratic scaling in most realistic environments. In degenerate cases, in which all of the targets are so densely clustered that they cannot be individually resolved, there is no way to avoid comparing each report with each track. When each report can be feasibly associated only with a constant number of tracks on average, subquadratic scaling is achievable. This will become clear later in the chapter. Even with the new methods, multiple-target tracking remains a complex task that strains the capacity of the largest and fastest supercomputers. However, the new methods have brought important problem instances within reach.

4.1.1   Keeping Track

The modern need for tracking algorithms began with the development of radar during World War II. By the 1950s, radar was a relatively mature technology. Systems were installed aboard military ships and aircrafts and at airports. The tracking of radar targets, however, was still performed manually by drawing lines through blips on a display screen. The first attempts to automate the tracking process were modeled closely on human performance. For the single-target case, the resulting algorithm was straightforward—the computer accumulated a series of positions from radar reports and estimated the velocity of the target to predict its future position.

Even single-target tracking presented certain challenges related to the uncertainty inherent in position measurements. The first problem involves deciding how to represent this uncertainty. A crude approach is to define an error radius surrounding the position

---

* The material in this chapter updates and supplements material that first appeared in *American Scientist*.[1]

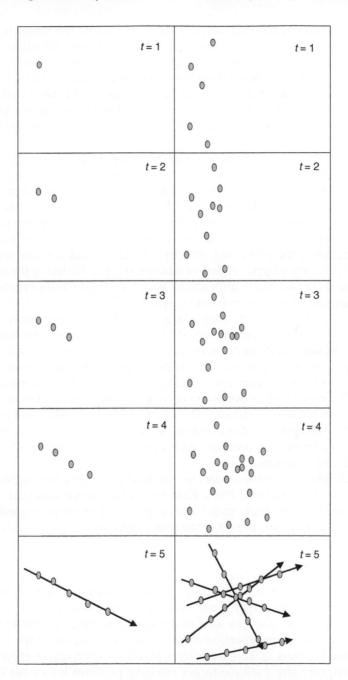

FIGURE 4.1

The information available for plotting a track consists of position reports (shown as dots) from a sensor such as a radar system. In tracking a single target (left), one can accumulate a series of reports and then fit a line or curve corresponding to those data points to estimate the object's trajectory. With multiple targets (right), there is no obvious way to determine which object has generated each report. Here, five reports appear initially at time step $t = 1$, then five more are received at $t = 2$. Neither the human eye nor a computer can easily distinguish which of the later dots goes with which of the earlier ones. (In fact, the problem is even more difficult given that the reports at $t = 2$ could be newly detected targets that are not correlated with the previous five reports.) As additional reports arrive, coherent tracks begin to emerge. The tracks from which these reports were derived are shown in the lower panels at $t = 5$. Here and in subsequent figures, all targets are assumed to have constant velocity in two dimensions. The problem is considerably more difficult for ballistic or maneuvering trajectories in three dimensions.

estimate. This practice implies that the probability of finding the target is uniformly distributed throughout the volume of a three-dimensional sphere. Unfortunately, this simple approach is far from optimal. The error region associated with many sensors is highly nonspherical; radar, for example, tends to provide accurate range information but has relatively poorer radial resolution. Furthermore, one would expect the actual position of the target to be closer on average to the mean position estimate than to the perimeter of the error volume, which suggests, in turn, that the probability density should be greater near the center.

The second difficulty in handling uncertainty is determining how to interpolate the actual trajectory of the target from multiple measurements, each with its own error allowance. For targets known to have constant velocity (e.g., they travel in a straight line at constant speed), there are methods for calculating projectile straight-line path that best fits, by some measure, the series of past positions. A desirable property of this approach is that it should always converge on the correct path—as the number of reports increases, the difference between the estimated velocity and the actual velocity should approach zero. However, retaining all past reports of a target and recalculating the entire trajectory every time a new report arrives is impractical. Such a method would eventually exceed all constraints on computation time and storage space.

A near-optimal method for addressing a large class of tracking problems was developed in 1960 by Kalman.[2] His approach, referred to as *Kalman filtering*, involves the recursive fusion of noisy measurements to produce an accurate estimate of the state of a system of interest. A key feature of the Kalman filter is its representation of state estimates in terms of mean vectors and error covariance matrices, where a covariance matrix provides an estimate (usually a conservative overestimate) of the second moment of the error distribution associated with the mean estimate. The square root of the estimated covariance gives an estimate of the standard deviation. If the sequence of measurement errors is statistically independent, the Kalman filter produces a sequence of conservative fused estimates with diminishing error covariances.

Kalman's work had a dramatic impact on the field of target tracking in particular and data fusion in general. By the mid-1960s, Kalman filtering was a standard methodology. It has become as central to multiple-target tracking as it has been to single-target tracking; however, it addresses only one aspect of the overall problem.

4.1.2 Nearest Neighbors

What multiple targets add to the tracking problem is the need to assign each incoming position report to a specific target track. The earliest mechanism for classifying reports was the nearest-neighbor rule. The idea of the rule is to estimate each object's position at the time of a new position report, and then assign that report to the object nearest to such estimate (see Figure 4.2). This intuitively plausible approach is especially attractive because it decomposes the multiple-target tracking problem into a set of single-target problems.

The nearest-neighbor rule is straightforward to apply when all tracks and reports are represented as points; however, there is no clear means for defining what constitutes *nearest neighbors* among tracks and reports with different error covariances. For example, if a sensor has an error variance of 1 cm, then the probability that measurements 10 cm apart are from the same object is $P(10^{-20})$, whereas measurements having a variance of 10 cm could be 20–30 cm apart and feasibly correspond to the same object. Therefore, the appropriate measure of distance must reflect the relative uncertainties in the mean estimates.

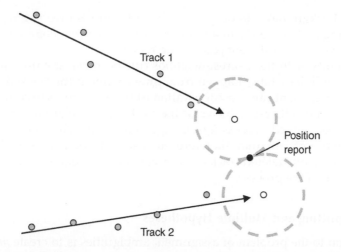

FIGURE 4.2
The nearest-neighbor rule is perhaps the simplest approach for determining which tracked object produced a given sensor report. When a new position report arrives, all existing tracks are projected forward to the time of the new measurement. (In this diagram, earlier target positions are indicated by dots and projected positions by circles; the new position report is labeled.) Then, the distance from the report to each projected position is calculated, and the report is associated with the nearest track. More generally, the distance calculation is computed to reflect the relative uncertainties (covariances) associated with each track and report. In the situation depicted in the diagram, the report would be assigned to track 1, based purely on its Euclidean proximity to the report. If this assignment is erroneous, the subsequent tracking process will be adversely affected.

The most widely used measure of the correlation between two mean and covariance pairs $\{x_1, P_1\}$, which are assumed to be Gaussian-distributed random variables, is[3,4]

$$P_{\text{association}}(x_1, x_2) = \frac{1}{\sqrt{2\pi|(P_2 + P_2)|}} \exp\left(-\frac{1}{2}(x_1 - x_2)(P_1 + P_2)^{-1}(x_1 - x_2)^{\mathrm{T}}\right) \qquad (4.1)$$

which reflects the probability that x_1 is a realization of x_2 or, symmetrically, the probability that x_2 is a realization of x_1. If this quantity is above a given threshold—called a gate—then the two estimates are considered to be feasibly correlated. If the assumption of Gaussian distribution does not hold exactly—and it generally does not—then this measure is heuristically assumed (or hoped) to yield results that are at least good enough to be used for ranking purposes (i.e., to say confidently that one measurement is more likely than the other to be associated with a given track). If this assumption approximately holds, then the gate will tend to discriminate high- and low-probability associations. Accordingly, the nearest-neighbor rule can be redefined to state that a report should be assigned to the track with which it has the highest association ranking. In this way, a multiple-target problem can still be decomposed into a set of single-target problems.

The nearest-neighbor rule has strong intuitive appeal, but doubts and difficulties connected with it soon emerged. For example, early implementers of the method discovered problems in creating initial tracks for multiple targets. In the case of a single target, two reports can be accumulated to derive a velocity estimate, from which a track can be created. For multiple targets, however, there is no obvious way to deduce such initial velocities. The first two reports received could represent successive positions of a single object or the initial detection of two distinct objects. Every subsequent report could be the continuation of a known track or the start of a new one. To make matters worse, almost every sensor produces some background rate of spurious reports, which give rise to spurious tracks. Thus, the tracking system needs an additional mechanism to recognize and delete tracks that do not receive any subsequent confirming reports.

Another difficulty with the nearest-neighbor rule becomes apparent when reports are misclassified, as will inevitably happen from time to time if the tracked objects are close together. A misassignment can cause the Kalman-filtering process to converge very slowly, or fail to converge altogether, in which case the track cannot be predicted. Moreover, tracks updated with misassigned reports (or not updated at all) will tend to correlate poorly with subsequent reports and may, therefore, be mistaken as spurious by the track-deletion mechanism. Mistakenly deleted tracks then necessitate subsequent track initiations and a possible repetition of the process.

4.1.3   Track Splitting and Multiple Hypotheses

A robust solution to the problem of assignment ambiguities is to create *multiple-hypothesis tracks*. Under this scheme, the tracking system does not have to commit immediately or irrevocably to a single assignment of each report. If a report is highly correlated with more than one track, an updated copy of each track can be created; subsequent reports can be used to determine which assignment is correct. As more reports come in, the track associated with the correct assignment will rapidly converge on the true target trajectory, whereas the falsely updated tracks are less likely to be correlated with subsequent reports.

This basic technique is called track splitting.[3,5] One of its worrisome consequences is a proliferation in the number of tracks upon which a program must keep tabs. The proliferation can be controlled with the same track-deletion mechanism used in the nearest-neighbor algorithm, which scans through all the tracks from time to time and eliminates those that have a low probability of association with recent reports. A more sophisticated approach to track splitting, called multiple-hypothesis tracking, maintains a history of track branching, so that as soon as one branch is confirmed, the alternative branches can be pruned away.

Track splitting in its various forms[6] is a widely applied strategy for handling the ambiguities inherent in correlating tracks with reports from multiple targets. It is also used to minimize the effects of spurious reports when tracking a single target. Nevertheless, some serious difficulties remain. First, track splitting does not completely decompose a multiple-target tracking problem into independent single-target problems, the way the nearest-neighbor strategy was intended to function. For example, two hypothesis tracks may lock onto the trajectory of a single object. Because both tracks are valid, the standard track-deletion mechanism cannot eliminate either of them. The deletion procedure has to be modified to detect redundant tracks and, therefore, cannot look at just one track at a time. This coupling between multiple tracks is theoretically troubling; however, experience has shown that it can be managed in practice at low computational cost.

The second problem is the difficulty of deciding when a position report and a projected track are correlated closely enough to justify creating a new hypothesis track. If the correlation threshold is set too high, correct assignments may be missed so often as to prevent convergence of the Kalman filter. If the threshold is too low, the number of hypotheses could grow exponentially. The usual practice is to set the threshold low enough to ensure convergence, and then add another mechanism to limit the rate of hypothesis generation. A simple strategy is to select the n hypothesis candidates with the highest probabilities of association, where n is the maximum number of hypotheses that computational resource constraints will allow. This "greedy" method often yields good performance.

Even with these enhancements, the tracking algorithm makes such prodigious demands on computing resources that large problems remain beyond practical reach. Monitoring the computation to see how much time is spent on various subtasks shows that calculating probabilities of association is, by far, the biggest expense. The program gets bogged down projecting target tracks to the time of a position report and calculating association probabilities. Because this is the critical section of the algorithm, further effort has focused on improving performance in this area.

4.1.4 Gating

The various calculations involved in estimating a probability of association are numerically intensive and inherently time-consuming. Thus, one approach to speed up the tracking procedure is to streamline or fine-tune these calculations—to encode them more efficiently without changing their fundamental nature. An obvious example is to calculate

$$\text{dist}^2(x_1, x_2) = (x_1 - x_2)(P_1 + P_2)^{-1}(x_1 - x_2)^{\text{T}} \tag{4.2}$$

rather than the full probability of association. This measure is proportional to the logarithm of the probability of association and is commonly referred to as the Mahalanobis distance or log-likelihood measure.[4] Applying a suitably chosen threshold to this quantity yields a method for obtaining the same set of feasible pairs, while avoiding a large number of numerically intensive calculations.

An approach for further reducing the number of computations is to minimize the number of log-likelihood calculations by performing a simpler preliminary screening of tracks and sensor reports. Only if a track–report pair passes this computationally inexpensive feasibility check is there a need to complete the log-likelihood calculation. Multiple gating tests also can be created for successively weeding out infeasible pairs, so that each gate involves more calculations but is applied to considerably fewer pairs than the previous gate.

Several geometric tests could serve as gating criteria. For example, if each track is updated, on average, every 5 s, and the targets are known to have a maximum speed of 10 km/s, a track and report more than 50 km apart are not likely to be correlated. A larger distance may be required to take into account the uncertainty measures associated with both the tracks and the reports.

Simple gating strategies can successfully reduce the numerical overhead of the correlation process and increase the number of targets that can be tracked in real time. Unfortunately, the benefits of simple gating diminish as the number of targets increases. Specifically, implementers of gating algorithms have found that increasing the number of targets by a factor of 20 often increases the computational burden by a factor of more than 100. Moreover, the largest percentage of computation time is still spent on the correlation process; although now the bulk of the demand is for simple distance calculations within the gating algorithm. This implies that the quadratic growth in the number of gating tests is more critical than the constant numerical overhead associated with the individual tests. In other words, simple gating can reduce the average cost of each comparison, but what is really needed is a method to reduce the sheer number of comparisons. Some structure must be imposed on the set of tracks that will allow correlated track–report pairs to be identified without requiring every report to be compared with every track.

The gating problem is difficult conceptually because it demands that most pairs of tracks and reports be excluded from consideration without ever being examined. At the same time, no track–report pair whose probability of association exceeds the correlation threshold can be disregarded. Until the 1980s, the consensus in the tracking literature was that these constraints were impossible to satisfy simultaneously. Consequently, the latter constraint was often sacrificed using methods that did allow some, but hopefully few, track–report pairs to be missed even though their probabilities of association exceeded the threshold. This seemingly reasonable compromise, however, has led to numerous *ad hoc* schemes that fail to adequately limit the number of either comparisons or missed correlations. Some approaches are susceptible to both problems.

Most of the *ad hoc* strategies depend heavily on the distribution of the targets. A common approach is to identify clusters of targets that are sufficiently separated so that reports from targets in one cluster will never have a significant probability of association with tracks from another cluster.[7] This allows the correlation process to determine from which cluster a particular report could have originated and then compare the report only to the tracks in that cluster. The problem with this approach is that the number of properly separated clusters depends on the distribution of the targets and, therefore, cannot be controlled by the clustering algorithm (Figure 4.3). If $O(n)$ tracks are partitioned into $O(n)$ clusters, each consisting of a constant number of tracks, or into a constant number of clusters of $O(n)$ tracks, the method still results in a computational cost that is proportional to the comparison of every report to every track. Unfortunately, most real-world tracking problems tend to be close to one of these extremes.

A gating strategy that avoids some of the distribution problems associated with clustering involves partitioning the space in which the targets reside into grid cells. Each track can then be assigned to a cell according to its mean projected position. In this way, the tracks that might be associated with a given report can be found by examining the tracks

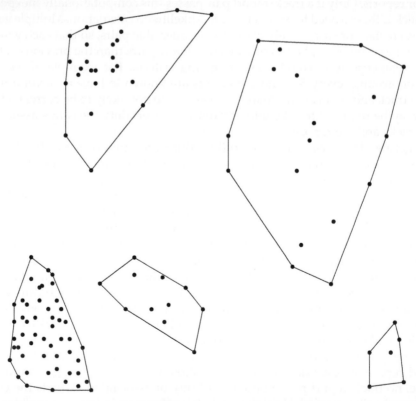

FIGURE 4.3
Clustering algorithms may produce spatially large clusters with few points and spatially small ones with many points.

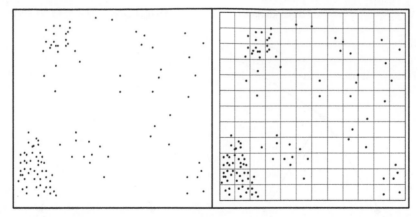

FIGURE 4.4
Grids may have a few cells with many points, whereas the remaining cells contain few or no points.

in only those cells that are within close proximity to the report's cell. The problem with this approach is that its performance depends heavily on the size of the grid cells, as well as on the distribution of the targets (Figure 4.4). If the grid cells are large and the targets are densely distributed in a small region, every track will be within a nearby cell. Conversely, if the grid cells are small, the algorithm may spend as much time examining cells (most of which may be empty) as would be required to simply examine each track.

4.1.5   Binary Search and *kd*-Trees

The deficiencies of grid methods suggest the need for a more flexible data structure. The main requirement imposed on the data structure has already been mentioned—it must allow all proximate track–report pairs to be identified without having to compare every report with every track (unless every track is within the prescribed proximity to every report).

A clue to how real-time gating might be accomplished comes from one of the best-known algorithms in computer science: binary search. Suppose one is given a sorted list of n numbers and asked to find out whether or not a specific number, q, is included in the list. The most obvious search method is simply to compare q with each number in sequence; in the worst case (when q is the last number or is not present at all), the search requires n comparisons. There is a much better way. Because the list is sorted, if q is found to be greater than a particular element of the list, one can exclude from further consideration not only that element but all those that precede it in the list. This principle is applied optimally in binary search. The algorithm is recursive—first compare q to the median value in the list of numbers (by definition, the median will be found in the middle of a sorted list). If q is equal to the median value, then stop, and report that the search was successful. If q is greater than the median value, then apply the same procedure recursively to the sublist greater than the median; otherwise apply it to the sublist less than the median (Figure 4.5). Eventually either q will be found—it will be equal to the median of some sublist—or a sublist will turn out to be empty, at which point the procedure terminates and reports that q is not present in the list.

The efficiency of this process can be analyzed as follows. At every step, half of the remaining elements in the list are eliminated from consideration. Thus, the total number of comparisons is equal to the number of halves, which in turn is $O(\log n)$. For example, if n is 1,000,000, then only 20 comparisons are needed to determine whether a given number is in the list.

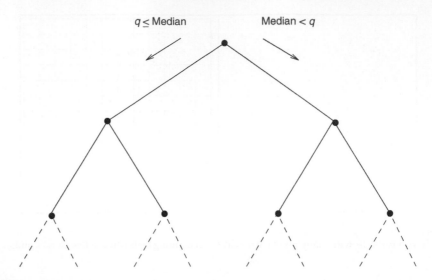

FIGURE 4.5
Each node in a binary search tree stores the median value of the elements in its subtree. Searching the tree requires a comparison at each node to determine whether the left or right subtree should be searched.

Binary search can also be used to find all elements of the list that are within a specified range of values (*min, max*). Specifically, it can be applied to find the position in the list of the largest element less than *min* and the position of the smallest element greater than *max*. The elements between these two positions then represent the desired set. Finding the positions associated with *min* and *max* requires $O(\log n)$ comparisons. Assuming that some operation will be carried out on each of the *m* elements of the solution set, the overall computation time for satisfying a range query scales as $O(\log n + m)$.

Extending binary search to multiple dimensions yields a *kd*-tree.[8] This data structure permits the fast retrieval of all three-dimensional points; for example, in a data set whose *x* coordinate is in the range (x_{min}, x_{max}), *y* coordinate is in the range (y_{min}, y_{max}), and *z* coordinate is in the range (z_{min}, z_{max}). The *kd*-tree for $k = 3$ is constructed as follows: The first step is to list the *x* coordinates of the points and choose the median value, then partition the volume by drawing a plane perpendicular to the *x*-axis through this point. The result is to create two subvolumes, one containing all the points whose *x* coordinates are less than the median and the other containing the points whose *x* coordinates are greater than the median. The same procedure is then applied recursively to the two subvolumes, except that now the partitioning planes are drawn perpendicular to the *y*-axis and they pass through points that have median values of the *y* coordinate. The next round uses the *z* coordinate, and then the procedure returns cyclically to the *x* coordinate. The recursion continues until the subvolumes are empty.*

Searching the subdivided volume for the presence of a specific point with given *x, y,* and *z* coordinates is a straightforward extension of standard binary search. As in the one-dimensional case, the search proceeds as a series of comparisons with median values, but now attention alternates among the three coordinates. First the *x* coordinates are compared,

---

* An alternative generalization of binary search to multiple dimensions is to partition the data set at each stage according to its distance from a selected set of points:[9,10,24–28] those that are less than the median distance comprise one branch of the tree, and those that are greater comprise the other. These data structures are very flexible because they offer the freedom to use an appropriate application-specific metric to partition the data set; however, they are also much more computationally intensive because of the number of distance calculations that must be performed.

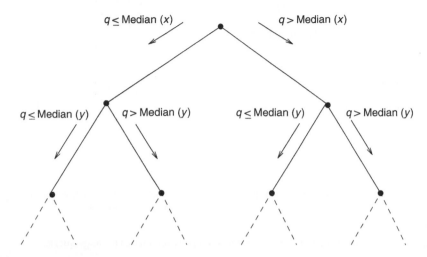

FIGURE 4.6
A *kd*-tree partitions on a different coordinate at each level in the tree. It is analogous to an ordinary binary search tree, except that each node stores the median of the multidimensional elements in its subtree projected onto one of the coordinate axes.

then the y, the z, and so on (Figure 4.6). In the end, either the chosen point will be found to lie on one of the median planes or the procedure will come to an empty subvolume.

Searching for all of the points that fall within a specified interval is somewhat more complicated. The search proceeds as follows: If x_{min} is less than the median value of x coordinate, the left subvolume must be examined. If x_{max} is greater than the median value of x coordinate, the right subvolume must be examined. At the next level of recursion, the comparison is done using y_{min} and y_{max}, then z_{min} and z_{max}.

A detailed analysis[11–13] of the algorithm reveals that for k dimensions (provided that k is greater than 1), the number of comparisons performed during the search can be as high as $O(n^{1-1/k} + m)$; thus, for three dimensions the search time is proportional to $O(n^{2/3} + m)$. In the task of matching n reports with n tracks, the range query must be repeated n times, so the search time scales as $O(n \times n^{2/3} + m)$ or $O(n^{5/3} + m)$. This scaling is better than quadratic scaling, but not nearly as good as the logarithmic scaling observed in the one-dimensional case, which works out for n range queries to be $O(n \log n + m)$. The reason for the penalty in searching a multidimensional tree is the possibility at each step that both subtrees will have to be searched without necessarily finding an element that satisfies the query. (In one dimension, a search of both subtrees implies that the median value satisfies the query.) In practice, however, this seldom happens, and the worst-case scaling is rarely seen. Moreover, for query ranges that are small relative to the extent of the data set—as they typically are in gating applications—the observed query time for *kd*-trees is consistent with $O(\log^{1+\varepsilon} n + n)$, where $\varepsilon > 0$.

4.2 Ternary Trees

The *kd*-tree is provably optimal for satisfying multidimensional range queries if one is constrained to using only linear (i.e., $O(n)$) storage.[12,13] Unfortunately, it is inadequate for gating purposes because the track estimates have spatial extent due to uncertainty in their

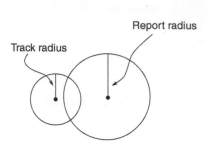
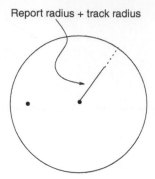

If the position uncertainties are thresholded, then
gating requires intersection detection.

If the largest track radius is added to all the report radii,
then the tracks can be treated as points.

FIGURE 4.7
Transferring uncertainty from tracks to reports reduces intersection queries to range queries.

exact position. In other words, a *kd*-tree would be able to identify all track points that fall within the observation uncertainty bounds. However, it would fail to return any imprecisely localized map item whose uncertainty region intersects the observation region, but whose mean position does not. Thus, the gating problem requires a data structure that stores sized objects and is able to retrieve those objects that intersect a given query region associated with an observation.

One approach for solving this problem is to shift all of the uncertainty associated with the tracks onto the reports.[14,15] The nature of this transfer is easy to understand in the simple case of a track and a report whose error ellipsoids are spherical and just touching. Reducing the radius of the track error sphere to zero, while increasing the radius of the report error sphere by an equal amount, leaves the enlarged report sphere just touching the point representing the track, so the track still falls within the gate of the report (Figure 4.7). Unfortunately, when this idea is applied to multiple tracks and reports, the query region for every report must be enlarged in all directions by an amount large enough to accommodate the largest error radius associated with any track. Techniques have been devised to find the minimum enlargement necessary to guarantee that every track correlated with a given report will be found[15]; however, many tracks with large error covariances can result in such large query regions that an intolerable number of uncorrelated tracks will also be found.

A solution that avoids the need to inflate the search volumes is to use a data structure that can satisfy ellipsoid intersection queries instead of range queries. One such data structure that has been applied in large-scale tracking applications is an enhanced form of *kd*-tree that stores coordinate-aligned boxes.[1,16] A box is defined as the smallest rectilinear shape, with sides parallel to the coordinate axes, that can entirely surround a given error ellipsoid (see Figure 4.8). Because the axes of the ellipse may not correspond to those of the coordinate system, the box may differ significantly in size and shape from the ellipse it encloses. The problem of determining optimal approximating boxes is presented in reference 29.

An enhanced form of the *kd*-tree is needed for searches in which one range of coordinate values is compared with another range, rather than the simpler case in which a range is compared with a single point. A binary tree will not serve this purpose because it is not possible to say that one interval is entirely greater than or less than another when they intersect. What is needed is a ternary tree, with three descendants per node (Figure 4.9). At each stage in a search of the tree, the maximum value of one interval is compared with the

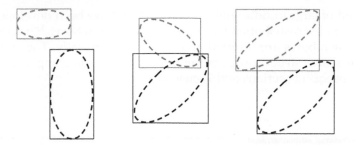

FIGURE 4.8
The intersection of error boxes offers a preliminary indication that a track and a report probably correspond to the same object. A more definitive test of correlation requires a computation to determine the extent to which the error ellipses (or their higher-dimensional analogs) overlap, but such computations can be too time-consuming when applied to many thousands of track-report pairs. Comparing bounding boxes is more computationally efficient; if they do not intersect, an assumption can be made that the track and report do not correspond to the same object. However, intersection does not necessarily imply that they do correspond to the same object. False positives must be weeded out in subsequent processing.

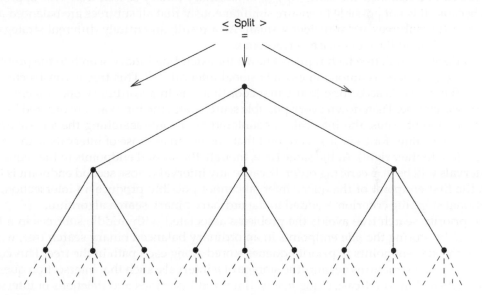

FIGURE 4.9
Structure of a ternary tree. In a ternary tree, the boxes in the left subtree fall on one side of the partitioning (split) plane, the boxes in the right subtree fall to the other side of the plane, and the boxes in the middle subtree are strictly cut by the plane.

minimum of the other, and vice versa. These comparisons can potentially eliminate either the left subtree or the right subtree. In either case, examining the middle subtree—the one made up of nodes representing boxes that might intersect the query interval—is necessary. However, because all of the boxes in a middle subtree intersect the plane defined by the split value, the dimensionality of the subtree can be reduced by 1, causing subsequent searches to be more efficient.

The middle subtree represents obligatory search effort; therefore, one goal is to minimize the number of boxes that straddle the split value. However, if most of the nodes fall to the left or right of the split value, then few nodes will be eliminated from the search, and query performance will be degraded. Thus, a trade-off must be made between the effects of unbalance and of large middle subtrees. Techniques have been developed for adapting ternary trees to exploit distribution features of a given set of boxes,[16] but they cannot easily be applied when boxes are inserted and deleted dynamically. The ability to dynamically update the search structure can be very important in some applications; this topic is addressed in subsequent sections of this chapter.

4.3   Priority kd-Trees

The ternary tree represents a very intuitive approach to extending the kd-tree for the storage of boxes. The idea is that, in one dimension, if a balanced tree is constructed from the minimum values of each interval, then the only problematic cases are those intervals whose *min* endpoints are less than a split value while their *max* endpoints are greater. Thus, if these cases can be handled separately (i.e., in separate subtrees), then the rest of the tree can be searched the same way as an ordinary binary search tree. This approach fails because it is not possible to ensure simultaneously that all subtrees are balanced and that the extra subtrees are sufficiently small. As a result, an entirely different strategy is required to bound the worst-case performance.

The priority search tree technique is known for extending binary search to the problem of finding intersections among one-dimensional intervals.[18,19] This tree is constructed by sorting the intervals according to the first coordinate as in an ordinary one-dimensional binary search tree. Then down every possible search path, the intervals are ordered by the second endpoint. Thus, the intervals encountered by always searching the left subarray will all have values for their first endpoint that are less than those of intervals with larger indices (i.e., to their right). At the same time, though, the second endpoints in the sequence of intervals will be in ascending order. Because any interval whose second endpoint is less than the first endpoint of the query interval cannot possibly produce an intersection, an additional stopping criterion is added to the ordinary binary search algorithm.

The priority search tree avoids the problems associated with middle subtrees in a ternary tree by storing the *min* endpoints in an ordinary balanced binary search tree, while storing the *max* endpoints in priority queues stored along each path in the tree. This combination of data structures permits the storage of n intervals, such that intersection queries can be satisfied in worst-case $O(\log n + m)$ time, and insertions and deletions of intervals can be performed in worst-case $O(\log n)$ time. Thus, the priority search tree generalizes binary search on points to the case of intervals, without any penalty in terms of errors. Unfortunately, the priority search tree is defined purely for intervals in one dimension.

The kd-tree can store multidimensional points, but not multidimensional ranges, whereas the priority search tree can store one-dimensional ranges, but not multiple dimensions. The question that arises is whether the kd-tree can be extended to store boxes efficiently, or whether the priority search tree can be extended to accommodate the analog of intervals in higher dimensions (i.e., boxes). The answer to the question is "yes" for both data structures, and the solution is, in fact, a combination of the two.

A priority kd-tree[20] is defined as follows: Given a set S of k-dimensional box intervals (lo_i, hi_i), $1 < i < k$, a priority kd-tree consists of a kd-tree constructed from the *lo* endpoints of the intervals with a priority set containing up to k items stored at each node (Figure 4.10).*

---

* Other data structures have been independently called *priority kd-trees* in the literature, but they are different data structures designed for different purposes.

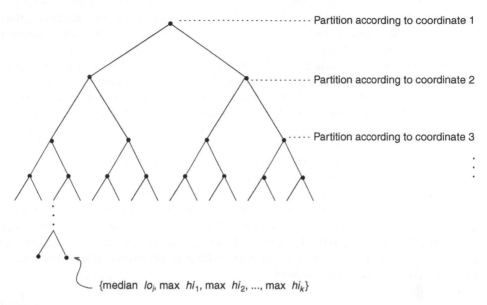

FIGURE 4.10
Structure of a priority kd-tree. The priority kd-tree stores multidimensional boxes, instead of vectors. A box is defined by an interval (lo_i, hi_i) for each coordinate i. The partitioning is applied to the lo coordinates analogously to an ordinary kd-tree. The principal difference is that the maximum hi value for each coordinate is stored at each node. These hi values function analogously to the priority fields of a priority search tree. In searching a priority kd-tree, the query box is compared to each of the stored values at each visited node. If the node partitions on coordinate i, then the search proceeds to the left subtree if lo_i is less than the median lo_i associated with the node. If hi_i is greater than the median lo_i, then the right subtree must be searched. The search can be terminated, however, if for any j, lo_j of the query box is greater than the hi_j stored at the node.

The items stored at each node are the minimum set so that the union of the hi endpoints in each coordinate includes a value greater than the corresponding hi endpoint of any interval of any item in the subtree. Searching the tree proceeds exactly as for all ordinary priority search trees, except that the intervals compared at each level in the tree cycle through the k dimensions as in a search of a kd-tree.

The priority kd-tree can be used to efficiently satisfy box intersection queries. Just as important, however, is the fact that it can be adapted to accommodate the dynamic insertion and deletion of boxes in optimal $O(\log n)$ time by replacing the kd-tree structure with a divided kd-tree structure.[21] The difference between the divided kd-tree and an ordinary kd-tree is that the divided variant constructs a d-layered tree in which each layer partitions the data structure according to only one of the d coordinates. In three dimensions, for example, the first layer would partition on the x coordinate, the next layer on y, and the last layer on z. The number of levels per layer/coordinate is determined so as to minimize query time complexity. The reason for stratifying the tree into layers for the different coordinates is to allow updates within the different layers to be treated just like updates in ordinary one-dimensional binary trees.

Associating priority fields with the different layers results in a dynamic variant of the priority kd-tree, which is referred to as a *layered box tree*. Note that the i priority fields, for coordinates $1, \ldots, i$, need to be maintained at level i. This data structure has been proven[17] to be maintainable at a cost of $O(\log n)$ time per insertion or deletion and can satisfy box intersection queries $O(n^{1-1/k} \log^{1/k} n + m)$, where m is the number of boxes in S that intersect a given query box. Application of an improved variant[22] of the divided kd-tree improves the query complexity to $O(n^{1-1/k} + m)$, which is optimal.

The priority *kd*-tree is optimal among the class of linear-sized data structures, that is, those using only $O(n)$ storage, but asymptotically better $O(\log^k n + m)$ query complexity is possible if $O(n \log^{k-1} n)$ storage is used.[12,13] However, the extremely complex structure, called a range-segment tree, requires $O(\log^k n)$ update time, and the query performance is $O(\log^k n + m)$. Unfortunately, this query complexity holds in the average case, as well as in the worst case, so it can be expected to provide superior query performance in practice only when *n* is extremely large. For realistic distributions of objects, however, it may never provide better query performance practice. Whether or not that is the case, the range-segment tree is almost never used in practice because the values of $n^{1-1/k}$ and $\log^k n$ are comparable even for *n* as large as 1,000,000, and for data sets of that size, the storage for the range-segment tree is multiplied by a factor of $\log^2(1,000,000) = 400$.

Another data structure, called the DE-tree,[23] has recently been described in the literature. It has suboptimal worst-case theoretical performance but practical performance that seems to exceed all other known data structures in a wide variety of experimental contexts. The DE-tree is of particular interest in target tracking because it can efficiently identify intersections between boxes and rays, which is necessary when tracking with bearings-only (i.e., line-of-sight) sensors.

4.3.1 Applying the Results

The method in which multidimensional search structures are applied in a tracking algorithm can be summarized as follows: tracks are recorded by storing the information—such as current positions, velocities, and accelerations—that a Kalman filter needs to estimate the future position of each candidate target. When a new batch of position reports arrives, the existing tracks are projected forward to the time of the reports. An error ellipsoid is calculated for each track and each report, and a box is constructed around each ellipsoid. The boxes representing the track projections are organized into a multidimensional tree. Each box representing a report becomes the subject of a complete tree search; the result of the search is the set of all track boxes that intersect the given report box. Track–report pairs whose boxes do not intersect are excluded from all further consideration. Next the set of track–report pairs whose boxes do overlap is examined more closely to see whether the inscribed error ellipsoids also overlap. Whenever this calculation indicates a correlation, the track is projected to the time of the new report. Tracks that consistently fail to be associated with any report are eventually deleted; reports that cannot be associated with any existing track initiate new tracks.

The above-mentioned approach for multiple-target tracking ignores a plethora of intricate theoretical and practical details. Unfortunately, such details must eventually be addressed, and the SDI forced a generation of tracking, data fusion, and sensor system researchers to face all of the thorny issues and constraints of a real-world problem of immense scale. The goal was to develop a space-based system to defend against a full-scale missile attack against the United States. Two of the most critical problems were the design and deployment of sensors to detect the launch of missiles at the earliest possible moment in their 20 min midcourse flight, and the design and deployment of weapons systems capable of destroying the detected missiles. Although an automatic tracking facility would clearly be an integral component of any SDI system, it was not generally considered a "high-risk" technology. Tracking, especially of aircraft, had been widely studied for more than 30 years, so the tracking of nonmaneuvering ballistic missiles seemed to be a relatively simple engineering exercise. The principal constraint imposed by SDI was that the tracking be precise enough to predict a missile's future position to within a few meters, so that it could be destroyed by a high-energy laser or a particle-beam weapon.

The high-precision tracking requirement led to the development of highly detailed models of ballistic motion that took into account the effects of atmospheric drag and various gravitational perturbations over the earth. By far the most significant source of error in the tracking process, however, resulted from the limited resolution of existing sensors. This fact reinforced the widely held belief that the main obstacle to effective tracking was the relatively poor quality of sensor reports. The impact of large numbers of targets seemed manageable, just build larger, faster computers. Although many in the research community thought otherwise, the prevailing attitude among funding agencies was that if 100 objects could be tracked in real time, then little difficulty would be involved in building a machine that would be 100 times faster—or simply having 100 machines run in parallel—to handle 10,000 objects.

Among the challenges facing the SDI program, multiple-target tracking seemed far simpler than what would be required to further improve sensor resolution. This belief led to the awarding of contracts to build tracking systems in which the emphasis was placed on high precision at any cost in terms of computational efficiency. These systems did prove valuable for determining bounds on how accurately a single cluster of three to seven missiles could be tracked in an SDI environment, but ultimately pressures mounted to scale up to more realistic numbers. In one case, a tracker that had been tested on five missiles was scaled up to track 100, causing the processing time to increase from a couple of hours to almost a month of nonstop computation for a simulated 20 min scenario. The bulk of the computations were later determined to have involved the correlation step, where reports were compared against hypothesis tracks.

In response to a heightened interest in scaling issues, some researchers began to develop and study prototype systems based on efficient search structures. One of these systems demonstrated that 65–100 missiles could be tracked in real time on a late-1980s personal workstation. These results were based on the assumption that a good-resolution radar report would be received every 5 s for every missile, which is unrealistic in the context of SDI; nevertheless, the demonstration did provide convincing evidence that SDI trackers could be adapted to avoid quadratic scaling. A tracker that had been installed at the SDI National Testbed in Colorado Springs achieved significant performance improvements after a tree-based search structure was installed in its correlation routine; the new algorithm was superior for as few as 40 missiles. Stand-alone tests showed that the search component could process 5,000–10,000 range queries in real time on a modest computer workstation of the time. These results suggested that the problem of correlating vast numbers of tracks and reports had been solved. Unfortunately, a new difficulty was soon discovered.

The academic formulation of the problem adopts the simplifying assumption that all position reports arrive in batches, with all the reports in a batch corresponding to measurements taken at the same instant of all of the targets. A real distributed sensor system would not work this way; reports would arrive in a continuing stream and would be distributed over time. To determine the probability that a given track and report correspond to the same object, the track must be projected to the measurement time of the report. If every track has to be projected to the measurement time of every report, the combinatorial advantages of the tree-search algorithm are lost.

A simple way to avoid the projection of each track to the time of every report is to increase the search radius in the gating algorithm to account for the maximum distance an object could travel during the maximum time difference between any track and report. For example, if the maximum speed of a missile is 10 km/s, and the maximum time difference between any report and track is 5 s, then 50 km would have to be added to each search radius to ensure that no correlations are missed. For boxes used to approximate ellipsoids, this means that each side of the box must be increased by 100 km.

As estimates of what constitutes a realistic SDI scenario became more accurate, members of the tracking community learned that successive reports of a particular target often would be separated by as much as 30–40 s. To account for such large time differences would require boxes so immense that the number of spurious returns would negate the benefits of efficient search. Demands for a sensor configuration that would report on every target at intervals of 5–10 s were considered unreasonable for a variety of practical reasons. The use of sophisticated correlation algorithms seemed to have finally reached its limit. Several heuristic "fixes" were considered, but none solved the problem.

A detailed scaling analysis of the problem ultimately pointed the way to a solution. Simply accumulate sensor reports until the difference between the measurement time of the current report and the earliest report exceeds a threshold. A search structure is then constructed from this set of reports, the tracks are projected to the mean time of the reports, and the correlation process is performed with the maximum time difference being no more than half of the chosen time-difference threshold. The subtle aspect of this deceptively simple approach is the selection of the threshold. If it is too small, every track will be projected to the measurement time of every report. If it is too large, every report will fall within the search volume of every track. A formula has been derived that, with only modest assumptions about the distribution of targets, ensures the optimal trade-off between these two extremes.

Although empirical results confirm that the track file projection approach essentially solves the time difference problem in most practical applications, significant improvements are possible. For example, the fact that different tracks are updated at different times suggests that projecting all of the tracks at the same points in time may be wasteful. An alternative approach might take a track updated with a report at time t_i and construct a search volume sufficiently large to guarantee that the track gates with any report of the target arriving during the subsequent s seconds, where s is a parameter similar to the threshold used for triggering track file projections. This is accomplished by determining the region of space the target could conceivably traverse based on its kinematic state and error covariance. The box circumscribing this search volume can then be maintained in the search structure until time $t_i + s$, at which point it becomes stale and must be replaced with a search volume that is valid from time $t_i + s$ to time $t_i + 2s$. However, if before becoming stale it is updated with a report at time t_j, $t_i < t_j < t_i + s$, then it must be replaced with a search volume that is valid from time t_j to time $t_j + s$.

The benefit of the enhanced approach is that each track is projected only at the times when it is updated or when all extended period has passed without an update (which could possibly signal the need to delete the track). To apply the approach, however, two conditions must be satisfied. First, there must be a mechanism for identifying when a track volume has become stale and needs to be recomputed. It is, of course, not possible to examine every track upon the receipt of each report because the scaling of the algorithm would be undermined. The solution is to maintain a priority queue of the times at which the different track volumes will become invalid. A priority queue is a data structure that can be updated efficiently and supports the retrieval of the minimum of n values in $O(\log n)$ time. At the time a report is received, the priority queue is queried to determine which, if any, of the track volumes have become stale. New search volumes are constructed for the identified tracks, and the times at which they will become invalid are updated in the priority queue.

The second condition that must be satisfied for the enhanced approach is a capability to incrementally update the search structure as tracks are added, updated, recomputed, or deleted. The need for such a capability was hinted at in the discussion of dynamic search structures. Because the layered box tree supports insertions and deletions in $O(\log n)$ time, the update of a track's search volume can be efficiently accommodated. The track's associated box is deleted from the tree, an updated box is computed, and then the result is inserted back into the tree. In summary, the cost for processing each report involves updates of the search structure and the priority queue, at $O(\log n)$ cost, plus the cost of determining the set of tracks with which the report could be feasibly associated.

4.4   Conclusion

The correlation of reports with tracks numbering in thousands can now be performed in real time on a personal computer. More research on large-scale correlation is needed, but work has already begun on implementing efficient correlation modules that can be incorporated into existing tracking systems. Ironically, by hiding the intricate details and complexities of the correlation process, these modules give the appearance that multiple-target tracking involves little more than the concurrent processing of several single-target problems. Thus, a paradigm with deep historical roots in the field of target tracking is at least partially preserved.

Note that the techniques described in this chapter are applicable only to a very restricted class of tracking problems. Other problems, such as the tracking of military forces, demand more sophisticated approaches. Not only does the mean position of a military force change, its shape also changes. Moreover, reports of its position are really only reports of the positions of its parts, and various parts may be moving in different directions at any given instant. Filtering out the local deviations in motion to determine the net motion of the whole is beyond the capabilities of a simple Kalman filter. Other difficult tracking problems include the tracking of weather phenomena and soil erosion. The history of multiple-target tracking suggests that, in addition to new mathematical techniques, new algorithmic techniques will certainly be required for any practical solution to these problems.

Acknowledgment

The author gratefully acknowledges support from the Naval Research Laboratory, Washington, DC.

References

1. Uhlmann, J.K., Algorithms for multiple-target tracking, *American Scientist*, 80(2): 128–141, 1992.
2. Kalman, R.E., A new approach to linear filtering and prediction problems, *ASME, Basic Engineering*, 82: 34–45, 1960.
3. Blackman, S.S., *Multiple-Target Tracking with Radar Applications*, Artech House, Norwood, MA, 1986.
4. Bar Shalom, Y. and Fortmann, T.E., *Tracking and Data Association*, Academic Press, New York, NY, 1988.
5. Blackman, S. and Popoli R., *Design and Analysis of Modern Tracking Systems*, Artech House, Norwood, MA, 1999.

6. Bar Shalom, Y. and Li, X.R., *Multitarget-Multisensor Tracking: Principles and Techniques*, YBS Press, New York, NY, 1995.

7. Uhlmann, J.K., Zuniga, M.R., and Picone, J.M., Efficient approaches for report/cluster correlation in multi-target tracking systems, *NRL Report 9281*, US Government, Washington, DC, 1990.

8. Bentley, J., Multidimensional binary search trees for associative searching, *Communications of the ACM*, 18(9): 509–517, 1975.

9. Yianilos, P.N., Data structures and algorithms for nearest neighbor search in general metric spaces, in *SODA*, Austin, TX, pp. 311–321, 1993.

10. Uhlmann, J.K., Implementing metric trees to satisfy general proximity/similarity queries, *NRL Code 5570 Technical Report 9192*, 1992.

11. Lee, D.T. and Wong, C.K., Worse-case analysis for region and partial region searches in multidimensional binary search trees and quad trees, *Acta Informatica*, 9(1): 23–29, 1977.

12. Preparata, F. and Shamos, M., *Computational Geometry*, Springer-Verlag, New York, NY, 1985.

13. Mehlhorn, K., *Multidimensional Searching and Computational Geometry*, Vol. 3, Springer-Verlag, Berlin, 1984.

14. Uhlmann, J.K. and Zuniga, M.R., Results of an efficient gating algorithm for large-scale tracking scenarios, *Naval Research Reviews*, 43(1): 24–29, 1991.

15. Zuniga, M.R., Picone, J.M., and Uhlmann, J.K., Efficient algorithm for unproved gating combinatorics in multiple-target tracking, submitted to *IEEE Transactions on Aerospace and Electronic Systems*, 1990.

16. Uhlmann, J.K., Adaptive partitioning strategies for ternary tree structures, *Pattern Recognition Letters*, 12(9): 537–541, 1991.

17. Boroujerdi, A. and Uhlmann, J.K., Large-scale intersection detection using layered box trees, *AIT-DSS Report*, US Government, 1998.

18. McCreight, E.M., Priority search trees, *SIAM Journal of Computing*, 14(2): 257–276, 1985.

19. Wood, D., *Data, Structures, Algorithms, and Performance*, Addison-Wesley Longman, Boston, MA, 1993.

20. Uhlmann, J.K., Dynamic map building and localization for autonomous vehicles, *Engineering Sciences Report*, Oxford University, London, UK, 1994.

21. van Kreveld, M. and Overmars, M., Divided *k-d* trees, *Algorithmica*, 6: 840–858, 1991.

22. Kanth, K.V.R. and Singh, A., Optimal dynamic range searching in non-replicating index structures, in *Lecture Notes in Computer Science*, Vol. 1540, pp. 257–276, Springer, Berlin, 1999.

23. Zuniga, M.R. and Uhlmann, J.K., Ray queries with wide object isolation with DE-trees, *Journal of Graphics Tools*, 11(3): 19–37, 2006.

24. Ramasubramanian, V. and Paliwal, K., An efficient approximation-elimination algorithm for fast nearest-neighbour search on a spherical distance coordinate formulation, *Pattern Recognition Letters*, 13(7): 471–480, 1992.

25. Vidal, E., An algorithm for finding nearest neighbours in (approximately) constant average time complexity, *Pattern Recognition Letters*, 4(3): 145–157, 1986.

26. Vidal, E., Rulot, H., Casacuberta, F., and Benedi, J.M., On the use of a metric-space search algorithm (AESA) for fast DTW-based recognition of isolated words, *IEEE Transactions on Acoustics, Speech, and Signal Processing*, 36(5): 651–660, 1988.

27. Uhlmann, J.K., Metric trees, *Applied Mathematics Letters*, 4(5): 61–62, 1991.

28. Uhlmann, J.K., Satisfying general proximity/similarity queries with metric trees, *Information Processing Letters*, 40: 175–170, 1991.

29. Collins, J.B. and Uhlmann, J.K., Efficient gating in data association for multivariate Gaussian distributions, *IEEE Transactions on Aerospace and Electronic Systems*, 28(3): 909–916, 1990.

5

Principles and Practice of Image and Spatial Data Fusion*

Ed Waltz and Tim Waltz

CONTENTS

5.1 Introduction

The joint use of imagery and spatial data from different imaging, mapping, or other spatial sensors has the potential to provide significant performance improvements over single-sensor detection, classification, and situation assessment functions. The terms *imagery*

* Adapted from The Principles and Practice of Image and Spatial Data Fusion, in *Proceedings of the 8th National Data Fusion Conference*, Dallas, Texas, March 15–17, pp. 257–278, 1995.

fusion and *spatial data fusion* have been applied to describe a variety of combining operations for a wide range of image enhancement and understanding applications. Surveillance, robotic machine vision, and automatic target cueing (ATC) are among the application
areas that have explored the potential benefits of multiple-sensor imagery. This chapter
provides a framework for defining and describing the functions of image data fusion in
the context of the Joint Directors of Laboratories (JDL) data fusion model. The chapter also
describes representative methods and applications.

   Sensor fusion and *data fusion* have become the *de facto* terms to describe the general abductive or deductive combination processes by which diverse sets of related data are joined
or merged to produce a product that is greater than the individual parts. A range of mathematical operators have been applied to perform this process for a wide range of applications. Two areas that have received increasing research attention over the past decade
are the processing of imagery (two-dimensional [2D] information) and spatial data (three-
dimensional [3D] representations of real-world surfaces and objects that are imaged).
These processes combine multiple data views into a composite set that incorporates the
best attributes of all contributors. The most common product is a spatial (3D) model, or
virtual world, which represents the best estimate of the real world as derived from all
sensors.

5.2 Motivations for Combining Image and Spatial Data

A diverse range of applications have employed image data fusion to improve imaging
and automatic detection or classification performance over that of single imaging sensors. Table 5.1 summarizes representative and recent research and development in six key
application areas.

   Satellite and airborne imagery used for military intelligence, photogrammetric, earth
resources, and environmental assessments can be enhanced by combining registered data
from different sensors to refine the spatial or spectral resolution of a composite image
product. Registered imagery from different passes (multitemporal) and different sensors (multispectral and multiresolution) can be combined to produce composite imagery
with spectral and spatial characteristics equal to, or better than, that of the individual
contributors.

   Composite SPOT™ and LANDSAT™ satellite imagery and 3D terrain relief composites
of military regions demonstrate current military applications of such data for mission planning purposes.[1-3] The Joint National Intelligence Development Staff (JNIDS) pioneered the
development of workstation-based systems to combine a variety of image and nonimage
sources for intelligence analysts[4] who perform:

- *Registration.* Spatial alignment of overlapping images and maps to a common coordinate system
- *Mosaicking.* Registration of nonoverlapping, adjacent image sections to create a
  composite of a larger area
- *3D mensuration-estimation.* Calibrated measurement of the spatial dimensions of
  objects within in-image data

TABLE 5.1

Representative Range of Activities Applying Spatial and Imagery Fusion

	Activities	Sponsors
Satellite/airborne imaging		
Multiresolution image sharpening	Multiple algorithms, tools in commercial packages	United States, commercial vendors
Terrain visualization	Battlefield visualization, mission planning	Army, Air Force
Planetary visualization-exploration	Planetary mapping missions	NASA
Mapping, charting, and geodesy (MC & G)		
Geographic information system (GIS) generation from multiple sources	Terrain feature extraction, rapid map generation	DARPA, Army, Air Force
Earth environment information system	Earth observing system, data integration system	NASA
Military automatic target recognition (ATR)		
Battlefield surveillance	Various MMW/LADAR/FLIR	Army
Battlefield seekers	Millimeter wave (MMW)/ forward-looking IR (FLIR)	Army, Air Force
IMINT correlation	Single Intel IMINT correlation	DARPA
IMINT-SIGINT/MTI correlation	Dynamic database	DARPA
Industrial robotics		
3D multisensor inspection	Product line inspection	Commercial
Nondestructive inspection	Image fusion analysis	Air Force, commercial
Medical imaging		
Human body visualization, diagnosis	Tomography, magnetic resonance imaging, 3D fusion	Various research and development (R&D) hospitals

Similar image functions have been incorporated into a variety of image processing systems, from tactical image systems such as the premier Joint Service Image Processing System (JSIPS) to UNIX- and PC-based commercial image processing systems. Military services and the National Geospatial Intelligence Agency (NGA) are performing cross intelligence (i.e., IMINT and other intelligence sources) data fusion research to link signals and human reports to spatial data.[5]

When the fusion process extends beyond imagery to include other spatial data sets, such as digital terrain data, demographic data, and complete geographic information system (GIS) data layers, numerous mapping applications may benefit. Military intelligence preparation of the battlefield (IPB) functions (e.g., area delimitation and transportation network identification), as well as wide area terrain database generation (e.g., precision GIS mapping), are complex mapping problems that require fusion to automate processes that are largely manual. One area of ambitious research in spatial data fusion is the U.S. Army Topographic Engineering Center's (TEC) efforts to develop automatic terrain feature generation techniques based on a wide range of source data, including imagery, map data, and remotely sensed terrain data.[6] On the broadest scale, NGA's Global Geospatial Information and Services (GGIS) vision includes spatial data fusion as a core functional element.[7] NGA's Mapping, Charting, and Geodesy Utility Software package (MUSE), for example, combines vector and raster data to display base maps with overlays of a variety of data to support geographic analysis and mission planning.

Real-time ATC/ATR (ATR—automatic target recognition) for military applications has turned to multiple-sensor solutions to expand spectral diversity and target feature dimensionality, seeking to achieve high probabilities of correct detection or identification at acceptable false alarm rates. Forward-looking infrared (FLIR), imaging millimeter wave (MMW), and light amplification for detection and ranging (LADAR) sensors are the most promising suite capable of providing the diversity needed for reliable discrimination in battlefield applications. In addition, some applications seek to combine the real-time imagery to present an enhanced image to the human operator for driving, control, and warning, as well as manual target recognition.

Industrial robotic applications for fusion include the use of 3D imaging and tactile sensors to provide sufficient image understanding to permit robotic manipulation of objects. These applications emphasize automatic object position understanding rather than recognition (e.g., the target recognition) that is, by nature, noncooperative.[8]

Transportation applications combine MMW and electrooptical imaging sensors to provide collision avoidance warning by sensing vehicles whose relative rates and locations pose a collision threat.

Medical applications fuse information from a variety of imaging sensors to provide a complete 3D model or enhanced 2D image of the human body for diagnostic purposes. The United Medical and Dental Schools of Guy's and St. Thomas' Hospital (London, U.K.) have demonstrated methods for registering and combining magnetic resonance (MR), positron emission tomography (PET), and computer tomography (CT) into composites to aid surgery.[9]

5.3 Defining Image and Spatial Data Fusion

In this chapter, image and spatial data fusion are distinguished as subsets of the more general data fusion problem that is typically aimed at associating and combining 3D data about sparse point objects located in space. Targets on a battlefield, aircraft in airspace, ships on the ocean surface, or submarines in the 3D ocean volume are common examples of targets represented as point objects in a 3D space model.

Image data fusion, however, is involved with associating and combining complete, spatially filled sets of data in 2D (images) or 3D (terrain or high-resolution spatial representations of real objects). Herein lies the distinction: image and spatial data fusion requires data representing every point on a surface or in space to be fused, rather than selected points of interest.

The more general problem is described in detail in the introductory texts by Waltz and Llinas[10] and Hall,[11] while the progress in image and spatial data fusion is reported over a wide range of the technical literature, as cited in this chapter.

The taxonomy in Figure 5.1 distinguishes the data properties and objectives that distinguish four categories of fusion applications.

In all the image and spatial applications cited in the text above, the common thread of the fusion function is its emphasis on the following distinguishing functions:

- Registration involves spatial and temporal alignment of physical items within imagery or spatial data sets and is a prerequisite for further operations. It can

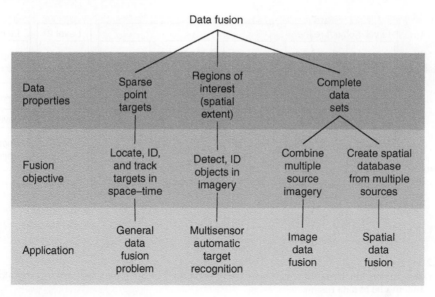

FIGURE 5.1
Data fusion application taxonomy.

occur at the raw image level (i.e., any pixel in one image may be referenced with known accuracy to a pixel or pixels in another image, or to a coordinate in a map) or at higher levels, relating objects rather than individual pixels. Of importance to every approach to combining spatial data is the accuracy with which the data layers have been spatially aligned relative to each other or to a common coordinate system (e.g., geolocation or geocoding of earth imagery to an earth projection). Registration can be performed by traditional internal image-to-image correlation techniques (when the images are from sensors with similar phenomena and are highly correlated)[12] or by external techniques.[13] External methods apply in-image control knowledge or as-sensed information that permits accurate modeling and estimation of the true location of each pixel in 2D or 3D space.

- The combination function operates on multiple, registered layers of data to derive composite products using mathematical operators to perform integration; mosaicking; spatial or spectral refinement; spatial, spectral, or temporal (change) detection; or classification.

- Reasoning is the process by which intelligent, often iterative search operations are performed between the layers of data to assess the meaning of the entire scene at the highest level of abstraction and of individual items, events, and data contained in the layers.

The image and spatial data fusion functions can be placed in the JDL data fusion model context to describe the architecture of a system that employs imagery data from multiple sensors and spatial data (e.g., maps and solid models) to perform detection, classification, and assessment of the meaning of information contained in the scenery of interest.

Figure 5.2 compares the JDL general model[14] with a specific multisensor ATR image data fusion functional flow to show how the more abstract model can be related to a specific

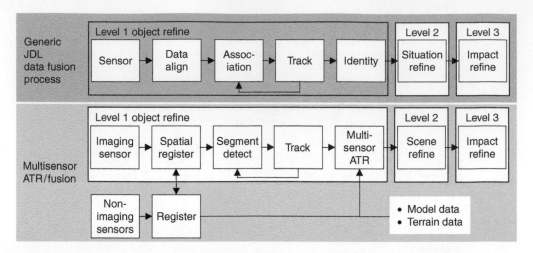

FIGURE 5.2
Image of a data fusion functional flow can be directly compared to the Joint Directors of Laboratories data fusion subpanel model of data fusion.

imagery fusion application. The level 1 processing steps can be directly related to image counterparts as follows:

- *Alignment.* The alignment of data into a common time, space, and spectral reference frame involves spatial transformations to warp image data to a common coordinate system (e.g., projection to an earth reference model or 3D space). At this point, nonimaging data that can be spatially referenced (perhaps not to a point, but often to a region with a specified uncertainty) can then be associated with the image data.

- *Association.* New data can be correlated with the previous data to detect and segment (select) targets on the basis of motion (temporal change) or behavior (spatial change). In time-sequenced data sets, target objects at time t are associated with target objects at time $t - 1$ to discriminate newly appearing targets, moved targets, and disappearing targets.

- *Tracking.* When objects are tracked in dynamic imagery, the dynamics of target motion are modeled and used to predict the future location of targets (at time $t + 1$) for comparison with new sensor observations.

- *Identification.* The data for segmented targets are combined from multiple sensors (at any one of several levels) to provide an assignment of the target to one or more of several target classes.

Level 2 and 3 processing deals with the aggregate of targets in the scene and other characteristics of the scene to derive an assessment of the meaning of data in the scene or spatial data set.

In the following sections, the primary image and spatial data fusion application areas are described to demonstrate the basic principles of fusion and the state of the practice in each area (Figure 5.3).

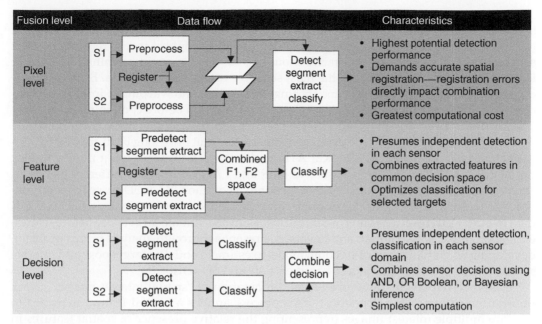

FIGURE 5.3
Three basic levels of fusion are provided to the multisensor automatic target recognition designer as the most logical alternative points in the data chain for combining data.

5.4 Three Classic Levels of Combination for Multisensor Automatic Target Recognition Data Fusion

Since the late 1970s, the ATR literature has adopted three levels of image data fusion as the basic design alternatives offered to the system designer. The terminology was adopted to describe the point in the traditional ATR processing chain at which registration and combination of different sensor data occurred. These functions can occur at multiple levels, as described later in this chapter. First, a brief overview of the basic alternatives and representative research and development results is presented. (Broad overviews of the developments in ATR in general, with specific comments on data fusion, are available in other literature.[15-17])

5.4.1 Pixel-Level Fusion

At the lowest level, *pixel-level fusion* uses the registered pixel data from all image sets to perform detection and discrimination functions. This level has the potential to achieve the greatest signal detection performance (if registration errors can be contained) at the highest computational expense. At this level, detection decisions (pertaining to the presence or absence of a target object) are based on the information from all sensors by evaluating the spatial and spectral data from all layers of the registered image data. A subset of this level of fusion is *segment-level fusion*, in which basic detection decisions are made independently in each sensor domain, but the segmentation of image regions is performed by evaluation of the registered data layers.

Fusion at the pixel level involves accurate registration of the different sensor images before applying a combination operator to each set of registered pixels (which correspond to associated measurements in each sensor domain at the highest spatial resolution of the sensors). Spatial registration accuracies should be subpixel to avoid combination of unrelated data, making this approach the most sensitive to registration errors. Because image data may not be sampled at the same spacing, resampling and warping of images are generally required to achieve the necessary level of registration before combining pixel data.

In the most direct 2D image applications of this approach, coregistered pixel data may be classified on a pixel-by-pixel basis using approaches that have long been applied to multispectral data classification.[18] Typical ATR applications, however, pose a more complex problem when dissimilar sensors, such as FLIR and LADAR, image in different planes. In such cases, the sensor data must be projected into a common 2D or 3D space for combination. Gonzalez and Williams, for example, have described a process for using 3D LADAR data to infer FLIR pixel locations in three dimensions to estimate target pose before feature extraction (FE).[19] Schwickerath and Beveridge present a thorough analysis of this problem, developing an eight-degree-of-freedom model to estimate both the target pose and relative sensor registration (coregistration) based on a 2D and 3D sensor.[20]

Delanoy et al. demonstrated pixel-level combination of "spatial interest images" using Boolean and fuzzy logic operators.[21] This process applies a spatial feature extractor to develop multiple interest images (representing the relative presence of spatial features in each pixel), before combining the interest images into a single detection image. Similarly, Hamilton and Kipp describe a probe-based technique that uses spatial templates to transform the direct image into probed images that enhance target features for comparison with reference templates.[22,23] Using a limited set of television and FLIR imagery, Duane compared pixel-level and feature-level fusion to quantify the relative improvement attributable to the pixel-level approach with well-registered imagery sets.[24]

5.4.2 Feature-Level Fusion

At the intermediate level, feature-level fusion combines the features of objects that are detected and segmented in the individual sensor domains. This level presumes independent detectability of objects in all the sensor domains. The features for each object are independently extracted in each domain; these features create a common feature space for object classification.

Such feature-level fusion reduces the demand on registration, allowing each sensor channel to segment the target region and extract features without regard to the other sensor's choice of target boundary. The features are merged into a common decision space only after a spatial association is made to determine that the features were extracted from objects whose centroids were spatially associated.

During the early 1990s, the Army evaluated a wide range of feature-level fusion algorithms for combining FLIR, MMW, and LADAR data for detecting battlefield targets under the multisensor feature-level fusion (MSFLF) program of the OSD multisensor-aided targeting initiative. Early results demonstrated marginal gains over single-sensor performance and reinforced the importance of careful selection of complementary features to specifically reduce single-sensor ambiguities.[25]

At the feature level of fusion, researchers have developed model-based (or model-driven) alternatives to the traditional statistical methods, which are inherently data driven. Model-based approaches maintain target and sensing models that predict all possible views (and target configurations) for comparison with extracted features rather than using a more limited set of real signature data for comparison.[26] The application of model-based approaches

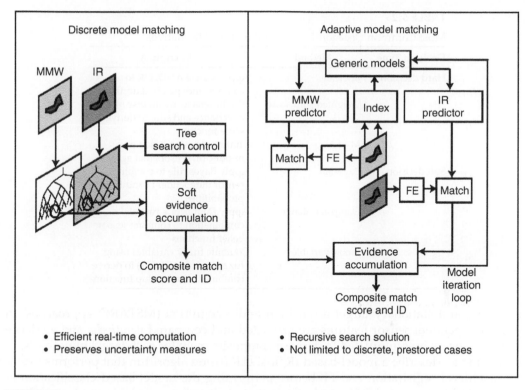

FIGURE 5.4

Two model-based sensor alternatives demonstrate the use of a prestored hierarchy of model-based templates or an online iterative model that predicts features based on estimated target pose.

to multiple-sensor ATR offers several alternative implementations, two of which are described in Figure 5.4. The adaptive model matching approach performs FE and comparison (match) with predicted features for the estimated target pose. The process iteratively searches to find the best model match for the extracted features.

5.4.2.1 Discrete Model Matching Approach

A multisensor model-based matching approach described by Hamilton and Kipp[27] develops a relational tree structure (hierarchy) of 2D silhouette templates. These templates capture the spatial structure of the most basic all-aspect target "blob" (at the top or root node), down to individual target hypotheses at specific poses and configurations. This predefined search tree is developed on the basis of model data for each sensor, and the ATR process compares segmented data to the tree, computing a composite score at each node to determine the path to the most likely hypotheses. At each node, the evidence is accumulated by applying an operator (e.g., weighted sum, Bayesian combination, etc.) to combine the score for each sensor domain.

5.4.2.2 Adaptive Model Matching Approach

Rather than using prestored templates, this approach implements the sensor or the target modeling capability within the ATR algorithm to dynamically predict features for direct comparison. Figure 5.4 illustrates a two-sensor extension of the one-sensor, model-based ATR paradigm (e.g., automatic radar air-to-ground target acquisition program (ARAGTAP)[28]

TABLE 5.2

Most Common Decision-Level Combination Alternatives

Decision Type	Method	Description
Hard decision	Boolean	Apply logical AND, OR to combine independent decisions
	Weighted sum score	Weight sensors by inverse of covariance and sum to derive score function
	M-of-N	Confirm decision based on m-out-of-n sensors that agree
Soft decision	Bayesian	Apply Bayes rule to combine sensor independent conditional probabilities
	Dempster-Shafer	Apply Dempster's rule of combination to combine sensor belief functions
	Fuzzy variable	Combine fuzzy variables using fuzzy logic (AND, OR) to derive combined membership function

or moving and stationary target acquisition and recognition (MSTAR)[29] approaches) in which independent sensor features are predicted and compared *iteratively*, and evidence from the sensors is accumulated to derive a composite score for each target hypothesis.

Larson et al. describe a model-based IR/LADAR fusion algorithm that performs extensive pixel-level registration and FE before performing the model-based classification at the extracted feature level.[30] Similarly, Corbett et al. describe a model-based feature-level classifier that uses IR and MMW models to predict features for military vehicles.[31] Both of these follow the adaptive generation approach.

5.4.3   Decision-Level Fusion

Fusion at the decision level (also called postdecision or postdetection fusion) combines the decisions of independent sensor detection or classification paths by Boolean (AND, OR) operators or by a heuristic score (e.g., M-of-N, maximum vote, or weighted sum). Two methods of making classification decisions exist: hard decisions (single, optimum choice) and soft decisions, in which decision uncertainty in each sensor chain is maintained and combined with a composite measure of uncertainty (Table 5.2).

The relative performance of alternative combination rules and independent sensor thresholds can be optimally selected using distribution data for the features used by each sensor.[32] In decision-level fusion, each path must independently detect the presence of a candidate target and perform a classification on the candidate. These detections and classifications (the sensor decisions) are combined into a fused decision. This approach inherently assumes that the signals and signatures in each independent sensor chain are sufficient to perform independent detection before the sensor decisions are combined. This approach is much less sensitive to spatial misregistration than all others and permits accurate association of detected targets to occur with registration errors over an order of magnitude larger than for pixel-level fusion. Lee and Vleet have shown procedures for estimating the registration error between sensors to minimize the mean square registration error and optimize the association of objects in dissimilar images for decision-level fusion.[33]

Decision-level fusion of MMW and IR sensors has long been considered a prime candidate for achieving the level of detection performance required for autonomous precision-guided munitions.[34] Results of an independent two-sensor (MMW and IR) analysis on military targets demonstrated the relative improvement of two-sensor decision-level fusion over either independent sensor.[35–37] A summary of ATR comparison methods was compiled by Diehl et al.[38] These studies demonstrated the critical sensitivity of performance gains to the relative performance of each contributing sensor and the independence of the sensed phenomena.

5.4.4 Multiple-Level Fusion

In addition to the three classic levels of fusion, other alternatives or combinations have been advanced. At a level even higher than the decision level, some researchers have defined scene-level methods in which target detections from a low-resolution sensor are used to cue a search-and-confirm action by a higher-resolution sensor. Menon and Kolodzy described such a system that uses FLIR detections to cue the analysis of high-spatial resolution laser radar data using a nearest-neighbor neural network classifier.[39] Maren describes a scene structure method that combines information from hierarchical structures developed independently by each sensor by decomposing the scene into element representations.[40] Others have developed hybrid, multilevel techniques that partition the detection problem to a high level (e.g., decision level) and the classification to a lower level. Aboutalib et al. described a hybrid algorithm that performs decision-level combination for detection (with detection threshold feedback) and feature-level classification for air target identification in IR and TV imagery.[41]

Other researchers have proposed multilevel ATR architectures, which perform fusion at all levels, carrying out an appropriate degree of combination at each level based on the ability of the combined information to contribute to an overall fusion objective. Chu and Aggarwal describe such a system that integrates pixel-level to scene-level algorithms.[42] Eggleston has long promoted such a knowledge-based ATR approach that combines data at three levels, using many partially redundant combination stages to reduce the errors of any single unreliable rule.[43,44] The three levels in this approach are:

1. *Low level.* Pixel-level combinations are performed when image enhancement can aid higher-level combinations. The higher levels adaptively control this fine grain combination.

2. *Intermediate symbolic level.* Symbolic representations (tokens) of attributes or features for segmented regions (image events) are combined using a symbolic level of description.

3. *High level.* The scene or context level of information is evaluated to determine the meaning of the overall scene, by considering all intermediate-level representations to derive a situation assessment. For example, this level may determine that a scene contains a brigade-sized military unit forming for attack. The derived situation can be used to adapt lower levels of processing to refine the high-level hypotheses.

Bowman and DeYoung described an architecture that uses neural networks at all levels of the conventional ATR processing chain to achieve pixel-level performances of up to .99 probability of correct identification for battlefield targets using pixel-level neural network fusion of UV, visible, and MMW imagery.[45]

Pixel-, feature-, and decision-level fusion designs have focused on combining imagery for the purposes of detecting and classifying specific targets. The emphasis is on limiting processing by combining only the most likely regions of target data content at the minimum necessary level to achieve the desired detection or classification performance. This differs significantly from the next category of image fusion designs, in which all data must be combined to form a new spatial data product that contains the best composite properties of all contributing sources of information.

5.5   Image Data Fusion for Enhancement of Imagery Data

Both still and moving image data can be combined from multiple sources to enhance desired features, combine multiresolution or differing sensor look geometries, mosaic multiple views, and reduce uncorrelated noise.

5.5.1   Multiresolution Imagery

One area of enhancement has been in the application of "band sharpening" or "multiresolution image fusion" algorithms to combine differing resolution satellite imagery. The result is a composite product that enhances the spatial boundaries in lower-resolution multispectral data using higher-resolution panchromatic or synthetic aperture radar (SAR) data.

Veridian-ERIM International has applied its Sparkle algorithm to the band sharpening problem, demonstrating the enhancement of lower-resolution SPOT multispectral imagery (20 m ground sample distance (GSD)) with higher-resolution airborne SAR (3 m GSD) and panchromatic photography (1 m) to sharpen the multispectral data. Radar backscatter features are overlayed on the composite to reveal important characteristics of the ground features and materials. The composite image preserves the spatial resolution of the panchromatic data, the spectral content of the multispectral layers, and the radar reflectivity of the SAR.

Vrabel has reported the relative performance of a variety of band sharpening algorithms, concluding that Veridian ERIM International's Sparkle algorithm and a color normalization (CN) technique provided the greatest GSD enhancement and overall utility.[46] Additional comparisons and applications of band sharpening techniques have been published in the literature.[47–50]

Imagery can also be mosaicked by combining overlapping images into a common block, using classical photogrammetric techniques (bundle adjustment) that use absolute ground control points and tie points (common points in overlapped regions) to derive mapping polynomials. The data may then be forward resampled from the input images to the output projection or backward resampled by projecting the location of each output pixel onto each source image to extract pixels for resampling.[51] The latter approach permits spatial deconvolution functions to be applied in the resampling process. Radiometric feathering of the data in transition regions may also be necessary to provide a gradual transition after overall balancing of the radiometric dynamic range of the mosaicked image is performed.[52] Such mosaicking fusion processes have also been applied to 3D data to create composite digital elevation models (DEMs) of terrain.[53]

5.5.2   Dynamic Imagery

In some applications, the goal is to combine different types of real-time video imagery to provide the clearest possible composite video image for a human operator. The David Sarnoff Research Center has applied wavelet encoding methods to selectively combine IR and visible video data into a composite video image that preserves the most desired characteristics (e.g., edges, lines, and boundaries) from each data set.[54] The Center later extended the technique to combine multitemporal and moving images into composite mosaic scenes that preserve the "best" data to create a current scene at the best possible resolution at any point in the scene.[55,56]

5.5.3 Three-Dimensional Imagery

3D perspectives of the earth's surface are a special class of image data fusion products that have been developed by draping orthorectified images of the earth's surface over digital terrain models. The 3D model can be viewed from arbitrary static perspectives, or a dynamic fly-through, which provides a visualization of the area for mission planners, pilots, or land planners.

Off-nadir regions of aerial or spaceborne imagery include a horizontal displacement error that is a function of the elevation of the terrain. A DEM is used to correct for these displacements to accurately overlay each image pixel on the corresponding post (i.e., terrain grid coordinate). Photogrammetric orthorectification functions[57] include the following steps to combine the data:

- *DEM preparation.* The DEM is transformed to the desired map projection for the final composite product.
- *Transform derivation.* Platform, sensor, and the DEM are used to derive mapping polynomials that will remove the horizontal displacements caused by terrain relief, placing each input image pixel at the proper location on the DEM grid.
- *Resampling.* The input imagery is resampled into the desired output map grid.
- *Output file creation.* The resampled image data (x, y, and pixel values) and DEM (x, y, and z) are merged into a file with other georeferenced data, if available.
- *Output product creation.* 2D image maps may be created with map grid lines, or 3D visualization perspectives can be created for viewing the terrain data from arbitrary viewing angles.

The basic functions necessary to perform registration and combination are provided in an increasing number of commercial image processing software packages (Table 5.3), permitting users to fuse static image data for a variety of applications.

TABLE 5.3

Basic Image Data Fusion Functions Provided in Several Commercial Image Processing Software Packages

	Function	Description
Registration	Sensor-platform modeling	Model sensor-imaging geometry; derive correction transforms (e.g., polynomials) from collection parameters (e.g., ephemeris, pointing, and earth model)
	Ground control point (GCP) calibration	Locate known GCPs and derive correction transforms
	Warp to polynomial	Spatially transform (warp) imagery to register pixels to regular grid or to a digital terrain model
	Orthorectify to digital terrain model	
	Resample imagery	Resample warped imagery to create fixed pixel-sized image
Combination	Mosaic imagery	Register adjacent and overlapped imagery; resample to common pixel grid
	Edge feathering	Combine overlapping imagery data to create smooth (feathered) magnitude transitions between two image components
	Band sharpening	Enhance spatial boundaries (high-frequency content) in lower-resolution band data using higher-resolution registered imagery data in a different band

5.6   Spatial Data Fusion Applications

Robotic and transportation applications include a wide range of applications similar to military applications. Robotics applications include relatively short-range, high-resolution imaging of cooperative target objects (e.g., an assembly component to be picked up and accurately placed) with the primary objectives of position determination and inspection. Transportation applications include longer-range sensing of vehicles for highway control and multiple-sensor situation awareness within a vehicle to provide semiautonomous navigation, collision avoidance, and control.

The results of research in these areas are chronicled in a variety of sources, beginning with the 1987 Workshop on Spatial Reasoning and MultiSensor Fusion,[58] and many subsequent SPIE conferences.[59-63]

5.6.1   Spatial Data Fusion: Combining Image and Nonimage Data to Create Spatial Information Systems

One of the most sophisticated image fusion applications combines diverse sets of imagery (2D), spatially referenced nonimage data sets, and 3D spatial data sets into a composite spatial data information system. The most active area of research and development in this category of fusion problems is the development of GIS by combining earth imagery, maps, demographic and infrastructure, or facilities mapping (geospatial) data into a common spatially referenced database.

Applications for such capabilities exist in three areas. In civil government, the need for land and resource management has prompted intense interest in establishing GISs at all levels of government. The U.S. Federal Geographic Data Committee is tasked with the development of a National Spatial Data Infrastructure (NSDI), which establishes standards for organizing the vast amount of geospatial data currently available at the national level and coordinating the integration of future data.[64]

Commercial applications for geospatial data include land management, resources exploration, civil engineering, transportation network management, and automated mapping or facilities management for utilities.

The military application of such spatial databases is the IPB,[65] which consists of developing a spatial database containing all terrain, transportation, groundcover, man-made structures, and other features available for use in real-time situation assessment for command and control. The Defense Advanced Research Projects Agency (DARPA) Terrain Feature Generator is one example of a major spatial database and fusion function defined to automate the functions of IPB and geospatial database creation from diverse sensor sources and maps.[66]

Realization of efficient, affordable systems capable of accommodating the volume of spatial data required for large regions and performing reasoning that produces accurate and insightful information depends on two critical technology areas:

1. *Spatial data structure.* Efficient, linked data structures are required to handle the wide variety of vector, raster, and nonspatial data sources. Hundreds of point, lineal, and areal features must be accommodated. Data volumes are measured in terabytes and short access times are demanded for even broad searches.

2. *Spatial reasoning.* The ability to reason in the context of dynamically changing spatial data is required to assess the meaning of the data. The reasoning process

must perform the following kinds of operations to make assessments about the data:

a. Spatial measurements (e.g., geometric, topological, proximity, and statistics)

b. Spatial modeling

c. Spatial combination and inference operations, in uncertainty

d. Spatial aggregation of related entities

e. Multivariate spatial queries

Antony surveyed the alternatives for representing spatial and spatially referenced semantic knowledge[67] and published the first comprehensive data fusion text[68] that specifically focused on spatial reasoning for combining spatial data.

5.6.2 Mapping, Charting, and Geodesy Applications

The use of remotely sensed image data to create image maps and generate GIS base maps has long been recognized as a means of automating map generation and updating to achieve currency as well as accuracy.[69–71] The following features characterize integrated geospatial systems:

- *Currency.* Remote sensing inputs enable continuous update with change detection and monitoring of the information in the database.

- *Integration.* Spatial data in a variety of formats (e.g., raster and vector data) is integrated with metadata and other spatially referenced data, such as text, numerical, tabular, and hypertext formats. Multiresolution and multiscale spatial data coexist, are linked, and share a common reference (i.e., map projection).

- *Access.* The database permits spatial query access for multiple user disciplines. All data is traceable, and the data accuracy, uncertainty, and entry time are annotated.

- *Display.* Spatial visualization and query tools provide maximum human insight into the data content using display overlays and 3D capability.

Ambitious examples of such geospatial systems include the DARPA Terrain Feature Generator, the European ESPRIT II MultiSource Image Processing System (MuSIP),[72,73] and NASA's Earth Observing Systems Data and Information System (EOSDIS).[74]

Figure 5.5 illustrates the most basic functional flow of such a system, partitioning the data integration (i.e., database generation) function from the scene assessment function. The integration function spatially registers and links all data to a common spatial reference and also combines some data sets by mosaicking, creating composite layers, and extracting features to create feature layers. During the integration step, higher-level spatial reasoning is required to resolve conflicting data and to create derivative layers from extracted features. The output of this step is a registered, refined, and traceable spatial database.

The next step is scene assessment, which can be performed for a variety of application functions (e.g., further FE, target detection, quantitative assessment, or creation of vector layers) by a variety of user disciplines. This stage extracts information in the context of the scene, and is generally query driven.

Table 5.4 summarizes the major kinds of registration, combination, and reasoning functions that are performed, illustrating the increasing levels of complexity in each level of spatial processing. Faust described the general principles for building such a geospatial

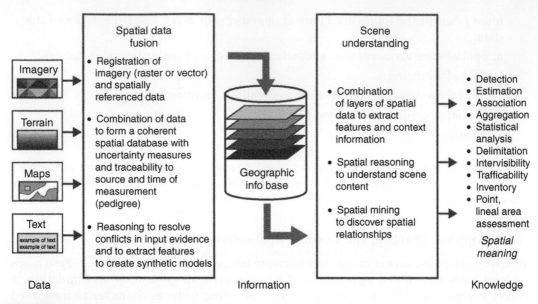

FIGURE 5.5
The spatial data fusion process flow includes the generation of a spatial database and the assessment of spatial information in the database by multiple users.

TABLE 5.4

Spatial Data Fusion Functions

	Increasing Complexity and Processing		
	Registration	**Combination**	**Reasoning**
Data fusion functions	Image registration Image-to-terrain registration	Multiresolution image sharpening Multispectral classification of registered imagery	Image-to-image cross layer searches Feature finding: extraction by roaming across layers to increase detection, recognition, and confidence
	Orthorectification Image mosaicking, including radiometric balancing and feathering	Image-to-image cueing Spatial detection via multiple layers of image data	Context evaluation Image-to-nonimage cueing (e.g., IMINT to SIGINT)
	Multitemporal change detection	Feature extraction using multilayer data	Area delimitation
Examples	Coherent radar imagery change detection	Multispectral image sharpening using panchromatic image	Area delimitation to search for critical target
	SPOT imagery mosaicking		Automated map feature extraction
	LANDSAT magnitude change detection	3D scene creation from multiple spatial sources	Automated map feature updating

Note: Spatial data fusion functions include a wide variety of registration, combination, and reasoning processes and algorithms.

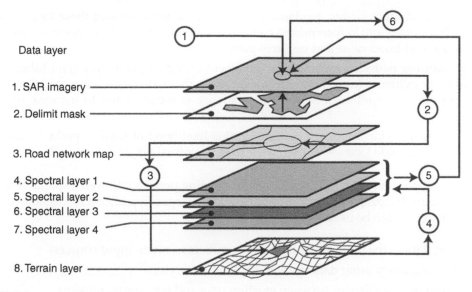

FIGURE 5.6
Target search example uses multiple layers of spatial data and applies iterative spatial reasoning to evaluate alternative hypotheses while accumulating evidence for each candidate target.

database, the hierarchy of functions, and the concept for a blackboard architecture expert system to implement the functions described earlier.[75]

5.6.2.1 Representative Military Example

The spatial reasoning process can be illustrated by a hypothetical military example that follows the process an image or intelligence analyst might follow in search of critical mobile targets (CMTs). Consider the layers of a spatial database illustrated in Figure 5.6, in which recent unmanned air vehicle (UAV) SAR data (the top data layer) has been registered to all other layers, and the following process is performed (process steps correspond to path numbers on the figure):

1. A target cueing algorithm searches the SAR imagery for candidate CMT targets, identifying potential targets in areas within the allowable area of a predefined delimitation mask (data layer 2).*

2. Location of a candidate target is used to determine the distance to transportation networks (which are located in the map, data layer 3) and to hypothesize feasible paths from the network to the hide site.

3. The terrain model (data layer 8) is inspected along all paths to determine the feasibility that the CMT could traverse the path. Infeasible path hypotheses are pruned.

4. Remaining feasible paths (on the basis of slope) are then inspected using the multispectral data (data layers 4, 5, 6, and 7). A multispectral classification algorithm is scanned over the feasible paths to assess ground load-bearing strength, vegetation

---

* This mask is a *derived* layer produced by a spatial reasoning process in the scene generation stage to delimit the entire search region to only those allowable regions in which a target may reside.

cover, and other factors. Evidence is accumulated for slope and these factors (for each feasible path) to determine a composite path likelihood. Evidence is combined into a likelihood value and unlikely paths are pruned.

5. Remaining paths are inspected in the recent SAR data (data layer 1) for other significant evidence (e.g., support vehicles along the path, recent clear cut) that can support the hypothesis. Supportive evidence is accumulated to increase likelihood values.

6. Composite evidence (target likelihood plus likelihood of feasible paths to candidate target hide location) is then used to make a final target detection decision.

In the example presented in Figure 5.6, the reasoning process followed a spatial search to accumulate (or discount) evidence about a candidate target. In addition to target detection, similar processes can be used to:

- Insert data in the database (e.g., resolve conflicts between input sources)
- Refine accuracy using data from multiple sources, etc.
- Monitor subtle changes between existing data and new measurements
- Evaluate hypotheses about future actions (e.g., trafficability of paths, likelihood of flooding given rainfall conditions, and economy of construction alternatives)

5.6.2.2 Representative Crime Mapping Examples

The widespread availability of desktop GIS systems has allowed local crime analysis units to develop new ways of mapping geocoded crime data and visualizing spatial-temporal patterns of criminal activity to better understand patterns of activity and underlying causal factors.[76] This crime mapping process requires only that law enforcement systems geocode the following categories of source information in databases for subsequent analysis:

- *Calls for service.* Date/time group, call number, call category, associated incident report identifier, address, and latitude and longitude of location a police unit was sent
- *Reported crimes.* Date/time group for crime event, case number, crime category, address, and latitude and longitude of the crime location
- *Arrests.* Date/time group of arrest, case number, arrested person information, charge category, address, and latitude and longitude of the arrest location
- *Routes.* Known routes of travel (to-from crime scene, to-from drug deliveries, route between car-stolen and car-recovered, etc.) derived from arrest, investigation, and interrogation records

The spatial data fusion process is relatively simple, registering the geocoded event data to GIS layers for visualization and statistical analysis. Analysts can visualize and study the spatial distributions of this data (crimes, as well as addresses of victims, suspects, travel routes to-from crime scenes, and other related evidence) to

- Discover patterns of spatial association, including spatial clusters or hot spots—areas that have a greater than average number of criminal or disorder events, or areas where people have a higher than average risk of victimization.[77]

The attributes of the hot spots (e.g., spatial properties for each block such as number of buildings, number of families, per capita income, distance to police station, etc.) may be used to predict other vulnerable areas for similar criminal activities.

- Discover spatial-temporal patterns that can be correlated to behavior tempos or profiles (e.g., a drug addict's tempo of burglaries, purchases, and rest).
- Identify trends and detect changes in behavioral patterns to investigate root cause factors (e.g., changes in demographics, movement of people, economics, etc.).
- Identify unusual event locations or spatial outliers.

Crime mapping is particularly useful in the analysis of a crime series—criminal offenses that are thought to share the same causal factor (usually a single offender or group of offenders) given their descriptive, behavior, spatial, or temporal commonality.[78] Crime series are generally related to the category of "crimes of design" that are conducted within certain constraints and therefore may have an observable pattern, rather than "crimes of opportunity" that occur when the criminal, victim and circumstances converge at a particular time and place.

The crime data can be presented in a variety of thematic map formats (Table 5.5) for visual analysis, and the analyst can compute statistical properties of selected events, such as the spatial distribution of events, including measures like the mean center, center of minimum distance, or ellipses of standard deviation. Statistics can be computed for the properties of distances between events including nearest-neighbor measures or Ripley's *K* statistic distance measure. Clustering methods are applied to spatial and temporal data to locate highly correlated cluster of similar events (hot spots) and to identify locations with spatial attributes that make them vulnerable to similar crime patterns.[79] (This predictive analysis may be used to increase surveillance or set up decoys where criminals are expected to sustain a pattern.)

TABLE 5.5

Typical Thematic Crime Map Formats

Map Type	Description	Law Enforcement Use
Dot or pin map	Individual event locations are plotted as dots; symbol coding identifies the event (crime) type at each location	Identify general crime activity patterns, trends, vulnerable locations For a specific crime series, identify candidate spatial patterns
Statistical map	Proportional symbols (e.g., pin sizes, pie charts, or histograms) are used to display quantitative data at locations or in areas	Identify crime densities and relative rates of events by area to manage police coverage
Choropleth map	Display discrete distributions of data within defined boundaries (police beats, precincts, districts, or census blocks)	Define hot spot areas that share the same level of risk
Isoline map	Display contour lines that bound areas of common attributes (e.g., crime rate) and show gradients between bounded areas	Inform law enforcement personnel about high incident areas to increase field contacts and surveillance in an area

5.7 Spatial Data Fusion in GEOINT

The term *geospatial intelligence* (GEOINT) refers to the exploitation and analysis of imagery and geospatial information to describe, assess, and visually depict physical features and geographically referenced activities on the Earth. GEOINT consists of three elements: imagery, imagery-derived intelligence, and geospatial information.[80] The previously mentioned processes of spatial data fusion are at the core of GEOINT processing and analysis functions, enabling the registration, correlation, and spatial reasoning over many sources of intelligence data.

The three elements of GEOINT bring together different types of data that when integrated give a complete spatially coherent product capable of being used in a more detailed analysis. The first element, imagery, refers to any product that depicts features, objects, or activity, natural or man-made with the positional data from the same time. Imagery data can be collected by a large variety of platforms including satellite, airborne, and unmanned platforms; it is a crucial element that provides the initial capability for analysis. The second component of GEOINT is imagery intelligence, a derived result of the interpretation and analysis of imagery, adding context to imagery products. The third element is geospatial information that identifies locations and characteristics of features of the earth. This data includes two categories:

1. Static information that is collected through remote sensing, mapping, and surveying; it is often derived from existing maps, charts, or related products.
2. Dynamic information that is provided by objects being tracked by radar, a variety of tagging mechanisms, or self-reporting techniques such as blue force tracking systems that report the location of personnel and vehicles.

The capability to visualize registered geospatial in three dimensions increases situational awareness and adds to the context of GEOINT analysis. This capability allows reconstruction of scenes and dynamic activities using advanced modeling and simulation techniques. These techniques allow the creation of 3D fly-through products that can then be enhanced with the addition of information gathered from other intelligence disciplines; the fourth dimension of time and movement, provided by dynamic information, can be added to create dynamic and interactive products. These GEOINT products apply several of the fusion methods described in earlier sections to provide analysts with an accurate simulation of a site or an activity for analysis, mission training, or targeting.

To provide accurate, timely, and applicable intelligence, NGA has adopted a systematic process to apply spatial data fusion to analyze intelligence problems and produce standard GEOINT products. The four-step geospatial preparation of the environment (GPE) process (Figure 5.7) was adapted from the military's IPB process to meet a broader spectrum of analysis including civilian and nontraditional threat issues.[81] The process is described as a cycle, though in practice the steps need not be performed sequentially in a linear process; the cycle simply provides the GEOINT analyst a template to follow when attempting to solve an intelligence problem.

The first step defines the environment, based on the mission requirements provided by the intelligence customer. The analyst gathers all pertinent information about the mission location, and determines applicable boundaries, features, and characteristics. This first grouping of information provides the foundation layer(s) for the GEOINT product, including essential features that change rarely or slowly.

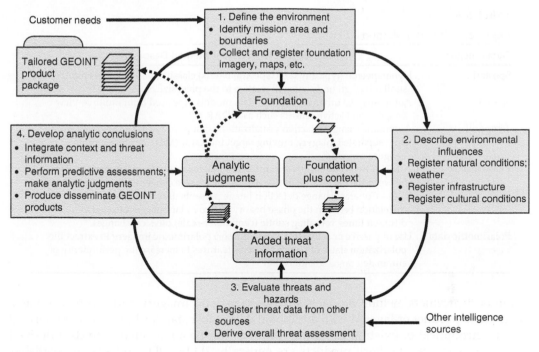

FIGURE 5.7
Geospatial preparation of the environment process flow and products.

The second step is to describe the environmentally related influence. In this step it is important to provide descriptive information about the location being analyzed, including all existing natural conditions such as cultural influences, infrastructure, and political environment. The analyst must also consider other factors that could potentially effect operations in the area, including weather; vegetation; roads; facilities; population; language; and social, ethnic, religious, and political factors. This information is then registered to the data layer(s) prepared in the first step.

The third step evaluates threats and hazards to the mission. This requires gathering available threat data from multiple intelligence disciplines related to the location, including details of the adversary, their forces, doctrine capabilities, and intent. Any information that provides background or insight on the threats for the location is closely investigated, registered, and added to the layers of information from the last two steps. In many cases the estimated geolocation of nonspatial data requires analysts to describe the uncertainty in spatial information and inferences; this requires collaboration with other entities within the intelligence community.

The last step of the GPE is to develop an analytic conclusion. After all the layers of information gathered have been integrated, registered, and considered, the analyst must make analytic judgments regarding operational considerations, such as physical terrain obstacles to vehicle movement, human terrain sensitivities to psychological operations (PSYOP) activities, line-of-sight restrictions to sensor placements, etc. The emphasis in this stage of analysis is on deriving proactive and predictive assessments; these assessments may include predicted effects of next courses of action, estimated impact statements, and assessments.

Most GEOINT products can be categorized as either standard or specialized; however, since GEOINT products are generally tailored to meet specific issues they cannot always be easily categorized.

TABLE 5.6

Representative AGI Methods

Data Category	Example AGI Exploitation Method Description
Spectral	Hyperspectral (typically >100 spectral bands) classification of targets hidden in similar backgrounds or small relative to the pixel size of the image.[83]
Spatial	Automated 3D urban area spatial data models produced from multiple-view imagery, and active sensors such as LIDAR.[84]
Temporal	Automatic change detection visualizations that portray changes derived from time-separated imagery, moving target indicator (MTI), or other motion imagery sources.[84]
Phase history	2CMV—Two-color multiview difference images show the instantaneous change between the backscatter in two SAR scenes where coherence is generally low.[85] CCD—Coherent change detection image presents the complex signal correlation between the phase history of two radar images collected at different times, revealing subtle (centimeter-scale) physical changes.[86]
Polarimetric data	Use of passive or active (e.g., SAR or laser) polarimetric imagery to extract the polarization states of targets to use as features to increase the performance of automated target identification

Standard products include products such as maps and imagery and can be used as a stand-alone product or layered with additional data. Most standard products are derived from electrooptical or existing geospatial data but can be augmented with data derived from other sources. Standard products are generally 2D but 3D products are available; they make up the bulk of the GEOINT requirements and may include in-depth analysis, depending on the consumer's requirements.

Specialized products take standard products to the next level by providing additional capability to tailor them for more specific situations. These products typically use data from a wider variety of geospatial sources and even data from other intelligence disciplines. Specialized products incorporate data from many more technically advanced sensors and typically incorporate more complex registration and exploitation techniques. One of the complex exploitation techniques commonly used in specialized products is Advanced Geospatial Intelligence (AGI), which includes all types of information technically derived from the processing, exploitation, and nonliteral analysis (to include integration or fusion) of spectral, spatial, temporal, phase history, and polarimetric data (Table 5.6). These types of data can be collected on stationary and moving targets by electrooptical, infrared, radar, and related sensors (both active and passive). AGI also includes both ancillary data needed for data processing or exploitation and signature information (to include development, validation, simulation, data archival, and dissemination).[82]

5.8 Summary

The fusion of image and spatial data is an important process that promises to achieve new levels of performance and integration with GISs for a wide variety of military, intelligence, and commercial application areas. By combining registered data from multiple sensors or views, and performing intelligent reasoning on the integrated data sets, fusion systems are beginning to significantly improve the performance of current-generation ATR, single-sensor imaging, and geospatial data systems.

There remain significant challenges to translate the state-of-the-art manual and limited semiautomated capabilities to large-scale production. A recent study by the National Research Council of image and spatial data fusion in GEOINT concluded, "Yet analysis methods have not evolved to integrate multiple sources of data rapidly to create actionable intelligence. Nor do today's means of information dissemination, indexing, and preservation suit this new agenda or future needs."[87] Among the challenges for research identified in the study were developing ontologies for tagging GEOINT objects, image data fusion across space, time, spectrum and scale, and spatiotemporal database management systems to support fusion across all spatial sources. The report summarized that "Data acquired from multiple sensors carry varying granularity, geometric type, time stamps, and registered footprints. Data fusion rectifies coordinate positions to establish which features have not changed over time to focus on what has changed. The fusion, however, involves confronting several hard problems including spatial and temporal conflation, dealing with differential accuracy and resolutions, creating the ontologies and architectures necessary for interoperability, and managing uncertainty with metadata."[88]

References

1. Composite photo of Kuwait City in *Aerospace and Defense Science,* Spring 1991.
2. *Aviation Week and Space Technology,* May 2, 1994, 62.
3. Composite Multispectral and 3-D Terrain View of Haiti in *Aviation Week and Space Technology,* 49, October 17, 1994.
4. R. Ropelewski, Team helps analysts cope with data flood, *Signal,* 47(12), 40–45, 1993.
5. Intelligence and imagery exploitation, Solicitation BAA 94-09-KXPX, *Commerce Business Daily,* April 12, 1994.
6. Terrain feature generation testbed for war breaker intelligence and planning, Solicitation BAA 94-03, *Commerce Business Daily,* July 28, 1994; Terrain visualization and feature extraction, Solicitation BAA 94-01, *Commerce Business Daily,* July 25, 1994.
7. *Global Geospace Information and Services (GGIS),* Defense Mapping Agency, Version 1.0, 36–42, August 1994.
8. M.A. Abidi and R.C. Gonzales, Eds., *Data Fusion in Robotics and Machine Intelligence,* Academic Press, Boston, 1992.
9. Hill, D.L.G. et al., Accurate frameless registration of MR and CT images of the head: Applications in surgery and radiotherapy planning, *Radiology,* 191(2), 447–454, 1994.
10. E.L. Waltz and J. Llinas, *Multisensor Data Fusion,* Artech House, Norwood, MA, 1990.
11. D.L. Hall, *Mathematical Techniques in Multisensor Data Fusion,* Artech House, Norwood, MA, 1992.
12. W.K. Pratt, Correlation techniques of image registration, *IEEE Trans. AES,* 10(13), 353–358, 1974.
13. L. Gottsfield Brown, A survey of image registration techniques, *Computing Surveys,* Vol. 29, 325–376, 1992; see also B.R. Bruce, *Registration for Tactical Imagery: An Updated Taxonomy,* Defence Science and Technology Organisation, Australia, DSTO-TR-1855, April 2006; and J. Ronald, D.M. Booth, P.G. Perry, and N.J. Redding, *A Review of Registration Capabilities in the Analyst's Detection Support System,* Defence Science and Technology Organisation, Australia, DSTO-TR-1632, March 2005.
14. F.E. White, Jr., Data Fusion Subpanel Report, *Proc. Fifth Joint Service Data Fusion Symp.,* Vol. I, 335–361, October 1991.
15. B. Bhanu, Automatic target recognition: State-of-the-art survey, *IEEE Trans. AES,* 22(4), 364–379, 1986.

16. B. Bhanu and T.L. Jones, Image understanding research for automatic target recognition, *IEEE AES Systems Magazine*, 15–23, October 1993.

17. W.G. Pemberton, M.S. Dotterweich, and L.B. Hawkins, An overview of ATR fusion techniques, *Proc. Tri-Service Data Fusion Symp.*, 115–123, June 1987.

18. L. Lazofson and T.J. Kuzma, Scene classification and segmentation using multispectral sensor fusion implemented with neural networks, *Proc. 6th Nat'l. Sensor Symp.*, Vol. I, 135–142, August 1993.

19. V.M. Gonzales and P.K. Williams, Summary of progress in FLIR/LADAR fusion for target identification at Rockwell, *Proc. Image Understanding Workshop*, ARPA, Vol. I, 495–499, November 1994.

20. A.N.A. Schwickerath and J.R. Beveridge, Object to multisensor coregistration with eight degrees of freedom, *Proc. Image Understanding Workshop*, ARPA, Vol. I, 481–490, November 1994.

21. R. Delanoy, J. Verly, and D. Dudgeon, Pixel-level fusion using "interest" images, *Proc. 4th National Sensor Symp.*, Vol. I, 29, August 1991.

22. M.K. Hamilton and T.A. Kipp, Model-based multi-sensor fusion, *Proc. IEEE Asilomar Circuits and Systems Conference*, Vol. 1, 283–289, November 1993.

23. T.A. Kipp and M.K. Hamilton, Model-based automatic target recognition, *4th Joint Automatic Target Recognition Systems and Technology Conf.*, November 1994.

24. G. Duane, Pixel-level sensor fusion for improved object recognition, *Proc. SPIE Sensor Fusion*, Vol. 931, 180–185, 1988.

25. D. Reago et al., Multi-sensor feature level fusion, *4th Nat'l. Sensor Symp.*, Vol. I, 230, August 1991.

26. E. Keydel, Model-based ATR, *Tutorial Briefing*, Environmental Research Institute of Michigan, February 1995.

27. M.K. Hamilton and T.A. Kipp, ARTM: Model-based mutisensor fusion, *Proc. Joint NATO AC/243 Symp. on Multisensors and Sensor Data Fusion*, Vol. 1, November 1993.

28. D.A. Analt, S.D. Raney, and B. Severson, An angle and distance constrained matcher with parallel implementations for model based vision, *Proc. SPIE Conf. on Robotics and Automation*, Boston, MA, October 1991.

29. Model-Driven Automatic Target Recognition Report, ARPA/SAIC System Architecture Study Group, October 14, 1994.

30. J. Larson, L. Hung, and P. Williams, FLIR/laser radar fused model-based target recognition, *4th Nat'l. Sensor Symp.*, Vol. I, 139–154, August 1991.

31. F. Corbett et al., Fused ATR algorithm development for ground to ground engagement, *Proc. 6th Nat'l. Sensor Symp.*, Vol. I, 143–155, August 1993.

32. J.D. Silk, J. Nicholl, and D. Sparrow, Modeling the performance of fused sensor ATRs, *Proc. 4th Nat'l. Sensor Symp.*, Vol. I, 323–335, August 1991.

33. R.H. Lee and W.B. Van Vleet, Registration error between dissimilar sensors, *Proc. SPIE Sensor Fusion*, Vol. 931, 109–114, 1988.

34. J.A. Hoschette and C.R. Seashore, IR and MMW sensor fusion for precision guided munitions, *Proc. SPIE Sensor Fusion*, Vol. 931, 124–130, 1988.

35. D. Lai and R. McCoy, A radar-IR target recognizer, *Proc. 4th Nat'l. Sensor Symp.*, Vol. I, 137, August 1991.

36. M.C. Roggemann et al., An approach to multiple sensor target detection, Sensor Fusion II, *Proc. SPIE*, Vol. 1100, 42–50, March 1989.

37. K. Siejko et al., Dual mode sensor fusion performance optimization, *Proc. 6th Nat'l. Sensor Symp.*, Vol. I, 71–89, August 1993.

38. V. Diehl, F. Shields, and A. Hauter, Testing of multi-sensor automatic target recognition and fusion systems, *Proc. 6th Nat'l. Sensor Fusion Symp.*, Vol. I, 45–69, August 1993.

39. M. Menon and P. Kolodzy, Active/passive IR scene enhancement by Markov random field sensor fusion, *Proc. 4th Nat'l. Sensor Symp.*, Vol. I, 155, August 1991.

40. A.J. Maren, A hierarchical data structure representation for fusing multisensor information, Sensor Fusion II, *Proc. SPIE*, Vol. 1100, 162–178, March 1989.

41. A. Omar Aboutalib, L. Tran, and C.-Y. Hu, Fusion of passive imaging sensors for target acquisition and identification, *Proc. 5th Nat'l. Sensor Symp.*, Vol. I, 151, June 1992.

42. C.-C. Chu and J.K. Aggarwal, Image interpretation using multiple sensing modalities, *IEEE Trans. Pattern Analysis Machine Intell.*, 14(8), 840–847, 1992.
43. P.A. Eggleston and C.A. Kohl, Symbolic fusion of MMW and IR imagery, *Proc. SPIE Sensor Fusion*, Vol. 931, 20–27, 1988.
44. P.A. Eggleston, *Algorithm Development Support Tools for Machine Vision*, Amerinex Artificial Intelligence, Inc. (n.d., received February 1995).
45. C. Bowman and M. DeYoung, Multispectral neural network camouflaged vehicle detection using flight test images, *Proc. World Conf. on Neural Networks*, June 1994.
46. J. Vrabel, MSI band sharpening design trade study, *Presented at 7th Joint Service Data Fusion Symp.*, October 1994.
47. P.S. Chavez, Jr. et al., Comparison of three different methods to merge multiresolution and multispectral data: LANDSAT™ and SPOT Panchromatic, *Photogramm. Eng. Remote Sensing*, 57(3), 295–303, 1991.
48. K. Edwards and P.A. Davis, The use of intensity-hue-saturation transformation for producing color-shaded relief images, *Photogramm. Eng. Remote Sensing*, 60(11), 1369–1374, 1994.
49. R. Tenney and A. Willsky, Multiresolution image fusion, DTIC Report AD-B162322L, January 31, 1992.
50. B.N. Haack and E. Terrance Slonecker, Merging spaceborne radar and thematic mapper digital data for locating villages in Sudan, *Photogramm. Eng. Remote Sensing*, 60(10), 1253–1257, 1994.
51. C.C. Chesa, R.L. Stephenson, and W.A. Tyler, Precision mapping of spaceborne remotely sensed imagery, *Geodetical Info Magazine*, 8(3), 64–67, 1994.
52. R.M. Reinhold, *Arc Digital Raster Imagery (ADRI) Program*, Air Force Spatial data Technology Workshop, Environmental Research Institute of Michigan, July 1991.
53. I.S. Kweon and T. Kanade, High resolution terrain map from multiple sensor data, *IEEE Trans. Pattern Analysis Machine Intell.*, 14(2), 278–292, 1992.
54. P.J. Burt, Pattern selective fusion of IR and visible images using pyramid transforms, *Proc. 5th Nat'l. Sensor Symp.*, Vol. I, 313–325, June 1992.
55. M. Hansen et al., Real-time scene stabilization and mosaic construction, *Proc. of the 2nd IGEE Workshop on Application of Computer Vision*, ARPA, Vol. I, 54–62, December 1994.
56. P.J. Burt and P. Anandan, Image stabilization by registration to a reference mosaic, *Proc. Image Understanding Workshop*, ARPA, Vol. I, 425–434, November 1994.
57. C.C. Chiesa and W.A. Tyler, Data fusion of off-Nadir SPOT panchromatic images with other digital data sources, *Proc. 1990 ACSM-ASPRS Annual Convention*, Denver, 86–98, March 1990.
58. A. Kak and S.-S. Chen (Eds.), *Proc. Spatial Reasoning and Multi-Sensor Fusion Workshop*, AAAI, October 1987.
59. P.S. Shenker (Ed.), Sensor fusion: Spatial reasoning and scene interpretation, *SPIE*, Vol. 1003, November 1988.
60. P.S. Shenker, Sensor fusion III: 3-D perception and recognition, *SPIE*, Vol. 1383, November 1990.
61. P.S. Shenker, Sensor fusion IV: Control paradigms and data structures, *SPIE*, Vol. 1611, November 1991.
62. P.S. Shenker, Sensor fusion V, *SPIE*, Vol. 1828, November 1992.
63. P.S. Shenker, Sensor fusion VI, *SPIE*, Vol. 2059, November 1993.
64. *Condensed Standards for Digital Geographic Metadata*, Federal Geographic Data Committee, Washington, DC, June 8, 1994.
65. *Intelligence Preparation of the Battlefield*, FM-34-130, HQ Dept. of the Army, May 1989.
66. *Development and Integration of the Terrain Feature Generator (TFG)*, Solicitation DACA76-94-R-0009, Commerce Business Daily Issue PSA-1087, May 3, 1994.
67. R.T. Antony, Eight canonical forms of fusion: A proposed model of the data fusion process, *Proc. of 1991 Joint Service Data Fusion Symp.*, Vol. III, October 1991.
68. R.T. Antony, *Principles of Data Fusion Automation*, Artech House, Norwood, MA, 1995.
69. R.L. Shelton and J.E. Estes, Integration of remote sensing and geographic information systems, *Proc. 13th Int'l. Symp. Remote Sensing of the Environment*, Environmental Research Institute of Michigan, April 1979, 463–483.

70. J.E. Estes and J.L. Star, Remote sensing and GIS integration: Towards a prioritized research agenda, *Proc. 25th Int'l Symp. on Remote Sensing and Change*, Graz Austria, Vol. I, 448–464, April 1993.

71. J.L. Star (Ed.), *The Integration of Remote Sensing and Geographic Information Systems*, American Society for Photogrammetry and Remote Sensing, Annapolis, MD, 1991.

72. G. Sawyer et al., MuSIP multi-sensor image processing system, *Image Vis. Comput.*, 10(9), 589–609, 1992.

73. D.C. Mason et al., Spatial database manager for a multi-source image understanding system, *Image Vis. Comput.*, 11(1), 25–34, 1993.

74. N.D. Gershon and C. Grant Miller, Dealing with the data deluge, *IEEE Spectr.*, 30(7), 28–32, 1993.

75. N.L. Faust, Design concept for database building, *Proc. of SPIE*, March 1988.

76. K.W. Harries, *Mapping Crime: Principle and Practice*, U.S. Department of Justice, Office of Justice Programs Report NCJ 178919, National Institute of Justice, December 1999; see also Advanced crime mapping topics, *Proc. of Invitational Advanced Crime Mapping Symp.*, Crime Mapping & Analysis Program (CMAP) of the National Law Enforcement & Corrections Technology Center (NLECTC), Denver, CO, 2001; M. Velasco and R. Boba, *Manual of Crime Analysis Map Production*, 97-CK-WXK-004, Office of Community Oriented Policing Services, November 2000.

77. J.E. Eck, S. Chainey, J.G. Cameron, M. Leitner, and R.E. Wilson, *Mapping Crime: Understanding Hot Spots*, U.S. Department of Justice, Office of Justice Programs, National Institute of Justice, NCJ, 209–393, August 2005.

78. Advanced crime mapping topics, *Proc. of Invitational Advanced Crime Mapping Symp.*, Crime Mapping & Analysis Program (CMAP) of the National Law Enforcement & Corrections Technology Center (NLECTC), Denver, CO, June 2001.

79. D. Helms, *The Use of Dynamic Spatio-Temporal Analytical Techniques to Resolve Emergent Crime Series*, Las Vegas Metropolitan Police Department, 1999.

80. Geospatial Intelligence (GEOINT) Basic Doctrine, National System for Geospatial Intelligence, Publication 1-0, National Geospatial-Intelligence Agency, September 2005. The term GEOINT is defined in Title 10 U.S. Code §467.

81. Ibid., pp. 12–13.

82. Ibid., p. 45.

83. J.A. Benediktsson and I. Kanellopoulos, Classification of multisource and hyperspectral data based on decision fusion, *IEEE Trans. Geosci. Remote Sensing*, 37(3), 1367–1377, 1999.

84. *Tools and Techniques for Advanced Exploitation of Geospatial Intelligence*, White Paper, Intergraph Corp., 2006.

85. CCDMap™ Automated Coherent Change Detection Mosaic System, Product Brochure, Vexcel Corp., Boulder CO, 2003.

86. M. Preiss and N.J.S. Stacy, *Coherent Change Detection: Theoretical Description and Experimental Results*, Defence Science and Technology Organisation, DSTO-TR-1851, August 2006.

87. Priorities for GEOINT Research at the National Geospatial-Intelligence Agency, Mapping Science Committee, National Research Council, 2006, p. 2.

88. Ibid., p. 76.

6

Data Registration

Richard R. Brooks and Lynne Grewe

CONTENTS

6.1   Introduction

Sensor fusion refers to the use of multiple sensor readings to infer a single piece of information. Inputs may be received from a single sensor over a period of time. They may also be received from multiple sensors of the same or different types. This is a very important technology for radiology and remote sensing applications. In medical imaging alone, the number of papers published per year has increased from 20 in 1990 to about 140 in 2001.[1] Registering elastic images and finding theoretical bounds for image registration are particularly active research topics.

Registration inputs may be raw data, extracted features, or higher-level decisions. This process provides increased robustness and accuracy in machine perception. This is conceptually similar to the use of repeated experiments to establish parameter values using statistics.[2] Several reference books have been published on sensor fusion.[3-6]

One decomposition of the sensor fusion process is shown in Figure 6.1. Sensor readings are gathered, preprocessed, compared, and combined, arriving at the final result. An essential preprocessing step for comparing readings from independent physical sensors is transforming all input data into a common coordinate system. This is referred to as *data registration*. In this chapter, we describe data registration, provide a review of the existing methods, and discuss some recent results.

Data registration transformation is often assumed to be known *a priori* partially because the problem is not trivial. Traditional methods are based on methods developed

FIGURE 6.1
Decomposition of sensor fusion process. (From Keller, Y. and Averbuch, A., *IEEE Trans. Patt. Anal. Mach. Intell.* 28, 794, 2006. With permission.)

by cartographers. These methods have a number of drawbacks and often make invalid assumptions concerning the input data.

Although data input includes raw sensor readings, features extracted from sensor data, and higher-level information, registration is a preprocessing stage and, therefore, is usually applied only to either raw data or extracted features. Sensor readings can have one to n dimensions. The number of dimensions will not necessarily be an integer. Most techniques deal with data of two or three dimensions; however, same approaches can be trivially applied to one-dimensional (1D) readings. Depending on the sensing modalities used, occlusion may be a problem with data in more than two dimensions, leading to obscurity of data by the relative position of objects in the environment. The specific case studies presented in this chapter use image data in two dimensions and range data in 2½ dimensions.

This chapter is organized as follows. Section 6.2 gives a formal definition of image registration. Section 6.3 provides a brief survey of existing methods. Section 6.4 discusses meta-heuristic techniques that have been used for image registration. This includes objective functions for sensor readings with various types of noise. Section 6.5 discusses a multiresolution implementation of image registration. Section 6.6 discusses human-assisted registration. Section 6.7 looks at registration of warped images. In Section 6.8, we discuss problems faced by researchers registering data from multiple sensing modes. The theoretical bounds that have been established for the registration process are given in Section 6.9. Section 6.10 concludes with a brief summary discussion.

6.2 Registration Problem

Competitive multiple sensor networks consist of a large number of physical sensors providing readings that are at least partially redundant. The first step in fusing multiple sensor readings is registering them to a common frame of reference.[7] *Registration* refers to finding the correct mapping of one image onto another. When an inaccurate estimate of the registration is known, finding the exact registration is referred to as *refined registration*. Another survey of image registration can be found in Ref. 8.

As shown in Figure 6.2, the general image registration problem is, given two N-dimensional sensor readings, find the function F that best maps the reading from sensor 2, $S_2(x_1, \ldots, x_n)$ onto the reading from sensor 1, $S_1(x_1, \ldots, x_n)$. Ideally, $F(S_2(x_1, \ldots, x_n)) = S_1(x_1, \ldots, x_n)$. Because all sensor readings contain some amount of measurement error or noise, the ideal case rarely occurs.

Many processes require that data from one image, called the *observed image*, be compared with or mapped to another image, called the *reference image*. Hence, a wide range of critical applications depends on image registration.

Perhaps the largest amount of image registration research is focused on medical imaging. One application is sensor fusion to combine outputs from several medical imaging

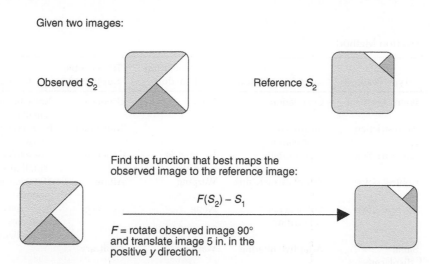

Given two images:

Observed S_2 Reference S_2

Find the function that best maps the
observed image to the reference image:

$$F(S_2) - S_1$$

F = rotate observed image 90°
and translate image 5 in. in the
positive y direction.

FIGURE 6.2
Registration is finding the mapping function $F(S_2)$.

technologies such as positron emission tomography (PET) and magnetic resonance imagery (MRI) to form a more complete image of internal organs.[9] Registered images are then used for medical diagnosis of illness[10] and automated control of radiation therapy.[11] Similar applications of registered and fused images are common[12] in military applications (e.g., terrain "footprints"),[13] remote sensing applications, and robotics. A novel application is registering portions of images to estimate motion. Descriptions of motion can then be used to construct intermediate images in television transmissions. Jain and Jain describe the applications of this to bandwidth reduction in video communications.[14] These are some of the more recent applications that rely on accurate image registration. Methods of image registration have been studied since the beginning of the field of cartography.

6.3 Review of Existing Research

This section discusses the current state of research concerning image registration. Image registration is a basic problem in image processing, and a large number of methods have been proposed.

Table 6.1 summarizes the features of representative image registration methods discussed in this section. The discussion is followed by a detailed discussion of the established methodologies and algorithms currently in use. Each is explored in more detail in the remainder of the section.

The traditional method of registering two images is an extension of methods used in cartography. A number of control points are found in both images. The control points are matched, and this match is used to deduce equations that interpolate all points in the new image to corresponding points in the reference image.[15,16]

Several algorithms exist for each phase of this process. Control points must be unique and easily identified in both images. Control points have been explicitly placed in the image by the experimenter,[11] and edges have been defined by intensity changes,[17,18] specific points peculiar to a given image,[8,19] line intersections, center of gravity of closed regions,

TABLE 6.1

Sensor Registration Methods

Algorithm	Data Type/Features	Matching Method	Interpolation Function	Transforms Supported	Comments
Andrus	Boundary maps	Correlation	None	Gruence	Noise intolerant, small rotations
Barnea	No restriction	Improved correlation	None	Translation	No rotation, scaling noise, rubber sheet
Barrow	No restriction	Hill climbing	Parametric chamfer	Gruence	Noise tolerant, small displacement
Bentoutou	Control points	Template matching	Warping	Affine	Matches different modes
Brooks–Iyengar	No restriction	Elitist genetic algorithm	None	Gruence	Noise tolerant, tolerates periodicity
Caner	Small displacements	Adaptive filtering	Warping	Arbitrary	
Cox	Line segments	Hill climbing	None	Gruence	Matches using small number of features
Dai[67]	Chain code of edges of closed contours	Mapping invariant features	Least squares	Polynomial transform	Centers of gravity for closed contours are superimposed to give the automatic registration
Davis	Specific shapes	Relaxation	None	Affine	Matches shapes
Goshtasby, 1986	Control points	Various	Piecewise linear	Rubber sheet	Fits images using mapped points
Goshtasby, 1987	Control points	Various	Piecewise cubic	Rubber sheet	Fits images using mapped points
Goshtasby, 1988	Control points	Various	Lease squares	Rubber sheet	Fits images using mapped points
Han	Person silhouettes	Hierarchical Genetic Algorithm	None	Translation	Uses centroids of silhouettes in color, infrared images.
Irani	Oriented edge maps	Regression and correlation	Square of directional derivative	Affine	Different modalities, multiresolution approach
Jain	Subimages	Hill climbing	None	Translation	Small translations, no rotation, no noise
Kybic	2D to 3D images	Gradient descent	B-splines	Arbitrary	
Li	Control points and intensity values	Newton's method to optimize one parameter at a time	None	Affine	
Mandara	Control points	Classic G.A.S.A.	Bilinear	Rubber sheet	Fits four fixed points using error fitness
Michel	Internal clock, temporal multimodal events	Correlation	None	Temporal	Temporal registration—matches audio, video frames
Mitiche	Control points	Least squares	None	Affine	Uses control points
Oghabian	Control points	Sequential search	Least squares	Rubber sheet	Assumes small displacement
Penney	2D to 3D	Multiple similarity measures	None	Translation, rotation, 3D perspective	

(continued)

TABLE 6.1   (Continued)

Algorithm	Data Type/Features	Matching Method	Interpolation Function	Transforms Supported	Comments
Petrushin	Motion and color features	Correlation	None	Temporal, translation	Temporal registration—loose synchronization first, refined with features
Pinz	Control points	Tree search	None	Affine	Difficulty with local minima
Silva	Range images/surfaces	Genetic algorithm	None	Rotation and translation	Use a small gene pool at first. At the end, the gene pool size increases
Simonson	Edges in local regions	Hypothesis testing	None	Translations	
Stockman	Control points	Cluster	None	Affine	Assumes landmarks, periodicity problem
Wang	Pixels then control points	Correlation, warping	Thin-plate spline/warping	Affine then elastic	Two-stage approach, local refinement uses initial global affine transform
Wong	Intensity differences	Exhaustive search	None	Affine	Uses edges, intense computation
Xia	2D to 3D	Newton's method	B-splines	Warping	
Yao	Different sensing modes/control points	Genetic algorithm, squared gradient magnitude	Pixel migration	Arbitrary	Matches radically different sensing modes
Zokai	Template matching	Levenberg–Marquadt optimization (Newton's method)	None	Affine with limited perspective	Templates are log-polar transforms of circular regions

or points of high curvature.[15] The type of control point that should be used primarily depends on the application and contents of the image. For example, in medical image processing, the contents of the image and approximate poses are generally known *a priori*.

When registering images from different sensing modes, features that are invariant to sensor differences are sought.[18] Registration of data not explicitly represented in the same data space like sound and imagery is complicated, and a common "spatial" representation must be constructed.[20–22] Petrushin et al. have used "events" as "control points" in registration.[20] In their system, 32 web cameras, a higher-quality video camera, and a fingerprint reader are the sensors used for indoor surveillance. Gradient maxima have been used to find small sets of pixels to use as candidate control points.[23,24] In a subapplication of person-tracking, the event is associated with the movement of a person from one "location" to another. A location is predefined as a spatial area. This is what is registered between the sensors for the purpose of identification and tracking. Han and Bhanu[25] have also used movement queues to determine "control points" for matching. In particular, "human silhouettes" representing humans moving are extracted from synchronous color and infrared image sequences. The two-dimensional (2D) to three-dimensional (3D) multimodal registration problem[26,27] is fairly common in radiology but is done using radiometry values instead of control points.

Similarly, many methods have been proposed for matching control points in the observed image to the control points in the reference image. The obvious method is to correlate a template of the observed image.[28,29] Another widely used approach is to calculate the transformation matrix, which describes the mapping with the least square error.[12,19,30] Other standard computational methods, such as relaxation and hill climbing, have also been used.[14,31,32] Pinz et al. have used a hill climbing algorithm to match images and note the difficulty posed by local minima in the search space; to overcome this, they run a number of attempts in parallel with different initial conditions.[33] Inglada and Giros have compared many similarity measures for comparing readings of different modes.[34] Some measures consider the variance of radiometry values at each sample point. Other measures are based on entropy and mutual information content. Different metrics perform better for different combinations of sensors.

Some interesting methods have been implemented that consider all possible transformations. Stockman et al. have constructed vectors between all pairs of control points in an image.[13] For each vector in each image, an affine transformation matrix is computed that converts the vector from the observed image to one of the vectors from the reference image. These transformations are then plotted, and the region containing the largest number of correspondences is assumed to contain the correct transformation.[13] This method is computationally expensive because it considers the power set of control points in each image. Wong and Hall have matched scenes by extracting edges or intensity differences and constructing a tree of all possible matches that fall below a given error threshold.[17] They have reduced the amount of computation needed by stopping all computation concerning a potential matching once the error threshold is exceeded; however, this method remains computationally intensive. Dai and Khorram have extracted affine transform invariant features based on the central moments of regions found in remote sensing images.[35] Regions are defined by zero-crossing points. Similarly, Yang and Cohen have described a moments-based method for registering images using affine transformations given sets of control points.[36]

Registration of multisensor data to a 3D scene, given a knowledge of the contents of the scene, is discussed by Chellappa et al.[37] The use of an extended Kalman filter (EKF) to register moving sensors in a sensor fusion problem is discussed by Zhou et al.[38] Mandara and Fitzpatrick have implemented a very interesting approach[10] using simulated annealing and genetic algorithm heuristics to find good matches between two images. They have found a rubber sheet transformation that fits two images by using linear interpolation around four control points, and assumed that the images match approximately at the beginning. A similar approach has been espoused by Matsopoulos et al.[39] Silva et al. have used a modified genetic algorithm, where they have modified the size of the gene pool over time.[40] During the first few generations, most mappings are terrible. By considering only a few samples at that stage, the system saves a lot of computation. During the final stages, the gene pool size is increased to allow a finer grain search for the optimal solution. Yao and Goh have registered images from different sensing modes by using a genetic algorithm constrained to search only in subspaces of the problem space that could contain the solution.[24]

A number of researchers have used multiresolution methods to prune the search space considered by their algorithms. Mandara and Fitzpatrick[10] have used a multiresolution approach to reduce the size of their initial search space for registering medical images using simulated annealing and genetic algorithms. This work has influenced Oghabian and Todd-Pokropek, who have similarly reduced their search space when registering brain images with small displacements.[9] Pinz et al. have adjusted both multiresolution scale space and step size to reduce the computational complexity of a hill climbing registration method.[33] A hierarchical genetic algorithm has been used by Han and Bhanu to achieve registration of color and infrared image sequences.[25] These researchers believe that by starting with low-resolution images or feature spaces, they can reject large numbers of possible matches and find the correct match by progressively increasing the resolution. Zokai and Wolberg have done Levenberg–Marquadt optimization (similar to Newton's algorithm) to find optimal correspondences in the state space.[41] Note that in images with a strong periodic component, a number of low-resolution matches may be feasible. In such cases, the multiresolution approach would be unable to prune the search space and, instead, will increase the computational load. Another problem with a common multiresolution approach, the wavelet transform, is its sensitivity to translation.[42] Kybic and Unser have used splines as basis functions instead of wavelets.[27]

A number of methods have been proposed for fitting the entire image around the control points once an appropriate match has been found. Simple linear interpolation is computationally straightforward.[10] Goshtasby has explored using a weighted least-squares approach,[30] constructing piecewise linear interpolation functions within triangles defined by the control points,[15] and developing piecewise cubic interpolation functions.[43] These methods create nonaffine rubber sheet transformation functions to attempt to reduce the image distortion caused by either errors in control point matching or differences in the sensors that constructed the image. Similarly, Wang et al. have presented a system that looks at merging 2D medical images.[44] In a two-stage approach, the intensity values are used to determine the best global affine transformation between the two images. This first step yields good results because much of the material, like bone, that may be imaged can be matched between two images using an affine transform. However, as there are also small local tissue deformations that are not rigid, a refinement of this initial registration takes place locally using the thin-plate spine interpolation method with landmark control points selected as a function of the image properties.[44] This problem is discussed further in Section 6.7.

Several algorithms exist for image registration. The algorithms described have some common drawbacks. The matching algorithms assume that (a) a small number of distinct features can be matched,[13,19,45] (b) specific shapes are to be matched,[46] (c) no rotation exists, or (d) the relative displacement is small.[9,10,14,19,28,32] Refer to Table 6.1 for a summary of many of these points.

Choosing a small number of control points is not a trivial problem and has a number of inherent drawbacks. For example, the control point found may be a product of measurement noise. When two readings have more than a trivial relative displacement, control points in one image may not exist in the other image. This requires considering the power set of the control points. When an image contains periodic components, control points may not define a unique mapping of the observed image to the reference image. Additional problems exist. The use of multiresolution cannot always trim the search space and, if the image is dominated by periodic elements, it would only increase the computational complexity of an algorithm.[9,10,33]

Many algorithms attempt to minimize the square error over the image; however, this does not consider the influence of noise in the image.[9,10] Most of the existing methods are sensitive to noise.[8,19,28] Section 6.4 discusses meta-heuristics-based methods, which try to overcome these drawbacks. Section 6.5 discusses a multiresolution approach.

Although much of the discussion of this chapter concentrates on the issue of spatial registration, temporal registration for some applications is important. Michel and Stanford[21] have developed a system that uses off-the-shelf video cameras and microphones for a smart-meeting application. Video and sound frames from multiple sources must be synchronized and this is done in a two-part approach. Because of the lower quality of sensor components there can be a lot of temporal drift and some sensors do not possess internal clocks; therefore, a master network external clock is used and each sensor/capture node is synchronized to this. This brings capture nodes within a few milliseconds of each other temporally. Next, a multimodal event is generated for a more refined synchronization between the captured video and the audio streams. They combine a "clap" sound with a strobe flash to create a multimodal event that is then found in each video and audio stream. The various streams are correlated temporally using this multimodal event trigger.

Petrushin et al.[20] have provided another example of a system performing temporal registration/synchronization where they have a number of video sensors that can run at different sampling rates. Because of this, they have defined the concept of a "tick," which represents a unit of time large enough to contain at least one frame for the slowest video capture but also hopefully small enough to detect changes necessary for their application, including people tracking. Using ticks, a kind of "loose" synchronization is achieved. Petrushin et al.[20] have used self-organizing maps to cluster tick data for the detection of events such as people moving in a scene.

6.4   Registration Using Meta-Heuristics

This section discusses research on automatically finding a *gruence* (i.e., translation and rotation) registering two overlapping images. Results from this research have previously been presented in a number of sources.[3,47,48,50,66]

This approach attempts to correctly calibrate two 2D sensor readings with identical geometries. These assumptions about the sensors can be made without a loss of generality because

- A method that works for two readings can be extended to register any number of readings sequentially.
- The majority of sensors work in one or two dimensions. Extensions of calibration methods to more dimensions are desirable, but not imperative.
- Calibration of two sensors presupposes known sensor geometry. If geometries are known, a function can be derived that maps the readings as if the geometries were identical when a registration is given.

This approach finds gruences because these functions best represent the most common class of problems. The approach that has been used can be directly extended to include the class of all affine transformations by adding scaling transformations.[49] It does not consider "rubber sheet" transformations that warp the contents of the image because these transformations mainly correct local effects after use of an affine transformation correctly matches the images.[16] It assumes that any rubber sheet deformations of the sensor image are known and corrected before the mapping function is applied, or that their effects over the image intersections are negligible.

The computational examples used pertain to two sensors returning 2D gray scale data from the same environment. The amount of noise and the relative positions of the two sensors are not known. Sensor 2 is translated and rotated by an unknown amount with relation to sensor 1.

If the size or content of the overlapping areas is known, a correlation using the contents of the overlap on the two images could find the point where they overlap directly. Use of central moments could also find relative rotation of the readings. However, when the size or content of the areas is unavailable, this approach is impossible.

In this work, the two sensors have identical geometric characteristics. They return readings covering a circular region, and these readings overlap. Both sensors' readings contain noise. What is not known, however, is the relative positions of the two sensors. Sensor 2 is translated and rotated by an unknown amount with relation to sensor 1.

The best way to solve this problem depends on the nature of the terrain being observed. If unique landmarks can be identified in both the images, those points can be used as control points. Depending on the number of landmarks available, minor adjustments may be needed to fit the readings exactly. Goshtasby's methods could be used at that point.[15,30,43]

Hence, the problem to be solved is, given noisy gray scale data readings from sensor 1 and sensor 2, finding the optimal set of parameters (x-displacement, y-displacement, and angle of rotation) that defines the center of the sensor 2 image relative to the center of the sensor 1 image. These parameters would provide the optimal mapping of sensor 2 readings to the readings from sensor 1. This can be done using meta-heuristics for optimization. Brooks and Iyengar have described implementations of genetic algorithms, simulated annealing, and tabu search for this problem.[3] Chen et al. have applied TRUST, a subenergy tunneling approach from Oak Ridge National Laboratories.[48]

To measure optimality, a fitness function can be used. The fitness function provides a numerical measure of the goodness of a proposed answer to the registration problem. Brooks and Iyengar have derived a fitness function for sensor readings corrupted with Gaussian noise:[3]

$$\frac{\sum(read_1(x,y)read_2(x',y'))^2}{K(W)^2} = \frac{\sum(gray_1(x,y)gray_2(x',y'))^2 - \sum(noise_1(x,y)noise_2(x',y'))^2}{K(W)^2}$$

(6.1)

where w is a point in the search space; $K(W)$ is the number of pixels in the overlap for w; (x',y') is the point corresponding to (x,y); $read_1(x,y)read_2(x',y')$ is the pixel value returned by sensor 1 (2) at point $(x,y)(x',y')$; $gray_1(x,y)gray_2(x',y')$ is the noiseless value for sensor 1 (2) at $(x,y)(x',y')$; $noise_1(x,y)noise_2(x',y')$ is the noise in the sensor 1 (2) reading at $(x,y)(x',y')$.

The equation is derived by separating the sensor reading into information and additive noise components. This means the fitness function is made up of two components: (a) lack of fit and (b) stochastic noise. The lack of fit component has a unique minimum when the two images have the same gray scale values in the overlap (i.e., when they are correctly registered). The noise component follows a χ^2 distribution, whose expected value is proportional to the number of pixels in the region where the two sensor readings intersect. Dividing the difference squared by the cardinality of the overlap makes the expected value of the noise factor constant. Dividing by the cardinality squared favors large intersections. For a more detailed explanation of this derivation, see Ref. 3.

Other noise models simply modify the fitness function. Another common noise model addresses salt-and-pepper noise typically caused by either malfunctioning pixels in electronic cameras or dust in optical systems. In this model, the correct gray scale value in a picture is replaced by a value of 0 (255) with an unknown probability $p(q)$. An appropriate fitness function for this type of noise is

$$\begin{aligned}&read_1(x,y) \neq 0\\&read_1(x,y) \neq 255\\&read_2(x',y') \neq 0\\&read_2(x',y') \neq 255\end{aligned} \quad \sum \frac{(read_1(x,y) - read_2(x',y'))^2}{K(W)}$$

(6.2)

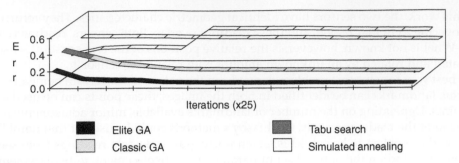

FIGURE 6.3
Fitness function results with variance 1.

A similar function can be derived for uniform noise by using the expected value $E[(U_1 - U_2)^2]$ of the squared difference of two uniform variables U_1 and U_2. An appropriate fitness function is then given by

$$\sum \frac{(read_1(x, y) - read_2(x', y'))^2}{E[(U_1 - U_2)^2]K(W)} \tag{6.3}$$

Figure 6.3 shows the best fitness function value found by simulated annealing, elitist genetic algorithms, classic genetic algorithms, and tabu search versus the number of iterations performed. In the study by Brooks and Iyengar, elitist genetic algorithms outperformed the other methods attempted. Further work by Chen et al. indicates that TRUST is more efficient than the elitist genetic algorithms.[48] These studies show that optimization techniques can work well on the problem, even in the presence of large amounts of noise. This is surprising because the fitness functions take the difference of noise-corrupted data—essentially a derivative. Derivatives are sensitive to noise. Further inspection of the fitness functions explains this surprising result. Summing over the area of intersection is equivalent to integrating over the area of intersection. Implicitly, integrating counteracts the derivative's magnification of noise.

Matsopoulos et al. have used affine, bilinear, and projective transformations to register medical images of the retina.[39] The techniques tested include genetic algorithms, simulated annealing, and the downhill simplex method. They use image correlation as a fitness function. For their application, much preprocessing is necessary, which removes sensor noise. Their results indicate the superiority of genetic algorithms for automated image registration. This is consistent with Brooks' results.[3,50]

6.5   Wavelet-Based Registration of Range Images

This section uses range sensor readings. More details have been provided by Grewe and Brooks.[51] Range images consist of pixels with values corresponding to range or depth rather than photometric information. The range image represents a perspective of a 3D world. The registration approach described herein can be trivially applied to other kinds of images, including 1D readings. If desired, the approaches described by Brooks[3,50] and

Chen et al.[48] can be directly extended to include the class of all affine transformations by adding scaling transformations. This section discusses an approach for finding these transformations.

The approach uses a multiresolution technique, the wavelet transform, to extract features used to register images. Other researchers have also applied wavelets to this problem, including using locally maximum wavelet coefficient values as features from two images.[52] The centroids of these features are used to compute the translation offset between the two images. A principal components analysis is then performed and the eigenvectors of the covariance matrix provide an orthogonal reference system for computing the rotation between the two images. (This use of a simple centroid difference is subject to difficulties when the scenes only partially overlap and, hence, contain many other features.)

In another example, the wavelet transform is used to obtain a complexity index for two images.[53] The complexity measure is used to determine the amount of compression appropriate for the image. Compression is then performed, yielding a small number of control points. The images, made up of control points for rotations, are tested to determine the best fit.

The system described by Grewe and Brooks[51] is similar to that in some of the previous works discussed. Similar to DeVore et al.,[53] Grewe and Brooks have used wavelets to compress the amount of data used in registration. Unlike previous wavelet-based systems prescribed by Sharman et al.[52] and DeVore et al.,[53] Grewe and Brooks[51] have capitalized on the hierarchical nature of the wavelet domain to further reduce the amount of data used in registration. Options exist to perform a hierarchical search or simply to perform registration inside one wavelet decomposition level. Other system options include specifying an initial registration estimate, if known, and the choice of the wavelet decomposition level in which to perform or start registration. At higher decomposition levels, the amount of data is significantly reduced, but the resulting registration would be approximate. At lower decomposition levels, the amount of data is reduced to a lesser extent, but the resulting registration is more exact. This allows the user to choose between accuracy and speed as necessary.

Figure 6.4 shows a block diagram of the system. It consists of a number of phases, beginning with the transformation of the range image data to the wavelet domain. Registration

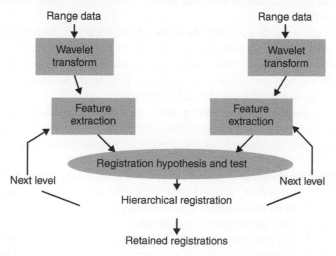

FIGURE 6.4
Block diagram of WaveReg system.

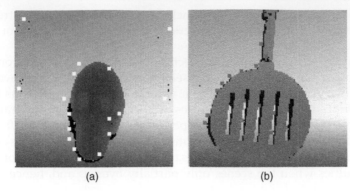

(a)   (b)

FIGURE 6.5
Features detected, approximate location indicated by white squares: (a) for wavelet level 2 and (b) for wavelet level 1.

can be performed on only one decomposition level of this space to reduce registration complexity. Alternately, a hierarchical registration across multiple levels would extract features from a wavelet decomposition level as a function of a number of user-selected parameters that determine the amount of compression desired in the level. Matching features from the two range images are used to hypothesize the transformation between the two images and evaluated. The "best" transformations are retained. This process is explained in the following paragraphs.

First, a Daubechies-4 wavelet transform is applied to each range image. The wavelet data is compressed by thresholding the data to eliminate low magnitude wavelet coefficients. The wavelet transform produces a series of 3D edge maps at different resolutions. A maximal wavelet value indicates a relatively sharp change in depth.

Features, special points of interest in the wavelet domain, are simply points of maximum value in the current wavelet decomposition level under examination. These points are selected so that no two points are close to each other. The minimum distance is scaled with the changing wavelet level under examination. Figure 6.5 shows features detected for different range scenes at different wavelet levels. Notice how these correspond to points of sharp change in depth.

Using a small number of feature points allows this approach to overcome the wavelets transform's sensitivity to translation. Stone et al.[42] have proposed another method for overcoming the sensitivity to translation. They have noted that the low-pass portions of the wavelet transform are less sensitive to translation and that coarse-to-fine registration of images using the wavelet transform should be robust.

The next stage involves hypothesizing correspondences between features extracted from the two unregistered range images. Each hypothesis represents a possible registration and is subsequently evaluated for its goodness. Registrations are compared and the best retained.

Hypothesis formation begins at a default wavelet decomposition level. Registrations retained at this level are further "refined" at the next lower level, L-1. This process continues until the lowest level in the wavelet space is reached.

For each hypothesis, the corresponding geometric transformation relating the matched features is calculated, and the remaining features from one range image are transformed into the other's space. This greatly reduces the computation involved in hypothesis evaluation in comparison to those systems that perform nonfeature-based registration. Next, features that are not part of the hypothesis are compared. Two features match if they are

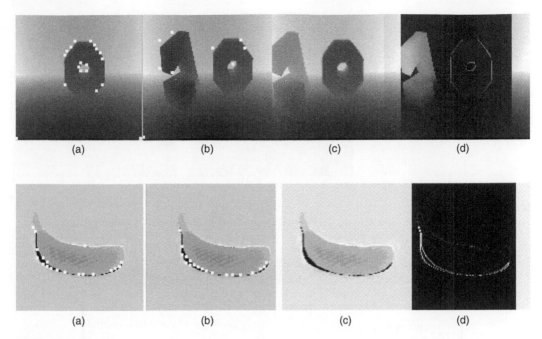

FIGURE 6.6
(a) Features extracted level 1, image 1, (b) features extracted level 1, image 2, (c) merged via averaging registered images, and (d) merged via subtraction of registered images.

close in value and location. Hypotheses are ranked by the number of features matched and how closely the features match. Examples are given in Figure 6.6.

6.6 Registration Assistance/Preprocessing

All the registration techniques discussed herein operate on the basic premise that there is identical content in the data sets being compared. However, the difficulty in registration pertains to the fact that the content is the same semantically, but often not numerically. For example, sensor readings taken at different times of the day can lead to lighting changes that can significantly alter the underlying data values. Also, weather changes can lead to significant changes in data sets. Registration of these kinds of data sets can be improved by first preprocessing the data. Figure 6.7 shows some preliminary work by Grewe[54] on the process of altering one image to appear more like another image in terms of photometric values. Such systems may improve registration systems of the future.

In some systems, we are dealing with very different kinds of data and registration can require operator assistance. Grewe et al.[22] have developed a system for aiding decisions and communication in disaster relief scenarios. The system, named DiRecT, takes as input different kinds of data; some such as images and maps follow the spatial grid pattern of images. However, DiRecT also takes in intelligence data from the field, which represent reports of victims, emergency situations, and facilities found by field workers. In this case, DiRecT requires the operator to gather information about the placement of these items in

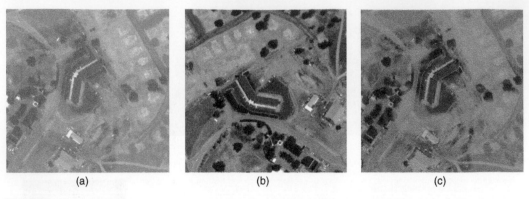

FIGURE 6.7
(a) Image 1, (b) image 2, and (c) new image 1 corrected to appear more like image 2 in photometric content.

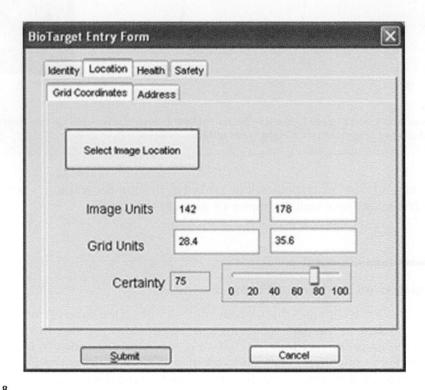

FIGURE 6.8
DiRecT GUI, which will direct operator to select location in current grid of, in this case a victim, or biotarget, along with a measure of certainty of this information. Different kinds of disaster items would have varying interfaces for registration in the scene. Victims are simply placed by location, orientation is not considered unlike other kinds of items.

the field, itself represented by a grid, from the reporter. Figure 6.8 indicates the process of how this can be done relative to a grid backdrop that may be visually aided by the display of a corresponding map if one is available. The operator not only fixes the hypothetical location using a mouse but also gives a measure of certainty about the location. This uncertainty in the registration process is visually represented in different controllable ways using uncertainty visualization techniques as shown in Figure 6.9.

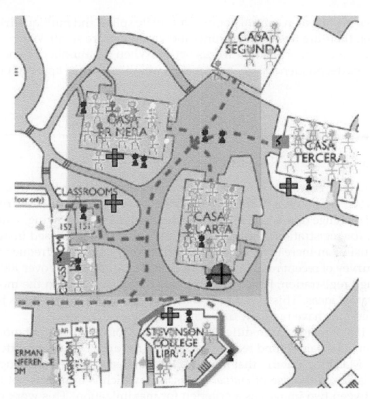

FIGURE 6.9
An image scene alterable in DiRecT interface.

6.7 Registration Using Elastic Transformations

Many registration problems assume that the images only undergo rigid body transformations. Unfortunately, this assumption is rarely true. Most technologies either rely on lenses that warp the data or have local imperfections that also complicate the mapping process. This is why elastic transformations are often needed to accurately register images. The approach from Bentoutou et al., which we discuss in Section 6.8, is a good example of how elastic transforms can be integrated into a general registration approach.[23] The rest of this section details how warping transforms registered images.

Kybic and Unser[27] and Xia and Liu[55] have used B-splines as basis functions for registering images. In both cases, a variant of Newton's algorithm uses gradient descent to optimize the mapping of the observed image to the reference image. Xia and Liu[55] varies from Kybic and Unser[27] in that the B-splines find a "super-curve," which maps two affine transforms.

Caner et al. have registered images containing small displacements with warping transforms.[56] To derive the elastic mapping, they have converted the 2D problem into a 1D problem by using a space-filling Hilbert curve to scan the image. Adaptive filtering techniques compensate for smooth changes in the image. This approach has been verified both by recovering a watermark from an intentionally distorted image and by compensating for lens distortions in photographs.

Probably the most useful results in this domain can be found in Ref. 57, which compares nonrigid registration techniques for problems defined by the number of control points and variation in spacing between control points. Thin-plate spline and multiquadric approaches work well when there are few control points and variations are small. When variations are large, piecewise linear interpolation works better. When the number of control points is large and their correspondences are inaccurate, a weighted means approach is preferable.

6.8   Multimodal Image Registration

Multimodal image registration is important for many applications. Real-time military and intelligence systems rely on a number of complementary sensing modes from a number of platforms. The range of medical imaging technologies keeps growing, which leads to a growing need for registration. Similarly, more and more satellite-based imaging services are available, using an increasing number of technologies and data frequencies. We have conducted a survey of recent radiology research literature and found over 100 publications a year on image registration; hence, we limit our discussion here to the most important and representative ideas. This problem is challenging because pixel values in the images being registered may have radically different meanings.

For registration of images from different kinds of sensors and different modalities, careful attention must be paid to find features or spaces that are invariant to differences in the modalities.[18,58] Matching criteria that measure invariant similarities are sought.[18,59] Pluim et al.[59] have discussed the use of *mutual information*, which is a measure of the statistical correlation between two images as a criterion for maximization. This work not only uses mutual information, which can work well for images from different modalities, but also includes a spatial component in matching via the inclusion of image gradient information, which in general is more invariant to differences in modalities than raw data. Similarly, Keller and Averbuch[58] have discussed some of the problems of using mutual information, including the problem of finding local, not global, maxima. They have suggested that as images become lower in resolution, or contain limited information or reduced overlap, they also become less statistically dependent and, hence, the maximization of mutual information becomes ill-posed. Keller and Averbuch[58] have developed a matching criterion that does not use the intensities of both images at the same time to reduce some of these effects. First, they have examined both images (I_1, I_2) and, using metrics related to sharpness, have determined which of the two images is of better quality (in focus, not blurred), I_2. The pixels corresponding to the locations of higher-gradient values of the poorer-quality image, I_1, are recorded. Iteratively, these pixels are aligned to the better-quality image, I_2, using only the pixel values from I_2. Specifically, the maximization of the magnitude of the intensity gradient of the matched pixels in I_2 is iteratively sought. Thus, there is at no step a direct comparison of I_1 and I_2 intensity values that supports invariance of multimodal differences.

One particularly important problem is mapping 2D images of varying modalities to 3D computed tomography data sets. Penney et al.[26] have matched (2D) fluoroscopy images to (3D) computed tomography images using translation, rotation, 3D perspective, and position on film. They have considered the following similarity measures:

- Normalized cross-correlation
- Entropy of the difference image
- Mutual information
- Gradient correlation

- Pattern intensity
- Gradient difference

They have compared their results using synthetic images derived from existing images. Ground truth has been derived by inserting fiducial artifacts in the images. The noise factors they have considered included adding soft tissue and stents. Their search algorithm has kept the optimal perspective known from the ground truth, introduced small errors in other parameters, and then used a greedy search by one parameter at a time. Their results found mutual information to be the least accurate measure. Correlation measures were sensitive to thin line structures. Entropy measures were not sensitive to thin lines, but fail when slowly varying changes exist. Pattern intensity and gradient difference approaches have been found to be most applicable to medical registration problems.

Kybic and Unser[27] have used B-spline basis functions to deform 2D images so as to best match a 3D reference image as determined by a sum-square of differences metric. In their work, splines have been shown to be computationally more efficient than wavelet basis functions. Because warping could make almost any image fit another one, they assume some operator intervention. The optimization algorithms they use include gradient descent and a variation of Newton's algorithm. This approach was tested on a number of real and synthetic data sets.

Another medical imaging problem involving mapping sensing modes of the same dimensionality, such as ultrasound and magnetic resonance, can be found in Yao and Goh.[24] There is no clear correspondence between the intensity values of the two sensing modes. It is also unlikely that all the control points in one mode exist in the other mode. To counter this, a "pixel migration" approach is considered. The control points chosen are gradient maxima, and the fitness function they use is the sum of the squared gradient magnitude (SSG). Unfortunately, the SSG maximum is not necessarily the best solution. To counter this, they partition the search space into feasible and unfeasible regions and restrict the search to feasible regions. A genetic algorithm is then used to find the optimal matching within the feasible regions. Electrooptical, infrared, and synthetic aperture radar (SAR) images are matched in examples.

Other multimodal registration problems have been addressed for remote sensing applications. Inglada and Giros[34] have performed subpixel accurate registration of images from different sensing modalities using rigid body transformations. This is problematic because the different sensing modes have radically different radiometry values. Correlation metrics are not applicable to this problem. The authors[34] have presented a number of similarity measures that are very useful for this problem, including the following:

- Normalized standard deviation—the variance of the differences between pixel values of two images will be minimized when the two are registered correctly
- Correlation ratio—which is similar conceptually to normalized standard deviation but provides worse results
- A normalized version of the χ^2-test—measures the statistical dependence between two images (should be maximal when registration is correct)
- The Kolmogorov distance of the relative entropy between the two images
- The mutual information content of the two images
- Many additional entropy-related metrics
- A cluster rewards algorithm that measures the dispersion of the marginal histogram of the registered image

They have tested these methods by registering optical and SAR remote sensing images.

In the study by Bentoutou et al.,[23] SAR and Systeme Pour l'Observation de la Terre [French remote sensing satellite] (SPOT) optical remote sensing images have been registered. They have considered translation rotation and scaling (i.e., rigid body) transformations. Their algorithm has been the following:

- Control points found by edge detection, retaining points with large gradients. A strength measure computed by finding local maxima as a function of the determinant of the autocorrelation function is used to reduce the number of control points.

- Central moments of the set of control functions are used to derive image templates that are made up of invariant features. Matches are found from the minimum distance between control points in the invariant space.

- An inverse mapping is found for the least-squares affine transform that best fits the two pairs of strongest control points.

- An interpolation algorithm finally warps the image to perform subpixel registration.

The approaches presented are typical of the state of this field, and provide insights that can be used in many related applications.

Related to registration preprocessing is the topic of sensor calibration. The meaning of calibration changes with the sensor and even its use in an application. In some ways, calibration is similar to registration, but here the goal is not correlation between sensors but from a sensor to some known ground truth. Whitehouse and Culler[60] have discussed the problems of calibration of multiple sensors. They have specifically addressed the application of localization of a target, but discussed the use in other multisensor network applications. The calibration problem is framed as a general parameter estimation problem. However, rather than calibrating each sensor independently, they have chosen for each sensor the parameters that optimize the entire sensor networks response. This, although not necessarily optimal for each sensor, eliminates the work of observing each sensor if a sensor network response can be formulated and observed. Also, it reduces the task of collecting ground truth for each sensor to a task that collects ground truth for the network response. Interestingly, in this work[60] and specifically for their application of simple localization, they have reduced individual sensor calibration and multisensor registration to one of "network calibration."

6.9 Theoretical Bounds

In practice, the images being registered are always less than ideal. Data are lost in the sampling process and the presence of noise further obscures the object being observed. Because noise processes are due to random processes following statistical laws, it is natural to phrase the image registration problem as a statistical decision problem. For example, the registration fitness functions given in Section 6.4 (see Equations 6.1 through 6.3) include random variables and are suited to statistical analysis. Recent research has established theoretical bounds to image registration using both the confidence intervals (CIs) and the Cramer–Rao lower bound (CRLB).

Two recent approaches use CIs to determine the theoretical bounds of their registration approach. Simonson et al.[61] have used statistical hypothesis testing for determining when to accept or reject candidate registrations. They try to find translations that match pairs of "chip" regions and calculate joint confidence regions for these correspondences. If the statistics do not adequately support the hypothesis, the registration is rejected. The statistical foundation of their approach helps make it noise tolerant. Wang et al. have used the residual sum of squares statistic to determine CIs for their registration results.[62] This information is valuable because it provides the application with not only an answer but also a metric for how accurate the result is likely to be.

A larger body of research has found the CRLB for a number of registration approaches. Unlike the CI, which states that the true answer would be within a known envelope for a given percentage of the data samples, CRLB expresses the best performance that could ever be obtained for a specific problem. This information is useful to researchers because it tells them objectively how close their approach is to the best possible performance.

Li and Leung[63] have registered images using a fitness function containing both control points and intensity values. They have searched for the best affine match between the observed and the reference images by varying one parameter at a time while keeping all the others fixed. This process iterates until a good solution is found. The CRLB they have derived for this process shows that the combination of control points and intensity values can perform better than either set of features alone.

Robinson and Milanfar[64] have taken a similar approach to finding the theoretical bounds of registration. They look at translations of images corrupted with Gaussian white noise. The CRLB of this process provides important insights. For periodic signals, registration quality is independent of the translation distance. It depends only on the image contents. However, translation registration accuracy depends on the angle of translation. Unfortunately, unbiased estimators do not generally exist for image registration. The CRLB for gradient-based translation algorithms is given in Ref. 64.

Yetik and Nehorai[65] have carried this idea further by deriving the CRLB for 2D affine transforms, 3D affine transforms, and warping transformations using control points. For translations, accuracy depends only on the number of points used, not on their placement or the amount of translation. These results are consistent with those of Robinson and Milanfar.[64] For rotations, it is best to use points far away from the rotation axis. Affine transformation registration accuracy is bounded by noise and the points but is independent of transformation parameters. For warping transformations, accuracy depends on the warping parameters. The bounds get to be large as the amount of warping increases.

6.10 Conclusion

Addressing the data registration problem is an essential preprocessing step in multisensor fusion. Data from multiple sensors must be transformed onto a common coordinate system. This chapter has provided a survey of existing methods, including methods for finding registrations and applying registrations to data after they have been found. In addition, example approaches have been described in detail.

Brooks and Iyengar[3] and Chen et al.[48] have detailed meta-heuristic-based optimization methods that can be applied to raw data. Of these methods, TRUST, a new meta-heuristic from Oak Ridge National Laboratories, is the most promising. Fitness functions have been given for readings corrupted with Gaussian, uniform, and salt-and-pepper noise. Because these methods use raw data, they are computationally intensive.

Grewe and Brooks[51] have presented a wavelet-based approach to registering range data. Features are extracted from the wavelet domain. A feedback approach is then applied to search for good registrations. Use of the wavelet domain compresses the amount of data that must be considered, providing increased computational efficiency. Drawbacks to using feature-based methods have also been discussed in this chapter.

Acknowledgments

The study has been sponsored by the Defense Advance Research Projects Agency (DARPA) and Air Force Research Laboratory, Air Force Materiel Command, USAF, under agreement number F30602-99-2-0520 (Reactive Sensor Network). The U.S. Government is authorized to reproduce and distribute reprints for governmental purposes notwithstanding any copyright annotation thereon. The views and conclusions that have been contained herein are those of the authors and should not be interpreted as necessarily representing the official policies or endorsements, either expressed or implied, of the DARPA, the Air Force Research Laboratory, or the U.S. Government.

References

1. J. P. W. Pluim and J. M. Fitzpatrick, "Image registration" introduction to special issue, *IEEE Transactions on Medical Imaging*, 22(11), 1341–1342, 2003.
2. D. C. Montgomery, *Design and Analysis of Experiments, 4th Edition*, Wiley, New York, NY, 1997.
3. R. R. Brooks and S. S. Iyengar, *Multi-Sensor Fusion: Fundamentals and Applications with Software*, Prentice Hall, PTR, Upper Saddle River, NJ, 1998.
4. B. V. Dasarathy, *Decision Fusion*, IEEE Computer Society Press, Los Alamitos, CA, 1994.
5. D. L. Hall, *Mathematical Techniques in Multi-sensor Fusion*, Artech House, Norwood, MA, 1992.
6. R. Blum and Z. Liu, *Multi-Sensor Image Fusion and Its Applications*, CRC Press, Boca Raton, FL, 2006.
7. R. C. Gonzalez and R. E. Woods, *Digital Image Processing*, Addison-Wesley, Reading, PA, 302, 1993.
8. L. G. Brown, A survey of image registration techniques, *ACM Computing Surveys*, 24(4), 325–376, 1992.
9. M. A. Oghabian and A. Todd-Pokropek, Registration of brain images bay a multiresolution sequential method, in *Information Processing in Medical Imaging*, Springer, Berlin, pp. 165–174, 1991.
10. V. R. Mandara and J. M. Fitzpatrick, Adaptive search space scaling in digital image registration, *IEEE Transactions on Medical Imaging*, 8(3), 251–262, 1989.
11. C. A. Palazzini, K. K. Tan, and D. N. Levin, Interactive 3D patient-image registration, *Information Processing in Medical Imaging*, Springer, Berlin, pp. 132–141, 1991.
12. P. Van Wie and M. Stein, A LANDSAT digital image rectification system, *IEEE Transactions on Geoscience Electronics*, GE-15, 130–137, 1977.
13. G. Stockman, S. Kopstein, and S. Benett, Matching images to models for registration and object detection via clustering, *IEEE Transactions on Pattern Analysis and Machine Intelligence*, 4(3), 229–241, 1982.
14. J. R. Jain and A. K. Jain, Displacement measurement and its application in interface image coding, *IEEE Transactions on Communications*, COM-29(12), 1799–1808, 1981.
15. A. Goshtasby, Piecewise linear mapping functions for image registration, *Pattern Recognition*, 19(6), 459–466, 1986.
16. G. Wolberg, *Digital Image Warping*, IEEE Computer Society Press, Los Alamitos, CA, 1964.
17. R. Y. Wong and E. L. Hall, Performance comparison of scene matching techniques, *IEEE Transaction on Pattern Analysis and Machine Intelligence*, PAMI-1(3), 325–330, 1979.
18. M. Irani and P. Anandan, Robust multisensor image alignment, in *IEEE Proceedings of International Conference on Computer Vision*, 959–966, 1998.
19. A. Mitiche and K. Aggarwal, Contour registration by shape specific points for shape matching comparison, *Vision, Graphics and Image Processing*, 22, 396–408, 1983.

20. V. Petrushin, G. Wei, R. Ghani, and A. Gershman, Multiple sensor integration for indoor surveillance, in *Proceedings of the 6th International Workshop on Multimedia Data Mining: Mining Integrated Media and Complex Data MDM'05*, 53–60, 2005.

21. M. Michel and V. Stanford, Synchronizing multimodal data streams acquired using commodity hardware, in *Proceedings of the 4th ACM International Workshop on Video Surveillance and Sensor Networks VSSN'06*, 3–8, 2006.

22. L. Grewe, S. Krishnagiri, and J. Cristobal, DiRecT: A disaster recovery system, *Signal Processing, Sensor Fusion, and Target Recognition XIII*, Proceedings of SPIE, Vol. 5429, 480–489, 2004.

23. Y. Bentoutou, N. Taleb, K. Kpalma, and J. Ronsin, An automatic image registration for applications in remote sensing, *IEEE Transactions in Geoscience and Remote Sensing*, 43(9), 2127–2137, 2005.

24. J. Yao and K. L. Goh, A refined algorithm for multisensor image registration based on pixel migration, *IEEE Transactions on Image Processing*, 15(7), 1839–1847, 2006.

25. J. Han and B. Bhanu, Hierarchical multi-sensor image registration using evolutionary computation, *Proceedings of the 2005 Conference on Genetic and Evolutionary Computation*, 2045–2052, 2005.

26. G. P. Penney, J. Weese, J. A. Little, P. Desmedt, D. L. G. Hill, and D. J. Hawkes, A comparison of similarity measures for use in 2D–3D medical image registration, *IEEE Transactions on Medical Imaging*, 17(4), 586–595, 1998.

27. J. Kybic and M. Unser, Fast parametric image registration, *IEEE Transactions on Image Processing*, 12(11), 1427–1442, 2003.

28. J. Andrus and C. Campbell, Digital image registration using boundary maps, *IEEE Transactions on Computers*, 19, 935–940, 1975.

29. D. Barnea and H. Silverman, A class of algorithms for fast digital image registration, *IEEE Transaction on Computers*, C-21(2), 179–186, 1972.

30. A. Goshtasby, Image registration by local approximation methods, *Image and Vision Computing*, 6(4), 255–261, 1988.

31. B. R. Horn and B. L. Bachman, Using synthetic images with surface models, *Communications of the ACM*, 21, 914–924, 1977.

32. H. G. Barrow, J. M. Tennenbaum, R. C. Bolles, and H. C. Wolf, Parametric correspondence and chamfer matching: Two new techniques for image matching, in *Proceedings of International Joint Conference on Artificial Intelligence*, 659–663, 1977.

33. A. Pinz, M. Prontl, and H. Ganster, Affine matching of intermediate symbolic presentations, in *CAIP'95 Proceedings LNCS 970*, Hlavac and Sara (Eds.), Springer-Verlag, New York, NY, pp. 359–367, 1995.

34. J. Inglada and A. Giros, On the possibility of automatic multisensor image registration, *IEEE Transactions on Geoscience and Remote Sensing*, 42(10), 2104–2102, 2004.

35. X. Dai and S. Khorram, A feature-based image registration algorithm using improved chaincode representation combined with invariant moments, *IEEE Transactions on Geoscience and Remote Sensing*, 37(5), 2351–2362, 1999.

36. Z. Yang and F. S. Cohen, Cross-weighted moments and affine invariants for image registration and matching, *IEEE Transactions on Pattern Analysis and Machine Intelligence*, 21(8), 804–814, 1999.

37. R. Chellappa, Q. Zheng, P. Burlina, C. Shekhar, and K. B. Eom, On the positioning of multisensory imagery for exploitation and target recognition, *Proceedings of IEEE*, 85(1), 120–138, 1997.

38. Y. Zhou, H. Leung, and E. Bosse, Registration of mobile sensors using the parallelized extended Kalman filter, *Optical Engineering*, 36(3), 780–788, 1997.

39. G. K. Matsopoulos, N. A. Mouravliansky, K. K. Delibasis, and K. S. Nikita, Automatic retinal image registration scheme using global optimization techniques, *IEEE Transactions on Information Technology in Biomedicine*, 3(1), 47–68, 1999.

40. L. Silva, O. R. P. Bellon, and K. L. Boyer, Precision range image registration using a robust surface interpenetration measure and enhanced genetic algorithms, *IEEE Transactions on Pattern Analysis and Machine Intelligence*, 27(5), 762–776, 2005.

41. S. Zokai and G. Wolberg, Image registration using log-polar mappings for recovery of large-scale similarity and projective transformations, *IEEE Transactions on Image Processing*, 14(10), 1422–1434, 2005.

42. H. S. Stone, J. Lemoigne, and M. McGuire, The translation sensitivity of wavelet-based registration, *IEEE Transactions on Pattern Analysis and Machine Intelligence*, 21(10), 1074–1081, 1999.
43. A. Goshtasby, Piecewise cubic mapping functions for image registration, *Pattern Recognition*, 20(5), 525–535, 1987.
44. X. Wang, D. Feng, and H. Hong, Novel elastic registration for 2D medical and gel protein images, in *Proceedings of the First Asia-Pacific Bioinformatics Conference on Bioinformatics 2003*, Vol. 19, 223–22, 2003.
45. H. S. Baird, *Model-Based Image Matching Using Location*, MIT Press, Boston, MA, 1985.
46. L. S. Davis, Shape matching using relaxation techniques, *IEEE Transaction on Pattern Analysis and Machine Intelligence*, 1(1), 60–72, 1979.
47. R. R. Brooks and S. S. Iyengar, Self-calibration of a noisy multiple sensor system with genetic algorithms, self-calibrated intelligent optical sensors and systems, SPIE, Bellingham, WA, in *Proceedings of SPIE International Symposium Intelligent Systems and Advanced Manufacturing*, 1995.
48. Y. Chen, R. R. Brooks, S. S. Iyengar, N. S. V. Rao, and J. Barhen, Efficient global optimization for image registration, *IEEE Transactions on Knowledge and Data Engineering*, 14(1), 79–92, 2002.
49. F. S. Hill, *Computer Graphics*, Prentice Hall, Englewood Cliffs, NJ, 1990.
50. R. R. Brooks, *Robust Sensor Fusion Algorithms: Calibration and Cost Minimization*, PhD dissertation in Computer Science, Louisiana State University, Baton Rouge, LA, 1996.
51. L. Grewe and R. R. Brooks, Efficient registration in the compressed domain, in *Wavelet Applications VI, SPIE Proceedings*, H. Szu (Ed.), Vol. 3723, AeroSense 1999.
52. R. Sharman, J. M. Tyler, and O. S. Pianykh, Wavelet-based registration and compression of sets of images, *SPIE Proceedings*, Vol. 3078, 497–505, 1997.
53. R. A. DeVore, W. Shao, J. F. Pierce, E. Kaymaz, B. T. Lerner, and W. J. Campbell, Using nonlinear wavelet compression to enhance image registration, *SPIE Proceedings*, Vol. 3078, 539–551, 1997.
54. L. Grewe, *Image Correction*, Technical Report, CSUMB, 2000.
55. M. Xia and B. Liu, Image Registration by "Super-Curves", *IEEE Transactions on Image Processing*, 13(5), 720–732, 1994.
56. G. Caner, A. M. Tekalp, G. Sharma, and W. Heinzelman, Local image registration by adaptive filtering, *IEEE Transactions on Image Processing*, 15(10), 3053–3065, 2006.
57. L. Zagorchev and A. Goshtasby, A comparative study of transformation functions for nonrigid image registration, *IEEE Transactions on Image Processing*, 15(3), 529–538, 2006.
58. Y. Keller and A. Averbuch, Multisensor image registration via implicit similarity, *IEEE Transactions on Pattern Analysis and Machine Intelligence Archive*, 28, 794–801, 2006.
59. J. Pluim, J. Maintz, and M. Viergever, Image registration by maximization of combined mutual information and gradient information, *IEEE Transactions on Medical Imaging*, 19(8), 809–814, 2000.
60. K. Whitehouse and D. Culler, Macro-calibration in sensor/actuator networks, *Mobile Networks and Applications*, 8, 463–472, 2003.
61. K. M. Simonson, S. M. Drescher Jr., and F. R. Tanner, A statistics-based approach to binary image registration with uncertainty analysis, *IEEE Transactions on Pattern Analysis and Machine Intelligence*, 29(1), 112–125, 2007.
62. H. S. Wang, D. Feng, E. Yeh, and S. C. Huang, Objective assessment of image registration results using statistical confidence intervals, *IEEE Transactions on Nuclear Science*, 48(1), 106–110, 2001.
63. W. Li and H. Leung, A maximum likelihood approach for image registration using control point and intensity, *IEEE Transactions on Image Processing*, 13(8), 1115–1127, 2004.
64. D. Robinson and P. Milanfar, Fundamental performance limits in image registration, *IEEE Transactions on Image Processing*, 13(9), 1185–1199, 2004.
65. I. S. Yetik and A. Nehorai, Performance bounds on image registration, *IEEE Transactions on Signal Processing*, 54(5), 1737–1749, 2006.
66. R. R. Brooks, S. S. Iyengar, and J. Chen, Automatic correlation and calibration of noisy sensor readings using elite genetic algorithms, *Artificial Intelligence*, 84(1), 339–354, 1996.
67. X. Dai and S. Khorram, A feature-based image registration algorithm using improved chain-code representation combined with invariant moments, *IEEE Transactions on Geoscience and Remote Sensing*, 37(5), 2351–2362, 1999.

7

Data Fusion Automation: A Top-Down Perspective

Richard Antony

CONTENTS

7.1 Introduction

This chapter offers a conceptual-level view of the data fusion process and discusses key principles associated with both data analysis and information combination. The discussion begins with a high-level view of data fusion requirements and analysis options. Although the discussion focuses on tactical situation awareness development, a much wider range of applications exists for this technology.

After motivating the concepts behind effective information combination and decision making through the use of a number of simple metaphors, the chapter

- Presents a top-down view of the data fusion process
- Discusses the inherent complexities of combining uncertain, erroneous, and fragmentary information
- Identifies key requirements for achieving practical and effective machine-based reasoning

7.1.1   Biological Fusion Metaphor

Sensory fusion in biological systems provides a natural metaphor for studying artificial data fusion systems.[1] As with any good metaphor, consideration of a simpler or more familiar phenomenon can provide valuable insight into the study of more complex or less familiar processes.

Even the most primitive animals sense their environment, develop some level of situation awareness, and react to the acquired information. By assisting in the acquisition of food and the avoidance of threats, situation awareness directly supports survival of the species. A barn owl, for instance, fuses visual and auditory information to help accurately locate mice under very low light conditions, whereas a mouse responds to threatening visual and auditory cues to avoid becoming the owl's dinner.

In general, natural selection has tended to favor the development of more capable senses (sensors) and more effective utilization of the derived information (exploitation and fusion). Color vision in humans, for instance, is believed to have been a natural adaptation that permitted apes to find ripe fruit in trees. Situation awareness in animals can rely on a single, highly developed sense, or on multiple, often less capable, senses. A hawk depends principally on a highly acute visual search and tracking capability, whereas a shark relies primarily on its sense of smell when hunting. Sexual attraction can depend primarily on sight (plumage), smell (pheromones), or sound (mating call). For humans, sight is arguably the most vital sense, with hearing a close second. Dogs, however, rely more heavily on the senses of smell and hearing, with vision playing the role of a secondary information source.

Sensory input in biological organisms typically supports both sensory cueing and situation awareness development. Sounds cue the visual sense to the presence and the general direction of a potentially relevant event. Information gained by the aural sense (i.e., direction, speed, and tentative object classification) is then combined (fused) with the information gathered by the visual system to produce a more complete, higher-confidence, or higher-level situation awareness.

In many cases, sensory fusion can be critical to successful decision making. Food that looks appetizing (sight) might be extremely salty (taste), spoiled (smell), or too hot (touch). At the other extreme, sensory fusion might be unnecessary if the various senses provide redundant, as well as highly reliable, information. Bacon frying in a pan, for instance, need not be seen, smelled, and tasted because any of the senses taken alone would suffice.

Although it might seem prudent to simply ignore apparently redundant information, such information can help sort out conflicts, both intentional (deception) and unintentional (confusion). Whereas single-source deception is reasonably straightforward to perpetrate, deception across multiple senses (sensor modalities) is considerably more difficult. Successful hunting and fishing depend, to a large degree, on effective multisource deception. Duck hunters use both decoys and mating calls to simultaneously provide deceptive visual and auditory information. Because deer can sense danger through the sense of smell, sound, and sight, the shrewd hunter must mask his scent (or stay down-wind), make little or no noise, and remain motionless if the deer looks in his direction. In nonadversarial applications, data fusion provides a natural means of resolving unintentional conflicts that arise due to uncertainty in both the measurement and the decision spaces.

Sensory fusion need not be restricted to the familiar five senses of sight, sound, smell, taste, and touch. Internal signals, such as acidity of the stomach, coupled with visual cues, olfactory cues, or both, can trigger hunger pains. The fusion of vision, inner-ear balance information, and muscle feedback signals facilitates motor control. Measurement and signature intelligence (MASINT) in tactical applications likewise focuses on a wide range of "nontraditional" information classes.

7.1.2 Puzzle-Solving Metaphor

Because situation awareness development requires the production and maintenance of an adequate multiple level-of-abstraction picture of a (dynamic) situation, the data fusion process can be compared to assembling a jigsaw puzzle for which no picture of the completed scene exists. Although assembling puzzles that contain hundreds of pieces (*information fragments*) can challenge an individual's skill and patience, the production of a comprehensive situational picture, created by fusing disparate and fragmentary sensor-derived information, typically represents a far more challenging task. Despite the fact that a jigsaw puzzle represents a static product, the assembly (fusion) strategy evolves over time once patterns begin to develop. In tactical situation awareness applications, time represents a key problem dimension, with both the information fragments and the fused picture constantly evolving.

The partially completed puzzle (*fused situation awareness product*) illustrated in Figure 7.1 contains numerous aggregate objects (i.e., forest and meadow), each composed of simpler objects (i.e., trees and ground cover). Each of these objects, in turn, has been assembled from multiple puzzle pieces, some representing a section of bark on a single tree trunk, whereas

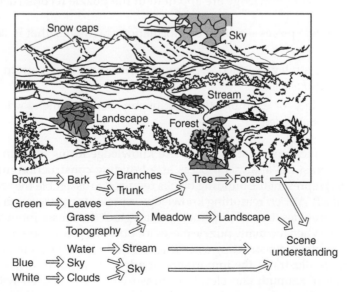

FIGURE 7.1
Puzzle-solving metaphor example.

others might be a grassy area associated with a meadow. In general, sensor-derived information (individual puzzle pieces) can run the gamut from providing just color and texture to depicting higher level-of-abstraction objects such as trees and buildings.

At the outset of the puzzle construction process, problem solving necessarily relies on rather general problem-solving strategies (e.g., locate border pieces). Because there exists little context to direct either puzzle piece selection or placement, simple, brute-force pattern matching strategies are used at the early stages of the process. A blue-colored piece, for example, might represent either sky or water with little basis for distinguishing between the two interpretations during the early stages of the analysis. If the pieces came from a previously opened box, additional complications arise because some pieces may be missing (*no available sensor data*) whereas pieces from another puzzle may have inadvertently been dumped into the box (*irrelevant or erroneous sensor reports*). Once portions of the scene begin to fill in, the assembly process becomes much more goal-directed.

Slipping a single puzzle piece into the correct spot reduces entropy and supports higher level-of-abstraction scene interpretation at the same time. As regions of the puzzle begin to take form, identifiable features in the scene emerge (e.g., trees, grass, and cliffs), permitting higher-level interpretations to be developed (e.g., forest, meadows, and mountains). By supporting the placement of the individual pieces, as well as the goal-driven search (*sensor resource management*) for specific pieces, *context* provided by the developing scene (*situation awareness product*) helps further focus the construction process (*fusion process optimization*).

Just as duplicate or erroneous pieces significantly complicate puzzle assembly, redundant and irrelevant sensor-derived information similarly burdens machine-based situation development. Thus, goal-directed information collection offers a twofold benefit: critical information requirements are satisfied, and the collection (and subsequent analysis) of unnecessary information is minimized. Although numerous puzzle pieces may be yet unplaced (*undetected objects*) and some pieces might actually be missing (*information not collectible by the available sensor suite*), a reasonably comprehensive, multiple level-of-abstraction understanding of the overall scene (*situation awareness*) gradually emerges.

Three broad classes of knowledge are apparent in the puzzle reconstruction metaphor:

1. Individual puzzle pieces—collected information fragments, that is, *sensor-derived knowledge*
2. Puzzle-solving strategies, such as edge piece-finding and pattern matching—*a priori reasoning knowledge*
3. World knowledge, for example, the relationship between meadows and grass—*domain context knowledge*

To investigate the critical role that each of these knowledge forms plays in fusion product development, we recast the analysis process as a building construction metaphor. Puzzle pieces (*sensor input*) represent the building blocks required to assemble the scene (*fused situation awareness product*). *A priori* reasoning knowledge represents construction knowledge and skills, and context provides the nails and mortar that "glue" the sensor input together to form a coherent product. When too many puzzle pieces (or building blocks) are missing (*inadequate sensor-derived information*), constructing a scene (or a building) becomes impossible.

A simple example illustrates the importance of adequate input information. Figure 7.2a illustrates a cluster of azimuth and elevation measurements associated with two separate groups of air targets. Given the spatial overlap between the data sets, reliable target-to-group assignment may not be possible, regardless of the selected analysis paradigm or the extent of

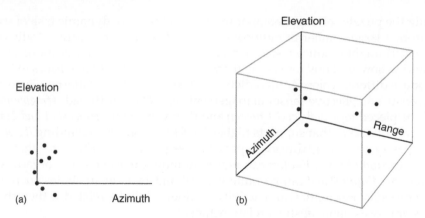

FIGURE 7.2
(a) Two-dimensional measurements and (b) the corresponding three-dimensional measurement space.

algorithm training. However, with the addition of range measurements (*increased measurement space dimensionality*), two easily separable clusters become readily apparent (Figure 7.2b). Because the information content of the original 2D data set was fundamentally inadequate, even sophisticated clustering algorithms would be unable to discriminate between the two target groups. However, with the addition of the third measurement dimension, even a crude clustering algorithm easily accomplishes the required decision task.

A spectrum of problem-solving paradigms can be used to implement *reasoning knowledge* (e.g., rules, procedures, and statistic-based algorithms), evidence combination strategies (e.g., Bayes, Dempster–Shafer, and fuzzy set theory), and decision-making approaches (e.g., rule instantiation and parametric algorithms). In general, the process of solving a complex puzzle (or performing automated situation awareness) benefits from both bottom-up (deductive-based) and top-down (goal-directed) reasoning that exploits relationships among the hierarchy of domain entities (i.e., primitive, composite, aggregate, and organizational).

In the puzzle-solving metaphor, *context knowledge* refers to relevant domain knowledge not explicitly contained within a puzzle piece (*nonsensor-derived knowledge*). Humans routinely apply a wide range of contextual knowledge during analysis and decision making.* For example, context-sensitive evaluation of Figure 7.1 permits the conjecture that the picture is a summer scene taken somewhere in the western United States. The season and location are deduced from the presence of deciduous trees in full leaf (summer) in the foreground and jagged snow-capped mountain peaks in the distance (western United States). In a similar fashion, the exploitation of context knowledge in automated fusion systems can promote more robust interpretations of sensor-derived information.

During both puzzle assembly and automated situation development, determining when an *adequate* situation representation has been achieved can be difficult. In the puzzle reconstruction problem, although the general landscape characteristics might be evident, missing puzzle pieces could depict denizens of the woodland community that can be hypothesized, but for which no compelling evidence yet exists. Individual puzzle pieces might contain partial or ambiguous information. For example, the presence of a section of log wall in the evolving scene suggests the possibility of a log cabin. However, additional evidence is required to validate this hypothesis.

---

* This fact partially accounts for the disparity in performance between manual and automated approaches to data fusion.

Recasting the puzzle-solving metaphor in terms of the more dynamic task of weaving a tapestry from a large number of multicolored threads can be instructive. Individual sensors (e.g., *radar*) might produce just a portion of a single thread (red portions for moving vehicles and yellow portions for stationary vehicles), whereas other sensors (e.g., *imagery*) might produce different colored "threadlets" or even higher-level patterns. Given adequate sensor diversity and effective sensor management (*level 4 fusion*), threads (*fragmentary object tracks*) can be pieced together (*level 1 fusion*) and then woven (*fusion levels 1 and 2*) to generate a "picture" of a scene that supports higher-level situation understanding (*level 3 fusion*).

In the tapestry metaphor, threads represent a sequence of *entity states* that implicitly occur as a function of time. *Entities* can represent individuals, vehicles, organizations, or abstract objects. *Entity states* can be quite general and include attributes such as status, location, emissions, kinematics, and activity. Entities might be related if they share common states or possess more abstract relationships.

7.1.3   Command and Control Metaphor

The game of chess provides an appropriate metaphor for military command and control (C^2), as well as an abstract metaphor for any system that senses and reacts to its environment. Both chess players and battlefield commanders require a clear picture of the "playing field" to properly evaluate the options available to them and their opponents. In both chess and C^2, opposing players command numerous individual resources (i.e., pieces or units) that possess a range of characteristics and capabilities. Resources and strategies vary over time. Groups of chess pieces are analogous to higher-level organizations on the battlefield. The chessboard represents domain constraints to movement that are similar to constraints posed by terrain, weather, logistics, and other features of the military problem domain. Player-specific strategies are analogous to tactics, whereas legal moves represent established doctrine. In both domains, the overall objective of an opponent may be known, whereas specific tactics and subgoals must be deduced.

Despite a chess player's complete knowledge of the chessboard (*all domain constraints*), the location of all pieces (*own and opponent-force locations*), all legal moves (*own and opponent-force doctrine*), and his ability to exercise direct control over all of his own assets, chess remains a highly challenging game. Tactical situation development possesses numerous complicating factors that make it far more of a challenge.

First, battlefield commanders rarely possess a complete or fully accurate picture of their own forces let alone those of their adversaries. If forced to deal with incomplete and inaccurate force structure knowledge, as well as location uncertainty, chess players would be reduced to guessing the location and composition of an adversary's pieces, akin to playing "Battleship," the popular children's game.

Second, individual sensors provide only limited observables, coverage, resolution, and accuracy. Thus, the analysis of individual sensor reports tends to lead to ambiguous and rather local interpretations. Third, domain constraints in tactical situation awareness are considerably more complex than the well-structured (and level) playing field in chess. Fourth, doctrinal knowledge in the tactical domain tends to be more difficult to exploit effectively and far less reliable than its counterpart in chess.

Application-motivated metaphors can be useful in terms of extending current fusion applications. For example, data fusion seems destined to play a significant role in the development of future "smart highway" control systems. The underpinning of such a system is a sophisticated control capability that optimally resolves a range of conflicting requirements, such as (1) to expedite the movement of both local and long-distance traffic, (2) to ensure maximum safety for all vehicles, and (3) to create the minimum environmental impact. The actors in the metaphor are drivers (or automated vehicle control systems), the rules of the game are the "rules of the road," and domain constraints are the road network and traffic control means. Players possess individualized objectives and tactics; road characteristics and vehicle performance capabilities provide physical constraints on the problem solution.

7.1.4 Evidence Combination

Because reliance on a single information source can lead to ambiguous, uncertain, and inaccurate situation awareness, data fusion seeks to overcome such limitations by synergistically combining relevant (and available) information sources leading to the generation of a potentially more consistent, accurate, comprehensive, and global situation awareness. A famous poem by John Godfrey Saxe[2] written more than a century ago, aptly demonstrates both the need for and the challenge of effectively combining fragmentary information.

The poem describes an attempt by six blind men to gain a first-hand understanding of an elephant. The first man happens to approach the elephant from the side and surmises that an elephant must be similar to a wall. The second man touches the tusk and imagines an elephant to be like a spear. The third man approaches the trunk and decides an elephant is rather like a snake. The fourth man reaches out and touches a leg and decides that an elephant must be similar to a tree. The fifth man chances to touch an ear and imagines an elephant must be like a fan. The sixth man grabs the tail and concludes an elephant is similar to a rope. Although each man's assessment is entirely consistent within his own limited sensory space and myopic frame of reference, a true picture of an elephant emerges only after these six observations are effectively integrated (*fused*).

Among other insights, the puzzle-solving metaphor presented earlier illustrated that (1) complex dependencies can exist between the information fragments and the completed situation description and (2) determining whether an individual puzzle piece actually belongs to the scene being assembled can be difficult. Even when the collected information is known to be relevant, based on strictly local interpretations, it might not be possible to determine whether a given blue-colored piece represents sky, water, or some other feature class.

During criminal investigations, observations are assembled, investigators hunt for clues, and motives are evaluated. Tactical data fusion involves a similar approach. Although the existence of a single strand of hair might appear insignificant at the outset of a criminal investigation, it might prove to be the key piece of evidence that discriminates among several suspects. Similarly, seemingly irrelevant pieces of sensor-derived information might ultimately link observations with motives, or provide other significant benefits.

Thus, both the information content (*information measure*) of a given piece of relevant data and its relationship to the overall fusion task are vital considerations. As a direct consequence, the development of a simple information theoretic framework for the data fusion process appears problematic. Assessing the utility of information in a given application effectively demands a top-down, holistic analysis approach.

7.1.5 Information Requirements

Because no widely accepted theory exists for determining when adequate information has been assembled to support a given fusion task, empirical measures of performance must generally be relied upon to evaluate the effectiveness of both individual fusion algorithms and an overall fusion system. In general, data fusion performance can be enhanced by

- Technical improvements in sensor measurements (i.e., longer range, higher resolution, improved signal-to-noise ratio, better accuracy, higher reliability)
- Increasing measurement space dimensionality by employing heterogeneous sensors that provide at least partially independent information
- Spatially distributing sensors to provide improved coverage, perspective, and measurement reliability
- Incorporation of relevant nonsensor-derived domain knowledge to constrain information combination and the decision-making process

In general, effective data fusion automation requires the development of robust, context-sensitive algorithms that are practical to implement. *Robust* performance argues for the use of all potentially relevant sensor-derived information sources, reasoning knowledge, and maximal utilization of relevant nonsensor-derived information. However, to be *practical* to implement and efficient enough to employ in an operational setting, the algorithms may need to compromise fusion performance quality. Consequently, system developers must quantify or otherwise assess the value of these various information sources in light of system requirements, moderated by programmatic, budgetary, and performance constraints (e.g., decision time-line and hardware capability). The interplay between achieving optimal algorithm robustness and context sensitivity on the one hand, and a practical implementation on the other, is a fundamental tension associated with virtually any form of machine-based reasoning directed at solving complex, real-world problems.

7.1.6   Problem Dimensionality

Effective situational awareness (with or without intentional deception) generally benefits from the collection and analysis of a wide range of observables. As a result of the dynamic nature of many problem domains, observables can change with time and, in some cases, may require continuous monitoring. In a tactical application, objects of interest can be stationary (fixed or currently nonmoving), quasi-stationary (highly localized motion), or in motion. Individual objects possess characteristics that constrain their behavior. Objects emit different forms of electromagnetic energy that vary with time and can indicate the state of the object. Object emissions include *intentional* or active emissions, such as radar, communications, and data link signals, as well as *unintentional* or passive emissions, such as acoustic, magnetic, or thermal signatures generated by internal heat sources or environmental loading. Patterns of physical objects and their behavior provide indications of organization, tactics, and intent, as can patterns of emissions, both active and passive. For example, a sequence of signals emitted from a surface-to-air missile radar system that changes from search, to lock-on, to launch clearly indicates hostile intent.

Because a single sensor modality is incapable of measuring all relevant information dimensions, multiple sensor classes may be needed to detect, track, classify, and infer the likely intent of a host of objects from submarines and surface vessels, to land-, air-, and space-based objects. Certain sensor classes lend themselves to surveillance applications, providing both wide-area and long-range coverage, plus readily automated target detection. Examples of such sensor classes include signals intelligence (SIGINT) for collecting active emissions, moving target indication (MTI) radar for detecting and tracking moving targets against a high clutter background, and synthetic aperture radar (SAR) for detecting stationary targets. Appropriately cued, other sensor classes that possess narrower fields of view and typically operate at a much shorter range, can provide information that supports refined analysis. Geospatial and other intelligence databases can provide the static domain context for interpreting target-sensed data, whereas environmental sensors can generate dynamic estimates of the current atmospheric conditions.

7.1.7 Commensurate and Noncommensurate Data

Although the fusion of *similar* (commensurate) information would seem to be more straightforward than the fusion of *dissimilar* (noncommensurate) information, that is not necessarily the case. Three examples are offered to highlight the varying degrees of difficulty associated with combining multiple-source data. First, consider the relative simplicity of fusing registered electronic intelligence (ELINT) data and real-time SAR imagery. Although these sensors measure dramatically different information dimensions, both sources provide reasonably wide area coverage, relatively good geolocation, and highly complementary information. As a consequence, the fusion process tends to be straightforward. Even when an ELINT sensor provides little more than target line-of-bearing, overlaying the two data sets might be instructive. If the line-of-bearing intercepts a single piece of equipment in the SAR image, the radar system class, as well as its precise location, would be known. This information, in turn, can support the identification of other nearby objects in the SAR image (e.g., missile launchers normally associated with track-while-scan radar).

At the other end of the spectrum, the fusion of information from two or more identical sensors can present a significant challenge. Consider, for example, fusing data sets obtained from spatially separated forward-looking infrared (FLIR) radars. Although FLIR imagery provides good azimuth and elevation resolution, it does not directly measure range. Because the range and view angles to targets will be different for multiple sensors, combining such data sets demands some form of range estimation, as well as sophisticated registration and normalization techniques.

Finally, consider the fusion of two bore-sited sensors: light-intensified and FLIR. The former device amplifies low-intensity optical images to enhance night vision. When coupled with the human's natural ability to separate moving objects from the relatively stationary background, such devices permit visualization of the environment and detection of both stationary and moving objects. However, such devices offer limited capability for the detection of stationary personnel and equipment located in deep shadows or under extremely low ambient light levels (e.g., heavy cloud cover, no moon, or inside buildings). FLIR devices, however, detect thermal radiation from objects. Consequently, these devices support the detection of humans, vehicles, and operating equipment based on their higher temperature relative to the background. Consequently, with bore-sighted sensors, pixel-by-pixel combination of the two separate images provides a highly effective night vision capability.

7.2 Biologically Motivated Fusion Process Model

A hierarchically organized *functional-level* model of data fusion is presented in Chapter 3. In contrast, this section focuses on a *process-level* model. Whereas the functional model describes what analysis functions or processes need to be performed, a process-level model describes how this analysis is accomplished.

The goal of data fusion, as well as most other forms of data processing, is to turn data into useful information. In perhaps the simplest possible view, all the required information is assumed to be present within a set of sensor measurements. Thus, the role of data fusion is extraction of information embedded in a data set (separating the wheat from the chaff). In this case, fusion algorithms can be characterized as a function of

- Observables
- Current situation description (e.g., target track files and current situation description)
- *A priori* declarative knowledge (e.g., distribution functions, templates, constraint sets, filters, and decision threshold values)

As shown in Figure 7.3a, the fusion process output provides updates to the *situation description*, as well as feedback to the *reasoning knowledge base* to support knowledge refinement (learning).

Signal processing, statistical hypothesis testing, target localization performed by intersecting two independently derived error ellipses, and target identification based on correlation of an image with a set of rigid templates are simple examples of such a fusion model. In general, this "information extraction" view of data fusion makes a number of unstated, simplifying assumptions including the existence of

- Adequate information content in the sensor observables
- Adequate sensor update rates
- Homogeneous sensor data
- Relatively small number of readily distinguishable targets

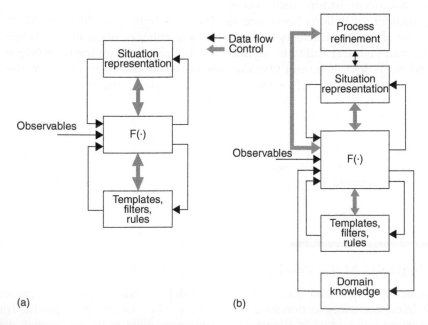

(a)   (b)

FIGURE 7.3
(a) Basic fusion process model and (b) generalized process model.

- Relatively high resolution sensors
- High-reliability sensors
- Full sensor coverage of the area of interest
- Stationary, Gaussian random interference

When such assumptions are appropriate, data analysis tends to be straightforward and an "information extraction" fusion model is adequate. Rigid template-match paradigms typically perform well when a set of observables closely matches a single template and are uncorrelated with the balance of the templates. Track association algorithms perform well against a small number of moving, widely spaced targets provided the radar provides relatively high update rates. The combination of similar features is often more straightforward than the combination of disparate features. When the sensor data possesses adequate information content, high confidence analysis is possible. High signal-to-noise ratios tend to enhance signal detection. High-resolution sensors reduce ambiguity and uncertainty with respect to feature measurements (e.g., location and frequency). High-reliability sensors maximize sensor availability. Adequate sensor coverage provides a "complete" view of the areas of interest. Statistic-based reasoning is generally simplified when signal interference can be modeled as a Gaussian random process.

Typical applications where such assumptions are realistic include:

- Track assignment in low-target-density environments or for ballistic targets that obey well-established physical laws of motion
- Classification of military organizations based on associated radio types
- Detection of signals and targets exhibiting high signal-to-background ratio

However, numerous real-world data fusion tasks exhibit one or more of the following complexities:

- Large number of target and nontarget entities (e.g., garbage trucks may be nearly indistinguishable from armored personnel carriers)
- Within-class variability of individual targets (e.g., hatch open vs. hatch closed)
- Low data rates (exacerbating track association problems)
- Disparate sensor classes (numeric and symbolic observables can be difficult to combine)
- Inadequate sensor coverage (i.e., inadequate number of sensors, obscuration due to terrain and foliage, radio frequency interference, weather, or countermeasures)
- Inadequate set of sensor observables (e.g., inadequate input space dimensionality)
- Inadequate sensor resolution
- Registration and measurement errors
- Inadequate *a priori* statistical knowledge (e.g., unknown prior and conditional probabilities, multimodal density functions, or non-Gaussian and nonstationary statistics)
- Processing and communication latencies
- High level-of-abstraction analysis product required (beyond platform location and identification)

- Complex collection environment (i.e., multipath, diffraction, or atmospheric attenuation)
- Purposefully deceptive behavior

When such complexities exist, sensor-derived information tends to be incomplete, ambiguous, erroneous, and difficult to combine or abstract. Thus, a data fusion process that relies on rigid composition among (1) the observables, (2) the current situation description, and (3) a set of rigid templates or filters tends to be fundamentally inadequate.

As stated earlier, rather than simply "extracting" information from sensor-derived data, effective data fusion requires the combination, consolidation, organization, and abstraction of information. Such analysis can enhance the fusion product, its confidence, and its ultimate utility in at least four ways:

1. Existing sensors can be improved to provide better resolution, accuracy, sensitivity, and reliability.
2. Additional similar sensors can be employed to improve the coverage or confidence in the domain observables.
3. Dissimilar sensors can be used to increase the dimensionality of the observation space, permitting the measurement of at least partially independent target attributes (a radar can offer excellent range and azimuth resolution, whereas an ELINT sensor can provide target identification).
4. Additional domain knowledge and context constraints can be utilized.

Whereas the first three recommendations effectively *increase* the information content or dimensionality of the observables, the latter effectively *reduces* the decision space dimensionality by constraining the possible decision states.

If observables are considered to represent *explicit* knowledge (i.e., knowledge that is explicitly provided by the sensors), then context knowledge can be considered *implicit* (or nonsensor-derived) knowledge. Although fusion analysts routinely use both forms, automated fusion approaches have traditionally relied almost exclusively on the former.

As an example of the utility of implicit domain knowledge, consider the extrapolation of the track of a ground-based vehicle that has been observed moving along the relatively straight-line path shown in Figure 7.4. Although the target is a wheeled vehicle traveling along a road with a hairpin curve just beyond the last detection point, a purely statistical-based tracker would likely extend the track through the hill (the reason for the curve in the road) and into the lake on the other side. To address such problems, modern ground-based trackers typically accommodate multiple model approaches that include road-following strategies.

Additional complications remain, however, including potentially large numbers of ground vehicles, nonresolvable individual vehicles, terrain and vegetation masking, and infrequent target update rates. The application of relevant domain knowledge (e.g., mobility, observability, vehicle class behavior, and vehicle group behavior) provides at least some help in managing these additional complications.

In addition to demonstrating the value of *reasoning in context*, the road-following target tracking problem illustrates the critical role *paradigm selection* has in the algorithm development process. Rather than demonstrating the failure of a statistical-based tracker, the above-mentioned example illustrates its misapplication. Applying a purely statistical approach to this problem assumes (perhaps unwittingly) that domain constraints are either irrelevant or insignificant. However, in this application, the domain constraints proved far stronger than those provided by a strictly statistical-based motion model.

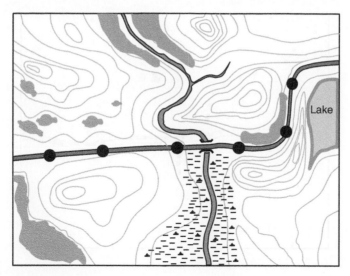

FIGURE 7.4
Road-following target tracking model.

Paradigm selection, in fact, must be viewed as a key component of successful data fusion automation. Consequently, algorithm developers must ensure that both the capability and limitations of a selected problem-solving paradigm are appropriately matched to the requirements of the fusion task they are attempting to automate.

Traditional trackers associate new detections based on a fairly rigid evaluation criterion. Even state-of-the-art trackers tend to associate detections on the basis of, at most, a small number of discrete behavior models. Analysts, however, typically accommodate a broader range of behavior that might be highly context-sensitive. Whether the analysis is manually performed or automated, uncertainty in both the measurement space and the reasoning process must be effectively dealt with. If additional information becomes available to help resolve residual uncertainty, confidences in supported hypotheses are increased whereas false and under-supported hypotheses must be pruned.

To illustrate the importance of both context-sensitive reasoning and paradigm selection, consider the problem of analyzing the radar detections from multiple closely spaced targets, some with potentially crossing trajectories, as illustrated in Figure 7.5.

To effectively exploit the background context depicted in this figure, such knowledge must be readily accessible and a means of exploiting it must be in place. For this data set, an automated tracker could potentially infer that (1) tracks 1–3 appear to be following a road, (2) tracks 4 and 5 are most consistent with minimum terrain gradient following motion model, and (3) track 6 is inconsistent with any ground-based vehicle behavior model. By evaluating track updates from targets 1–3 with respect to road-following model, estimated vehicle speeds and observed intertarget spacing (assuming individual targets are resolvable) can be used to deduce that targets 1–3 are wheeled vehicles traveling in a convoy along a secondary road. On the basis of the maximum observed vehicle speeds and the worst-case surface conditions along their trajectories, tracks 4 and 5 can be deduced to be tracked vehicles. Finally, the relatively high speed and the rugged terrain suggest that track 6 is most consistent with a low-flying airborne target. Given that the velocity of target 6 is too low for it to be a fixed-wing aircraft, the target can be deduced to be a helicopter. Thus, rather than simply "connecting the dots," a context-sensitive tracker can potentially enhance the level of abstraction and the confidence of its output.

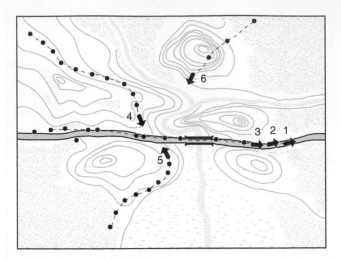

FIGURE 7.5
Example of the fusion of multiple-target tracks over time.

TABLE 7.1

Mapping between Sensor Classes and Target States

	Sensor Classes								
Target States	**MTI Radar**	**SAR**	**Laser Radar**	**Communications Intelligence (COMINT)**	**ELINT**	**FLIR**	**Optical**	**Acoustic**	**Measurement and Signature**
Moving/emitting	•		•	•	•	•		•	•
Moving/nonemitting	•		•			•		•	
Nonmoving/emitting		•		•	•	•	•		•
Nonmoving/nonemitting		•				•	•		

Behavior can change over time. Targets might be moving at one instant and stationary at another. Entities might be communicating during one interval and silent during another. Thus, entities capable of movement can be grouped into four mutually exclusive states: (1) moving, nonemitting; (2) moving, emitting; (3) nonmoving, nonemitting; and (4) nonmoving, emitting. Because entities switch between these four states over time, data fusion inherently entails a recursive analysis process. Table 7.1 shows the mapping between these four target states and a wide range of sensor classes. The table reveals that tracking entities through such state changes virtually demands the use of multiple-source sensor data.

Individual targets can potentially exhibit complex patterns of behavior that may help discriminate among certain object classes and identify activities of interest. Consider the scenario depicted in Figure 7.6, showing the movement of a tactical erectable missile launcher (TEL) between time t_0 and time t_6. At t_0, the vehicle is in a location that makes it difficult to detect. At t_1, the vehicle is moving along a dirt road at velocity v_1. At time t_2, the vehicle continues along the road and begins communicating with its support elements. At time t_3, the vehicle is traveling off road at velocity v_3 along a minimum terrain gradient path. At time t_4, the target has stopped moving and begins to erect its launcher. At time t_5, just before launch, radar emissions begin. At time t_6, the vehicle is traveling to a new hide location at velocity v_6.

Table 7.2 identifies sensor classes that could contribute to detection and identification of the various target states. The "Potentially Contributing Sensors" column lists sensor

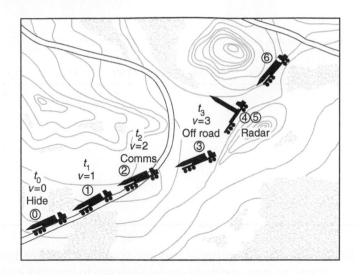

FIGURE 7.6
Dynamic target scenario showing sensor snapshots over time.

TABLE 7.2

Interpretation of Scenario Depicted in Figure 7.6

State (1)	State Class (2)	Velocity (3)	Emission (4)	Potentially Contributing Sensors (5)	Local Interpretation (6)	High-Level Interpretation (7)	Global Interpretation (8)
0	Nonmoving/ nonemitting	0		SAR, FLIR, imagery, video	Light foliage	Concealment	Hide
1	Moving/ nonemitting	v_1		MTI, FLIR, video	Road association	High-speed mobility	Move to launch
2	Moving/ emitting	v_2	Communication type 1	MTI, FLIR, video, COMINT	Road association C2 network active	High-speed mobility Coordination status	
3	Moving/ nonemitting	v_3		MTI, FLIR, video	Off road Good mobility	Minimum terrain gradient path Local goal seeking	
4	Nonmoving/ nonemitting	v_4		MTI, FLIR, imagery	Open, flat Good mobility Good visibility	Tactical activity or staging area	Launch pre-paration and launch
5	Nonmoving/ emitting	v_5	Communication types 1 and 2, radar	SAR, FLIR, imagery, video, SIGINT	Coordination status Prelaunch transmission	Launch indication	
6	Moving/ nonemitting	v_6		MTI, FLIR, video	High-speed travel	Rapid movement Road seeking	Move to hide

cross-cueing opportunities. At the lowest level of abstraction, observed behavior can be interpreted with respect to a highly local perspective, as indicated in column 6, "Local Interpretation." By assuming that the object is performing some higher-level behavior, progressively more global interpretations can be developed as indicated in columns 7 and 8.

TABLE 7.3

Mapping between Sensor Classes and Activities for a Bridging Operation

State	MTI Radar	SAR	COMINT	ELINT	FLIR	Optical	Acoustic
Engineers move to river bank	•		•			•	•
Construction activity		•	•	•	•	•	•
Forces move toward river bank	•		•	•		•	•
Forces move from opposite side of river	•		•			•	•

If individual battle space objects are organized into operational or functional-level units, observed behavior among groups of objects can be analyzed to generate higher-level situation awareness products. Table 7.3 categorizes the behavioral fragments of an engineer battalion engaged in a bridge-building operation and identifies sensors that could contribute to the recognition of each fragment.

Situation awareness development can thus be viewed as the recursive refinement of a composite multiple level-of-abstraction scene description. The generalized fusion process model shown in Figure 7.3b supports the effective combination of (1) domain observables, (2) *a priori* reasoning knowledge, and (3) the multiple level-of-abstraction/multiple-perspective fusion product. The process refinement loop controls both effective information combination and collection management. As we have seen, each element of the process model is potentially sensitive to implicit (nonsensor-derived) domain knowledge.

7.3 Fusion Process Model Extensions

Traditional multisource fusion algorithms have relied heavily on three key problem dimensions: *object, location*, and *time*. Table 7.4 characterizes all possible combinations of these three dimensions and offers an interpretation of each distinct "triple" class. If *time, location,* or both are unknown (or not applicable), these dimensions can be treated as a special case of *different*.

Table 7.4 provides a number of useful insights into the nature of the fusion process, its relationship to behavior analysis and data mining, as well as the role of context in machine-based reasoning.

Class 1 represents single-entity multisource *space–time association*. Fusing an ELINT report with SAR detections is a typical example. Linking a human intelligence (HUMINT) report with an ATM transaction that occurred at roughly the same time and in the general vicinity represents a more contemporary application.

Class 2 represents the *absence of change* (no change in imagery or stationary entities). When the measurement interval is relatively long, the scene or entity being observed may only be quasi-stationary (e.g., a vehicle could have moved and returned before the second observation).

Because physical objects cannot be at two different places at the same time, *class 3* represents an infeasible condition that can be used to detect inconsistent hypotheses. The

TABLE 7.4

Eight Fundamental Fusion Classes

Class	Object	Location	Time	Type of Analysis	Comment
1	Same	Same	Same	Fuse multiple information sources	Traditional space–time correlation
2	Same	Same	Different	Static object	Stationary object
3	Same	Different	Same	Infeasible condition	Contradictory data
4	Same	Different	Different	Object tracking (mining if extended time)	Traditional space–time tracking
5	Different	Same	Same	Associate colocated objects	Traditional space–time correlation
6	Different	Same	Different	Independent objects at the same location	Location could link objects
7	Different	Different	Same	Communications link (cell phone, chat, e-mail)	Nonspatial correlation
8	Different	Different	Different	Multiple different objects at different times	No spatialtemporal association

obvious generalization of this condition—that objects cannot move between two locations in less time than is realizable given the most optimistic mode of transit and the existing environmental conditions—provides a valuable constraint that can be exploited by both hypothesis generation, as well as truth maintenance systems.

Class 4 represents traditional bottom-up *space–time tracking*. When "tracks" evolve over relatively long periods of time (e.g., hours, days, or months), data mining-like approaches tend to be more appropriate than data-driven algorithms. Support to forensic analysis may require extensive archival storage, as well as appropriate data indexing, to support efficient data retrieval.

Class 5 represents *multiple object coincidence* (two or more objects discovered to be at approximately the same place at the same time). Examples include (1) electro optical (EO) detection of several side-by-side vehicles and (2) an ELINT report on a radar system and a COMINT report associated with a nearby command vehicle. In a more contemporary application, a HUMINT report on *individual A* could be weakly associated with *individual B* who placed a cell phone call from the same general vicinity and at approximately the same time.

Class 6 represents *association by location independent of time*. Individuals who visit the same safe house or mosque at different times can potentially be linked through independent relationships with these buildings (key locations). Because class 6 association does not rely on time coincidence of events, exploitation inherently involves data mining-like operations rather than more traditional bottom-up fusion of "streaming" sensor data. As with forensic track development (class 4), the temporal "distance" between events and event *ordering* may be more important than the absolute time.

By contrast, *class 7* represents object linking when there exists *no spatial coincidence* among entities. In this case other dimensions are exploited to establish links between entities whose activities appear to be "temporally connected." Examples include phone records, online chat, and financial transactions.

Finally, *class 8* represents the case when *no spatiotemporal link* exists between two or more entities. These links can be either *explicit* or *implicit*. Explicit links include associations among individuals derived from e-mail communications, counterintelligence reports, records showing joint ownership of property, known familial relationships, and organizational memberships. Implicit links imply more subtle associations including both indirect and purposely hidden relationships (e.g., as in money laundering operations).

7.3.1 All-Source

Because individual data sources provide only partial characterizations of observed behavior, fusing data from multiple sources offers the promise of enhanced situational awareness. Although the term *all-source* has been in common usage for decades, only recently have automated fusion approaches included unstructured text input. We loosely classify all human-generated information as HUMINT. Although the addition of HUMINT potentially adds valuable problem dimensions, it presents new challenges, as well.

Whereas traditional sensors generally provide some degree of "continuous monitoring" and possess well-understood technical capabilities in terms of coverage, resolution, environmental sensitivity, and probability of detection, HUMINT tends to be highly asynchronous and opportunistic. For applications that rely heavily on HUMINT (e.g., tracking terrorists in urban environments where traditional intelligence, surveillance, and reconnaissance (ISR) sensors offer limited support), effective integration of HUMINT with traditional sensors requires that both the strengths and weaknesses of all data sources be fully understood.

HUMINT provides potentially unequaled entity description and problem domain insights. At the same time, the information content tends to be far more subjective than traditional sensor reports. Whereas conventional sensor data primarily feeds level 1, HUMINT can have relevancy across all levels of the Joint Director of Laboratories (JDL) fusion model. With traditional sensors, data latency tends to be small. By comparison, HUMINT information can be hours (e.g., CNN reports) or even days old (e.g., counterintelligence reports) by the time it reaches an automated fusion system. HUMINT is clearly not "just another Intel sensor."

A comprehensive discussion of the many challenges associated with incorporating HUMINT into the all-source fusion process is beyond the scope of this chapter. Robust machine-based natural language understanding, however, tops the list of issues. The technical issues associated with natural language processing are well documented in the literature. Related challenges include such considerations as effective exploitation of cultural, regional, and religious biases, extraction of hidden meaning, and assessment of information validity. For our present purposes, we sidestep the many technical challenges and focus instead on the pragmatic aspects of incorporating HUMINT into the all-source fusion process.

Whereas conventional sensors typically produce highly structured outputs, HUMINT tends to be partially structured at best. To fuse human-generated level 1 "messages" with traditional multisource data, unstructured text must first be "normalized." At a minimum, this implies that an entity's triple attribute set (i.e., *name, location, time*) must be extractable along with other relevant attributes (e.g., action, direct and indirect objects, explicitly indicated relationships).

To facilitate *organization discovery* (level 2) and higher level *situation understanding*-oriented processes (level 3), preprocessing may likewise be required to create a more standardized information base that lends itself to effective machine-based information exploitation.

7.3.2 Entity Tracking

Stripped to its essence, a track is little more than a temporal sequence of states derived through physical observation, detection of relevant transactions, indirect evidence, or some other means. At least conceptually, tracks can be developed for individual entities (tanks, trucks, individuals, bank accounts), composites (tank companies, battalions, and Al Queda cells), as well as more abstract concepts such as social movements, political mood, and financing methods for illicit operations.

Historically, tracking has focused on detecting and following the movement of aircraft, ships, and ground-based vehicles using a variety of active and passive sensing systems. When kinematics plays a major role and data latencies are relatively low, state-of-the-art multiple hypothesis trackers perform reasonably well.

In contrast to traditional space–time tracking applications, following individuals in urban settings is likely to involve at least some high-latency data sources. Depending on the mix of sources, traditional kinematics-based trackers will play a diminished or non-existent role. Figure 7.7 shows the track of an individual based on HUMINT and SIGINT and Figure 7.8 demonstrates how time-late information modified that track (the smaller the error ellipse, the higher the location accuracy of the source data).

Fusion systems that rely on HUMINT must effectively manage uncertainty across all sensing modalities. Whereas traditional sensor systems (radar, EO/infrared (IR), unattended ground sensors (UGS)) tend to have relatively well-defined operating characteristics, human-generated information can be far harder to characterize due to ambiguity associated with natural language, perceptual coloring, and nonsystematic errors. Perceptual coloring includes personal biases, training limitations, and experience base. Reporting errors can be either unintentional or deliberately introduced (e.g., interrogations of individuals having questionable motives).

FIGURE 7.7
Track of an high-value individual based on a combination of eyewitness reports and cell phone intercepts.

FIGURE 7.8
Track of an high-value individual after insertion of a time-late message (note retracted link).

7.3.3 Track Coincidence

Relationships between entities can be either *explicit* or *implicit*. A cell phone call between two individuals or an observation that places an individual in a vehicle with a known person of interest establishes an *explicit* link. *Implicit* links are relationships that must be deduced. Fusing multisource tracks provides an important mechanism for discovering implicit links between entities. For example, *two* individuals found to be (1) in the vicinity of each other at approximately the same time, (2) eating in the same restaurant at overlapping times, (3) staying in the same building on different days, and (4) driving a vehicle owned by another individual may be connected. In general, implicit links tend to have lower confidence than explicit links.

7.3.4 Key Locations

Any location that possesses particular significance (historical, tactical, strategic) can be considered a *key location*. Embassies, police stations, suspected safe houses, isolated sections of town, and an abandoned warehouse might all be key locations in a certain application. Specific homes, businesses, religious institutions, neighborhoods, districts, or regions might also qualify. Any location associated with high-value individuals (HVIs) could automatically be added to the key location database. Because key locations need not be a literal feature (e.g., "the region between the park and Mosque C"), accommodations must be made for handling both crisp and fuzzy spatial location descriptions.

Key locations can be rated based on their estimated level of significance. Certain locations might be important enough to fire imbedded triggers whenever certain individuals or classes of tracks intercept these regions. Any track or activity that associates (however loosely) with such locations might raise the interest level of the track, justifying a more detailed analysis or even operator intervention.

7.3.5 Behavior

Ideally, tracks capture at least a partial representation of entity *behavior*. Interpreting/understanding behavior can involve something as simple as associating a vehicle's trajectory with a road network or as complex as predicting an entity's goal state. Fusing individual tracks helps link previously unassociated entities, as well as identify higher level-of-abstraction entities and patterns.

Fusing multiple entity tracks helps uncover more complicated relationships (e.g., construction and emplacement of an improvised explosive device (IED)). The feasibility of assessing "apparent coordination" among a group of entities tends to increase with the number of problem dimensions. Behavior patterns of interest are likely to involve apparent coordination and coincidence patterns among individuals, each serving specific functional roles within an organization. In general, exploitation of additional information sources potentially leads to a more robust recognition of both general, as well as specific, behaviors.

Treating tracks from a more global perspective, one that spans levels 1–3 can lead to a more robust behavioral understanding. Suppose, for example, the first portion of a track of a military vehicle represents "advancing to contact" while the next portion of the same track involves little movement but a great deal of communications or other activities (e.g., coordinating activity, firing weapons). Viewed as disconnected (local) activities, no intrinsic information in the track states explains these distinctly different behavior modes. Only when viewed at a higher level of abstraction (in conjunction with levels 2 and 3) can the two separate lower-level "behavior" patterns be "understood" as phases of battle.

As a second example, consider the track of an individual that begins when he leaves his home in the morning. The man is seen at a mosque several hours later, and then at the market in the early afternoon. The man finally returns home around 5 p.m. This particular individual may have visited many other locations and engaged in numerous unobserved activities. On the basis of the currently available information, however, this is all we know.

Given our limited knowledge, the man's known "track" might have a vast number of interpretations. Assuming intentional behavior, we are left to presume that this series of states has a definite (albeit unknown) purpose. When and if more information becomes available to "fill in the blanks," the information can be added to the existing track file and exploited by the higher-level fusion algorithms.

If significant portions of the man's track (pattern of activity) are repeated at predictable intervals (e.g., daily), variations in his routine can be detected. In some cases, such anomalies might be of more interest than the individual's nominal behavior. Variations could include missing a regular prayer meeting or remaining at a particular location for a much shorter or much longer time than usual. A track may warrant additional scrutiny if the individual happens to be an HVI, the mosque he visited is a suspected meeting place for insurgents, and he remained there long after the noon prayer meeting ended.

Similarly, if an individual who normally sends just a few e-mails per day suddenly writes dozens of them, this change in "expected" behavior might be significant. If an individual that normally remains within a few miles of his home suddenly appears in a different portion of the city or in a completely different city, this anomaly might justify a more detailed analysis or even trigger an alert.

7.3.6 Context

Humans, often subconsciously, incorporate a wide range of contextual knowledge during the analysis of sensor-derived data. To achieve comparable robustness, automated approaches must incorporate such "external" knowledge, as well. Consider the following message:

> "Individual A met Individual B at XYZ" where XYZ represents specific global positioning system (GPS) coordinates

Although the message clearly links *individual A* and *individual B*, it fails to reveal that (1) the meeting took place in front of Mosque C, (2) it happened in a section of town where there has recently been considerable sectarian violence, and (3) Mosque C is a suspected insurgent-controlled facility, all of which places the sensor-derived information within a larger framework. Global information system (GIS) systems, organization charts, existing databases, and numerous other data sources maintain domain knowledge that, if properly exploited, supports more effective use of sensor-derived, as well human-generated information.

Applying relevant *a priori* domain knowledge potentially increases information content (and thereby the dimensionality of the decision-making process) by either adding relevant supporting features or constraining the decision process. Interpreting the above-mentioned message "out of context" (i.e., without knowing about the mosque or the recent sectarian violence in the area) is analogous to solving a set of linear equations when there exist more unknowns than equations. In both cases, the problem is under-constrained.

As mentioned earlier, adding HUMINT to the fusion process introduces a number of challenges. Dealing with semantic descriptions of *location* is one such challenge. To correlate on location, the "location" descriptions must be comparable to those provided by conventional sensors.

The example shown in Figure 7.9 highlights the need to translate from semantic location descriptions to more formal "mathematical" representations. In this case, all eight

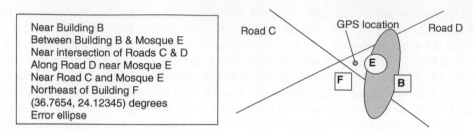

FIGURE 7.9
Sample data set to illustrate various semantic location descriptions of the same general region.

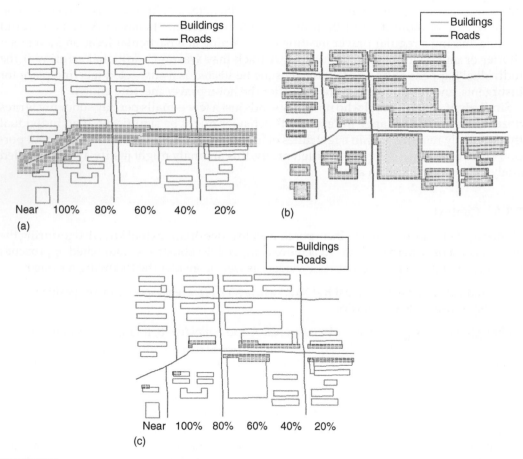

FIGURE 7.10
(a) Near road 1, (b) all buildings, and (c) fuzzy intersection of (a) and (b). Dark gray regions represent 100% and light gray regions represent 80% solutions.

locations listed in the left-hand box effectively refer to the same general region of space. Thus, all are semantically "similar," something that would be difficult to deduce using a strict semantic-based reasoning approach.

Chapter 24 discusses "semantic location translation" in more detail. For the present, we merely illustrate that it is feasible to relate crisp (Boolean) and semantic (fuzzy) spatial representations. Figure 7.10 depicts the product of intersecting *near a named road* (fuzzy feature) with *all buildings* (Boolean features).

7.3.7 HUMINT and the JDL Fusion Model

Whereas traditional sensors provide a relatively low level-of-abstraction input to the fusion process (before entity extraction), humans intuitively combine and abstract sensory and nonsensory information, generating distilled results that potentially fall across all levels of the JDL fusion model. For example, a text-based message might (1) report on the current location of a suspected terrorist (level 1 input), (2) describe the organization to which an individual is associated (level 2 input), or (3) indicate a specific threat an individual or an organization might pose (level 3 input).

In more traditional fusion applications, raw sensor data must undergo level 1 processing before data exists to *link* and *organize* in level 2. Once HUMINT becomes a viable information source, however, vast stores of both text-based open source (e.g., TV, radio, newspapers, and web sites) and classified data (e.g., messages, reports, databases) exist that can be directly mined by higher-level fusion processes to help discover organizations and develop threat awareness products. Thus, once HUMINT is included as a data source, mixed-initiative bottom-up (feed forward) and top-down (feedback) reasoning across all levels of the JDL fusion model becomes feasible.

In a level 2 counterinsurgency application, for instance, both classified (messages, reports, studies, briefings) and open source information (Internet, newspapers, TV, radio) may provide direct support to higher-level fusion applications. Relevant information about specific terrorist organizations might come from reports, web sites, and media sources. Whereas some organizations remain relatively static over time, the loose organizational structures favored by many terrorist organizations tend to evolve, as well as dramatically change, over time.

Full automation of organization construction and maintenance based on digesting a variety of unstructured text sources represents an extremely challenging problem. During the early stages of the analysis process, documents are typically read line by line. Syntactic and lexical analysis algorithms tag parts of speech and assign some level of meaning to individual sentences. Word sense disambiguation and syntactic ambiguity require consideration of implicit sentence context (ranging from fixed and narrowly defined to very general).

Extracting valid relationships between tagged words found in different sentences of the same document presents many challenges. Because different sources might express the same information in different ways or may contain conflicting information, combining information from different information sources presents a far greater challenge. Even when all available information sources are fully exploited, it might not be possible to resolve all uncertainty. In the short term, at least, it is safe to say there must be a human in the loop to help guide automated processes, draw their own conclusions, and remove erroneous deductions.

7.3.8 Context Support Extensions

Existing context-sensitive fusion algorithms tend to be highly specialized and tuned to a specific application. To provide the appropriate contextual information, developers must tap an array of information sources (databases, GIS, and other knowledge repositories), as well as write code for required intermediate processing. Rather than tightly coupling context support to an application, building general services that support a broader range of applications appears to have considerable merit. Given the increasing importance of context in data fusion applications, adding a Context Support module to the existing JDL fusion model would both acknowledge the importance of nonsensor-derived information to the fusion process, as well as likely foster the development of reusable functionality that is not tied to a specific program, developer, or data source.

7.4 Observations

This chapter concludes with five general observations pertaining to data fusion automation.

7.4.1 Observation 1

For an effective automation of key elements of the fusion process, it may be necessary to emulate more of the strengths of human analysts. In general, humans

- Are adept at model-based reasoning (which supports robustness and extensibility)
- Naturally employ domain knowledge to augment formally supplied information (which supports context sensitivity)
- Update or modify existing beliefs to accommodate new information as it becomes available (which supports dynamic reasoning)
- Intuitively differentiate between context-sensitive and context-insensitive knowledge (which supports maintainability)
- Control the analysis process in a highly focused, often top-down fashion (which enhances efficiency)

7.4.2 Observation 2

Global phenomena naturally require global analysis. Analysis of local phenomena can benefit from a global perspective as well. The target track assignment process, for instance, is typically treated as a strictly local analysis problem. Track assignment is typically based on recent, highly local behavior (often assuming a Markoff process). For ground-based objects, a vehicle's historical trajectory and its maximum performance capabilities provide rather weak constraints on future target motion.

Although applying nearby domain constraints could adequately explain the local behavior of an object (e.g., constant velocity travel along a relatively straight, level road), a more global viewpoint is required to interpret global behavior. Figure 7.11 demonstrates local (i.e., concealment, minimum terrain gradient, and road seeking), medium-level (i.e., river-crossing and road-following), and global (i.e., reinforce at unit) interpretations of a target's trajectory over space and time. The development and maintenance of such a multiple level-of-abstraction perspective is critical to achieving robust automated situation awareness.

7.4.3 Observation 3

A historical view of fusion approaches can be instructive. Production systems have historically been found to perform better against static, well-behaved, finite-state diagnostic-like problems than against problems possessing complex dependencies and exhibiting dynamic, time-varying behavior. In general, any system that relies on rigid, single level-of-abstraction control would be expected to exhibit such characteristics. Despite this fact,

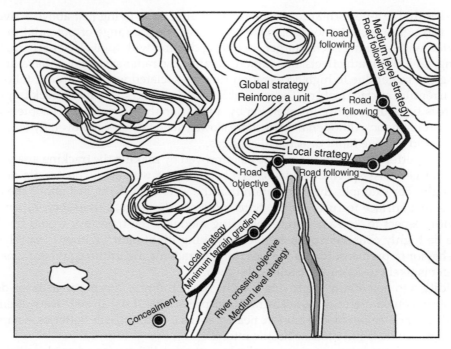

FIGURE 7.11
Multiple level-of-abstraction situation understanding.

during the early 1990s, expert systems were routinely applied to dynamic, highly context-sensitive problems, often with disappointing results.

The lesson to be learned from this and other similar examples is that both the strengths and limitations of a selected problem-solving paradigm must be fully understood by the algorithm developer from the outset. Interestingly, when an appropriately constrained task was successfully automated using an expert system, developers frequently discovered that the now well-understood problem could be more efficiently implemented using a different paradigm.

When an expert system proved inadequate for a given problem, artificial neural systems were often seen as an alternative or even preferred approach. Neural networks require no programming, suggesting that the paradigm would be ideal for handling ill-defined or poorly understood problems. Whereas expert systems could have real-time performance problems, artificial neural systems promised high-performance hardware implementations. In addition, the adaptive nature of the neural net learning process seemed to match real world, dynamically evolving problem-solving requirements. However, most artificial neural systems operate more like a statistical or fuzzy pattern recognizer than a sophisticated reasoning system capable of generalization, reasoning by analogy, and abstract inference. In addition, network training can be problematic for realistic applications that involve complex behaviors evolving over time.

7.4.4   Observation 4

Traditional air target radar trackers employed a single statistic-based algorithm regardless of whether an aircraft was flying at an altitude of 20 km or just above tree-top level. Likewise, the algorithms were generally insensitive with regard to whether the target happened to be a high-performance fighter aircraft or a relatively low speed helicopter. Suppose a nonfriendly high-performance reconnaissance aircraft is flying just above a river that snakes through a mountainous region. There exists a wide range of potential problems associated with tracking such a target, including dealing with high clutter return, terrain masking, and multipath effects. In addition, an airborne radar system may have difficulty tracking the target due to high acceleration turns associated with an aircraft following a highly irregular surface feature. The inevitable track loss and subsequent track fragmentation errors typically require intervention by a radar analyst. Tracking helicopters can be equally problematic. Although they fly more slowly, such targets can hover, fly below tree-top level, and execute rapid directional changes.

Tracking performance can potentially be improved by making the tracking analysis sensitive to target class-specific (model-based) behavior, as well as to constraints posed by the domain. For example, the recognition that the aircraft is flying just above the terrain suggests that surface features are likely to influence the target's trajectory. When evaluated with respect to "terrain feature-following models," the trajectory would be discovered to be highly consistent with a "river-following flight path." Rather than relying on past behavior to predict future target positions, a tracking algorithm could anticipate that the target is likely to continue to follow the river.

In addition to potentially improving tracking performance, interpreting sensor-derived data within context permits more abstract interpretations. If the aircraft were attempting to avoid radar detection by one or more nearby surface-to-air missile batteries, a nap of the earth flight profile could indicate hostile intent.

Even more global interpretations can be hypothesized. Suppose a broader view of the "situation picture" reveals another unidentified aircraft operating in the vicinity of the river-following target. By evaluating the apparent coordination between the two aircraft, the organization and mission of the target group can be conjectured. For example, if the second aircraft begins jamming friendly communication channels just as the first aircraft reaches friendly airspace, the second aircraft's role can be inferred to be "standoff protection for the primary collection or weapon delivery aircraft." The effective utilization of relevant domain knowledge and physical domain constraints offers the potential for developing both more effective and higher level-of-abstraction interpretations of sensor-derived information.

7.4.5　Observation 5

Indications and warnings, as well as many other forms of expectation-based analysis, have traditionally relied on relatively rigid doctrinal and tactical knowledge. Contemporary data fusion applications, however, often must support intelligence applications where flexible, ill-defined, and highly creative tactics and doctrine are employed. Consequently, the credibility of any analysis that relies on rigid expectation-based behavior needs to be carefully scrutinized. Although the lack of strong, reliable *a priori* knowledge handicaps all forms of expectation-based reasoning, the effective application of relevant logical, physical, and logistical context at least partially compensates for the lack of more traditional problem domain constraints.

Acknowledgment

The authors acknowledge CERDEC I2WD, Fort Monmouth, NJ for providing support for both the preparation and the subsequent revision of this chapter.

References

1. Antony, R. T., *Principles of Data Fusion Automation*, Artech House Inc., Boston, MA, 1995.
2. Saxe, J. G., The blind man and the elephant, *The Poetical Works of John Godfrey Saxe*, Houghton, Mifflin and Company, Boston, MA, 1882.

8

Overview of Distributed Decision Fusion

Martin E. Liggins*

CONTENTS

8.1 Introduction

The concept of distributed detection fusion[†] or decision fusion has evidenced wide interest over the past 25 years, beginning with applications ranging from multimode signal detection and distributed multisensor decision theory to multiple change detection algorithms for surveillance imaging systems to subordinate team decision managers or decision makers (DMs) within a command and control formulation. Recent applications have involved team-decision concepts for wireless networks and particle swarm theory for biometric security systems.

Distributed detection fusion problems are often represented as optimization problems involving a composite collection of local DMs (involving sensors, algorithms, control processes, and persons) feeding a centralized (global) DM. As we embrace the ever-growing challenge of network-centric systems, the need and the opportunity now exist to communicate between any number of local DMs, providing distributed interaction between local DMs within a region or between global managers. With this emphasis on network-centric development, demand for optimal decision techniques would need to grow as well.

---

* The author's affiliation with the MITRE Corporation is provided only for identification purposes and is not intended to convey or imply MITRE's concurrence with, or support for, the positions, opinions, or viewpoints expressed by the author.

[†] It is important to point out that this work focuses on multiple sensor detection fusion. Other forms of decision fusion involving higher cognitive-based decisions such as situation awareness or impact assessment are discussed throughout this handbook.

This chapter provides the basic concepts of distributed detection fusion, beginning with the concepts of classical detection theory portrayed by Van Trees.[1] We will discuss the most basic architecture—a parallel fusion architecture—and in doing so, lay out the means to develop and design system decision fusion rules. Finally, we will provide a discussion on fusion rules and the assumptions placed on local DMs.

The structure of this chapter is as follows: Section 8.2 provides the fundamental concepts of detection fusion theory, beginning with a single DM. We build on this approach to extend to multiple DMs operating within a parallel fusion system in Section 8.3, based on the concepts of team-decision theory. Section 8.4 provides an analysis of some of the common fusion rules, how they are constructed, and their performance.

8.2 Single Node Detection Fundamentals

In developing an optimal detection approach for distributed DMs, we consider the binary hypothesis testing problem for N local DMs. Consider each DM having to select between two hypotheses, H_1 and H_0, with known prior probabilities for each hypothesis. These hypotheses represent observations, indicating acceptance (H_1) or rejection (H_0) of the existence of, say, a target versus noise, correct versus false target classification, or detecting relevant anomaly detections over that of random activity.

Consider Figure 8.1. When sensors monitor their environment, either they observe an entity of interest, Z_1, and the local DM declares H_1 (e.g., detect the target), or they detect noise (no target), with the local DM declaring H_0 with some statistical basis. It is assumed that these choices represent the total observation space as the union of the observations, $Z = Z_1 \cup Z_0$, with no overlap between the observation subspaces, that is, $Z_1 \cap Z_0 = \phi$.

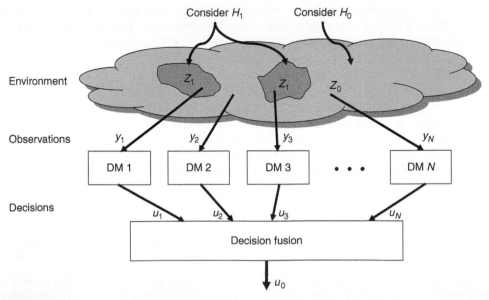

FIGURE 8.1
Distributed detection fusion concept.

To develop the concepts of minimum risk associated with making a decision on a target, consider an individual DM generating a decision, u_i, representing one of the two hypotheses based on its observation, y_i:

$H_0 : y_i = n_i$ (noise only)

$H_1 : y_i = r + n_i$ (target plus noise)

The noise and signals observed by each DM are considered to be statistically independent from any other DM. Decisions characterized by each DM, u_i, are dependent on both the observation and a threshold. We can denote these decisions as follows:

$$u_i = \gamma(y_i) = \begin{cases} 1 & \text{detector } i \text{ decides } H_1 \\ 0 & \text{detector } i \text{ decides } H_0 \end{cases}$$

This detection performance is characterized by the local DM based on a probability conditioned on choosing the true hypothesis, $P(u_i|H_j)$. Bayes formulation attributes the cost to each possible choice with decision rules designed to minimize the cost of making a decision, that is, minimizing the expected risk, R.

For a single DM, we can express this risk as follows:

$$R = \sum_{i=0}^{1} \sum_{j=0}^{1} C_{ij} P_j P(H_i|H_j) = \sum_{i=0}^{1} \sum_{j=0}^{1} C_{ij} P_j \int_{Z_i} p(y|H_j) dy$$

where C_{ij} is the cost of declaring hypothesis i, given that hypothesis j is true and P_j is the prior probability representing the choice that hypothesis j is true based on the observation y. The conditional probability represents the probability that integrating over a region, Z_i, will generate a correct response for a target.

Expanding over the summation gives

$$R = C_{00} P_0 \int_{Z_0} p(y|H_0) dy + C_{10} P_0 \int_{Z_1} p(y|H_0) dy + C_{01} P_1 \int_{Z_0} p(y|H_1) dy + C_{11} P_1 \int_{Z_1} p(y|H_1) dy$$

$$= (C_{00} P_0 + C_{01} P_1) \int_{Z_0} p(y|H_0) dy + (C_{10} P_0 + C_{11} P_1) \int_{Z - Z_0} p(y|H_1) dy$$

which reduces to

$$R = (C_{10} P_0 + C_{11} P_1) + \int_{Z_0} \{ P_1(C_{01} - C_{11}) p(y|H_1) dy - P_0(C_{10} - C_{00}) p(y|H_0) dy \} \tag{8.1}$$

by taking advantage of the fact that Z represents the total observation space, with unity probability. Minimizing risk is equivalent to minimizing the integrand, that is, assigning y to H_1 if the integrand is positive or to H_0 for a negative integrand. Alternatively, we can write the integrand in the form of a likelihood ratio as follows:

$$\Lambda(y) = \frac{p(y|H_1)}{p(y|H_0)} \underset{H_0}{\overset{H_1}{\gtrless}} \frac{P_0(C_{10} - C_{00})}{P_1(C_{01} - C_{11})} = \tau \tag{8.2}$$

where the left-hand side indicates the likelihood ratio and the right-hand side the detection threshold.

We can rewrite the risk in terms of the following well-known probabilities:

- Probability of a false alarm, $P_F = \int_{Z_1} p(y|H_0)dy$
- Probability of a detection, $P_D = \int_{Z_1} p(y|H_1)dy$
- Probability of miss detection, $P_M = \int_{Z_0} p(y|H_1)dy = 1 - P_D$.

Also note that the prior probability of H_1 can be written as $P_1 = 1 - P_0$.
The risk becomes

$$R = P_0 C_{00} + (1 - P_0) C_{01} + P_0(C_{10} - C_{00}) P_F - (1 - P_0)(C_{01} - C_{11}) P_D \qquad (8.3)$$

which can be reduced to

$$R = C + C_F P_F - C_D P_D \qquad (8.4)$$

where we define the costs to be $C_F = P_0(C_{10} - C_{00})$, $C_D = (1 - P_0)(C_{01} - C_{11})$, and $C = P_0 C_{00} + (1 - P_0)C_{01}$, assuming the cost of an incorrect decision to be higher than the cost of a correct decision, that is, $C_{10} > C_{00}$ and $C_{01} > C_{11}$.

8.3 Parallel Fusion Network

Now consider a number of DMs within the parallel decision fusion network shown in Figure 8.1. Extending the Bayesian minimum risk concepts to a set of distributed detectors, each DM provides a localized decision and transmits results to a centralized fusion system, where they are combined into a composite result. The fusion system combines the local results by an appropriate fusion rule, and provides a final decision, u_0, representing the viability that the global hypothesis represents the true hypothesis. Fused optimization of the results depends on an evaluation of the transmitted decisions from each of the N DMs and developing the appropriate fusion rule to define how individual local decisions are combined.

Varshney[2-5] provides an excellent derivation. We follow a similar approach. We define the overall fused probabilities as follows:

- Fused probability of a false alarm, $P_F = \sum_u P(u_0 = 1|u)P(u|H_0)$
- Fused probability of a detection, $P_D = \sum_u P(u_0 = 1|u)P(u|H_1)$

where $\sum_u$ is the summation over the N DMs. Substituting these system probabilities into Equation 8.4 gives the overall risk as follows:

$$R = C + C_F \sum_u P(u_0 = 1|u)P(u|H_0) - C_D \sum_u P(u_0 = 1|u)P(u|H_1) \qquad (8.5)$$

Tsitsiklis[6-8] points out that decentralized detection problems of this type fall within the class of team-decision problems. Once a decision rule is optimized for each local DM, the ensemble set of decisions transmitted to the fusion center represents a realization from all DMs. That is, each DM transmits $u_i = \gamma(y_i)$ contributing to a global ensemble set $u_0 = \gamma_0(u_1, u_2, u_3, \ldots, u_N)$ to be evaluated at the fusion center. For conditionally independent conditions, observations of one DM are unaffected by the choice of the other DMs. Tsitsiklis shows that once the individual strategies are optimized, the overall system fusion can be represented by a person-by-person optimization approach.

Person-by-person optimization can be performed in two steps:

1. Likelihood rules are generated for each local DM, holding all other DMs fixed, including the fusion decision.
2. Fusion rule is obtained in a similar way, assuming all local DMs are fixed.

8.3.1 Optimizing Local Decision Managers (Step 1)

For the individual DMs, assume all other DMs are fixed to an optimal setting, except for the kth DM. Then from Equation 8.5, we can expand the effects of the other DMs on u_k as follows:

$$R = C + \sum_{\substack{u \\ u \neq u_k}} \left\{ \begin{array}{l} P(u_0 = 1 | u \text{ fixed}, u_k = 1)[C_F P(u_k = 1, u \text{ fixed} | H_0) - C_D P(u_k = 1, u \text{ fixed} | H_1)] \\ + P(u_0 = 1 | u \text{ fixed}, u_k = 0)[C_F P(u_k = 0, u \text{ fixed} | H_0) - C_D P(u_k = 0, u \text{ fixed} | H_1)] \end{array} \right\}$$

(8.6)

From the total probability, $P(u_k = 1 | H_j) + P(u_k = 0 | H_j) = 1$, we can separate out the effect of the kth detector as follows:

$$R = \overbrace{C + \sum_{\substack{u \\ u \neq u_k}} \{P(u_0 = 1 | u \text{ fixed}, u_k = 0)[C_F P(u | H_0) - C_D P(u | H_1)]\}}^{C_k}$$

$$+ \sum_{\substack{u \\ u \neq u_k}} \left\{ \begin{array}{l} \overbrace{[P(u_0 = 1 | u \text{ fixed}, u_k = 1) - P(u_0 = 1 | u \text{ fixed}, u_k = 0)]}^{\alpha(u)} \\ \underbrace{[C_F P(u_k = 1 | H_0) - C_D P(u_k = 1 | H_1)]}_{k\text{th detector}} \end{array} \right\}$$

(8.7)

where the terms in the first summation remain fixed and can be absorbed into the constant, whereas the first bracket in the second summation does not directly affect u_k and can be treated as the weighted influence, $\alpha(u)$, on the kth detector.

Expanding to show the impact of the observations, we have $P(u | H_j) = \int_Y P(u | Y)p(Y | H_j)dY$. Then the risk becomes

$$R = C_k + \sum_{\substack{u \\ u \neq u_k}} \{\alpha(u)[C_F P(u_k = 1 | H_0) - C_D P(u_k = 1 | H_1)]\}$$

$$= C_k + \sum_{\substack{u \\ u \neq u_k}} \left\{ \alpha(u) \left[C_F \int_Y P(u_k = 1 | y_k)P(u | Y)p(Y | H_0)dY - C_D \int_Y P(u_k = 1 | y_k)P(u | Y)p(Y | H_1)dY \right] \right\}$$

The conditional probability of the kth detector can be pulled out of the integrand because of the independence of the observations and formulated as a likelihood ratio. The optimal decision (i.e., the kth detector) is represented as follows:

$$\Lambda(y) = \frac{p(y_k|H_1)}{p(y_k|H_0)} \mathop{\gtrless}_{u_k=0}^{u_k=1} \frac{\sum_{\substack{u \\ u \neq u_k}} C_F \int_Y \alpha(u)P(u|y)p(y|H_0)dY}{\sum_{\substack{u \\ u \neq u_k}} C_D \int_Y \alpha(u)P(u|y)p(y|H_1)dY}$$

For conditional independence, this reduces to the threshold test as follows:

$$\frac{p(y_k|H_1)}{p(y_k|H_0)} \mathop{\gtrless}_{u_k=0}^{u_k=1} \frac{\sum_{\substack{u \\ u \neq u_k}} C_F \prod_{\substack{i=1 \\ i \neq k}}^{N} P(u_i|H_0)}{\sum_{\substack{u \\ u \neq u_k}} C_D \prod_{\substack{i=1 \\ i \neq k}}^{N} P(u_i|H_1)} = \tau_i \qquad (8.8)$$

Tenney and Sandell[9] in their groundbreaking paper, point out that even for identical local detectors, the thresholds τ_i are coupled to the other detectors, as shown in Equation 8.8. That is, $\tau_k = f(\tau_1, \tau_2, \ldots, \tau_N)_{i \neq k}$ for each detector. Even if we have two identical detectors, the thresholds will not be the same. The impact of this is important. Individually, sensor thresholds may be required to maintain a high threshold setting to keep false alarms low, reducing the opportunity to detect low signal targets. By not requiring all sensors to keep the same high threshold, the fusion system can take advantage of differing threshold settings to ensure that increased false alarms from the lower-threshold sensors are "fused out" of the final decision while ensuring an opportunity to detect difficult targets.

8.3.2 Optimizing Fusion Rules (Step 2)

The number of possible combinations that could be received from N local DMs is 2^N, assuming a binary decision. In a similar way, the fusion center uses these results to develop an overall system assessment of the most appropriate hypothesis. The goal is to continue the team-decision process to minimize the overall average risk at the fusion center. Referring back to Equation 8.5, the overall system risk becomes

$$R = C + \sum_u P(u_0 = 1|u)[C_F P(u|H_0) - C_D P(u|H_1)]$$

Choose one decision, $\tilde{u}$, as one possible realization for the set of local DMs, then the fused risk becomes

$$R = C_k(u) + P(u_0 = 1|\tilde{u})[C_F P(\tilde{u}|H_0) - C_D P(\tilde{u}|H_1)]$$

where

$$C_k(u) = C + \sum_{\substack{u \\ u \neq \tilde{u}}} P(u_0 = 1|u)[C_F P(u|H_0) - C_D P(u|H_1)]$$

The risk is minimized for the fusion rule when

$$P(u_0 = 1|\tilde{u}) = \begin{cases} 1 & \text{if } [C_F P(\tilde{u}|H_0) - C_D P(\tilde{u}|H_1)] \leq 0 \\ 0 & \text{otherwise} \end{cases}$$

This provides the fusion rule:

$$\frac{P(\tilde{u}_i|H_1)}{P(\tilde{u}_i|H_0)} \overset{u_0=1}{\underset{u_0=0}{\lessgtr}} \frac{C_F}{C_D}$$

Assuming conditional independence, we can write the overall fused system likelihood as a product of the individual local decisions:

$$\prod_{i=1}^{N} \frac{P(u_i|H_1)}{P(u_i|H_0)} \overset{u_0=1}{\underset{u_0=0}{\lessgtr}} \frac{C_F}{C_D} \tag{8.9}$$

Thus, Equation 8.8 provides 2^N simultaneous fusion likelihoods that work to achieve system optimization for each local DM.

Other architectures expand on these principles. For instance, Hashemi and Rhodes,[10] Veeravalli et al.,[11] and Hussain[12] discuss a series of transmissions over time using similar person-by-person sequential detections that use stopping criteria such as Wald's sequential probability ratio tests. Tandem network topologies[13,14] are discussed by transmitting from one DM to the neighboring manager, then the combined decisions are sent to the next neighbor, and so on. Multistage and asymmetric fusion strategies are discussed in Refs 15 and 16.

8.4 Fusion rules

The combined team-decision solution, involves solving the N local DM thresholds in Equation 8.8 simultaneously with the 2^N fusion likelihoods in Equation 8.9, coupling the local and fused solutions. Thus, to achieve system optimization, the decision fusion process optimizes the thresholds for each local DM by holding the other DMs fixed, and repeating the process until all local DM thresholds are determined. The resulting thresholds are coupled to the global likelihood and solved simultaneously. Their simultaneous solution provides the basis for determining the set of fusion rules, a total set of 2^{2^N} rules. Selection of the appropriate fusion rule is determined by an evaluation of all possible decisions.

Table 8.1 shows the explosive growth required to develop a complete set of fusion rules—for instance, four DMs give rise to more than 65,500 rules. However, an examination of the rule sets shows that a great many of the rules can be eliminated through the principle of monotonicity,[2,17–20] while even further reduction can be made with some knowledge of the DMs.

The assumption of monotonicity implies that for every DM, $P_{D_i} > P_{F_i}$, and as such not all fusion rules satisfy threshold optimality. For example, consider a condition where K sensors select the primary hypothesis, H_1. Under a subsequent test, if $M > K$ sensors select hypothesis H_1, where the set of M sensors includes the set of K sensors, then it is expected that the global decision provides a higher likelihood for the M sensors than for the K sensors. Choosing the fusion rule with the largest likelihood value ensures an optimal result. Alternatively, as Table 8.1 shows, any fusion rule that contradicts this process can be removed, significantly saving on the number of necessary rules to evaluate.

Figure 8.2 shows a simple but common example of a complete set of fusion rules generated for two DMs. We can gain considerable insight by examining these rules further. For example, fusion rule d1 rejects H_1 regardless of the decisions of the two detectors (d1 = 0

TABLE 8.1

Fusion Rules Based on Number of DMs

Number of DMs	Set of Fusion Rules	Number of Relevant Rules after Applying Monotonicity
1	4	3
2	16	6
3	256	20
4	65,536	168

DMs		Fusion rules																
u_1	u_2	d1	d2	d3	d4	d5	d6	d7	d8	d9	d10	d11	d12	d13	d14	d15	d16
0	0	0	0	0	0	0	0	0	0	1	1	1	1	1	1	1	1
0	1	0	0	0	0	1	1	1	1	0	0	0	0	1	1	1	1
1	0	0	0	1	1	0	0	1	1	0	0	1	1	0	0	1	1
1	1	0	1	0	1	0	1	0	1	0	1	0	1	0	1	0	1

FIGURE 8.2
Global fusion rules based on two local decision makers.

for all combinations of u_1 and u_2), whereas fusion rule d16 accepts H_1 under any condition (d16 = 1 for all combinations of u_1 and u_2). Rules d9 through d16 declare H_1 when both local detectors reject H_1. Also rules d3, d5, d7, d9, d11, d13, and d15 all reject H_1 even though both detectors declare H_1 (i.e., $u_1 = 1$, $u_2 = 1$). Thus monotonicity maintains that we keep six rules: $\gamma(u) = \{d1, d2, d4, d6, d8, d16\}$.

Figure 8.3 shows the final fusion rule selection using two local detectors. Note that the fusion rules can be reduced even further, since d1 and d16 just bound the set of rules and are not practical. We are left with four rules: the well-known AND and OR rules, as well as rules for relying on one or the other single DM.

Much of the literature has emphasized this particular set, preferring to study the impact of AND and OR rules for their particular development. However, the case for three detectors shows a great deal more flexibility in the rule set.

Figure 8.4 shows the set of fusion rules for three detectors after we apply monotonicity, reducing the number of rules from 256 to 20. Two important additional rules become evident: "Majority Logic" and "Sensor Dominance." For the former rule, the global decision rule requires at least two or more DMs to agree on H_1 before declaring H_1 at the fusion center. For Sensor Dominance, we assume that a single detector, say u_2, operates with sufficient quality to be accepted by the fusion system whenever it declares H_1; however, when u_2 rejects the hypothesis, the other two DMs must both have positive decision before making a global declaration of H_1. This condition could occur, for example, when a high quality imaging radar detects a target by itself, while two poor resolution sensors (detectors u_1 and u_3) must have consensus for the fusion system to declare H_1.

As such, we can use this approach to eliminate some of these rules on the basis of the quality of performance of the local DMs and their sensors. For the example just discussed, we would not necessarily keep rule d34 (u_1 dominates) or rule d88 (u_3 dominates), since their performance quality is poorer than u_2. To assess the rule set further, we would need to examine the impact of each individual detector performance on the environment and set the system fusion rules accordingly.

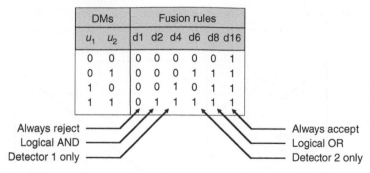

FIGURE 8.3
Final fusion rule selection for two decision makers.

Detector	Monotonic fused detection			
u_1 u_2 u_3	d1 d2 d4 d6 d8	d16 d18 d20 d22 d24	d34 d52 d56 d64 d86	d88 d96 d120 d128 d256
0 0 0	0 0 0 0 0	0 0 0 0 0	0 0 0 0 0	0 0 0 0 1
0 0 1	0 0 0 0 0	0 0 0 0 0	0 0 0 0 1	1 1 1 1 1
0 1 0	0 0 0 0 0	0 0 0 0 0	0 1 1 1 0	0 0 1 1 1
0 1 1	0 0 0 0 0	0 1 1 1 1	1 1 1 1 1	1 1 1 1 1
1 0 0	0 0 0 0 0	1 0 0 0 0	1 0 0 1 0	0 1 0 1 1
1 0 1	0 0 0 1 1	1 0 0 1 1	1 0 1 1 1	1 1 1 1 1
1 1 0	0 0 1 0 1	1 0 1 0 1	1 1 1 1 0	1 1 1 1 1
1 1 1	0 1 1 1 1	1 1 1 1 1	1 1 1 1 1	1 1 1 1 1

AND rule   Majority Logic   OR rule

u_2 Dominates, (u_1, u_3) must

FIGURE 8.4
Global decision rules for three detectors, applying monotonicity.

Functionally, each rule applies a different set of constraints on the basis of individual detection performance to provide the fusion system with a complete set of representative rules. Selecting the rule with the highest likelihood ensures improved target detection while reducing false alarms. If we have knowledge of the individual sensors, we can select those fusion rules that best fit their individual characteristics. Figure 8.5 shows a representative set of fusion performance curves that compare performance from three detectors against that of an individual detector. For convenience, we have assumed that all three detectors perform equally well. We can use the graph to examine the effects of the fusion rules for both fused system probability of detection (P_D) as well as system probability of false alarm (P_F). That is, for probability values ≥ 0.5 along the x-axis, we can attribute these to individual detector probability of detection (P_{D_i}), and those probability values ≤ 0.5 can represent individual detector probability of false alarm (P_{F_i}). The same can be considered for the fusion system performance along the y-axis. For instance, consider the two vertical lines at $P_{D_i} = 0.88$ and $P_{F_i} = 0.10$. The P_D line intersects the AND rule giving a fused value of 0.6815; intersection with the OR rule occurs at $P_{D_i} = 1 - (1 - 0.88)^3$ for a system value of

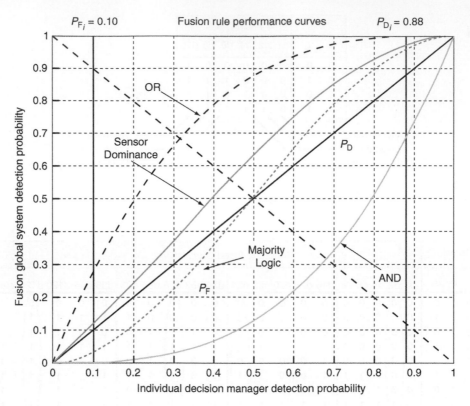

FIGURE 8.5
Fusion system performance in terms of individual detector performance.

0.9983. Similarly for P_{F_i}, the AND rule gives system $P_F = 0.001$ and the OR rule gives $P_F = 0.2710$. From this, it appears that the OR rule can significantly outperform all other rules for P_D, but at the cost of higher false alarms, whereas the AND fusion rule appears to be an excellent means of reducing P_F, but provides poor overall system performance.

Now let us consider each of the four fusion rules shown in Figure 8.5 in more detail:

1. The AND rule relaxes constraints on the individual detectors, allowing each detector to pass a large number of both detections and false alarms, relying on the fusion process to improve performance by eliminating most false reports. In this way, relaxed individual performance conditions will ensure improved target reportability from each detector, even though a larger set of false reports may also be developed.

2. The OR rule provides the reversed effect, relying on individual detectors to operate at higher detection thresholds to significantly eliminate false alarms for the fused system. This rule assumes the detectors are quite adept at distinguishing target reports from false alarms. If the detectors produce too many false reports, this rule can be detrimental. As such, a higher set of thresholds are applied to each detection source.

3. The Majority Logic rule falls somewhere between AND and OR performance curves, exhibiting some of the best characteristics from both strategies. It is based on the assumption that not as many false alarms would be generated by the individual

detectors as in the AND case, whereas the potential to distinguish target reports from false alarms is not as stringent as in the OR case.

4. The Sensor Dominance rule presents an interesting case, relaxing constraints only for the single, dominant detector, expecting a lower number of false alarms for that detector, but relying on the fused element to reduce false reports from the other two poorer performing detectors. Depending on the degree of improvement of the dominant detector over the other two, system performance can improve, but false alarm performance is determined by the performance of the individual dominant detector. This rule relies on the assumption that the dominant detector is of very high quality.

Recently, particle swarm optimization theory has been suggested as an alternative to person-by-person optimization[21,22] when we consider the explosive growth of rules. Here, multiple classifiers are fused at the decision level, and the particle swarm optimization techniques determine the optimal local decision thresholds for each classifier and the fusion rule, replacing the person-by-person strategy discussed in Section 8.3. Other recent trends have focused on the application of distributed fusion to wireless sensor networks.[23]

8.5 Summary

We have presented an introduction into the details involving the development of distributed detection fusion, beginning with the fundamentals of classical Bayesian likelihoods and extending these concepts to a parallel distributed fusion detection system, building on the concepts of system fusion rules to combine the local detectors. Each fusion rule applies a coupled decision statistic that adjusts detection performance thresholds for each DM on the basis of the combined performance of the other local detectors. We have also discussed how global fusion rules perform in terms of individual (local) detection performance, pointing out how detections and false alarms are treated on the basis of different fusion rules.

References

1. Van Trees, H.L., *Detection, Estimation, and Modulation Theory,* Vol. 1, Wiley, NY, 1968.
2. Varshney, P.K., *Distributed Detection and Data Fusion,* Springer-Verlag, NY, 1997.
3. Hoballah, I.Y. and P.K. Varshney, Distributed Bayesian signal detection, *IEEE Transactions on Information Theory,* 35(5), 995–1000, 1989.
4. Barkat, M. and P.K. Varshney, Decentralized CFAR signal detection, *IEEE Transactions on Aerospace and Electronic Systems,* 25(2), 141–149, 1989.
5. Hashlamoun, W.A. and P.K. Varshney, Further results on distributed Bayesian signal detection, *IEEE Transactions on Information Theory,* 39(5), 1660–1661, 1993.
6. Tsitsiklis, J.N., Decentralized detection, in *Advances in Statistical Signal Processing,* Vol. 2, JAI Press, Greenwich, CT, 1993.
7. Irving, W.W. and J.N. Tsitsiklis, Some properties of optimal thresholding in decentralized detection, *IEEE Transactions on Automatic Control,* 39(4), 835–838, 1994.
8. Tay, W.P., J.N. Tsitsiklis, and M.Z. Win, *Data Fusion Trees for Detection: Does Architecture Matter,* Manuscript to be Submitted, November 2006.

9. Tenney, R.R. and N.R. Sandell, Jr., Detection with distributed sensors, *IEEE Transactions on Aerospace and Electronic Systems,* 17(4), 501–509, 1981.

10. Hashemi, H.R. and I.B. Rhodes, Decentralized sequential detection, *IEEE Transactions on Information Theory,* 35(3), 509–520, 1989.

11. Veeravalli, V.V., T. Basar, and H.P. Poor, Decentralized sequential detection with a fusion center performing the sequential test, *IEEE Transactions on Information Theory,* 39(2), 433–442, 1993.

12. Hussain, A.M., Multisensor distributed sequential detection, *IEEE Transactions on Aerospace and Electronic Systems,* 30(3), 698–708, 1994.

13. Barkat, M. and P.K. Varshney, Adaptive cell-averaging CFAR detection in distributed sensor networks, *IEEE Transactions on Aerospace and Electronic Systems,* 27(3), 424–429, 1991.

14. Tang, Z.B. and K.R. Pattipati, Optimization of detection networks: part I—tandem structures, *IEEE Transactions, Systems, Man and Cybernetics,* 21(5), 1044–1059, 1991.

15. Dasarathy, B.V., Decision fusion strategies for target detection with a three-sensor suite, *Sensor Fusion: Architectures, Algorithms, and Applications,* Dasarathy, B.V. (ed.), Proceedings of the SPIE Aerosense, Vol. 3067, pp. 14–25, Orlando, FL, April 1997.

16. Dasarathy, B.V., Asymmetric fusion strategies for target detection in multisensor environments, *Sensor Fusion: Architectures, Algorithms, and Applications,* Dasarathy, B.V. (ed.), Proceedings of the SPIE Aerosense, Vol. 3067, pp. 26–37, Orlando, FL, April 1997.

17. Thomopoulos, S.C.A., R. Viswanathan, and D.K. Bougoulias, Optimal distributed decision fusion, *IEEE Transactions on Aerospace and Electronic Systems,* 25(5), 761–765, 1989.

18. Liggins, M.E. and M.A. Nebrich, Adaptive multi-image decision fusion, *Signal Processing, Sensor Fusion, and Target Recognition IX,* Kadar, I. (ed.), Proceedings of the SPIE Aerosense, Vol. 4052, pp. 213–228, Orlando, FL, April 2000.

19. Liggins, M.E. An evaluation of CFAR effects on adaptive Boolean decision fusion performance for SAR/EO change detection, *Signal Processing, Sensor Fusion, and Target Recognition IX,* Kadar, I. (ed.), Proceedings of the SPIE Aerosense, Vol. 4380, pp. 406–416, Orlando, FL, April 2001.

20. Liggins, M.E., Extensions to adaptive Boolean decision fusion, *Signal Processing, Sensor Fusion, and Target Recognition IX,* Kadar, I. (ed.), Proceedings of the SPIE Aerosense, Vol. 4729, pp. 288–296, Orlando, FL, April 2002.

21. Veeramachaneni, K., L. Osadciw, and P.K. Varshney, An adaptive multimodal biometric management algorithm, *IEEE Transactions, Systems, Man, and Cybernetics,* 35(3), 344–356, 2005.

22. Veeramachaneni, K. and L. Osadciw, Adaptive multimodal biometric fusion algorithm using particle swarm, *MultiSensor, Multisource Information Fusion: Architectures, Algorithms, and Applications 2003,* Dasarathy, B. (ed.), Proceedings of the SPIE Aerosense, Vol. 5099, Orlando, FL, April 2003.

23. Chen, B., Tong, L., and Varshney, P.K., Channel-aware distributed detection in wireless sensor networks, *IEEE Signal Processing Magazine,* 23(4), 16–26, 2006.

9

Introduction to Particle Filtering: The Next Stage in Tracking

Martin E. Liggins* and Kuo-Chu Chang

CONTENTS

9.1 Introduction

Tracking development has progressed slowly in terms of its ability to track under more stressing conditions. Rudolph Kalman's[1] innovative paper on linear filtering and prediction problems led the way to provide one of the first adaptive and optimal approaches for realistic target tracking. As a result of the importance of this work, Kalman was awarded the Kyoto Prize ($340,000), considered to be the Japanese equivalent to the Nobel Prize.[2] Since that time, the Kalman filter has become the cornerstone for most of the major tracking developments to date.[3–7] Kalman filters represent the class of optimal Bayesian estimators that operate under conditions involving linear-Gaussian uncertainty. As long as the target dynamics operates within a polynomial motion model and sensor measurements are related linearly to the target state with Gaussian noise, the approach is an optimal estimator. Although the success of the Kalman filter led to numerous military and civilian applications, it has its problems[8]: to name a few, there is marginal stability of the numerical solution of the Riccati equation, small round-off errors accumulate and degrade

---

* The author's affiliation with the MITRE Corporation is provided only for identification purposes and is not intended to convey or imply MITRE's concurrence with, or support for, the positions, opinions, or viewpoints expressed by the author.

performance, and conversion of non-Cartesian sensor measurements introduces coupling biases within the Cartesian Kalman state. This did not stop researchers from developing methods to expand Kalman filter operating bounds to the larger nonlinear problem classes.

The extended Kalman filter (EKF) has provided the means to adapt these linear principles to conditions involving nonlinear dynamics or nonlinear measurements by employing partial derivatives as a means to linearize the systems. The Kalman filter retains its basic form, but partial derivatives are applied either to the state transition matrix or the measurement matrix to develop a minimum variance estimate of the target state. In essence, this provides a "piecewise linear" application to the nonlinear tracking problem.

Still certain problem classes are not addressed by these methods; the most prominent problem involves conditions where sensor measurements provide incomplete observations (such as range-only, bearing-only, or time difference of arrival tracking) or where tracking models drastically change (for instance, road-following and move-stop-move tracking). Recently, problems of this class have been successfully addressed by techniques such as the unscented* (transform) Kalman filter (UKF) and particle filtering. While the EKF provides an approximation to optimal nonlinear estimation, the UKF has been shown to provide good performance in propagating the state mean and covariance through small-scale nonlinear transformations. The basic premise is that a set of *sigma points* (generally $2n + 1$, n being the state space size) represent the target state estimate probability density—one particle for the mean and $2n$ particles to bound the covariance of the track estimate. However, these samples are not randomly selected but they are deterministic. Yet they are sufficient to capture third-order moments. Other techniques also exist that extend the Kalman filter to nonlinear dynamic conditions, such as the modified gain Kalman filter, the Gauss–Hermite quadrature filter—a version similar to the UKF.

Particle filtering extends the notions of the unscented transform, using sequential Monte Carlo sum (SMS) methods to represent the state estimate for nonlinear non-Gaussian problem classes. The full potential of particle filtering has not been explored. While many of the concepts were introduced in the 1960s and 1970s, interest did not develop until recently.[9] This chapter explores some of the concepts that lead to the development of particle filters, and the details behind their implementation. An example is provided to demonstrate some of the strengths and weaknesses of this approach. Particle filtering is an important contribution to the world of dynamic target estimation, but like the other approaches, it is designed to satisfy a specific class of tracking problems. The Kalman filter provides an optimal solution for linear-Gaussian problem classes, whereas the particle filter is designed to address nonlinear non-Gaussian conditions, but it is not an optimal filter. Still, it represents a major step toward satisfying the needs of working in nonlinear tracking environments.

9.2 Target State Filtering Problem

The target state model can be represented by a discrete-time stochastic model

$$x_{k+1} = f_k(x_k, u_k, v_k) \tag{9.1}$$

---

* The unscented Kalman filter is described in detail in Chapter 15. As such, it will not be discussed here.

where u_k represents the external deterministic control, v_k represents the random process noise that aids in capturing any mismatch of the target model with the true state dynamics, and the time index k represents current time, whereas $k + 1$ provides a prediction to the next time step. The function $f(\cdot)$ indicates the possible nonlinear relationship of the target dynamic behavior. Likewise, the measurement equation could also be nonlinear as indicated by $h(\cdot)$ in the following equation:

$$z_k = h_k(x_k, u_k, w_k) \tag{9.2}$$

where w_k represents the measurement noise.

We are interested in generating filtered track estimates of x_k from a sequence of sensor measurements. Using the measurement history, $\mathbf{Z}^k \equiv \{z_i, I = 1, 2, ...,k\}$, we can develop the *posterior* probability density function (pdf), $p(x_k|\mathbf{Z}^k)$. The posterior pdf provides the means to determine the conditional predicted target state, based on Bayes' theorem.

9.2.1 Chapman–Komolgorov Equation

Given the posterior pdf $p(x_k|\mathbf{Z}^k)$, we want to obtain a recursive relationship relating the prior pdf at time $k + 1$. The Chapman–Komolgorov equation represents a one-step transition density for a Markov random sequence,[10] an optimal estimator. As $\mathbf{Z}^k$ represents the cumulative information that leads to the current state estimate, the posterior density summarizes the target past in a probabilistic sense. Starting with Bayes' theorem, we can write,

$$p(x_{k+1}|\mathbf{Z}^{k+1}) = p(x_{k+1}|\mathbf{Z}^k, z_{k+1}) = \frac{p(z_{k+1}|x_{k+1}, \mathbf{Z}^k)p(x_{k+1}|\mathbf{Z}^k)}{p(z_{k+1}|\mathbf{Z}^k)}$$

or more compactly as

$$p(x_{k+1}|\mathbf{Z}^{k+1}) = \frac{1}{c}p(z_{k+1}|x_{k+1})p(x_{k+1}|\mathbf{Z}^k) \tag{9.3}$$

where

$$c = p(z_{k+1}|\mathbf{Z}^{k+1}) = \int p(z_{k+1}|x_{k+1})p(x_{k+1}|\mathbf{Z}^k)dx_k$$

The first step is to show the relationship between the prediction pdf $p(x_{k+1}|\mathbf{Z}^k)$ and the prior update $p(x_k|\mathbf{Z}^k)$. That is, we can expand the second term of Equation 9.3 as

$$p(x_{k+1}|\mathbf{Z}^k) = \frac{p(x_{k+1}, \mathbf{Z}^k)}{p(\mathbf{Z}^k)} = \int \frac{p(x_{k+1}, x_k, \mathbf{Z}^k)}{p(\mathbf{Z}^k)}dx_k = \int \frac{p(x_{k+1}|x_k, \mathbf{Z}^k)p(x_k, \mathbf{Z}^k)}{p(\mathbf{Z}^k)}dx_k$$

or simply recognizing the joint pdf as $p(x_k, \mathbf{Z}^k) = p(x_k|\mathbf{Z}^k)p(\mathbf{Z}^k)$ (assuming the process is Markov), we get

$$p(x_{k+1}|\mathbf{Z}^k) = \int p(x_{k+1}|x_k)p(x_k|\mathbf{Z}^k)dx_k$$

Equation 9.3 becomes the Chapman–Komolgorov equation:

$$p(x_{k+1}|\mathbf{Z}^{k+1}) = \frac{1}{c}p(z_{k+1}|x_{k+1})\int p(x_{k+1}|x_k)p(x_k|\mathbf{Z}^k)dx_k \tag{9.4}$$

Note that the observation likelihood $p(z_{k+1}|x_{k+1})$ can be taken outside the integral of the Chapman—Komolgorov (Equation 9.4). The terms inside the integral include the dynamic transition of the target to the next state prediction $p(x_{k+1}|x_k)$, and the prior state $p(x_k|\mathbf{Z}^k)$. The conditional pdf $p(x_k|\mathbf{Z}^k)$ has the property of being an information state. The optimal estimator then consists of a recursive formulation of the conditional pdf. In general, the process is computationally intensive.

The Chapman–Komolgorov equation forms the basis for developing track estimation. If the system is linear, and all noises are Gaussian, then the functional recursion becomes the recursion for the conditional mean and covariance (Kalman filter), in which case the covariance estimation errors (accuracy) are independent of measurements and can be computed off-line.

For the linear case, the target dynamic equation can generally be written as

$$x_{k+1} = F_k x_k + G_k u_k + v_k$$

where v_k is assumed be a zero-mean white Gaussian noise with covariance Q_k. The mean and covariance of the one-step prediction track state estimate then become, respectively,

$$\hat{x}_{k+1|k} = F_k \hat{x}_{k|k} + G_k u_k$$

and

$$P_{k+1|k} = F_k P_{k|k} F_k' + Q_k$$

while the measurement equation can be written as

$$z_k = H_k x_k + w_k$$

where w_k is the measurement noise generally assumed to be zero-mean white Gaussian with covariance R.

The nonlinear case, however, shows the accuracy to be measurement dependent. The EKF represents an attempt to apply Kalman filtering principles to conditions of nonlinearity, either in the target dynamic or the measurement model (or both) by applying a first-order Taylor series expansion. In essence, it attempts to linearize the nonlinear functions of the state or measurement models. Taking Equations 9.1 and 9.2, we can expand the state estimate to include the first-order derivative and higher-order terms (HOT) as (ignoring the external control)

$$\hat{x}_{k+1|k} = f(k, \hat{x}_{k|k}) + f_X(k)(x_k - \hat{x}_{k|k}) + \text{HOT} + v_k$$

where $f_X(k)$ represents the Jacobian matrix

$$f_X(k) \equiv [\nabla_X f(x, k)']' \,|_{x=x_{k|k}}$$

$$= \begin{pmatrix} \dfrac{\partial f_1}{\partial x_1} & \cdots & \dfrac{\partial f_1}{\partial x_n} \\ \vdots & \vdots & \vdots \\ \dfrac{\partial f_m}{\partial x_1} & \cdots & \dfrac{\partial f_m}{\partial x_n} \end{pmatrix} = F(k)$$

Then the prediction state and the associated covariance become, respectively,

$$\hat{x}_{k+1|k} = f(k, \hat{x}_{k|k})$$

and

$$P_{k+1|k} = F(k)P_{k|k}F(k)' + Q_k$$

Likewise, nonlinearities in the measurement prediction can be written as

$$z_{k+1} = h(k+1, \hat{x}_{k+1|k}) + h_X(k+1)(x_{k+1} - \hat{x}_{k+1|k}) + \text{HOT} + w_k$$

where

$$h_X(k+1) \equiv [\nabla_X h(x, k+1)']' \,|_{x=x_{k+1|k}} = H(k+1)$$

The prediction measurement and the prediction measurement covariance become, respectively,

$$\hat{z}_{k+1|k} = h(x_{k+1}, \hat{x}_{k+1|k})$$

and

$$S_{k+1} = H(k+1)P_{k+1|k}H(k+1)' + R_{k+1}$$

A complete description can be found in Refs 5 and 11. Some of the key issues with using an EKF include the need to evaluate a Jacobian matrix for the state and measurement equations to approximate nonlinear calculations, and covariance computations are no longer decoupled from the state estimate calculations.

Limitations with the EKF could involve introducing (1) biases through the linearization of the nonlinear conditions, (2) reduced accuracy in the prediction state, or (3) complications due to human error in calculating Jacobian matrices. In addition, there is no guarantee of reduced error even if we include second-order terms from the Taylor series expansion. Finally, extensive simulations may be required to test the consistency of the filter.

As the EKF uses the first-order terms as an approximation, errors will "creep" into the system if neglected higher-order terms begin to dominate.

Note that the EKF always approximates the $p(x_k|Z^k)$ as Gaussian, which is often considered to be piecewise continuous between updates. As such, the principles of the EKF assume that the posterior densities hold for the Gaussian $p(x_k|Z^k) = N(x_k, \hat{x}_{k|k}, P_{k|k})$.

The approach provides a fast and simple method to apply higher-order moments to solve non-Gaussian posterior densities, as long as the nonlinearity is not too severe. Other methods, such as the UKF, extend these techniques to include accuracy up to third order.[*]

Particle filters represent the next level, designed to characterize non-Gaussian process noise in the target state or measurements through a sequence of particles (point masses) of

---

[*] See Chapter 15 for a more detailed discussion.

the density that estimates the state in terms of a sequential Monte Carlo process. Before we proceed, we need to examine the notion of importance sampling.

9.2.2 Monte Carlo Integration and Importance Sampling

Monte Carlo integration involves developing a numerical integration for a multidimensional integral that represents a particular function, which draws from samples of x over a specified set of dimensions, n_x.

$$I = \int g(x)dx \quad \text{where } x \in \Re^{n_x}$$

We need to factor $g(x)$ into two functions, $g(x) = f(x)\pi(x)$, such that $\pi(x)$ represents a pdf with probabilistic properties $\pi(x) \geq 0$ and $\int \pi(x)dx = 1$. Then we can use these properties to approximate the integral

$$I = \int f(x)\pi(x)dx \tag{9.5}$$

in terms of its sample moments. The sample mean would be expressed as $E[I_N] = (1/N)\sum_{i=1}^{N} f(x^i)$ by sampling N independent samples from $f(x)$, where x^i is the ith sample. The variance (second central moment) of $f(x)$ is $\sigma^2 = \int (f(x) - I)^2 \pi(x)dx$. In the limiting sense, convergence occurs based on the central limit theorem showing that the characteristic function approaches a normal distribution, that is, $\lim_{N \to \infty} \sqrt{N}(E[I_N] - I) \sim N(0, \sigma^2)$.

For the Bayesian tracking problem, the pdf $\pi(x)$ represents the posterior density. Ideally, we want to generate samples from this density $\pi(x)$; in reality, we may know little about the density. However, suppose we can only approximate $\pi(x)$ with a similar density $q(x)$, termed an *importance density*. We want to ensure that a sample drawn from $q(x)$ will have a good chance of coming from the true posterior pdf $\pi(x)$. An acceptable distribution would contain the true pdf, but it need not fully contain the entire pdf as long as the samples rejected (outside of $q(x)$) would have a high probability of not affecting the target state. Ideally, a good approximation density would be the one proportional to the true posterior. If the ratio $\pi(x)/q(x)$ is bounded by some constant, say M, then we can accept every sample drawn with probability 1.

We can rewrite Equation 9.5 in terms of this approximation as

$$I = \int f(x)\pi(x)dx = \int f(x)\frac{\pi(x)}{q(x)}q(x)dx$$

To illustrate the problem, consider the pdf $\pi(x)$ in Figure 9.1 to be a Gaussian mixture that represents the true target conditions. Then, assume that the true pdf has the form:

$$\pi(x) = 0.5N(x; 0, 1.0) + 0.5N(x; 1.25, 0.25)$$

where each Gaussian density has a uniform probability of acceptance.* We can let the initial importance function be a uniform density that expands the original true pdf to have a size, say $U(-2, 2)$. Because the importance density supports (covers) the interval containing most of the true pdf $\pi(x)$, samples from this estimate will eventually converge to $\pi(x)$ (central limit theorem). If, however, we could use the function $q^*(x)$ in Figure 9.1, convergence

---

* $N(x; m, \sigma^2)$ represents a Gaussian density on x, with a mean m and a standard deviation σ.

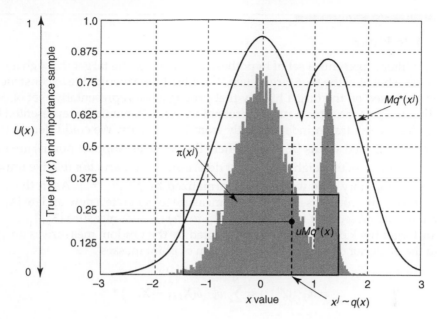

FIGURE 9.1
Conceptual example of importance sampling.

would be more rapid and the ratio $\tilde{w}(x^i) \propto (\pi(x^i)/Mq(x^i))$ would ensure a sample $q(x)$ has a probability of acceptance as the ratio of the higher curve to the lower density.[12]

This would give a sample mean estimate as

$$E[I_N] = \frac{1}{N}\sum_{i=1}^{N} f(x^i)\tilde{w}(x^i) \tag{9.6}$$

where $\tilde{w}(x^i) = \pi(x^i)/q(x^i)$ represents the importance sampling weights. Either $q(x^i)$ or $q^*(x^i)$ would be adequate, but $q^*(x^i)$ would ensure a quicker convergence. However, it is customary to use a uniform density when initiating a particle filter. Note that the density $q(x)$ need not sum to 1, but it must have a positive finite solution. To compensate for this, we can normalize $\tilde{w}(x^i)$ by simply replacing Equation 9.6 with

$$E[I_N] = \frac{1}{N}\sum_{i=1}^{N} f(x^i)\left\{\frac{\tilde{w}(x^i)}{(1/N)\sum_{j=1}^{N}\tilde{w}(x^j)}\right\} = \frac{1}{N}\sum_{i=1}^{N} f(x^i)w(x^i)$$

where $\tilde{w}(x^i)$ represents importance sampling weights that support (cover) the posterior density of the Bayesian track, $\pi(x)$.

If we compare results between the true and the estimated mixtures, as long as the importance function supports the posterior density, the means will be relatively consistent. Similarly, the sampled variance approaches the desired posteriori density, $\pi(x)$, as[13]

$$\lim_{N\to\infty} \sqrt{N}(E[I_N] - I) \to N(0, E[f(x) - I]^2 w(x))$$

which is consistent with sampling the original distribution: $\sigma_{f_n} \sim \sigma_I/\sqrt{N}$.

9.3 Particle Filter

The particle filter represents a set of hypothesized states for the target through a finite set of particles as a belief of the true target state.[9,14–28] Ideally, we would like to construct a posterior density based on the target history that propagates a representative set of particles through the current time. This posterior probability $p(\mathbf{X}_k|\mathbf{Z}^k)$ would be represented by a set of samples (particles) drawn independently from this density. We could denote this random measure (collected set) as $\{\mathbf{X}_k^j, w_k^j\}_{j=1}^N$ where the particles (support points) are given by $\{\mathbf{X}_k^j\}_{j=1}^N$, along with a set of weights, $\{w_k^j\}$, that determine each particle's relative importance to the target's trajectory. The weights are normalized by $\sum_{j=1}^N w_k^j = 1$. As in the standard tracking problem, we could use this random measure to generate a new update $\{\mathbf{X}_{k+1}^j\}_{j=1}^N$ by applying the next subsequent measurement z_{k+1} to the current measure $\{\mathbf{X}_k^j\}_{j=1}^N$.

The true target track density can be represented by the random measure as an approximation to the posterior density by a discrete sum of point masses*:

$$p(\mathbf{X}_{k+1}|\mathbf{Z}^{k+1}) \approx \sum_{j=1}^N w_{k+1}^j \delta(\mathbf{X}_{k+1} - \mathbf{X}_{k+1}^j) * \tag{9.7}$$

Unfortunately, as we saw in Section 9.2, little may be known of this density. We will have to rely on developing a set of particles from a more convenient density—the *importance density*[†] $q(\mathbf{X}_{k+1}|\mathbf{Z}^{k+1})$. If we sample the particles $\{x_k^j, w_k^j\}_{j=1}^N$ from this density rather than the true posterior, we need to modify the weights accordingly,

$$w_{k+1}^j \propto \frac{p(\mathbf{X}_{k+1}^j|\mathbf{Z}^{k+1})}{q(\mathbf{X}_{k+1}^j|\mathbf{Z}^{k+1})} \tag{9.8}$$

for the posterior density as

$$p(\mathbf{X}_{k+1}|\mathbf{Z}^{k+1}) \approx \sum_{j=1}^N \frac{p(\mathbf{X}_{k+1}^j|\mathbf{Z}^{k+1})}{q(\mathbf{X}_{k+1}^j|\mathbf{Z}^{k+1})} \delta(\mathbf{X}_{k+1} - \mathbf{X}_{k+1}^j) \tag{9.9}$$

However, to be effective for target tracking, we need to develop a recursive expression for the weights in Equation 9.8 that represent a product of the current state and the previous history.

Thus, for the denominator of the weights in Equation 9.8, we want the recursive relationship:

$$q(\mathbf{X}_{k+1}|\mathbf{Z}^{k+1}) = q(x_{k+1}|\mathbf{X}_k, \mathbf{Z}^{k+1})q(\mathbf{X}_k|\mathbf{Z}^k) \tag{9.10}$$

For the numerator, we want to adapt the pdf correspondingly as

$$p(\mathbf{X}_{k+1}^j|\mathbf{Z}^{k+1}) = \frac{p(z_{k+1}|\mathbf{X}_{k+1}^j, \mathbf{Z}^k)p(\mathbf{X}_{k+1}^j|\mathbf{Z}^k)}{p(z_{k+1}|\mathbf{Z}^k)} \tag{9.11}$$

---

* $\delta(\mathbf{X}_{k+1} - \mathbf{X}_{k+1}^j)$ is the Dirac delta function.
† Also known as sequential importance sampling.

where the target trajectory history becomes, using the Markov assumption

$$p(\mathbf{X}_{k+1}^j|\mathbf{Z}^k) = p(x_{k+1}^j|\mathbf{X}_k^j, \mathbf{Z}^k)p(\mathbf{X}_k^j|\mathbf{Z}^k)$$

Then the pdf (Equation 9.11) becomes

$$p(\mathbf{X}_{k+1}^j|\mathbf{Z}^{k+1}) = \frac{p(z_{k+1}|x_{k+1}^j)p(x_{k+1}^j|x_k^j)}{p(z_{k+1}|\mathbf{Z}^k)}p(\mathbf{X}_k^j|\mathbf{Z}^k)$$

$$= \frac{1}{c}p(z_{k+1}|x_{k+1}^j)p(x_{k+1}^j|x_k^j)p(\mathbf{X}_k^j|\mathbf{Z}^k)$$

with c being the normalization constant based on the measurement history.
  Thus, the weights in Equation 9.8 can now be rewritten as

$$w_{k+1}^j = \frac{1}{c}\frac{p(z_{k+1}|x_{k+1}^j)p(x_{k+1}^j|x_k^j)}{q(x_{k+1}^j|\mathbf{X}_k^j, \mathbf{Z}^{k+1})}\frac{p(\mathbf{X}_k^j|\mathbf{Z}^k)}{q(\mathbf{X}_k^j|\mathbf{Z}^k)}$$

$$= \frac{1}{c}\frac{p(z_{k+1}|x_{k+1}^j)p(x_{k+1}^j|x_k^j)}{q(x_{k+1}^j|\mathbf{X}_k^j, \mathbf{Z}^{k+1})}w_k^j$$

By the Markov property, we need only take the previous sample particles drawn from $q(x_k^j|\mathbf{Z}^k)$ and apply the current measurement z_{k+1}. We can augment the weights to deal with the last state x_k^j and the current measurement z_{k+1} as

$$w_{k+1}^j \propto \frac{p(z_{k+1}|x_{k+1}^j)p(x_{k+1}^j|x_k^j)}{q(x_{k+1}^j|x_k^j, z_{k+1})}w_k^j \tag{9.12}$$

We now have a good approximation to the posterior density of the true target state, Equation 9.7, as

$$p(x_{k+1}|z^{k+1}) \approx \sum_{j=1}^N w_{k+1}^j\delta(x_{k+1} - x_{k+1}^j) \tag{9.13}$$

using the weights given in Equation 9.12. If the importance function provides an effectively bounded support, that is, $q(x) = \{x:p(x) > 0\}$, then as $N \rightarrow \infty$, Equation 9.13 converges to the true pdf Equation 9.7.

9.4 Resampling

Ideally, the development thus far should be sufficient to provide an effective particle filtering system. However, it turns out that over time the variances of the weights increase significantly. This will often result in only a small percentage of particles gaining sufficient weight insuring that they continue to be selected over again. Depending on the quality of the match of the importance function, any substantive deviation from the true pdf will provide conditions for divergence. If the measurements appear in the tail of the prior, this will lead to potential failure of the particle filter. Eventually, a single particle will dominate while the remaining particles will develop negligible weights, leading to an ineffective representation for the pdf estimate.

Depending on the degree of dimensionality, this *degeneracy effect* may take only a small number of updates to occur. One concern with higher dimensionality is that it will become more difficult to select a suitable $q(x)$. A large number of particles are necessary to properly converge, but in general, convergence is not guaranteed. As such, another concern involves inefficiency. Evaluating each of N samples when most of them have negligible weights is computationally inefficient.

A balance is necessary to maintain a sufficiently large particle set for effective importance sampling, while at the same time, managing the effects of degeneracy that severely decrease the number of effective particles.

Resampling leads to an effective solution by recreating a set of new particles whenever the need arises. Examining Figure 9.2, we can get a better perspective of the problem. Step A shows a set of 12 uniformly generated samples $\{x^j_{k+1}, N^{-1}\}$ that approximates the *prediction density* $p(x_{k+1}|\mathbf{Z}^k)$. These samples provide a uniformly weighted random measure of the particle set derived from the previously updated estimate of the target trajectory. As a new measurement is received, z_{k+1} (step B), we determine the *likelihood function*, $p(z_{k+1}|x_{k+1})$, to compute a new set of particle weights. These weights, w^j_{k+1}, from Equation 9.12, provide a new weighted random measure, $\{x^j_{k+1}, w^j_{k+1}\}$ (step C), that gives more value to those particles that match the measurement, effectively selecting the most suitable particles—*update function*, $p(x_{k+1}|\mathbf{Z}^{k+1})$.

At this step, we can see where the key concern resides regarding a practical implementation for the particle filter. Assuming that the five most heavily weighted particles dominate the particle set, all the particle diversity lies in two compact regions.

Repeating the cycle for the next time index will only increase the possibility for track failure, leading to the same particles being chosen over again. This problem is referred to

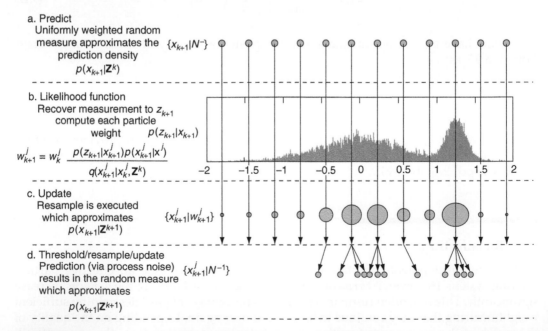

FIGURE 9.2
Graphical representation of particle filter process cycle.

as *particle impoverishment*. The impact is the effect that the track system is trying to predict target dynamics with very small process noise, and could cause divergence between the true path and the predicted estimate.

Resampling/update (step D) focuses on representing the random measure from the most representative particles with a new set $\{x_{k+1}^{j*}, N^{-1}\}$, providing a uniformly resampled estimate of the update density. This brings "new life" to the particle set, insuring that the sample size is large enough to maintain diversity. For instance, in Figure 9.2, the larger weighted particles are used to resample the original set of 12 particles. Note that this "regenerates" the particle set, clustering them around the location of the heavily weighted particles.

To prevent the particle size from becoming too small, we need to develop a metric that assesses when the sample set is beginning to contain too many particles of insufficient weight, before the set becomes too small. An effective metric discussed in Ristic et al.[9] and devised by Liu et al.[29] sets an effectiveness condition that can be compared to a threshold N_{THRESH}. This metric

$$1 < N_{\text{Eff}} = \frac{1}{\sum_{j=1}^{N}(w_{k+1}^{j})^{2}} < N_{\text{THRESH}} < N$$

sets the resampling step into place by regenerating N samples over the span of a number of higher-valued importance weights before their number becomes too restrictive, while eliminating samples with very low weights. A common threshold is to set $N_{\text{THRESH}} = (2/3)N$. The new sample set generates N uniformly weighted samples, providing a "reapproximation" of the updated function, $p(x_{k+1}|\mathbf{Z}^{k+1})$. We add track process noise to the resampled set to reintroduce some diversity to the particle set. This new random measure $\{x_{k+1}^{j}, N^{-1}\}$ approximates the new predictive density $p(x_{k+1}|\mathbf{Z}^{k+1})$, and the cycle repeats (step D). This particle filtering cycle is referred to as a *forward sequential importance sampling* and can be represented by the following sequence:

Sampling $\rightarrow$ importance weights $\rightarrow$ selection $\rightarrow$ resampling/propagation
(Step A)   (Step B)   (Step C)   (Step D)

According to Ristic et al.,[9] resampling improves against degeneracy, but it also introduces another key problem—sample impoverishment.

9.5 Markov Chain Monte Carlo

Sample impoverishment involves the lack of particle diversity with repeated applications of the recycling step. Over time, selected particles will begin to dominate, gaining heavier weights, and could produce multiple generations of offspring through the selection and resampling process. This continued selection of the same set of particles causes a lack of diversity among the particles. In the extreme case, we could see a single particle giving rise to the entire next generation of particles. This is most pronounced when process noise is too small to contain any deviations in actual target motion.

Markov Chain Monte Carlo (MCMC) techniques are an effective counter to this problem, having been instrumental in solving statistical physics problems in the 1950s,[30] mainly involving spatial analysis.

The MCMC approach relies on the generation of samples based on the transition process that impacts the target state space through the Markov chain. There are a number of advantages when one considers the discrete-time Markov Chain process. The primary property is the conditional effect of containing the entire history of the distribution of the target state in the previous state. That is,

$$p(x_{k+1}^j | x_k^j) = p(x_{k+1}^j | x_k^j, x_{k-1}^j \dots, x_0^j)$$

for each particle.

To converge to a stationary distribution, the Markov chain needs to satisfy the following[31]: (1) It must be irreducible. That is, every state is reachable from every other state (probabilistic connectivity). From the tracking problem, every state defines a dynamic model—linear acceleration models, turning models, move-stop-move models, etc. (2) The chain needs to be aperiodic, which prevents the states from oscillating periodically. (3) The chain must be positive recurrent. The discrete-time Markov chain is guaranteed to return to any state infinitely often. That is, no state becomes prohibited during the track process. Finally, an irreducible, aperiodic Markov chain in the limiting condition always exists and is independent of the initial probability distribution.[32]

As an example, consider the two-state problem,[11] where

$$T = [p_{ij}] = \begin{bmatrix} 0.9 & 0.1 \\ 0.33 & 0.67 \end{bmatrix}$$

Here the term p_{ij} is the Markov probability of transitioning from state i to state j. This state problem can be interpreted as in Figure 9.3.

The rationale is that targets will travel between two points in the shortest time ($p_{11} = 0.9$). The maneuver probability is based on an expected time within a turn. The transition probability from nonmaneuver to maneuver mode should occur relatively infrequently ($p_{12} = 0.1$) and the return to a nonmaneuver state should occur as timely as possible ($p_{21} = 0.33$).

On the basis of the properties of the Markov chain, the probability of starting from any initial point will stabilize to the distribution $p(x) = (0.767, 0.233)$ and the target distribution will represent the invariant (balanced) density of the chain. The transition kernel

$$T(\mathbf{X}^*|\mathbf{X})p(\mathbf{X}) = T(\mathbf{X}|\mathbf{X}^*)p(\mathbf{X}^*)$$

satisfies the balance condition, meaning that transition from one state to another is feasible.

A number of MCMC methods have been designed over the years. The Metropolis–Hastings (M-H) is the most popular and widely used.[30] Gibbs sampling is another, being a special case of M-H, that is, it represents the simplest of the Markov chain simulation algorithms. The Gibbs is most effective when we can sample directly from the conditional posterior distribution. The M-H is effective when we cannot sample directly from these distributions.[33]

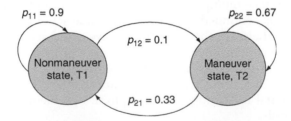

FIGURE 9.3
Markov state transition.

9.6 Metropolis–Hastings

Recall that resampling creates conditions of sample impoverishment due to the selection of the same particles over repeated cycles.[9] To prevent this problem, each cycle needs to undergo a test to determine whether the current state model is sufficient, x_k^i, or that a possible change of state has taken place, x_k^{i*}, according to the proposed distribution, $T(x_k^{i*} \mid x_k^i)$.

The test involves first drawing the new sample x_k^{i*} from $T(x_k^{i*} \mid x_k^i)$ and accepting the new state $x_{k+1}^i = x_k^{i*}$ with probability[9]

$$\alpha = \min\left\{1, \ \frac{p(\mathbf{X}_k^{j*} \mid \mathbf{Z}^k) T(x_k^j \mid x_k^{j*})}{p(\mathbf{X}_k^j \mid \mathbf{Z}^k) T(x_k^{j*} \mid x_k^j)}\right\} \tag{9.14}$$

only if $\alpha \geq u$, where u is a uniform random number between 0 and 1. The particles are excited based on either the representation of the zero-mean Gaussian density of the new state $T(x_k^{j*} \mid x_k^j)$ or the current state $T(x_k^j \mid x_k^{j*})$.

The impact of the M-H process is to "allow" the particles to jump according to the transition matrix, preventing particles from becoming too attached to the current target state model. This has a similar effect of genetic algorithms that use "natural selection" to search for an acceptable solution. Here we search for the most appropriate state.

Figure 9.4 provides a description of the M-H process. We are expanding Figure 9.2 for ease of description, with the M-H–modified process (step D*) replacing the "step D" from the figure, whenever the M-H test has passed.

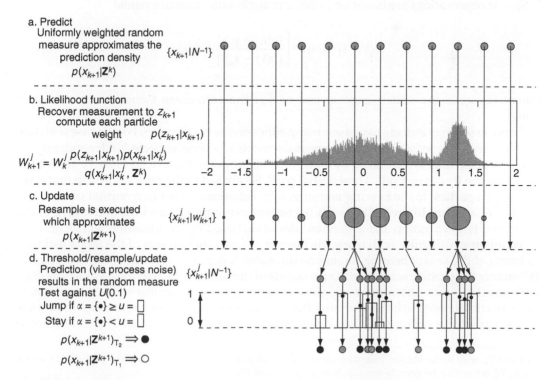

FIGURE 9.4
Graphical representation of Metropolis–Hastings in particle filter process cycle.

In this case, let the box shown in step D be the uniformly selected random value to be used to test against each particle shown. Let the "dot" {●} refer to the acceptance value α. When $\alpha \geq u$, we accept the new state* $x^i_{k+1} = (x^i_k)_{T2}$ whenever the "dot" lies above the box u. Then, only those particles exceeding the randomly selected u will be given the new state and accepted for update in T2; all other particles are rejected, and maintain the old state, T1.

The M-H process greatly impacts the efficiency of the particle filter. If the jump size is set too small, convergence will be slow; if the jump size is set too large, the particles are rejected and the algorithm maintains its current state.

9.7 Particle Filtering Example

We will consider a bearings-only tracking example for a target state with sampling time T, and operating with a dynamic state in two dimensions: 2D position and 2D velocity. The sampling period T is defined as $T = (t_{k+1} - t_k)$; the target state is represented as $x = f(x, \dot{x}, y, \dot{y})$, where the position and velocity terms, respectively, are

$$x(k+1) = x(k) + \dot{x}(k)dt$$

and

$$\dot{x}(k+1) = \dot{x}(k) + v(k)$$

where $v(k)$ is the zero-mean Gaussian process noise.

Sensor observations are based on nonlinear angle-only measurements[†]

$$z(k+1) = \tan^{-1}\left[\frac{(y(k) - y_s)}{(x(k) - x_s)}\right] + w(k) \qquad (9.15)$$

where (x_s, y_s) is the sensor location and $w(k)$ is the zero-mean Gaussian measurement noise.

Scenario settings include two cases: using either one or two sensors. For the case with two sensors, only one sensor can provide measurements at a time, collected independently.

The particle filter is initialized by a uniform distribution of particles of a size allowed to subsume the true target pdf. These initial particles are fed to the prediction step that compares each particle to the bearing measurement statistically. That is, a normal pdf value is used with a mean difference between the sensor measurement and the arctangent of the estimated target position and a standard deviation based on the measurement error.

An inherent difficulty with using the normal pdf as a comparison of the target estimate in terms of the sensor bearing angle measurement is dealing with outlier measurements. Whenever the particle set exceeds three standard deviations, the statistical values produce very small weights. Often these weights are close to zero. When this effect occurs, a new set of particles needs to be regenerated from the previous time index to attempt to provide a new particle set representatively closer to the bearing measurement. This is attempted

---

[*] T1 and T2 refer to the Markov transition states. In this case, when $\alpha \geq u$, we "move" the particle to the new state T2, otherwise the particle remains in its present state, T1.

[†] The model is only valid for small bearing errors where noise is additive, but is a good representation to demonstrate the effects of nonlinear measurements.

up to a number of times, until an appropriate set of weights are determined, or the measurement is rejected (at the last attempt, currently set as the 20th) and the target propagates based on its dynamic model.

Next, the particle set is resampled based on either the surviving weighted particles or a new uniform weighting.

Finally, a M-H sampling algorithm is used when conditions indicate a possibility that the true target path deviates significantly from the estimated target dynamic model. We use the measurement $z(k + 1)$ to calculate the weights and resample the particles at time k. A move is performed, if necessary. We choose the new dynamic model by calculating $T(x_k^{i*} \mid x_k^i) = N[x^*(k), \sigma^2]$ for both dimensions, $x(k)$ and $y(k)$. The dynamic model change is accepted based on Equation 9.14 as follows:

$$\alpha = \min\left\{1, \ \frac{T(x_k^{i*} \mid x_k^i)}{T(x_k^i \mid x_k^{i*})}\right\}$$

This could continue up to N steps. Currently N is set to 1.

9.8 Provide a Set of Performance Evaluations

Figure 9.5 shows a particle filtering bearing-only scenario for both one and two sensors. A target track is initiated for a target heading southeast and abruptly turns southwest approximately half-way through the scenario. The sensors are bearing-only, providing accurate angle information on the target, but contain no range observations. As such, the single-sensor example must rely on the accuracy of particle filter target dynamic estimates to represent true target dynamics, whereas, for the two-sensor case, target estimates can take advantage of sensor diversity to build a better track estimate.

The particle filter is a standard bootstrap filter* that uses a Gaussian pdf (importance density) to choose the particle weights (Equation 9.12) by comparing the predicted particle estimate with the sensor measurement (Equation 9.15). The filter then tests to see if the weights are close to zero (outside the range of the measurement). If this occurs, the algorithm attempts to try to randomly readjust the estimate back to the measurement (attempting this up to 20 times) from the previous time step, which if reached, the new weights are applied to the update. If not, the dynamic model propagates the particles. Regardless, the particles are updated according to Equation 9.13. The weights are then used to repropagate the surviving particles after resampling. Finally, a comparison is made between the resampled particles and an update based on the dynamic model using the M-H (Equation 9.14). The resultant particle set is used to update the filter.

Sample tracking results are shown in Figure 9.6 for both the single-sensor and the two-sensor conditions.

Note the track results for the single-sensor case. In a number of instances, the track estimate tends to veer from the true target track, but particle spread brings it back into line when track angle estimates become incongruous with bearing measurements. This is quite evident during the maneuver, where the M-H was necessary to adjust track estimates to meet new dynamic conditions. For the two-sensor case, however, the algorithm can characterize angular mismatch much sooner due to the geographic diversity of the

---

* Also known as a sequential importance resampling filter.

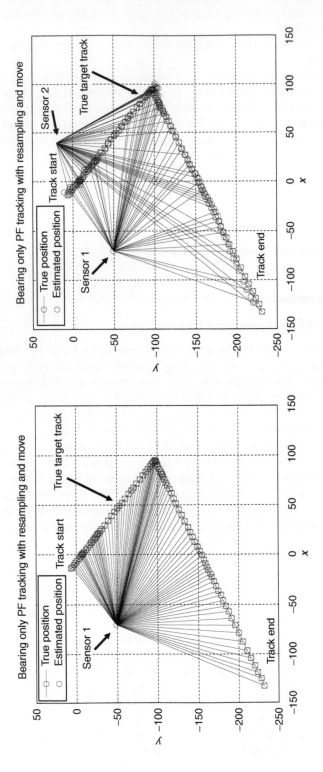

FIGURE 9.5
Particle filtering scenario involving a single sensor and two sensors.

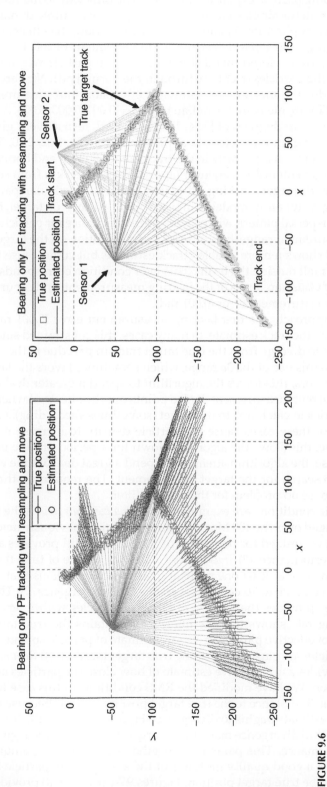

FIGURE 9.6

Sample performance involving a single sensor and two sensors.

two sensors, keeping particle expansion in check. The only way to maintain a significantly smaller track error in the single-sensor is to develop more accurate dynamic model error.

For each test, we applied 100 Monte Carlo runs. We show the effects of different sized particle sets ranging from 100 to 1000 particles in increments of 100. Four different sensor bearing errors, σ_z, are explored that distinctly affect the particle weights, and ultimately the dispersion of the particles over time through the normal pdf, $N(z-\arctan(y/x), \sigma_z)$. The $\arctan(y/x)$ provides the mean-target estimate conversion into measurement space. The bearing errors (STD) are measured in radian terms as [0.001, 0.003, 0.005, 0.010], corresponding to angular errors, in degrees, of [0.34, 1.03, 1.72, 3.44], respectively, giving 3σ accuracy bounds corresponding, in kilometers, to approximately [0.54, 1.62, 2.70, 5.40] errors when the target is nearest to sensor 1, early in the track (about 90 km) and bounds twice as large at the time the target enters the maneuver (approximately 180 km from sensor 1).

One problem with target tracking, in general, and in particular particle filtering, is building the target dynamic model to meet a wide variety of target track conditions. A wide particle dispersal pattern will provide a greater chance for particles to "fall" within the bearing measurement range, but causes a wider variation on the target track estimate. A tighter pattern shows a more "stable" track estimate, but could cause track divergence when all particles fall outside the beam. This later condition corresponds to particles falling outside the 3σ Gaussian limits of the bearing measurement, and occurs many times for the first bearing measurement error (0.001 rad).

By using this approach, the first bearing measurement error (0.001 rad) enforces very tight constraints on the track estimate. In many cases, this constrained smaller error causes the track estimate to diverge from the true target track in part due to the algorithm setting and in part due to this use of the 3σ range, which sometimes favors the larger sensor error cases. For small errors, this forces the algorithm to spend a greater deal of time trying to catch up to the current measurement. It also points out the effects of when the algorithm's track estimate "tries to catch up" to the target (sometimes succeeding), or when it "loses" the target. Both of these cases cause the particle density to increase its "search" space. In the former case, this slows the algorithm down for a period of time until it does catch up; in the later case, the algorithm attempts to spend a great deal of time with each particle (up to 20 tries) to search for the target measurement. The remaining three bearing measurement errors pose no problem for the particle filter.

Four parametric conditions are examined: (1) *normalized run time* is the time to complete a track run, averaged over the 100 Monte Carlo runs. This is obviously machine-dependent and has not been optimized for performance. However, it still provides a reasonable estimate of overall performance. (2) *Tracking rate* is the percentage of time that the true target position is within 3σ of at least one particle. This indicator tells us that the particle filter has not deviated so much as to cause irreversible track divergence. (3) The *average number of effective particles* provides the average number of particles within 3σ of the true target position. Tracking rate shows that the filter error contains the true target, whereas the average number of effective particles gives an estimate of particle spread by telling us how many of the particles are clustered near the true target location. (4) *Root mean square (RMS) tracking error* provides a quantitative estimate of how close the particles actually are to the true target position. We have modified the RMS conditions in this case to show *only those particles* within the 3σ distance to the true target position. This is because under conditions where the target estimate begins to diverge (see Figure 9.6), using the total number of particles may show RMS divergence many times larger than as indicated, giving a misleading estimate of performance. This parameter, together with the average number of effective particles, provides a good quality indicator of the ability of the particle track estimate to remain "close" to the true target position. Figures 9.7 through 9.10 provide these results.

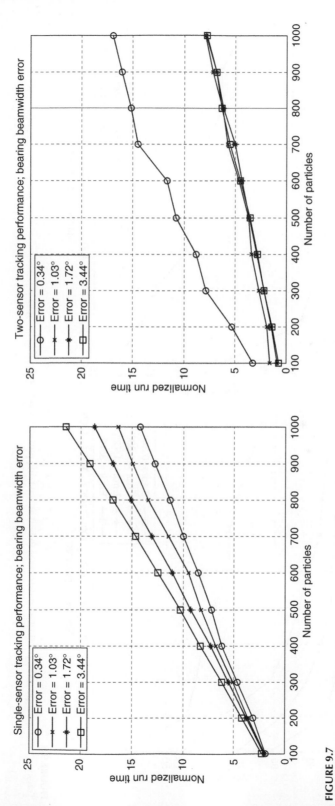

FIGURE 9.7
Normalized run time for single-sensor and two-sensor cases.

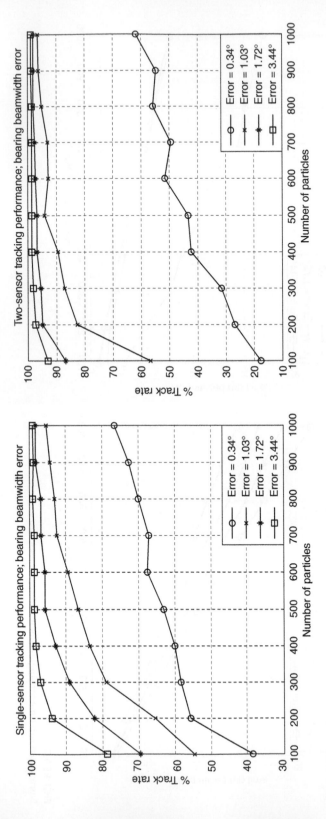

FIGURE 9.8
Tracking rate for single-sensor and two-sensor cases (when at least one particle is within 3σ of the target).

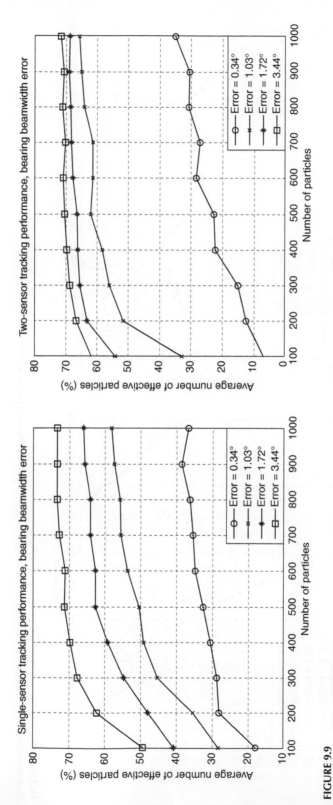

FIGURE 9.9

Average number of effective particles for single-sensor and two-sensor cases (average number of particles within 3σ of the target).

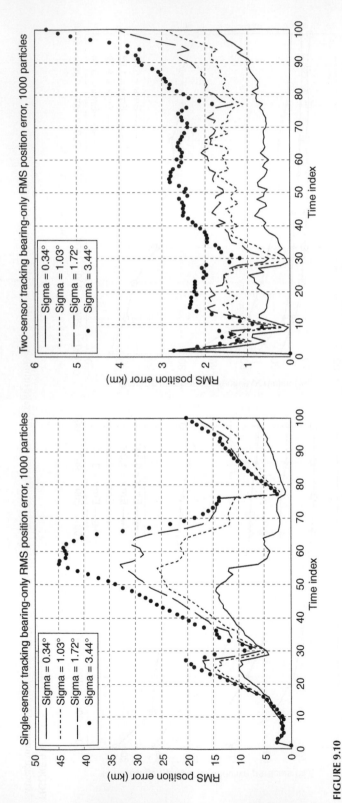

FIGURE 9.10
Root mean square tracking error for single-sensor and two-sensor cases (when the particles are within 3σ error).

Figure 9.7 provides the normalized run time for the single-sensor and two-sensor cases. For the two-sensor case, note that normalized run time performance is consistently good for the higher bearing errors, corresponding to good track estimate. For the smallest error (0.34°), however, we have a high rate of track divergence causing the particles to often fall outside the bearing range, which gives high run times (as much as three times processing speed). For the bearing errors in the single-sensor case, processing speeds are dominated by the algorithm's "tests" to ensure that the particles stay within bounded measurement limits (see Figure 9.6). As the particle density increases, processing speeds increase linearly.

The tracking rate is shown in Figure 9.8. For both single- and two-sensor cases, the small bearing error gives poor performance, due to a higher rate of track divergence. However, for each of the other bearing errors, high track rate can be seen to converge, starting with a particle set of between 200 and 300 particles, giving a strong indication that the track will be maintained.

Figure 9.9 shows the average number of effective particles that fall within 3σ of the target for both cases. Again, the small (0.34°) beam pointing error shows that track divergence has difficulty maintaining the track giving the poorest performance, whereas the other three bearing errors tend to converge quicker for increasing numbers of particles. As expected, the two-sensor case shows rapid convergence, demonstrating that for particle densities as low as 200, at least 50% of the particles are near the true target position; however, for the single-sensor condition, good performance requires a density of at least 400 particles to ensure this 50% condition.

Finally, Figure 9.10 shows the RMS tracking error only for those particles within 3σ of the target. For the single-sensor case, tracking errors are relatively small, except during the target-turn, which occurred around time index 50. While the smaller pointing error (0.34°) shows small RMS errors, this is to be expected as this occurs only for those particles that fall within 3σ of the target—close to the true target position. However, together with the average number of effective particles (Figure 9.9), we see that only 30% of these particles on average actually remain close to the true target. For larger bearing errors, a greater number of particles (and a larger spread) fall within the acceptable error bound giving rise to somewhat higher RMS error—at least 50% of the particles being within the 3σ bound. The two-sensor case shows a significant improvement over the single-sensor case, as expected.

9.9   Summary

This chapter has provided an overview of the development of particle filtering from recognition of the limits of the Kalman filter that led to the EKF, to the conception of particle filtering. Discussion of the development led to the use of importance sampling to characterize acceptable density support functions that reasonably encompass the true target pdf even though the target pdf may not be known. Problems inherent to particle filter development and use were also discussed. Finally, an example scenario was provided that shows some of the overall conditions expected in particle filter tracking along with metrics that help to assess tracking performance. Many articles presenting particle filters also provide excellent pseudocode adequate for developing particle filter simulations. For instance, as Ristic et al.[9] builds the theoretical development of the particle filter, sample pseudocode descriptions are provided, with each subsequent section building upon the previous pseudocode descriptions.

Additional efforts continue to expand the capabilities of the particle filters. For instance, in an upcoming IEEE issue[34] methods are provided to apply particle filters to dynamic Bayesian networks to address nonlinearity and hybrid state conditions in sequential inferencing for complex dynamic systems.

Additionally, particle filtering plays a role in fault diagnostics for autonomous operation of mobile robots,[35] and recent developments have provided reconfigurable adaptation of a suite of particle filters for domain-specific processing,[36] just to name a few examples.

References

1. R. Kalman, A new approach to linear filtering and prediction problems, *Transactions of AME, Journal of Basic Engineering*, 82, 35–45, 1960.
2. E. Brookner, *Tracking and Kalman Filtering Made Easy*, Wiley, New York, NY, 1998.
3. Y. Bar-Shalom (Ed.) *Multitarget-Multisensor Tracking: Advanced Applications*, Vol. I, Artech House, Norwood, MA, 1990.
4. Y. Bar-Shalom (Ed.) *Multitarget-Multisensor Tracking: Applications and Advances*, Vol. II, Artech House, Norwood, MA, 1992.
5. Y. Bar-Shalom and W. Blair, Eds., *Multitarget-Multisensor Tracking: Applications and Advances*, Vol. III, Artech House, Norwood, MA, 2000.
6. S. Blackman, *Multiple-Target Tracking with Radar Applications*, Artech House, Norwood, MA, 1986.
7. S. Blackman and R. Popoli, *Design and Analysis of Modern Tracking Systems*, Artech House, Norwood, MA, 1999.
8. M. Grewal and A. Andrews, *Kalman Filtering: Theory and Practice*, Prentice Hall, Upper Saddle River, NJ, 1993.
9. B. Ristic, S. Arulampalam, and N. Gordon, *Beyond the Kalman Filter: Particle Filters for Tracking Applications*, Artech House, Boston, MA, 2004.
10. H. Stark and J. Woods, *Probability, Random Processes, and Estimation Theory for Engineers*, Prentice Hall, Upper Saddle River, NJ, 1986.
11. Y. Bar-Shalom and X.R. Li, *Multitarget Multisensor Tracking: Principles and Techniques*, YBS Publishing, Storrs, CT, 1995.
12. A. Gelman, J. Carlin, H. Stern, and D. Rubin, *Bayesian Data Analysis*, Chapman and Hall/CRC, Boca Raton, FL, 2004.
13. E. Dudewicz and S. Mishra, *Modern Mathematical Statistics*, Wiley Series in Probability & Mathematical Statistics, Wiley, New York, NY, 1988. [The work is discussed in Chapter 6.3, *Central Limit Theorem, Limit Laws, Applications.*]
14. M. Sanjeev Arulampalam, S. Maskell, N. Gordon, and T. Clapp, A tutorial on particle filters for online non-linear/non-Gaussian Bayesian tracking, *IEEE Transactions on Signal Processing*, 50(2), 174–188, 2002.
15. C. Andrieu, A. Doucet, S. Singh, and V. Tadic, Particle methods for change detection, system identification, and control, *Proceedings of the IEEE, Special Issue on Sequential State Estimation*, 92(3), 423–438, 2002.
16. C. Kwok, D. Fox, and M. Meila, Real-time particle filters, *Proceedings of the IEEE, Special Issue on Sequential State Estimation*, 92(3), 469–484, 2002.
17. C. Agate and K. Sullivan, Road-constrained target tracking and identification using a particle filter, *Signal and Data Processing of Small Targets 2003*, Proceedings of SPIE, Vol. 5204, 2003.
18. M. Rutten, N. Gordon, S. Maskell, Particle-based track-before-detect in Rayleigh noise, *Signal and Data Processing of Small Targets 2004*, Proceedings of SPIE, Vol. 5428, 2004.
19. M. Ulmke, On multitarget track extraction and maintenance using sequential Monte Carlo methods, *Signal and Data Processing of Small Targets 2005*, Proceedings of SPIE, Vol. 5913, 2005.
20. E. Blasch, A. Rice, and C. Yang, Nonlinear track evaluation using absolute and relative metrics, *Signal and Data Processing of Small Targets 2006*, Proceedings of SPIE, Vol. 6236, 2006.

21. H. Kamel and W. Badawy, Comparison between smoothing and auxiliary particle filter in tracking a maneuvering target in a multiple sensor network, *Acquisition, Tracking, and Pointing XX*, SPIE, Vol. 5810, 2005.

22. T. Clemons III and K.C. Chang, Estimation filters for missile tracking with airborne laser, *Acquisition, Tracking, and Pointing XX*, SPIE, Vol. 6238, 2006.

23. F. Gustafsson, F. Gunnarsson, N. Bergman, U. Forssell, J. Jansson, R. Karlsson, and P.J. Nordlund, Particle filters for position, navigation, and tracking, *IEEE Transactions on Signal Processing, Special Issue on Monte Carlo Methods for Statistical Signal Processing*, 50(2), 425–437, 2002.

24. D. Crisan and A. Doucet, A survey of convergence results on particle filtering methods for practitioners, *IEEE Transactions on Signal Processing*, 50(3), 736–746, 2002.

25. S. Arulampalam and B. Ristic, Comparison of the particle filter with range-parameterized and modified polar EKFs for angle-only tracking, *Signal and Data Processing of Small Targets 2000*, Proceedings of SPIE, Vol. 4048, 2000.

26. K. Ezal and C. Agate, Tracking and interception of ground-based RF sources using autonomous guided munitions with passive-bearings-only sensors and tracking algorithms, *Acquisition, Tracking, and Pointing XVIII*, Proceedings of SPIE, Vol. 5430, 2004.

27. A. Marrs, S. Maskell, and Y. Bar-Shalom, Expected likelihood for tracking in clutter with particle filters, *Signal and Data Processing of Small Targets 2002*, Proceedings of SPIE, Vol. 4728, 2002.

28. R. Van der Merwe, A. Doucet, N. De Freitas, and E. Wan, The unscented particle filter, Technical Report, CUED/F-INFENG/TR 380, Cambridge University Engineering Department, August 2000.

29. J. Liu and R. Chen, Sequential Monte Carlo methods for dynamical systems, *Journal of the American Statistical Association*, 93, 1032–1044, 1998.

30. C. Andrieu, N. Freitas, A. Doucet, and M. Jordan, *Introduction to Markov Chain Monte Carlo for Machine Learning*, Kluwer Academic Publishers, The Netherlands, 2001.

31. W. Gilks, S. Richardson, and D. Spiegelhalter, *Markov Chain Monte Carlo in Practice*, Chapman and Hall, London, UK, 1996.

32. W. Steward, *Introduction to Numerical Solutions of Markov Chains*, Princeton University Press, Princeton, NJ, 1994.

33. A. Gelman, J. Carlin, H. Stern, and D. Rubin, *Bayesian Data Analysis*, Chapman and Hall/CRC, Boca Raton, FL 2004.

34. H. Chen and K.C. Chang, *K*-nearest neighbor particle filters for dynamic hybrid Bayesian networks, *IEEE Transactions on Aerospace and Electronic Systems*, 44(2), 2008.

35. N. de Freitas, R. Dearden, F. Hutter, R. Morales-Menendez, J. Mutch, and D. Poole, Diagnosis by a waiter and a Mars explorer, *Proceedings of the IEEE, Special Issue on Sequential State Estimation*, 92(3), 455–468, 2002.

36. S. Hong, J. Lee, A. Athalye, P, Djuric, and W.D. Cho, Design methodology for domain specific parameterizable particle filter realizations, *IEEE Transactions on Circuits and Systems I: Regular Papers*, 54(9), 1987–2000, 2007.

10

Target Tracking Using Probabilistic Data Association-Based Techniques with Applications to Sonar, Radar, and EO Sensors

T. Kirubarajan and Yaakov Bar-Shalom

CONTENTS

10.1   Introduction

In tracking targets with less-than-unity probability of detection in the presence of false alarms (FA) (clutter), data association—deciding which of the received multiple measurements to use to update each track—is crucial. A number of algorithms have been developed to solve this problem.[1-4] Two simple solutions are the strongest neighbor filter (SNF) and the nearest neighbor filter (NNF). In the SNF, the signal with the highest intensity among the validated measurements (in a gate) is used for track update and the others are discarded. In the NNF, the measurement closest to the predicted measurement is used. Although these simple techniques work reasonably well with benign targets in sparse scenarios, they begin to fail as the FA rate increases or with low observable (LO) (low probability of target detection) maneuvering targets.[5,6] Instead of using only one measurement among the received ones and discarding the others, an alternative approach is to use all of the validated measurements with different weights (probabilities), known as probabilistic data association (PDA).[3] The standard PDA and its numerous improved versions have been shown to be very effective in tracking a single target in clutter.[6,7]

   Data association becomes more difficult with multiple targets where the tracks compete for measurements. Here, in addition to a track validating multiple measurements as in the single target case, a measurement itself can be validated by multiple tracks (i.e., contention occurs among tracks for measurements). Many algorithms exist to handle this contention. The joint probabilistic data association (JPDA) algorithm is used to track multiple targets by evaluating the measurement-to-track association probabilities and combining them to find the state estimate.[3] The multiple-hypothesis tracking (MHT) is a more powerful (but much more complex) algorithm that handles the multitarget tracking problem by evaluating the likelihood that there is a target given a sequence of measurements.[4] In the tracking benchmark problem[8] designed to compare the performance of different algorithms for tracking highly maneuvering targets in the presence of electronic countermeasures (ECM), the PDA-based estimator, in conjunction with the interacting multiple model (IMM) estimator, yielded one of the best solutions. Its performance was comparable to that of the MHT algorithm.[6,9]

   This chapter presents an overview of the PDA technique and its application for different target-tracking scenarios. Section 10.2 summarizes the PDA technique. Section 10.3 describes the use of the PDA technique for tracking LO targets with passive sonar measurements. This target motion analysis (TMA) is an application of the PDA technique, in conjunction with the maximum likelihood (ML) approach for target motion parameter estimation via a batch procedure. Section 10.4 presents the use of the PDA technique for tracking highly maneuvering targets and for radar resource management. It illustrates the application of the PDA technique for recursive state estimation using the IMM estimator with probabilistic data association filter (IMMPDAF). Section 10.5 presents a state-of-the-art sliding-window (which can also expand and contract) parameter estimator using the

PDA approach for tracking the state of a maneuvering target using measurements from an electrooptical (EO) sensor. This, while still a batch procedure, offers the flexibility of varying the batches depending on the estimation results.

10.2 Probabilistic Data Association

The PDA algorithm calculates in real time the probability that each validated measurement is attributable to the target of interest. This probabilistic (Bayesian) information is used in a tracking filter, the PDA filter (PDAF), which accounts for the measurement origin uncertainty.

10.2.1 Assumptions

The following assumptions are made to obtain the recursive PDAF state estimator (tracker)

- There is only one target of interest whose state evolves according to a dynamic equation driven by process noise.
- The track has been initialized.
- The past information about the target is summarized approximately by

$$p[x(k)|Z^{k-1}] = N[x(k); \hat{x}(k|k-1), P(k|k-1)] \tag{10.1}$$

where $N[k(k); \hat{x}(k|k-1), P(k|k-1)]$ denotes the normal probability density function (pdf) with argument $x(k)$, mean $\hat{x}(k|k-1)$, and covariance matrix $P(k|k-1)$. This assumption of the PDAF is similar to the generalized pseudo-Bayesian (GPB1) approach,[10] where a single "lumped" state estimate is a quasi-sufficient statistic.

- At each time, a validation region as in Ref. 3 are set up (Equation 10.4).
- Among the possibly several validated measurements, at most one of them can be target-originated—if the target was detected and the corresponding measurement fell into the validation region.
- The remaining measurements are assumed to be FA or clutter and are modeled as independent identically distributed (iid) measurements with uniform spatial distribution.
- The target detections occur independently over time with known probability PD.

These assumptions enable a state estimation scheme to be obtained, which is almost as simple as the Kalman filter, but much more effective in clutter.

10.2.2 PDAF Approach

The PDAF approach uses a decomposition of the estimation with respect to the origin of each element of the latest set of validated measurements, denoted as

$$Z(k) = \{z_i(k)\}_{i=1}^{m(k)} \tag{10.2}$$

where $z_i(k)$ is the ith validated measurement and $m(k)$ is the number of measurements in the validation region at time k.

The cumulative set (sequence) of measurements* is

$$Z^k = \{Z(j)\}_{j=1}^k \qquad (10.3)$$

10.2.3 Measurement Validation

From the Gaussian assumption (Equation 10.1), the validation region is the elliptical region

$$V(k, \gamma) = \{Z : [z - \hat{z}(k|k-1)]'S(k)^{-1}[z - \hat{z}(k|k-1)] \leq \gamma\} \qquad (10.4)$$

where γ is the gate threshold and

$$S(k) = H(k)P(k|k-1)H(k)' + R(k) \qquad (10.5)$$

is the covariance of the innovation corresponding to the true measurement. The volume of the validation region (Equation 10.4) is

$$V(k) = c_{n_z} |\gamma S(k)|^{1/2} = c_{n_z} \gamma^{n_z/2} |S(k)|^{1/2} \qquad (10.6)$$

where the coefficient c_{n_z} depends on the dimension of the measurement (it is the volume of the n_z-dimensional unit hypersphere: $c_1 = 2, c_2 = \pi, c_3 = 4\pi/3$, etc.).

10.2.4 State Estimation

In view of the assumptions listed, the association events

$$\theta_i(k) = \begin{cases} \{Z_i(k) \text{ is the target-originated measurement}\} & i = 1, \ldots, m(k) \\ \{\text{none of the measurements is target originated}\} & i = 0 \end{cases} \qquad (10.7)$$

are mutually exclusive and exhaustive for $m(k) \geq 1$.

Using the total probability theorem[10] with regard to the events given in Equation 10.7, the conditional mean of the state at time k can be written as

$$\hat{x}(k|k) = E[x(k)|Z^k]$$

$$= \sum_{i=0}^{m(k)} E[x(k)|\theta_i(k), Z^k] P\{\theta_i(k)|Z^k\}$$

$$= \sum_{i=0}^{m(k)} \hat{x}_i(k|k)\beta_i(k) \qquad (10.8)$$

---

* When the running index is a time argument, a sequence exists; otherwise it is a set where the order is not relevant. The context should indicate which is the case.

where $\hat{x}_i(k|k)$ is the updated state conditioned on the event that the ith validated measurement is correct, and

$$\beta_i(k) \triangleq P\{\theta_i(k)|Z_k\} \tag{10.9}$$

is the conditional probability of this event—the association probability, obtained from the PDA procedure presented in Section 10.2.7.

The estimate conditioned on measurement i being correct is

$$\hat{x}_i(k|k) = \hat{x}(k|k-1) + W(k)v_i(k) \qquad i = 1,\dots,m(k) \tag{10.10}$$

where the corresponding innovation is

$$v_i(k) = z_i(k) - \hat{z}(k|k-1) \tag{10.11}$$

The gain $W(k)$ is the same as in the standard filter

$$W(k) = P(k|k-1)H(k)'S(k)^{-1} \tag{10.12}$$

because conditioned on $\theta_i(k)$, there is no measurement origin uncertainty.

For $i = 0$ (i.e., if none of the measurements is correct) or $m(k) = 0$ (i.e., there is no validated measurement)

$$\hat{x}_0(k|k) = \hat{x}(k|k-1) \tag{10.13}$$

10.2.5 State and Covariance Update

Combining Equations 10.10 and 10.13 with Equation 10.8 yields the state update equation of the PDAF

$$\hat{x}(k|k) = \hat{x}(k|k-1) + W(k)v(k) \tag{10.14}$$

where the combined innovation is

$$v(k) = \sum_{i=1}^{m(k)} \beta_i(k)v_i(k) \tag{10.15}$$

The covariance associated with the updated state is

$$P(k|k) = \beta_0(k)P(k|k-1) + [1 - \beta_0(k)]P^c(k|k-1) + \tilde{P}(k) \tag{10.16}$$

where the covariance of the state updated with the correct measurement is[3]

$$P^c(k|k) = P(k|k-1) - W(k)S(k)W(k)' \tag{10.17}$$

and the spread of the innovations term (similar to the spread of the means term in a mixture)[10] is

$$\tilde{P}(k) \triangleq W(k)\left[\sum_{i=1}^{m(k)} \beta_i(k)v_i(k)v_i(k)' - v(k)v(k)'\right]W(k)' \tag{10.18}$$

10.2.6   Prediction Equations

The prediction of the state and measurement to $k + 1$ is done as in the standard filter, that is,

$$\hat{x}(k + 1|k) = F(k)\hat{x}(k|k) \tag{10.19}$$

$$\hat{z}(k + 1|k) = H(k + 1)\hat{x}(k + 1|k) \tag{10.20}$$

The covariance of the predicted state is, similarly,

$$P(k + 1|k) = F(k)P(k|k)F(k)' + Q(k) \tag{10.21}$$

where $P(k|k)$ is given by Equation 10.16.

The innovation covariance (for the correct measurement) is, again, as in the standard filter

$$S(k + 1) = H(k + 1)P(k + 1|k)H(k + 1)' + R(k + 1) \tag{10.22}$$

10.2.7   Probabilistic Data Association

To evaluate the association probabilities, the conditioning is broken down into the past data Z^{k-1} and the latest data $Z(k)$. A probabilistic inference can be made on both the number of measurements in the validation region (from the clutter density, if known) and on their location, expressed as

$$\beta_i(k) = P\{\theta_i(k)|Z^k\} = P\{\theta_i(k)|Z(k), m(k), Z^{k-1}\} \tag{10.23}$$

Using Bayes' formula, the Equation 10.23 is rewritten as

$$\beta_i(k) = \frac{1}{c}p\big[Z(k)|\theta_i(k), m(k), Z^{k-1}\big]p\{\theta_i(k)|m(k), Z^{k-1}\} \quad i = 0, \ldots, m(k) \tag{10.24}$$

The joint density of the validated measurements conditioned on $\theta_i(k)$, $i \neq 0$, is the product of

- The (assumed) Gaussian pdf of the correct (target-originated) measurements
- The pdf of the incorrect measurements, which are assumed to be uniform in the validation region whose volume $V(k)$ is given in Equation 10.6

The pdf of the correct measurement (with the P_G factor that accounts for restricting the normal density to the validation gate) is

$$p[z_i(k)|\theta_i(k), m(k), Z^{k-1}] = P_G^{-1}N[z_i(k); z(k|k - 1), S(k)] = P_G^{-1}N[v_i(k); 0, S(k)] \tag{10.25}$$

The pdf from Equation 10.24 is then

$$p[Z(k)|\theta_i(k), m(k), Z^{k-1}] = \left\{V(k)^{-m(k)+1}P_G^{-1}N[v_i(k); 0, S(k)]\right. \tag{10.26}$$

The probabilities of the association events conditioned only on the number of validated measurements are

$$p[Z(k)|\theta_i(k), m(k), Z^{k-1}] = \begin{cases} V(k)^{-m(k)+1}P_G^{-1}N[v_i(k); 0, S(k)] & i = 1, \ldots, m(k) \\ V(k)^{-m(k)} & i = 0 \end{cases} \tag{10.27}$$

where $\mu_F(m)$ is the probability mass function (pmf) of the number of false measurements (FA or clutter) in the validation region.

Two models can be used for the pmf $\mu_F(m)$ in a volume of interest V

1. A Poisson model with a certain spacial density λ

$$\mu_F(m) = e^{-\lambda V}\frac{(\lambda V)^m}{m!} \tag{10.28}$$

2. A diffuse prior model[3]

$$\mu_F(m) = \mu_F(m-1) = \delta \tag{10.29}$$

where the constant δ is irrelevant since it cancels out.

Using the (parametric) Poisson model in Equation 10.27 yields

$$\gamma_i[m(k)] = \begin{cases} P_DP_G[P_DP_Gm(k) + (1 - P_DP_G)\lambda V(k)]^{-1} & i = 1, \ldots, m(k) \\ (1 - P_DP_F)\lambda V(k)[P_DP_Gm(k) + (1 - P_DP_G)\lambda V(k)]^{-1} & i = 0 \end{cases} \tag{10.30}$$

The (nonparametric) diffuse prior (Equation 10.29) yields

$$\gamma_i[m(k)] = \begin{cases} \dfrac{1}{m(k)}P_DP_G & i = 1, \ldots, m(k) \\ (1 - P_DP_G) & i = 0 \end{cases} \tag{10.31}$$

The nonparametric model (Equation 10.31) can be obtained from Equation 10.30 by setting

$$\lambda = \frac{m(k)}{V(k)} \tag{10.32}$$

that is, replacing the Poisson parameter with the sample spatial density of the validated measurements. The volume $V(k)$ of the elliptical (i.e., Gaussian-based) validation region is given in Equation 10.6.

10.2.8 Parametric PDA

Using Equations 10.30 and 10.26 with the explicit expression of the Gaussian pdf in Equation 10.24 yields, after some cancellations, the final equations of the parametric PDA with the Poisson clutter model

$$\beta_i(k) = \begin{cases} \dfrac{e_i}{b + \sum_{j=1}^{m(k)} e_j} & i = 1,\ldots,m(k) \\[4mm] \dfrac{b}{b + \sum_{j=1}^{m(k)} e_j} & i = 0 \end{cases} \tag{10.33}$$

where

$$e_i \triangleq e^{-(1/2)v_i(k)'S(k)^{-1}v_i(k)} \tag{10.34}$$

$$b \triangleq \lambda|2\pi S(k)|^{1/2}\,\frac{1 - P_D P_G}{P_D} \tag{10.35}$$

Equation (10.35) can be rewritten as

$$b = \left(\frac{2\pi}{\lambda}\right)^{n_z/2} \lambda V(k) c_{n_z}^{-1}\frac{1 - P_D P_G}{P_D} \tag{10.36}$$

10.2.9 Nonparametric PDA

The nonparametric PDA is the same as the parametric PDA except for replacing $\lambda V(k)$ in Equation 10.36 by $m(k)$—this obviates the need to know λ.

10.3 Low Observable TMA Using the ML-PDA Approach with Features

This section considers the problem of TMA—estimation of the trajectory parameters of a constant velocity target—with a passive sonar, which does not provide full target position measurements. The methodology presented here applies equally to any *target motion characterized by a deterministic equation*, in which case the initial conditions (a finite dimensional parameter vector) characterize in full the *entire motion*. In this case the (batch) ML parameter estimation can be used; this method is more powerful than state estimation when the target motion is deterministic (it does not have to be linear). Furthermore, the ML-PDA approach makes no approximation, unlike the PDAF in Equation 10.1.

10.3.1 Amplitude Information Feature

The standard TMA consists of estimating the target's position and its constant velocity from bearings-only (wideband sonar) measurements corrupted by noise.[10] Narrowband passive sonar tracking, where frequency measurements are also available, has been studied.[11] The advantages of narrowband sonar are that it does not require a maneuver of the platform for observability, and it greatly enhances the accuracy of the estimates. However, not all passive sonars have frequency information available. In both cases, the intensity of the signal at the output of the signal processor, which is referred to as *measurement amplitude* or *amplitude information* (AI), is used implicitly to determine whether there is a valid measurement. This is usually done by comparing it with the detection threshold, which is a design parameter.

This section shows that the measurement amplitude carries valuable information and that its use in the estimation process increases the observability even though the AI cannot be correlated to the target state directly. Also, superior global convergence properties are obtained.

The pdf of the envelope detector output (i.e., the AI) a when the signal is due to noise only is denoted as $p_0(a)$ and the corresponding pdf when the signal originated from the target is $p_1(a)$. If the signal-to-noise ratio (SNR—this is the SNR in a resolution cell, to be denoted later as SNR_c) is d, the density functions of noise-only and target-originated measurements can be written as

$$p_0(a) = a\exp\left(-\frac{a^2}{2}\right) \quad a \ge 0 \tag{10.37}$$

$$p_1(a) = \frac{a}{1+d}\exp\left(-\frac{a^2}{2(1+d)}\right) \quad a \ge 0 \tag{10.38}$$

respectively. This is a Rayleigh fading amplitude (Swerling I) model believed to be the most appropriate for shallow water passive sonar.

A suitable threshold, denoted by τ, is used to declare a detection. The probability of detection and the probability of FA are denoted by P_D and P_{FA}, respectively. Both P_D and P_{FA} can be evaluated from the pdfs of the measurements. Clearly, to increase P_D, the threshold τ must be lowered; however, this also increases P_{FA}. Therefore, depending on the SNR, τ must be selected to satisfy two conflicting requirements.*

The density functions given above correspond to the signal at the envelope detector output. Those corresponding to the output of the threshold detector are

$$\rho_0^\tau(a) = \frac{1}{P_{FA}}p_0(a) = \frac{1}{P_{FA}}a\exp\left(-\frac{a^2}{2}\right) \quad a > \tau \tag{10.39}$$

$$\rho_1^\tau(a) = \frac{1}{P_D}p_1(a) = \frac{1}{P_D}\frac{a}{1+d}a\exp\left(-\frac{a^2}{2(1+d)}\right) \quad a > \tau \tag{10.40}$$

---

* For other probabilistic models of the detection process, different SNR values correspond to the same P_D, P_{FA} pair. Compared to the Rician model receiver operating characteristic (ROC) curve, the Rayleigh model ROC curve requires a higher SNR for the same pair (P_D, P_{FA}), that is, the Rayleigh model considered here is pessimistic.

where $p_0^{\tau}(a)$ is the pdf of the validated measurements that are caused by noise only, and $p_1^{\tau}(a)$ is the pdf of those that originated from the target. In the following, a is the amplitude of the candidate measurements. The amplitude likelihood ratio, ρ, is defined as

$$\rho = \frac{p_1^{\tau}(a)}{p_0^{\tau}(a)} \tag{10.41}$$

10.3.2 Target Models

Assume that n sets of measurements, made at times $t = t_1, t_2, \ldots, t_n$, are available.

For bearings-only estimation, the target motion is defined by the four-dimensional parameter vector

$$x \equiv [\xi(t_0), \eta(t_0), \dot{\xi}, \dot{\eta}] \tag{10.42}$$

where $\xi(t_0)$ and $\eta(t_0)$ are the distances of the target in the east and north directions, respectively, from the origin at the reference time t_0. The corresponding velocities, assumed constant, are $\dot{\xi}$ and $\dot{\eta}$, respectively. This assumes deterministic target motion (i.e., no process noise[10]). Any other deterministic motion (e.g., constant acceleration) can be handled within the same framework.

The state of the platform at t_i ($i = 1, \ldots, n$) is defined by

$$x_p(t_i) \equiv [\xi_p(t_i), \eta_p(t_i), \dot{\xi}_p(t_i), \dot{\eta}_p(t_i)]' \tag{10.43}$$

The relative position components in the east and north directions of the target with respect to the platform at t_i are defined by $r_\xi(t_i, x)$ and $r_\eta(t_i, x)$, respectively. Similarly, $v_\xi(t_i, x)$ and $v_\eta(t_i, x)$ define the relative velocity components. The true bearing of the target from the platform at t_i is given by

$$\theta_i(x) \triangleq \tan^{-1} \left[\frac{r_\xi(t_i, x)}{r_\eta(t_i, x)} \right] \tag{10.44}$$

The range of possible bearing measurements is

$$U_\theta \triangleq [\theta_1, \theta_2] \subset [0, 2\pi] \tag{10.45}$$

The set of measurements at t_i is denoted by

$$Z(i) \triangleq \{z_j(i)\}_{j=1}^{m_i} \tag{10.46}$$

where m_i is the number of measurements at t_i, and the pair of bearing and amplitude measurements $z_j(i)$, is defined by

$$z_j(i) \triangleq [\beta_{ij} a_{ij}]' \tag{10.47}$$

The cumulative set of measurements during the entire period is

$$Z^n \triangleq \{Z(i)\}_{i=1}^n \tag{10.48}$$

The following additional assumptions about the statistical characteristics of the measurements are also made:[11]

1. The measurements at two different sampling instants are conditionally independent, that is,

$$p[Z(i_1), Z(i_2)|x] = p[Z(i_1)|x]p[Z(i_2)|x] \quad \forall i_1 \neq i_2 \qquad (10.49)$$

where $p[\cdot]$ is the pdf.

2. A measurement that originated from the target at a particular sampling instant is received by the sensor only once during the corresponding scan with probability P_D and is corrupted by zero-mean Gaussian noise of known variance. That is,

$$\beta_{ij} = \theta_i(x) + \epsilon_{ij} \qquad (10.50)$$

where $\epsilon_{ij} \sim N[0, \sigma_\theta^2]$ is the bearing measurement noise. Owing to the presence of false measurements, the index of the true measurement is not known.

3. The false bearing measurements are distributed uniformly in the surveillance region, that is,

$$\beta_{ij} \sim U[\theta_1, \theta_2] \qquad (10.51)$$

4. The number of false measurements at a sampling instant is generated according to Poisson law with a known expected number of false measurements in the surveillance region. This is determined by the detection threshold at the sensor (exact equations are given in Section 10.3.5).

For narrowband sonar (with frequency measurements) the target motion model is defined by the five-dimensional vector

$$x \equiv [\xi(t_1), \eta(t_1), \dot{\xi}, \dot{\eta}, \gamma] \qquad (10.52)$$

where γ is the unknown emitted frequency assumed constant. Owing to the relative motion between the target and platform at t_i, this frequency will be Doppler shifted at the platform. The (noise-free) shifted frequency, denoted by $\gamma_i(x)$, is given by

$$\gamma_i(x) = \gamma \left[1 - \frac{v_\xi(t_i, x)\sin\theta_i(x) + v_\eta(t_i, x)\cos\theta_i(x)}{c} \right] \qquad (10.53)$$

where c is the velocity of sound in the medium. If the bandwidth of the signal processor in the sonar is $[\Omega_1, \Omega_2]$, the measurements can lie anywhere within this range. As in the case of bearing measurements, we assume that an operator is able to select a frequency subregion $[\Gamma_1, \Gamma_2]$ for scanning. In addition to the bearing surveillance region given in Equation 10.45, the region for frequency is defined as

$$U_\gamma \triangleq [\Gamma_1, \Gamma_2] \subset [\Omega_1, \Omega_2] \qquad (10.54)$$

The noisy frequency measurements are denoted by f_i and the measurement vector is

$$z_j(i) \triangleq [\beta_{ij}, f_{ij}, a_{ij}]'$$ (10.55)

As for the statistical assumptions, those related to the conditional independence of measurements (assumption 1) and the number of false measurements (assumption 4) are still valid. The equations relating the number of FA in the surveillance region to detection threshold are given in Section 10.3.5.

The noisy bearing measurements satisfy Equation 10.50 and the noisy frequency measurements f_{ij} satisfy

$$f_{ij} = \gamma_i(x) + v_{ij}$$ (10.56)

where $v_{ij} \sim N[0, \sigma_\gamma^2]$ is the frequency measurement noise.

It is also assumed that these two measurement noise components are conditionally independent. That is,

$$p(\epsilon_{ij}, v_{ij}|x) = p(\epsilon_{ij}|x)p(v_{ij}|x)$$ (10.57)

The measurements resulting from noise only are assumed to be uniformly distributed in the entire surveillance region.

10.3.3 Maximum Likelihood Estimator Combined with PDA: The ML-PDA

In this section, we present the derivation and implementation of the ML estimator combined with the PDA technique for both bearings-only tracking and narrowband sonar tracking. If there are m_i detections at t_i, one has the following mutually exclusive and exhaustive events:[3]

$$\varepsilon_j(i) \triangleq \begin{cases} \{\text{measurement } z_j(i) \text{ is from the target}\} & j = 1, \ldots, m_i \\ \{\text{all measurements are false}\} & j = 0 \end{cases}$$ (10.58)

The pdf of the measurements corresponding to the above events can be written as

$$p[Z(i)|e_j(i), x] = \begin{cases} u^{1-m_i} p(\beta_{ij}) \rho_{ij} \prod_{j=1}^{m_i} p_0^\tau(a_{ij}) & j = 1, \ldots, m_i \\ u^{-m_i} \prod_{j=1}^{m_i} p_0^\tau(a_{ij}) & j = 0 \end{cases}$$ (10.59)

where $u = U_\theta$ is the area of the surveillance region.

Using the total probability theorem, the likelihood function of the set of measurements at t_i can be expressed as

$$p[Z(i)|x] = u^{-m_i}(1 - P_D)\prod_{j=1}^{m_i} p_0^\tau(a_{ij})\mu_F(m_i) + \frac{u^{1-m_i}P_D\mu_F(m_i - 1)}{m_i}\prod_{j=1}^{m_i} p_0^\tau(a_{ij})\sum_{j=1}^{m_i} p(b_{ij})\rho_{ij}$$

$$= u^{-m_i}(1 - P_D)\prod_{j=1}^{m_i} p_0^\tau(a_{ij})\mu_F(m_i) + \frac{u^{1-m_i}P_D\mu_F(m_i - 1)}{m_i}\prod_{j=1}^{m_i} p_0^\tau(a_{ij})$$

$$\cdot \sum_{j=1}^{m_i} \frac{1}{\sqrt{2\pi}\sigma_\theta}\exp\left(-\frac{1}{2}\left[\frac{\beta_{ij} - \theta_i(x)}{\sigma_\theta}\right]^2\right)\rho_{ij}$$ (10.60)

$\mu_F(m_i)$ is the Poisson pmf of the number of false measurements at t_i. Dividing the above by $p[Z(i)|\varepsilon_0(i), x]$ yields the *dimensionless* likelihood ratio $\Phi_i[Z(i), x]$ at t_i. Then

$$\Phi_i[Z(i), x] = \frac{p[Z(i), x]}{p[Z(i) \mid \varepsilon_0(i), x]}$$

$$= (1 - P_D) + \frac{P_D}{\lambda} \sum_{j=1}^{m_i} \frac{1}{\sqrt{2\pi\sigma_\theta}} \rho_{ij} \exp\left(-\frac{1}{2}\left[\frac{\beta_{ij} - \theta_i(x)}{\sigma_\theta}\right]^2\right) \tag{10.61}$$

where λ is the expected number of FA per unit area. Alternately, the log-likelihood ratio at t_i can be defined as

$$\phi_i[Z(i), x] = \ln\left[(1 - P_D) + \frac{P_D}{\lambda} \sum_{j=1}^{m_i} \frac{1}{\sqrt{2\pi\sigma_\theta}} \rho_{ij} \exp\left(-\frac{1}{2}\left[\frac{\beta_{ij} - \theta_i(x)}{\sigma_\theta}\right]^2\right)\right] \tag{10.62}$$

Using conditional independence of measurements, the likelihood function of the entire set of measurements can be written in terms of the individual likelihood functions as

$$p[Z^n|x] = \prod_{i=1}^{n} p[Z(i)|x] \tag{10.63}$$

Then the dimensionless likelihood ratio for the entire data is given by

$$\Phi[Z^n, x] = \prod_{i=1}^{n} \Phi_i[Z(i), x] \tag{10.64}$$

From the above, the total log-likelihood ratio $\Phi_i[Z(i), x]t_i$ can be expressed as

$$\Phi[Z^n, x] = \sum_{i=1}^{n} \ln\left[(1 - P_D) + \frac{P_D}{\lambda} \sum_{j=1}^{m_i} \frac{1}{\sqrt{2\pi\sigma_\theta}} \rho_{ij} \exp\left(-\frac{1}{2}\left[\frac{\beta_{ij} - \theta_i(x)}{\sigma_\theta}\right]^2\right)\right] \tag{10.65}$$

The maximum likelihood estimate (MLE) is obtained by finding the state $x = \hat{x}$ that maximizes the total log-likelihood function. In deriving the likelihood function, the gate probability mass, which is the probability that a target-originated measurement falls within the surveillance region, is assumed to be 1. The operator selects the appropriate region.

Arguments similar to those given earlier can be used to derive the MLE when frequency measurements are also available. Defining $\varepsilon_j(i)$ as in Equation 10.58, the pdf of the measurements is

$$p[Z(i)|\varepsilon_j(i), x] = \begin{cases} u^{1-m_i} p(\beta_{ij})p(f_{ij})p_{ij} \displaystyle\prod_{j=1}^{m_i} p_0^\varsigma(a_{ij}) & j = 1, \ldots, m_i \\[3ex] u^{-m_i} \displaystyle\prod_{j=1}^{m_i} p_0^\tau(a_{ij}) & j = 0 \end{cases} \tag{10.66}$$

where $u = U_\theta\, U_\gamma$ is the volume of the surveillance region.

After some lengthy manipulations, the total log-likelihood function is obtained as

$$\phi[Z^n, x] = \sum \ln\left[(1 - P_D) + \frac{P_D}{\lambda}\sum_{j=1}^{m_i}\frac{\rho_{ij}}{2\pi\sigma_\theta\sigma_\gamma}\exp\left(-\frac{1}{2}\left[\frac{\beta_{ij} - \theta_i(x)}{\sigma_\theta}\right]^2 - \frac{1}{2}\left[\frac{f_{ij} - \gamma_i(x)}{\sigma_\gamma}\right]^2\right)\right]$$

(10.67)

For narrowband sonar, the MLE is found by maximizing Equation 10.67.

This section demonstrated the essence of the use of the PDA—all the measurements are accounted for and the likelihood function is evaluated using the total probability theorem, similar to Equation 10.8. However, since Equation 10.67 is exact (for the parameter estimation formulation), there is no need for the approximation in Equation 10.1, which is necessary in the PDAF for state estimation.

The same ML-PDA approach is applicable to the estimation of the trajectory of an exoatmospheric ballistic missile.[12,13] The modification of this fixed-batch ML-PDA estimator to a flexible (sliding/expanding/contracting) procedure is discussed in Section 10.5 and demonstrated with an actual EO data example.

10.3.4   Cramér–Rao Lower Bound for the Estimate

For an unbiased estimate, the Cramér–Rao lower bound (CRLB) is given by

$$E\{(x - \hat{x})(x - \hat{x})'\} \geq J^{-1}$$

(10.68)

where J is the Fisher information matrix (FIM) given by

$$J = E\{[\nabla_x \ln p(Z^n|x)][\nabla_x \ln p(Z^n|x)]'\}\,|_{x=x_{\text{true}}}$$

(10.69)

Only in simulations will the true value of the state parameter be available. In practice CRLB is evaluated at the estimate.

As expounded in Ref. 14, the FIM J is given in the present ML-PDA approach for the bearings-only case, *wideband sonar*, by

$$J = q_2(P_D, \lambda v_g, g)\sum_{i=1}^{n}\frac{1}{\sigma_\theta^2}[\nabla_x\theta_i(x)][\nabla_x\theta_i(x)]'$$

(10.70)

where $q_2(P_D, \lambda v_g, g)$ is the *information reduction* factor that accounts for the loss of information resulting from the presence of false measurements and less-than-unity probability of detection,[3] and the expected number of FA per unit volume is denoted by λ.

In deriving Equation 10.70, only the bearing measurements that fall within the validation region

$$V_g^i(x) \triangleq \left\{\beta_{ij} : \frac{|\beta_{ij} - \theta_i(x)|}{\sigma_\theta} \leq g\right\}$$

(10.71)

at t_i were considered. The validation region volume (g-sigma region), v_g, is given by

$$v_g = 2\sigma_\theta g$$

(10.72)

The information reduction factor $q_2(P_D, \lambda v_g, g)$ for the present two-dimensional measurement situation (bearing and amplitude) is given by

$$q_2(P_D, \lambda v_g, g) = \frac{1}{1+d} \sqrt{\frac{2}{\pi}} \sum_{m=1}^{\infty} \frac{\mu_f(m-1)}{(gP_{FA})^{m-1}} I_2(m, P_D, g) \qquad (10.73)$$

where $I_2(m, P_D, g)$ is a $2m$-fold integral given in Ref. 14 where numerical values of $q_2(P_D, \lambda v_g, g)$ for different combinations of P_D and λv_g are also presented. The derivation of the integral is based on Bar-Shalom and Li.[3] In this implementation, $g = 5$ was selected. Knowing P_D and λv_g, P_{FA} can be determined by using

$$\lambda v_g = P_{FA} \frac{v_g}{V_c} \qquad (10.74)$$

where V_c is the resolution cell volume of the signal processor (discussed in more detail in Section 10.3.5). Finally, d, the SNR, can be calculated from P_D and λv_g.

The rationale for the term *information reduction factor* follows from the fact that the FIM for zero FA probability and unity target detection probability, J_0, is given by Ref. 10.

$$J_0 = \sum_{i=1}^{n} \frac{1}{\sigma_\theta^2} [\nabla_x \theta_i(x)][\nabla_x \theta_i(x)]' \qquad (10.75)$$

Equations 10.70 and 10.75 clearly show that $q_2(P_D, \lambda v_g, g)$, which is always less than or equal to unity, represents the loss of information due to clutter.

For *narrowband sonar* (bearing and frequency measurements), the FIM is given by

$$J = q_2(P_D, \lambda v_g, g) \sum_{i=1}^{n} \left\{ \frac{1}{\sigma_\theta^2} [\nabla_x \theta_i(x)][\nabla_x \theta_i(x)]' + \frac{1}{\sigma_\theta^2} [\nabla_x \gamma_i(x)][\nabla_x \gamma_i(x)]' \right\} \qquad (10.76)$$

where $q_2(P_D, \lambda v_g, g)$ for this three-dimensional measurement (bearing, frequency, and amplitude) case is evaluated[14] using

$$q_2(P_D, \lambda v_g, g) = \frac{1}{1+d} \sum \frac{2^{m-1} \mu_F(m-1)}{(g^2 P_{FA})^{m-1}} I_2(m, P_D, g) \qquad (10.77)$$

The expression for $I_2(m, P_D, g)$ and the numerical values for $q_2(P_D, \lambda v_g, g)$ are also given by Kirubarajan and Bar-Shalom.[14]

For narrowband sonar, the validation region is defined by

$$V_g^i(x) \triangleq \left\{ (\beta_{ij}, f_{ij}): \left[\frac{\beta_{ij} - \theta_i(x)}{\sigma_\theta} \right]^2 + \left[\frac{f_{ij} - \theta_i(x)}{\sigma\gamma} \right]^2 \right\} \leq g^2 \qquad (10.78)$$

and the volume of the validation region, v_g, is

$$v_g = \sigma^\theta \sigma^\gamma g^2 \qquad (10.79)$$

10.3.5   Results

Both the bearings-only and the narrowband sonar problems with AI were implemented to track a target moving at constant velocity. The results for the narrowband case are given in the following text, accompanied by a discussion of the advantages of using AI by comparing the performances of the estimators with and without AI.

In narrowband signal processing, different bands in the frequency domain are defined by an appropriate cell resolution and a center frequency about which these bands are located. The received signal is sampled and filtered in these bands before applying FFT and beamforming. Then the angle of arrival is estimated using a suitable algorithm.[15] As explained earlier, the received signal is registered as a valid measurement only if it exceeds the threshold τ. The threshold value, together with the SNR, determines the probability of detection and the probability of FA.

The signal processor was assumed to consist of the frequency band (500 Hz, 1000 Hz) with a 2048-point FFT. This results in a frequency cell whose size is given by

$$C_\gamma = \frac{500}{2048} \approx 0.25 \text{ Hz} \tag{10.80}$$

Regarding azimuth measurements, the sonar is assumed to have 60 equal beams, resulting in an azimuth cell C_θ with size

$$C_\theta = \frac{180°}{60} = 3.0° \tag{10.81}$$

Assuming uniform distribution in a cell, the frequency and azimuth measurement standard deviations are given by*

$$\sigma_\gamma = \frac{0.25}{\sqrt{12}} = 0.0722 \text{ Hz} \tag{10.82}$$

$$\sigma_\theta = \frac{3.0}{\sqrt{12}} = 0.866° \tag{10.83}$$

The SNR_c in a cell† was taken as 6.1 dB and $P_D = 0.5$. The estimator is not very sensitive to an incorrect P_D. This is verified by running the estimator with an incorrect P_D on the data generated with a different P_D. Differences up to 0.15 are tolerated by the estimator. The corresponding SNR in a 1 Hz bandwidth SNR_1 is 0.1 dB. These values give

$$\tau = 2.64 \tag{10.84}$$

$$P_{FA} = 0.306 \tag{10.85}$$

From P_{FA}, the expected number of FA per unit volume, denoted by λ, can be calculated using

$$P_{FA} = \lambda C_\theta C_\gamma \tag{10.86}$$

---

* The "uniform" factor corresponds to the worst case. In practice, σ_θ and σ_γ are functions of the 3 dB bandwidth and the SNR.
† The commonly used SNR, designated here as SNR_1, is signal strength divided by the noise power in a 1 Hz bandwidth. SNR_c is signal strength divided by the noise power in a resolution cell. The relationship between them, for $C_\gamma = 0.25$ Hz is $SNR_c = SNR_1 - 6$ dB. SNR_c is believed to be the more meaningful SNR because it determines the ROC curve.

Substituting the values for C_θ and λ gives

$$\lambda = \frac{0.0306}{3.0 \times 0.25} = 0.0407/\text{deg} \cdot \text{Hz} \tag{10.87}$$

The surveillance regions for azimuth and frequency, denoted by U_θ and U_y, respectively, are taken as

$$U_\theta = [-20°, 20°] \tag{10.88}$$

$$U_y = [747\,\text{Hz}, 753\,\text{Hz}] \tag{10.89}$$

The expected number of FA in the entire surveillance region and that in the validation gate V_g can be calculated. These values are 9.8 and 0.2, respectively, where the validation gate is restricted to $g = 5$. These values mean that, for every true measurement that originated from the target, there are about 10 FA that exceed the threshold.

The estimated tracks were validated using the hypothesis testing procedure described in Ref. 14. The track acceptance test was carried out with a miss probability of 5%.

To check the performance of the estimator, simulations were carried out with clutter only (i.e., without a target) and also with a target present; measurements were generated accordingly. Simulations were done in batches of 100 runs.

When there was no target, irrespective of the initial guess, the estimated track was always rejected. This corroborates the accuracy of the validation algorithm given by Kirubarajan and Bar-Shalom.[14]

For the set of simulations with a target, the following scenario was selected: the target moves at a speed of 10 m/s heading west and 5 m/s heading north starting from (5,000 m, 35,000 m). The signal frequency is 750 Hz. The target parameter is $x = [5,000\,\text{m}, 35,000\,\text{m}, -10\,\text{m/s}, 5\,\text{m/s}, 750\,\text{Hz}]$. The motion of the platform consisted of two velocity legs in the northwest direction during the first half, and in the northeast direction during the second half of the simulation period with a constant speed of 7.1 m/s. Measurements were taken at regular intervals of 30 s. The observation period was 900 s. Figure 10.1 shows the scenario including the target true trajectory (solid line), platform trajectory (dashed line), and the

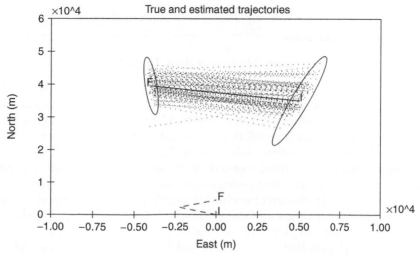

FIGURE 10.1
Estimated tracks from 100 runs for narrowband sonar with amplitude information.

TABLE 10.1

Results of 100 Monte Carlo Runs for Narrowband Sonar with AI ($SNR_c = 6.1$ dB)

Unit	x_{true}	x_{init}	$\bar{x}$	σ_{CRLB}	$\hat{\sigma}$
M	5,000	−12,000 to 12,000	4,991	667	821
M	35,000	49,000 to 50,000	35,423	5,576	5,588
m/s	−10	−16 to 5	−9.96	0.85	0.96
m/s	5	−4 to 9	4.87	4.89	4.99
Hz	750	747 to 751	749.52	2.371	2.531

95% probability regions of the position estimates at the initial and final sampling instants based on the CRLB (Equation 10.76). The initial and the final positions of the trajectories are marked by I and F, respectively. The purpose of the probability region is to verify the validity of the CRLB as the actual parameter estimate covariance matrix from a number of Monte Carlo runs.[4]

Figure 10.1 shows the 100 tracks formed from the estimates. Note that in all but 6 runs (i.e., 94 runs) the estimated trajectory endpoints fall in the corresponding 95% uncertainty ellipses.

Table 10.1 gives the numerical results from 100 runs. Here $\bar{x}$ is the average of the estimates, $\hat{\sigma}$ the variance of the estimates evaluated from 100 runs, and σ_{CRLB} the theoretical CRLB derived in Section 10.3.4. The range of initial guesses found by rough grid search to start off the estimator are given by x_{init}. The efficiency of the estimator was verified using the normalized estimation error squared (NEES)[10] defined by

$$\epsilon_x \triangleq (x - \hat{x})'J(x - \hat{x}) \tag{10.90}$$

where $\bar{x}$ is the estimate and J is the FIM (Equation 10.76). Assuming approximately Gaussian estimation error, the NEES is χ^2 distributed with n degrees of freedom where n is the number of estimated parameters. For the 94 accepted tracks, the NEES obtained was 5.46, which lies within the 95% confidence region [4.39; 5.65]. Also note that each component of $\bar{x}$ is within $2\hat{\sigma}/\sqrt{100}$ of the corresponding component of x_{true}.

10.4   IMMPDAF for Tracking Maneuvering Targets

Target tracking is a problem that has been well studied and documented. Some specific problems of interest in the single-target, single-sensor case are tracking maneuvering targets,[10] tracking in the presence of clutter,[3] and ECM. In addition to these tracking issues, a complete tracking system for a sophisticated electronically steered antenna radar has to consider radar scheduling, waveform selection, and detection threshold selection.

Although many researchers have worked on these issues and many algorithms are available, there had been no standard problem comparing the performances of the various algorithms. Rectifying this, the first benchmark problem[16] was developed, focusing only on tracking a maneuvering target and pointing/scheduling phased array radar. Of all the algorithms considered for this problem, the IMM estimator yielded the best performance.[17] The second benchmark problem[9] included FA and ECM—specifically, a standoff jammer (SOJ) and range gate pull off (RGPO)—as well as several possible radar waveforms (from which the resource allocator has to select one at every revisit time). Preliminary results for this problem showed that the IMM and MHT algorithms were the best solutions.[6,9] For the problem considered, the MHT algorithm yielded similar results as the IMMPDAF modules;[3] although the MHT algorithm was one to two orders of magnitude costlier computationally (as many as 40 hypotheses were needed*). The benchmark problem of Ref. 18 was upgraded in Ref. 8 for the radar resource allocator/manager to select the operating constant FA rate (CFAR), and included the effects of the SOJ on the direction of arrival (DOA) measurements; also the SOJ power was increased to present a more challenging benchmark problem. Although in Ref. 18 the primary performance criterion for the tracking algorithm was minimization of radar energy, the primary performance was changed in Ref. 8 to minimization of a weighted combination of radar time and energy.

This section presents the IMMPDAF technique for automatic track formation, maintenance, and termination. The coordinate selection for tracking and radar scheduling/pointing, and the models used for mode-matched filtering (the modules inside the IMM estimator) are also discussed. These cover the target tracking aspects of the solution to the benchmark problem. These are based on the benchmark problem tracking and sensor resource management.[6,8]

10.4.1   Coordinate Selection

For target tracking in track-dwell mode of the radar, the number of detections at scan k (time t_k) is denoted by m_k. The mth detection report $\bar{\zeta}_m(t_k)$ ($m = 1, 2, ..., m_k$) consists of a time stamp t_k, range r_m, bearing b_m, elevation e_m, AI ρ_m given by the SNR, and the standard deviations of bearing and elevation measurements, σ_m^b and σ_m^e, respectively. Thus,

$$\bar{\zeta}_m(t_k) = [t_k, r_m, b_m, e_m, \rho_m, \sigma_m^b, \sigma_m^e]'$$ (10.91)

where the overbar indicates that this is in the radar's spherical coordinate system.

The AI is used only to declare detections and select the radar waveform for the next scan. Since the use of AI, for example, as in Ref. 17, can be counterproductive in discounting RGPO measurements, which generally have higher SNR than target-originated measurements, AI is not utilized in the estimation process itself. Using the AI would require a separate model for the RGPO intensity, which cannot be estimated in real time due to its short duration and variability.[17]

For target tracking, the measurements are converted from spherical coordinates to Cartesian coordinates, and then the IMMPDAF is used on these converted measurements. This conversion avoids the use of extended Kalman filters and makes the problem linear.[4] The converted measurement report $\zeta_m(t_k)$ corresponding to $\bar{\zeta}_m(t_k)$ is given by[6]

$$\zeta_m(t_k) = [t_k, x_m, y_m, z_m, \rho_m, R_m]$$ (10.92)

where x_m, y_m, z_m, and R_m are the three position measurements in the Cartesian frame and their covariance matrix, respectively. The converted values are

$$x_m = r_m \cos(b_m)\cos(e_m)$$ (10.93)

$$y_m = r_m \sin(b_m)\cos(e_m)$$ (10.94)

---

* The more recent IMM-MHT (as opposed to Kalman filter-based MHT) requires six to eight hypotheses.

$$z_m = r_m \sin(e_m) \tag{10.95}$$

$$R_m = T_m \cdot \mathrm{diag}[(\sigma_k^r)^2, (\sigma_m^b)^2, (\sigma_m^e)^2] \cdot T_m' \tag{10.96}$$

where σ_k^r is the standard deviation of range measurements at scan k and T_m the spherical-to-Cartesian transformation matrix given by

$$T_m = \begin{bmatrix} \cos(b_m)\cos(e_m) & -r_m \sin(b_m)\cos(e_m) & -r_m \cos(b_m)\cos(e_m) \\ \sin(b_m)\cos(e_m) & r_m \cos(b_m)\cos(e_m) & -r_m \sin(b_m)\cos(e_m) \\ \sin(e_m) & 0 & r_m \cos(e_m) \end{bmatrix} \tag{10.97}$$

For the scenarios considered here, this transformation is practically unbiased and there is no need for the debiasing procedure of Ref. 4.

10.4.2 Track Formation

In the presence of FA, track formation is crucial. Incorrect track initiation will result in target loss. In Ref. 3, an automatic track formation/deletion algorithm in the presence of clutter is presented based on the IMM algorithm. In the present benchmark problem, a noisy measurement corresponding to the target of interest is given in the first scan.* Forming new tracks for each validated measurement (based on a velocity gate) at subsequent scans, as suggested in Ref. 3 and as implemented in Ref. 6, is expensive in terms of both radar energy and computational load. In this implementation, track formation is simplified and handled as follows.

Scan 1 ($t = 0$ s): As defined by the benchmark problem, there is only one (target-originated, noisy) measurement. The position component of this measurement is used as the starting point for the estimated track.

Scan 2 ($t = 0.1$ s): The beam is pointed at the location of the first measurement. This yields, possibly, more than one measurement and these measurements are gated using the maximum possible velocity of the targets to avoid the formation of impossible tracks. This validation region volume, which is centered on the initial measurement, is given by

$$V_{xyz} = \left[2\left(\dot{x}_{\max}\delta_2 + 3\sqrt{R_{m2}^x}\right)\right]\left[2\left(\dot{y}_{\max}\delta_2 + 3\sqrt{R_{m2}^y}\right)\right]\left[2\left(\dot{z}_{\max}\delta_2 + 3\sqrt{R_{m2}^z}\right)\right] \tag{10.98}$$

where $\delta_2 = 0.1$ s is the sampling interval and $\dot{x}_{\max}\,\delta_2$, $\dot{y}_{\max}\,\delta_2$, and $\dot{z}_{\max}\,\delta_2$ are the maximum speeds in the X, Y, and Z directions, respectively; R_{m2}^x, R_{m2}^y, and R_{m2}^z are the variances of position measurements in these directions obtained from the diagonal components of Equation 10.96. The maximum speed in each direction is assumed to be 500 m/s.

The measurement in the first scan and the measurement with the highest SNR in the second scan are used to form a track with the two-point initialization technique.[10] The track splitting used in Refs 3 and 6 was found unnecessary—the strongest validated measurement was adequate. This technique yields the position and velocity estimates and the associated covariance matrices in all three coordinates.

Scan 3 ($t = 0.2$ s): The pointing direction for the radar is given by the predicted position at $t = 0.2$ s using the estimates at scan 2. An IMMPDA filter with three models discussed in

---

* Assuming that this is a search pulse without (monopulse) split-beam processing, the angular errors are uniformly distributed in the beam.

the sequel is initialized with the estimates and covariance matrices obtained at the second scan. The acceleration component for the third-order model is assumed zero with variance $(a_{max})^2$, where $a_{max} = 70$ m/s^2 is the maximum expected acceleration of the target.

From scan 3 onward, the track is maintained using the IMMPDAF as described in Section 10.4.3. To maintain a high SNR for the target-originated measurement during track formation, a high-energy waveform is used. Also, scan 3 dwells are used to ensure target detection. This simplified approach cannot be used if the target-originated measurement is not given at the first scan. In that case, the track formation technique in Ref. 3 can be used.

Immediate revisit with sampling interval 0.1 s is carried out during track formation because the initial velocity of the target is not known—in the first scan only the position is measured and there is no *a priori* velocity. This means that in the second scan the radar must be pointed at the first scan position, assuming zero velocity. Waiting longer to obtain the second measurement could result in the loss of the target-originated measurement due to incorrect pointing. Also, to make the IMM mode probabilities converge to the correct values as quickly as possible, the target is revisited at a high rate.

10.4.3   Track Maintenance

The true state of the target at t_k is

$$x(t_k) = [x(t_k)\dot{x}(t_k)\ddot{x}(t_k)y(t_k)\dot{y}(t_k)\ddot{y}(t_k)z(t_k)\dot{z}(t_k)\ddot{z}(t_k)]'$$

where $x(t_k)$, $y(t_k)$, and $z(t_k)$ are the positions, $\dot{x}(t_k)$, $\dot{y}(t_k)$, and $\dot{z}(t_k)$ are the velocities, and $\ddot{x}(t_k)$, $\ddot{y}(t_k)$, and $\ddot{z}(t_k)$ are the accelerations of the target in the corresponding coordinates, respectively. The measurement vector consists of the Cartesian position components at t_k and is denoted by $z(t_k)$.

Assuming that the target motion is linear in the Cartesian coordinate system, the true state of the target can be written as

$$x(t_k) = F(\delta_k)x(t_{k-1}) + \Gamma(\delta_k)v(t_{k-1}) \tag{10.99}$$

and the target-originated measurement is related to the state according to

$$z(t_k) = Hx(t_k) + w(t_k) \tag{10.100}$$

where $\delta_k = t - t_{k-1}$. The white Gaussian noise sequences $v(t_k)$ and $w(t_k)$ are independent and their covariances are $Q(\delta_k)$ and $R(t_k)$, respectively.

With the assumptions from Equations 10.99 and 10.100, the predicted state $\hat{x}(t_k^-)$ at time t_k is

$$\hat{x}(t_k^-) = F(\delta_k) + \hat{x}(t_{k-1}) \tag{10.101}$$

and the predicted measurement is

$$\hat{z}(t_k^-) = H\hat{x}(t_k^-) \tag{10.102}$$

with associated innovation covariance

$$S(t_k) = HP(t_k^-)H' + R(t_k) \tag{10.103}$$

where $P(t_k^-)$ is the predicted state covariance to be defined in Equation 10.117 and $R(t_k)$ is the (expected) measurement noise covariance.

10.4.3.1 Probabilistic Data Association

During track maintenance, each measurement at scan t_k is validated against the established track. This is achieved by setting up a validation region centered around the predicted measurement at t_k^-. The validation region is

$$[z(t_k) - \hat{z}(t_k^-)]'S(t_k)^{-1}[z(t_k) - \hat{z}(t_k^-)] \leq \gamma \tag{10.104}$$

where $S(t_k)$ is the expected covariance of the innovation corresponding to the correct measurement and $\gamma = 16$ (0.9989 probability mass[3]) is the gate size. The appropriate covariance matrix for Equation 10.104 is discussed later in this section.

The set of measurements validated for the track at t_k is

$$Z(k) = \{z_m(t_k), \quad m = 1, 2, ..., m_k\} \tag{10.105}$$

where m_k is the number of measurements validated and associated with the track. Also, the cumulative set of validated measurements up to and including scan k is denoted by z_1^k. All unvalidated measurements are discarded.

With these m_k validated measurements at t_k, one has the following mutually exclusive and exhaustive events:

$$\varepsilon_m(t_k) \triangleq \begin{cases} \{\text{measurement } z_m(t_k) \text{ is from the target}\} & m = 1, ..., m_k \\ \{\text{all measurements are false}\} & m = 0 \end{cases} \tag{10.106}$$

Using the nonparametric version of the PDAF,[4] the validated measurements are associated probabilistically to the track. The combined target state estimate is obtained as

$$\hat{x}(t_k) = \sum_{m=0}^{m_k} \beta_m(t_k)\hat{x}_m(t_k) \tag{10.107}$$

where $\beta_m(t_k)$ is the probability that the mth validated measurement is correct and $\hat{x}_m(t_k)$ is the updated state conditioned on that event. The conditionally updated states are given by

$$\hat{x}_m(t_k) = \hat{x}(t_k^-) + W_m(t_k)v_m(t_k) \quad m = 1, 2, ..., m_k \tag{10.108}$$

where $W_m(t_k)$ is the filter gain and $v_m(t_k) = z_m(t_k) - \hat{z}_m(t_k^-)$ the innovation associated with the mth validated measurement. The gain, which depends on the measurement noise covariance, is

$$W_m(t_k) = P(t_k^-)H'[HP(t_k^-)H' + R_m(t_k)]^{-1} \triangleq P(t_k^-)H'S_m(t_k)^{-1} \tag{10.109}$$

where $R_m(t_k)$ depends on the observed SNR for measurement m.[8]

The association event probabilities $\beta_m(t_k)$ are given by

$$\beta_m(t_k) = P\{\varepsilon_m(t_k)|z_1^k\} \tag{10.110}$$

$$= \begin{cases} \dfrac{e(m)}{b + \sum_{j=1}^{m_k} e(j)} & m = 1, 2, ..., m_k \\[4mm] \dfrac{b}{b + \sum_{j=1}^{m_k} e(j)} & m = 0 \end{cases} \tag{10.111}$$

where

$$e(m) = N[v_m(t_k); 0, S_m(t_k)] \tag{10.112}$$

$$b = m_k \frac{1 - P_D}{P_D V(t_k)} \tag{10.113}$$

and P_D is the probability of detection of a target-originated measurement. The probability that a target-originated measurement, if detected, falls within the validation gate is assumed to be unity. Also, $N[v; 0, S]$ denotes the normal pdf with argument v, mean zero, and covariance matrix S. The common validation volume $V(t_k)$ is the union of the validation volumes $V_m(t_k)$ used to validate the individual measurements associated with the target $V(t_k)$ and is given by

$$V_m(t_k) = \gamma^{n_z/2} V_{n_z} |S_m(t_k)|^{1/2} \tag{10.114}$$

where V_{n_z} is the volume of the unit hypersphere of dimension n_z, based on the dimension of the measurement $\mathbf{z}$. For the three-dimensional position measurements $V_{n_z} = 4\pi/3$.[3]
The state estimate is updated as

$$\hat{x}(t_k) = \hat{x}(t_k^-) + \sum_{m=1}^{m_k} \beta_m(t_k) W_m(t_k) v_m(t_k) \tag{10.115}$$

and the associated covariance matrix is updated as

$$P(t_k) = P(t_k^-) - \sum_{m=1}^{m_k} \beta_m(t_k) W_m(t_k) S_m(t_k) W_m(t_k)' + \tilde{P}(t_k) \tag{10.116}$$

where

$$P(t_k^-) = F(\delta_k) P(t_{k-1}) F(\delta_k)' + \Gamma(\delta_k) Q(\delta_k) \Gamma(\delta_k)' \tag{10.117}$$

is the predicted state covariance and the term

$$\tilde{P}(t_k) = \sum_{m=1}^{m_k} \beta_m(t_k)[W_m(t_k)v_m(t_k)][W_m(t_k)v_m(t_k)]'$$

$$- \left[\sum_{m=1}^{m_k} \beta_m(t_k) W_m(t_k) v_m(t_k) \right] \left[\sum_{m=1}^{m_k} \beta_m(t_k) W_m(t_k) v_m(t_k) \right]' \tag{10.118}$$

is analogous to the *spread of the innovations* in the standard PDA.[3] Monopulse processing results in different accuracies (standard deviations) for different measurements within the same dwell. This accounts for the difference in the preceding equations from the standard PDA, where the measurement accuracies are assumed to be the same for all of the validated measurements.

To initialize the filter at $k = 3$, the following estimates are used:[10]

$$
\mathbf{x}(t_2) =
\begin{bmatrix}
\hat{x}(t_2) \\
\dot{\hat{x}}(t_2) \\
\ddot{\hat{x}}(t_2) \\
\hat{y}(t_2) \\
\dot{\hat{y}}(t_2) \\
\ddot{\hat{y}}(t_2) \\
\hat{z}(t_2) \\
\dot{\hat{z}}(t_2) \\
\ddot{\hat{z}}(t_2)
\end{bmatrix}
=
\begin{bmatrix}
\mathbf{z}_h^x(t_2) \\
\dfrac{\mathbf{z}_h^x(t_2) - \mathbf{z}^x(t_1)}{\delta_2} \\
0 \\
\mathbf{z}_h^y(t_2) \\
\dfrac{\mathbf{z}_h^y(t_2) - \mathbf{z}^y(t_1)}{\delta_2} \\
0 \\
\mathbf{z}_h^z(t_2) \\
\dfrac{\mathbf{z}_h^z(t_2) - \mathbf{z}^z(t_1)}{\delta_2} \\
0
\end{bmatrix}
\tag{10.119}
$$

where h is the index corresponding to the validated measurement with the highest SNR in the second scan, and the superscripts x, y, and z denote the components in the corresponding directions, respectively. The associated covariance matrix can be derived[10] using the measurement covariance R_h and the maximum target acceleration a_{max}. If the two-point differencing results in a velocity component that exceeds the corresponding maximum speed, it is replaced by that speed. Similarly, the covariance terms corresponding to the velocity components are upper bounded by the corresponding maximum values.

10.4.3.2 IMM Estimator Combined with PDA Technique

In the IMM estimator it is assumed that at any time the target trajectory evolves according to one of a finite number of models, which differ in their noise levels and structures.[10] By probabilistically combining the estimates of the filters, typically Kalman, matched to these modes, an overall estimate is found. In the IMMPDAF the Kalman filter is replaced with the PDAF (given in Section 10.4.3.1 for mode-conditioned filtering of the states), which handles the data association.

Let r be the number of mode-matched filters used, $M(t_k)$ the index of the mode in effect in the semiopen interval (t_{k-1}, t_k) and $\mu_j(t_k)$ be the probability that mode j ($j = 1, 2, \ldots, r$) is in effect in this interval. Thus,

$$
\mu_j(t_k) = P\{M(t_k) = j | Z_1^k\}
\tag{10.120}
$$

The mode transition probability is defined as

$$
p_{ij} = P\{M(t_k) = j | M(t_{k-1}) = i\}
\tag{10.121}
$$

The state estimates and their covariance matrix at t_k conditioned on the jth mode are denoted by $\hat{x}_j(t_k)$ and $P_j(t_k)$, respectively.

The steps of the IMMPDAF are as follows.[3]

Step 1: Mode interaction or mixing. The mode-conditioned state estimate and the associated covariances from the previous iteration are mixed to obtain the initial condition for the mode-matched filters. The initial condition in cycle k for the PDAF matched to the jth mode is computed using

$$\hat{x}_{0j}(t_{k-1}) = \sum_{i=1}^{r} \hat{x}_i(t_{k-1})\mu_{i|j}(t_{k-1}) \tag{10.122}$$

where

$$\mu_{i|j}(t_{k-1}) = P\{M(t_{k-1}) = i | M(t_{k-1}) = j, Z_1^{k-1}\} = \frac{p_{ij}\mu_i(t_{k-1})}{\sum_{l=1}^{r} p_{lj}\mu_l(t_{k-1})} \qquad i, j = 1, 2, \ldots, r \tag{10.123}$$

are the mixing probabilities. The covariance matrix associated with Equation 10.122 is given by

$$P_{0j}(t_{k-1}) = \sum_{i=1}^{r} \mu_{i|j}(t_{k-1})\{P_i(t_{k-1}) + [\hat{x}_i(t_k) - \hat{x}_{0j}(t_{k-1})][\hat{x}_i(t_k) - \hat{x}_{0j}(t_{k-1})]'\} \tag{10.124}$$

Step 2: Mode-conditioned filtering. A PDAF is used for each mode to calculate the mode-conditioned state estimates and covariances. In addition, we evaluate the likelihood function $\Lambda_j(t_k)$ of each mode at t_k using the Gaussian-uniform mixture:

$$\Lambda_j(t_k) \triangleq p[Z(k)|M(t_k) = j, Z_1^{k-1}]$$

$$= V(t_k)^{-m_l}(1 - P_D) + V(t_k)^{1-m_k}\frac{P_D}{m_k}\sum_{m=1}^{m_k} e_j(m) \tag{10.125}$$

$$= \frac{P_D V(t_k)^{1-m_k}}{m_k}\left(b + \sum_{m=1}^{m_k} e_j(m)\right) \tag{10.126}$$

where $e_j(m)$ is defined in Equation 10.112 and b in Equation 10.113. Note that the likelihood function, as a pdf, has a physical dimension that depends on m_k. Since ratios of these likelihood functions are to be calculated, they all must have the same dimension, that is, the same m_k. Thus, a common validation region (Equation 10.104) is vital for all the models in the IMMPDAF. Typically, the "largest" innovation covariance matrix corresponding to "noisiest" model covers the others and, therefore, this can be used in Equations 10.104 and 10.114.

Step 3: Mode update. The mode probabilities are updated based on the likelihood of each mode using

$$\mu_j(t_k) = \frac{\Lambda_j(t_k)\sum_{l=1}^{r} p_{lj}\mu_l(t_{k-1})}{\sum_{l=1}^{r}\sum_{l=1}^{r}\Lambda_i(t_k)p_{lj}\mu_l(t_{k-1})} \tag{10.127}$$

Step 4: State combination. The mode-conditioned estimates and covariances are combined to find the overall estimate $\hat{x}(t_k)$ and its covariance matrix $P(t_k)$, as follows:

$$\hat{x}(t_k) = \sum_{j=1}^{r} \mu_j(t_k)\hat{x}(t_k) \tag{10.128}$$

$$P(t_k) = \sum_{j=1}^{r} \mu_j(t_k)\{P_j(t_k) + [\hat{x}_j(t_k) - \hat{x}(t_k)][\hat{x}_j(t_k) - \hat{x}(t_k)]'\} \tag{10.129}$$

10.4.3.3 Models in the IMM Estimator

The selection of the model structures and their parameters is one of the critical aspects of the implementation of IMMPDAF. Designing a good set of filters requires *a priori* knowledge about the target motion, usually in the form of maximum accelerations and sojourn times in various motion modes.[10] The tracks considered in the benchmark problem span a wide variety of motion modes—from benign constant velocity motions to maneuvers up to 7g. To handle all possible motion modes and to handle automatic track formation and termination, the following models are used.

Benign motion model (M^1). This second-order model with low noise level (to be given later) has a probability of target detection P_D given by the target's expected SNR and corresponds to the nonmaneuvering intervals of the target trajectory. For this model the process noise is, typically, assumed to model air turbulence.

Maneuver model (M^2). This second-order model with high noise level corresponds to ongoing maneuvers. For this white noise acceleration model, the process noise standard deviation σ_{v2} is obtained using

$$\sigma_{v2} = \alpha a_{\max} \tag{10.130}$$

where $a_{\max}$ is the maximum acceleration in the corresponding modes and $0.5 < \alpha \le 1$.[10]

Maneuver detection model (M^3). This is a third-order (Wiener process acceleration) model with high noise level. For highly maneuvering targets, like military attack aircraft, this model is useful for detecting the onset and termination of maneuvers. For civilian air traffic surveillance,[19] this model is not necessary.

For a Wiener process acceleration model, the standard deviation σ_{v3} is chosen using

$$\sigma_{v3} = \min\{\beta\Delta_a\delta, a_{\max}\} \tag{10.131}$$

where Δ_a is the maximum acceleration increment per unit time (jerk), δ the sampling interval, and $0.5 < \alpha \le 1$.[3]

For the targets under consideration, $a_{\max} = 70$ m/s^2 and $\Delta_a = 35$ m/s^3. Using these values, the process noise standard deviations were taken as

$$\sigma_{v1} = 3\,\text{m/s}^2 \quad \text{(for nonmaneuvering intervals)}$$

$$\sigma_{v2} = 35\,\text{m/s}^2 \quad \text{(for maneuvering intervals)}$$

$$\sigma_{v3} = \min\{35\delta, 70\} \quad \text{(for maneuver start/termination)}$$

In addition to the process noise levels, the elements of the Markov chain transition matrix between the modes, defined in Equation 10.121, are also design parameters. Their selection depends on the sojourn time in each motion mode. The transition probability depends on the expected sojourn time via

$$\tau_i = \frac{\delta}{1 - p_{ii}} \tag{10.132}$$

where τ_i is the expected sojourn time of the ith mode, p_{ii} the probability of transition from ith mode to the same mode, and δ the sampling interval.[10]

For the above models, p_{ii}, $I = 1, 2, 3$ are calculated using

$$p_{ii} = \min\left\{ u_{i,\max}\left(l_i, 1 - \frac{\delta}{\tau_i} \right) \right\} \tag{10.133}$$

$l_i = 0.1$ and $u_i = 0.9$ are the lower and upper limits, respectively, for the ith model transition probability.

The expected sojourn times of 15, 4, and 2 s, are assumed for modes M^1, M^2, and M^3, respectively. The selection of the off-diagonal elements of the Markov transition matrix depends on the switching characteristics among the various modes and is done as follows:

$$p_{12} = 0.1(1 - p_{11}) \qquad p_{13} = 0.9(1 - p_{11})$$

$$p_{21} = 0.1(1 - p_{22}) \qquad p_{23} = 0.9(1 - p_{22})$$

$$p_{31} = 0.3(1 - p_{33}) \qquad p_{32} = 0.7(1 - p_{33})$$

The x, y, and z components of target dynamics are uncoupled, and the same process noise is used in each coordinate.

10.4.4 Track Termination

According to the benchmark problem, a track is declared lost if the estimation error is greater than the two-way beam width in angles or 1.5 range gates in range. In addition to this problem-specific criterion, the IMMPDAF declares (on its own) track loss if the track is not updated for 100 s. Alternatively, one can include a "no target" model,[3] which is useful for automatic track termination, in the IMM mode set. In a more general tracking problem, where the true target state is not known, the "no target" mode probability or the track update interval would serve as the criterion for track termination, and the IMMPDAF would provide a unified framework for track formation, maintenance, and termination.

10.4.5 Simulation Results

This section presents the simulation results obtained using the algorithms described earlier. The computational requirements and root-mean-square errors (RMSE) are given.

The tracking algorithm using the IMMPDAF is tested on the following six benchmark tracks (the tracking algorithm does not know the type of the target under track—the parameters are selected to handle any target):

Target 1: A large military cargo aircraft with maneuvers up to 3g

Target 2: A Learjet or commercial aircraft that is smaller and more maneuverable than target 1 with maneuvers up to 4g

TABLE 10.2

Performance of IMMPDAF in the Presence of False Alarms, Range Gate Pull Off, and the Standoff Jammer

Target	Time Length (s)	Max. Acceleration (m/s^2)	Max. Density (%)	Sample Period (s)	Average Power (W)	Positive RMSE (m)	Velocity RMSE (m/s)	Average Load (kFLOPS)	Lost Tracks (%)
1	165	31	25	2.65	8.9	98.1	61.3	22.2	1
2	150	39	28.5	2.39	5.0	97.2	68.5	24.3	0
3	145	42	20	2.38	10.9	142.1	101.2	24.6	1
4	184	58	20	2.34	3.0	26.5	25.9	24.3	0
5	182	68	38	2.33	18.4	148.1	110.7	27.1	2
6	188	70	35	2.52	12.4	98.6	71.4	24.6	1
Average	—	—	—	2.48	8.3	—	—	24.5	—

Target 3: A high-speed medium bomber with maneuvers up to 4g

Target 4: Another medium bomber with good maneuverability up to 6g

Targets 5 and 6: Fighter or attack aircraft with very high maneuverability up to 7g

In Table 10.2, the performance measures and their averages of the IMMPDAF (in the presence of FA, RGPO, and SOJ[6,8]) are given. The averages are obtained by adding the corresponding performance metrics of the six targets (with those of target 1 added twice) and dividing the sum by 7. In the table, the maneuver density is the percentage of the total time that the target acceleration exceeds 0.5g. The average floating point operation (FLOP) count per second was obtained by dividing the total number of FLOPs by the target track length. This is the computational requirement for target and jammer tracking, neutralizing techniques for ECM, and adaptive parameter selection for the estimator, that is, it excludes the computational load for radar emulation.

The average FLOP requirement is 25 kFLOPS, which can be compared with the FLOP rate of 78 MFLOPS of a Pentium® processor running at 133 MHz. (The FLOP count is obtained using the built-in MATLAB function flops. Note that these counts, which are given in terms of thousands of floating point operations per second [kFLOPS] or millions of floating point operations per second [MFLOPS], are rather pessimistic—the actual FLOP requirement would be considerably lower.) Thus, the real-time implementation of the complete tracking system is possible. With the average revisit interval of 2.5 s, the FLOP requirement of the IMMPDAF is 62.5 kFLOP/radar cycle. With the revisit time calculations taking about the same amount of computation as a cycle of the IMMPDAF, but running at half the rate of the Kalman filter (which runs at constant rate), the IMMPDAF with adaptive revisit time is about 10 times costlier computationally than a Kalman filter. Owing to its ability to save radar resources, which are much more expensive than computational resources, the IMMPDAF is a viable alternative to the Kalman filter, which is the standard "workhorse" in many current tracking systems. (Some systems still use the $\alpha - \beta$ filter as their "work mule.")

10.5 Flexible-Window ML-PDA Estimator for Tracking Low Observable Targets

One difficulty with the ML-PDA approach of Section 10.3, which uses a set of scans of measurements as a batch, is the incorporation of noninformative scans when the target is not present in the surveillance region for some consecutive scans. For example, if the target appears within the surveillance region of the sensor after the first few scans, the estimator can be misled by the pure clutter in those scans—the earlier scans contain no relevant information, and the incorporation of these into the estimator not only increases the amount of processing (without adding any more information), but also results in less accurate estimates or even track rejection. Also, a target could disappear from the surveillance region for a while during tracking and reappear sometime later. Again, these intervening scans contain little or no information about the target and can potentially mislead the tracker.

In addition, the standard ML-PDA estimator assumes that the target SNR, the target velocity, and the density of FA over the entire tracking period remain constant. In practice, this may not be the case, and then the standard ML-PDA estimator will not yield the desired results. For example, the average target SNR may vary significantly as the target gets closer to or moves away from the sensor. In addition, the target might change its course and speed intermittently over time. For EO sensors, depending on the time of the day and weather, the number of FA may vary as well.

Because of these concerns, an estimator capable of handling time-varying SNR (with online adaptation), FA density, and slowly evolving course and speed is needed. Although a recursive estimator like the IMMPDA is a candidate, to operate under low SNR conditions in heavy clutter, a batch estimator is still preferred. In this section, the above-mentioned problems are addressed by introducing an estimator that uses the ML-PDA with AI adaptively in a sliding-window fashion,[20] rather than using all the measurements in a single batch as the standard ML-PDA estimator does.[14] The initial time and the length of this sliding window are adjusted adaptively based on the information content in the measurements in the window. Thus, scans with little or no information content are eliminated and the window is moved over to scans with "informative" measurements.

This algorithm is also effective when the target is temporarily lost and reappears later. In contrast, recursive algorithms will diverge in this situation and may require an expensive track reinitiation. The standard batch estimator will be oblivious to the disappearance and may lose the whole track. This section demonstrates the performance of the adaptive sliding-window ML-PDA estimator on a real scenario with heavy clutter for tracking a fast-moving aircraft using an EO sensor.

10.5.1 Scenario

The adaptive ML-PDA algorithm was tested on an actual scenario consisting of 78 frames of long wave infrared (LWIR) IR data collected during the Laptex data collection, which occurred in July 1996 at Crete, Greece. The sequence contains a single target—a fast-moving Mirage F1 fighter jet. The 920×480 pixel frames taken at a rate of 1 Hz were registered to compensate for frame-to-frame line-of-sight (LOS) jitter. Figure 10.2 shows the last frame in the Mirage F1 sequence.

A sample detection list for the Mirage F1 sequence obtained at the end of preprocessing is shown in Figure 10.3. Each "×" in the figure represents a detection above the threshold.

10.5.2 Formulation of ML-PDA Estimator

This section describes the target models used by the estimator in the tracking algorithm and the statistical assumptions made by the algorithm. The ML-PDA estimator for these models is introduced, and the CRLB for the estimator and the hypothesis test used to validate the track are presented.

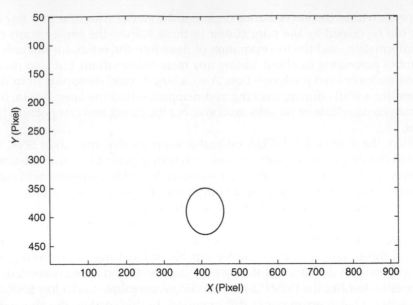

FIGURE 10.2
The last frame in the Mirage F1 sequence.

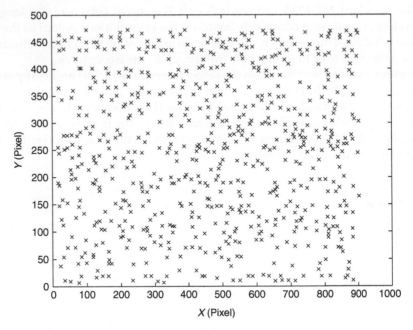

FIGURE 10.3
Detection list corresponding to the frame in Figure 10.2.

10.5.2.1 Target Models

The ML-PDA tracking algorithm is used on the detection lists after the data preprocessing phase. It is assumed that there are n detection lists obtained at $t = t_1, t_2, ..., t_n$. The ith detection list, where $1 \leq i \leq n$, consists of m_i detections at pixel positions (x_{ij}, y_{ij}) along the X and Y directions. In addition to locations, the signal strength or amplitude, a_{ij}, of the jth detection in the *i*th list, where $1 \le j \le m$, is also known. Thus, assuming constant velocity over a number of scans, the problem can be formulated as a two-dimensional scenario in space with the target motion defined by the four-dimensional vector

$$\mathbf{x} \equiv [\xi(t_0), \eta(t_0), \dot{\xi}, \dot{\eta}] \tag{10.134}$$

where $\xi(t_0)$ and $\eta(t_0)$ are the horizontal and vertical pixel positions of the target, respectively, from the origin at the reference time t_0. The corresponding velocities along these directions are assumed constant at $\dot{\xi}(t_0)$ pixel/s and $\dot{\eta}(t_0)$ pixel/s, respectively.

The set of measurements in list *i* at time t_i is denoted by

$$Z(i) = \left\{ z_j(i) \right\}_{j=1}^{m_i} \tag{10.135}$$

where m_i is the number of measurements at t_i. The measurement vector $z_j(i)$ is denoted by

$$z_j(i) \triangleq [x_{ij}, y_{ij}, a_{ij}]' \tag{10.136}$$

where x_{ij} and y_{ij} are observed *X* and *Y* positions, respectively.

The cumulative set of measurements made in scans t_1 through t_n is given by

$$Z^n = \{Z(i)_{i=1}^n\} \tag{10.137}$$

A measurement can either originate from a true target or from a spurious source. In the former case, each measurement is assumed to have been received only once in each scan with a detection probability P_D and to have been corrupted by zero-mean additive Gaussian noise of known variance, that is,

$$x_{ij} = \xi(t_i) + \epsilon_{ij} \tag{10.138}$$

$$y_{ij} = \eta(t_i) + v_{ij} \tag{10.139}$$

where ϵ_{ij} and v_{ij} are the zero-mean Gaussian noise components with variances σ_1^2 and σ_2^2 along the *X* and *Y* directions, respectively.

Thus, the joint pdf of the position components of z_{ij} is given by

$$p(z_{ij}) = \frac{1}{2\pi\sigma_1\sigma_2} \exp\left(-\frac{1}{2}\left[\frac{x_{ij} - \xi(t_i)}{\sigma_1} \right]^2 - \frac{1}{2}\left[\frac{y_{ij} - \eta(t_i)}{\sigma_2} \right]^2 \right) \tag{10.140}$$

The FA are assumed to be distributed uniformly in the surveillance region and their number at any sampling instant obeys the Poisson pmf

$$\mu_F(m_i) = \frac{(\lambda U)^{m_i} e^{-\lambda U}}{m_i!} \tag{10.141}$$

where *U* is the area of surveillance and λ the expected number of FA per unit of this area. Kirubarajan and Bar-Shalom[14] have shown that the performance of the ML-PDA estimator can be improved by using AI of the received signal in the estimation process itself, in addition to thresholding. After the signal has been passed through the matched filter, an envelope detector can be used to obtain the amplitude of the signal. The noise at the matched filter is assumed to be narrowband Gaussian. When this is fed through the envelope detector, the output is Rayleigh distributed. Given the detection threshold, τ, the probability of detection P_D and the probability of FA P_{FA} are

$$P_D \triangleq P \quad \text{(the target-oriented measurement exceeds the threshold } \tau) \qquad (10.142)$$

and

$$P_{FA} \triangleq P \quad \text{(a measurement caused by noise only exceeds the threshold } \tau) \quad (10.143)$$

where $P(\cdot)$ is the probability of an event.

The pdfs at the output of the threshold detector, which corresponds to signals from the target and FA are denoted by $p_1^\tau(a)$ and $p_0^\tau(a)$, respectively. Then the amplitude likelihood ratio, ρ, can then be written as[3]

$$\rho = \frac{p_1^\tau(a)}{p_0^\tau(a)} = \frac{P_{FA}}{P_D(1+d)} \exp\left(\frac{a^2 d}{1(1+d)}\right) \qquad (10.144)$$

where τ is the detection threshold.

10.5.2.2 Maximum Likelihood-Probabilistic Data Association Estimator

This section focuses on the ML estimator combined with the PDA approach. If there are m_i detections at t_i, one has the following mutually exclusive and exhaustive events:[3]

$$\varepsilon_j(i) \triangleq \begin{cases} \{\text{measurement } z_j(i) \text{ is from the target}\} & j = 1, \ldots, m \\ \{\text{all measurements are false}\} & j = 0 \end{cases} \qquad (10.145)$$

The pdf of the measurements corresponding to the above events can be written as[3]

$$p[Z(i)|\varepsilon_j(i), \mathbf{x}] = \begin{cases} U^{1-m_i} p(z_{ij}) \rho_{ij} \prod_{j=1}^{m_i} p_0^\tau(a_{ij}) & j = 1, \ldots, m_i \\[2ex] U^{1-m_i} \prod_{j=1}^{m_i} p_0^\tau(a_{ij}) & j = 0 \end{cases} \qquad (10.146)$$

Using the total probability theorem,

$$p[Z(i)|\mathbf{x}] = \sum_{j=0}^{m_i} p[Z(i)|\varepsilon_j(i), \mathbf{x}] p[\varepsilon_j(i), \mathbf{x}]$$

$$= \sum_{j=0}^{m_i} p[Z(i)|\varepsilon_j(i), \mathbf{x}] p[\varepsilon_j(i)] \qquad (10.147)$$

the above can be written explicitly as

$$p[Z(i)|\mathbf{x}] = U^{-m_i}(1 - P_D)\prod_{j=1}^{m_i} p_0^{\tau}(a_{ij})\mu_F(m_i) + \frac{U^{1-m_i}P_D\mu_F(m_i - 1)}{m_i}\prod_{j=1}^{m_i} p_0^{\tau}(a_{ij})\sum_{j=1}^{m_i} p(z_{ij})\rho_{ij}$$

$$= U^{-m_i}(1 - P_D)\prod_{j=1}^{m_i} p_0^{\tau}(a_{ij})\mu_F(m_i) + \frac{U^{1-m_i}P_D\mu_F(m_i - 1)}{2\pi\sigma_1\sigma_2 m_i}\prod_{j=1}^{m_i} p_0^{\tau}(a_{ij})$$

$$\cdot \sum_{j=1}^{m_i} \rho_{ij}\exp\left(-\frac{1}{2}\left[\frac{x_{ij} - \xi(t_i)}{\sigma_1}\right]^2 - \frac{1}{2}\left[\frac{y_{ij} - \eta(t_i)}{\sigma_2}\right]^2\right) \qquad (10.148)$$

To obtain the likelihood ratio, $\Phi[Z(i), \mathbf{x}]$, at t_i, divide Equation 10.148 by $p[Z(i)|\varepsilon_0(i), \mathbf{x}]$

$$\Phi[Z(i), \mathbf{x}] = \frac{p[Z(i), \mathbf{x}]}{p[Z(i)|\varepsilon_0(i), \mathbf{x}]}$$

$$= (1 - P_D) + \frac{P_D}{2\pi\lambda\sigma_1\sigma_2}\sum \rho_{ij}\exp\left(-\frac{1}{2}\left[\frac{x_{ij} - \xi(t_i)}{\sigma_1}\right]^2 - \frac{1}{2}\left[\frac{y_{ij} - \eta(t_i)}{\sigma_2}\right]^2\right) \qquad (10.149)$$

Assuming that measurements at different sampling instants are conditionally independent, that is,

$$p[Z^n|\mathbf{x}] = \prod p[Z(i)|\mathbf{x}] \qquad (10.150)$$

the total likelihood ratio[3] for the entire data set is given by

$$\Phi[Z^n, \mathbf{x}] = \prod_{i=1}^{n} \Phi_i[Z(i), \mathbf{x}] \qquad (10.151)$$

Then, the total log-likelihood ratio $\Phi[Z^n, \mathbf{x}]$, expressed in terms of the individual log-likelihood ratios $\phi[Z(i), \mathbf{x}]$ at sampling time instants t_i, becomes

$$\phi[Z^n, \mathbf{x}] = \sum_{i=1}^{n} \phi_i[Z(i), \mathbf{x}]$$

$$= \sum \ln\left[(1 - P_D) + \frac{P_D}{2\pi\lambda\sigma_1\sigma_2}\sum_{j=1}^{m_i} \rho_{ij}\exp\left(-\frac{1}{2}\left[\frac{x_{ij} - \xi(t_i)}{\sigma_1}\right]^2 - \frac{1}{2}\left[\frac{y_{ij} - \eta(t_i)}{\sigma_2}\right]^2\right)\right] \qquad (10.152)$$

The MLE is obtained by finding the vector $\mathbf{x} = \hat{\mathbf{x}}$ that maximizes the total log-likelihood ratio given in Equation 10.152. This maximization is performed using a quasi-Newton (variable metric) method. This can also be accomplished by minimizing the negative log-likelihood function. In our implementation of the MLE, the Davidon-Fletcher-Powell variant of the variable metric method is used. This method is a conjugate gradient technique that finds the minimum value of the function iteratively.[21] However, the negative log-likelihood function may have several local minima; that is, it has multiple modes. Owing to this property, if the search is initiated too far away from the global minimum, the line search algorithm may converge to a local minimum. To remedy this, a multipass approach is used as in Ref. 14.

10.5.3 Adaptive ML-PDA

Often, the measurement process begins before the target becomes visible—that is, the target enters the surveillance region of the sensor some time after the sensor started to record measurements. In addition, the target may disappear from the surveillance region for a certain period of time before reappearing. During these periods of blackout, the received measurements are purely noise-only, and the scans of data contain no information about the target under track. Incorporating these scans into a tracker reduces its accuracy and efficiency. Thus, detecting and rejecting these scans is important to ensure the fidelity of the estimator. This subsection presents a method that uses the ML-PDA algorithm in a sliding-window fashion. In this case, the algorithm uses only a subset of the data at a time rather than all of the frames at once, to eliminate the use of scans that have no target. The initial time and the length of the sliding window are adjusted adaptively based on the information content of the data—the smallest window, and thus the fewest number of scans, required to identify the target is determined online and adapted over time.

The key steps in the adaptive ML-PDA estimator are as follows:

1. Start with a window of minimum size.

2. Run the ML-PDA estimator within this window and carry out the validation test on the estimates.

3. If the estimate is accepted (i.e., if the test is passed), and if the window is of minimum size, accept the window. The next window is the present window advanced by one scan. Go to step 2.

4. If the estimate is accepted, and if the window is greater than minimum size, try a shorter window by removing the initial scan. Go to step 2 and accept the window only if estimates are better than those from the previous window.

5. If the test fails and if the window is of minimum size, increase the window length to include one more scan of measurements and, thus, increase the information content in the window. Go to step 2.

6. If the test fails and if the window is greater than minimum size, eliminate the first scan, which could contain pure noise only. Go to step 2.

7. Stop when all scans are used.

The algorithm is described as follows. To specify the exact steps in the estimator, the following variables are defined:

```
W = current window length
Wmin = minimum window length
Z(ti) = scan (set) of measurements at time ti
With these definitions, the algorithm is
BEGIN PROCEDURE Adaptive ML-PDA estimator (Wmin, Z(t1), Z(tn))
    i = 1—Initialize the window at the first scan.
    W = Wmin—Initially, use a window of minimum size.
    WHILE (i + W < n)—Repeat until the last scan at tn.
        Do grid search for initial estimates by numerical search on
            Z(ti), Z(ti+1),..., Z(ti+W)
        Apply ML-PDA estimator on the measurements in Z(ti), Z(ti+1),..., Z(ti+W)
        Validate the estimates
        IF the estimates are rejected
            IF (W > Wmin)—Check if we can reduce the window size.
```

```
                    i = i + 1—Eliminate the initial scan that might be due to
                        noise only.
          ELSEIF (W = W_min)
                    W = W + 1—Expand window size to include an additional scan.
          ENDIF
       ENDIF
       IF the estimates are accepted
          IF (W > W_min)—Check if we can reduce the window size.
                    Try a shorter window by removing the initial scan and
                        check if estimates are better, i = i + 1
          ENDIF
          IF estimates for shorter window are NOT better OR (W = W_min)
                    Accept estimates and try next window, i = i + 1
          ENDIF
       ENDIF
  END WHILE
END PROCEDURE
```

To illustrate the adaptive algorithm, consider a scenario where a sensor records 10 scans of measurements over a surveillance region. The target, however, appears in this region (i.e., its intensity exceeds the threshold) only after the second scan (i.e., from the third scan onward). This case is illustrated in Figure 10.4. The first two scans are useless because they contain only noise.

Consider the smallest window size required for a detection to be 5. Then the algorithm will evolve as shown in Figure 10.5. First, for the sake of illustration, assume that a single "noisy" scan present in the data set is sufficient to cause the MLE to fail the hypothesis test for track acceptance. The algorithm tries to expand the window to include an additional scan if a track detection is not made. This is done because an additional scan of data may bring enough additional information to detect the target track. The algorithm next tries to cut down the window size by removing the initial scans. This is done to check whether a better estimate can be obtained without this scan. If this initial scan is noise-only, then it degrades the accuracy of the estimate. If a better estimate is found (i.e., a more accurate estimate) without this scan, the latter is eliminated. Thus, as in the above-mentioned example, the algorithm expands at the front (most recent scan used) and contracts at the rear end of the window to find the best window that produces the strongest detection, on the basis of the validation test.

10.5.4 Results

10.5.4.1 Estimation Results

The Mirage F1 data set consists of 78 scans or frames of LWIR IR data. The target appears late in this scenario and moves toward the sensor. There are about 600 detections per frame. In this implementation the parameters shown in Table 10.3 were chosen.

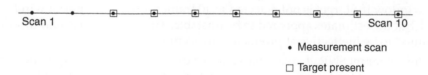

FIGURE 10.4
Scenario with a target being present for only a partial time during observation.

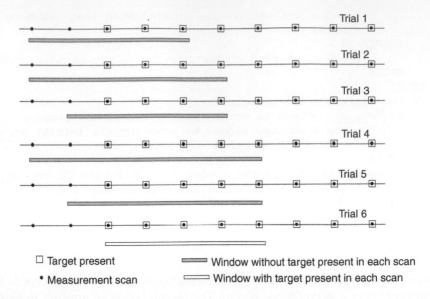

FIGURE 10.5
Adaptive ML-PDA algorithm applied to the scenario illustrated in Section 10.5.3.

TABLE 10.3

Parameters Used in the ML-PDA Algorithm for the Mirage F1 Jet

Parameter	Value
σ_1	1.25
σ_2	1.25
Minimum window size, W	10
Initial target SNR, d_0	9.5
P_{DC}	.70
α	0.85
π_m	5%
$\bar{v}$	5.0
$\bar{\sigma}_v$	0.15
K	4

The choice of these parameters is as follows:

- σ_1 and σ_2 are, as in Equation 10.140, the standard deviations along the horizontal and vertical axes, respectively. The value of 1.25 for both variables models the results of the preprocessing.

- The minimum window size, W_{min}, was chosen to be 10. The algorithm will expand this window if a target is not detected in 10 frames. Initially a shorter window was used, but the estimates appeared to be unstable. Therefore, fewer than 10 scans are assumed to be ineffective at producing an accurate estimate.

- The initial target SNR, d_0, was chosen as 9.5 dB because the average SNR of all the detections over the frames is approximately 9.0 dB. However, in most frames, random spikes were noted. In the first frame, where a target is unlikely to be present, a single spike of 15.0 dB is noted. These spikes, however, cannot and should not be modeled as the target SNR.

- A constant probability of detection (P_{DC}) of .70 was chosen. A value that is too high would bring down the detection threshold and increase P_{FA}

- α is the parameter used to update the estimated target SNR with an α filter. A high value is chosen for the purpose of detecting a distant target that approaches the sensor over time and to account for the presence of occasional spikes of noise. Thus, the estimated SNR is less dependent on a detection that could originate from a noisy source and, thus, set the bar too high for future detections.

- π_m is the miss probability.

- $\bar{v}$ and $\bar{\sigma}_v$ are used in the multipass approach of the optimization algorithm.[11,14]

- The number of passes $\mathbf{K}$ in the multipass approach of the optimization algorithm was chosen as 4.

Figure 10.6 further clarifies the detection process by depicting the windows where the target has been detected.

From the above results, note the following.

The first detection uses 22 scans and occurs at scan 28. This occurs because the initial scans have low-information content as the target appears late in the frame of surveillance.

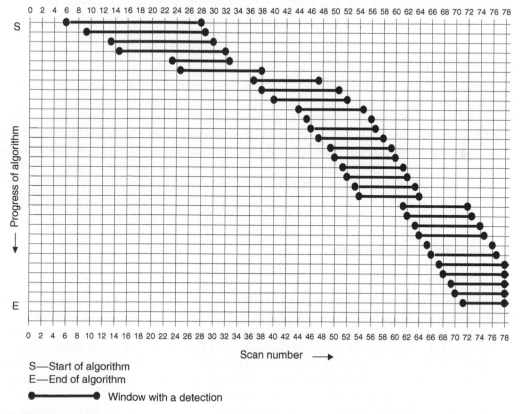

S—Start of algorithm
E—End of algorithm

●————● Window with a detection

FIGURE 10.6
Progress of the algorithm showing windows with detections.

The IMM-MHT algorithm[22] required 38 scans for a detection, whereas the IMMPDA[23] required 39 scans. Some spurious detections were noticed at earlier scans, but these were rejected.

The next few detection windows produce similar target estimates. This is because a large number of scans repeat themselves in these windows.

After the initial detections, there is a "jump" in the scan number at which a detection is made. In addition, the estimates, particularly the velocity estimates, deteriorate. This could indicate that either the target has suddenly disappeared (became less visible) from the region of surveillance or the target made a maneuver.

From scan 44 onward, the algorithm stabilizes for several next windows. At scan 52, however, there is another jump in detection windows. This is also followed by a drop in the estimated target SNR, as explained above. This, however, indicates that the algorithm can adjust itself and restart after a target has become suddenly invisible. Recursive algorithms will diverge in this case.

From scan 54 onward, the algorithm stabilizes, as indicated by the estimates. Also, a detection is made for every increasing window because the target has come closer to the sensor and, thus, is more visible. This is noted by the sharp rise in the estimated target SNR after scan 54.

The above results provide an understanding of the target's behavior. The results suggest that the Mirage F1 fighter jet appears late in the area of surveillance and moves toward the sensor. However, initially it remains quite invisible and possibly undergoes maneuvers. As it approaches the sensor, it becomes more and more visible and, thus, easier to detect.

10.5.4.2 Computational Load

The adaptive ML-PDA tracker took 442 s, including the time for data input/output, on a Pentium III processor running at 550 MHz to process the 78 scans of the Mirage F1 data. This translates into about 5.67 s/frame (or 5.67 s running time for one-second interval data), including input/output time. A more efficient implementation on a dedicated processor can easily make the algorithm real-time capable on a similar processor. Also, by parallelizing the initial grid search, which required more than 90% of the time, the adaptive ML-PDA estimator can be made even more efficient.

10.6 Summary

This chapter presented the use of the PDA technique for different tracking problems. Specifically, the PDA approach was used for parameter estimation as well as recursive state estimation. As an example of parameter estimation, track formation of a LO target using a nonlinear ML estimator in conjunction with the PDA technique with passive (sonar) measurements was presented. The use of the PDA technique in conjunction with the IMM estimator, resulting in the IMMPDAF, was presented as an example of recursive estimation on a radar-tracking problem in the presence of ECM. Also presented was an adaptive sliding-window PDA-based estimator that retains the advantages of the batch (parameter) estimator while being capable of tracking the motion of maneuvering targets. This was illustrated on an EO surveillance problem. These applications demonstrate the usefulness of the PDA approach for a wide variety of real tracking problems.

References

1. Bar-Shalom, Y. (Ed.), *Multitarget-Multisensor Tracking: Advanced Applications*, Vol. I, Artech House Inc., Dedham, MA, 1990. Reprinted by YBS Publishing, 1998.
2. Bar-Shalom, Y. (Ed.), *Multitarget-Multisensor Tracking: Applications and Advances*, Vol. II, Artech House Inc., Dedham, MA, 1992. Reprinted by YBS Publishing, 1998.
3. Bar-Shalom, Y. and Li, X. R., *Multitarget-Multisensor Tracking: Principles and Techniques*, YBS Publishing, Storrs, CT, 1995.
4. Blackman, S. S. and Popoli, R., *Design and Analysis of Modern Tracking Systems*, Artech House Inc., Dedham, MA, 1999.
5. Feo, M., Graziano, A., Miglioli, R., and Farina, A., IMMJPDA vs. MHT and Kalman filter with NN correlation: Performance comparison, *IEE Proceedings on Radar, Sonar and Navigation (Part F)*, 144(2), 49–56, 1997.
6. Kirubarajan, T., Bar-Shalom, Y., Blair, W. D., and Watson, G. A., IMMPDA solution to benchmark for radar resource allocation and tracking in the presence of ECM, *IEEE Transactions on Aerospace and Electronic Systems*, 34(3), 1023–1036, 1998.
7. Lerro, D. and Bar-Shalom, Y., Interacting multiple model tracking with target amplitude feature, *IEEE Transactions on Aerospace and Electronic Systems*, AES-29(2), 494–509, 1993.
8. Blair, W. D., Watson, G. A., Kirubarajan, T., and Bar-Shalom, Y., Benchmark for radar resource allocation and tracking in the presence of ECM, *IEEE Transactions on Aerospace and Electronic Systems*, 34(3), 1015–1022, 1998.
9. Blackman, S. S., Dempster, R. J., Busch, M. T., and Popoli, R. F., IMM/MHT solution to radar benchmark tracking problem, *IEEE Transactions on Aerospace and Electronic Systems*, 35(2), 730–738, 1999.
10. Bar-Shalom, Y. and Li, X. R., *Estimation and Tracking: Principles, Techniques and Software*, Artech House Inc., Dedham, MA, 1993. Reprinted by YBS Publishing, 1998.
11. Jauffret, C. and Bar-Shalom, Y., Track formation with bearing and frequency measurements in clutter, *IEEE Transactions on Aerospace and Electronic Systems*, AES-26, 999–1010, 1990.
12. Kirubarajan, T., Wang, Y., and Bar-Shalom, Y., Passive ranging of a low observable ballistic missile in a gravitational field using a single sensor, *Proceedings of the 2nd International Conference on Information Fusion*, July 1999.
13. Sivananthan, S., Kirubarajan, T., and Bar-Shalom, Y., A radar power multiplier algorithm for acquisition of LO ballistic missiles using an ESA radar, *Proceedings of the IEEE Aerospace Conference*, March 1999.
14. Kirubarajan, T. and Bar-Shalom, Y., Target motion analysis in clutter for passive sonar using amplitude information, *IEEE Transactions on Aerospace and Electronic Systems*, 32(4), 1367–1384, 1996.
15. Nielsen, R. O., *Sonar Signal Processing*, Artech House Inc., Boston, MA, 1991.
16. Blair, W. D., Watson, G. A., Hoffman, S. A., and Gentry, G. L., Benchmark problem for beam pointing control of phased array radar against maneuvering targets, *Proceedings of the American Control Conference*, June 1995.
17. Blair, W. D. and Watson, G. A., IMM lgorithm for solution to benchmark problem for tracking maneuvering targets, *Proceedings of SPIE Acquisition, Tracking and Pointing Conference*, April 1994.
18. Blair, W. D., Watson, G. A., Hoffman, S. A., and Gentry, G. L., Benchmark problem for beam pointing control of phased array radar against maneuvering targets in the presence of ECM and false alarms, *Proceedings of the American Control Conference*, June 1995.

19. Yeddanapudi, M., Bar-Shalom, Y., and Pattipati, K. R., IMM estimation for multitarget-multisensor air traffic surveillance, *Proceedings of IEEE*, 85(1), 80–96, 1997.
20. Chummun, M. R., Kirubarajan, T., and Bar-Shalom, Y., An adaptive early-detection ML-PDA estimator for LO targets with EO sensors, *Proceedings of SPIE Conference on Signal and Data Processing of Small Targets*, 4048, 345–356, April 2000.
21. Press, W. H., Teukolsky, S. A., Vetterling, W. T., and Flannery, B. P., *Numerical Recipes in C*, Cambridge University Press, Cambridge, U.K., 1992.
22. Roszkowski, S. H., Common database for tracker comparison, *Proceedings of SPIE Conference on Signal and Data Processing of Small Targets*, 3373, 95–102, April 1998.
23. Lerro, D. and Bar-Shalom, Y., IR Target detection and clutter reduction using the interacting multiple model estimator, *Proceedings of SPIE Conference on Signal and Data Processing of Small Targets*, 3373, April 1998.

11

Introduction to the Combinatorics of Optimal and Approximate Data Association

Jeffrey K. Uhlmann

CONTENTS

11.1 Introduction

Applying filtering algorithms to track the states of multiple targets first requires the correlation of the tracked objects with their corresponding sensor observations. A variety of probabilistic measures can be applied to each track estimate to determine independently how likely it is to have produced the current observation; however, such measures are useful only in practice for eliminating obviously infeasible candidates. Chapter 4 uses these measures to construct gates for efficiently reducing the number of feasible candidates to a number that can be accommodated within real-time computing constraints. Subsequent elimination of candidates can then be effected by measures that consider the joint relationships among the remaining track and report pairs.

After the gating step has been completed, a determination must be made concerning which feasible associations between observations and tracks are most likely to be correct. In a dense environment, however, resolving the ambiguities among various possible assignments of sensor reports to tracks may be impossible. The general approach proposed in the literature for handling such ambiguities is to maintain a set of *hypothesized* associations in the hope that some would be eliminated by future observations.[19–21] The key challenge is somehow to bound the overall computational cost by limiting the proliferation of pairs under consideration, which may increase in number at a geometric rate.

If n observations are made of n targets, one can evaluate an independent estimate of the probability that a given observation is associated with a given beacon. An independently computed probability, however, may be a poor indicator of how likely a particular observation is associated with a particular track because it does not consider the extent to which other observations may also be correlated with that item. More sophisticated methods generally require a massive batch calculation (i.e., they are a function of all n beacon estimates and $O(n)$ observations of the beacons) where approximately one observation of each target is assumed.[2,3] Beyond the fact that a real-time tracking system must process each sensor report as soon as it is obtained, the most informative joint measures of association scale exponentially with n and are, therefore, completely useless in practice.

This chapter examines some of the combinatorial issues associated with the batch data association problem arising in tracking and correlation applications. A procedure is developed that addresses a large class of data association problems involving the calculation of permanents of submatrices of the original association matrix. This procedure yields what is termed as the joint assignment matrix (JAM), which can be used to optimally rank associations for hypothesis selection. Because the computational cost of the permanent scales exponentially with the size of the matrix, improved algorithms are developed both for calculating the exact JAM and for generating approximations to it. Empirical results suggest that at least one of the approximations is suitable for real-time hypothesis generation in large-scale tracking and correlation applications. Novel theoretical results include an improved upper bound on the calculation of the JAM and new upper bound inequalities for the permanent of general nonnegative matrices. One of these inequalities is an improvement over the best previously known inequality.

11.2 Background

The batch data association problem[2,4–7] can be defined as follows.

Given predictions about the states of n objects at a future time t, n measurements of that set of objects at time t, and a function to compute a probability of association between each prediction and measurement pair, calculate a new probability of association for a prediction and a measurement that is conditioned on the knowledge that the mapping from the set of predictions to the set of measurements is one-to-one.

For real-time applications, the data association problem is usually defined only in terms of estimates maintained at the current timestep of the filtering process. A more general problem can be defined that considers observations over a series of timesteps. Both problems are intractable, but the single-timestep variant appears to be more amenable to efficient approximation schemes.

As an example of the difference between these two measures of the probability of association, consider an indicator function that provides a binary measure of whether or not a given measurement is compatible with a given prediction. If the measure considers each prediction/measurement pair independently of all other predictions and measurements, then it could very well indicate that several measurements are feasible realizations of a single prediction. However, if one of the measurements has only that prediction as a feasible association, then from the constraint that the assignment is one-to-one, the measurement and

TABLE 11.1

Assignment Example

Consider the effect of the assignment constraint on the following matrix of feasible associations:

	R_1	R_2	R_3	R_4
T_1	0	0	1	1
T_2	1	1	0	1
T_3	0	1	0	1
T_4	0	0	1	1

Every track has more than one report with which it could be feasibly assigned. The report R_1, however, can only be assigned to T_2. Given the one-to-one assignment constraint, R_1 clearly must have originated from track T_2. Making this assignment leaves us with the following options for the remaining tracks and reports:

	R_1	R_2	R_3	R_4
T_1	—	0	1	1
T_2	1	—	—	—
T_3	—	1	0	1
T_4	—	0	1	1

The possibility that R_2 originated from T_2 has been eliminated; therefore, R_2 could only have originated from the remaining candidate, T_3. This leaves us with the following options:

	R_1	R_2	R_3	R_4
T_1	—	—	1	1
T_2	1	—	—	—
T_3	—	1	—	—
T_4	—	—	1	1

where two equally feasible options now exist for assigning tracks T_1 and T_4 to reports R_3 and R_4. From this ambiguity only the following options can be concluded:

	R_1	R_2	R_3	R_4
T_1	—	—	0.5	0.5
T_2	1	—	—	—
T_3	—	1	—	—
T_4	—	—	0.5	0.5

the prediction must correspond to the same object. Furthermore, it can then be concluded, based on the same constraint, that the other measurements absolutely are not associated with that prediction. Successive eliminations of candidate pairs would hopefully yield a set of precisely n feasible candidates that must represent the correct assignment; however, this is rare in real-world applications. The example presented in Table 11.1 demonstrates the process.

The above-mentioned association problem arises in a number of practical tracking and surveillance applications. A typical example is the following: a surveillance aircraft flies over a region of ocean and reports the positions of various ships. Several hours later another aircraft repeats the mission. The problem then arises how to identify which reports in the second pass are associated with which reports from the earlier pass. The information available here includes known kinematic constraints (e.g., maximum speed) and the time difference between the various reports. If each ship is assumed to be traveling at a speed in the interval $[v_{min}, v_{max}]$, then the indicator function can identify feasible pairs simply by determining which reports from the second pass fall within the radial

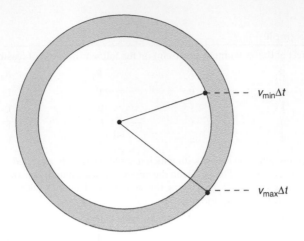

FIGURE 11.1
The inner circle represents the possible positions of the ship if it travels at minimum speed, whereas the outer circle represents its possible positions at maximum speed.

interval $[v_{min}\Delta t, v_{max}\Delta t]$ about reports from the first pass (see Figure 11.1). This problem is called track initiation[2,4,3] because its solution provides a full position and velocity estimate, referred to as a *track*, which can then permit proper tracking.

After the tracking process has been initiated, the association problem arises again at the arrival of each batch of new reports. Thus, data association in this case is not a "one shot" problem; it is a problem of associating series of reports over a period of time to identify distinct trajectories. This means that attempting to remove ambiguity entirely at each step is not necessary; it is possible to retain a set of pairs in the hope that some would be eliminated at future steps. The maintenance of tentative tracks, referred to as *hypotheses*, is often termed as *track splitting*. Track splitting can be implemented in several ways, ranging from methods that simply add extra tracks to the track set with no logical structure to indicate which were formed from common reports, to methods that construct a complete "family tree" of tracks so that confirmation of a single leaf can lead to the pruning of large branches. No matter what the method is, the critical problem is to determine which pairs to keep and which to discard.

11.3   Most Probable Assignments

One way to deal with ambiguities arising in the joint analysis of the prior probabilities of association is to determine which of the *a priori* $n!$ possible assignments is most probable. In this case, "most probable" means the assignment that maximizes the product of the prior probabilities of its component pairs. In other words, it is the assignment, σ_i, that maximizes

$$\prod_i a_{i\sigma_i} \tag{11.1}$$

where a_{ij} is the matrix element giving the probability that track i is associated with report j. Unfortunately, this approach seems to require the evaluation and examination of $n!$ products. There exists, however, a corpus of work on the closely related problem of optimal assignment (also known as maximum-weighted bipartite matching). The optimal assignment problem seeks the assignment that maximizes the *sum* of the values of its component pairs. In other words, it maximizes

$$\sum_i a_{i\sigma_i} \tag{11.2}$$

This can be accomplished in $O(n^3)$ time.[8] Thus, the solution to the maximum product problem can be obtained with an optimal assignment algorithm simply by using the logs of the prior probabilities, with the log of zero replaced with an appropriately large negative number. The optimal assignment approach eliminates ambiguity by assuming that the best assignment is always the correct assignment. Thus, it never maintains more than n tracks. The deficiency with this approach is that unless there are very few ambiguous pairs, many assignments will have almost the same probability. For example, if two proximate reports are almost equally correlated with each of two tracks, then swapping their indices in the optimal assignment will generate a new but only slightly less probable assignment. In fact, in most nontrivial applications, the best assignment has a very low probability of being correct.

The optimal assignment method can be viewed as the best choice of track–report pairs if the criterion is to maximize the probability of having *all* the pairs correct. Another reasonable optimality criterion would seek the set of n pairs that maximizes the *expected number* of pairs that are correct. To illustrate the difference between these criteria, consider the case in which two proximate reports are almost equally correlated with two tracks, and no others. The optimal assignment criterion would demand that the two reports be assigned to distinct tracks, whereas the other criterion would permit all four pairs to be kept as part of the n selected pairs. With all four possible pairs retained, the second criterion ensures that at least two are correct. The two pairs selected by the optimal assignment criterion, however, have almost a .5 probability of both being incorrect.

11.4   Optimal Approach

A strategy for generating multiple hypotheses within the context of the optimal assignment approach is to identify the k best assignments and take the union of their respective pairings.[5,9,10] The assumption is that pairs in the most likely assignments are most likely to be correct. Intuition also would suggest that a pair common to all the k best assignments stands a far better chance of being correct than its prior probability of association might indicate. Generalizing this intuition leads to the exact characterization of the probabilities of association under the assignment constraint. Specifically, the probability that report R_i is associated with track T_j is simply the sum of the probabilities of all the assignments containing the pair, normalized by the sum of the probabilities of all assignments:[5]

$$p(a_{ij}|\text{Assignment Constraint}) = \frac{1}{\sum_\sigma \prod_k a_{k\sigma_k}} \sum_{(\sigma|\sigma_i=j)} \prod_m a_{m\sigma_m} \tag{11.3}$$

For example, suppose the following matrix is given containing the prior probabilities of association for two tracks with two reports:

$$\begin{array}{ccc} & R_1 & R_2 \\ T_1 & .3 & .7 \\ T_2 & .5 & .4 \end{array} \tag{11.4}$$

Given the assignment constraint, the correct pair of associations must be either (T_1, R_1) and (T_2, R_2), or (T_1, R_2) and (T_2, R_1). To assess the joint probabilities of association, Equation 11.3 must be applied to each entry of the matrix to obtain:

$$\frac{1}{(.3)(.4)+(.5)(.7)} \times \begin{vmatrix} (.3)(.4) & (.5)(.7) \\ (.5)(.7) & (.3)(.4) \end{vmatrix} = \begin{vmatrix} .255 & .745 \\ .745 & .255 \end{vmatrix} \tag{11.5}$$

Notice that the resulting matrix is doubly stochastic (i.e., it has rows and columns all of which sum to unity), as one would expect. (The numbers in the matrix in Equation 11.5 have been rounded, but still sum appropriately. One can verify that the true values do as well.) Notice also that the diagonal elements are equal. This is the case for any 2×2 matrix because elements of either diagonal can only occur jointly in an assignment; therefore, one element of a diagonal cannot be more or less likely than the other. The matrix of probabilities generated from the assignment constraint is JAM.

Now consider the following matrix:

$$\begin{array}{cc} & R_1 \quad R_2 \\ T_1 & .9 \quad .5 \\ T_2 & .4 \quad .1 \end{array} \tag{11.6}$$

which, given the assignment constraint, leads to the following:

$$\begin{array}{cc} & R_1 \quad R_2 \\ T_1 & .31 \quad .69 \\ T_2 & .69 \quad .31 \end{array} \tag{11.7}$$

This demonstrates how significant the difference can be between the prior and the joint probabilities. In particular, the pair (T_1, R_1) has a prior probability of .9, which is extremely high. Considered jointly with the other measurements, however, its probability drops to only .31.

A more extreme example is the following:

$$\begin{array}{cc} & R_1 \quad R_2 \\ T_1 & .99 \quad .01 \\ T_2 & .01 \quad 0 \end{array} \tag{11.8}$$

which leads to

$$\begin{array}{cc} & R_1 \quad R_2 \\ T_1 & 0 \quad 1.0 \\ T_2 & 1.0 \quad 0 \end{array} \tag{11.9}$$

where the fact that T_2 cannot be associated with R_2 implies that there is only one feasible assignment. Examples like this show why a "greedy" selection of hypotheses based only on the independently assessed prior probabilities of association can lead to highly suboptimal results.

The examples that have been presented here have considered only the ideal case where the actual number of targets is known and the number of tracks and reports equals that number. However, the association matrix can easily be augmented* to include rows and columns to account for the cases in which some reports are spurious and some targets are not detected. Specifically, a given track can have a probability of being associated with each report as well as a probability of not being associated with any of them. Similarly, a given report has a probability of association with each of the tracks and a probability that it is a false alarm not associated with any target. Sometimes a report that is not associated with any tracks signifies a newly detected target. If the actual number of targets is not known, a combined probability of false alarm and probability of new target must be used. Estimates of probabilities of detection, probabilities of false alarms, and probabilities of new detections are difficult to determine because of complex dependencies on the type of sensor used, the environment, the density of targets, and a multitude of other factors whose effects are almost never known. In practice, such probabilities are often lumped together into one tunable parameter (e.g., a "fiddle factor").

11.5 Computational Considerations

A closer examination of Equation 11.3 reveals that the normalizing factor is a quantity that resembles the determinant of the matrix, but without the alternating ±1 factors. In fact, the determinant of a matrix is just the sum of products over all even permutations minus the sum of products over all odd permutations. The normalizing quantity of Equation 11.3, however, is the sum of products over all permutations. This latter quantity is called the permanent[6,11,1] of a matrix, and it is often defined as follows:

$$\text{per}(A) = \sum_{\sigma} a_{1\sigma_1} a_{2\sigma_2} \cdots a_{n\sigma_n} \tag{11.10}$$

where the summation extends over all permutations Q of the integers $1, 2, \ldots, n$. The Laplace expansion of the determinant also applies for the permanent:

$$\text{per}(A) = \sum_{j=1}^{n} a_{ij} \cdot \text{per}(A_{\overline{ij}})$$

$$= \sum_{i=1}^{n} a_{ij} \cdot \text{per}(A_{\overline{ij}}) \tag{11.11}$$

where $A_{\overline{ij}}$ is the submatrix obtained by removing row i and column j. This formulation provides a straightforward mechanism for evaluating the permanent, but it is not efficient. As in the case of the determinant, expanding by Laplacians requires $O(n \ldots n!)$ computations. The unfortunate fact about the permanent is that although the determinant can be evaluated by other means in $O(n^3)$ time, effort exponential in n seems to be necessary to evaluate the permanent.[12]

If Equation 11.3 were evaluated without the normalizing coefficient for every element of the matrix, determining the normalizing quantity by simply computing the sum of any

---

* Remember the use of an association "matrix" is purely for notational convenience. In large applications, such a matrix would not be generated in its entirety; rather, a sparse graph representation would be created by identifying and evaluating only those entries with probabilities above a given threshold.

row or column would seem possible as the result must be doubly stochastic. In fact, this is the case. Unfortunately, Equation 11.3 can be rewritten, using Equation 11.11, as

$$p(a_{ij}|\text{assignment constraint}) = a_{ij} \cdot \frac{\text{per}(A_{\overline{ij}})}{\text{per}(A)} \tag{11.12}$$

where the evaluation of permanents seems unavoidable. Knowing that such evaluations are intractable, the question then is how to deal with computational issues.

The most efficient method for evaluating the permanent is attributable to Ryser.[6] Let A be an $n \times n$ matrix, let A_r denote a submatrix of A obtained by deleting r columns, let $\prod(A_r)$ denote the product of the row sums of A_r, and let $\sum\prod(A_r)$ denote the sum of the products $\prod(A_r)$ taken over all possible A_r. Then,

$$\text{per}(A) = \prod(A) - \sum\prod(A_1) + \sum\prod(A_2) - \cdots + (-1)^{n-1}\sum\prod(A_{n-1}) \tag{11.13}$$

This formulation involves only $O(2^n)$, rather than $O(n!)$, products. This may not seem like a significant improvement, but for n in the range of 10–15, the reduction in compute time is from hours to minutes on a typical workstation. For n greater than 35, however, scaling effects thwart any attempt at evaluating the permanent. Thus, the goal must be to reduce the coefficients as much as possible to permit the optimal solution for small matrices in real time, and to develop approximation schemes for handling large matrices in real time.

11.6 Efficient Computation of Joint Assignment Matrix

Equation 11.13 is known to permit the permanent of a matrix to be computed in $O(n^2 \ldots 2^n)$ time. From Equation 11.12, then, one can conclude that the JAM is computable in $O(n^4 \ldots 2^n)$ time (i.e., the amount of time required to compute the permanent of A_{ij} for each of the n^2 elements a_{ij}). However, this bound can be improved by showing that the JAM can be computed in $O(n^3 \ldots 2^n)$ time, and that the time can be further reduced to $O(n^2 \ldots 2^n)$.

First, the permanent of a general matrix can be computed in $O(n \ldots 2^n)$ time. This is accomplished by eliminating the most computationally expensive step in direct implementations of Ryser's method—the $O(n^2)$ calculation of the row sums at each of the 2^n iterations. Specifically, each term of Ryser's expression of the permanent is just a product of the row sums with one of the 2^n subsets of columns removed. A direct implementation therefore requires the summation of the elements of each row that are not in one of the removed columns. Thus, $O(n)$ elements are summed for each of the n rows at a total cost of $O(n^2)$ arithmetic operations per term. To reduce the cost required to update the row sums, Nijenhuis and Wilf showed that the terms in Ryser's expression can be ordered so that only one column is changed from one term to the next.[13] At each step the algorithm updates the row sums by either adding or subtracting the column element corresponding to the change. Thus, the total update time is only $O(n)$. This change improves the computational complexity from $O(n^2 \ldots 2^n)$ to $O(n \ldots 2^n)$.

The above-mentioned algorithm for evaluating the permanent of a matrix in $O(n \ldots 2^n)$ time can be adapted for the case in which a row and a column of a matrix are assumed removed. This permits the evaluation of the permanents of the submatrices associated with each of the n^2 elements, as required by Equation 11.12, to calculate the JAM in $O(n^3 \ldots 2^n)$ time. This is an improvement over the $O(n^4 \ldots 2^n)$ scaling obtained by the direct application of Ryser's formula (Equation 11.13) to calculate the permanent of each submatrix. The scaling can, however, be reduced even further. Specifically, note that the permanent of the submatrix associated with element a_{ij} is the sum of all $((n-1)2^{(n-1)})/(n-1)$ terms in Ryser's formula that do not involve row i or column j. In other words, one can eliminate a factor of $(n-1)$ by factoring the products of the $(n-1)$ row sums common to the submatrix permanents of each element in the same row. This factorization leads to an optimal JAM algorithm with complexity $O(n^2 \ldots 2^n)$.[14]

An optimized version of this algorithm can permit the evaluation of 12×12 joint assignment matrices in well under a second. Because the algorithm is highly parallelizable—the $O(2^n)$ iterative steps can be divided into k subproblems for simultaneous processing on k processors—solutions to problems of size n in the range of 20–25 should be computable in real time with $O(n^2)$ processors. Although not practical, the quantities computed at each iteration could also be computed in parallel on $n \ldots 2^n$ processors to achieve an $O(n^2)$ sequential scaling. This might permit the real-time solution of somewhat larger problems but at an exorbitant cost. Thus, the processing of $n \ldots n$ matrices, for $n > 25$, will require approximations.

Figure 11.2 provides actual empirical results (on a ca. 1994 workstation) showing that the algorithm is suitable for real-time applications for $n < 10$, and it is practical for offline applications for n as large as 25. The overall scaling has terms of $n^2 2^n$ and 2^n, but the test results and the algorithm itself clearly demonstrate that the coefficient on the $n^2 2^n$ term is small relative to the 2^n term.

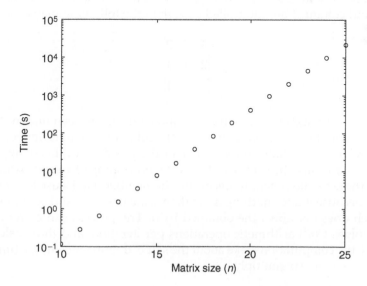

FIGURE 11.2
Tests of the new joint assignment matrix algorithm reveal the expected exponentially scaling computation time.

11.7 Crude Permanent Approximations

In the late 1980s, several researchers identified the importance of determining the number of feasible assignments in sparse association matrices arising in tracking applications. In this section, a recently developed approach for approximating the JAM via permanent inequalities is described, which yields surprisingly good results within time roughly proportional to the number of feasible track–report pairs.

Equation 11.12 shows that an approximation to the permanent would lead to a direct approximation of Equation 11.3. Unfortunately, research into approximating the permanent has emphasized the case in which the association matrix A has only 0–1 entries.[15] Moreover, the methods for approximating the permanent, even in this restricted case, still scale exponentially for reasonable estimates.[15-17] Even "unreasonable" estimates for the general permanent, however, may be sufficient to produce a reasonable estimate of a conditional probability matrix. This is possible because the solution matrix has additional structure that may permit the filtering of noise from poorly estimated permanents. The fact that the resulting matrix should be doubly stochastic, for example, suggests that the normalization of the rows or columns (i.e., dividing each row or column by the sum of its elements) should improve the estimate.

One of the most important properties of permanents relating to the doubly stochastic property of the JAM is the following: *Multiplying a row or column by a scalar c has the effect of multiplying the permanent of the matrix by the same factor.*[1,6,11] This fact verifies that the multiplication of a row or column by some c also multiplies the permanent of any submatrix by c. This implies that the multiplication of any combination of rows or columns of a matrix by any values (other than zero) has no effect on the JAM. This is because the factors applied to the various rows and columns cancel in the ratio of permanents in Equation 11.12. Therefore, the rows and columns of a matrix can be normalized in any manner before attempting to approximate the JAM.

To see why a conditioning step could help, consider a 3×3 matrix with all elements equal to 1. If row 1 and column 1 are multiplied by 2, then the following matrix is obtained:

$$\begin{vmatrix} 4 & 2 & 2 \\ 2 & 1 & 1 \\ 2 & 1 & 1 \end{vmatrix} \tag{11.14}$$

where the effect of the scaling of the first row and column has been undone.* This kind of preconditioning could be expected to improve the reliability of an estimator. For example, it seems to provide more reliable information for the greedy selection of hypotheses. This iterative process could also be useful for the postconditioning of an approximate JAM (e.g., to ensure that the estimate is doubly stochastic). Remember that its use for preconditioning is permissible because it does nothing more than scale the rows and columns during each iteration, which does not affect the obtained JAM. The process can be repeated for $O(n)$ iterations,† involves $O(n^2)$ arithmetic operations per iteration, and thus scales as $O(n^3)$. In absolute terms, the computations take about the same amount of compute time as required to perform an $n \times n$ matrix multiply.

---

* A physical interpretation of the original matrix in sensing applications is that each row and column corresponds to a set of measurements, which is scaled by some factor due to different sensor calibrations or models.
† This iterative algorithm for producing a doubly stochastic matrix is best known in the literature as the Sinkhorn iteration, and many of its properties have been extensively analyzed.

11.8 Approximations Based on Permanent Inequalities

The possibility has been discussed of using crude estimators of the permanent, combined with the knowledge about the structure of the JAM, to obtain better approximations. Along these lines, four upper bound inequalities, $\varepsilon_1 - \varepsilon_4$, for the permanent are examined for use as crude estimators. The first inequality, ε_1, is a well-known result:

$$\text{per}(A) \le \prod_{i=1}^{n} r_i$$

(11.15)

where r_i is the sum of the elements in row i. This inequality holds for nonnegative matrices because it sums over all products that do not contain more than one element from the same row. In other words, it sums over a set larger than that of the actual permanent because it includes products with more than one element from the same column. For example, the above-mentioned inequality applied to a 2×2 matrix would give $a_{11}a_{21} + a_{11}a_{22} + a_{12}a_{21} + a_{12}a_{22}$, rather than the sum over one-to-one matchings $a_{11}a_{22} + a_{12}a_{21}$. In fact, the product of the row sums is the first term in Ryser's equation. This suggests that the evaluation of the first k terms should yield a better approximation. Unfortunately, the required computation scales exponentially in the number of evaluated terms, thus making only the first two or three terms practical for approximation purposes. All the inequalities in this section are based on the first term of Ryser's equation applied to a specially conditioned matrix. They can all, therefore, be improved by the use of additional terms, noting that an odd number of terms yields an upper bound, whereas an even number yields a lower bound.

This inequality also can be applied to the columns to achieve a potentially better upper bound. This would sum over products that do not contain more than one element from the same column. Thus, the following bound can be placed on the permanent:[1,6,11]

$$\text{per}(A) \le \min\left\{\prod_{i=1}^{n} r_i, \prod_{i=1}^{n} c_i\right\}$$

(11.16)

Although taking the minimum of the product of the row sums and the product of the column sums tends to yield a better bound on the permanent, there is no indication that it is better than always computing the bound with respect to the rows. This is because the goal is to estimate the permanent of a submatrix for every element of the matrix, as required by Equation 11.12, to generate an estimate of the JAM. If all of the estimates are of the same "quality" (i.e., are all too large by approximately the same factor), then some amount of postconditioning could yield good results. If the estimates vary considerably in their quality, however, postconditioning might not provide significant improvement.

The second inequality considered, ε_2, is the Jurkat–Ryser upper bound:[11,18] Given a non-negative $n \times n$ matrix $A = [a_{ij}]$ with row sums $r_1, r_2, \dots, r_n$ and column sums $c_1, c_2, \dots, c_n$, where the row and column sums are indeed so that $r_k \le r_{k+1}$ and $c_k \le c_{k+1}$ for all k, then:

$$\text{per}(A) \le \prod_{i=1}^{n} \min\{r_i, c_i\}$$

(11.17)

For cases in which there is at least one k, such that $r_k \ne c_k$, this upper bound is less than that obtained from the product of row or column sums. This has been the best of all known upper bounds since it was discovered in 1966. The next two inequalities, ε_3 and ε_4, are the new results.[14]

ε_3 is defined as

$$\text{per}(A) \le \prod_{i=1}^{n} \sum_{j=1}^{n} \frac{a_{ij}c_i}{c_j} \tag{11.18}$$

This inequality is obtained by normalizing the columns, computing the product of the row sums of the resulting matrix, and then multiplying that product by the product of the original column sums. In other words,

$$\text{per}(A) \le (c_1 c_2 \ldots c_n) \prod_{i=1}^{n} \sum_{j=1}^{n} \frac{a_{ij}}{c_j} \tag{11.19}$$

$$= \prod_{i=1}^{n} c_i \cdot \sum_{j=1}^{n} \frac{a_{ij}}{c_j} \tag{11.20}$$

$$= \prod_{i=1}^{n} \sum_{j=1}^{n} c_i \frac{a_{ij}}{c_j} \tag{11.21}$$

Note that the first summation is just the row sums after the columns have been normalized.

This new inequality is interesting because it seems to be the first general upper bound to be discovered, which is, in some cases, superior to Jurkat–Ryser. For example, consider the following matrix:

$$\begin{vmatrix} 1 & 2 \\ 2 & 3 \end{vmatrix} \tag{11.22}$$

Jurkat–Ryser (ε_2) yields an upper bound of 15, whereas the estimator ε_3 yields a bound of 13.9. Although ε_3 is not generally superior, it is considered for the same reasons as the estimator ε_1.

The fourth inequality considered, ε_4, is obtained from ε_3 via ε_2:

$$\text{per}(A) \le \prod_{i=1}^{n} \min \left\{ \sum_{j=1}^{n} \frac{a_{ij}c_i}{c_j}, \ c_i \right\} \tag{11.23}$$

The inequality is derived by first normalizing the columns. Then applying Jurkat–Ryser with all column sums equal to unity yields the following:

$$\text{per}(A) \le (c_1 c_2 \ldots c_n) \prod_{i=1}^{n} \min \left\{ \sum_{j=1}^{n} \frac{a_{ij}}{c_j}, 1 \right\} \tag{11.24}$$

$$= \prod_{i=1}^{n} c_i \cdot \min\left\{\sum_{j=1}^{n} \frac{a_{ij}}{c_j}, 1\right\} \tag{11.25}$$

$$= \prod_{i=1}^{n} \min\left\{\sum_{j=1}^{n} \frac{a_{ij}c_i}{c_j}, c_i\right\} \tag{11.26}$$

where the first summation simply represents the row sums after the columns have been normalized.

Similar to ε_1 and ε_3, this inequality can be applied with respect to the rows or columns—whichever yields the better bound. In the case of ε_4, this is critical because one case usually provides a bound that is smaller than the other and is smaller than that obtained from Jurkat–Ryser.

In the example matrix (Equation 11.22), the ε_4 inequality yields an upper bound of 11, an improvement over the other three estimates. Small-scale tests of the four inequalities on matrices of uniform deviates suggest that ε_3 almost always provides better bounds than ε_1, ε_2 almost always provides better bounds than ε_3, and ε_4 virtually always (more than 99% of the time) produces superior bounds to the other three inequalities. In addition to producing relatively tighter upper bound estimates in this restricted case, inequality ε_4 should be more versatile analytically than Jurkat–Ryser because it does not involve a reindexing of the rows and columns.

11.9 Comparisons of Different Approaches

Several of the JAM approximation methods described in this chapter have been compared on matrices containing

1. Uniformly and independently generated association probabilities
2. Independently generated binary (i.e., 0–1) indicators of feasible association
3. Probabilities of association between two three-dimensional (3D) sets of correlated objects

The third group of matrices was generated from n tracks with uniformly distributed means and equal covariances by sampling the Gaussian defined by each track covariance to generate n reports. A probability of association was then calculated for each track–report pair.

The first two classes of matrices are examined to evaluate the generality of the various methods. These matrices have no special structure to be exploited. Matrices from the third class, however, contain structure typical of association matrices arising in tracking and correlation applications. Performance on these matrices should be indicative of performance in real-world data association problems, whereas performance in the first two classes should reveal the general robustness of the approximation schemes.

The approximation methods considered are the following:

1. ε_1, the simplest of the general upper bound inequalities on the permanent. Two variants are considered

    a. The upper bound taken with respect to the rows

    b. The upper bound taken with respect to the rows or columns, whichever is less

    Scaling: $O(n^3)$

2. ε_2, the Jurkat–Ryser inequality

    Scaling: $O(n^3 \log n)$

3. ε_3 with two variants:

    a. The upper bound taken with respect to the rows

    b. The upper bound taken with respect to the rows or columns, whichever is less

    Scaling: $O(n^3)$ (see the appendix)

4. ε_4 with two variants

    a. The upper bound taken with respect to the rows

    b. The upper bound taken with respect to the rows or columns, whichever is less

    Scaling: $O(n^3)$

5. Iterative renormalization alone

    Scaling: $O(n^3)$

6. The standard greedy method[2] that assumes the prior association probabilities are accurate (i.e., performs no processing of the association matrix)

    Scaling: $O(n^2 \log n)$

7. The one-sided normalization method that normalizes only the rows (or columns) of the association matrix

    Scaling: $O(n^2 \log n)$

The four ε estimators include the $O(n^3)$ cost of pre- and postprocessing via iterative renormalization.

The quality of hypotheses generated by the greedy method $O(n^2 \log n)$ is also compared to those generated optimally via the JAM. The extent to which the greedy method is improved by first normalizing the rows (or columns) of the association matrix is also examined. The latter method is relevant for real-time data association applications in which the set of reports (or tracks) cannot be processed in batch.

Tables 11.2–11.4 give the results of the various schemes when applied to different classes of $n \times n$ association matrices. The n best associations for each scheme are evaluated via the true JAM to determine the expected number of correct associations. The ratio of the expected number of correct associations for each approximation method and the optimal method yields the percentages in Table 11.2.* For example, an entry of 50% implies that the expected number of correct associations is half of what would be obtained from the JAM.

Table 11.2 provides a comparison of the schemes on matrices of size 20×20. Matrices of this size are near the limit of practical computability. For example, the JAM computations

---

* The percentages are averages over enough trials to provide an accuracy of at least five decimal places in all cases except tests involving 5×5 0–1 matrices. The battery of tests for the 5×5 0–1 matrices produced some instances in which no assignments existed. The undefined results for these cases were not included in the averages, so the precision may be slightly less.

TABLE 11.2

Results of Tests of Several Joint Assignment Matrix Approximation Methods on 20×20 Association Matrices (% Optimal)

Method	Uniform Matrices	0–1 Matrices	3D Spatial Matrices
ε_{1r}	99.9996186	99.9851376	99.9995588
ε_{1c}	99.9996660	99.9883342	100.0000000
Jurkat–Ryser (ε_2)	99.9232465	99.4156865	99.8275153
ε_{3r}	99.9996660	99.9992264	99.9999930
ε_{3c}	99.9997517	99.9992264	100.0000000
ε_{4r}	99.9996660	99.9992264	99.9999930
ε_{4c}	99.9997517	99.9992264	100.0000000
Iterative normalized	99.9875369	99.9615623	99.9698123
Standard greedy	84.7953351	55.9995049	86.3762418
One-sided normalized	93.6698728	82.5180206	98.0243181

Note: Each entry in the table represents the ratio of the expected number of correct associations made by the approximate method and the expected number for the optimal JAM. The subscripts r and c on the ε_{1-4} methods denote the application of the method to the rows and to the columns, respectively.

TABLE 11.3

Results of Tests of the Joint Assignment Matrix Approximation on Variously Sized Association Matrices Generated from Uniform Random Deviates

Method	5×5 Matrices	10×10 Matrices	15×15 Matrices	20×20 Matrices
ε_{1r}	99.9465167	99.9883438	99.9996111	99.9996186
ε_{1c}	99.9465167	99.9920086	99.9996111	99.9996660
Jurkat–Ryser (ε_2)	99.8645867	99.7493972	99.8606475	99.9232465
ε_{3r}	99.9465167	99.9965856	99.9996111	99.9996660
ε_{3c}	99.9465167	99.9965856	99.9997695	99.9997517
ε_{4r}	99.9465167	99.9965856	99.9996111	99.9996660
ε_{4c}	99.9465167	99.9965856	99.9997695	99.9997517
Iterative normalized	99.4492256	99.8650315	99.9646233	99.9875369
Standard greedy	80.3063296	80.5927739	84.2186048	84.7953351
One-sided normalized	90.6688891	90.7567223	93.2058342	93.6698728

for a 35×35 matrix would demand more than a hundred years of nonstop computing on current high-speed workstations.

The most interesting information that has been provided by Table 11.2 is that the inequality-based schemes are all more than 99% of optimal, with the approximation based on the Jurkat–Ryser inequality performing worst. The ε_3 and ε_4 methods performed best and always yielded identical results on the doubly stochastic matrices obtained by preprocessing. Preprocessing also improved the greedy scheme by 10–20%. Tables 11.3 and 11.4 show the effect of matrix size on each of the methods.

In almost all cases, the inequality-based approximations seemed to improve with matrix size. The obvious exception is the case of 5×5 0–1 matrices—the approximations are perfect because there tends to be only one feasible assignment for randomly generated 5×5 0–1 matrices. Surprisingly, the one-sided normalization approach is 80–90% optimal, yet scales as $O(p \log p)$, where p is the number of feasible pairings.

The one-sided approach is the only practical choice for large-scale applications that do not permit batch processing of sensor reports because the other approaches require the generation of the (preferably sparse) assignment matrix. The one-sided normalization approach requires only the set of tracks with which it gates (i.e., one row of the assignment

TABLE 11.4

Results of Tests of the Joint Assignment Matrix Approximation on Variously Sized Association
Matrices Generated from Uniform 0–1 Random Deviates

Method	5 × 5 Matrices	10 × 10 Matrices	15 × 15 Matrices	20 × 20 Matrices
ε_{1r}	100.0000000	99.9063047	99.9606670	99.9851376
ε_{1c}	100.0000000	99.9137304	99.9754337	99.9883342
Jurkat–Ryser (ε_2)	100.0000000	99.7915349	99.6542955	99.4156865
ε_{3r}	100.0000000	99.9471028	99.9947939	99.9992264
ε_{3c}	100.0000000	99.9503096	99.9949549	99.9992264
ε_{4r}	100.0000000	99.9471028	99.9947939	99.9992264
ε_{4c}	100.0000000	99.9503096	99.9949549	99.9992264
Iterative normalized	100.0000000	99.7709328	99.9256354	99.9615623
Standard greedy	56.1658957	55.7201435	53.8121279	55.9995049
One-sided normalized	72.6976451	77.6314664	83.6890193	82.5180206

matrix). Therefore, it permits the sequential processing of measurements. Of the methods compared here, it is the only one that satisfies online constraints in which each observation must be processed at the time it is received.

To summarize, the JAM approximation schemes based on the new permanent inequalities appear to yield near-optimal results. An examination of the matrices produced by these methods reveals a standard deviation of less than 3×10^{-5} from the optimal JAM computed via permanents. The comparison of expected numbers of correct assignments given in the Tables 11.2–11.4, however, is the most revealing in terms of applications to multiple-target tracking. Specifically, the recursive formulation of the tracking process leads to highly nonlinear dependencies on the quality of the hypothesis generation scheme. In a dense tracking environment, a deviation of less than 1% in the expected number of correct associations can make the difference between convergence and divergence of the overall process. The next section considers applications to large-scale problems.

11.10   Large-Scale Data Association

This section examines the performance of the one-sided normalization approach. The evidence provided in the Section 11.9 indicates that the estimator ε_3 yields probabilities of association conditioned on the assignment constraint that are very near optimal. Therefore, ε_3 can be used as a baseline of comparison for the one-sided estimator for problems that are too large to apply the optimal approach. The results in the previous section demonstrate that the one-sided approach yields relatively poor estimates when compared to the optimal and near-optimal methods. However, because the latter approaches cannot be applied online to process each observation as it arrives, the one-sided approach is the only feasible alternative. The goal, therefore, is to demonstrate only that its estimates do not diminish in quality as the size of the problem increases. This is necessary to ensure that a system that is tuned and tested on problems of a given size behaves predictably when applied to larger problems.

In the best-case limit, as the amount of ambiguity goes to zero, any reasonable approach to data association should perform acceptably. In the worst-case limit, as all probabilities converge to the same value, no approach can perform any better than a simple random selection of hypotheses. The worst-case situation in which information can be exploited is

TABLE 11.5

Example of an Association Matrix Generated from a Uniform Random Number Generator

.266	.057	.052	.136	.227	.020	.059
.051	.023	.208	.134	.199	.135	.058
.031	.267	.215	.191	.117	.227	.002
.071	.057	.243	.029	.230	.281	.046
.020	.249	.166	.148	.095	.178	.121
.208	.215	.064	.268	.067	.180	.039
.018	.073	.126	.062	.125	.141	.188

TABLE 11.6

True Probabilities of Association Conditioned on the Assignment Constraint

.502	.049	.037	.130	.191	.014	.077
.075	.026	.244	.167	.243	.136	.109
.034	.310	.182	.193	.093	.186	.003
.088	.055	.244	.027	.237	.278	.071
.022	.285	.139	.144	.076	.144	.191
.259	.200	.043	.279	.048	.124	.048
.021	.074	.111	.060	.113	.119	.501

when the probabilities of association appear to be uncorrelated random deviates. In such a case, only higher-order estimation of the joint probabilities of association can provide useful discriminating information. In Table 11.5 is an example of an association matrix that was generated from a uniform random number generator. This example matrix was chosen from among the 10 that were generated because it produced the most illustrative JAM. The uniform deviate entries have been divided by n so that they are of magnitude comparable to that of actual association probabilities.

Applying the optimal JAM algorithm yields the true probabilities of association conditioned on the assignment constraint, as shown in Table 11.6.

A cursory examination of the differences between corresponding entries in the two matrices demonstrates that a significant amount of information has been extracted by considering higher-order correlations. For example, the last entries in the first two rows of the association matrix are .059 and .058—differing by less than 2%—yet their respective JAM estimates are .077 and .109, a difference of almost 30%. Remarkably, despite the fact that the entries in the first matrix have been generated from uniform deviates, the first entry in the first row and the last entry in the last row of the resulting JAM represent hypotheses that each have a better than 50% chance of being correct.

To determine whether the performance of the one-sided hypothesis selection approach suffers as the problem size is increased, its hypotheses have been compared with those of estimator ε_3 on $n \times n$, for n in the range of 10–100, association matrices generated from uniform random deviates. Figure 11.3 shows that the number of correct associations for the one-sided approach seems to approach the same number as ε_3 as n increases. This may be somewhat misleading, however, because the expected number of correct associations out of n hypotheses selected from $n \times n$ possible candidates would tend to decrease as n increases. More specifically, the ratio of correct associations to number of hypotheses (proportional to n) would tend to zero for all methods if the prior probabilities of association are generated at random.

A better measure of performance is how many hypotheses are necessary for a given method to ensure that a fixed number are expected to be correct. This can be determined from the JAM by summing its entries corresponding to a set of hypotheses. Because each

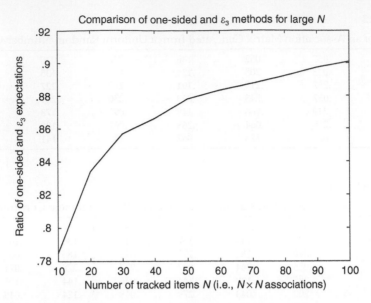

FIGURE 11.3
The performance of the one-sided approach relative to ε_3 improves with increasing N. This is somewhat misleading, however, because the expected number of correct assignments goes to zero in the limit $N \to \infty$ for both approaches.

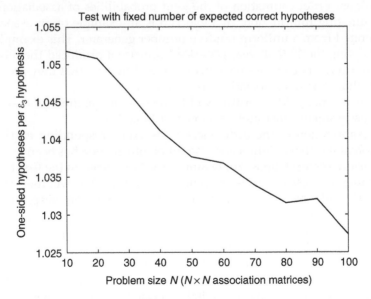

FIGURE 11.4
This test plots the number of hypotheses required by the one-sided approach to achieve some fixed expected number of correct assignments. Again, it appears that the one-sided approach performs comparably to ε_3 as N increases.

entry contains the expectation that a particular track–report pair corresponds to the same object, the sum of the entries gives the expected number of correct assignments. To apply this measure, it is necessary to fix a number of hypotheses that must be correct, independent of n, and determine the ratio of the number of hypotheses required for the one-sided approach to that required by ε_3. Figure 11.4 also demonstrates that the one-sided method seems to approach the same performance as ε_3 as n increases in a highly ambiguous environment. In conclusion, the performance of the one-sided approach scales robustly even in highly ambiguous environments.

The good performance of the one-sided approach—which superficially appears to be little more than a crude heuristic—is rather surprising given the amount of information it fails to exploit from the association matrix. In particular, it uses information only from the rows (or columns) taken independently, thus making no use of information provided by the columns (or rows). Because the tests described in the previous text have all rows and columns of the association matrix scaled randomly, but uniformly, the worst-case performance of the one-sided approach may not have been seen. In tests in which the columns have been independently scaled by vastly different values, results generated from the one-sided approach show little improvement over those of the greedy method. (The performances of the optimal and near-optimal methods, of course, are not affected.) In practice, a system in which all probabilities are scaled by the same value (e.g., as a result of using a single sensor) should not be affected by this limitation of the one-sided approach. In multisensor applications, however, association probabilities must be generated consistently. If a particular sensor is not modeled properly, and its observations produce track–report pairs with consistently low or high probabilities of association, then the hypotheses generated from these pairs by the one-sided approach would be ranked consistently low or high.

11.11   Generalizations

The combinatorial analysis of the assignment problem in Sections 11.9 and 11.10 has considered only the case of a single "snapshot" of sensor observations. In actual tracking applications, however, the goal is to establish tracks from a sequence of snapshots. Mathematically, the definition of a permanent can be easily generalized to apply not only to assignment matrices but also to tensor extensions. Specifically, the generalized permanent sums over all assignments of observations at timestep k, for each possible assignment of observations at timestep $k - 1$, continuing recursively down to the base case of all possible assignments of the first batch of observations to the initial set of tracks. (This multidimensional assignment problem is described more fully in Chapter 13.) Although generalizing the crude permanent approximations for application to the multidimensional assignment problem is straightforward, the computation time scales geometrically with exponent k. This is vastly better than the superexponential scaling required for the optimal approach, but it is not practical for large values of n and k unless the gating process yields a very sparse association matrix (see Chapter 4).

11.12   Conclusions

This chapter has discussed some of the combinatorial issues arising in the batch data association problem. It has described the optimal solution for a large class of data association problems involving the calculation of permanents of submatrices of the original association matrix. This procedure yields the JAM, which can be used to optimally rank associations for hypothesis selection. Because the computational cost of the permanent scales exponentially in the size of the matrix, improved algorithms have been developed both for calculating the exact JAM and for generating approximations to it. Empirical results suggest that the approximations are suitable for hypothesis generation in large-scale tracking and correlation applications. New theoretical results include an improved upper bound on the calculation of the JAM and new upper bound inequalities, ε_3 and ε_4, for the permanent of general nonnegative matrices.

The principal conclusion that can be drawn from this chapter is that the ambiguities introduced by a dense environment are extremely difficult and computationally expensive to resolve. Although this chapter has examined the most general possible case in which tracking has to be performed in an environment that is dense with indistinct (other than position) targets, there is little doubt that the constraints imposed by most real-world applications would necessitate some sacrifice of this generality.

Acknowledgments

The author gratefully acknowledges support from the University of Oxford, UK, and the Naval Research Laboratory, Washington, DC.

Appendix: Algorithm for Data Association Experiment

The following algorithm demonstrates how $O(n^3)$ scaling can be obtained for approximating the JAM of an $n \times n$ association matrix using inequality ε_3. A straightforward implementation that scales as $O(n^4)$ can be obtained easily; the key to removing a factor of n comes from the fact that from a precomputed product of row sums, *rprod*, the product of all row sums excluding row i is just *rprod* divided by row sum i. In other words, performing an $O(n)$ step of explicitly computing the product of all row sums except row i is not necessary. Some care must be taken to accommodate row sums that are zero, but the following pseudocode shows that there is little extra overhead incurred by the more efficient implementation. Similar techniques lead to the advertised scaling for the other inequality approaches. (Note, however, that sums of logarithms should be used in place of explicit products to ensure numerical stability in actual implementations.)

$E3r(M, P, n)$

> r and c are vectors of length n corresponding to the row and column sums, respectively. *rn* is a vector of length n of normalized row sums.
>
> for $i = 1$ to n: $r_i \leftarrow c_i \leftarrow 0.0$
>
> Apply iterative renormalization to M
>
> for $i = 1$ to n:
>
>> for $j = 1$ to n:
>>
>>> $r_i \leftarrow r_i + M_{ij}$
>>>
>>> $c_i \leftarrow c_i + M_{ij}$
>>
>> end
>
> end

```
for i = 1 to n:
    for l = 1 to n:
        rn_l ← 0.0
        for k = 1 to n, if (c_k - M_ik) > 0.0
            then rn_l ← rn_l + M_ik/(c_k - M_ik)
    end
    for j = 1 to n:
        nprod = 1.0
        for k = 1 to n, if k ≠ j
            then nprod ← nprod* (c_k - M_ik)
        rprod ← cprod ← 1.0
        for k = 1 to n, if k ≠ 1
            if (c_j - M_ij) > 0.0
                then rprod ← rprod* (rn_k - M_kj)/(c_j - M_ij)
            else rprod ← rprod* rn_k
            end
        rprod ← rprod* nprod
        P_ij ← M_ij* rprod
    end
end
Apply iterative renormalization to P
end.
```

References

1. Minc, H., *Permanents, Encyclopedia of Mathematics and its Applications*, 4th edition, Vol. 6, Part F, Addison-Wesley, New York, NY, 1978.
2. Blackman, S., *Multiple-Target Tracing with Radar Applications*, Artech House, Dedham, MA, 1986.
3. Blackman, S. and Popoli, R., *Design and Analysis of Modern Tracking Systems*, Artech House, Norwood, MA, 1999.
4. Bar-Shalom, Y. and Fortmann, T.E., *Tracking and Data Association*, Academic Press, New York, NY, 1988.
5. Collins, J.B. and Uhlmann, J.K., Efficient gating in data association for multivariate Gaussian distributions, *IEEE Transactions on Aerospace and Electronic Systems*, 28(3), 909–916, 1990.
6. Ryser, H.J., *Combinatorial Mathematics*, 14, Carus Mathematical Monograph Series, Mathematical Association of America, 1963.
7. Uhlmann, J.K., Algorithms for multiple-target tracking, *American Scientist*, 80(2), 128–141, 1992.
8. Papadimitrious, C.H. and Steiglitz, K., *Combinatorial Optimization Algorithms and Complexity*, Prentice Hall, Englewood Cliffs, NJ, 1982.
9. Cox, I.J. and Miller, M.L., On finding ranked assignments with application to multitarget tracking and motion correspondence, *IEEE Transactions on Aerospace and Electronic Systems*, 31, 486–489, 1995.

10. Nagarajan, V., Chidambara, M.R., and Sharma, R.M., New approach to improved detection and tracking in track-while-scan radars, Part 2: Detection, track initiation, and association, *IEEE Proceedings*, 134(1), 89–92, 1987.

11. Brualdi, R.A. and Ryser, H.J., *Combinatorial Matrix Theory*, Cambridge University Press, Cambridge, UK, 1992.

12. Valiant, L.G., The complexity of computing the permanent, *Theoretical Computer Science*, 8, 189–201, 1979.

13. Nijenhuis, A. and Wilf, H.S., *Combinatorial Algorithms*, 2nd edition, Academic Press, New York, NY, 1978.

14. Uhlmann, J.K., *Dynamic Localization and Map Building: New Theoretical Foundations*, PhD thesis, University of Oxford, 1995.

15. Karmarkar, N. et al., A Monte Carlo algorithm for estimating the permanent, *SIAM Journal on Computing*, 22(2), 284–293, 1993.

16. Jerrum, M. and Vazirani, U., *A Mildly Exponential Approximation Algorithm for the Permanent*, Technical Report, Department of Computer Science, University of Edinburgh, Scotland, 1991. ECS-LFCS-91–179.

17. Jerrum, M. and Sinclair, A., Approximating the permanent, *SIAM Journal on Computing*, 18(6), 1149–1178, 1989.

18. Jurkat, W.B. and Ryser, H.J., Matrix factorizations of determinants and permanents, *Journal of Algebra*, 3, 1–27, 1966.

19. Reid, D.B., An algorithm for tracking multiple targets, *IEEE Transactions on Automatic Control*, AC-24(6), 843–854, 1979.

20. Cox, I.J. and Leonard, J.J., Modeling a dynamic environment using a Bayesian multiple hypothesis approach, *Artificial Intelligence*, 66(2), 311–344, 1994.

21. Bar-Shalom, Y. and Li, X.R., *Multitarget-Multisensor Tracking: Principles and Techniques*, YBS Press, Storrs, CT, 1995.

12

Bayesian Approach to Multiple-Target Tracking*

Lawrence D. Stone

CONTENTS

---

* This chapter is based on Stone, L.D., Barlow, C.A., and Corwin, T.L., *Bayesian Multiple Target Tracking*, Artech House, Norwood, MA, 1999. www.artechhouse.com.

12.1   Introduction

This chapter views the multiple-target tracking problem as a Bayesian inference problem and highlights the benefits of this approach. The goal of this chapter is to provide the reader with some insights and perhaps a new view of the multiple-target tracking. It is not designed to provide the reader with a set of algorithms for multiple-target tracking.

The chapter begins with a Bayesian formulation of the single-target tracking problem and then extends this formulation to multiple targets. It then discusses some of the interesting consequences of this formulation, including

- A mathematical formulation of the multiple-target tracking problem with a minimum of complications and formalisms

- The emergence of likelihood functions as a generalization of the notion of contact and as the basic currency for valuing and combining information from disparate sensors

- A general Bayesian formula for calculating association probabilities
- A method, called unified tracking, for performing multiple-target tracking when the notions of contact and association are not meaningful
- A delineation of the relationship between multiple-hypothesis tracking (MHT) and unified tracking
- A Bayesian track-before-detect methodology called likelihood ratio detection and tracking

12.1.1 Definition of Bayesian Approach

To appreciate the discussion in this chapter, the reader must first understand the concept of Bayesian tracking. For a tracking system to be considered Bayesian, it must have the following characteristics:

- *Prior distribution.* There must be a prior distribution on the state of the targets. If the targets are moving, the prior distribution must include a probabilistic description of the motion characteristics of the targets. Usually the prior is given in terms of a stochastic process for the motion of the targets.
- *Likelihood functions.* The information in sensor measurements, observations, or contacts must be characterized by likelihood functions.
- *Posterior distribution.* The basic output of a Bayesian tracker is a posterior probability distribution on the (joint) state of the target(s). The posterior at time t is computed by combining the motion-updated prior at time t with the likelihood function for the observation(s) received at time t.

These are the basics: prior, likelihood functions, and posterior. If these are not present, the tracker is not Bayesian. The recursions given in this chapter for performing Bayesian tracking are all *recipes* for calculating priors, likelihood functions, and posteriors.

12.1.2 Relationship to Kalman Filtering

Kalman filtering resulted from viewing tracking as a least-squares problem and finding a recursive method of solving that problem. One can think of many standard tracking solutions as methods for minimizing mean squared errors. Chapters 1–3 of Ref. 1 give an excellent discussion on tracking from this point of view. One can also view Kalman filtering as Bayesian tracking. To do this, one starts with a prior that is Gaussian in the appropriate state space with a *very large* covariance matrix. Contacts are measurements that are linear functions of the target state with Gaussian measurement errors. These are interpreted as Gaussian likelihood functions and combined with motion-updated priors to produce posterior distributions on target state. Because the priors are Gaussian and the likelihood functions are Gaussian, the posteriors are also Gaussian. When doing the algebra, one finds that the mean and covariance of the posterior Gaussian are identical to the mean and covariance of the least-squares solution produced by the Kalman filter. The difference is that from the Bayesian point of view, the mean and covariance matrices represent posterior Gaussian distributions on a target state. Plots of the mean and characteristic ellipses are simply shorthand representations of these distributions.

Bayesian tracking is not simply an alternate way of viewing Kalman filtering. Its real value is demonstrated when some of the assumptions required for Kalman filtering are not satisfied. Suppose the prior distribution on target motion is not Gaussian, or the measurements are not linear functions of the target state, or the measurement error is not Gaussian. Suppose that multiple sensors are involved and are quite different. Perhaps they produce measurements that are not even in the target state space. This can happen if, for example, one of the measurements is the observed signal-to-noise ratio at a sensor. Suppose that one has to deal with measurements that are not even contacts (e.g., measurements that are so weak that they fall below the threshold at which one would call a contact). Tracking problems involving these situations do not fit well into the mean squared error paradigm or the Kalman filter assumptions. One can often stretch the limits of Kalman filtering by using linear approximations to nonlinear measurement relations or by other nonlinear extensions. Often these extensions work very well. However, there does come a point where these extensions fail. That is where Bayesian filtering can be used to tackle these more difficult problems. With the advent of high-powered and inexpensive computers, the numerical hurdles to implementing Bayesian approaches are often easily surmounted. At the very least, knowing how to formulate the solution from the Bayesian point of view will allow one to understand and choose wisely the approximations needed to put the problem into a more tractable form.

12.2   Bayesian Formulation of Single-Target Tracking Problem

This section presents a Bayesian formulation of single-target tracking and a basic recursion for performing single-target tracking.

12.2.1   Bayesian Filtering

Bayesian filtering is based on the mathematical theory of probabilistic filtering described by Jazwinski.[2] Bayesian filtering is the application of Bayesian inference to the problem of tracking a single target. This section considers the situation where the target motion is modeled in continuous time, but the observations are received at discrete, possibly random, times. This is called continuous-discrete filtering by Jazwinski.

12.2.2   Problem Definition

The single-target tracking problem assumes that there is one target present in the state space; as a result, the problem becomes one of estimating the state of that target.

12.2.2.1   Target State Space

Let S be the state space of the target. Typically, the target state will be a vector of components. Usually some of these components are kinematic and include position, velocity, and possibly acceleration. Note that there may be constraints on the components, such as a maximum speed for the velocity component. There can be additional components that may be related to the identity or other features of the target. For example, if one of the components specifies target type, then that may also specify information such as radiated noise levels at various frequencies and motion characteristics (e.g., maximum speeds). To use the recursion presented in this section, there are additional requirements on the target state space. The state space must be rich enough that (1) the target's motion is Markovian in the chosen state space and (2) the sensor likelihood functions depend only on the state of the target at the time of the observation. The sensor likelihood functions depend on the characteristics of the sensor, such as its position and measurement error distribution that are assumed to be known. If they are not known, they need to be determined by experimental or theoretical means.

12.2.2.2 Prior Information

Let $X(t)$ be the (unknown) target state at time t. We start the problem at time 0 and are interested in estimating $X(t)$ for $t \geq 0$. The prior information about the target is represented by a stochastic process $\{X(t); t \geq 0\}$. Sample paths of this process correspond to possible target paths through the state space, S. The state space S has a measure associated with it. If S is discrete, this measure is a discrete measure. If S is continuous (e.g., if S is equal to the plane), this measure is represented by a density. The measure on S can be a mixture or product of discrete and continuous measures. Integration with respect to this measure will be indicated by ds. If the measure is discrete, then the integration becomes summation.

12.2.2.3 Sensors

There is a set of sensors that report observations at an ordered, discrete sequence of (possibly random) times. These sensors may be of different types and report different information. The set can include radar, sonar, infrared, visual, and other types of sensors. The sensors may report only when they have a contact or are on a regular basis. Observations from sensor j take values in the measurement space H_j. Each sensor may have a different measurement space. The probability distribution of each sensor's response conditioned on the value of the target state s is assumed to be known. This relationship is captured in the likelihood function for that sensor. The relationship between the sensor response and the target state s may be linear or nonlinear, and the probability distribution representing measurement error may be Gaussian or non-Gaussian.

12.2.2.4 Likelihood Functions

Suppose that by time t observations have been obtained at the set of times $0 \leq t_1 \leq \ldots \leq t_K \leq t$. To allow for the possibility that more than one sensor observation may be received at a given time, let Y_k be the set of sensor observations received at time t_k. Let y_k denote a value of the random variable Y_k. Assume that the likelihood function can be computed as

$$L_k(y_k|s) = \mathbf{Pr}\{Y_k = y_k|X(t_k) = s\} \quad \text{for } s \in S \tag{12.1}$$

The computation in Equation 12.1 can account for correlation among sensor responses. If the distribution of the set of sensor observations at time t_k is independent of the given target state, then $L_k(y_k|s)$ is computed by taking the product of the probability (density) functions for each observation. If they are correlated, then one must use the joint density function for the observations conditioned on target state to compute $L_k(y_k|s)$.

Let $\mathbf{Y}(t) = (Y_1, Y_2, \ldots, Y_K)$ and $\mathbf{y} = (y_1, \ldots, y_K)$. Define $L(\mathbf{y}|s_1, \ldots, s_K) = \mathbf{Pr}\{\mathbf{Y}(t) = \mathbf{y}|X(t_1) = s_1, \ldots, X(t_K) = s_K\}$.

Assume

$$\mathbf{Pr}\{\mathbf{Y}(t) = \mathbf{y}|X(u) = s(u), \quad 0 \leq u \leq t\} = L(\mathbf{y}|s(t_1), \ldots, s(t_K)) \tag{12.2}$$

Equation 12.2 means that the likelihood of the data $\mathbf{Y}(t)$ received through time t depends only on the target states at the times $\{t_1, \ldots, t_K\}$ and not on the whole target path.

12.2.2.5 Posterior

Define $q(s_1, \ldots, s_K) = \mathbf{Pr}\{X(t_1) = s_1, \ldots, X(t_K) = s_K\}$ to be the prior probability (density) that the process $\{X(t); t \geq 0\}$ passes through the states $s_1, \ldots, s_K$ at times $t_1, \ldots, t_K$. Let $p(t_K, s_K) = \mathbf{Pr}\{X(t_K) = s_K | \mathbf{Y}(t_K) = \mathbf{y}\}$. Note that the dependence of p on $\mathbf{y}$ has been suppressed. The function $p(t_K, \cdot)$ is the posterior distribution on $X(t_K)$ given $\mathbf{Y}(t_K) = \mathbf{y}$. In mathematical terms, the problem is to compute this posterior distribution. Recall that from the point of view of Bayesian inference, the posterior distribution on target state represents our knowledge of the target state. All estimates of target state derive from this posterior.

12.2.3 Computing the Posterior

Compute the posterior by the use of Bayes' theorem as follows:

$$p(t_k, s_K) = \frac{\mathrm{Pr}\{\mathbf{Y}(t_K) = \mathbf{y} \quad \text{and} \quad X(t_K) = s_K\}}{\mathrm{Pr}\{\mathbf{Y}(t_K) = \mathbf{y}\}}$$

$$= \frac{\int L(\mathbf{y}|s_1, \ldots, s_K) q(s_1, s_2, \ldots, s_K) ds_1 ds_2 \cdots ds_{K-1}}{\int L(\mathbf{y}|s_1, \ldots, s_K) q(s_1, s_2, \ldots, s_K) ds_1 ds_2 \cdots ds_K} \qquad (12.3)$$

Computing $p(t_K, s_K)$ can be quite difficult. The method of computation depends on the functional forms of q and L. The two most common ways are batch computation and a recursive method.

12.2.3.1 Recursive Method

Two additional assumptions about q and L permit recursive computation of $p(t_K, s_K)$. First, the stochastic process $\{X(t; t \geq 0\}$ must be Markovian on the state space S. Second, for $i \neq j$, the distribution of $Y(t_i)$ must be independent of $Y(t_j)$ given $(X(t_1) = s_1, \ldots, X(t_K) = s_K)$ so that

$$L(\mathbf{y}|s_1, \ldots, s_K) = \prod_{k=1}^{K} L_k(y_k|s_k) \qquad (12.4)$$

The assumption in Equation 12.4 means that the sensor responses (or observations) at time t_k depend only on the target state at the time t_k. This is not automatically true. For example, if the target state space is only position and the observation is a velocity measurement, this observation will depend on the target state over some time interval near t_k. The remedy in this case is to add velocity to the target state space. There are other observations, such as failure of a sonar sensor to detect an underwater target over a period of time, for which the remedy is not so easy or obvious. This observation may depend on the whole past history of target positions and, perhaps, velocities.

Define the transition function $q_k(s_k|s_{k-1}) = \mathbf{Pr}\{X(t_k) = s_k | X(t_{k-1}) = s_{k-1}\}$ for $k \geq 1$, and let q_0 be the probability (density) function for $X(0)$. By the Markov assumption

$$q(s_1, \ldots, s_K) = \int_S \prod_{k=1}^{K} q_k(s_k|s_{k-1}) q_0(s_0) ds_0 \qquad (12.5)$$

12.2.3.2 Single-Target Recursion

Applying Equations 12.4 and 12.5 to 12.3 results in the basic recursion for single-target tracking.

Basic recursion for single-target tracking

$$\text{Initialize distribution: } p(t_0, s_0) = q_0(s_0) \quad \text{for } s_0 \in S \tag{12.6}$$

For $k \geq 1$ and $s_k \in S$,

$$\text{Perform motion update: } p^-(t_k, s_k) = \int q_k(s_k|s_{k-1}) p(t_{k-1}, s_{k-1}) ds_{k-1} \tag{12.7}$$

Compute likelihood function: L_k from the observation $Y_k = y_k$

$$\text{Perform information update: } p(t_k, s_k) = \frac{1}{C} L_k(y_k|s_k) p^-(t_k, s_k) \tag{12.8}$$

The motion update in Equation 12.7 accounts for the transition of the target state from time t_{k-1} to t_k. Transitions can represent not only the physical motion of the target, but also the changes in other state variables. The information update in Equation 12.8 is accomplished by pointwise multiplication of $p^-(t_k, s_k)$ by the likelihood function $L_k(y_k|s_k)$. Likelihood functions replace and generalize the notion of contacts in this view of tracking as a Bayesian inference process. Likelihood functions can represent sensor information such as detections, no detections, Gaussian contacts, bearing observations, measured signal-to-noise ratios, and observed frequencies of a signal. Likelihood functions can also represent and incorporate information in situations where the notion of a contact is not meaningful. Subjective information also can be incorporated by using likelihood functions. Examples of likelihood functions are provided in Section 12.2.4. If there has been no observation at time t_k, then there is no information update, only a motion update.

The above recursion does not require the observations to be linear functions of the target state. It does not require the measurement errors or the probability distributions on target state to be Gaussian. Except in special circumstances, this recursion must be computed numerically. Today's high-powered scientific workstations can compute and display tracking solutions for complex nonlinear trackers. To do this, discretize the state space and use a Markov chain model for target motion so that Equation 12.7 is computed through the use of discrete transition probabilities. The likelihood functions are also computed on the discrete state space. A numerical implementation of a discrete Bayesian tracker is described in Section 3.3 of Ref. 3.

12.2.4 Likelihood Functions

The use of likelihood functions to represent information is at the heart of Bayesian tracking. In the classical view of tracking, contacts are obtained from sensors that provide estimates of (some components of) the target state at a given time with a specified measurement error. In the classic Kalman filter formulation, a measurement (contact) Y_k at time t_k satisfies the measurement equation

$$Y_k = \mathbf{M}_k X(t_k) + \varepsilon_k \tag{12.9}$$

where Y_k is an r-dimensional real column vector, $X(t_k)$ an l-dimensional real column vector, $\mathbf{M}_k$ an $r \times l$ matrix, and $\varepsilon_k \sim N(0, \Sigma_k)$.

Note that $\sim N(\mu, \Sigma)$ means "has a Normal (Gaussian) distribution with mean μ and covariance Σ." In this case, the measurement is a linear function of the target state and the measurement error is Gaussian. This can be expressed in terms of a likelihood function as follows. Let $L_G(y|x) = \Pr\{Y_k = y | X(t_k) = x\}$. Then

$$L_G(y|x) = (2\pi)^{-r/2} \left| \det \sum\nolimits_k \right|^{-1/2} \exp\left(-\frac{1}{2}(y - \mathbf{M}_k x)^T \sum\nolimits^{-1} (y - \mathbf{M}_k x) \right) \qquad (12.10)$$

Note that the measurement y is data that is known and fixed. The target state x is unknown and varies, so that the likelihood function is a function of the target state variable x. Equation 12.10 looks the same as a standard elliptical contact, or estimate of target state, expressed in the form of multivariate normal distribution, commonly used in Kalman filters. There is a difference, but it is obscured by the symmetrical positions of y and $\mathbf{M}_k x$ in the Gaussian density in Equation 12.10. A likelihood function does not represent an estimate of the target state. It looks at the situation in reverse. For each value of target state x, it calculates the probability (density) of obtaining the measurement y given that the target is in state x. In most cases, likelihood functions are not probability (density) functions on the target state space. They need not integrate to one over the target state space. In fact, the likelihood function in Equation 12.10 is a probability density on the target state space only when Y_k is l-dimensional and $\mathbf{M}_k$ is an $l \times l$ matrix.

Suppose one wants to incorporate into a Kalman filter information such as a bearing measurement, speed measurement, range estimate, or the fact that a sensor did or did not detect the target. Each of these is a nonlinear function of the normal Cartesian target state. Separately, a bearing measurement, speed measurement, and range estimate can be handled by forming linear approximations and assuming Gaussian measurement errors or by switching to special non-Cartesian coordinate systems in which the measurements are linear and hopefully the measurement errors are Gaussian. In combining all this information into one tracker, the approximations and the use of disparate coordinate systems become more problematic and dubious. In contrast, the use of likelihood functions to incorporate all this information (and any other information that can be put into the form of a likelihood function) is quite straightforward, no matter how disparate the sensors or their measurement spaces. Section 12.2.4.1 provides a simple example of this process involving a line-of-bearing measurement and a detection.

12.2.4.1   Line-of-Bearing Plus Detection Likelihood Functions

Suppose that there is a sensor located in the plane at (70, 0) and that it has produced a detection. For this sensor the probability of detection is a function, $P_d(r)$, of the range r from the sensor. Take the case of an underwater sensor such as an array of acoustic hydrophones and a situation where the propagation conditions produce convergence zones of high detection performance that alternate with ranges of poor detection performance. The observation (measurement) in this case is $Y = 1$ for detection and 0 for no detection. The likelihood function for detection is $L_d(1|x) = P_d(r(x))$, where $r(x)$ is the range from the state x to the sensor. Figure 12.1 shows the likelihood function for this observation.

Suppose that, in addition to the detection, there is a bearing measurement of 135° (measured counterclockwise from the x_1-axis) with a Gaussian measurement error having mean 0 and standard deviation 15°. Figure 12.2 shows the likelihood function for this

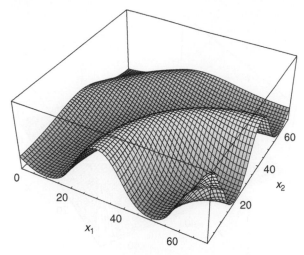

FIGURE 12.1
Detection likelihood function for a sensor at (70, 0).

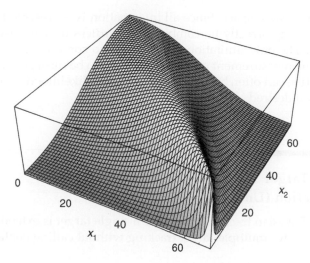

FIGURE 12.2
Bearing likelihood function for a sensor at (70, 0).

observation. Notice that, although the measurement error is Gaussian in bearing, it does not produce a Gaussian likelihood function on the target state space. Furthermore, this likelihood function would integrate to infinity over the whole state space. The information from these two likelihood functions is combined by pointwise multiplication. Figure 12.3 shows the likelihood function that results from this combination.

12.2.4.2 Combining Information Using Likelihood Functions

Although the example of combining likelihood functions presented in Section 12.2.4.1 is simple, it illustrates the power of using likelihood functions to represent and combine information. A likelihood function converts the information in a measurement into a

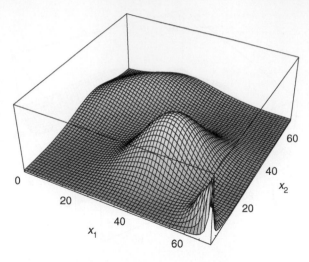

FIGURE 12.3
Combination of bearing and detection likelihood functions.

function on the target state space. Since all information is represented on the same state space, it can easily and correctly be combined, regardless of how disparate the sources of the information. The only limitation is the ability to compute the likelihood function corresponding to the measurement or the information to be incorporated. As an example, subjective information can often be put into the form of a likelihood function and incorporated into a tracker if desired.

12.3 Multiple-Target Tracking without Contacts or Association (Unified Tracking)

In this section, the Bayesian tracking model for a single target is extended to multiple targets in a way that allows multiple-target tracking without calling contacts or performing data association.

12.3.1 Multiple-Target Motion Model

In Section 12.2, the prior knowledge about the single target's state and its motion through the target state space S were represented in terms of a stochastic process $\{X(t); t \geq 0\}$ where $X(t)$ is the target state at time t. This motion model is now generalized to multiple targets.

Begin the multiple-target tracking problem at time $t = 0$. The total number of targets is unknown but bounded by $\bar{N}$, which is known. We assume a known bound on the number of targets because it allows us to simplify the presentation and produces no restriction in practice. Designate a region, R, which defines the boundary of the tracking problem. Activity outside of R has no importance. For example, we might be interested in targets having only a certain range of speeds or contained within a certain geographic region.

Add an additional state φ to the target state space S. If a target is not in the region R, it is considered to be in state φ. Let $S^{+} = S \cup \{\varphi\}$ be the extended state space for a single target and $\mathbf{S}^{+} = S^{+} \times \cdots \times S^{+}$ be the joint target state space where the product is taken $\bar{N}$ times.

12.3.1.1 Multiple-Target Motion Process

Prior knowledge about the targets and their *movements* through the state space $\mathbf{S}^+$ is expressed as a stochastic process $\mathbf{X} = \{X(t); t \geq 0\}$. Specifically, let $\mathbf{X}(t) = (X_1(t), \ldots, X_N(t))$ be the state of the system at time t where $X_n(t) \in S^+$ is the state of target n at time t. The term *state of the system* is used to mean the joint state of all the targets. The value of the random variable $X_n(t)$ indicates whether target n is present in R and, if so, in what state. The number of components of $\mathbf{X}(t)$ with states not equal to ϕ at time t gives the number of targets present in R at time t. Assume that the stochastic process $\mathbf{X}$ is Markovian in the state space $\mathbf{S}^+$ and that the process has an associated transition function. Let $q_k(s_k|s_{k-1}) = \mathbf{Pr}\{\mathbf{X}(t_k) = \mathbf{s}_k|\mathbf{X}(t_{k-1}) = s_{k-1}\}$ for $k \geq 1$, and let q_0 be the probability (density) function for $\mathbf{X}(0)$. By the Markov assumption

$$\mathbf{Pr}\{\mathbf{X}(t_1) = \mathbf{s}_1, \ldots, \mathbf{X}(t_K) = \mathbf{s}_K\} = \int \prod_{k=1}^{K} q_k(\mathbf{s}_k|\mathbf{s}_{k-1})q_0(\mathbf{s}_0)ds_0 \qquad (12.11)$$

The state space $\mathbf{S}^+$ of the Markov process $\mathbf{X}$ has a measure associated with it. If the process $\mathbf{S}^+$ is a discrete space Markov chain, then the measure is discrete and integration becomes summation. If the space is continuous, then functions such as transition functions become densities on $\mathbf{S}^+$ with respect to that measure. If $\mathbf{S}^+$ has both continuous and discrete components, then the measure will be the product or mixture of discrete and continuous measures. The symbol $d\mathbf{s}$ will be used to indicate integration with respect to the measure on $\mathbf{S}^+$, whether it is discrete or not. When the measure is discrete, the integrals become summations. Similarly, the notation $\mathbf{Pr}$ indicates either probability or probability density as appropriate.

12.3.2 Multiple-Target Likelihood Functions

There is a set of sensors that report observations at a discrete sequence of possibly random times. These sensors may be of different types and may report different information. The sensors may report only when they have a contact or on a regular basis. Let $Z(t, j)$ be an observation from sensor j at time t. Observations from sensor j take values in the measurement space H_j. Each sensor may have a different measurement space.

For each sensor j, assume that one can compute

$$\mathbf{Pr}\{Z(t, j) = z|\mathbf{X}(t) = \mathbf{s}\} \quad \text{for } z \in H_j \text{ and } \mathbf{s} \in \mathbf{S}^+ \qquad (12.12)$$

To compute the probabilities in Equation 12.12, one must know the distribution of the sensor response conditioned on the value of the state $\mathbf{s}$. In contrast to Section 12.2, the likelihood functions in this section can depend on the joint state of all the targets. The relationship between the observation and the state $\mathbf{s}$ may be linear or nonlinear, and the probability distribution may be Gaussian or non-Gaussian.

Suppose that by time t, observations have been obtained at the set of discrete times $0 \leq t_1 \leq \ldots \leq t_K \leq t$. To allow for the possibility of receiving more than one sensor observation at a given time, let Y_k be the set of sensor observations received at time t_k. Let y_k denote a value of the random variable Y_k. Extend Equation 12.12 to assume that the following computation can be made

$$L_k(y_k|\mathbf{s}) = \mathbf{Pr}\{Y_k = y_k|\mathbf{X}(t_k) = \mathbf{s}\} \quad \text{for } \mathbf{s} \in \mathbf{S}^+ \qquad (12.13)$$

$L_k(y_k|\cdot)$ is called the *likelihood function* for the observation $Y_k = y_k$. The computation in Equation 12.13 can account for correlation among sensor responses if required.

Let $\mathbf{Y}(t) = (Y_1, Y_2, ..., Y_K)$ and $\mathbf{y} = (y_1, ..., y_K)$. Define $L(\mathbf{y}|\mathbf{s}_1, ..., \mathbf{s}_K) = \Pr\{\mathbf{Y}(t) = \mathbf{y}|\mathbf{X}(t_1) = \mathbf{s}_1, ..., \mathbf{X}(t_K) = \mathbf{s}_K\}$.

In parallel with Section 12.2, assume that

$$\Pr\{\mathbf{Y}(t) = \mathbf{y}|\mathbf{X}(u) = \mathbf{s}(u), 0 \le u \le t\} = L(\mathbf{y}|\mathbf{s}(t_1), ..., \mathbf{s}(t_k)) \tag{12.14}$$

and

$$L(\mathbf{y}|\mathbf{s}_1, ..., \mathbf{s}_K) = \prod_{k=1}^{K} L_k(y_k|\mathbf{s}_k) \tag{12.15}$$

Equation 12.14 assumes that the distribution of the sensor response at the times $\{t_k, k = 1, ..., K\}$ depends only on the system states at those times. Equation 12.15 assumes independence of the sensor response distributions across the observation times. The effect of both assumptions is to assume that the sensor response at time t_k depends only on the system state at that time.

12.3.3 Posterior Distribution

For unified tracking, the tracking problem is equivalent to computing the posterior distribution on $\mathbf{X}(t)$ given $\mathbf{Y}(t)$. The posterior distribution of $\mathbf{X}(t)$ represents our knowledge of the number of targets present and their state at time t given $\mathbf{Y}(t)$. From this distribution point, estimates can be computed, when appropriate, such as maximum *a posteriori* probability estimates or means. Define $q(\mathbf{s}_1, ..., \mathbf{s}_K) = \Pr\{\mathbf{X}(t_1) = \mathbf{s}_1, ..., \mathbf{X}(t_K) = \mathbf{s}_K\}$ to be the prior probability (density) that the process $\mathbf{X}$ passes through the states $\mathbf{s}_1, ..., \mathbf{s}_K$ at times $t_1, ..., t_K$. Let q_0 be the probability (density) function for $\mathbf{X}(0)$. By the Markov assumption

$$q(\mathbf{s}_1, ..., \mathbf{s}_K) = \int \prod_{k=1}^{K} q_k(\mathbf{s}_k|\mathbf{s}_{k-1})q_0(\mathbf{s}_0)d\mathbf{s}_0 \tag{12.16}$$

Let $p(t, \mathbf{s}) = \Pr\{\mathbf{X}(t) = \mathbf{s}|\mathbf{Y}(t)\}$. The function $p(t, \cdot)$ gives the posterior distribution on $\mathbf{X}(t)$ given $\mathbf{Y}(t)$. By Bayes' theorem,

$$\begin{aligned} p(t_K, \mathbf{s}_K) &= \frac{\Pr\{\mathbf{Y}(t_K) = \mathbf{y} \quad \text{and} \quad \mathbf{X}(t_K) = \mathbf{s}_K\}}{\Pr\{\mathbf{Y}(t_K) = \mathbf{y}\}} \\ &= \frac{\int L(\mathbf{y}|\mathbf{s}_1, ..., \mathbf{s}_K)q(\mathbf{s}_1, \mathbf{s}_2, ..., \mathbf{s}_K)d\mathbf{s}_1 \cdots d\mathbf{s}_{K-1}}{\int L(\mathbf{y}|\mathbf{s}_1, ..., \mathbf{s}_K)q(\mathbf{s}_1, \mathbf{s}_2, ..., \mathbf{s}_K)d\mathbf{s}_1 \cdots d\mathbf{s}_K} \end{aligned} \tag{12.17}$$

12.3.4 Unified Tracking Recursion

Substituting Equations 12.15 and 12.16 into Equations 12.17 gives

$$\begin{aligned} p(t_K, \mathbf{s}_K) &= \frac{1}{C'} \int \prod_{k=1}^{K} L_k(y_k|\mathbf{s}_k) \prod_{k=1}^{K} q_k(\mathbf{s}_k|\mathbf{s}_{k-1})q_0(\mathbf{s}_0)d\mathbf{s}_0 \cdots d\mathbf{s}_{K-1} \\ &= \frac{1}{C'} L_K(y_K|\mathbf{s}_K) \int q_K(\mathbf{s}_K|\mathbf{s}_{K-1}) \\ &\quad \times \left[\int \prod_{k=1}^{K-1} L_k(y_k|\mathbf{s}_k)q_k(\mathbf{s}_k|\mathbf{s}_{k-1})q_0(\mathbf{s}_0)\,d\mathbf{s}_0 \cdots d\mathbf{s}_{K-2} \right] d\mathbf{s}_{K-1} \end{aligned}$$

and

$$p(t_K, \mathbf{s}_K) = \frac{1}{C} L_K(y_K|\mathbf{s}_K) \int q_K(\mathbf{s}_K|\mathbf{s}_{K-1}) p(t_{K-1}, \mathbf{s}_{K-1}) d\mathbf{s}_{K-1} \qquad (12.18)$$

where C and C' normalize $p(t_K, \cdot)$ to be a probability distribution. Equation 12.18 provides a recursive method of computing $p(t_K, \cdot)$. Specifically,

Unified tracking recursion

$$\text{Initialize distribution: } p(t_0, \mathbf{s}_0) = q_0(\mathbf{s}_0) \quad \text{for } \mathbf{s}_0 \in \mathbf{S}^+ \qquad (12.19)$$

For $k \geq 1$ and $\mathbf{s}_k \in \mathbf{S}^+$,

$$\text{Perform motion update: } p^-(t_k, \mathbf{s}_k) = \int q_k(\mathbf{s}_k|\mathbf{s}_{k-1}) p(t_{k-1}, \mathbf{s}_{k-1}) d\mathbf{s}_{k-1} \qquad (12.20)$$

Compute likelihood function: L_k from the observation $Y_k = y_k$

$$\text{Perform information update: } p(t_k, \mathbf{s}_k) = \frac{1}{C} L_k(y_k|\mathbf{s}_k) p^-(t_k, \mathbf{s}_k) \qquad (12.21)$$

12.3.4.1 Multiple-Target Tracking without Contacts or Association

The unified tracking recursion appears deceptively simple. The difficult part is performing the calculations in the joint state space of the $\bar{N}$ targets. Having done this, the combination of the likelihood functions defined on the joint state space with the joint distribution function of the targets automatically accounts for all possible association hypotheses without requiring explicit identification of these hypotheses. Section 12.4 demonstrates that this recursion produces the same joint posterior distribution as MHT does when the conditions for MHT are satisfied. However, the unified tracking recursion goes beyond MHT. One can use this recursion to perform multiple-target tracking when the notions of contact and association (notions required by MHT) are not meaningful. Examples of this are given in Section 5.3 of Ref. 3. Another example by Finn[4] applies to tracking two aircraft targets with a monopulse radar when the aircrafts become so close together in bearing that their signals become unresolved. They merge inextricably at the radar receiver.

12.3.4.1.1 Merged Measurements

The problem tackled by Finn[4] is an example of the difficulties caused by merged measurements. A typical example of merged measurements is when a sensor's received signal is the sum of the signals from all the targets present. This can be the case with a passive acoustic sensor. Fortunately, in many cases the signals are separated in space or frequency so that they can be treated as separate signals. In some cases, two targets are so close in space (and radiated frequency) that it is impossible to distinguish which component of the received signal is due to which target. This is a case when the notion of associating a contact to a target is not well defined. Unified tracking will handle this problem correctly, but the computational load may be too onerous. In this case an MHT algorithm with special approximations could be used to provide an approximate but computationally feasible solution. See, for example, Ref. 5.

Section 12.4 presents the assumptions that allow contact association and multiple-target tracking to be performed by using MHT.

12.3.4.2 Summary of Assumptions for Unified Tracking Recursion

In summary, the assumptions required for the validity of the unified tracking recursion are

1. The number of targets is unknown but bounded by $\bar{N}$.
2. $S^+ = S \cup \{\phi\}$ is the extended state space for a single target where ϕ indicates the target is not present. $X_n(t) \in S^+$ is the state of the nth target at time t.
3. $\mathbf{X}(t) = (X_1(t), \ldots, X_N(t))$ is the state of the system at time t, and $\mathbf{X} = \{\mathbf{X}(t); t \geq 0\}$ is the stochastic process describing the evolution of the system over time. The process, $\mathbf{X}$, is Markov in the state space $\mathbf{S}^+ = S^+ \times \cdots \times S^+$ where the product is taken $\bar{N}$ times.
4. Observations occur at discrete (possibly random) times, $0 \leq t_1 \leq t_2 \ldots$. Let $Y_k = y_k$ be the observation at time t_k, and let $\mathbf{Y}(t_K) = \mathbf{y}_K = (y_1, \ldots, y_K)$ be the first K observations. Then the following is true

$$\mathbf{Pr}\{\mathbf{Y}(t_K) = \mathbf{y}_K | \mathbf{X}(u) = \mathbf{s}(u), 0 \leq u \leq t_K\}$$

$$= \mathbf{Pr}\{\mathbf{Y}(t_K) = \mathbf{y}_K | \mathbf{X}(t_K) = \mathbf{s}(t_K), k = 1, \ldots, K\}$$

$$= \prod_{k=1}^{K} L_k(y_k | \mathbf{s}(t_K))$$

12.4 Multiple-Hypothesis Tracking

In classical multiple-target tracking, the problem is divided into two steps: (1) association and (2) estimation. Step 1 associates contacts with targets. Step 2 uses the contacts associated with each target to produce an estimate of that target's state. Complications arise when there is more than one reasonable way to associate contacts with targets. The classical approach to this problem is to form association hypotheses and to use MHT, which is the subject of this section. In this approach, alternative hypotheses are formed to explain the source of the observations. Each hypothesis assigns observations to targets or false alarms. For each hypothesis, MHT computes the probability that it is correct. This is also the probability that the target state estimates that result from this hypothesis are correct. Most MHT algorithms display only the estimates of target state associated with the highest probability hypothesis.

The model used for the MHT problem is a generalization of the one given by Reid[6] and Mori et al.[7] Section 12.4.3.3 presents the recursion for general MHT. This recursion applies to problems that are nonlinear and non-Gaussian as well as to standard linear-Gaussian situations. In this general case, the distributions on target state may fail to be independent of one another (even when conditioned on an association hypothesis) and may require a joint state space representation. This recursion includes a conceptually simple Bayesian method of computing association probabilities. Section 12.4.4 discusses the case where the target distributions (conditioned on an association hypothesis) are independent of one another. Section 12.4.4.2 presents the independent MHT recursion that holds when these independence conditions are satisfied. Note that not all tracking situations satisfy these independence conditions.

Numerous books and articles on multiple-target tracking examine in detail the many variations and approaches to this problem. Many of these discuss the practical aspects of implementing multiple-target trackers and compare approaches. See, for example, Refs 1 and 6–13. With the exception of Ref. 7 these references focus primarily on the linear-Gaussian case.

In addition to the full or classical MHT as defined by Reid[6] and Mori et al.,[7] a number of approximations are in common use for finding solutions to tracking problems. Examples include joint probabilistic data association[9] and probabilistic MHT.[14] Rather than solve the full MHT, Poore[15] attempts to find the data association hypothesis (or the *n* hypotheses) with the highest likelihood. The tracks formed from this hypothesis then become the solution. Poore does this by providing a window of scans in which contacts are free to float among hypotheses. The window has a constant width and always includes the latest scan. Eventually contacts from older scans fall outside the window and become assigned to a single hypothesis. This type of hypothesis management is often combined with a nonlinear extension of Kalman filtering called an interactive multiple model Kalman filter.[16]

Section 12.4.1 presents a description of general MHT. Note that general MHT requires many more definitions and assumptions than unified tracking.

12.4.1 Contacts, Scans, and Association Hypotheses

This discussion of MHT assumes that sensor responses are limited to contacts.

12.4.1.1 Contacts

A *contact* is an observation that consists of a called detection and a measurement. In practice, a detection is called when the signal-to-noise ratio at the sensor crosses a predefined threshold. The measurement associated with a detection is often an estimated position for the object generating the contact. Limiting the sensor responses to contacts restricts responses to those in which the signal level of the target, as seen at the sensor, is high enough to call a contact. Section 12.6 demonstrates how tracking can be performed without this assumption being satisfied.

12.4.1.2 Scans

This discussion further limits the class of allowable observations to scans. The observation Y_k at time t_k is a *scan* if it consists of a set C_k of contacts such that each contact is associated with at most one target, and each target generates at most one contact (i.e., there are no merged or split measurements). Some of these contacts may be false alarms, and some targets in R might not be detected on a given scan.

More than one sensor group can report a scan at the same time. In this case, the contact reports from each sensor group are treated as separate scans with the same reporting time. As a result, $t_{k+1} = t_k$. A scan can also consist of a single contact report.

12.4.1.3 Data Association Hypotheses

To define a data association hypothesis, h, let

C_j = set of contacts of the *j*th scan

$H(k)$ = set of all contacts reported in the first *k* scans

Note that

$$H(k) = \bigcup_{j=1}^{k} C_j$$

A data association hypothesis, h, on $K(k)$ is a mapping

$$h: K(k) \rightarrow \{0, 1, \ldots, \bar{N}\}$$

such that $h(c) = n > 0$ means contact c is associated to target n, $h(c) = 0$ means contact c is associated to a false alarm, and no two contacts from the same scan are associated to the same target.

Let $H(k)$ = set of all data association hypotheses on $H(k)$. A hypothesis h on $K(k)$ partitions $K(k)$ into sets $U(n)$ for $n = 0, 1, \ldots, \bar{N}$ where $U(n)$ is the set of contacts associated to target n for $n > 0$ and $U(0)$ is the set of contacts associated to false alarms.

12.4.1.4 Scan Association Hypotheses

Decomposing a data association hypothesis h into scan association hypotheses is convenient. For each scan Y_k, let M_k = the number of contacts in scan k, Γ_k = the set of all functions $\gamma: \{1, \ldots, M_k\} \rightarrow \{0, \ldots, \bar{N}\}$ such that no two contacts are assigned to the same positive number. If $\gamma(m) = 0$, then contact m is associated to a false alarm. If $\gamma(m) = n > 0$, then contact m is associated to target n.

A function $\gamma \Gamma_k$ is called a *scan association hypothesis* for the kth scan, and Γ_k is the set of scan association hypotheses for the kth scan. For each contact, a scan association hypothesis specifies which target generated the contact or that the contact was due to a false alarm.

Consider a data association hypothesis $h_K \in H(K)$. Think of h_K as being composed of K scan association hypotheses $\{\gamma_1, \ldots, \gamma_K\}$ where γ_k is the association hypothesis for the kth scan of contacts. The hypothesis $h_K \in H(K)$ is the extension of the hypothesis $h_{K-1} = \{\gamma_1, \ldots, \gamma_{K-1}\} \in H(K-1)$. That is, h_K is composed of h_{K-1} plus γ_K. This can be written as $h_K = h_{K-1} \wedge \gamma_K$.

12.4.2 Scan and Data Association Likelihood Functions

The correctness of the scan association hypothesis γ is equivalent to the occurrence of the event "the targets to which γ associates contacts generate those contacts." Calculating association probabilities requires the ability to calculate the probability of a scan association hypothesis being correct. In particular, we must be able to calculate the probability of the event $\{\gamma \wedge Y_K = y_K\}$, where $\{\gamma \wedge Y_K = y_K\}$ denotes the conjunction or intersection of the events γ and $Y_k = y_k$.

12.4.2.1 Scan Association Likelihood Function

Assume that for each scan association hypothesis γ, one can calculate the scan association likelihood function

$$l_k(\gamma \wedge Y_k = y_k | \mathbf{s}_k) = \mathbf{Pr}\{\gamma \wedge Y_k = y_k | \mathbf{X}(t_k) = \mathbf{s}_k\}$$

$$= \mathbf{Pr}\{Y_k = y_k | \gamma \wedge \mathbf{X}(t_k) = \mathbf{s}_k\} \mathbf{Pr}\{\gamma | \mathbf{X}(t_k) = \mathbf{s}_k\} \quad \text{for } \mathbf{s}_k \in \mathbf{S}^+ \quad (12.22)$$

The factor $\mathbf{Pr}\{\gamma|\mathbf{X}(t_k) = \mathbf{s}_k\}$ is the prior probability that the scan association γ is the correct one. We normally assume that this probability does not depend on the system state $\mathbf{s}_k$, so that one may write

$$l_k(\gamma \wedge Y_k = y_k|\mathbf{s}_k) = \mathbf{Pr}\{Y_k = y_k|\gamma \wedge \mathbf{X}(t_k) = \mathbf{s}_k\}\mathbf{Pr}\{\gamma\} \quad \text{for } \mathbf{s}_k \in \mathbf{S}^+ \tag{12.23}$$

Note that $l_k(\gamma \wedge Y_K = y_k|\cdot)$ is not, strictly speaking, a likelihood function because γ is not an observation. Nevertheless, it is called a likelihood function because it behaves like one. The likelihood function for the observation $Y_k = y_k$ is

$$L_k(y_k|\mathbf{s}_k) = \mathbf{Pr}\{Y_k = y_k|\mathbf{X}(t_k) = \mathbf{s}_k\} = \sum_{\gamma \in \Gamma_k} l_k(\gamma \wedge Y_k = y_k|\mathbf{s}_k) \quad \text{for } \mathbf{s}_k \in \mathbf{S}^+ \tag{12.24}$$

12.4.2.1.1 Scan Association Likelihood Function Example

Consider a tracking problem where detections, measurements, and false alarms are generated according to the following model. The target state, s, is composed of an l-dimensional position component, z, and an l-dimensional velocity component, v, in a Cartesian coordinate space, so that $s = (z, v)$. The region of interest, R, is finite and has volume V in the l-dimensional position component of the target state space. There are at most $\bar{N}$ targets in R.

Detections and measurements. If a target is located at z, then the probability of its being detected on a scan is $P_d(z)$. If a target is detected then a measurement Y is obtained where $Y = z + \varepsilon$ and $\varepsilon \sim N(0, \Sigma)$. Let $\eta(y, z, \Sigma)$ be the density function for a $N(z, \Sigma)$ random variable evaluated at y. Detections and measurements occur independently for all targets.

False alarms. For each scan, false alarms occur as a Poisson process in the position space with density ρ. Let Φ be the number of false alarms in a scan, then

$$\Pr\{\Phi = j\} = \frac{(\rho V)^j}{j!} \quad \text{for } j = 0, 1, \ldots$$

Scan. Suppose that a scan of M measurements is received $\mathbf{y} = (y_1, \ldots, y_M)$ and γ is a scan association. Then γ specifies which contacts are false and which are true. In particular, if $\gamma(m) = n > 0$, measurement m is associated to target n. If $\gamma(m) = 0$, measurement m is associated to a false target. No target is associated with more than one contact. Let

$\phi(\gamma)$ = the number of contacts associated to false alarms

$I(\gamma) = \{n:\gamma$ associates no contact in the scan to target $n\}$ = the set of targets that have no contacts associated to them by γ

Scan Association Likelihood Function. Assume that the prior probability is the same for all scan associations, so that for some constant G, $\mathbf{Pr}\{\gamma\} = G$ for all γ. The scan association likelihood function is

$$l(\gamma \wedge \mathbf{y}|\mathbf{s} = (z, v)) = \frac{G(\rho V)^{\varphi(\gamma)}}{\varphi(\gamma)! V^{\varphi(\gamma)}} e^{-\rho V} \prod_{\{m:\gamma(m)>0\}} P_D(z_{\gamma(m)})\eta\left(y_m, z_{\gamma(m)}, \Sigma\right) \prod_{n \in I(\gamma)} (1 - P_D(z_n)) \tag{12.25}$$

12.4.2.2   Data Association Likelihood Function

Recall that $\mathbf{Y}(t_K) = \mathbf{y}_K$ is the set of observations (contacts) contained in the first K scans and $H(K)$ is the set of data association hypotheses defined on these scans. For $h \in H(K)$, $\Pr\{h \wedge \mathbf{Y}(t_K) = \mathbf{y}_K | \mathbf{X}(u) = \mathbf{s}_u, 0 \leq u \leq t_K\}$ is the likelihood of $\{h \wedge \mathbf{Y}(t_K) = \mathbf{y}_K\}$, given $\{\mathbf{X}(u) = \mathbf{s}_u, 0 \leq u \leq t_K\}$. Technically, this is not a likelihood function either, but it is convenient and suggestive to use this terminology. As with the observation likelihood functions, assume that

$$\Pr\{h \wedge \mathbf{Y}(t_K) = \mathbf{y}_K | \mathbf{X}(u) = \mathbf{s}(u), 0 \leq u \leq t_K\}$$

$$= \Pr\{h \wedge \mathbf{Y}(t_K) = \mathbf{y}_K | \mathbf{X}(t_k) = \mathbf{s}(t_k), k = 1, \ldots, K\} \qquad (12.26)$$

In addition, assuming that the scan association likelihoods are independent, the data association likelihood function becomes

$$\Pr\{h \wedge \mathbf{Y}(t_K) = \mathbf{y}_K | \mathbf{X}(t_k) = \mathbf{s}_k, k = 1, \ldots, K\} = \prod_{k=1}^{K} l_k(\gamma_k \wedge Y_k = y_k | \mathbf{s}_k) \qquad (12.27)$$

where $\mathbf{y}_K = (y_1, \ldots, y_K)$ and $h = \{\gamma_1, \ldots, \gamma_k\}$.

12.4.3   General Multiple-Hypothesis Tracking

Conceptually, MHT proceeds as follows. It calculates the posterior distribution on the system state at time t_K, given that data association hypothesis h is true, and the probability, $\alpha(h)$, that hypothesis h is true for each $h \in H(K)$. That is, it computes

$$p(t_K, \mathbf{s}_K | h) \equiv \Pr\{\mathbf{X}(t_K) = \mathbf{s}_K | h \wedge \mathbf{Y}(t_K) = \mathbf{y}_K\} \qquad (12.28)$$

and

$$\alpha(h) \equiv \Pr\{h | \mathbf{Y}(t_K) = \mathbf{y}_K\} = \frac{\Pr\{h \wedge \mathbf{Y}(t_K) = \mathbf{y}_K\}}{\Pr\{\mathbf{Y}(t_K) = \mathbf{y}_K\}} \quad \text{for each } h \in H(K) \qquad (12.29)$$

Next, MHT can compute the Bayesian posterior on system state by

$$p(t_K, \mathbf{s}_K) = \sum_{h \in H(K)} \alpha(h) p(t_K, \mathbf{s}_K | h) \qquad (12.30)$$

Subsequent sections show how to compute $p(t_K, \mathbf{s}_K | h)$ and $\alpha(h)$ in a joint recursion.

A number of difficulties are associated with calculating the posterior distribution in Equation 12.30. First, the number of data association hypotheses grows exponentially as the number of contacts increases. Second, the representation in Equation 12.30 is on the joint $\bar{N}$-fold target state space, a state space that is dauntingly large for most values of $\bar{N}$. Even when the size of the joint state space is not a problem, displaying and understanding the joint distribution is difficult.

Most MHT algorithms overcome these problems by limiting the number of hypotheses carried, displaying the distribution for only a small number of the highest probability hypotheses—perhaps only the highest. Finally, for a given hypothesis, they display the marginal distribution on each target, rather than the joint distribution. (Note, specifying a data association hypothesis specifies the number of targets present in R.) Most MHT implementations make the linear-Gaussian assumptions that produce Gaussian distributions for the posterior on a target state. The marginal distribution on a two-dimensional target position can then be represented by an ellipse. It is usually these ellipses, one for each target, that are displayed by an MHT to represent the tracks corresponding to a hypothesis.

12.4.3.1 Conditional Target Distributions

Distributions conditioned on the truth of a hypothesis are called *conditional target distributions*. The distribution $p(t_K, \cdot|h)$ in Equation 12.28 is an example of a conditional joint target state distribution. These distributions are always conditioned on the data received (e.g., $\mathbf{Y}(t_K) = \mathbf{y}_K$), but this conditioning does not appear in our notation, $p(t_K, \mathbf{s}_K|h)$.

Let $h_K = \{\gamma_1, \ldots, \gamma_K\}$, then

$$p(t_K, \mathbf{s}_K|h_K) = \mathbf{Pr}\{\mathbf{X}(t_K) = \mathbf{s}_K|h_K \wedge \mathbf{Y}(t_K) = \mathbf{y}_K\}$$

$$= \frac{\mathbf{Pr}\{(\mathbf{X}(t_K) = \mathbf{s}_K) \wedge h_K \wedge (\mathbf{Y}(t_K) = \mathbf{y}_K)\}}{\mathbf{Pr}\{h_K \wedge \mathbf{Y}(t_K) = \mathbf{y}_K\}}$$

and by Equation 12.11 and the data association likelihood function in Equation 12.27,

$$\mathbf{Pr}\{(\mathbf{X}(t_K) = \mathbf{s}_K) \wedge h_K \wedge (\mathbf{Y}(t_k) = \mathbf{y}_K)\}$$

$$= \int \left\{ \prod_{k=1}^{K} l_k(\gamma_k \wedge Y_k = y_k|\mathbf{s}_k) \prod_{k=1}^{K} q_k(\mathbf{s}_k|\mathbf{s}_{k-1}) q_0(\mathbf{s}_0) \right\} d\mathbf{s}_0 d\mathbf{s}_1 \cdots d\mathbf{s}_{k-1} \tag{12.31}$$

Thus,

$$p(t_K, \mathbf{s}_K|h_K) = \frac{1}{C(h_K)} \int \left\{ \prod_{k=1}^{K} l_k(\gamma_k \wedge Y_k = y_k|\mathbf{s}_k) q_k(\mathbf{s}_k|\mathbf{s}_{k-1}) \right\} q_0(\mathbf{s}_0) d\mathbf{s}_0 d\mathbf{s}_1 \cdots d\mathbf{s}_{k-1} \tag{12.32}$$

where $C(h_K)$ is the normalizing factor that makes $p(t_K, \cdot|h_K)$ a probability distribution. Of course

$$C(h_K) = \int \left\{ \prod_{k=1}^{K} l_k(\gamma_k \wedge Y_k = y_k|\mathbf{s}_k) q_k(\mathbf{s}_k|\mathbf{s}_{k-1}) \right\} q_0(\mathbf{s}_0) d\mathbf{s}_0 d\mathbf{s}_1 \cdots d\mathbf{s}_k$$

$$= \mathbf{Pr}\{h_K \wedge \mathbf{Y}(t_K) = \mathbf{y}_K\} \tag{12.33}$$

12.4.3.2 Association Probabilities

Sections 4.2.1 and 4.2.2 of Ref. 3 show that

$$\alpha(h) = \mathbf{Pr}\{h|\mathbf{Y}(t_K) = \mathbf{y}_K\} = \frac{\mathbf{Pr}\{h \wedge \mathbf{Y}(t_K) = \mathbf{y}_K\}}{\mathbf{Pr}\{\mathbf{Y}(t_K) = \mathbf{y}_K\}} = \frac{C(h)}{\sum_{h' \in H(K)} C(h')} \quad \text{for } h \in H(K) \tag{12.34}$$

12.4.3.3   General MHT Recursion

Section 4.2.3 of Ref. 3 provides the following general MHT recursion for calculating conditional target distributions and hypothesis probabilities.

General MHT recursion

1. *Intialize:* Let $H(0) = \{h_0\}$, where h_0 is the hypothesis with no associations. Set

$$\alpha(h_0) = 1 \quad \text{and} \quad p(t_0, \mathbf{s}_0 | h_0) = q_0(\mathbf{s}_0) \quad \text{for } \mathbf{s}_0 \in \mathbf{S}^+$$

2. *Compute conditional target distributions:* For $k = 1, 2, \ldots$, compute

$$p^-(t_k, \mathbf{s}_k | h_{k-1}) = \int q_k(\mathbf{s}_k | \mathbf{s}_{k-1}) p(t_{k-1}, \mathbf{s}_{k-1} | h_{k-1}) d\mathbf{s}_{k-1} \quad \text{for } h_{k-1} \in H(k-1) \tag{12.35}$$

For $h_k = h_{k-1} \wedge \gamma_k \in H(k)$, compute

$$\beta(h_k) = \alpha(h_{k-1}) \int l_k(\gamma_k \wedge Y_k = y_k | \mathbf{s}_k) p^-(t_k, \mathbf{s}_k | h_{k-1}) d\mathbf{s}_k \tag{12.36}$$

$$p(t_k, \mathbf{s}_k | h_k) = \frac{\alpha(h_{k-1})}{\beta(h_k)} l_k(\gamma_k \wedge Y_k = y_k | \mathbf{s}_k) p^-(t_k, \mathbf{s}_k | h_{k-1}) \quad \text{for } \mathbf{s}_k \in \mathbf{S}^+ \tag{12.37}$$

3. *Compute association probabilities:* For $k = 1, 2, \ldots$, compute

$$\alpha(h_k) = \frac{\beta(h_k)}{\sum_{h \in H(k)} \beta(h)} \quad \text{for } h_K \in H(K) \tag{12.38}$$

12.4.3.4   Summary of Assumptions for General MHT Recursion

In summary, the assumptions required for the validity of the general MHT recursion are

1. The number of targets is unknown but bounded by $\bar{N}$.
2. $S^+ = S \cup \{\phi\}$ is the extended state space for a single target, where ϕ indicates that the target is not present.
3. $X_n(t) \in S^+$ is the state of the nth target at time t.
4. $\mathbf{X}(t) = (X)_1(t), \ldots, X_{\bar{N}}(t))$ is the state of the system at time t, and $\mathbf{X} = \{\mathbf{X}(t); t \geq 0\}$ is the stochastic process describing the evolution of the system over time. The process, $\mathbf{X}$, is Markov in the state space $\mathbf{S}^+ = S^+ \times \cdots \times S^+$ where the product is taken $\bar{N}$ times.
5. Observations occur as contacts in scans. Scans are received at discrete (possibly random) times $0 \leq t_1 \leq t_2 \ldots$. Let $Y_k = y_k$ be the scan (observation) at time t_k, and let $\mathbf{Y}(t_k) = y_K \equiv (y_1, \ldots, y_k)$ be the set of contacts contained in the first K scans. Then, for each data association hypothesis $h \in H(K)$, the following is true:

$$\mathbf{Pr}\{h \wedge \mathbf{Y}(t_K) = \mathbf{y}_K | \mathbf{X}(u) = \mathbf{s}(u), 0 \leq u \leq t_K\}$$

$$= \mathbf{Pr}\{h \wedge \mathbf{Y}(t_K) = \mathbf{y}_K | \mathbf{X}(t_K) = \mathbf{s}(t_K), k = 1, \ldots, K\}$$

6. For each scan association hypothesis γ at time t_k, there is a scan association likelihood function

$$l_k(\gamma \wedge Y_k = y_k | \mathbf{s}_k) = \mathbf{Pr}\{\gamma \wedge Y_k = y_k | \mathbf{X}(t_k) = \mathbf{s}_k\}$$

7. Each data association hypothesis, $h \in H(K)$, is composed of scan association hypotheses so that $h = \{\gamma_1, \ldots, \gamma_K\}$ where γ_k is a scan association hypothesis for scan k.

8. The likelihood function for the data association hypothesis $h = \{\gamma_1, \ldots, \gamma_K\}$ satisfies

$$\mathbf{Pr}\{h \wedge \mathbf{Y}(t_K) = \mathbf{y}_K | \mathbf{X}(u) = \mathbf{s}(u), 0 \le u \le t_K\} = \prod_{k=1}^{K} l_k(\gamma_k \wedge Y_k = y_k | \mathbf{s}(t_K))$$

12.4.4 Independent Multiple-Hypothesis Tracking

The decomposition of the system state distribution into a sum of conditional target distributions is most useful when the conditional distributions are the product of independent single-target distributions. This section presents a set of conditions under which this happens and restates the basic MHT recursion for this case.

12.4.4.1 Conditionally Independent Scan Association Likelihood Functions

Prior to this section no special assumptions were made about the scan association likelihood function $l_k(\gamma \wedge Y_k = y_k | \mathbf{s}_k) = \mathbf{Pr}\{\gamma \wedge Y_k = y_k | \mathbf{X}(t_k) = \mathbf{s}_k\}$ for $\mathbf{s}_k \in \mathbf{S}^+$. In many cases, however, the joint likelihood of a scan observation and a data association hypothesis satisfies an independence assumption when conditioned on a system state.

The likelihood of a scan observation $Y_k = y_k$ obtained at time t_k is *conditionally independent*, if and only if, for all scan association hypotheses $\gamma \in \Gamma_k$,

$$l_k(\gamma \wedge Y_k = y_k | \mathbf{s}_k) = \mathbf{Pr}\{\gamma \wedge Y_k = y_k | \mathbf{X}(t_k) = \mathbf{s}_k\}$$

$$= g_0^\gamma(y_k) \prod_{n=1}^{\bar{N}} g_n^\gamma(y_k, x_n) \quad \text{for } \mathbf{s}_k = (x_1, \ldots, x_{\bar{N}}) \tag{12.39}$$

for some functions g_n^γ, $n = 0, \ldots, \bar{N}$, where g_0^γ can depend on the scan data but not $\mathbf{s}_k$.

Equation 12.39 shows that conditional independence means that the probability of the joint event $\{\gamma \wedge Y_k = y_k\}$, conditioned on $\mathbf{X}(t_k) = (x_1, \ldots, x_{\bar{N}})$, factors into a product of functions that each depend on the state of only one target. This type of factorization occurs when the component of the response due to each target is independent of all other targets. As an example, the scan association likelihood in Equation 12.25 is conditionally independent. This can be verified by setting

$$g_0^\gamma(\mathbf{y}) = \frac{G(\rho V)^{\varphi(\gamma)}}{\varphi(\gamma)! V^{\varphi(\gamma)}} = \frac{G \rho^{\varphi(\gamma)}}{\varphi(\gamma)!}$$

and for $n = 1, \ldots, \bar{N}$

$$g_n^\gamma(\mathbf{y}, z_n) = \begin{cases} P_d(z_n) \eta(y_m, z_n, n) & \text{if } \gamma(m) = n \text{ for some } m \\ 1 - P_d(z_n) & \text{if } \gamma(m) \ne n \text{ for all } m \end{cases}$$

Conditional independence implies that the likelihood function for the observation $Y_k = y_k$ is given by

$$L_k(Y_k = y_k|\mathbf{X}(t_k)) = (x_1, \ldots, x_{\bar{N}}))$$

$$= \sum_{\gamma \in \Gamma_k} g_0^\gamma(y_k) \prod_{n=1}^{\bar{N}} g_n^\gamma(y_k, x_n) \quad \text{for all } (x_1, \ldots, x_{\bar{N}}) \in \mathbf{S}^+ \qquad (12.40)$$

The assumption of conditional independence of the observation likelihood function is implicit in most multiple-target trackers. The notion of conditional independence of a likelihood function makes sense only when the notions of contact and association are meaningful. As noted in Section 12.3, there are cases in which these notions do not apply. For these cases, the scan association likelihood function will not satisfy Equation 12.39.

Under the assumption of conditional independence, the Independence Theorem, discussed in Section 12.4.4.1.1, says that conditioning on a data association hypothesis allows the multiple-target tracking problem to be decomposed into $\bar{N}$ independent single-target problems. In this case, conditioning on a hypothesis greatly simplifies the joint tracking problem. In particular, no joint state space representation of the target distributions is required when they are conditional on a data association hypothesis.

12.4.4.1.1　Independence Theorem

Suppose that (1) the assumptions of Section 12.4.3.4 hold, (2) the likelihood functions for all scan observations are conditionally independent, and (3) the prior target motion processes, $\{X_n(t); t \geq 0\}$ for $n = 1, \ldots, \bar{N}$ are mutually independent. Then the posterior system state distribution conditioned on the truth of a data association hypothesis is the product of independent distributions on the targets.

Proof. The proof of this theorem is given in Section 4.3.1 of Ref. 3.

Let $\mathbf{Y}(t) = \{Y_1, Y_2, \ldots, y_{K(t)}\}$ be scan observations that are received at times $0 \leq t_1 \leq, \ldots, \leq t_k \leq t$, where $K = K(t)$, and let $H(k)$ be the set of all data association hypotheses on the first k scans. Define $p_n(t_k, x_n|h) = \mathbf{Pr}\{X_n(t_k) = x_n|h\}$ for $x_n \in S^+$, $k = 1, \ldots, K$, and $n = 1, \ldots, \bar{N}$. Then by the independence theorem,

$$p(t_k, \mathbf{s}_k|h) = \prod_n p_n(t_k, x_n|h) \quad \text{for } h \in H(k) \text{ and } \mathbf{s}_k = (x_1, \ldots, x_{\bar{N}}) \in \mathbf{S}^+ \qquad (12.41)$$

Joint and Marginal Posteriors. From Equation 12.30 the full Bayesian posterior on the joint state space can be computed as follows:

$$p(t_K, \mathbf{s}_K) = \sum_{h \in H(K)} \alpha(h) p(t_K, \mathbf{s}_K|h)$$

$$= \sum_{h \in H(K)} \alpha(h) \prod_n p_n(t_K, x_n|h) \quad \text{for } \mathbf{s}_K = (x_1, \ldots, x_{\bar{N}}) \in \mathbf{S}^+ \qquad (12.42)$$

Marginal posteriors can be computed in a similar fashion. Let $\bar{p}_n(t_K, \cdot)$ be the marginal posterior on $X_n(t_K)$ for $n = 1, \ldots, \bar{N}$. Then

$$\bar{p}_n(t_K, x_n) = \int \left[\sum_{h \in H(K)} \alpha(h) \prod_{l=1}^{\bar{N}} p_l(t_K, x_l|h) \right] \prod_{l \neq n} dx_l$$

$$= \sum_{h \in H(K)} \alpha(h) p_n(t_K, x_n|h) \int \prod_{l \neq n} p_l(t_K, x_l|h) dx_l$$

$$= \sum_{h \in H(K)} \alpha(h) p_n(t_K, x_n|h) \quad \text{for } n = 1, \ldots, \bar{N}$$

Thus, the posterior marginal distribution on target n may be computed as the weighted sum over n of the posterior distribution for target n conditioned on h.

12.4.4.2 Independent MHT Recursion

Let $q_0(n, x) = \mathbf{Pr}\{X_n(0) = x\}$ and $q_k(x|n, x') = \mathbf{Pr}\{X_n(t_k) = x|X_n(t_{k-1}) = x'\}$. Under the assumptions of the independence theorem, the motion models for the targets are independent, and $q_n(\mathbf{s}_k|\mathbf{s}_{k-1}) = \prod_n q_k(x_n|n, x'_n)$ where $\mathbf{s}_k = (x_1, \dots, x_{\bar{N}})$ and $\mathbf{s}_{k-1} = (x'_1, \dots, x'_{\bar{N}})$. As a result, the transition density, $q_k(\mathbf{s}_k|\mathbf{s}_{k-1})$, factors just as the likelihood function does. This produces the independent MHT recursion.
Independent MHT recursion

1. *Intialize:* Let $H(0) = \{h_0\}$ where h_0 is the hypothesis with no associations. Set

$$\alpha(h_0) = 1 \quad \text{and} \quad p_n(t_0, \mathbf{s}_0|h_0) = q_0(n, \mathbf{s}_0) \quad \text{for } \mathbf{s}_0 \in S^+ \text{ and } n = 1, \dots, \bar{N}$$

2. *Compute conditional target distributions:* For $k = 1, 2, \dots$, do the following: For each $h_k \in H(k)$, find $h_{k-1} \in H(k-1)$ and $\gamma \in \Gamma_k$, such that $h_k = h_{k-1} \wedge \gamma$. Then compute

$$p_n^-(t_k, x|h_{k-1}) = \int_{S^+} q_k(x|n, x')p_n(t_{k-1}, x'|h_{k-1})dx' \quad \text{for } n = 1, \dots, \bar{N}$$

$$p_n(t_k, x|h_k) = \frac{1}{C(n, h_k)} g_n^\gamma(y_k, x)p_n^-(t_k, x \mid h_{k-1}) \quad \text{for } x \in S^+ \text{ and } n = 1, \dots, \bar{N}$$

$$p(t_k, \mathbf{s}_k|h_k) = \prod_n p_n(t_k, x_n|h_k) \quad \text{for } \mathbf{s}_k = (x_1, \dots, x_{\bar{N}}) \in S^+ \tag{12.43}$$

where $C(n, h_k)$ is the constant that makes $p_n(t_k, \cdot|h_k)$ a probability distribution.

3. *Compute association probabilities:* For $k = 1, 2, \dots$, and $h_k = h_{k-1} \wedge \gamma \in H(k)$ compute

$$\beta(h_k) = \alpha(h_{k-1})\int g_0^\gamma(y_k)\left[\prod_n g_n^\gamma(y_k, x_n)p_n^-(t_k, x_n|h_{k-1}) \right]dx_1 \dots dx_{\bar{N}}$$

$$= \alpha(h_{k-1})g_0^\gamma(y_k)\prod_n \int g_n^\gamma(y_k, x_n)p_n^-(t_k, x_n|h_{k-1})dx_n \tag{12.44}$$

Then

$$\alpha(h_k) = \frac{\beta(h_k)}{\sum\limits_{h'_k \in H(k)} \beta(h'_k)} \quad \text{for } h_K \in H(K) \tag{12.45}$$

In Equation 12.43, the independent MHT recursion performs a motion update of the probability distribution on target n given h_{k-1} and multiplies the result by $g_n^\gamma(y_k, x)$, which is the likelihood function of the measurement associated to target n by γ. When this product is normalized to a probability distribution, we obtain the posterior on target n given $h_k = h_{k-1} \wedge \gamma$. Note that these computations are all performed independently of the other targets. Only the computation of the association probabilities in Equations 12.44 and 12.45 requires interaction with the other targets and the likelihoods of the measurements associated to them. This is where the independent MHT obtains its power and simplicity. *Conditioned on a data association hypothesis, each target may be treated independently of all other targets.*

12.5 Relationship of Unified Tracking to MHT and Other Tracking Approaches

This section discusses the relationship of unified tracking to other tracking approaches such as general MHT.

12.5.1 General MHT Is a Special Case of Unified Tracking

Section 5.2.1 of Ref. 3 shows that the assumptions for general MHT that are given in Section 12.4.3.4 imply the validity of the assumptions for unified tracking given in Section 12.3.4.2. This means that whenever it is valid to perform general MHT, it is valid to perform unified tracking. In addition, Section 5.2.1 of Ref. 3 shows that when the assumptions for general MHT hold, MHT produces the same Bayesian posterior on the joint target state space as unified tracking does. Section 5.3.2 of Ref. 3 presents an example where the assumptions of unified tracking are satisfied, but those of general MHT are not. This example compares the results of running the general MHT algorithm to that obtained from unified tracking and shows that unified tracking produces superior results. This means that general MHT is a special case of unified tracking.

12.5.2 Relationship of Unified Tracking to Other Multiple-Target Tracking Algorithms

Bethel and Paras,[17] Kamen and Sastry,[18] Kastella,[19–21] Lanterman et al.,[22] Mahler,[23] and Washburn[24] have formulated versions of the multiple-target tracking problem in terms of computing a posterior distribution on the joint target state space. In these formulations the steps of data association and estimation are unified as shown in Section 12.3.

Kamen and Sastry,[18] Kastella,[19] and Washburn[24] assume that the number of targets is known and that the notions of contact and association are meaningful. They have additional restrictive assumptions. Washburn[24] assumes that all measurements take values in the same space. (This assumption appears to preclude sets of sensors that produce disparate types of observations.) Kamen and Sastry[18] and Kastella[19] assume that the measurements are position estimates with Gaussian errors. Kamen and Sastry[18] assume perfect detection capability. Kastella[20] considers a fixed but unknown number of targets. The model in Ref. 20 is limited to identical targets, a single sensor, and discrete time and space. Kastella[21] extends this to targets that are not identical. Bethel and Paras[17] require the notions of contact and association to be meaningful. They also impose a number of special assumptions, such as requiring that contacts be line-of-bearing and assuming that two targets cannot occupy the same cell at the same time.

Mahler's formulation (Section 3 of Ref. 23) uses a random set approach in which all measurements take values in the same space with a special topology. Mahler[23] does not provide an explicit method for handling unknown numbers of targets. Lanterman et al.[22] consider only observations that are camera images. They provide formulas for computing posterior distributions only in the case of stationary targets. They discuss the possibility of handling an unknown number of targets but do not provide an explicit procedure for doing so.

In Ref. 25, Mahler develops an approach to tracking that relies on random sets. The random sets are composed of finite numbers of contacts; therefore, this approach applies only to situations where there are distinguishable sensor responses that can clearly be called out as contacts or detections. To use random sets, one must specify a topology and a rather complex measure on the measurement space for the contacts. The approach, presented in Sections 6.1 and 6.2 of Ref. 25, requires that the measurement spaces be identical for all sensors. In contrast, the likelihood function approach presented in Section 12.3 of this chapter, which transforms sensor information into a function on the target state space, is simpler and appears to be more general. For example, likelihood functions and the tracking approach presented in Section 12.3 can accommodate situations in which sensor responses are not strong enough to call contacts.

The approach presented in Section 12.3 differs from previous work in the following important aspects:

- The unified tracking model applies when the number of targets is unknown and varies over time.

- Unified tracking applies when the notions of contact and data association are not meaningful.

- Unified tracking applies when the nature (e.g., measurement spaces) of the observations to be fused are disparate. It can correctly combine estimates of position, velocity, range, and bearing as well as frequency observations and signals from sonars, radars, and infrared (IR) sensors. Unified tracking can fuse any information that can be represented by a likelihood function.

- Unified tracking applies to a richer class of target motion models than are considered in the references cited in the foregoing text. It allows for targets that are not identical. It provides for space-and-time dependent motion models that can represent the movement of troops and vehicles through terrain and submarines and ships through waters near land.

12.5.3 Critique of Unified Tracking

The unified tracking approach to multiple-target tracking has great power and breadth, but it is computationally infeasible for problems involving even moderate numbers of targets. Some shrewd numerical approximation techniques are required to make more general use of this approach.

The approach does appear to be feasible for two targets as explained by Finn.[4] Koch and Van Keuk[26] also consider the problem of two targets and unresolved measurements. Their approach is similar to the unified tracking approach; however, they consider only probability distributions that are mixtures of Gaussian ones. In addition, the target motion model is Gaussian.

A possible approach to dealing with more than two targets is to develop a system that uses a more standard tracking method when targets are well separated and then switches to a unified tracker when targets cross or merge.

12.6 Likelihood Ratio Detection and Tracking

This section describes the problem of detection and tracking when there is, at most, one target present. This problem is most pressing when signal-to-noise ratios are low.

This will be the case when performing surveillance of a region of the ocean's surface hoping to detect a periscope in the clutter of ocean waves or when scanning the horizon with an infrared sensor trying to detect a cruise missile at the earliest possible moment. Both of these problems have two important features: (1) a target may or may not be present; and (2) if a target is present, it will not produce a signal strong enough to be detected on a single glimpse by the sensor.

Likelihood ratio detection and tracking is based on an extension of the single-target tracking methodology, presented in Section 12.2, to the case where there is either one or no target present. The methodology presented here unifies detection and tracking into one seamless process. Likelihood ratio detection and tracking allows both functions to be performed simultaneously and optimally.

12.6.1 Basic Definitions and Relations

Using the same basic assumptions as in Section 12.2, we specify a prior on the target's state at time 0 and a Markov process for the target's motion. A set of K observations or measurements $\mathbf{Y}(t) = (Y_1, \ldots, Y_K)$ are obtained in the time interval $[0, t]$. The observations are received at the discrete (possibly random) times $(t_1, \ldots, t_K)$ where $0 < t_1 \cdots \leq t_K \leq t$. The measurements obtained at these various times need not be made with the same sensor or even with sensors of the same type; the data from the various observations need not be of the same structure. Some observations may consist of a single number whereas others may consist of large arrays of numbers, such as the range and azimuth samples of an entire radar scan. However, we do assume that, conditioned on the target's path, the statistics of the observations made at any time by a sensor are *independent* of those made at other times or by other sensors.

The state space in which targets are detected and tracked depends on the particular problem. Characteristically, the target state is described by a vector, some of whose components refer to the spatial location of the target, some to its velocity, and perhaps some to higher-order properties such as acceleration. These components as well as others that might be important to the problem at hand, such as target orientation or target strength, can assume continuous values. Other elements that might be part of the state description may assume discrete values. Target class (type) and target configuration (such as periscope extended) are two examples.

As in Section 12.3, the target state space S is augmented with a null state to make $S^+ = S \cup \phi$. There is a probability (density) function, p, defined on S^+, such that $p(\phi) + \int_{s \in S} p(s)ds = 1$.

Both the state of the target $X(t) \in S^+$ and the information accumulated for estimating the state probability densities evolve with time t. The process of target detection and tracking consists of computing the posterior version of the function p as new observations are available and propagating it to reflect the temporal evolution implied by target dynamics. Target dynamics include the probability of target motion into and out of S as well as the probabilities of target state changes.

Following the notation used in Section 12.2 for single-target Bayesian filtering, let $p(t, s) = \mathbf{Pr}\{X(t) = s \mid \mathbf{Y}(t) = (Y(t_1), \ldots, Y(t_k))\}$ for $s \in S^+$ so that $p(t, \cdot)$ is the posterior distribution on $X(t)$ given all observations received through time t. This section assumes that the conditions that ensure the validity of the basic recursion for single-target tracking in Section 12.2 hold, so that $p(t, \cdot)$ can be computed in a recursive manner. Recall that $\bar{p}(t_k, s_k) = \int_{S^+} q(s_k \mid s_{k-1}) p(t_{k-1}, s_{k-1})ds_{k-1}$ or $s_k \in S^+$ is the posterior from time t_{k-1} updated for target motion to time t_k, the time of the kth observation. Recall also the definition of the *likelihood* function L_k. Specifically, for the observation $Y_k = y_k$

$$L_k(y_k \mid s) = \mathbf{Pr}\{Y_k = y_k \mid X(t_k) = s\} \tag{12.46}$$

where for each $s \in S^+$, $L_k (\cdot|s)$ is a probability (density) function on the measurement space H_k.

According to Bayes' rule,

$$p(t_k, s) = \frac{p^-(t_k, s)L_k(y_k|s)}{C(k)} \quad \text{for } s \in S$$

$$p(t_k, \phi) = \frac{p^-(t_k, \phi)L_k(y_k|\phi)}{C(k)} \tag{12.47}$$

In these equations, the denominator is the probability of obtaining the measurement $Y_k = y_k$, that is, $C(k) = \bar{p}(t_k, \phi) L_k(y_k|\phi) + \int_{s \in S} \bar{p}(t_k, s)L_k(y_k|s)ds$.

12.6.1.1 Likelihood Ratio

The ratio of the state probability (density) to the null state probability $p(\phi)$ is defined to be the *likelihood ratio (density)*, $\Lambda(s)$, that is,

$$\Lambda(s) = \frac{p(s)}{p(\phi)} \quad \text{for } s \in S \tag{12.48}$$

It would be more descriptive to call $\Lambda(s)$ the target likelihood ratio to distinguish it from the measurement likelihood ratio defined in Section 12.6.1.2. However, for simplicity, we use the term likelihood ratio for $\Lambda(s)$. The notation for Λ is consistent with that already adopted for the probability densities. Thus, the prior and posterior forms become

$$\Lambda^-(t, s) = \frac{p^-(t, s)}{p^-(t, \phi)} \quad \text{and} \quad \Lambda(t, s) = \frac{p(t, s)}{p(t, \phi)} \quad \text{for } s \in S \text{ and } t \geq 0 \tag{12.49}$$

The likelihood ratio density has the same dimensions as the state probability density. Furthermore, from the likelihood ratio density one may easily recover the state probability density as well as the probability of the null state. Since $\int_S L(t, s)ds = (1 - p(t, \phi))/p(t, \phi)$, it follows that

$$p(t, s) = \frac{\Lambda(t, s)}{1 + \int_S \Lambda(t, s')ds'} \quad \text{for } s \in S \quad p(t, \phi) = \frac{1}{1 + \int_S \Lambda(t, s')ds'} \tag{12.50}$$

12.6.1.2 Measurement Likelihood Ratio

The measurement likelihood ratio L_k for the observation Y_k is defined as

$$L_k(y|s) = \frac{L_k(y|s)}{L_k(y|\phi)} \quad \text{for } y \in H_k, s \in S \tag{12.51}$$

$L_k(y|s)$ is the ratio of the likelihood of receiving the observation $Y_k = y_k$ (given the target is in state s) to the likelihood of receiving $Y_k = y_k$ given no target present. As discussed by Van Trees,[27] the measurement likelihood ratio has long been recognized as part of the prescription for optimal receiver design. This section demonstrates that it plays an even larger role in the overall process of sensor fusion.

Measurement likelihood ratio functions are chosen for each sensor to reflect its salient properties, such as noise characterization and target effects. These functions contain all the sensor information that is required for making optimal Bayesian inferences from sensor measurements.

12.6.2   Likelihood Ratio Recursion

Under the assumptions for which the basic recursion for single-target tracking in Section 12.1 holds, the following recursion for calculating the likelihood ratio holds.

Likelihood ratio recursion

$$Initialize: p(t_0, s) = q_0(s) \quad for \ s \in S^+ \tag{12.52}$$

For $k \geq 1$ and $s \in S^+$,

$$Perform \ motion \ update: p^-(t_k, s) = \int_{S^+} q_k(s|s_{k-1})p(t_{k-1}, s_{k-1})ds_{k-1} \tag{12.53}$$

$$Calculate \ likelihood \ function: L_k(y_k|s) = \mathbf{Pr}\{Y_k = y_k|X(t_k) = s\} \tag{12.54}$$

$$Perform \ information \ update: p(t_k, s) = \frac{1}{C}L_k(y_k|s)p^-(t_k, s) \tag{12.55}$$

For $k \geq 1$,

$$Calculate \ likelihood \ ratio: \Lambda(t_k, s) = \frac{p(t_k, s)}{p(t_k, \phi)} \quad for \ s \in S \tag{12.56}$$

The constant, C, in Equation 12.55 is a normalizing factor that makes $p(t_k, \cdot)$ a probability (density) function.

12.6.2.1   Simplified Recursion

The recursion given in Equations 12.52–12.56 requires the computation of the full probability function $p(t_k, \cdot)$ using the basic recursion for single-target tracking discussed in Section 12.2. A simplified version of the likelihood ratio recursion has probability mass flowing from the state ϕ to S and from S to ϕ in such a fashion that

$$p^-(t_k, \phi) = q_k(\phi|\phi)p(t_{k-1}, \phi) + \int_S q_k(\phi|s)p(t_{k-1}, s)ds$$

$$= p(t_{k-1}, \phi) \tag{12.57}$$

Since

$$p^-(t_k, s_k) = q_k(s_k|\phi)p(t_{k-1}, \phi) + \int_S q_k(s_k|s)p(t_{k-1}, s)ds \quad for \ s_k \in S$$

we have

$$\Lambda^-(t_k, s_k) = \frac{q_k(s_k|\phi)p(t_{k-1}, \phi) + \int_S q_k(s_k|s)p(t_{k-1}, s)ds}{p^-(t_k, \phi)}$$

$$= \frac{q_k(s_k|\phi) + \int_S q_k(s_k|s)\Lambda(t_{k-1}, s)ds}{p^-(t_k, \phi)/p(t_{k-1}, \phi)}$$

From Equation 12.57 it follows that

$$\Lambda^-(t_k, s_k) = q_k(s_k|\phi) + \int_S q_k(s_k|s)\Lambda(t_{k-1}, s)ds \quad \text{for } s_k \in S \qquad (12.58)$$

Assuming Equation 12.57 holds, a simplified version of the basic likelihood ratio recursion can be written.
Simplified likelihood ratio recursion

$$\text{Initialize likelihood ratio: } \Lambda(t_0, s) = \frac{p(t_0, s)}{p(t_0, \phi)} \quad \text{for } s \in S \qquad (12.59)$$

For $k \geq 1$ and $s \in S$,

$$\text{Perform motion update: } \Lambda^-(t_k, s) = q_k(s|\phi) + \int_S q_k(s|s_{k-1})\Lambda(t_{k-1}, s_{k-1})ds_{k-1} \qquad (12.60)$$

$$\text{Calculate measurement likelihood ratio: } L_k(y|s) = \frac{L_k(y|s)}{L_k(y|\phi)} \qquad (12.61)$$

$$\text{Perform information update: } \Lambda(t_k, s) = L_k(y|s)\Lambda^-(t_k, s) \qquad (12.62)$$

The simplified recursion is a reasonable approximation to problems involving surveillance of a region that may or may not contain a target. Targets may enter and leave this region, but only one target is in the region at a time.

As a special case, consider the situation where no mass moves from state ϕ to S or from S to ϕ under the motion assumptions. In this case $q_k(s|\phi) = 0$ for all $s \in S$, and $p^-(t_k, \phi) = p(t_{k-1}, \phi)$ so that Equation 12.60 becomes

$$\Lambda^-(t_k, s) = \int_S q_k(s|s_{k-1})\Lambda(t_{k-1}, s_{k-1})ds_{k-1} \qquad (12.63)$$

12.6.3 Log-Likelihood Ratios

Frequently, it is more convenient to write Equation 12.62 in terms of natural logarithms. Doing so results in quantities that require less numerical range for their representation. Another advantage is that, frequently, the logarithm of the measurement likelihood ratio is a simpler function of the observations than is the actual measurement likelihood ratio itself. For example, when the measurement consists of an array of numbers, the measurement log-likelihood ratio often becomes a linear combination of those data, whereas the measurement likelihood ratio involves a product of powers of the data. In terms of logarithms, Equation 12.62 becomes

$$\ln \Lambda(t_k, s) = \ln \Lambda^-(t_k, s) + \ln L_k(y_k|s) \quad \text{for } s \in S \tag{12.64}$$

The following example is provided to impart an understanding of the practical differences between a formulation in terms of probabilities and a formulation in terms of the logarithm of the likelihood ratios. Suppose there are I discrete target states, corresponding to physical locations so that the target state $X \in \{s_1, s_2, \ldots, s_I\}$ when the target is present. The observation is a vector, $\mathbf{Y}$, that is formed from measurements corresponding to these spatial locations, so that $\mathbf{Y} = (Y(s_1), \ldots, Y(s_I))$, where in the absence of a target in state, s_i, the observation $Y(s_i)$ has a distribution with density function $\eta(\cdot, 0, 1)$, where $\eta(\cdot, \mu, \sigma^2)$ is the density function for a Gaussian distribution with mean μ and variance σ^2. The observations are independent of one another regardless of whether a target is present. When a target is present in the ith state, the mean for $Y(s_i)$ is shifted from 0 to a value r. To perform a Bayesian update, the likelihood function for the observation $\mathbf{Y} = \mathbf{y} = (y(s_1), \ldots, y(s_I))$ is computed as follows:

$$L(\mathbf{y}|s_i) = \eta(y(s_i), r, 1) \prod_{j \neq i} \eta(y(s_j), 0, 1)$$

$$= \exp\left(ry(s_i) - \frac{1}{2}r^2\right) \prod_{j=1}^{I} \eta(y(s_j), 0, 1)$$

Contrast this with the form of the measurement log-likelihood ratio for the same problem. For state i,

$$\ln L_k(\mathbf{y}|s_i) = ry(s_i) - \frac{1}{2}r^2$$

Fix s_i and consider $\ln L_k(\mathbf{Y}|s_i)$ as a random variable. That is, consider $\ln L_k(\mathbf{Y}|s_i)$ before making the observation. It has a Gaussian distribution with

$$\mathbf{E}[\ln L(\mathbf{Y}|s_i)|X = s_i] = +\frac{1}{2}r^2$$

$$\mathbf{E}[\ln L(\mathbf{Y}|s_i)|X = \phi] = -\frac{1}{2}r^2$$

$$\mathbf{Var}[\ln L(\mathbf{Y}|s_i)] = r^2$$

This reveals a characteristic result. Although the likelihood function for any given state requires examination and processing of all the data, the log-likelihood ratio for a given state commonly depends on only a small fraction of the data—frequently only a single datum. Typically, this will be the case when the observation **Y** is a vector of independent observations.

12.6.4 Declaring a Target Present

The likelihood ratio methodology allows the Bayesian posterior probability density to be computed, including the discrete probability that no target resides in S at a given time. It extracts all possible inferential content from the knowledge of the target dynamics, the *a priori* probability structure, and the evidence of the sensors. This probability information may be used in a number of ways to decide whether a target is present. The following offers a number of traditional methods for making this decision, all based on the integrated likelihood ratio. Define

$$p(t, 1) = \int_S p(t, s)ds = \mathbf{Pr}\{\text{target present in } S \text{ at time } t\}$$

Then

$$\bar{\Lambda}(t) = \frac{p(t, 1)}{p(t, \phi)}$$

is defined to be the *integrated likelihood ratio at time t*. It is the ratio of the probability of the target being present in S to the probability of the target not being present in S at time t.

12.6.4.1 Minimizing Bayes' Risk

To calculate Bayes' risk, costs must be assigned to the possible outcomes related to each decision (e.g., declaring a target present or not). Define the following costs:

$C(1|1)$ if target is declared to be present and it is present

$C(1|\phi)$ if target is declared to be present and it is not present

$C(\phi|1)$ if target is declared to be not present and it is present

$C(\phi|\phi)$ if target is declared to be not present and it is not present

Assume that it is always better to declare the correct state, that is,

$$C(1|1) < C(\phi|1) \quad \text{and} \quad C(\phi|\phi) < C(1|\phi)$$

The *Bayes' risk* of a decision is defined as the expected cost of making that decision. Specifically the Bayes' risk is

$p(t, 1)C(1|1) + p(t, \phi)C(1|\phi)$ for declaring a target present

$p(t, 1)C(\phi|1) + p(t, \phi)C(\phi|\phi)$ for declaring a target not present

One procedure for making a decision is to take that action which minimizes the Bayes' risk. Applying this criterion produces the following decision rule. Define the threshold

$$\Lambda_\text{T} = \frac{C(1|\phi) - C(\phi|\phi)}{C(\phi|1) - C(1|1)} \tag{12.65}$$

Then declare

> Target present if $\bar{\Lambda}(t) > \Lambda_T$
> Target not present if $\bar{\Lambda}(t) \leq \Lambda_T$

This demonstrates that the integrated likelihood ratio is a sufficient decision statistic for taking an action to declare a target present or not when the criterion of performance is the minimization of the Bayes' risk.

12.6.4.2 Target Declaration at a Given Confidence Level

Another approach is to declare a target present whenever its probability exceeds a desired confidence level, p_T. The integrated likelihood ratio is a sufficient decision statistic for this criterion as well. The prescription is to declare a target present or not according to whether the integrated likelihood ratio exceeds a threshold, this time given by $\Lambda_T = p_T(1 - p_T)$.

A special case of this is the *ideal receiver*, which is defined as the decision rule that minimizes the average number of classification errors. Specifically, if $C(1|1) = 0$, $C(\phi|\phi) = 0$, $C(1|\phi) = 1$, and $C(\phi|1) = 1$, then minimizing Bayes' risk is equivalent to minimizing the expected number of miscalls of target present or not present. Using Equation 12.65 this is accomplished by setting $\Lambda_T = 1$, which corresponds to a confidence level of $p_T = 1/2$.

12.6.4.3 Neyman-Pearson Criterion for Declaration

Another standard approach in the design of target detectors is to declare targets present according to a rule that produces a specified false alarm rate. Naturally, the target detection probability must still be acceptable at that rate of false alarms. In the ideal case, one computes the distribution of the likelihood ratio with and without the target present and sets the threshold accordingly. Using the Neyman-Pearson approach, a threshold, Λ_T, is identified such that calling a target present when the integrated likelihood ratio is above Λ_T produces the maximum probability of detection subject to the specified constraint on false alarm rate.

12.6.5 Track-Before-Detect

The process of likelihood ratio detection and tracking is often referred to as *track-before-detect*. This terminology recognizes that one is tracking a possible target (through computation of $P(t, \cdot)$) before calling the target present. The advantage of track-before-detect is that it can integrate sensor responses over time on a moving target to yield a detection in cases where the sensor response at any single time period is too low to call a detection. In likelihood ratio detection and tracking, a threshold is set and a detection is called when the likelihood ratio surface exceeds that threshold. The state at which the peak of the threshold crossing occurs is usually taken to be the state estimate, and one can convert the likelihood ratio surface into a probability distribution for the target state.

Section 6.2 of Ref. 3 presents an example of performing track-before-detect using the likelihood ratio detection and tracking approach on simulated data. Its performance is compared to a matched filter detector that is applied to the sensor responses at each time period. The example shows that, for a given threshold setting, the likelihood ratio detection methodology produces a 0.93 probability of detection at a specified false alarm rate. To obtain that same detection probability with the matched filter detector, one has to suffer a false alarm rate that is higher by a factor of 10.[18] As another example, Section 1.1.3 of Ref. 3 describes the application of likelihood ratio detection and tracking to detecting a periscope with radar.

References

1. Blackman, S.S., and Popoli, R., *Design and Analysis of Modern Tracking Systems,* Artech House, Boston, 1999.
2. Jazwinski, A.H., *Stochastic Processes and Filtering Theory,* Academic Press, New York, 1970.
3. Stone, L.D., Barlow, C.A., and Corwin, T.L., *Bayesian Multiple Target Tracking,* Artech House, Boston, 1999.
4. Finn, M.V., Unified data fusion applied to monopulse tracking, *Proc. IRIS 1999 Nat'l. Symp. Sensor and Data Fusion,* I, 47–61, 1999.
5. Mori, S., Kuo-Chu, C., and Chong, C-Y., Tracking aircraft by acoustic sensors, *Proc. 1987 Am. Control Conf.,* 1099–1105, 1987.
6. Reid, D.B., An algorithm for tracking multiple targets, *IEEE Trans. Automatic Control,* AC-24, 843–854, 1979.
7. Mori, S., Chong, C-Y., Tse, E., and Wishner, R.P., Tracking and classifying multiple targets without *a priori* identification, *IEEE Trans. Automatic Control,* AC-31, 401–409, 1986.
8. Antony, R.T., *Principles of Data Fusion Automation,* Artech House, Boston, 1995.
9. Bar-Shalom, Y., and Fortman, T.E., *Tracking and Data Association,* Academic Press, New York, 1988.
10. Bar-Shalom, Y., and Li, X.L., *Multitarget-Multisensor Tracking: Principles and Techniques,* Yaakov Bar-Shalom, Storrs, CT, 1995.
11. Blackman, S.S., *Multiple Target Tracking with Radar Applications,* Artech House, Boston, 1986.
12. Hall, D.L., *Mathematical Techniques in Multisensor Data Fusion,* Artech House, Boston, 1992.
13. Waltz, E., and Llinas, J., *Multisensor Data Fusion,* Artech House, Boston, 1990.
14. Streit, R.L., *Studies in Probabilistic Multi-Hypothesis Tracking and Related Topics,* Naval Undersea Warfare Center Publication SES-98-101, Newport, RI, 1998.
15. Poore, A.B., Multidimensional assignment formulation of data association problems arising from multitarget and multisensor tracking, *Computat. Optimization Appl.,* 3(1), 27–57, 1994.
16. Yeddanapudi, M., Bar-Shalom, Y., and Pattipati, K.R., IMM estimation for multitarget-multisensor air traffic surveillance, *Proc. IEEE,* 85, 80–94, 1997.
17. Bethel, R.E., and Paras, G.J., A PDF multisensor multitarget tracker, *IEEE Trans. Aerospace Electronic Systems,* 34, 153–168, 1998.
18. Kamen, E.W., and Sastry, C.R., Multiple target tracking using products of position measurements, *IEEE Trans. Aerospace Electronics Systems,* 29(2), 476–493, 1993.
19. Kastella, K., Event-averaged maximum likelihood estimation and mean-field theory in multitarget tracking, *IEEE Trans. Automatic Control,* AC-40, 1070–1074, 1995.
20. Kastella, K., Discrimination gain for sensor management in multitarget detection and tracking, *IEEE-SMC and IMACS Multiconference CESA,* Vol. 1, pp. 167–172, 1996.
21. Kastella, K., Joint multitarget probabilities for detection and tracking, *Proc. SPIE, Acquisition, Tracking and Pointing XI,* 3086, 122–128, 1997.
22. Lanterman, A.D., Miller, M.I., Snyder, D.L., and Miceli, W.J., Jump diffusion processes for the automated understanding of FLIR Scenes, *Proc. SPIE,* 2234, 416–427, 1994.
23. Mahler, R., Global optimal sensor allocation, *Proc. 9th Nat'l. Symp. Sensor Fusion,* 347–366, 1996.
24. Washburn, R.B., A random point process approach to multiobject tracking, *Proc. Am. Control Conf.,* 3, 1846–1852, 1987.
25. Goodman, I.R., Mahler, R.P.S., and Nguyen, H.T., *Mathematics of Data Fusion,* Kluwer Academic Publishers, Boston, 1997.
26. Koch, W., and Van Keuk, G., Multiple hypothesis track maintenance with possibly unresolved measurements, *IEEE Trans. Aerospace Electronics Systems,* 33(3), 883–892, 1997.
27. Van Trees, H.L., *Detection, Estimation, and Modulation Theory, Part I: Detection, Estimation, and Linear Modulation Theory,* Wiley, New York, 1968.

13

Data Association Using Multiple-Frame Assignments

Aubrey B. Poore, Suihua Lu, and Brian J. Suchomel

CONTENTS

13.1 Introduction

The ever-increasing demand in surveillance is to produce highly accurate target identification and estimation in real time, even for dense target scenarios and in regions of high track contention. Past surveillance sensor systems have relied on individual sensors to solve this problem; however, current and future needs far exceed single sensor capabilities. The use of multiple sensors, through more varied information, has the potential to greatly improve state estimation and track identification. Fusion of information from multiple sensors is part of a much broader subject called data or information fusion, which for surveillance applications is defined as *a multilevel, multifaceted process dealing with the detection, association, correlation, estimation, and combination of data and information from multiple sources to achieve refined state and identity estimation, and complete and timely assessments of situation and threat.*[1] (A comprehensive discussion can be found in Waltz and Llinas.[1]) Level 1 deals with single- and multisource information, involving tracking, correlation, alignment, and association by sampling the external environment with multiple sensors and exploiting other available sources. Numerical processes thus dominate level 1. Symbolic reasoning involving various techniques from artificial intelligence permeates levels 2 and 3.[1]

Within level 1 fusion, architectures for single- and multiple-platform tracking must also be considered. These are generally delineated into centralized, distributed, and hybrid architectures,[2–4] each with its advantages and disadvantages. The architecture most appropriate to the current development is that of a centralized tracking, wherein all measurements are sent to one location and processed with tracks being transmitted back to the different platforms. This architecture is optimal in that it is capable of producing the best track quality (e.g., purity and accuracy) and a consistent air picture.[3,4] Although this architecture is appropriate for single-platform tracking, it may be unacceptable for multiple-platform tracking for several reasons. For example, communication loading and the single-point failure problems are important shortcomings. However, this architecture does provide a baseline against which other architectures should be compared. The case of distributed data association is discussed further in Section 13.6.

The methods for centralized tracking follow two different approaches: single-frame and multiple-frame processing. The three basic methods in single-frame processing are nearest neighbor, joint probabilistic data association (JPDA), and global nearest neighbor. The nearest neighbor works well when the objects are all far apart and with infrequent clutter. The JPDA works well for a few targets in moderate clutter, but is computationally expensive and may corrupt the target recognition or discrimination information.[5] The global nearest neighbor approach is posed as a two-dimensional assignment problem (for which there are algorithms that solve the problem optimally in polynomial time) and it has been successful for cases of moderate target density and light clutter.

Deferred logic techniques consider several data sets or frames of data all at once in making data association decisions. At one extreme is batch processing in which all observations (from all time) are processed together. This method is too computationally intensive for real-time applications. The other extreme is sequential processing. This chapter examines deferred logic methods that fall between these two extremes. The most popular deferred logic method used to track large numbers of targets in low-to-moderate clutter is called multiple-hypothesis tracking (MHT) in which a tree of possibilities is built, likelihood scores are assigned to each track, an intricate pruning logic is developed, and the data association problem is solved using an explicit enumeration scheme. The use of these enumeration schemes to solve this nondeterministic polynomial (NP)-hard combinatorial optimization problem in real time is inevitably faulty in dense scenarios, because the time required to solve the problem optimally can grow exponentially with the size of the problem.

Over the past 10 years a new formulation and class of algorithms for data association have proven to be superior to all other deferred logic methods.[3,6–13] This formulation is based on multidimensional assignment problems and the algorithms, on Lagrangian relaxation. The use of combinatorial optimization in multitarget tracking dates back to the pioneering work of Morefield,[14] who used integer programming to solve set packing and covering problems arising from a data association problem. MHT has been popularized by the fundamental work of Reid.[15] These works are further discussed in Blackman and Popoli,[3] Bar-Shalom and Li,[5] and Waltz and Llinas,[1] all of which also serve as excellent introductions to the field of multitarget tracking and multisensor data fusion. Bar-Shalom and coworkers[16–20] have also formulated sensor fusion problems in terms of these multidimensional assignment problems and have developed algorithms, as discussed in Section 13.4.

The performance of any tracking system is dependent on a large number of components. Having one component that is superior to all others does not guarantee a superior tracking system. To address some of these other issues, this chapter provides a brief overview of the many issues involved in the design of a tracking system, placing the problem within the context of the more general surveillance and fusion problem in Section 13.2. The formulation of the problem is presented in Section 13.3, and an overview of the Lagrangian relaxation–based methods appears in Section 13.4. Section 13.5 contains a summary of some opportunities for future investigation.

13.2 Problem Background

A question that often arises is that of the difference between air traffic control and surveillance. In the former, planes, through their beacon codes, generally identify themselves so that observations can be associated with the correct plane; whereas in surveillance, the objects being tracked do not identify themselves, requiring a figure of merit to be derived for the association of a sequence of observations to a particular target. The term *target* is used rather than *object*. This chapter addresses surveillance needs and describes the use of likelihood ratios for track association.

The targets under consideration are classified as point or small targets; measurements or observations of these targets are in the form of kinematic information, such as range, azimuth, elevation, and range rate. Future sensor systems will provide additional features or attribute information.[3]

A general surveillance problem involves the use of multiple platforms, such as ships, planes, or stationary ground-based radar systems, on which one or more sensors are located for tracking multiple objects. Optimization problems permeate the field of surveillance, particularly in the collection and fusion of information. First, there are the problems of routing and scheduling surveillance platforms and then dynamically retasking the platforms as more information becomes available. For each platform, the scarce sensor resources must be allocated and managed to maximize the information returned. The second area, information fusion, is the subject of this chapter.

Many issues are involved in the design of a fusion system for multiple surveillance platforms, such as fusion architectures, communication links between sensor platforms, misalignment problems, tracking coordinate systems, motion models, likelihood ratios, filtering and estimation, and the data association problem of partitioning reports into tracks and false alarms. The recent book, *Design and Analysis of Modern Tracking Systems*, by Blackman and Popoli[3] presents an excellent overview of these topics and also puts forward an extensive list of other references.

One aspect of surveillance that is seldom discussed in the literature is the development of data structures required to put all of this information together efficiently. In reality, a tracking system is generally composed of a dynamic search tree that organizes this information and recycles the memory for real-time processing. However, the central problem is the data association problem.

To place the current data association problem within the context of the different architectures of multiple-platform tracking, a brief review of the architectures is helpful. The first architecture is *centralized fusion*, in which raw observations are sent from the multiple platforms to a central processing unit where they can be combined to give superior state estimation (compared to sensor-level fusion).[3] At the other extreme is *track fusion*, wherein each sensor forms tracks along with the corresponding statistical information from its own reports and then sends this preprocessed information to a processing unit that correlates the tracks. Once the correlation is complete, the tracks can be combined and the statistics can be modified appropriately. In reality, many sensor systems are *hybrids* of these two architectures, in which some preprocessed data and some raw data are used and switches between the two are possible. A discussion of the advantages and disadvantages of these architectures is presented in the Blackman and Popoli book.[3] The centralized and hybrid architectures are most applicable to the current data association problem.

13.3 Assignment Formulation of Some General Data Association Problems

The goal of this section is to formulate the data association problem for a large class of multiple-target tracking and sensor fusion applications as a multidimensional assignment problem. This development extracts the salient features and assumptions that occur in a large class of these problems and is a brief update to an earlier work.[21] A general class of data association problems was posed as set packing problems by Morefield[14] in 1977. Using an abstract view of Morefield's work to include set coverings, packings, and partitionings, this section proceeds to formulate the assignment problem.

In tracking, a common surveillance challenge is to estimate the past, current, or future state of a collection of targets (e.g., airplanes in the air, ships on the sea, or automobiles on the ground) from a sequence of scans of the surveillance region by one or more sensors. This work specifically addresses *small* targets[22] for which the sensors generally supply kinematic information such as range, azimuth, elevation, range rate, and some limited attribute or feature information.

Suppose that one or more sensors, either colocated or distributed, survey the surveillance region and produce a stream of observations (or measurements), each with a distinct time tag. These observations are then arranged into sets of observations called *frames of data*. Mathematically, let $Z(k) = \{z_{i_k}^k\}_{i_k=1}^{M_k}$ denote the kth frame of data, where each $z_{i_k}^k$ is a vector of noise-contaminated observations with an associated time tag $t_{i_k}^k$. The index k represents the frame number and i_k represents the i_kth observation in frame k. An observation in the frame of data $Z(k)$ may emanate from a true target or may be a false report.

This discussion assumes that each frame of data is a *proper frame*, in which each target is seen no more than once. For a rotating radar, one sweep or scan of the field of view generally constitutes a proper frame. For sensors such as electronically scanning phased array radar, wherein the sensor switches from surveillance to tracking mode, the partitioning of the data into proper frames of data is interesting as there are several choices. More efficient partitioning methods would be addressed in the forthcoming work.

The data association problem is that of determining which observations or sensor reports emanate from which targets and which reports are false alarms. The combinatorial optimization problem that governs a large number of data association problems in multitarget tracking and multisensor data fusion[1,3,5,6,11–14,17–20,23] is generally posed as

Maximize

$$\{P(\Gamma = \gamma | Z^N) | \lambda \in \Gamma^*\} \tag{13.1}$$

where Z^N represents N data sets (Equation 13.2), γ is a partition of indices of the data (Equations 13.3 and 13.4a through 13.4d), Γ^* the finite collection of all such partitions (Equations 13.4a through 13.4d), Γ a discrete random element defined on Γ^*, γ^0 a reference partition, and $P(\Gamma = \gamma | Z^N)$ the posterior probability of a partition γ being true given the data Z^N. Each of these terms must be defined. The objective then is to formulate a reasonably general class of these data association problems (Equation 13.1) as multidimensional assignment problems (Equation 13.15).

In the surveillance example, the data sets were observations of the objects in the surveillance region, including false reports. Including more general types of data, such as tracks and track-observation combinations, as well as observations, Reid[15] used the term *reports* for the contents of the data sets. Thus, let $Z(k)$ denote a data set of M_k reports $\{z_{i_k}^k\}_{i_k=1}^{M_k}$, and Z^N the cumulative data set of N such sets defined, respectively, by

$$Z(k) = \{z_{i_k}^k\}_{i_k=1}^{M_k} \quad \text{and} \quad Z^N = \{Z(1), \dots, Z(N)\} \tag{13.2}$$

In multisensor data fusion and multitarget tracking, the data sets $Z(k)$ may represent different classes of objects. For track initiation in multitarget tracking, the objects are observations that must be partitioned into tracks and false alarms. In this formulation of track maintenance (Section 13.4), one data set is comprised of tracks and the remaining data sets include observations that are assigned either to existing tracks, false observations, or observations of new tracks. In sensor-level tracking, the objects to be fused are tracks.[1] In centralized fusion,[1,3] the objects can all be observations that represent targets or false reports, and the problem is to determine which observations emanate from a common source.

The next task is to define what is meant by a partition of the cumulative data set, Z^N, in Equation 13.2. Because this definition is independent of the actual data in the cumulative data set Z^N, a partition of the indices in Z^N must first be defined. Let

$$I^N = \{I(1), I(2), \dots, I(N)\}, \quad \text{where } I(k) = \{i_k\}_{i_k=1}^{M_k} \tag{13.3}$$

denote the indices in the data sets (Equation 13.2). A partition γ of I^N and the collection of all such partitions Γ^* are defined by

$$\gamma = \{\gamma_1, \dots, \gamma_{n(\gamma)} | \gamma_i \neq \varnothing \quad \text{for each } i\} \tag{13.4a}$$

$$\gamma_i \cap \gamma_j = \varnothing \quad \text{for } i \neq j \tag{13.4b}$$

$$I^N = \cup_{j=1}^{n(\gamma)} \gamma_j \tag{13.4c}$$

$$\Gamma^* = \{\gamma | \gamma \text{ satisfies Equations 13.4a and 13.4b}\} \tag{13.4d}$$

Here, $\gamma_i \subset I^N$ in Equation 13.4a will be called a track, so that $n(\gamma)$ denotes the number of tracks (or elements) in the partition γ. A $\gamma \in \Gamma^*$ is called a *set partitioning* of the indices I^N, if the properties in Equations 13.4a through 13.4c are valid; a *set covering* of I^N, if the property in Equation 13.4b is omitted, but the other two properties in Equations 13.4a and 13.4c are retained; and a *set packing* if the property in Equation 13.4c is omitted, but Equations 13.4a and 13.4b are retained.[24] A partition $\gamma \in \Gamma^*$ of the index set I^N induces a partition of the data Z^N via

$$Z_\gamma = \{Z_{\gamma_1}, \dots, Z_{\gamma_{n(\gamma)}}\} \quad \text{where } Z_{\gamma_i} \subset Z^N \tag{13.5}$$

Clearly, $Z_{\gamma_i} \cap Z_{\gamma_j} = \varnothing$ for $i \neq j$ and $Z^N = \cup_{j=1}^{n(\gamma)} Z_{\gamma_j}$. Each Z_{γ_i} is considered to be a track of data. Note that a Z_{γ_i} need not have observations from each frame of data, $Z(k)$, but it must, by definition, have at least one observation.

Under several independence assumptions between tracks,[21] a probabilistic framework can be established in which

$$P(\Gamma = \gamma | Z^N) = \frac{C}{p(Z^N)} \prod_{\gamma_i \in \gamma} p\left(Z_{\gamma_i}\right) G(\gamma_i) \tag{13.6}$$

where C is a constant and G is a function. This completes the formulation of the general data association problem as presented in the works of Poore[21] and Morefield.[14]

The next objective is to refine this formulation in a way that is amenable to the assignment problem. For notational convenience in representing tracks, add a *zero index* to each of the index sets $I(k)$ ($k = 1, \dots, N$) in Equation 13.3 and a *dummy report* z_0^k to each of the data sets $Z(k)$ in Equation 13.2, and require that each

$$\gamma_i = (i_1, \dots, i_N) \tag{13.7}$$

$$Z_{\gamma_i} = Z_{i_1 \dots i_n} \equiv \left(z_{i_1}^1, \dots, z_{i_N}^N\right)$$

where i_k and $z_{i_k}^k$ can assume the values of 0 and z_0^k, respectively. The dummy report z_0^k serves several purposes in the representation of missing data, false reports, initiating tracks, and terminating tracks. If Z_{γ_i} is missing an actual report from the data set $Z(k)$, then $\gamma_i = (i_1, \dots, i_{k-1}, 0, i_{k+1}, \dots, i_N)$ and $Z_{\gamma_i} = \{z_{i_1}^1, \dots, z_{i_{k-1}}^{k-1}, z_0^k, z_{i_{k+1}}^{k+1}, \dots, z_{i_N}^N\}$. A false report $z_{i_k}^k (i_k > 0)$ is represented by $\gamma_i = (0, \dots, 0, i_k, 0, \dots, 0)$ and $Z_{\gamma_i} = \{z_0^1, \dots, z_0^{k-1}, z_{i_k}^k, z_0^{k+1}, \dots, z_0^N\}$, in which there is only one actual report. The partition γ^0 of the data in which all reports are declared to be false reports is defined by

$$Z_{\gamma^0} = \{Z_{0\dots0 i_k 0\dots0} \equiv \left(z_0^1, \dots, z_0^{k-1}, z_{i_k}^k, z_0^{k+1}, \dots, z_0^N\right) | i_k = 1, \dots, M_k \quad k = 1, \dots, N\} \tag{13.8}$$

If each data set, $Z(k)$, represents a "proper frame" of observations, a track that initiates on frame $m > 1$ will contain only the dummy report z_0^k from each of the data sets, $Z(k)$, for each $k = 1, \dots, m - 1$. Likewise, a track that terminates on frame m would have only the dummy report from each of the data sets for $k > m$. These representations are discussed further in Section 13.3 for both track initiation and track maintenance.

The use of the 0–1 variable

$$z_{i_1,\dots,i_N} = \begin{cases} 1 & \text{if } (i_1, \dots, i_N) \in \gamma \\ 0 & \text{otherwise} \end{cases} \tag{13.9}$$

yields an equivalent characterization of a partition (Equations 13.4a through 13.4d and 13.7) as a solution of the equations:

$$\sum_{i_1=0}^{M_1} \cdots \sum_{i_{k-1}=0}^{M_{k-1}} \sum_{i_{k+1}=0}^{M_{k+1}} \cdots \sum_{i_N=0}^{M_N} z_{i_1 \dots i_N} = 1 \tag{13.10}$$

With this characterization of a partition of the cumulative data set Z^N as a set of equality constraints (Equation 13.10), the multidimensional assignment problem can then be formulated.

Observe that for $\gamma_i = (i_1, \ldots, i_N)$, as in Equation 13.7, and the reference partition (Equation 13.8),

$$\frac{P(\Gamma = \gamma | Z^N)}{P(\Gamma = \gamma^0 | Z^N)} \equiv L_\gamma \equiv \prod_{(i_1, \ldots, i_N) \in \gamma} L_{i_1 \ldots i_N} \tag{13.11}$$

where

$$L_{i_1 \ldots i_N} = \frac{p(Z_{i_1 \ldots i_N}) G(Z_{i_1 \ldots i_N})}{\prod p(Z_{0 \ldots 0 i_k 0 \ldots 0}) G(Z_{0 \ldots 0 i_k 0 \ldots 0})} \tag{13.12}$$

Here, the index i_k in the denominator corresponds to the kth index of $Z_{i_1 \ldots i_N}$ in the numerator. Next, define

$$c_{i_1 \ldots i_N} = -\ln L_{i_1 \ldots i_N} \tag{13.13}$$

so that

$$-\ln \left[\frac{P(\gamma | Z^N)}{P(\gamma^0 | Z^N)} \right] = \sum_{(i_1, \ldots, i_N) \in \gamma} c_{i_1 \ldots i_N} \tag{13.14}$$

Thus, in view of the characterization of a partition (Equations 13.4a through 13.4d and 13.5) specialized by Equation 13.7 as a solution of Equation 13.10, the independence assumptions[21] and the expansion (Equation 13.6) problem (Equation 13.1) are equivalently characterized as the following N-dimensional assignment problem:
Minimize

$$\sum_{i_1=0}^{M_1} \cdots \sum_{i_N=0}^{M_N} c_{i_1 \ldots i_N} z_{i_1 \ldots i_N} \tag{13.15}$$

Subject to

$$\sum_{i_2=0}^{M_2} \cdots \sum_{i_N=0}^{M_N} z_{i_1 \ldots i_N} = 1 \quad i = 1, \ldots, M_1 \tag{13.16}$$

$$\sum_{i_1=0}^{M_1} \cdots \sum_{i_{k-1}=0}^{M_{k-1}} \sum_{i_{k+1}=0}^{M_{k+1}} \cdots \sum_{i_N=0}^{M_N} z_{i_1 \ldots i_N} = 1 \tag{13.17}$$

for $i_k = 1, \ldots, M_k$ and $k = 2, \ldots, N-1$

$$\sum_{i_1=0}^{M_1} \cdots \sum_{i_{N-1}=0}^{M_{N-1}} z_{i_1 \ldots i_N} = 1 \quad i_N = 1, \ldots, M_N \tag{13.18}$$

$$z_{i_1 \ldots i_N} \in \{0, 1\} \quad \text{for all } i_1, \ldots, i_N \tag{13.19}$$

where $c_{0 \ldots 0}$ is arbitrarily defined to be zero. Note that the definition of a partition and the 0–1 variable $z_{i_1 \ldots i_N}$ in Equation 13.9 imply $z_{0 \ldots 0} = 0$. (If $z_{0 \ldots 0}$ is not preassigned to zero and $c_{0 \ldots 0}$ is defined arbitrarily, then $z_{0 \ldots 0}$ is determined directly from the value of $c_{0 \ldots 0}$, because it does not enter the constraints other than being a 0–1 variable.) Also, each cost coefficient with exactly one nonzero index is zero (i.e., $c_{0...0 i_k 0...0} = 0$ for all $i_k = 1, ..., M_k$ and $k = 1, ..., N$) based on the use of the normalizing partition γ^0 in the likelihood ratio in Equations 13.11 and 13.12. Deriving the same problem formulation is possible when not assuming that the cost coefficients with exactly one nonzero index are zero;[21] however, other formulations can be reduced to the above-mentioned using the invariance theorem presented in Section 13.5.1.

Derivation of the assignment problem (Equation 13.15) leads to several pertinent remarks. The definition of a partition in Equations 13.4 and 13.5 implies that each actual report belongs to, at most, one track of reports Z_{γ_i} in a partition Z_γ of the cumulative data set. This can be modified to allow multiassignments of one, some, or all of the actual reports. The assignment problem changes accordingly. For example, if $z_{i_k}^k$ is to be assigned no more than, exactly, or no less than $n_{i_k}^k$ times, then the "=1" in the constraint (Equation 13.15) is changed to "$\leq$, =, $\geq n_{i_k}^k$," respectively. (This allows both set coverings and packings in the formulation.) In making these changes, one must pay careful attention to the independence assumptions.[21] Inequality constraint problems, in addition to the problem of unresolved closely spaced objects, are common elements of the sensor fusion multiresolution problem.

The likelihood ratio, $L_{i_1...i_N}$ is a complicated expression containing probabilities for detection, termination, model maneuvers, and density functions for an expected number of false alarms and initiating targets. The likelihood that an observation arises from a particular target is also included; this requires target dynamics to be estimated through the corresponding sequence of observations $\{z_{i_1}^1, ..., z_{i_k}^k, ..., z_{i_N}^N\}$. Filtering such sequences is the most time-consuming part of the problem formulation and considerably exceeds the time required to solve the data association problem. Derivations for appropriate likelihood ratios can be found in the works of Poore[21] and Blackman and Popoli.[3]

13.4   Multiple-Frame Track Initiation and Track Maintenance

A general expression has been developed for the data association problem arising from tracking. The underlying tracking application is a dynamic one in that information from one or more sensors continually arrives at the processing unit where the data is partitioned into frames of data for the assignment problem. Thus, the dimension of the assignment problem grows with the number of frames of data, N. Processing all the data at once, called batch processing, eventually becomes computationally unacceptable as the dimension N increases. To circumvent this problem, a sliding window can be used, wherein changes to data association decisions are limited to those within the window. The first sliding window (single pane) formulation was presented in 1992[25] and refined in 1997[26] to include a dual-pane window. The single-pane sliding window and its refinements are described in this section.

The moving window and the resulting *search tree* are truly the heart of a tracking system. The importance of the underlying data structures to the efficiency of a tracking system cannot be overemphasized; however, these data structures can be very complex and are not discussed here.

13.4.1   Track Initiation

The pure track initiation problem is to formulate and solve the assignment problem described in the previous section with an appropriate number of frames of data N.

The choice of the number of frames N is not trivial. Choices of $N = 4, 5, 6$ have worked well in many problems; however, a good research topic would involve the development of a method that adaptively chooses the number of frames on the basis of the problem complexity.

13.4.2 Track Maintenance Using a Sliding Window

The term *track maintenance* as used in this section includes three functions: (1) extending existing tracks, (2) terminating existing tracks, and (3) initiating new ones. Suppose that the observations on P frames of observations have been partitioned into tracks and false alarms. and that K new frames of observations are to be added. One approach to solving the resulting data association problem is to formulate it as a track initiation problem with $P + K$ frames. This is the previously mentioned *batch* approach.

The *deferred logic* approach adopted here would treat the track extension problem within the framework of a window, sliding over the frames of observations. The P frames are partitioned into two components: the first H frames, in which *hard* data association decisions are made, and the next S frames, in which *soft* decisions are made. The K new frames of observations are added, making the number of frames in the sliding window $N = S + K$, whereas the number of frames in which data association decisions are hard is $H = P - S$. Various sliding windows can be developed, including single-pane, double-pane, and multiple-pane windows. The intent of each of these is efficiency in solving the underlying tracking problem.

13.4.2.1 Single-Pane Sliding Window

Assuming $K = 1$, let M_0 denote the number of confirmed tracks (i.e., tracks that arise from the solution of the data association problem) on frame k constructed from a solution of the data association problem utilizing frames up to $k + N - 1$. Data association decisions are fixed on frames up to k. Now a new frame of data is added. Thus, frame k denotes a list of tracks and frames $k + 1$ to $k + N$ denote observations. For $i_0 = 1, \ldots, M_0$, the i_0th such track is denoted by T_{i_0} and the $(N + 1)$-tuple $\{T_{i_0}, z_{i_1}^1, \ldots, z_{i_N}^N\}$ denotes a track T_{i_0} plus a set of observations $\{z_{i_1}^1, \ldots, z_{i_N}^N\}$, actual or dummy, that are feasible with the track T_{i_0}. The $(N + 1)$-tuple $\{T_0, z_{i_1}^1, \ldots, z_{i_N}^N\}$ denotes a track that initiates in the sliding window. A false report in the sliding window is one in which all indices except one are zero in the $(N + 1)$-tuple $\{T_0, z_{i_1}^1, \ldots, z_{i_N}^N\}$.

The hypothesis about a partition $\gamma \in \Gamma^*$ being true is now conditioned on the truth of the M_0 tracks entering the N-scan window. (Thus, the assignments before reaching this sliding window are fixed.) The likelihood function is given by $L_\gamma = \Pi_{\{T_{i_0}, z_{i_1}, \ldots, z_{i_N} \in \gamma\}} L_{i_0 i_1 \ldots i_N}$, where $L_{i_0 i_1 \ldots i_N} = L_{T_{i_0}} L_{i_1 \ldots i_N}$, $L_{T_{i_0}}$ is the composite likelihood from the discarded frames just before the first scan in the window for $i_0 > 0$, $L_{T_0} = 1$, and $L_{i_1 \ldots i_N}$ is defined as in Equation 13.8 for the N-scan window. ($L_{T_0} = 1$ is used for any tracks that initiate in the sliding window.) Thus, the track extension problem can be formulated as Maximize $\{L_\gamma | \gamma \in \Gamma^*\}$. With the same convention as in Section 13.3, a feasible partition is one that is defined by the properties in Equations 13.4 and 13.7. Analogously, the definition of the 0–1 variable

$$z_{i_0 i_1 \ldots i_N} = \begin{cases} 1 & \text{if } \{T_{i_0}, z_{i_1}^1, \ldots, z_{i_N}^N\} \text{ is assigned as a unit} \\ 0 & \text{otherwise} \end{cases}$$

and the corresponding cost for the assignment of the sequence $\{T_{i_0}, z_{i_1}, \ldots, z_{i_N}\}$ to a track by $c_{i_0 i_1 \ldots i_N} = -\ln L_{i_0 i_1 \ldots i_N}$ yield the following multidimensional assignment formulation of the data association problem for track maintenance:

Minimize

$$\sum_{i_0=0}^{M_0} \cdots \sum_{i_N=0}^{M_N} c_{i_0 \ldots i_N} z_{i_0 \ldots i_N} \tag{13.20}$$

Subject to

$$\sum_{i_1=0}^{M_1} \cdots \sum_{i_N=0}^{M_N} z_{i_0 \ldots i_N} = 1 \quad i_0 = 1, \ldots, M_0 \tag{13.21}$$

$$\sum_{i_0=0}^{M_0} \sum_{i_2=0}^{M_2} \cdots \sum_{i_M=0}^{M_N} z_{i_{1_0} i_1 \ldots i_N} = 1 \quad i = 1, \ldots, M_1 \tag{13.22}$$

$$\sum_{i_0=0}^{M_0} \cdots \sum_{i_{k-1}=0}^{M_{k-1}} \sum_{i_{k+1}=0}^{M_{k+1}} \cdots \sum_{i_M=0}^{M_N} z_{i_0 \ldots i_N} = 1 \tag{13.23}$$

for $i_k = 1, \ldots, M_k$ and $k = 2, \ldots, N-1$

$$\sum_{i_0=0}^{M_0} \cdots \sum_{i_{N-1}=0}^{M_{N-1}} z_{i_0 \ldots i_N} = 1 \quad i_N = 1, \ldots, M_N \tag{13.24}$$

$$z_{i_0 \ldots i_N} \in \{0, 1\} \quad \text{for all } i_0, \ldots, i_N \tag{13.25}$$

Note that the association problem involving N frames of observations is an N-dimensional assignment problem for track initiation and an $(N + 1)$-dimensional assignment problem for track maintenance.

13.4.2.2   Double- and Multiple-Pane Window

In the single-pane window, the first frame contains a list of tracks and the remaining N frames contain observations. The assignment problem is of dimension $N + 1$, where N is the number of frames of observations in front of the existing tracks. The same window is being used to initiate new tracks and continue existing ones. A newer approach is based on the belief that once a track is well established, only a few frames are needed to benefit from the soft decisions on track continuation, whereas a longer window is needed for track initiation. Thus, if the current frame is numbered k with frames $k + 1, \ldots, k + N$ being for track continuation, one can go back to frames $k - M, \ldots, k$ and allow observations not attached to tracks that exist through frame k, to be used to initiate new tracks. Indeed, this concept works extremely well in practice and was initially proposed in the work of Poore and Drummond.[26] The next approach to evolve is that of a multiple-pane window, in which the position of hard data association decisions of observations to tracks can vary within the frames $k - 1, \ldots, k, \ldots, k + N$, depending on the difficulty of the problem as measured by track contention. The efficiency of this approach is yet to be determined.

13.5 Algorithms

The multidimensional assignment problem for data association is one of combinatorial optimization and is NP-hard,[27] even for the case $N = 3$. The data association problems arising in tracking are generally sparse, large-scale, and noisy with real-time needs. Given the noise in the problem, the objective is generally to solve the problem to within the noise level. Thus, heuristic methods are recommended; however, efficient branch and bound schemes are potentially applicable to this problem because they provide the baseline against which other heuristic algorithms are judged. Lagrangian relaxation methods have worked extremely well, probably due to the efficiency of nonsmooth optimization methods[28,29] and the fact that the relaxed problem is the two-dimensional assignment problem for which there are some very efficient algorithms, such as the auction[30] and Jonker–Volbenant (JV) algorithms.[31] Another advantage is that these relaxation methods provide both a lower and upper bound on the optimal solution and, thus, some measure of closeness to optimality. (This is obviously limited by the duality gap.)

This section surveys some of the algorithms that have been particularly successful in solving the data association problems arising from tracking. There are, however, many potential algorithms that could be used. For example, greedy randomized adaptive local search procedure (GRASP)[32–34] also has been used successfully.

13.5.1 Preprocessing

Two frequently used preprocessing techniques, *fine gating* and *problem decomposition*, are presented in this section. These two methods can substantially reduce the complexity of the multidimensional assignment problem.

13.5.1.1 Fine Gating

The term *fine gating* is used because this method is the last in a sequence of techniques used to reduce the unlikely paths in the layered graph or pairings of combinations of reports. This method is based on the following theorem.[21]

Theorem 1 (Invariance property)

Let $N > 1$ and $M_k > 0$ for $k = 1, \ldots, N$, and assume $\hat{c}_{0\ldots0} = 0$ and $u_0^k = 0$ for $k = 1, \ldots, N$. Then the minimizing solution and objective function value of the following multidimensional assignment problem are independent of any choice of $u_{i_k}^k = 1$ for $i_k = 1, \ldots, M_k$ and $k = 1, \ldots, N$.

Minimize

$$\sum_{i_1=0}^{M_1} \cdots \sum_{i_N=0}^{M_N} \left(\hat{c}_{i_1\ldots i_N} - \sum_{k=1}^{N} u_{i_k}^k \right) z_{i_0\ldots i_N} + \sum_{k=1}^{N} \sum_{i_k=0}^{M_K} u_{i_k}^k \tag{13.26}$$

Subject to

$$\sum_{i_2=0}^{M_2} \cdots \sum_{i_N=0}^{M_N} z_{i_1\ldots i_N} = 1 \quad i_1 = 1, \ldots, M_1 \tag{13.27}$$

$$\sum_{i_1=0}^{M_1} \cdots \sum_{i_{k-1}=0}^{M_{k-1}} \sum_{i_{k+1}=0}^{M_{k+1}} \cdots \sum_{i_N=0}^{M_N} z_{i_1\ldots i_N} = 1 \tag{13.28}$$

for $i_k = 1, ..., M_k$ and $k = 2, ..., N - 1$

$$\sum_{i_1=0}^{M_1} \cdots \sum_{i_{N-1}=0}^{M_{N-1}} z_{i_1...i_N} = 1 \quad i_N = 1, ..., M_N \tag{13.29}$$

$$z_{i_1...i_N} \in \{0, 1\} \quad \text{for all } i_1, ..., i_N \tag{13.30}$$

If $\hat{c}_{i_1...i_N} = c_{i_1...i_N}$ and $u_{i_k}^k c_{0...0i_k0...0}$ is identified in this theorem, then

$$c_{i_1...i_N} - \sum_{k=1}^{N} c_{0...0i_k0...0} > 0 \tag{13.31}$$

implies that the corresponding 0–1 variable $z_{i_1...i_N}$ and cost $c_{i_1...i_N}$ can be removed from the problem because a lower cost can be achieved with the use of the variables $z_{0...0i_k0...0} = 1$. (This does not mean that one should set $z_{0...0i_k0...0} = 1$ for $k = 1, ..., N$.)

In the special case in which all costs with exactly one nonzero index are zero, this test is equivalent to

$$c_{i_1...i_N} > 0 \tag{13.32}$$

13.5.1.2 Problem Decomposition

Decomposition of the multidimensional assignment problem into a sequence of disjoint problems can improve the solution quality and the speed of the algorithm, even on a serial machine. The following decomposition method, originally presented in the work of Poore et al.,[35] uses graph theoretic methods.

Decomposition of the multidimensional assignment problem is accomplished by determining the connected components of the associated layered graph. Let

$$\mathcal{Z} = \left\{ z_{i_1 i_2...i_N} \mid z_{i_1 i_2...i_N} \text{ is not preassigned to zero} \right\} \tag{13.33}$$

denote the set of assignable variables. Define an undirected graph $G(N, A)$ where the set of nodes is

$$N = \left\{ z_{i_k}^k \mid i_k = 1, ..., M_k; k = 1, ..., N \right\} \tag{13.34}$$

and the set of arcs is

$$A = \left\{ \begin{array}{l} \left(z_{j_k}^k, z_{j_l}^l \right) \mid k \neq 1, j_k \neq 0, j_l \neq 0, \\ \text{and there exists } z_{i_1 i_2...i_N} \in \mathcal{Z} \text{ such that } j_k = i_k \text{ and } j_l = i_l \end{array} \right\} \tag{13.35}$$

The nodes corresponding to zero index have not been included in this graph because two variables that have only the zero index in common can be assigned independently.

Connected components of the graph are easily found by constructing a spanning forest through a depth-first search. Furthermore, this procedure can be used at each level in the relaxation (i.e., applied to each assignment problem for $k = 3,..., N$). Note that the decomposition algorithm depends only on the problem structure (i.e., the feasibility of the variables) and not on the cost function.

As an aside, this decomposition often yields small problems that are best and more efficiently handled by a branch and bound or an explicit enumeration procedure to avoid the overhead associated with relaxation. The remaining components are solved by relaxation. However, extensive decomposition can be time-consuming and limiting the number of components to approximately ten is desirable, unless one is using a parallel machine.

13.5.2   Lagrangian Relaxation Algorithm for the Assignment Problem

This section presents Lagrangian relaxation algorithms for multidimensional assignment, which have proven to be computationally efficient and accurate for tracking purposes. The N-dimensional assignment problem has $M_1 + \cdots + M_N$ individual constraints, which can be grouped into N constraint sets. Let $u_k = (u_0^k, u_1^k, ...,u^k M_k)$, denote the $(M_k + 1)$-dimensional Lagrange multiplier vector associated with the kth constraint set, with $u_0^k = 0$ and $k = 1, ..., N$. The full set of multipliers is denoted by the vector $u = [u^1,..., u^N]$. The multidimensional assignment problem is relaxed to an n-dimensional assignment problem by incorporating $N - n$ constraint sets into the objective function. There are several choices of n. The case $n = 0$ yields the linear programming dual; $n = 2$ yields a two-dimensional assignment problem and has been highly successful in practice.

Although any constraint sets can be relaxed, sets $n + 1, ..., N$ are chosen for convenience. In the tracking problem using a sliding window, these are the correct sets given the data structures that arise from the construction of tracks.

The *relaxed problem* for multiplier vector u is given by

$$L_n(u^{n+1}, ..., u^N) = \tag{13.36}$$

Minimize

$$\sum_{i_1=0}^{M_1} \cdots \sum_{i_N=0}^{M_N} \left(c_{i_1...i_N} + \sum_{k=n+1}^{N} u_{i_k}^k \right) z_{i_1...i_N} - \sum_{k=n+1}^{N} \sum_{i_k=0}^{M_k} u_{i_k}^k \tag{13.37}$$

Subject to

$$\sum_{i_2=0}^{M_2} \cdots \sum_{i_N=0}^{M_N} z_{i_1...i_N} = 1 \quad i_1 = 1, ..., M_1 \tag{13.38}$$

$$\sum_{i_1=0}^{M_1} \sum_{i_3=0}^{M_3} \cdots \sum_{i_N=0}^{M_N} z_{i_1...i_N} = 1 \quad i_2 = 1, ..., M_2 \tag{13.39}$$

$$\vdots$$

$$\sum_{i_1=0}^{M_1} \cdots \sum_{i_{n-1}=0}^{M_{n-1}} \sum_{i_{n+1}=0}^{M_{n+1}} \cdots \sum_{i_N=0}^{M_N} z_{i_1...i_N} = 1 \quad i_n = 1, ..., M_n \tag{13.40}$$

$$z_{i_1...i_N} \in \{0, 1\} \quad \text{for all } \{i_1, ..., i_N\} \tag{13.41}$$

The above-mentioned problem can be reduced to an n-dimensional assignment problem using the transformation

$$x_{i_1 i_2 \ldots i_n} = \sum_{i_{n+1}=0}^{M_{n+1}} \cdots \sum_{i_N=0}^{M_N} z_{i_1 \ldots i_n i_{n+1} \ldots i_N} \quad \text{for all } i_1, i_2, \ldots, i_n \tag{13.42}$$

$$c_{i_1 i_2 \ldots i_n} = \text{Min} \left\{ c_{i_1 \ldots i_n} + \sum_{k=n+1}^{N} u_{i_k}^u \,\middle|\, \text{ for all } i_{n+1}, \ldots, i_N \right\} \tag{13.43}$$

$$c_{0 \ldots 0} = \sum_{i_{n+1}=0}^{M_{n+1}} \cdots \sum_{i_N=0}^{M_N} \text{Min} \left\{ 0, c_{0 \ldots 0 i_{n+1} \ldots i_N} + \sum_{k=n+1}^{N} u_{i_k}^u \right\} \tag{13.44}$$

Thus, the Lagrangian relaxation algorithm can be summarized as follows:

1. Solve the problem
   Maximize

$$L_n(u^{n+1}, \ldots, u^N) \tag{13.45}$$

2. Give an optimal or near-optimal solution $(u^{n+1}, \ldots, u^N)$; solve the above-mentioned assignment problem for this given multiplier vector. This produces an alignment of the first n indices. Let these be enumerated by $\{(i_1^j, \ldots, i_n^j)\}_j^J = 0$, where $(i_1^0, \ldots, i_n^0) = (0, \ldots, 0)$. Then, the variable and cost coefficient

$$x_{j i_{n+1} \ldots i_N} = z_{i_1^j \ldots i_n^j i_{n+1} \ldots i_N} \tag{13.46}$$

$$c_{j i_{n+1} \ldots i_N} = c_{i_1^j \ldots i_n^j i_{n+1} \ldots i_N} \tag{13.47}$$

satisfy the following $(N - n + 1)$-dimensional assignment problem.
Minimize

$$\sum_{j=0}^{J} \sum_{i_{n+1}=0}^{M_{n+1}} \cdots \sum_{i_N=0}^{M_N} c_{j i_{n+1} \ldots i_N} z_{j i_{n+1} \ldots i_N} \tag{13.48}$$

Subject to

$$\sum_{i_{n+1}=0}^{M_{n+1}} \cdots \sum_{i_N=0}^{M_N} z_{j i_{n+1} \ldots i_N} = 1 \quad j = 1, \ldots, J \tag{13.49}$$

$$\sum_{j=0}^{J} \sum_{i_{n+2}=0}^{M_3} \cdots \sum_{i_N=0}^{M_N} z_{ji_{n+1}\ldots i_N} = 1 \quad \text{for } i_{n+1} = 1, \ldots, M_{n+1} \tag{13.50}$$

$$\vdots$$

$$\sum_{j=0}^{J} \cdots \sum_{i_{N-1}=0}^{M_{N-1}} z_{ji_{n+1}\ldots i_N} = 1 \quad i_N = 1, \ldots, M_N \tag{13.51}$$

$$z_{ji_{n+1}\ldots i_N} \in \{0, 1\} \quad \text{for all } \{j, i_{n+1}, \ldots, i_N\} \tag{13.52}$$

Let the nonzero 0–1 variables in a solution be denoted by $z_{ji_{n+1}^j \ldots i_N^j} = 1$, then the solution of the original problem is $z_{i_1^j \ldots i_n^j i_{n+1}^j \ldots i_N^j} = 1$ with all remaining values of z being zero.

In summary, this algorithm is the result of the N-dimensional assignment problem being relaxed to an n-dimensional assignment problem by relaxing $N - n$ constraint sets. The problem of restoring feasibility is defined as an $(N - n + 1)$-dimensional problem.

Notice that the problem of maximizing $L_n(u^{n+1}, \ldots, u^N)$ is one of nonsmooth optimization. Bundle methods[36,37] have proven to be particularly successful for this purpose.

13.5.2.1 Class of Algorithms

Earlier work[7,11,38,39] involved relaxing an N-dimensional assignment problem to an N-dimensional one, which is NP-hard for $N > 2$, by relaxing one set of constraints. The corresponding dual function $L_n(u^{n+1}, \ldots, u^N)$ is piecewise linear and concave, but the evaluation of the function and subgradients, as needed by the nonsmooth maximization, requires an optimal solution of an NP-hard n-dimensional assignment problem when $n > 2$. To address the real-time needs, suboptimal solutions to the relaxed problem must be used; however, suboptimal solutions only provide approximate function and subgradient values. To moderate this difficulty, Poore and Rijavec[39] used a concave, piecewise affine merit function to provide guidance for the function values for the nonsmooth optimization phase. This approach computed approximate subgradients from good-quality feasible solutions obtained from multiple relaxation and recovery cycles executed at lower levels in a recursive fashion. (The number of cycles can be any fixed number greater than or equal to 1 or it can be chosen adaptively by allowing the nonsmooth optimization solver to converge to within user-defined tolerances.) Despite these approximations, the numerical performance of these previous algorithms has been quite good.[39]

A variation on the N-to-$(N - 1)$ relaxation algorithm using one cycle is attributable to Deb et al.[16] Similar approximate function and subgradient values are used at each level of the relaxation process. To moderate this difficulty, they modify the accelerated subgradient method of Shor[40] by further weighting the search direction in the direction of violated constraints, and they report improvement over the accelerated subgradient method. They do not, however, use problem decomposition and a merit function as in Poore and Rijavec's previous work.[39]

When relaxing an N-dimensional assignment problem to an n-dimensional one, the one case in which the aforementioned difficulties are resolved is for $n = 2$; this is the algorithm that is currently used in the Poore and Rijavec tracking system.

13.5.3   Algorithm Complexity

In the absence of a complexity analysis, computational experience shows that between 90 and 95% of all the computation time is consumed in the search to find the minimizer in the list of feasible arcs that are present in the assignment problem. Thus, the time required to solve an assignment problem appears to be computationally linear in the number of feasible arcs (i.e., tracks) in the assignment problem. Obviously, the gating techniques used in tracking to control the number of feasible tracks are fundamental to managing complexity.

13.5.4   Improvement Methods

The above-mentioned relaxation methods have been enormously successful in providing quality solutions to the assignment problem. Improvement techniques are fundamentally important. On the basis of the relaxation and a branch and bound framework,[10] significant improvements in the quality of the solution have been achieved for tracking problems. On difficult problems in tracking, the improvement can be significant. Although straight relaxation can produce solutions to within 3% of optimality, the improvement techniques can produce solutions to within 0.5% of optimal on at least one class of very difficult tracking problems. Given the ever-increasing need for more accurate solutions, further improvements, such as local search methods, should be developed.

13.6   Future Directions

Following the previous sections' overview of the problem formulation and presentation of highly successful algorithms, this section summarizes some of the open issues that should be addressed in the future.

13.6.1   Other Data Association Problems and Formulations

Several alternate formulations, such as more general set packings and coverings of the data, should be pursued. The current formulation is sufficiently general to include other important cases, such as multisensor resolution problems and unresolved closely spaced objects in which an observation is necessarily assigned to more than one target. The assigning of a report to more than one target can be accomplished within the context of the multidimensional assignment problems by using inequality constraints. The formulation, algorithms, and testing are yet to be systematically developed. Although allowing an observation to be assigned to more than one target is easy to model mathematically, allowing a target to be assigned to more than one observation is difficult and may introduce nonlinearity into the objective function due to a loss of the independence assumptions discussed in Section 13.3. Certainly, the original formulation of Morefield[14] is worth revisiting.

13.6.2   Frames of Data

The use of the multidimensional assignment problems to solve the central data association problem rests on the assumption that each target is seen at most once in each frame of data (i.e., the frame is a "proper" frame). For a mechanically rotating radar, this is reasonably easy to approximate as a sweep of the surveillance region. For electronically scanning sensors, which can switch from searching mode to tracking mode, the solution to this partitioning problem is less obvious. Although one such solution has been developed, the formulation and solution are not yet optimal.

13.6.3 Sliding Windows

The batch approach to the data association problem of partitioning the observations into tracks and false alarms is to add a frame of data (observations) to the existing frames and then formulate and solve the corresponding multidimensional assignment problem. The dimension of the assignment problem increases by the number of frames added. To avoid the intractability of this approach, a moving window approach was developed in 1992 wherein the data association problem is resolved over the window of a limited number of frames of data.[25] This was revised in 1996[26] to use different length windows for track continuation (maintenance) and initiation. This approach improves the efficiency of the multiframe processing and maintains the longer window needed for track initiation. Indeed, numerical experiments to date show it to be far more efficient than a single-pane window. One can easily imagine using different depths in the window for continuing tracks, depending on the complexity of the problem. The efficiency of this approach in practice is yet to be determined.

A fundamental question relates to determining the dimension of the assignment problem that is most appropriate for a particular tracking problem. The goal of future research will be the development of a method that adapts the dimension to the difficulty of the problem or to the need in the surveillance problem.

13.6.4 Algorithms

Several Lagrangian relaxation methods were outlined in the previous section. The method that has been most successful involves relaxation to a two-dimensional assignment problem, maximization of the resulting relaxed problem with respect to the multipliers, and restoration of feasibility to the original problem by formulating this recovery problem as a multidimensional assignment problem of one dimension lower than the original. The process is then repeated until a two-dimensional assignment problem is reached that can be solved optimally.

Such an algorithm can generally produce solutions that are accurate to within 3% of optimal on very difficult problems and optimal for easy problems. The use of an improvement algorithm based on branch and bound can considerably improve the performance to within 0.5% of optimal, at least on some classes of difficult tracking problems.[10] Other techniques, such as local search, should equally improve the solution quality.

Speed enhancements, using the decomposition and clustering discussed in the previous section and as presented in 1992[35] can improve the speed of the assignment solvers by an order of magnitude on large-scale and difficult problems. Further work on distributed and parallel computing should enhance both the solution quality and speed.

Another direction of work is the computation of *K*-near optimal solutions similar to *K*-best solutions for two-dimensional assignment problems.[10] The *K*-near optimal solutions provide important information about the reliability of the computed tracks, which is not available with *K*-best solutions.

Finally, other approaches to the assignment problem, such as GRASP,[32–34] have also been successful, but not to the extent that relaxation has been.

13.6.5   Network-Centric Multiple-Frame Assignments

Multiple-platform tracking, like single-platform, multiple-sensor tracking, also has the potential to significantly improve track estimation by providing geometric diversity and sensor variety. The architecture for data association methods discussed in the previous sections can be applied to multiple-platform tracking in a centralized manner, wherein all measurements are sent to one location for processing and then tracks are transmitted back to the different platforms. The centralized architecture is probably optimal in that it is capable of producing the best track quality (e.g., purity and accuracy) and a consistent air picture; however, it is unacceptable in many applications as a result of issues such as communication loading and single-point failure. Thus, a distributed architecture is needed for both estimation/fusion[2] and data association. Although much has been achieved in the area of distributed fusion, few efforts have been extended to distributed, multiple-frame, data association.

The objectives for a distributed, multiple-frame data association approach to multiple-platform tracking are to achieve a performance, approaching that of the centralized architecture and to achieve a consistent or single integrated air picture (SIAP) across multiple platforms while maintaining communications loads to within a practical limit. Achieving these objectives will require researchers to address a host of problems or topics, including (1) distributed data association and estimation; (2) SIAP; (3) management of communication loading by using techniques such as data pruning, data compression (e.g., tracklets,[4] push/request schemes, and target prioritization; (4) network topology of the communication architecture, including the design of new communication architectures and the incorporation of legacy systems; (5) types of information (e.g., measurements, tracks, tracklets) sent across the network; (6) sensor location and registration errors (i.e., "gridlock"); (7) pedigree problems; and (8) out-of-order, latent, and missing data caused by both sensor and communication problems.

The network-centric algorithm architecture of the Navy's Cooperative Engagement Capability and Joint Composite Tracking Network provides a consistent or single integrated air picture across multiple platforms. This approach limits the communications loads to within a practical limit.[4] This was designed with single-frame data association in mind and has not been extended to multiple-frame approaches. Thus, the development of a "network multiple-frame assignment" approach to data association remains an open and fundamentally important problem.

Acknowledgments

This work was partially supported by the Air Force Office of Scientific Research through AFOSR grant numbers F49620-97-1-0273 and F49620-00-1-0108 and by the Office of Naval Research through ONR grant number N00014-99-1-0118.

References

1. Waltz, E. and Llinas, J., *Multisensor Data Fusion*, Artech House, Boston, MA, 1990.
2. Drummond, O.E., A hybrid fusion algorithm architecture and tracklets, In *Proceedings of SPIE Conference: Signal and Data Processing of Small Targets*, Vol. 3136, San Diego, CA, pp. 485–502, 1997.

3. Blackman, S. and Popoli, R., *Design and Analysis of Modern Tracking Systems*, Artech House Inc., Norwood, MA, 1999.

4. Moore, J.R. and Blair, W.D., Practical aspects of multisensor tracking, In *Multitarget-Multisensor Tracking: Applications and Advances III*, Bar-Shalom, Y. and Blair, W.D. (Eds.) Artech House, Norwood, MA, 2000.

5. Bar-Shalom, Y. and Li, X.R., *Multitarget-Multisensor Tracking: Principles and Techniques*, OPAMP Tech. Books, Los Angeles, CA, 1995.

6. Poore, A.B. and Rijavec, N., Multitarget tracking and multidimensional assignment problems, In *Proceedings of 1991 SPIE Conference: Signal and Data Processing of Small Targets*, Vol. 1481, pp. 345–356, 1991.

7. Poore, A.B. and Rijavec, N., A Lagrangian relaxation algorithm for multi-dimensional assignment problems arising from multitarget tracking, *SIAM Journal on Optimization*, 3(3), 544–563, 1993.

8. Poore, A.B. and Robertson, A.J. III, A new class of Lagrangian relaxation based algorithms for a class of multidimensional assignment problems, *Computational Optimization and Applications*, 8(2), 129–150, 1997.

9. Shea, P.J. and Poore, A.B., Computational experiences with hot starts for a moving window implementation of track maintenance, In *Proceedings of 1998 SPIE Conference: Signal and Data Processing of Small Targets*, Vol. 3373, 1998.

10. Poore, A.B. and Yan, X., Use of K-near optimal solutions to improve data association in multi-frame processing, In *Proceedings of SPIE: Signal and Data Processing of Small Targets*, Drummond, O.E. (Ed.), SPIE, Bellingham, WA, Vol. 3809, pp. 435–443, 1999.

11. Barker, T.N., Persichetti, J.A., Poore, A.B. Jr., and Rijavec, N., Method and system for tracking multiple regional objects, U.S. Patent No. 5,406,289, issued on April 11, 1995.

12. Poore, A.B. Jr., Method and system for tracking multiple regional objects by multi-dimensional relaxation, U.S. Patent No. 5,537,119, issued on July 16, 1996.

13. Poore, A.B. Jr., Method and system for tracking multiple regional objects by multi-dimensional relaxation, CIP, U.S. Patent No. 5,959,574, issued on September 28, 1999.

14. Morefield, C.L., Application of 0–1 integer programming to multitarget tracking problems, *IEEE Transaction on Automatic Control*, 22(3), 302–312, 1977.

15. Reid, D.B., An algorithm for tracking multiple targets, *IEEE Transaction on Automatic Control*, 24(6), 843–854, 1979.

16. Deb, S., Pattipati, K.R., Bar-Shalom, Y., and Yeddanapudi, M., A generalized s-dimensional assignment algorithm for multisensor multitarget state estimation, *IEEE Transactions on Aerospace and Electronic Systems*, 33(2), 523, 1997.

17. Kirubarajan, T., Bar-Shalom, Y., and Pattipati, K.R., Multiassignment for tracking a large number of overlapping objects, In *Proceedings of SPIE Conference: Signal and Data Processing of Small Targets*, Vol. 3163, 1997.

18. Kirubarajan, T., Wang, H., Bar-Shalom, Y., and Pattipati, K.R., Efficient multisensor fusion using multidimensional assignment for multitarget tracking, In *Proceedings of SPIE Conference: Signal Processing, Sensor Fusion and Target Recognition*, Vol. 3374, pp. 14–25, 1998.

19. Popp, R., Pattipati, K., Bar-Shalom, Y., and Gassner, R., An adaptive m-best assignment algorithm and parallelization for multitarget tracking, In *Proceedings of 1998 IEEE Aerospace Conference*, Snowmass, CO, Vol. 5, pp. 71–84, 1998.

20. Chummun, M., Kirubarajan, T., Pattipati, K.R., and Bar-Shalom, Y., Efficient multidimensional data association for multisensor-multitarget tracking using clustering and assignment algorithms, In *Proceedings of 2nd International Conference on Information Fusion*, 1999.

21. Poore, A.B., Multidimensional assignment formulation of data association problems arising from multitarget tracking and multisensor data fusion, *Computational Optimization and Applications*, 3(1), 27–57, 1994.

22. Drummond, O.E., Target tracking, In *Wiley Encyclopedia of Electrical and Electronics Engineering*, Vol. 21, Wiley, New York, NY, pp. 377–391, 1999.

23. Sittler, R.W., An optimal data association problem in surveillance theory, *IEEE Transaction on Military Electronics*, 8(2), 125, 1964.

24. Nemhauser, G.L. and Wolsey, L.A., *Integer and Combinatorial Optimization*, Wiley, New York, NY, 1988.
25. Poore, A.B., Rijavec, N., and Barker, T., Data association for track initiation and extension using multiscan windows, In *Proceedings of SPIE Conference: Signal and Data Processing of Small Targets*, Vol. 1698, pp. 432–437, 1992.
26. Poore, A.B. and Drummond, O.E., Track initiation and maintenance using multidimensional assignment problems, In *Lecture Notes in Economics and Mathematical Systems*, Pardalos, P.M., Hearn, D., and Hager, W. (Eds.) Vol. 450, Springer-Verlag, New York, NY, p. 407, 1997.
27. Garvey, M. and Johnson, D., *Computers and Intractability: A Guide to the Theory of NP-Completeness*, W.H. Freeman & Co., San Francisco, CA, 1979.
28. Hiriart-Urruty, J.B. and Lemaréchal, C., *Convex Analysis and Minimization Algorithms I*, Springer-Verlag, Berlin, 1993.
29. Hiriart-Urruty, J.-B. and Lemaréchal, C., *Convex Analysis and Minimization Algorithms II*, Springer-Verlag, Berlin, 1993.
30. Bertsekas, D.P., *Network Optimization*, Athena Scientific, Belmont, MA, 1998.
31. Jonker, R. and Volgenant, A., A shortest augmenting path algorithm for dense and sparse linear assignment problems, *Computing*, 38(4), 325–340, 1987.
32. Murphey, R., Pardalos, P., and Pitsoulis, L., A GRASP for the multi-target multi-sensor tracking problem, In *Networks, Discrete Mathematics and Theoretical Computer Science Series*, Vol. 40, American Mathematical Society, pp. 277–302, 1998.
33. Murphey, R., Pardalos, P., and Pitsoulis, L., A parallel GRASP for the data association multidimensional assignment problem, In *Parallel Processing of Discrete Problems, IMA Volumes in Mathematics and its Applications*, Vol. 106, Springer-Verlag, New York, NY, pp. 159–180, 1998.
34. Robertson, A., A set of greedy randomized adaptive local search procedure (GRASP) implementations for the multidimensional assignment problem, *Computational Optimization and Applications*, 19(2), 145–164, 2001.
35. Poore, A.B., Rijavec, N., Barker, T., and Munger, M., Data association problems posed as multidimensional assignment problems: numerical simulations, In *Proceedings of SPIE Conference: Signal and Data Processing of Small Targets*, Drummond, O.E. (Ed.) Vol. 1954, pp. 564–571, 1993.
36. Schramm, H. and Zowe, J., A version of the bundle idea for minimizing a nonsmooth function: Conceptual idea, convergence analysis, numerical results, *SIAM Journal on Optimization*, 2, 121, 1992.
37. Kiwiel, K.C., Methods of descent for nondifferentiable optimization, In *Lecture Notes in Mathematics*, Vol. 1133, Springer-Verlag, Berlin, 1985.
38. Poore, A.B. and Rijavec, N., Multidimensional assignment problems, and Lagrangian relaxation, In *Proceedings of The SDI Panels on Tracking*, Issue 2, Institute for Defense Analyses, pp. 3–51 to 3–74, 1991.
39. Poore, A.B. and Rijavec, N., A numerical study of some data association problems arising in multitarget tracking, In *Large Scale Optimization: State of the Art*, Hager, W.W., Hearn, D.W., and Pardalos, P.M. (Eds.), Kluwer Academic Publishers B.V., Boston, MA, pp. 347–370, 1994.
40. Shor, N.Z., *Minimization Methods for Non-Differentiable Functions*, Springer-Verlag, New York, NY, 1985.

14

General Decentralized Data Fusion
with Covariance Intersection

Simon Julier and Jeffrey K. Uhlmann

CONTENTS

14.1   Introduction

Decentralized (or distributed) data fusion (DDF) is one of the most important areas of research in the field of control and estimation. The motivation for decentralization is that it provides a degree of scalability and robustness that cannot be achieved using traditional centralized architectures. In industrial applications, decentralization offers the possibility of producing plug-and-play systems in which sensors of different types and capabilities can be slotted in and out, optimizing trade-offs such as price, power consumption, and performance. This has significant implications for military systems as well because it can dramatically reduce the time required to incorporate new computational and sensing components into fighter aircraft, ships, and other types of platforms.

The benefits of decentralization are not limited to sensor fusion aboard a single platform; decentralization makes it possible for a network of platforms to exchange information and coordinate activities in a flexible and scalable fashion. Interplatform information propagation and fusion form the crux of the network centric warfare (NCW) vision for the

U.S. military. The goal of NCW is to equip all battlespace entities—aircraft, ships, and even individual human combatants—with communication and computing capabilities such that each becomes a node in a vast decentralized command and control network. The idea is that each entity can dynamically establish a communication link with any other entity to obtain the information it needs to perform its warfighting role.

Although the notion of decentralization has had strong intuitive appeal for several decades, achieving its anticipated benefits has proven extremely difficult. Specifically, implementers quickly discovered that if communication paths are not strictly controlled, pieces of information begin to propagate redundantly. When these pieces of information are reused (i.e., double-counted), the fused estimates produced at different nodes in the network become corrupted. Various approaches for avoiding this problem were examined, but none seemed completely satisfactory. In the mid-1990s, the redundant information problem was revealed to be far more than just a practical challenge; it is a manifestation of a fundamental theoretical limitation that could not be surmounted using traditional Bayesian control and estimation methods such as the Kalman filter.[1] In response to this situation, a new data fusion framework, based on covariance intersection (CI), was developed. The CI framework effectively supports all aspects of general DDF.

The structure of this chapter is as follows: Section 14.2 describes the DDF problem. The CI algorithm is described in Section 14.3. Section 14.4 demonstrates how CI supports DDF and describes one such distribution architecture. A simple example of a network with redundant links is presented in Section 14.5. Section 14.6 shows how to exploit known information about network connectivity and information proliferation within the CI framework. Section 14.7 describes recent proposed generalizations of CI to explicit probability distributions, and describes a tractable approximation for Gaussian mixture models. This chapter concludes with a brief discussion on other applications of CI.

14.2 Decentralized Data Fusion

A DDF system is a collection of processing nodes, connected by communication links (Figure 14.1), in which none of the nodes has knowledge about the overall network topology. Each node performs a specific computing task using information from nodes with which it is linked, but no *central* node exists that controls the network. There are many attractive properties of such decentralized systems,[2] including

- Decentralized systems are *reliable* in the sense that the loss of a subset of nodes and links does not necessarily prevent the rest of the system from functioning. In a centralized system, however, the failure of a common communication manager or a centralized controller can result in immediate catastrophic failure of the system.

- Decentralized systems are *flexible* in the sense that nodes can be added or deleted by making only local changes to the network. For example, the addition of a node simply involves the establishment of links to one or more nodes in the network. In a centralized system, however, the addition of a new node can change the topology in such a way that the overall control and communications structure require massive changes.

The most important class of decentralized networks involves nodes associated with sensors or other information sources (such as databases of *a priori* information). Information from distributed sources propagates through the network so that each node obtains the

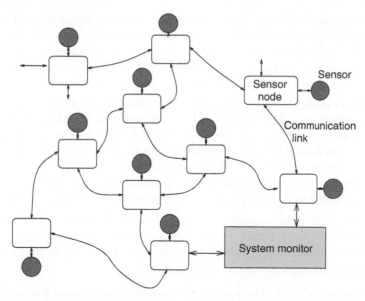

FIGURE 14.1
A distributed data fusion network. Each box represents a fusion node. Each node possesses 0 or more sensors and is connected to its neighboring nodes through a set of communication links.

data relevant to its own processing task. In a battle management application, for example, one node might be associated with the acquisition of information from reconnaissance photographs, another with ground-based reports of troop movements, and another with the monitoring of communication transmissions. Information from these nodes could then be transmitted to a node that estimates the position and movement of enemy troops. The information from this node could then be transmitted back to the reconnaissance photo node, which would use the estimated positions of troops to aid in the interpretation of ambiguous features in satellite photos.

In most applications, the information propagated through a network is converted into a form that provides the estimated state of some quantity of interest. In many cases, especially in industrial applications, the information is converted into means and covariances that can be combined within the framework of Kalman-type filters. A decentralized network for estimating the position of a vehicle, for example, could combine acceleration estimates from nodes measuring wheel speed, from laser gyros, and from pressure sensors on the accelerator pedal. If each independent node provides the mean and variance of its estimate of acceleration, fusing the estimates to obtain a better-filtered estimate is relatively easy.

The most serious problem arising in DDF networks is the effect of redundant information.[3] Specifically, pieces of information from multiple source cannot be combined within most filtering frameworks unless they are independent or have a known degree of correlation (i.e., known cross covariances). In the battle management example described in the preceding text, the effect of redundant information can be seen in the following scenario, sometimes referred to as *rumor propagation* or the *whispering in the hall* problem:

1. A photoreconnaissance node transmits information about potentially important features. This information then propagates through the network, changing form as it is combined with information at other nodes in the process.

2. The troop position estimation node eventually receives the information in some form and notes that one of the indicated features could possibly represent a mobilizing

tank battalion at position x. There are many other possible interpretations of the feature, but the possibility of a mobilizing tank battalion is deemed to be of such tactical importance that it warrants the transmission of a low confidence hypothesis (a *heads up* message). Again, the information can be synopsized, augmented, or otherwise transformed as it is relayed through a sequence of nodes.

3. The photoreconnaissance photo node receives the low confidence hypothesis that a tank battalion may have mobilized at position x. A check of available reconnaissance photos covering position x reveals a feature that is consistent with the hypothesis. Because the node is unaware that the hypothesis was based on that same photographic evidence, it assumes that the feature that it observes is an independent confirmation of the hypothesis. The node then transmits high confidence information that a feature at position x represents a mobilizing tank battalion.

4. The troop position node receives information from the photoreconnaissance node that a mobilizing tank battalion has been identified with high confidence. The troop position node regards this as confirmation of its early hypothesis and calls for an aggressive response to the mobilization.

The obvious problem is that the two nodes exchange redundant pieces of information and treat them as independent pieces of evidence in support of the hypothesis that a tank battalion has mobilized. The end result is that critical resources may be diverted in reaction to what is, in fact, a low probability hypothesis.

A similar situation can arise in a decentralized monitoring system for a chemical process:

1. A reaction vessel is fitted with a variety of sensors, including a pressure gauge.
2. Because the bulk temperature of the reaction cannot be measured directly, a node is added that uses pressure information, combined with a model of the reaction, to estimate temperature.
3. A new node is added to the system that uses information from the pressure and temperature nodes.

Clearly, the added node will always use redundant information from the pressure gauge. If the estimates of pressure and temperature are treated as independent, then the fact that their relationship is always exactly what is predicted by the model might lead to overconfidence in the stability of the system. This type of inadvertent use of redundant information arises commonly when attempts are made to decompose systems into functional modules. The following example is typical:

1. A vehicle navigation and control system maintains one Kalman filter for estimating position and a separate Kalman filter for maintaining the orientation of the vehicle.
2. Each filter uses the same sensor information.
3. The full vehicle state is determined (for prediction purposes) by combining the position and orientation estimates.
4. The predicted position covariance is computed essentially as a sum of the position and orientation covariances (after the estimates are transformed to a common vehicle coordinate frame).

The problem in this example is that the position and orientation errors are not independent. This means that the predicted position covariance will underestimate the actual position error. Obviously, such overly confident position estimates can lead to unsafe maneuvers.

To avoid the potentially disastrous consequences of redundant data on Kalman-type estimators, covariance information must be maintained. Unfortunately, maintaining consistent cross covariances in arbitrary decentralized networks is not possible locally.[1] In only a few special cases, such as tree and fully connected networks, can the proliferation of redundant information be avoided. These special topologies, however, fail to provide the reliability advantage because the failure of a single node or link results in either a disconnected network or one that is no longer able to avoid the effects of redundant information. Intuitively, the redundancy of information in a network is what provides reliability; therefore, if the difficulties with redundant information are avoided by eliminating redundancy, then reliability will also be eliminated.

The proof that cross covariance information cannot be consistently maintained in general decentralized networks seems to imply that the supposed benefits of decentralization are unattainable. However, the proof relies critically on the assumption that some knowledge of the degree of correlation is necessary to fuse pieces of information. This is certainly the case for all classical data fusion mechanisms (e.g., the Kalman filter and Bayesian nets), which are based on applications of Bayes' rule. Furthermore, independence assumptions are also implicit in many *ad hoc* schemes that compute averages over quantities with intrinsically correlated error components.*

The problems associated with assumed independence are often side stepped by artificially increasing the covariance of the combined estimate. This heuristic (or filter *tuning*) can prevent the filtering process from producing nonconservative estimates, but substantial empirical analysis and *tweaking* is required to determine the extent of increase in the covariances. Even with this empirical analysis, the integrity of the Kalman filter framework is compromised, and reliable results cannot be guaranteed. In many applications, such as in large decentralized signal/data fusion networks, the problem is much more acute and no amount of heuristic tweaking can avoid the limitations of the Kalman filter framework.[2] This is of enormous consequence, considering the general trend toward decentralization in complex military and industrial systems.

In summary, the only plausible way to simultaneously achieve robustness and consistency in a general decentralized network is to exploit a data fusion mechanism that does not require independence assumptions. Such a mechanism called CI satisfies this requirement.

14.3 Covariance Intersection

14.3.1 Problem Statement

Consider the following problem. Two pieces of information, labeled A and B, are to be fused together to yield an output, C. This is a very general type of data fusion problem. A and B could be two different sensor measurements (e.g., a batch estimation or track initialization problem), or A could be a prediction from a system model and B could be

---

* Dubious independence assumptions have permeated the literature over the decades and are now almost taken for granted. The fact is that statistical independence is an extremely rare property. Moreover, concluding that an approach will yield good approximations when *almost independent* is replaced with *assumed independent* in its analysis is usually erroneous.

sensor information (e.g., a recursive estimator similar to a Kalman filter). Both terms are corrupted by measurement noises and modeling errors; therefore, their values are known imprecisely and A and B are the random variables $\mathbf{a}$ and $\mathbf{b}$, respectively. Assume that the true statistics of these variables are unknown. The only available information are estimates of the means and covariances of $\mathbf{a}$ and $\mathbf{b}$ and the cross correlations between them, which are $\{\mathbf{a}, \mathbf{P}_{aa}\}$, $\{\mathbf{b}, \mathbf{P}_{bb}\}$, and $\mathbf{0}$, respectively.*

$$\bar{\mathbf{P}}_{aa} = E[\tilde{\mathbf{a}}\tilde{\mathbf{a}}^T], \; \bar{\mathbf{P}}_{ab} = E[\tilde{\mathbf{a}}\tilde{\mathbf{b}}^T], \; \bar{\mathbf{P}}_{bb} = E[\tilde{\mathbf{b}}\tilde{\mathbf{b}}^T] \tag{14.1}$$

where $\tilde{\mathbf{a}} \equiv \mathbf{a} - \bar{\mathbf{a}}$ and $\tilde{\mathbf{b}} \equiv \mathbf{b} - \bar{\mathbf{b}}$ are the true errors imposed by assuming that the means are $\mathbf{a}$ and $\mathbf{b}$. Note that the cross correlation matrix between the random variables, $\bar{\mathbf{P}}_{ab}$, is unknown and will not, in general, be $\mathbf{0}$.

The only constraint that we impose on the assumed estimate is consistency. In other words,

$$\mathbf{P}_{aa} - \bar{\mathbf{P}}_{aa} \geq 0$$
$$\mathbf{P}_{bb} - \bar{\mathbf{P}}_{bb} \geq 0 \tag{14.2}$$

This definition conforms to the standard definition of consistency.[4] The problem is to fuse the consistent estimates of A and B together to yield a new estimate C, $\{\mathbf{c}, \mathbf{P}_{cc}\}$, which is guaranteed to be consistent:

$$\mathbf{P}_{cc} - \bar{\mathbf{P}}_{cc} \geq 0 \tag{14.3}$$

where $\tilde{\mathbf{c}} \equiv \mathbf{c} - \bar{\mathbf{c}}$ and $\bar{\mathbf{P}}_{cc} = E[\tilde{\mathbf{c}}\tilde{\mathbf{c}}^T]$.

It is important to note that two estimates with different means do *not*, in general, supply additional information that can be used in the filtering algorithm.

14.3.2 Covariance Intersection Algorithm

In its generic form, the CI algorithm takes a convex combination of mean and covariance estimates expressed in information (or inverse covariance) space. The intuition behind this approach arises from a *geometric* interpretation of the Kalman filter equations. The general form of the Kalman filter equation can be written as

$$\bar{\mathbf{c}} = \mathbf{W}_a\bar{\mathbf{a}} + \mathbf{W}_b\bar{\mathbf{b}} \tag{14.4}$$

$$\mathbf{P}_{cc} = \mathbf{W}_a\mathbf{P}_{aa}\mathbf{W}_a^T + \mathbf{W}_a\mathbf{P}_{ab}\mathbf{W}_b^T + \mathbf{W}_b\mathbf{P}_{ba}\mathbf{W}_a^T + \mathbf{W}_b\mathbf{P}_{bb}\mathbf{W}_b^T \tag{14.5}$$

where the weights $\mathbf{W}_a$ and $\mathbf{W}_b$ are chosen to minimize the trace of $\mathbf{P}_{cc}$. This form reduces to the conventional Kalman filter if the estimates are independent ($\mathbf{P}_{ab} = 0$) and generalizes to the Kalman filter with colored noise when the correlations are known.

These equations have a powerful geometric interpretation: if one plots the covariance ellipses (for a covariance matrix $\mathbf{P}$ this is the locus of points $\{\mathbf{p}: \mathbf{p}^T\mathbf{P}^{-1}\mathbf{p} = c\}$ where c is a constant), $\mathbf{P}_{aa}$, $\mathbf{P}_{bb}$, and $\mathbf{P}_{cc}$ for all choices of $\mathbf{P}_{ab}$, then $\mathbf{P}_{cc}$ always lies within the *intersection* of $\mathbf{P}_{aa}$ and $\mathbf{P}_{bb}$. Figure 14.2 illustrates this for a number of different choices of $\mathbf{P}_{ab}$.

---

* Cross correlation can also be treated as a nonzero value. For brevity, we do not discuss this case here.

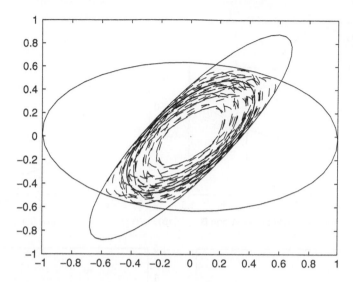

FIGURE 14.2

The shape of the updated covariance ellipse. The variances of $\mathbf{P}_{aa}$ and $\mathbf{P}_{bb}$ are the outer solid ellipses. Different values of $\mathbf{P}_{cc}$ that arise from different choices of $\mathbf{P}_{ab}$ are shown as dashed ellipses. The update with truly independent estimates is the inner solid ellipse.

This interpretation suggests the following approach: if $\mathbf{P}_{cc}$ lies within the intersection of $\mathbf{P}_{aa}$ and $\mathbf{P}_{bb}$ for any possible choice of $\mathbf{P}_{ab}$, then an update strategy that finds a $\mathbf{P}_{cc}$ that encloses the intersection region must be consistent even if there is no knowledge about $\mathbf{P}_{ab}$. The tighter the updated covariance encloses the intersection region, the more effectively the update uses the available information.*

The intersection is characterized by the convex combination of the covariances, and the CI algorithm is[5]

$$\mathbf{P}_{cc}^{-1} = \omega\mathbf{P}_{aa}^{-1} + (1 - \omega)\mathbf{P}_{bb}^{-1} \tag{14.6}$$

$$\mathbf{P}_{cc}^{-1}\mathbf{c} = \omega\mathbf{P}_{aa}^{-1}\mathbf{a} + (1 - \omega)\mathbf{P}_{bb}^{-1}\mathbf{b} \tag{14.7}$$

where $\omega \in [0, 1]$. Appendix 14.A proves that this update equation is consistent in the sense given by Equation 14.3 for all choices of $\mathbf{P}_{ab}$ and ω.

As illustrated in Figure 14.3, the free parameter ω manipulates the weights assigned to **a** and **b**. Different choices of ω can be used to optimize the update with respect to different performance criteria, such as minimizing the trace or the determinant of $\mathbf{P}_{cc}$. Cost functions, which are convex with respect to ω, have only one distinct optimum in the range $0 \leq \omega \leq 1$. Virtually any optimization strategy can be used, ranging from Newton-Raphson to sophisticated semidefinite and convex programming[6] techniques, which can minimize almost any norm. Appendix 14.B includes source code for optimizing ω for the fusion of two estimates.

Note that some measure of covariance size must be minimized at each update to guarantee nondivergence; otherwise, an updated estimate could be larger than the prior estimate.

---

* Note that the discussion of *intersection regions* and the plotting of particular covariance contours should not be interpreted in a way that confuses CI with ellipsoidal bounded region filters. CI does not exploit error bounds, only covariance information.

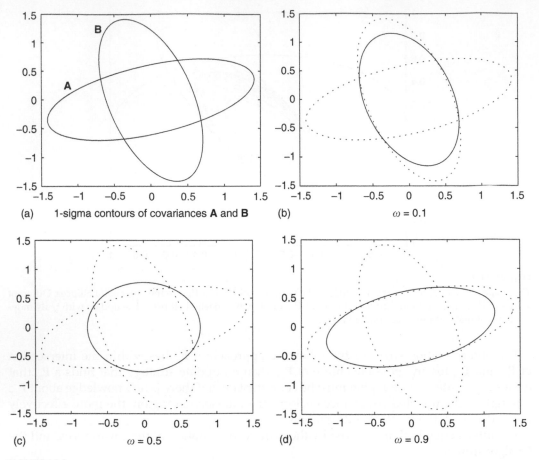

FIGURE 14.3
The value of ω determines the relative weights applied to each information term. (a) Shows the 1-sigma contours for 2D covariance matrices **A** and **B** and (b–d) show the updated covariance **C** (drawn in a solid line) for several different values of ω. For each value of ω, **C** passes through the intersection points of **A** and **B**.

For example, if one were always to use $\omega = 0.5$, then the updated estimate would simply be the Kalman updated estimate with the covariance inflated by a factor of 2. Thus, an update with an observation that has a very large covariance could result in an updated covariance close to twice the size of the prior estimate. In summary, the use of a fixed measure of covariance size with the CI equations leads to the nondivergent CI filter.

An example of the tightness of the CI update can be seen in Figure 14.4 for the case when the two prior covariances approach singularity:

$$\{\mathbf{a}, \mathbf{A}\} = \left\{ \begin{bmatrix} 1 \\ 0 \end{bmatrix}, \begin{bmatrix} 1.5 & 0.0 \\ 0.0 & \varepsilon \end{bmatrix} \right\} \tag{14.8}$$

$$\{\mathbf{b}, \mathbf{B}\} = \left\{ \begin{bmatrix} 0 \\ 1 \end{bmatrix}, \begin{bmatrix} \varepsilon & 0.0 \\ 0.0 & 1.0 \end{bmatrix} \right\} \tag{14.9}$$

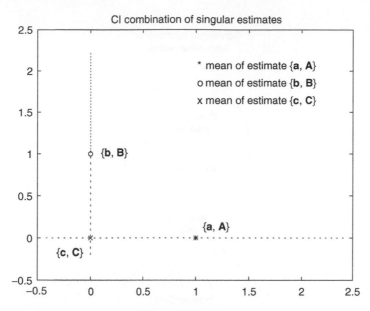

FIGURE 14.4

The CI update $\{c, C\}$ of two 2D estimates $\{a, A\}$ and $\{b, B\}$, where A and B are singular, defines the point of intersection of the colinear sigma contours of A and B.

The covariance of the combined estimate is proportional to ε, and the mean is centered on the intersection point of the 1D contours of the prior estimates. This makes sense intuitively because, if one estimate completely constrains one coordinate, and the other estimate completely constrains the other coordinate, there is only one possible update that can be consistent with both constraints.

CI can be generalized to an arbitrary number of $n > 2$ updates using the following equations:

$$P_{cc}^{-1} = \omega_1 P_{a_1 a_1}^{-1} + \cdots + \omega_n P_{a_n a_n}^{-1} \tag{14.10}$$

$$P_{cc}^{-1} c = \omega_1 P_{a_1 a_1}^{-1} a_1 + \cdots + \omega_n P_{a_n a_n}^{-1} a_n \tag{14.11}$$

where $\sum_{i=1}^{n} \omega_i = 1$. For this type of batch combination of large numbers of estimates, efficient codes such as the public domain MAXDET[7] and SPDSOL[8] are available. Some generalizations for low-dimensional spaces have been derived.

In summary, CI provides a general update algorithm that is capable of yielding an updated estimate even when the prediction and observation correlations are unknown.

14.4 Using Covariance Intersection for Distributed Data Fusion

Consider again the data fusion network that is illustrated in Figure 14.1. The network consists of N nodes whose connection topology is completely arbitrary (i.e., it might include loops and cycles) and can change dynamically. Each node has information only about its

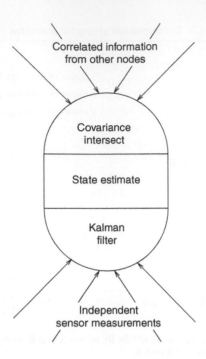

FIGURE 14.5
A canonical node in a general data fusion network that constructs its local state estimate using CI to combine information received from other nodes and a Kalman filter to incorporate independent sensor measurements.

local connection topology (e.g., the number of nodes with which it directly communicates and the type of data sent across each communication link). Assuming that the process and observation noises are independent, the only source of unmodeled correlations is the DDF system itself.

CI can be used to develop a DDF algorithm that directly exploits this structure. The basic idea is illustrated in Figure 14.5. Estimates that are propagated from other nodes are correlated to an unknown degree and fused with the state estimate using CI. Measurements taken locally are known to be independent and can be fused using the Kalman filter equations.

Using conventional notation,[9] the estimate at the ith node is $\hat{x}_i(k|k)$ with covariance $P_i(k|k)$. CI can be used to fuse the information that is propagated between the different nodes. Suppose that, at time step $k + 1$, node i locally measures the observation vector $z_i(k|k)$. A distributed fusion algorithm for propagating the estimate from time step k to time step $k + 1$ for node i is

1. Predict the state of node i at time $k + 1$ using the standard Kalman filter prediction equations.
2. Use the Kalman filter update equations to update the prediction with $z_i(k + 1)$. This update is the distributed estimate with mean $\hat{x}_i(k + 1|k + 1)$ and covariance $P_i(k + 1|k + 1)$. It is not the final estimate, because it does not include observations and estimates propagated from the other nodes in the network.
3. Node i propagates its distributed estimate to all of its neighbors.

4. Node i fuses its prediction $\hat{\mathbf{x}}_i(k + 1|k + 1)$ and $\mathbf{P}_i(k + 1|k)$ with the distributed estimates that it has received from all of its neighbors to yield the partial update with mean $\hat{\mathbf{x}}_i^+(k + 1|k + 1)$ and covariance $\mathbf{P}_i^+(k + 1|k + 1)$. Because these estimates are propagated from other nodes whose correlations are unknown, the CI algorithm is used. As explained earlier, if the node receives multiple estimates for the same time step, the batch form of CI is most efficient. Finally, node i uses the Kalman filter update equations to fuse $\mathbf{z}_i(k + 1)$ with its partial update to yield the new estimate $\hat{\mathbf{x}}_i(k + 1|k + 1)$ with covariance $\mathbf{P}_i(k + 1|k + 1)$. The Kalman filter equations are used because these are known to be independent of the prediction or data that has been distributed to the node from its neighbors. Therefore, CI is unnecessary. This concept is illustrated in Figure 14.5.

An implementation of this algorithm is given in Section 14.5. This algorithm has a number of important advantages. First, all nodes propagate their most accurate partial estimates to all other nodes without imposing any unrealistic requirements for perfectly robust communication. Communication paths may be uni- or bidirectional, there may be cycles in the network, and some estimates may be lost whereas others are propagated redundantly. Second, the update rates of the different filters do not need to be synchronized. Third, communications do not have to be guaranteed—a node can broadcast an estimate without relying on other nodes receiving it. Finally, each node can use a different observation model: one node may have a high accuracy model for one subset of variables of relevance to it, and another node may have a high accuracy model for a different subset of variables, but the propagation of their respective estimates allows nodes to construct fused estimates representing the union of the high accuracy information from both nodes.

The most important feature of this approach to DDF is that it is provably guaranteed to produce and maintain consistent estimates at the various nodes.* Section 14.5 demonstrates this consistency in a simple example.

14.5   Extended Example

Suppose the processing network, shown in Figure 14.6, is used to track the position, velocity, and acceleration of a 1D particle. The network is composed of four nodes. Node 1 measures the position of the particle only. Nodes 2 and 4 measure velocity and node 3 measures acceleration. The four nodes are arranged in a ring. From a practical standpoint, this configuration leads to a robust system with built-in redundancy: data can flow from one node to another through two different pathways. However, from a theoretical point of view, this configuration is extremely challenging. Because this configuration is neither fully connected nor tree connected, optimal data fusion algorithms exist only in the special case where the network topology and the states at each node are known.

---

* The fundamental feature of CI can be described as consistent estimates in, consistent estimates out. The Kalman filter, in contrast, can produce an inconsistent fused estimate from two consistent estimates if the assumption of independence is violated. The only way CI can yield an inconsistent estimate is if a sensor or model introduces an inconsistent estimate into the fusion process. In practice this means that some sort of fault-detection mechanism needs to be associated with potentially faulty sensors.

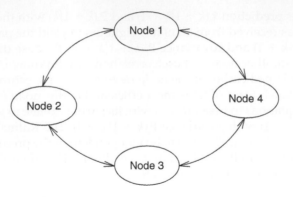

FIGURE 14.6
The network layout for our extended example.

TABLE 14.1

Sensor Information and Accuracy for Each
Node from Figure 14.6

Node	Measures	Variances
1	x	1
2	$\dot{x}$	2
3	$\ddot{x}$	0.25
4	$\dot{x}$	3

The particle moves using a nominal constant acceleration model with process noise injected into the jerk (derivative of acceleration). Assuming that the noise is sampled at the start of the time step and is held constant throughout the prediction step, the process model is

$$X_{(k+1)} = \mathbf{F}x_{(k)} + \mathbf{G}v_{(k+1)} \tag{14.12}$$

where

$$\mathbf{F} = \begin{bmatrix} 1 & \Delta T & \Delta T^2/2 \\ 0 & 1 & \Delta T \\ 0 & 0 & 1 \end{bmatrix} \quad \text{and} \quad \mathbf{G} = \begin{bmatrix} \Delta T^3/6 \\ \Delta T^2/2 \\ \Delta T \end{bmatrix}$$

$v(k)$ is an uncorrelated, zero-mean Gaussian noise with variance $\sigma^{2v} = 10$ and length of the time step $\Delta T = 0.1$ s.

The sensor information and the accuracy of each sensor are given in Table 14.1.

Assume, for the sake of simplicity, that the structure of the state space and the process models are the same for each node and are the same as the true system. However, this condition is not particularly restrictive and many of the techniques of model and system distribution that are used in optimal data distribution networks can be applied with CI.[10]

The state at each node is predicted using the process model:

$$\hat{\mathbf{x}}_i(k+1|k) = \mathbf{F}\hat{\mathbf{x}}_i(k|k)$$

$$\mathbf{P}_i(k+1|k) = \mathbf{F}\mathbf{P}_i(k+1|k)\mathbf{F}^T + \mathbf{Q}(k)$$

The partial estimates $\hat{\mathbf{x}}_i^*(k+1|k+1)$ and $\mathbf{P}_i^*(k+1|k+1)$ are calculated using the Kalman filter update equations. If $\mathbf{R}_i$ is the observation noise covariance on the ith sensor, and $\mathbf{H}_i$ is the observation matrix, then the partial estimates are

$$v_i(k+1) = \mathbf{z}_i(k+1) - \mathbf{H}_i\hat{\mathbf{x}}_i(k+1|k) \tag{14.13}$$

$$\mathbf{S}_i(k+1) = \mathbf{H}_i\mathbf{P}_i(k+1|k)\mathbf{H}_i^T + \mathbf{R}_i(k+1) \tag{14.14}$$

$$\mathbf{W}_i(k+1) = \mathbf{P}_i(k+1|k)\mathbf{H}_i^T\mathbf{S}_i^{-1}(k+1) \tag{14.15}$$

$$\hat{\mathbf{x}}_i^*(k+1|k+1) = \hat{\mathbf{x}}_i(k+1|k) + \mathbf{W}_i(k+1)v_i(k+1) \tag{14.16}$$

$$\mathbf{P}_i^*(k+1|k+1) = \mathbf{P}_i(k+1|k) - \mathbf{W}_i(k+1)\mathbf{S}_i(k+1)\mathbf{W}_i^T(k+1) \tag{14.17}$$

Examine three strategies for combining the information from the other nodes:

1. The nodes are disconnected. No information flows between the nodes and the final updates are given by

$$\hat{\mathbf{x}}_i(k+1|k+1) = \hat{\mathbf{x}}_i^*(k+1|k+1) \tag{14.18}$$

$$\mathbf{P}_i(k+1|k+1) = \mathbf{P}_i^*(k+1|k+1) \tag{14.19}$$

2. Assumed independence update. All nodes are assumed to operate independently of one another. Under this assumption, the Kalman filter update equations can be used in step 4 of the fusion strategy described in Section 14.4.
3. CI-based update. The update scheme described in Section 14.4 is used.

The performance of each of these strategies was assessed using a Monte Carlo of 100 runs.

The results from the first strategy (no data distribution) are shown in Figure 14.7. As expected, the system behaves poorly. Because each node operates in isolation, only node 1 (which measures x) is fully observable. The position variance increases without bound for the three remaining nodes. Similarly, the velocity is observable for nodes 1, 2, and 4, but it is not observable for node 3.

The results of the second strategy (all nodes are assumed independent) are shown in Figure 14.8. The effect of assumed independence observations is obvious: all of the estimates for all of the states in all of the nodes (apart from x for node 3) are inconsistent. This clearly illustrates the problem of double counting.

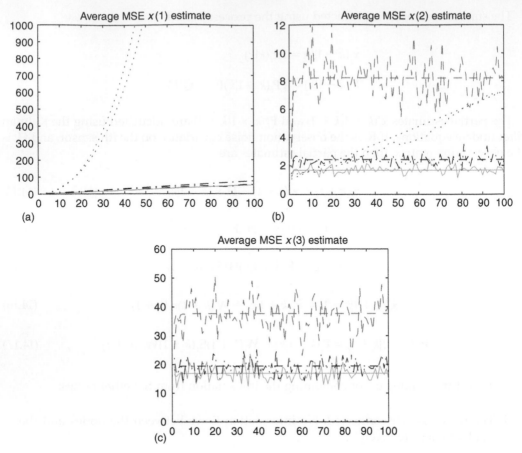

FIGURE 14.7
Disconnected nodes. (a) Mean squared error in x. (b) Mean squared error in $\dot{x}$. (c) Mean squared error in $\ddot{x}$. Mean squared errors and estimated covariances for all states in each of the four nodes. The curves for node 1 are solid, node 2 are dashed, node 3 are dotted, and node 4 are dash-dotted. The mean squared error is the rougher of the two lines for each node.

Finally, the results from the CI distribution scheme are shown in Figure 14.9. Unlike the other two approaches, all the nodes are consistent and observable. Furthermore, as the results in Table 14.2 indicate, the steady-state covariances of all of the states in all of the nodes are smaller than those for case 1. In other words, this example shows that this data distribution scheme successfully and usefully propagates data through an apparently degenerate data network.

This simple example is intended only to demonstrate the effects of redundancy in a general data distribution network. CI is not limited in its applicability to linear, time invariant systems. Furthermore, the statistics of the noise sources do not have to be unbiased and Gaussian. Rather, they only need to obey the consistency assumptions. Extensive experiments have shown that CI can be used with large numbers of platforms with nonlinear dynamics, nonlinear sensor models, and continuously changing network topologies (i.e., dynamic communication links).[11]

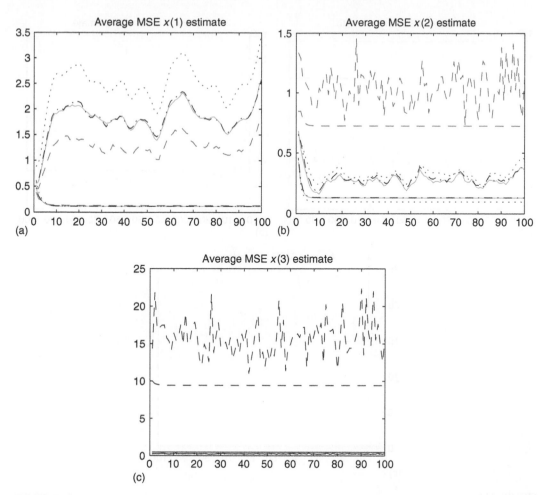

FIGURE 14.8
All nodes assumed independent. (a) Mean squared error in x. (b) Mean squared error in $\dot{x}$. (c) Mean squared error in $\ddot{x}$. Mean squared errors and estimated covariances for all states in each of the four nodes. The curves for node 1 are solid, node 2 are dashed, node 3 are dotted, and node 4 are dash-dotted. The mean squared error is the rougher of the two lines for each node.

14.6   Incorporating Known-Independent Information

CI and the Kalman filter are diametrically opposite in their treatment of covariance information: CI conservatively assumes that no estimate provides statistically independent information, and the Kalman filter assumes that every estimate provides statistically independent information. However, neither of these two extremes is representative of typical data fusion applications. This section demonstrates how the CI framework can be extended to subsume the generic CI filter and the Kalman filter and provide a completely general and optimal solution to the problem of maintaining and fusing consistent mean and covariance estimates.[12]

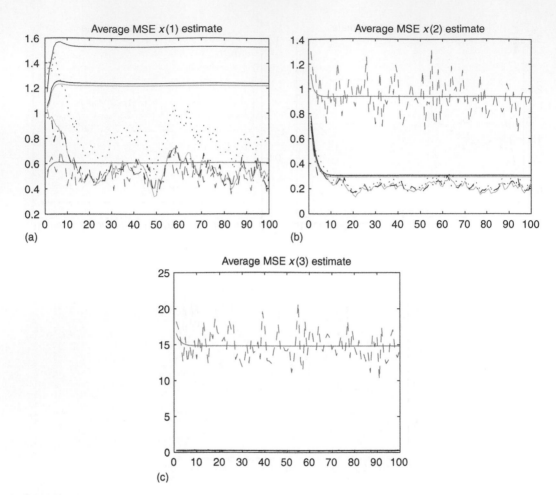

FIGURE 14.9

CI distribution scheme. (a) Mean squared error in x. (b) Mean squared error in $\dot{x}$. (c) Mean squared error in $\ddot{x}$. Mean squared errors and estimated covariances for all states in each of the four nodes. The curves for node 1 are solid, node 2 are dashed, node 3 are dotted, and node 4 are dash-dotted. The mean squared error is the rougher of the two lines for each node.

TABLE 14.2

The Diagonal Elements of the Covariance Matrices for Each Node at the End of 100 Time Steps for Each of the Consistent Distribution Schemes

Node	Scheme	σ_x^2	$\sigma_{\dot{x}}^2$	$\sigma_{\ddot{x}}^2$
1	NONE	0.8823	8.2081	37.6911
	CI	0.6055	0.9359	14.823
2	NONE	50.5716*	1.6750	16.8829
	CI	1.2186	0.2914	0.2945
3	NONE	77852.3*	7.2649*	0.2476
	CI	1.5325	0.3033	0.2457
4	NONE	75.207	2.4248	19.473
	CI	1.2395	0.3063	0.2952

Note: NONE—no distribution and CI—the CI algorithm. The asterisk denotes that a state is *unobservable* and its variance is increasing without bound.

The following equation provides a useful interpretation of the original CI result. Specifically, the estimates {a, A} and {b, B} are represented in terms of their joint covariance:

$$\left\{ \begin{bmatrix} a \\ b \end{bmatrix}, \begin{bmatrix} A & P_{ab} \\ P_{ab}^T & B \end{bmatrix} \right\}$$

(14.20)

where in most situations the cross covariance, P_{ab}, is unknown. The CI equations, however, support the conclusion that

$$\begin{bmatrix} A & P_{ab} \\ P_{ab}^T & B \end{bmatrix} \leq \begin{bmatrix} \omega A^{-1} & 0 \\ 0 & (1-\omega)B^{-1} \end{bmatrix}^{-1}$$

(14.21)

because CI must assume a joint covariance that is conservative with respect to the true joint covariance. Evaluating the inverse of the right-hand side (RHS) of the equation leads to the following consistent/conservative estimate for the joint system:

$$\left\{ \begin{bmatrix} a \\ b \end{bmatrix}, \begin{bmatrix} \frac{1}{\omega}A & 0 \\ 0 & \frac{1}{1-\omega}B \end{bmatrix} \right\}$$

(14.22)

From this result, the following generalization of CI can be derived:*

CI with independent error: Let $a = a_1 + a_2$ and $b = b_1 + b_2$, where a_1 and b_1 are correlated to an unknown degree, whereas the errors associated with a_2 and b_2 are completely independent of all others. Also, let the respective covariances of the components be A_1, A_2, B_1, and B_2. From the above results, a consistent joint system can be formed as follows:

$$\left\{ \begin{bmatrix} a_1 + a_2 \\ b_1 + b_2 \end{bmatrix}, \begin{bmatrix} \frac{1}{\omega}A_1 + A_2 & 0 \\ 0 & \frac{1}{1-\omega}B_1 + B_2 \end{bmatrix} \right\}$$

(14.23)

Letting $A = (1/\omega)A_1 + A_2$ and $B = [1/(1-\omega)]B_1 + B_2$ gives the following generalized CI equations:

$$C = \left[A^{-1} + B^{-1} \right]^{-1} = \left[\left(\tfrac{1}{\omega}A_1 + A_2 \right)^{-1} + \left(\tfrac{1}{1-\omega}B_1 + B_2 \right)^{-1} \right]^{-1}$$

(14.24)

$$c = [A^{-1}a + B^{-1}b]^{-1} = C\left[\left(\tfrac{1}{\omega}A_1 + A_2 \right)^{-1} a + \left(\tfrac{1}{1-\omega}B_1 + B_2 \right)^{-1} b \right]$$

(14.25)

where the known independence of the errors associated with a_2 and b_2 is exploited.

Although this generalization of CI exploits available knowledge about independent error components, further exploitation is impossible because the combined covariance C is formed from *both* independent and correlated error components. However, CI can be generalized even further to produce and maintain separate covariance components, C_1 and

---

* In the process, a consistent estimate of the covariance of $a + b$ is also obtained, where a and b have an unknown degree of correlation, as $(1/\omega)A + [1/(1-\omega)]B$. We refer to this operation as *covariance addition* (CA).

C_2, reflecting the correlated and known-independent error components, respectively. This generalization is referred to as *split CI*.

If we let $\tilde{a}_1$ and $\tilde{a}_2$ be the correlated and known-independent error components of a, with $\tilde{b}_1$ and $\tilde{b}_2$ similarly defined for b, then we can express the errors $\tilde{c}_1$ and $\tilde{c}_2$ in information (inverse covariance) form as

$$C^{-1}(\tilde{c}_1 + \tilde{c}_2) = A^{-1}(\tilde{a}_1 + \tilde{a}_2) + B^{-1}(\tilde{b}_1 + \tilde{b}_2) \tag{14.26}$$

from which the following can be obtained after premultiplying by C:

$$(\tilde{c}_1 + \tilde{c}_2) = C[A^{-1}(\tilde{a}_1 + \tilde{a}_2) + B^{-1}(\tilde{b}_1 + \tilde{b}_2)] \tag{14.27}$$

Squaring both sides, taking expectations, and collecting independent terms* yields:

$$C_2 = (A^{-1} + B^{-1})^{-1}(A^{-1}A_2A^{-1} + B^{-1}B_2B^{-1})(A^{-1} + B^{-1})^{-1} \tag{14.28}$$

where the nonindependent part can be obtained simply by subtracting the above result from the overall fused covariance $C = (A^{-1} + B^{-1})^{-1}$. In other words,

$$C_1 = (A^{-1} + B^{-1})^{-1} - C_2 \tag{14.29}$$

Split CI can also be expressed in batch form analogously to the batch form of original CI. Note that the CA equation can be generalized analogously to provide split CA capabilities.

The generalized and split variants of CI optimally exploit knowledge of statistical independence. This provides an extremely general filtering, control, and data fusion framework that completely subsumes the Kalman filter.

14.6.1 Example Revisited

The contribution of generalized CI can be demonstrated by revisiting the example described in Section 14.5. The scheme described earlier attempted to exploit information that is independent in the observations. However, it failed to exploit one potentially very valuable source of information—the fact that the distributed estimates $\hat{x}_i^*(k + 1|k + 1)$ with covariance $P_i^*(k + 1|k + 1)$ contain the observations taken at time step $k + 1$. Under the assumption that the measurement errors are uncorrelated, generalized CI can be exploited to significantly improve the performance of the information network. The distributed estimates are split into the (possibly) correlated and known-independent components, and generalized CI can be used to fuse the data remotely.

The estimate of node i at time step k is maintained in split form with mean $\hat{x}_i(k|k)$ and covariances $P_{i,1}(k|k)$ and $P_{i,2}(k|k)$. As explained in the following text, it is not possible to ensure that $P_{i,2}(k|k)$ will be independent of the distributed estimates that will be received at

---

* Recall that $A = (1/\omega)A_1 + A_2$ and $B = [1/(1 - \omega)]B_1 + B_2$.

time step k. Therefore, the prediction step combines the correlated and independent terms into the correlated term, and sets the independent term to 0:

$$\hat{\mathbf{x}}_i(k+1|k) = \mathbf{F}\hat{\mathbf{x}}_i(k|k)$$

$$\mathbf{P}_{i,1}(k+1|k) = \mathbf{F}(\mathbf{P}_{i,1}(k+1|k) + \mathbf{P}_{i,2}(k+1|k))\mathbf{F}^T + \mathbf{Q}(k) \qquad (14.30)$$

$$\mathbf{P}_{i,2}(k+1|k) = 0$$

The process noise is treated as a correlated noise component because each sensing node is tracking the same object. Therefore, the process noise that acts on each node is perfectly correlated with the process noise acting on all other nodes.

The split form of the distributed estimate is found by applying split CI to fuse the prediction with $\mathbf{z}_i(k+1)$. Because the prediction contains only correlated terms and the observation contains only independent terms ($\mathbf{A}_2 = 0$ and $\mathbf{B}_1 = 0$ in Equation 14.24), the optimized solution for this update occurs when $\omega = 1$. This is the same as calculating the normal Kalman filter update and explicitly partitioning the contributions of the predictions from the observations. Let $\mathbf{W}_i^*(k+1)$ be the weight used to calculate the distributed estimate. From Equation 14.30 its value is given by

$$S_i^* (k+1) = \mathbf{H}_i \mathbf{P}_{i,1}(k+1|k)\mathbf{H}_i^T + \mathbf{R}_i(k+1) \qquad (14.31)$$

$$W_i^* (k+1) = \mathbf{P}_{i,1}(k+1|k)\mathbf{H}_i^T S_i^* (k+1)^{-1} \qquad (14.32)$$

Note that the CA equation can be generalized analogously to provide split CA capabilities.

Taking outer products of the prediction and observation contribution terms, the correlated and independent terms of the distributed estimate are

$$\mathbf{P}_{i,1}^*(k+1|k+1) = \mathbf{X}(k) + 1\mathbf{P}_i(k+1|k)\mathbf{X}^T(k+1)$$

$$\mathbf{P}_{i,2}^*(k+1|k+1) = \mathbf{W}(k+1) + 1\mathbf{R}(k) + 1\mathbf{W}^T(k+1) \qquad (14.33)$$

where $\mathbf{X}(k+1) = \mathbf{I} - \mathbf{W}_i^*(k+1)\mathbf{H}(k+1)$.

The split distributed updates are propagated to all other nodes where they are fused with split CI to yield a split partial estimate with mean $\hat{\mathbf{x}}(k+1|k+1)$ and covariances $\mathbf{P}_{i,1}^+(k+1|k+1)$ and $\mathbf{P}_{i,2}^+(k+1|k+1)$. Split CI can now be used to incorporate $\mathbf{z}(k)$. However, because the observation contains no correlated terms ($\mathbf{B}_1 = 0$ in Equation 14.24), the optimal solution is always $\omega = 1$.

The effect of this algorithm can be seen in Figure 14.10 and in Table 14.3. As can be seen, the results of generalized CI are dramatic. The most strongly affected node is node 2, whose position variance is reduced almost by a factor of 3. The least affected node is node 1. This is not surprising, given that node 1 is fully observable. Even so, the variance on its position estimate is reduced by more than 25%.

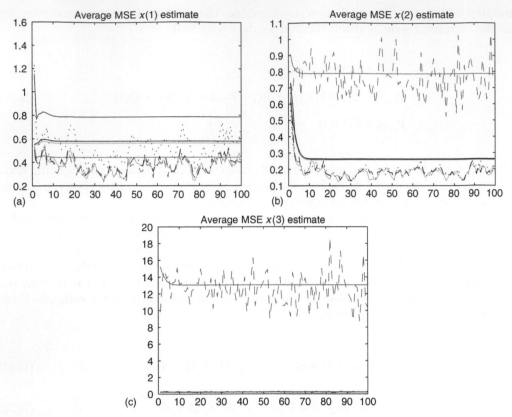

FIGURE 14.10

Mean squared errors and estimated covariances for all states in each of the four nodes. (a) Mean squared error in x. (b) Mean squared error in $\dot{x}$. (c) Mean squared error in $\ddot{x}$. The curves for node 1 are solid, node 2 are dashed, node 3 are dotted, and node 4 are dash-dotted. The mean squared error is the rougher of the two lines for each node.

TABLE 14.3

The Diagonal Elements of the Covariance Matrices for Each Node at the End of 100 Time Steps for Each of the Consistent Distribution Schemes

Node	Scheme	σ_x^2	$\sigma_{\dot{x}}^2$	$\sigma_{\ddot{x}}^2$
1	NONE	0.8823	8.2081	37.6911
	CI	0.6055	0.9359	14.823
	GCI	0.4406	0.7874	13.050
2	NONE	50.5716*	1.6750	16.8829
	CI	1.2186	0.2914	0.2945
	GCI	0.3603	0.2559	0.2470
3	NONE	77852.3*	7.2649*	0.2476
	CI	1.5325	0.3033	0.2457
	GCI	0.7861	0.2608	0.2453
4	NONE	75.207	2.4248	19.473
	CI	1.2395	0.3063	0.2952
	GCI	0.5785	0.2636	0.2466

Note: NONE—no distribution; CI—the CI algorithm; GCI—generalized CI algorithm, which is described in Section 14.6. The asterisk denotes that a state is *unobservable* and its variance is increasing without bound. The covariance used for the GCI values is $\mathbf{P}_i(k\,|\,k) = \mathbf{P}_{i,1}(k\,|\,k) + \mathbf{P}_{i,2}(k\,|\,k)$.

14.7 Conclusions

This chapter has considered the extremely important problem of data fusion in arbitrary data fusion networks. It has described a general data fusion/update technique that makes no assumptions about the independence of the estimates to be combined. The use of the CI framework to combine mean and covariance estimates without information about their degree of correlation provides a direct solution to the DDF problem.

However, the problem of unmodeled correlations reaches far beyond DDF and touches the heart of most types of tracking and estimation. Other application domains for which CI is highly relevant include

- *Fault-tolerant distributed data fusion*—It is essential in practice that DDF network be tolerant of corrupted information and fault conditions; otherwise, a single spurious piece of data could propagate and corrupt the entire system. The challenge is that it is not necessarily possible to identify which pieces of data in a network are corrupted. The use of CI with a related mechanism, CU, permits data that are inconsistent with the majority of information in a network to be eliminated automatically without ever having to be identified as spurious. In other words, the combination of CI and CU leads to fault tolerance as an emergent property of the system.[13]

- *Multiple model filtering*—Many systems switch behaviors in a complicated manner; therefore, a comprehensive model is difficult to derive. If multiple approximate models are available that capture different behavioral aspects with different degrees of fidelity, their estimates can be combined to achieve a better estimate. Because they are all modeling the same system, however, the different estimates are likely to be highly correlated.[14,15]

- *Simultaneous map building and localization for autonomous vehicles*—When a vehicle estimates the positions of landmarks in its environment while using those same landmarks to update its own position estimate, the vehicle and landmark position estimates become highly correlated.[5,16]

- *Track-to-track data fusion in multiple-target tracking systems*—When sensor observations are made in a dense target environment, there is ambiguity concerning which tracked target produced which observation. If two tracks are determined to correspond to the same target, assuming independence may not be possible when combining them, if they are derived from common observation information.[11,14]

- *Nonlinear filtering*—When nonlinear transformations are applied to observation estimates, correlated errors arise in the observation sequence. The same is true for time propagations of the system estimate. CI will ensure nondivergent nonlinear filtering if every covariance estimate is conservative. Nonlinear extensions of the Kalman filter are inherently flawed because they require independence regardless of whether the covariance estimates are conservative.[5,17–22]

Current approaches to these and many other problems attempt to circumvent troublesome correlations by heuristically adding *stabilizing noise* to updated estimates to ensure that they are conservative. The amount of noise is likely to be excessive to guarantee that no covariance components are underestimated. CI ensures the best possible estimate, given the amount of information available. The most important fact that must be emphasized is that the procedure makes no assumptions about either independence or the underlying distributions of the combined estimates. Consequently, CI likely will replace the Kalman filter in a wide variety of applications where independence assumptions are unrealistic.

Acknowledgments

The authors gratefully acknowledge the support of IDAK Industries for supporting the development of the full CI framework and the Office of Naval Research (Contract N000149WX20103) for supporting current experiments and applications of this framework. The authors also acknowledge support from RealityLab.com and the University of Oxford.

Appendix 14.A   Consistency of CI

This appendix proves that CI yields a consistent estimate for any value of ω and $\bar{P}_{ab}$, provided a and b are consistent.[23]

The CI algorithm calculates its mean using Equation 14.7. The actual error in this estimate is

$$\tilde{c} = P_{cc}\left\{\omega P_{aa}^{-1}\tilde{a} + (1 - \omega)P_{bb}^{-1}\tilde{b}\right\} \tag{14.34}$$

By taking outer products and expectations, the actual mean squared error, which calculates the mean using Equation 14.7, is

$$E[\tilde{c}\tilde{c}^{T}] = P_{cc}\left\{\omega^{2}P_{aa}^{-1}\bar{P}_{aa}P_{aa}^{-1} + \omega(1 - \omega)P_{aa}^{-1}\bar{P}_{ab}P_{bb}^{-1} + \omega(1 - \omega)P_{bb}^{-1}\bar{P}_{ba}P_{aa}^{-1} + (1 - \omega)^{2}P_{bb}^{-1}\bar{P}_{bb}P_{bb}^{-1}\right\}P_{cc} \tag{14.35}$$

Because $\bar{P}_{ab}$ is not known, the true value of the mean squared error cannot be calculated. However, CI implicitly calculates an upper bound of this quantity. If Equation 14.35 is substituted into Equation 14.3, the consistency condition can be written as

$$P_{cc} - P_{cc}\left\{\omega^{2}P_{aa}^{-1}\bar{P}_{aa}P_{aa}^{-1} + \omega(1 - \omega)P_{aa}^{-1}\bar{P}_{ab}P_{bb}^{-1} + \omega(1 - \omega)P_{bb}^{-1}\bar{P}_{ba}P_{aa}^{-1} + (1 - \omega)^{2}P_{bb}^{-1}\bar{P}_{bb}P_{bb}^{-1}\right\}P_{cc} \geq 0 \tag{14.36}$$

Pre- and postmultiplying both sides by P_{cc}^{-1} and collecting terms gives

$$P_{cc}^{-1} - \omega^{2}P_{aa}^{-1}\bar{P}_{aa}P_{aa}^{-1} - \omega(1 - \omega)P_{aa}^{-1}\bar{P}_{ab}P_{bb}^{-1} - \omega(1 - \omega)P_{bb}^{-1}\bar{P}_{ba}P_{aa}^{-1} - (1 - \omega)^{2}P_{bb}^{-1}\bar{P}_{bb}P_{bb}^{-1} \geq 0 \tag{14.37}$$

An upper bound on P_{cc}^{-1} can be found and expressed using P_{aa}, P_{bb}, $\bar{P}_{aa}$, and $\bar{P}_{bb}$. From the consistency condition for a,

$$P_{aa} - \bar{P}_{aa} \geq 0 \tag{14.38}$$

or, by pre- and postmultiplying by $\mathbf{P}_{aa}^{-1}$,

$$\mathbf{P}_{aa}^{-1} \geq \mathbf{P}_{aa}^{-1}\bar{\mathbf{P}}_{aa}\mathbf{P}_{aa}^{-1} \tag{14.39}$$

A similar condition exists for **b**, and substituting these results in Equation 14.6 leads to

$$\mathbf{P}_{cc}^{-1} = \omega\mathbf{P}_{aa}^{-1} + (1-\omega)\mathbf{P}_{bb}^{-1} \tag{14.40}$$

$$\geq \omega\mathbf{P}_{aa}^{-1}\bar{\mathbf{P}}_{aa}\mathbf{P}_{aa}^{-1} + (1-\omega)\mathbf{P}_{bb}^{-1}\bar{\mathbf{P}}_{bb}\mathbf{P}_{bb}^{-1} \tag{14.41}$$

Substituting this lower bound on $\mathbf{P}_{cc}^{-1}$ into Equation 14.37 leads to

$$\omega(1-\omega)\left(\mathbf{P}_{aa}^{-1}\bar{\mathbf{P}}_{aa}\mathbf{P}_{aa}^{-1} - \mathbf{P}_{aa}^{-1}\bar{\mathbf{P}}_{ab}\mathbf{P}_{bb}^{-1} - \mathbf{P}_{bb}^{-1}\bar{\mathbf{P}}_{ba}\mathbf{P}_{aa}^{-1}\bar{\mathbf{P}}_{bb}\mathbf{P}_{bb}^{-1}\right) \geq 0 \tag{14.42}$$

or

$$\omega(1-\omega)E\left[\left\{\mathbf{P}_{aa}^{-1}\tilde{\mathbf{a}} - \mathbf{P}_{bb}^{-1}\tilde{\mathbf{b}}\right\}\left\{\mathbf{P}_{aa}^{-1}\tilde{\mathbf{a}} - \mathbf{P}_{bb}^{-1}\tilde{\mathbf{b}}\right\}^{T}\right] \geq 0 \tag{14.43}$$

Clearly, the inequality must hold for all choices of $\bar{\mathbf{P}}_{ab}$ and $\omega \in [0, 1]$.

Appendix 14.B MATLAB Source Code

This appendix provides source code for performing the CI update in MATLAB.

14.B.1 Conventional CI

```
function [c,C,omega]=CI(a,A,b,B,H)
%
% function [c,C,omega]=CI(a,A,b,B,H)
%
% This function implements the CI algorithm and fuses two estimates
% (a,A) and (b,B) together to give a new estimate (c,C) and the value
% of omega, which minimizes the determinant of C. The observation
% matrix is H.
Ai=inv(A);
Bi=inv(B);
% Work out omega using the matlab constrained minimizer function
% fminbnd().
f=inline('1/det(Ai*omega+H''*Bi*H*(1-omega))', ...
 'omega', 'Ai', 'Bi', 'H');
omega=fminbnd(f,0,1,optimset('Display','off'),Ai,Bi,H);
% The unconstrained version of this optimization is:
% omega=fminsearch(f,0.5,optimset('Display','off'),Ai,Bi,H);
% omega=min(max(omega,0),1);
```

```
% New covariance
C=inv(Ai*omega+H'*Bi*H*(1-omega));
% New mean
nu=b-H*a;
W=(1-omega)*C*H'*Bi;
c=a+W*nu;
```

14.B.2 Split CI

```
function [c,C1,C2,omega]=SCI(a,A1,A2,b,B1,B2,H)
%
% function [c,C1,C2,omega]=SCI(a,A1,A2,b,B1,B2,H)
%
% This function implements the split CI algorithm and fuses two
% estimates (a,A1,A2) and (b,B1,B2) together to give a new estimate
% (c,C1,C2) and the value of omega which minimizes the determinant of
% (C1+C2). The observation matrix is H.
%
% Work out omega using the matlab constrained minimizer function
% fminbnd().
f=inline('1/det(omega*inv(A1+omega*A2)+(1-omega)*H''*inv(B1+(1-omega)*B2)*H)', ...
  'omega', 'A1', 'A2', 'B1', 'B2', 'H');
omega=fminbnd(f,0,1,optimset('Display','off'),A1,A2,B1,B2,H);
% The unconstrained version of this optimization is:
% omega=fminsearch(f,0.5,optimset('Display','off'),A1,A2,B1,B2,H);
% omega=min(max(omega,0),1);
Ai=omega*inv(A1+omega*A2);
HBi=(1-omega)*H'*inv(B1+(1-omega)*B2);
% New covariance
C=inv(Ai+HBi*H);
C2=C*(Ai*A2*Ai'+HBi*B2*HBi')*C;
C1=C-C2;
% New mean
nu=b-H*a;
W=C*HBi;
c=a+W*nu;
```

References

1. Utete, S.W., Network management in decentralised sensing systems, Ph.D. thesis, Robotics Research Group, Department of Engineering Science, University of Oxford, 1995.
2. Grime, S. and Durrant-Whyte, H., Data fusion in decentralized sensor fusion networks, *Control Eng. Pract.*, 2(5), 849–863, 1994.
3. Chong, C., Mori, S., and Chan, K., Distributed multitarget multisensor tracking, *Multitarget Multisensor Tracking: Advanced Application*, Artech House, Boston, 1990.
4. Jazwinski, A.H., *Stochastic Processes and Filtering Theory*, Academic Press, New York, 1970.
5. Uhlmann, J.K., Dynamic map building and localization for autonomous vehicles, Ph.D. thesis, University of Oxford, 1995/96.

6. Vandenberghe, L. and Boyd, S., Semidefinite programming, *SIAM Rev.*, 38(1), 49–95, 1996.
7. Wu, S.P., Vandenberghe, L., and Boyd, S., MAXDET: Software for determinant maximization problems, alpha version, Stanford University, April 1996.
8. Boyd, S. and Wu, S.P., *SDPSOL: User's Guide*, November 1995.
9. Bar-Shalom, Y. and Fortmann, T.E., *Tracking and Data Association*, Academic Press, New York, 1988.
10. Mutambara, A.G.O., *Decentralized Estimation and Control for Nonlinear Systems*, CRC Press, Boca Raton, FL, 1998.
11. Nicholson, D. and Deaves, R., Decentralized track fusion in dynamic networks, in *Proc. 2000 SPIE Aerosense Conf.*, Vol. 4048, pp. 452–460, 2000.
12. Julier, S.J. and Uhlmann, J.K., Generalized and split covariance intersection and addition, Technical Disclosure Report, Naval Research Laboratory, 1998.
13. Uhlmann, J.K., Covariance consistency methods for fault-tolerant distributed data fusion, *Inf. Fusion J.*, 4(3), 201–215, 2003.
14. Bar-Shalom, Y. and Li, X.R., *Multitarget-Multisensor Tracking: Principles and Techniques*, YBS Press, Storrs, CT, 1995.
15. Julier, S.J. and Durrant-Whyte, H., A horizontal model fusion paradigm, in *Proc. SPIE Aerosense Conf.*, Vol. 2738, pp. 37–48, 1996.
16. Uhlmann, J., Julier, S., and Csorba, M., Nondivergent simultaneous map building and localization using covariance intersection, in *Proc. 1997 SPIE Aerosense Conf.*, Vol. 3087, pp. 2–11, 1997.
17. Julier, S.J., Uhlmann, J.K., and Durrant-Whyte, H.F., A new approach for the nonlinear transformation of means and covariances in linear filters, *IEEE Trans. Automatic Control*, 45(3), 477–482, 2000.
18. Julier, S.J. Uhlmann, J.K., and Durrant-Whyte, H.F., A new approach for filtering nonlinear systems, in *Proc. American Control Conf.*, Seattle, WA, Vol. 3, pp. 1628–1632, 1995.
19. Julier, S.J. and Uhlmann, J.K., A new extension of the Kalman filter to nonlinear systems, in *Proc. AeroSense: 11th Internat'l. Symp. Aerospace/Defense Sensing, Simulation and Controls*, SPIE, Vol. 3068, pp. 182–193, 1997.
20. Julier, S.J. and Uhlmann, J.K., A consistent, debiased method for converting between polar and Cartesian coordinate systems, in *Proc. of AeroSense: 11th Internat'l. Symp. Aerospace/Defense Sensing, Simulation and Controls*, SPIE, Vol. 3086, pp. 110–121, 1997.
21. Juliers, S.J., A skewed approach to filtering, in *Proc. AeroSense: 12th Internat'l. Symp. Aerospace/ Defense Sensing, Simulation and Controls*, SPIE, Vol. 3373, pp. 271–282, 1998.
22. Julier, S.J. and Uhlmann, J.K., A General Method for Approximating Nonlinear Transformations of Probability Distributions, published on the Web at http://www.robots.ox.ac.uk/~siju, November 1996.
23. Julier, S.J. and Uhlmann, J.K., A non-divergent estimation algorithm in the presence of unknown correlations, *American Control Conf.*, Albuquerque, NM, 1997.

15

Data Fusion in Nonlinear Systems

Simon Julier and Jeffrey K. Uhlmann

CONTENTS

15.1 Introduction

The extended Kalman filter (EKF) has been one of the most widely used methods for tracking and estimation based on its apparent simplicity, optimality, tractability, and robustness. However, after more than 30 years of experience with it, the tracking and control community has concluded that the EKF is difficult to implement, difficult to tune, and only reliable for systems that are almost linear on the timescale of the update intervals. This chapter reviews the unscented transformation (UT), a mechanism for propagating mean and covariance information through nonlinear transformations, and describes its implications for data fusion.[1] This method is more accurate, is easier to implement, and uses the same order of calculations as the EKF. Furthermore, the UT permits the use of Kalman-type filters in applications where, traditionally, their use was not possible. For example, the UT can be used to rigorously integrate artificial intelligence (AI)-based systems with Kalman-based systems.

Performing data fusion requires estimates of the state of a system to be converted to a common representation. The mean and covariance representation is the *lingua franca* of modern systems engineering. In particular, the covariance intersection (CI)[2] and Kalman filter (KF)[3] algorithms provide mechanisms for fusing state estimates defined in terms of means and covariances, where each mean vector defines the nominal state of the system and its associated error covariance matrix defines a lower bound on the squared error. However, most data fusion applications require the fusion of mean and covariance estimates defining the state of a system in different coordinate frames. For example, a tracking system might maintain estimates in a global Cartesian coordinate frame, whereas observations of the tracked objects are generated in the local coordinate frames of various sensors. Therefore, a transformation must be applied to convert between the global coordinate frame and each local coordinate frame.

If the transformation between coordinate frames is linear, the linearity properties of the mean and covariance make the application of the transformation trivial. Unfortunately, most tracking sensors take measurements in a local polar or spherical coordinate frame (i.e., they measure range and bearings) that is not linearly transformable to a Cartesian coordinate frame. Rarely are the natural coordinate frames of two sensors linearly related. This fact constitutes a fundamental problem that arises in virtually all practical data fusion systems.

The UT, a mechanism that addresses the difficulties associated with converting mean and covariance estimates from one coordinate frame to another, can be applied to obtain mean and covariance estimates from systems that do not inherently produce estimates in that form. For example, this chapter describes how the UT can allow high-level AI and fuzzy control systems to be integrated seamlessly with low-level KF and CI systems.

The structure of this chapter is as follows. Section 15.2 describes the nonlinear transformation problem within the KF framework and analyzes the KF prediction problem in detail. The UT is introduced and its performance is analyzed in Section 15.3. Section 15.4 demonstrates the effectiveness of the UT with respect to a simple nonlinear transformation (polar to Cartesian coordinates with large bearing uncertainty) and a simple discontinuous system. Section 15.5 examines how the transformation can be embedded into a fully recursive estimator that incorporates process and observation noise. Section 15.6 discusses the use of the UT in a tracking example, and Section 15.7 describes its use with a complex process and observation model. Finally, Section 15.8 shows how the UT ties multiple levels of data fusion together into a single, consistent framework.

15.2 Estimation in Nonlinear Systems

15.2.1 Problem Statement

Minimum mean squared error (MMSE) estimators can be broadly classified into *linear* and *nonlinear* estimators. Of the linear estimators, by far the most widely used is the KF.[3]* Many researchers have attempted to develop suitable nonlinear MMSE estimators. However, the optimal solution requires that a complete description of the conditional probability density

---

* Researchers often (and incorrectly) claim that the KF can be applied only if the following two conditions hold: (a) all probability distributions are Gaussian and (b) the system equations are linear. The KF is, in fact, the minimum mean squared *linear* estimator that can be applied to *any* system with *any* distribution, provided the first two moments are known. However, it is the globally optimal estimator only under the special case that the distributions are all Gaussian.

be maintained,[4] and this exact description requires a potentially unbounded number of parameters. As a consequence, many suboptimal approximations have been proposed in the literature. Traditional methods are reviewed by Jazwinski[5] and Maybeck.[6] Recent algorithms have been proposed by Daum,[7] Gordon et al.,[8] and Kouritzin.[9] Despite the sophistication of these and other approaches, the EKF remains the most widely used estimator for nonlinear systems.[10,11] The EKF applies the KF to nonlinear systems by simply linearizing all the nonlinear models so that the traditional linear KF equations can be applied. However, in practice, the EKF has three well-known drawbacks:

1. Linearization can produce highly unstable filters if the assumption of local linearity is violated. Examples include estimating ballistic parameters of missiles[12–15] and some applications of computer vision.[16] As demonstrated later in this chapter, some extremely common transformations that are used in target tracking systems are susceptible to these problems.

2. Linearization can be applied only if the Jacobian matrix exists, and the Jacobian matrix exists only if the system is differentiable at the estimate. Although this constraint is satisfied by the dynamics of continuous physical systems, some systems do not satisfy this property. Examples include jump-linear systems, systems whose sensors are quantized, and expert systems that yield a finite set of discrete solutions.

3. Finally, the derivation of the Jacobian matrices is nontrivial in most applications and can often lead to significant implementation difficulties. In Dulimov,[17] for example, the derivation of a Jacobian requires six pages of dense algebra. Arguably, this has become less of a problem, given the widespread use of symbolic packages such as Mathematica[18] and Maple.[19] Nonetheless, the computational expense of calculating a Jacobian can be extremely high if the expressions for the terms are nontrivial.

Appreciating how the UT addresses these three problems requires an understanding of some of the mechanics of the KF and EKF.

Let the state of the system at a time step k be the state vector $\mathbf{x}(k)$. The KF propagates the first two moments of the distribution of $\mathbf{x}(k)$ recursively and has a distinctive *predictor–corrector* structure. Let $\hat{\mathbf{x}}(i|j)$ be the estimate of $\mathbf{x}(i)$ using the observation information up to and including time j, $\mathbf{Z}^j = [\mathbf{z}(1), \dots, \mathbf{z}(j)]$. The covariance of this estimate is $\mathbf{P}(i|j)$. Given an estimate $\hat{\mathbf{x}}(k|k)$, the filter first predicts what the future state of the system will be using the process model. Ideally; the predicted quantities are given by the expectations

$$\hat{\mathbf{x}}(k+1|k) = E[\mathbf{f}[\mathbf{x}(k), \mathbf{u}(k), \mathbf{v}(k), k]|\mathbf{Z}^k] \tag{15.1}$$

$$\mathbf{P}(k+1|k) = E[\{\mathbf{x}(k+1) - \hat{\mathbf{x}}(k+1|k)\}\{\mathbf{x}(k+1) - \hat{\mathbf{x}}(k+1|k)\}^T |\mathbf{Z}^k] \tag{15.2}$$

When $\mathbf{f}[\cdot]$ and $\mathbf{h}[\cdot]$ are nonlinear, the precise values of these statistics can be calculated only if the distribution of $\mathbf{x}(k)$ is perfectly known. However, this distribution has no general form, and a potentially unbounded number of parameters are required. Therefore, in most practical algorithms these expected values must be approximated.

The estimate $\hat{\mathbf{x}}(k+1|k+1)$ is found by updating the prediction with the current sensor measurement. In the KF, a linear update rule is specified and the weights are chosen to minimize the mean squared error of the estimate.

$$\hat{\mathbf{x}}(k+1|k+1) = \hat{\mathbf{x}}(k+1|k) + \mathbf{W}(k+1)v(k+1)$$

$$\mathbf{P}(k+1|k+1) = \mathbf{P}(k+1|k) - \mathbf{W}(k+1)\mathbf{P}_{vv}(k+1|k)\mathbf{W}^T(k+1)$$

$$v(k+1) = \mathbf{z}(k+1) - \hat{\mathbf{z}}(k+1|k) \tag{15.3}$$

$$\mathbf{W}(k+1) = \mathbf{P}_{xv}(k+1|k)\mathbf{P}_{vv}^{-1}(k+1|k)$$

Note that these equations are only a function of the predicted values of the first two moments of $\mathbf{x}(k)$ and $\mathbf{z}(k)$. Therefore, the problem of applying the KF to a nonlinear system is the ability to predict the first two moments of $\mathbf{x}(k)$ and $\mathbf{z}(k)$.

15.2.2 Transformation of Uncertainty

The problem of predicting the future state or observation of the system can be expressed in the following form. Suppose that $\mathbf{x}$ is a random variable with mean $\bar{\mathbf{x}}$ and covariance $\mathbf{P}_{xx}$. A second random variable, $\mathbf{y}$, is related to $\mathbf{x}$ through the nonlinear function

$$\mathbf{y} = \mathbf{f}[\mathbf{x}] \tag{15.4}$$

The mean $\bar{\mathbf{y}}$ and covariance $\mathbf{P}_{yy}$ of $\mathbf{y}$ must be calculated.

The statistics of $\mathbf{y}$ are calculated by (1) determining the density function of the transformed distribution and (2) evaluating the statistics from that distribution. In some special cases, exact, closed-form solutions exist (e.g., when $\mathbf{f}[\cdot]$ is linear or is one of the forms identified in Daum[7]). However; as explained in the preceeding text, most data fusion problems do not possess closed-form solutions and some kind of an approximation must be used. A common approach is to develop a transformation procedure from the Taylor series expansion of Equation 15.4 about $\bar{\mathbf{x}}$. This series can be expressed as

$$\mathbf{f}[\mathbf{x}] = \mathbf{f}[\bar{\mathbf{x}} + \delta\mathbf{x}]$$

$$= \mathbf{f}[\bar{\mathbf{x}}] + \nabla\mathbf{f}\delta\mathbf{x} + \frac{1}{2}\nabla^2\mathbf{f}\delta\mathbf{x}^2 + \frac{1}{3!}\nabla^3\mathbf{f}\delta\mathbf{x}^3 + \frac{1}{4!}\nabla^4\mathbf{f}\delta\mathbf{x}^4 + \cdots \tag{15.5}$$

where $\delta\mathbf{x}$ is a zero-mean Gaussian variable with covariance $\mathbf{P}_{xx}$ and $\nabla^n\mathbf{f}\delta\mathbf{x}^n$ is the appropriate nth-order term in the multidimensional Taylor series. The transformed mean and covariance are

$$\bar{\mathbf{y}} = \mathbf{f}[\bar{\mathbf{x}}] + \frac{1}{2}\nabla^2\mathbf{f}\mathbf{P}_{xx} + \frac{1}{2}\nabla^4\mathbf{f}E[\delta\mathbf{x}^4] + \cdots \tag{15.6}$$

$$\mathbf{P}_{yy} = \nabla\mathbf{f}\mathbf{P}_{xx}(\nabla\mathbf{f})^T + \frac{1}{2\times4!}\nabla^2\mathbf{f}(E[\delta\mathbf{x}^4] - E[\delta\mathbf{x}^2\mathbf{P}_{yy}] - E[\mathbf{P}_{yy}\delta\mathbf{x}^2] + \mathbf{P}_{yy}^2)(\nabla^2\mathbf{f})^T$$

$$+ \frac{1}{3!}\nabla^3\mathbf{f}E[\delta\mathbf{x}^4](\nabla\mathbf{f})^T + \cdots \tag{15.7}$$

In other words, the nth-order term in the series for $\bar{x}$ is a function of the nth-order moments of x multiplied by the nth-order derivatives of $f[\cdot]$ evaluated at $x = \bar{x}$. If the moments and derivatives can be evaluated correctly up to the nth order, the mean is correct up to the nth order as well. Similar comments hold for the covariance equation, although the structure of each term is more complicated. Since each term in the series is scaled by a progressively smaller and smaller term, the lowest-order terms in the series are likely to have the greatest impact. Therefore, the prediction procedure should be concentrated on evaluating the lower-order terms.

The EKF exploits linearization. Linearization assumes that the second- and higher-order terms of δx in Equation 15.5 can be neglected. Under this assumption,

$$\bar{y} = f[\bar{x}] \tag{15.8}$$

$$P_{yy} = \nabla f P_{xx} (\nabla f)^T \tag{15.9}$$

However, in many practical situations, linearization introduces significant biases or errors. These cases require more accurate prediction techniques.

15.3 Unscented Transformation

15.3.1 Basic Idea

The UT is a method for calculating the statistics of a random variable that undergoes a nonlinear transformation. This method is founded on the intuition that *it is easier to approximate a probability distribution than it is to approximate an arbitrary nonlinear function or transformation.*[20] The approach is illustrated in Figure 15.1: a set of points (*sigma points*) is chosen with sample mean and sample covariance of the nonlinear function is $\bar{x}$ and P_{xx}. The nonlinear function is applied to each point, in turn, to yield a cloud of transformed points; $\bar{y}$ and P_{yy} are the statistics of the transformed points.

Although this method bears a superficial resemblance to Monte Carlo–type methods, there is an extremely important and fundamental difference. The samples are not drawn at random; they are drawn according to a specific, deterministic algorithm. Since the problems of statistical convergence are not relevant, high-order information about the distribution can be captured using only a very small number of points. For an n-dimensional

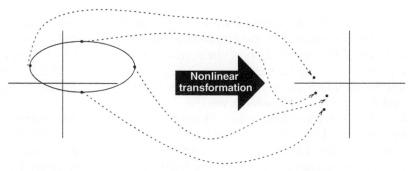

FIGURE 15.1
The principle of the unscented transformation.

space, only $n + 1$ points are needed to capture any given mean and covariance. If the distribution is known to be symmetric, $2n$ points are sufficient to capture the fact that the third-order and all higher-order odd moments are zero for any symmetric distribution.[20]

The set of sigma points, S, consists of l vectors and their appropriate weights, $S = \{i = 0, 0, ..., l - 1:X_i, W_i\}$. The weights W_i can be positive or negative but must obey the normalization condition

$$\sum_{i=0}^{l-1} W_i = 1 \tag{15.10}$$

Given these points, $\bar{y}$ and $\mathbf{P}_{yy}$ are calculated using the following procedure:

1. Instantiate each point through the function to yield the set of transformed sigma points,

$$\mathbf{y}_i = \mathbf{f}[X_i]$$

2. The mean is given by the weighted average of the transformed points,

$$\bar{y} = \sum_{i=0}^{l-1} W_i \mathbf{y}_i \tag{15.11}$$

3. The covariance is the weighted outer product of the transformed points,

$$\mathbf{P}_{yy} = \sum_{i=1}^{l-1} W_i \{\mathbf{y}_i - \bar{y}\}\{\mathbf{y}_i - \bar{y}\}^T \tag{15.12}$$

The crucial issue is to decide how many sigma points should be used, where they should be located, and what weights they should be assigned. The points should be chosen so that they capture the *most important* properties of $\mathbf{x}$. This can be formalized as follows. Let $P_x(\mathbf{x})$ be the density function of $\mathbf{x}$. The sigma points capture the necessary properties by obeying the condition

$$\mathbf{g}[S, p_x(\mathbf{x})] = 0$$

The decision about which properties of $\mathbf{x}$ are to be captured precisely and which are to be approximated is determined by the demands of the particular application in question. Here, the *moments* of the distribution of the sigma points are matched with those of $\mathbf{x}$. This is motivated by the Taylor series expansion, given in Section 15.2.2, which shows that matching the moments of $\mathbf{x}$ up to the nth-order means that Equations 15.11 and 15.12 captures $\bar{y}$ and $\mathbf{P}_{yy}$ up to the nth order as well.[21]

Note that the ÛT is distinct from other efforts published in the literature. First, some authors have considered the related problem of assuming that the distribution takes on a particular parameterized form, rather than an entire, arbitrary distribution. Kushner, for example, describes an approach whereby a distribution is approximated at each time step by a Gaussian.[22] However, the problem with this approach is that it does not address the fundamental problem of calculating the mean and covariance of the nonlinearly transformed distribution. Second, the UT bears some relationship to quadrature, which has been used to approximate the integrations implicit in statistical expectations. However, the UT avoids some of the difficulties associated with quadrature methods by approximating the unknown distribution. In fact, the UT is most closely related to perturbation analysis. In a 1989 article, Holtzmann introduced a noninfinitesimal perturbation for a scalar system.[23] Holtzmann's solution corresponds to that of the symmetric UT in the scalar case, but their respective generalizations (e.g., to higher dimensions) are not equivalent.

15.3.2 Example Set of Sigma Points

A set of sigma points can be constructed using the constraints that they capture the first three moments of a symmetric distribution: $g[S, p_x(\mathbf{x})] = [g_1[S, p_x(\mathbf{x})]g_2[S, p_x(\mathbf{x})]g_3[S, p_x(\mathbf{x})]]^T$ where

$$g_1[S, p_x(\mathbf{x})] = \sum_{i=0}^{p} W_i \mathbf{X}_i - \hat{\mathbf{x}} \tag{15.13}$$

$$g_2[S, p_x(\mathbf{x})] = \sum_{i=0}^{p} W_i (\mathbf{X}_i - \bar{\mathbf{x}})^2 - \mathbf{P}_{xx} \tag{15.14}$$

$$g_3[S, p_x(\mathbf{x})] = \sum_{i=0}^{p} W_i (\mathbf{X}_i - \bar{\mathbf{x}})^3 \tag{15.15}$$

The set is[21]

$$
\begin{aligned}
\mathbf{X}_0(k|k) &= \hat{\mathbf{x}}(k|k) \\
W_0 &= \frac{\kappa}{n + \kappa} \\
\mathbf{X}_i(k|k) &= \hat{\mathbf{x}}(k|k) + \left(\sqrt{(n + \kappa)\mathbf{P}(k|k)}\right)_i \\
W_i &= \frac{1}{2(n + \kappa)} \\
\mathbf{X}_{i+n}(k|k) &= \hat{\mathbf{x}}(k|k) - \left(\sqrt{(n + \kappa)\mathbf{P}(k|k)}\right)_i \\
W_{i+n} &= \frac{1}{2(n + \kappa)}
\end{aligned}
\tag{15.16}
$$

where κ is a real number, $\left(\sqrt{(n + \kappa)\mathbf{P}(k|k)}\right)_i$ is the ith row or column* of the matrix square root of $(n + \kappa)\mathbf{P}(k|k)$, and W_i is the weight associated with the ith point.

---

* If the matrix square root $\mathbf{A}$ of $\mathbf{P}$ is of the form $\mathbf{P} = \mathbf{A}^T\mathbf{A}$, then the sigma points are formed from the *rows* of $\mathbf{A}$. However, for a root of the form $\mathbf{P} = \mathbf{A}\mathbf{A}^T$, the *columns* of $\mathbf{A}$ are used.

15.3.3 Properties of the Unscented Transform

Despite its apparent similarity to other efforts described in the data fusion literature, the UT has a number of features that make it well suited for the problem of data fusion in practical problems:

- The UT can predict with the same accuracy as the second-order Gauss filter, but without the need to calculate Jacobians or Hessians. The reason is that the mean and covariance of x are captured precisely up to the second order, and the calculated values of the mean and covariance of y also are correct to the second order. This indicates that the mean is calculated to a higher order of accuracy than the EKF, whereas the covariance is calculated to the same order of accuracy.

- The computational cost of the algorithm is the same order of magnitude as the EKF. The most expensive operations are calculating the matrix square root and determining the outer product of the sigma points to calculate the predicted covariance. However, both operations are $O(n^3)$, which is the same cost as evaluating the $n \times n$ matrix multiplies needed to calculate the predicted covariance.*

- The algorithm naturally lends itself to a *black box* filtering library. The UT calculates the mean and covariance using standard vector and matrix operations and does not exploit details about the specific structure of the model.

- The algorithm can be used with distributions that are not continuous. Sigma points can straddle a discontinuity. Although this does not precisely capture the effect of the discontinuity, its effect is to spread the sigma points out such that the mean and covariance reflect the presence of the discontinuity.

- The UT can be readily extended to capture more information about the distribution. Because the UT captures the properties of the distribution, a number of refinements can be applied to improve greatly the performance of the algorithm. If only the first two moments are required, then $n + 1$ sigma points are sufficient. If the distribution is assumed or is known to be symmetric, then $n + 2$ sigma points are sufficient.[25] Therefore, the total number of calculations required for calculating the new covariance is $O(n^3)$, which is the same order as that required by the EKF. The transform has also been demonstrated to propagate successfully the fourth-order moment (or kurtosis) of a Gaussian distribution[26] and the third-order moments (or skew) of an arbitrary distribution.[27] In one dimension, Tenne and Singh developed the higher-order UT which can capture the first 12 moments of a Gaussian using only seven points.[28]

15.4 Uses of the Transformation

This section demonstrates the effectiveness of the UT with respect to two nonlinear systems that represent important classes of problems encountered in the data fusion literature—coordinate conversions and discontinuous systems.

---

* The matrix square root should be calculated using numerically efficient and stable methods such as the Cholesky decomposition.[24]

15.4.1 Polar to Cartesian Coordinates

One of the most important transformations in target tracking is the conversion from polar to Cartesian coordinates. This transformation is known to be highly susceptible to linearization errors. Lerro and Bar-Shalom, for example, show that the linearized conversion can become inconsistent when the standard deviation in the bearing estimate is less than a degree.[29] This subsection illustrates the use of the UT on a coordinate conversion problem with extremely high angular uncertainty.

Suppose a mobile autonomous vehicle detects targets in its environment using a range-optimized sonar sensor. The sensor returns polar information (range, r, and bearing, θ), which is converted to estimate Cartesian coordinates. The transformation is

$$\begin{pmatrix} x \\ y \end{pmatrix} = \begin{pmatrix} r\cos\theta \\ r\sin\theta \end{pmatrix} \quad \text{with} \quad \nabla\mathbf{F} = \begin{bmatrix} \cos\theta & -r\sin\theta \\ \sin\theta & r\cos\theta \end{bmatrix}$$

The real location of the target is $(0, 1)$. The difficulty with this transformation arises from the physical properties of the sonar. Fairly good range accuracy (with 2 cm standard deviation) is traded off to give a very poor bearing measurement (standard deviation of 15°).[30] The large bearing uncertainty causes the assumption of local linearity to be violated.

To appreciate the errors that can be caused by linearization, compare its values for the statistics of (x, y) with those of the true statistics calculated by Monte Carlo simulation. Owing to the slow convergence of random sampling methods, an extremely large number of samples (3.5×10^6) were used to ensure that accurate estimates of the true statistics were obtained. The results are shown in Figure 15.2a. This figure shows the mean and 1σ contours calculated by each method. The 1σ contour is the locus of points $\{\mathbf{y}:(\mathbf{y} - \bar{\mathbf{y}})\mathbf{P}_y^{-1}(\mathbf{y} - \bar{\mathbf{y}}) = 1\}$ and is a graphical representation of the size and orientation of $\mathbf{P}_{yy}$. The figure demonstrates that the linearized transformation is biased and inconsistent. This is most pronounced along the y-axis, where linearization estimates that the position is 1 m, whereas in reality it is 96.7 cm. In this example, linearization errors effectively introduce an error that is over 1.5 times the standard deviation of the range measurement. Since it is a bias that arises from the transformation process itself, the same error with the same sign will be committed each time a coordinate transformation takes place. Even if there were no bias, the transformation would still be inconsistent because its ellipse is not sufficiently extended along the y-axis.

In practice, this inconsistency can be resolved by introducing additional stabilizing noise that increases the size of the transformed covariance. This is one possible explanation of why EKFs are difficult to tune—sufficient noise must be introduced to offset the defects of linearization. However, introducing stabilizing noise is an undesirable solution because the estimate remains biased and there is no general guarantee that the transformed estimate remains consistent or efficient.

The performance benefits of using the UT can be seen in Figure 15.2b, which shows the means and 1σ contours determined by the different methods. The mismatch between the UT mean and the true mean is extremely small (6×10^{-4}). The transformation is consistent, ensuring that the filter does not diverge. As a result, there is no need to introduce artificial noise terms that would degrade performance even when the angular uncertainty is extremely high.

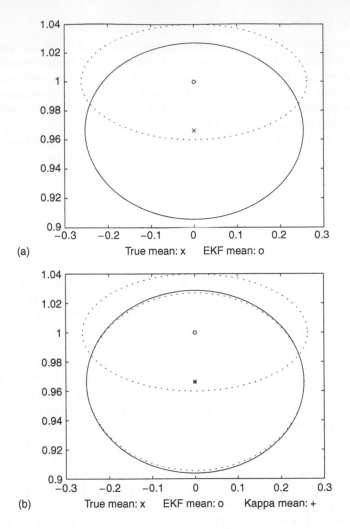

FIGURE 15.2
The mean and standard deviation ellipses for the true statistics, those calculated through linearization and those calculated by the unscented transformation. (a) Results from linearization. The true mean is at ×, and the uncertainty ellipse is solid. Linearization calculates the mean at o, and the uncertainty ellipse is dashed. (b) Results from the UT. The true mean is at ×, and the uncertainty ellipse is dotted. The UT mean is at + (overlapping the position of the true mean) and is the solid ellipse. The linearized mean is at o, and its ellipse is also dotted.

15.4.2 Discontinuous Transformation

Consider the behavior of a two-dimensional particle whose state consists of its position $\mathbf{x}(k) = [x(k), y(k)]^T$. The projectile is initially released at time 1 and travels at a constant and known speed, v_x, in the x direction. The objective is to estimate the mean position and covariance of the position at time 2, $[x(2), y(2)]^T$, where $\Delta T \equiv t_2 - t_1$. The problem is made difficult by the fact that the path of the projectile is obstructed by a wall that lies in the *bottom right quarter-plane* ($x \geq 0, y \leq 0$). If the projectile hits the wall, a perfectly elastic collision will occur, and the projectile will be reflected back at the same velocity as it traveled forward. This situation is illustrated in Figure 15.3a, which also shows the covariance ellipse of the initial distribution.

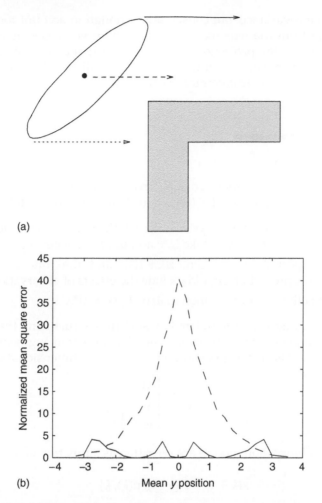

(a)

(b)

FIGURE 15.3
A discontinuous system example: a particle can either strike a wall and rebound, or continue to move in a straight line. The experimental results show the effect of using different start values for y.

The process model for this system is

$$x(2) = \begin{cases} x(1) + \Delta T v_x & y(1) \geq 0 \\ x(1) - \Delta T v_x & y(1) < 0 \end{cases} \tag{15.17}$$

$$y(2) = y(1) \tag{15.18}$$

At time 1, the particle starts in the left half-plane ($x \leq 0$) with position $[x(1), y(1)]^T$. The error in this estimate is Gaussian, has zero mean, and has covariance $\mathbf{P}(1|1)$. Linearized about this start condition, the system appears to be a simple constant velocity linear model.

The true conditional mean and covariance was determined using Monte Carlo simulation for different choices of the initial mean of y. The mean squared error calculated by the EKF and by the UT for different values is shown in Figure 15.3b. The UT estimates the mean very closely, suffering only small spikes as the translated sigma points successively pass the wall. Further analysis shows that the covariance for the filter is only slightly larger than the true covariance, but conservative enough to account for the deviation of its estimated mean from the true mean. The EKF, however, bases its entire estimate of the conditional mean on the projection of the prior mean; therefore, its estimates bear no resemblance to the true mean, except when most of the distribution either hits or misses the wall and the effect of the discontinuity is minimized.

15.5   Unscented Filter

The UT can be used as the cornerstone of a recursive Kalman-type of estimator. The transformation processes that occur in a KF (Equation 15.3) consist of the following steps:

1. Predict the new state of the system, $\hat{x}(k+1|k)$, and its associated covariance, $P(k+1|k)$. This prediction must take into account the effects of process noise.
2. Predict the expected observation, $\hat{z}(k+1|k)$, and the innovation covariance, $P_{vv}(k+1|k)$. This prediction should include the effects of observation noise.
3. Finally, predict the cross-correlation matrix, $P_{xz}(k+1|k)$.

These steps can be easily accommodated by slightly restructuring the state vector and process and observation models. The most general formulation augments the state vector with the process and noise terms to give an $n^a = n + q + r$ dimensional vector,

$$x^a(k) = \begin{bmatrix} x(k) \\ v(k) \\ w(k) \end{bmatrix} \tag{15.19}$$

The process and observation models are rewritten as a function of $x^a(k)$,

$$x(k+1) = f^a[x^a(k), u(k), k]$$
$$z(k+1) = h^a[x^a(k+1), u(k), k] \tag{15.20}$$

and the UT uses sigma points that are drawn from

$$\hat{x}^a(k|k) = \begin{pmatrix} \hat{x}(k|k) \\ 0_{q\times 1} \\ 0_{m\times 1} \end{pmatrix} \quad \text{and} \quad P^a(k|k) = \begin{bmatrix} P(k|k) & 0 & 0 \\ 0 & Q(k) & 0 \\ 0 & 0 & R(k) \end{bmatrix} \tag{15.21}$$

The matrices on the leading diagonal are the covariances, and the off-diagonal subblocks are the correlations between the state errors and the process noises.* Although this method

* If correlations exist between the noise terms, Equation 15.21 can be generalized to draw the sigma points from the covariance matrix

$$P^a(k|k) = \begin{bmatrix} P(k|k) & P_{xv}(k|k) & P_{xw}(k|k) \\ P_{vx}(k|k) & Q(k) & P_{vw}(k|k) \\ P_{wx}(k|k) & P_{wv}(k|k) & R(k) \end{bmatrix}$$

Such correlation structures commonly arise in algorithms such as the Schmidt–Kalman filter.[31]

requires the use of additional sigma points, it incorporates the noises into the predicted state with the same level of accuracy as the propagated estimation errors. In other words, the estimate is correct to the second order and no Jacobians, Hessians, or other numerical approximations must be calculated.

The full unscented filter is summarized in Table 15.1. However, recall that this is the most general form of the UF and many optimizations can be made. For example, if the process model is linear, but the observation model is not, the normal linear KF prediction equations can be used to calculate $\hat{x}(k + 1|k)$ and $P(k + 1|k)$. The sigma points would be drawn from the prediction distribution and would only be used to calculate $\hat{z}(k + 1|k)$, $P_{xv}(k + 1|k)$, and $P_{vv}(k + 1|k)$.

The following two sections describe the application of the unscented filter to two case studies. The first demonstrates the accuracy of the recursive filter, and the second considers the problem of an extremely involved process model.

TABLE 15.1

A General Formulation of The Kalman Filter Using the Unscented Transformation

---

1. The set of sigma points are created by applying a sigma point selection algorithm (e.g., Equation 15.16) to the augmented system given by Equation 15.21
2. The transformed set is given by instantiating each point through the process model

$X_i(k + 1|k) = f[X_i(k|k), u(k), k]$

3. The predicted mean is computed as

$$\hat{x}(k + 1|k) = \sum_{i=0}^{2n^a} W_i X_i(k + 1|k)$$

4. And the predicted covariance is computed as

$$P(k + 1|k) \sum_{i=0}^{2n^a} \{W_i(X_i(k + 1|k) - \hat{x}(k + 1|k))\}\{X_i(k + 1|k) - \hat{x}(k + 1|k)\}^T$$

5. Instantiate each of the prediction points through the observation model,

$Z_i(k + 1|k) = h[X_i(k + 1|k), u(k), k]$

6. The predicted observation is calculated by

$$\hat{z}(k + 1|k) = \sum_{i=1}^{2n^a} W_i Z(k + 1|k)$$

7. The innovation covariance is

$$P_{vv}(k + 1|k) = \sum_{i=0}^{2n^a} W_i\{Z_i(k|k - 1) - \hat{z}(k + 1|k)\}\{Z_i(k|k - 1) - \hat{z}(k + 1|k)\}^T$$

8. The cross-correlation matrix is determined by

$$P_{xz}(k + 1|k) = \sum_{i=0}^{2n^a} W_i\{X_i(k|k - 1) - \hat{x}(k + 1|k)\}\{Z_i(k|k - 1) - \hat{z}(k + 1|k)\}^T$$

9. Finally, the update can be performed using the normal Kalman filter equations:

$\hat{x}(k + 1|k + 1) = \hat{x}(k + 1|k) + W(k + 1)v(k + 1)$

$P(k + 1|k + 1) = P(k + 1|k) - W(k + 1)P_{vv}(k + 1|k)W^T(k + 1)$

$v(k + 1) = z(k + 1) - \hat{z}(k + 1|k)$

$W(k + 1) = P_{xv}(k + 1|k)P_{vv}^{-1}(k + 1|k)$

Note: As explained in the text, there is a significant scope for optimizing this algorithm.

15.6 Case Study: Using the UF with Linearization Errors

This section considers the problem that is illustrated in Figure 15.4: a vehicle entering the atmosphere at high altitude and at very high speed. The position of the body is to be tracked by a radar that accurately measures range and bearing. This type of problem has been identified by a number of authors[12–15] as being particularly stressful for filters and trackers, based on the strongly nonlinear nature of three types of forces that act on the vehicle. The most dominant is aerodynamic drag, which is a function of vehicle speed and has a substantial nonlinear variation in altitude. The second type of force is gravity, which accelerates the vehicle toward the center of the earth. The final type of force is random buffeting. The effect of these forces gives a trajectory of the form shown in Figure 15.4. Initially the trajectory is almost ballistic; however, as the density of the atmosphere increases, drag effects become important and the vehicle rapidly decelerates until its motion is almost vertical. The tracking problem is made more difficult by the fact that the drag properties of the vehicle could be only very crudely known.

In summary, the tracking system should be able to track an object that experiences a set of complicated, highly nonlinear forces. These depend on the current position and velocity of the vehicle, as well as on certain characteristics that are not precisely known. The filter's state space consists of the position of the body (x_1 and x_2), its velocity (x_3 and x_4), and a parameter of its aerodynamic properties (x_5). The vehicle state dynamics are

$$\dot{x}_1(k) = x_3(k)$$

$$\dot{x}_2(k) = x_4(k)$$

$$\dot{x}_3(k) = D(x)x_3(k) + G(x)x_1(k) + v_1(k) \qquad (15.22)$$

$$\dot{x}_4(k) = D(x)x_4(k) + G(x)x_2(k) + v_2(k)$$

$$\dot{x}_5(k) = v_3(k)$$

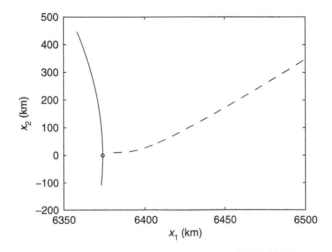

FIGURE 15.4
The reentry problem. The dashed line is the sample vehicle trajectory, and the solid line is a portion of the Earth's surface. The position of the radar is marked by an o.

where $D(k)$ is the drag-related force term, $G(k)$ is the gravity-related force term, and $v_1(k)$, $v_2(k)$, and $v_3(k)$ are the process noise terms. Defining $R(k) = \sqrt{x_1^2(k) + x_2^2(k)}$ as the distance from the center of the Earth and $V(k) = \sqrt{x_3^2(k) + x_4^2(k)}$ as absolute vehicle speed, the drag and gravitational terms are

$$D(k) = -\beta(k)\exp\left\{\frac{[R_0 - R(k)]}{H_0}\right\}V(k) \quad \text{and} \quad G(k) = -\frac{Gm_0}{r^3(k)}$$

where $\beta(k) = \beta_0 \exp x_5(k)$.

For this example, the parameter values are $\beta_0 = -0.59783$, $H_0 = 13.406$, $Gm_0 = 3.9860 \times 10^5$, and $R_0 = 6374$, and they reflect typical environmental and vehicle characteristics.[14] The parameterization of the ballistic coefficient, $\beta(k)$, reflects the uncertainty in vehicle characteristics.[13] β_0 is the ballistic coefficient of a *typical vehicle*, and it is scaled by $\exp x_5(k)$ to ensure that its value is always positive. This is vital for filter stability.

The motion of the vehicle is measured by a radar located at (x_r, y_r). It can measure range r and bearing θ at a frequency of 10 Hz, where

$$r_r(k) = \sqrt{(x_1(k) - x_r)^2 + (x_2(k) - y_r)^2} + w_1(k)$$

$$\theta(k) = \tan^{-1}\left(\frac{x_2(k) - y_r}{x_1(k) - x_r}\right) + w_2(k)$$

$w_1(k)$ and $w_2(k)$ are zero-mean uncorrelated noise processes with variances of 1 m and 17 mrad, respectively.[32] The high update rate and extreme accuracy of the sensor results in a large quantity of extremely high quality data for the filter.

The true initial conditions for the vehicle are

$$\mathbf{x}(0) = \begin{pmatrix} 6500.4 \\ 349.14 \\ -1.8093 \\ -6.7967 \\ 0.6932 \end{pmatrix} \quad \text{and} \quad \mathbf{P}(0) = \begin{bmatrix} 10^{-6} & 0 & 0 & 0 & 0 \\ 0 & 10^{-6} & 0 & 0 & 0 \\ 0 & 0 & 10^{-6} & 0 & 0 \\ 0 & 0 & 0 & 10^{-6} & 0 \\ 0 & 0 & 0 & 0 & 0 \end{bmatrix}$$

In other words, the vehicle's coefficient is twice the nominal coefficient.

The vehicle is buffeted by random accelerations,

$$\mathbf{Q}(k) = \begin{bmatrix} 2.4064 \times 10^{-5} & 0 & 0 \\ 0 & 2.4064 \times 10^{-5} & 0 \\ 0 & 0 & 0 \end{bmatrix}$$

The initial conditions assumed by the filter are

$$\hat{\mathbf{x}}(0|0) = \begin{pmatrix} 6500.4 \\ 349.14 \\ -1.8093 \\ -6.7967 \\ 0 \end{pmatrix} \quad \text{and} \quad \mathbf{P}(0|0) = \begin{bmatrix} 10^{-6} & 0 & 0 & 0 & 0 \\ 0 & 10^{-6} & 0 & 0 & 0 \\ 0 & 0 & 10^{-6} & 0 & 0 \\ 0 & 0 & 0 & 10^{-6} & 0 \\ 0 & 0 & 0 & 0 & 1 \end{bmatrix}$$

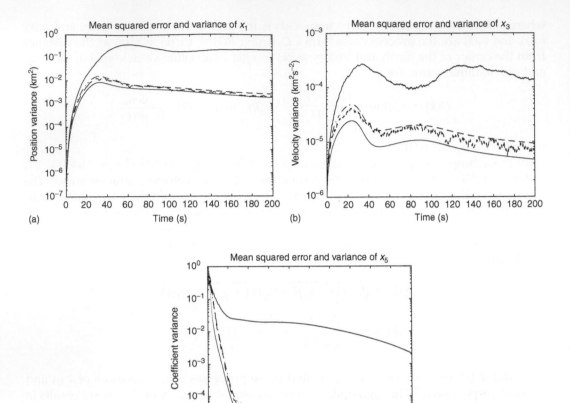

FIGURE 15.5

The mean squared errors and estimated covariances calculated by an EKF and an unscented filter. (a) Results for x_1. (b) Results for x_3. (c) Results for x_5. In all of the graphs, the solid line is the mean squared error calculated by the EKF and the dotted line is its estimated covariance. The dashed line is the unscented mean squared error, and the dot-dashed line is its estimated covariance. In all diagrams, the EKF estimate is inconsistent but the UT estimate is not.

The filter uses the nominal initial condition and, to offset for the uncertainty, the variance on the initial estimate is 1.

Both filters were implemented in discrete time, and observations were taken at a frequency of 10 Hz. However, as a result of the intense nonlinearities of the vehicle dynamics equations, the Euler approximation of Equation 15.22 was valid only for small time steps. The integration step was set to be 50 ms, which meant that two predictions were made per update. For the unscented filter, each sigma point was applied through the dynamics equations twice. For the EKF, an initial prediction step and relinearization had to be performed before the second step.

The performance of each filter is shown in Figure 15.5. This figure plots the estimated mean squared estimation error (the diagonal elements of $\mathbf{P}(k|k)$) against actual mean squared estimation error (which is evaluated using 100 Monte Carlo simulations). Only x_1, x_3, and x_5 are shown. The results for x_2 are similar to that for x_1, and the x_4 and x_3 results are the same. In all cases, the unscented filter estimates its mean squared error very accurately, maximizing the confidence of the filter estimates. The EKF, however, is highly inconsistent; the peak mean squared error in x_1 is 0.4 km^2, whereas its estimated covariance is over 100 times smaller. Similarly, the peak mean squared velocity error is 3.4×10^{-4} km^2 s^{-2}, which is more than five times the true mean squared error. Finally, x_5 is highly biased, and this bias decreases only slowly over time. This poor performance shows that, even with regular and high-quality updates, linearization errors inherent in the EKF can be sufficient to cause inconsistent estimates.

15.7   Case Study: Using the UF with a High-Order Nonlinear System

In many tracking applications, obtaining large quantities of accurate sensor data is difficult. For example, an air traffic control system might measure the location of an aircraft only once every few seconds. When information is scarce, the accuracy of the process model becomes extremely important for two reasons. First, if the control system must obtain an estimate of the target state more frequently than the tracker updates, a predicted tracker position must be used. Different models can have a significant impact on the quality of that prediction.[33] Second, to optimize the performance of the tracker, the limited data from the sensors must be exploited to the greatest degree possible. Within the KF framework, this can only be achieved by developing the most accurate process model that is practical. However, such models can be high order and nonlinear. The UF greatly simplifies the development and refinement of such models.

This section demonstrates the ease with which the UF can be applied to a prototype KF-based localization system for a conventional road vehicle. The road vehicle, shown in Figure 15.6, undertakes maneuvers at speeds in excess of 45 mph (15 m s^{-1}). The position of the vehicle is to be estimated with submeter accuracy. This problem is made difficult by the paucity of sensor information. Only the following sources are available: an inertial navigation system (which is polled only at 10 Hz), a set of encoders (also polled at 10 Hz), and a bearing-only sensor (rotation rate of 4 Hz) that measures bearing to a set of beacons. Because of the low quality of sensor information, the vehicle localization system can meet the performance requirements only through the use of an accurate process model. The model that was developed is nonlinear and incorporates kinematics, dynamics, and slip due to tire deformation. It also contains a large number of process noise terms. This model is extremely cumbersome to work with, but the UF obviates the need to calculate Jacobians, greatly simplifying its use.

The model of vehicle motion is developed from the two-dimensional *fundamental bicycle*, which is shown in Figure 15.6.[34-36] This approximation, which is conventional for vehicle ride and handling analysis, assumes that the vehicle consists of front and rear virtual wheels.* The vehicle body is the line *FGR* with the front axle affixed at *F* and the rear axle fixed at *R*. The center of mass of the vehicle is located at *G*, a distance *a* behind the front axle and *b* in front of the rear axle. The length of the wheel base is $B = a + b$. The wheels can slip laterally with slip angles α_f and α_r, respectively. The control inputs are the steer angle, δ, and angular speed, ω, of the front virtual wheel.

---

* Each virtual wheel lumps together the kinematic and dynamic properties of the pairs of wheels at the front and rear axles.

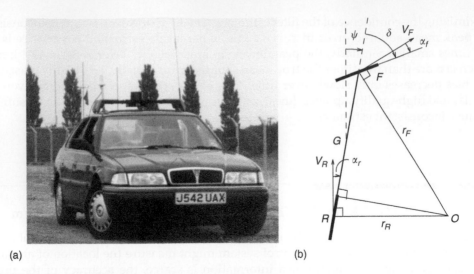

(a)   (b)

FIGURE 15.6
The actual experimental vehicle and the *fundamental bicycle model* representation used in the design of the vehicle process model. (a) The host vehicle at the test site with sensor suite. (b) Vehicle kinematics.

The filter estimates the position of F, (X_F, Y_F) the orientation of FGR, ψ, and the effective radius of the front wheel, R, (defined as the ratio of vehicle velocity to the rotation rate of the front virtual wheel). The speed of the front wheel is V_F, and the path curvature is ρ_F. From the kinematics, the velocity of F is

$$\dot{X}_F = V_F \cos[\psi + \delta - \alpha_f] \qquad \rho_F = \frac{\sin[\delta - \alpha_f] + \cos[\delta - \alpha_f]\tan\alpha_r}{B}$$

$$\dot{Y}_F = V_F \cos[\psi + \delta - \alpha_f] \qquad \text{where } V_F = R\omega\cos\alpha_f$$

$$\dot{\psi} = \rho_F V_F - \dot{\delta} - \dot{\alpha}_f$$

$$\dot{R} = 0$$

The slip angle (α_f) plays an extremely important role in determining the path taken by the vehicle and a model for determining the slip angle is highly desirable. The slip angle is derived from the properties of the tires. Specifically, tires behave as if they are linear torsional springs. The slip ankle on each wheel is proportional to the force that acts on the wheel[35]

$$\alpha_f = \frac{F_{y_f}}{C_{\alpha_f}} \qquad \alpha_r = \frac{F_{y_r}}{C_{\alpha_r}} \tag{15.23}$$

C_{α_f} and C_{α_r} are the front and rear wheel lateral stiffness coefficients (which are imprecisely known). The front and rear lateral forces, F_{y_f} and F_{y_r}, are calculated under the assumption that the vehicle has reached a steady state; at any instant in time the forces are such that the vehicle moves along an arc with constant radius and constant angular speed.[37] Resolving moments parallel and perpendicular to OG, and taking moments about G, the following simultaneous nonlinear equations must be solved:

$$\rho_G m V_G^2 = F_{x_f} \sin[\delta - \beta] + F_{y_f} \cos[\delta - \beta] + F_{y_r} \cos\beta, \quad \beta = \tan^{-1}\left(\frac{b\tan[\delta - \alpha_f] - \alpha\tan\alpha_r}{B}\right)$$

$$0 = F_{x_f} \cos[\delta - \beta] - F_{y_f} \sin[\delta - \beta] + F_{y_r} \sin\beta, \quad \rho_G = \cos\beta\,\frac{\tan[\delta - \alpha_f] + \tan\alpha_r}{B}$$

$$0 = F_{x_f} \sin\delta + aF_{y_f} \cos\delta - F_{y_r}, \quad\quad V_G = V_F \cos[\delta - \alpha_f]\sec\beta$$

m is the mass of the vehicle, V_G and ρ_G are the speed and path curvature of G, respectively, and β is the attitude angle (illustrated in Figure 15.6). These equations are solved using a conventional numerical solver[24] to give the tire forces. Through Equation 15.23, these determine the slip angles and, hence, the path of the vehicle. Since $C_{\alpha f}$ and $C_{\alpha r}$ must account for modeling errors (such as the inaccuracies of a linear force–slip angle relationship), these were treated as states and their values were estimated.

As this section has shown, a comprehensive vehicle model is extremely complicated. The state space consists of six highly interconnected states. The model is made even more complicated by the fact that that it possesses 12 process noise terms. Therefore, 18 terms must be propagated through the nonlinear process model. The observation models are also very complex. (The derivation and debugging of such Jacobians proved to be extremely difficult.) However, the UF greatly simplified the implementation, tuning, and testing of the filter. An example of the performance of the final navigation system is shown in Figure 15.7. Figure 15.7a shows a *figure of eight* route that was planned for the vehicle. This path is highly dynamic (with continuous and rapid changes in both vehicle speed and steer angle) and contains a number of well-defined landmarks (which were used to validate the algorithm). There is extremely good agreement between the estimated and the actual paths, and the covariance estimate (0.25 m^2 in position) exceeds the performance requirements.

15.8 Multilevel Sensor Fusion

This section discusses how the UT can be used in systems that do not inherently use a mean and covariance description to describe their state. Because the UT can be applied to such systems, it can be used as a consistent framework for multilevel data fusion. The problem of data fusion has been decomposed into a set of hierarchical domains.[38] The lowest levels, level 0 and level 1 (object refinement), are concerned with quantitative data fusion problems such as the calculation of a target track. Level 2 (situation refinement) and level 3 (threat refinement) apply various high-level data fusion and pattern recognition algorithms to attempt to glean strategic and tactical information from these tracks.

The difficulty lies in the fundamental differences in the representation and use of information. On the one hand, the low-level tracking filter provides only mean and covariance information. It does not specify an exact kinematic state from which an expert system could attempt to infer a tactical state. On the other hand, an expert system may be able to predict accurately the behavior of a pilot under a range of situations. However, the system does not define a rigorous low-level framework for fusing its predictions with raw sensor information to obtain high-precision estimates suitable for reliable tracking. The practical

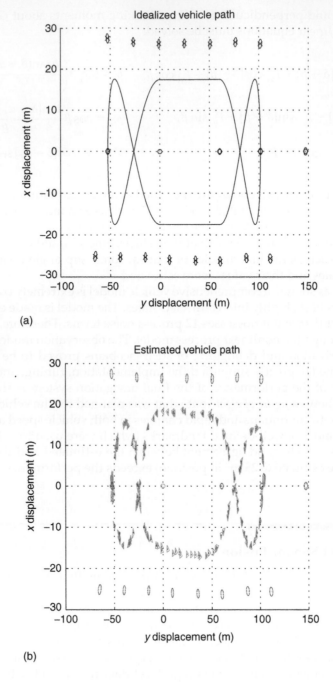

FIGURE 15.7
The positions of the beacons can be seen in (a) and (b) as the row of ellipses at the top and bottom of the figures.

solution to this problem has been to take the output of standard control and estimation routines, discretize them into a more symbolic form (e.g., *slow* or *fast*), and process them with an expert/fuzzy rule base. The results of such processing are then converted into forms that can be processed by conventional process technology.

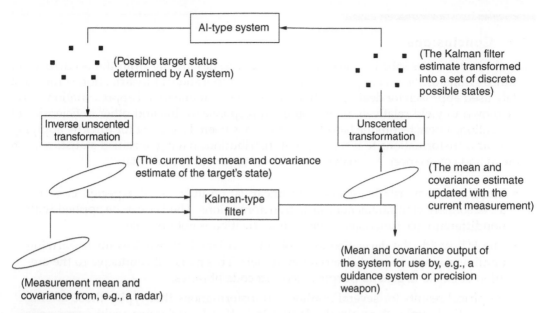

FIGURE 15.8
A possible framework for multilevel information fusion using the unscented transformation.

One approach for resolving this problem, illustrated in Figure 15.8, is to combine the different data fusion algorithms together into a single, composite data fusion algorithm that takes noise-corrupted raw sensor data and provides the inferred high-level state. From the perspective of the track estimator, the higher-level fusion rules are considered to be arbitrary, nonlinear transformations. From the perspective of the higher-level data fusion algorithms, the UT converts the output from the low-level tracker into a set of vectors. Each vector is treated as a possible kinematic state, which is processed by the higher-level fusion algorithms. In other words, the low-level tracking algorithms do not need to understand the concept of higher-level constructs, such as maneuvers, whereas the higher-level algorithms do not need to understand or produce probabilistic information.

Consider the problem of tracking an aircraft. The aircraft model consists of two components—a kinematic model, which describes the trajectory of the aircraft for a given set of pilot inputs, and an expert system, which attempts to infer current pilot intentions and predict future pilot inputs. The location of the aircraft is measured using a tracking system, such as a radar.

Some sigma points might imply that the aircraft is making a rapid acceleration, some might indicate a moderate acceleration, and yet others might imply that there is no discernible acceleration. Each of the state vectors produced from the UT can be processed individually by the expert system to predict various possible future states of the aircraft. For some of the state vectors, the expert system will signal air evasive maneuvers and predict the future position of the aircraft accordingly. Other vectors, however, will not signal a change of tactical state and the expert system will predict that the aircraft will maintain its current speed and bearing. The second step of the UT consists of computing the mean and covariance of the set of predicted state vectors from the expert system. This mean and covariance gives the predicted state of the aircraft in a form that can then be fed back to the low-level filter. The important observation to be made is that this mean and covariance reflect the probability that the aircraft will maneuver *even though the expert system did not produce any probabilistic information and the low-level filter knows nothing about maneuvers.*

15.9 Conclusions

This chapter has described some of the important issues arising from the occurrence of nonlinear transformations in practical data fusion applications. Linearization is the most widely used approach for dealing with nonlinearities, but linearized approximations have been shown to yield relatively poor results. In response to this and other deficiencies of linearization, a new technique based on the UT has been developed for directly applying nonlinear transformations to discrete point distributions having specified statistics (such as mean and covariance). An analysis of the new approach reveals that

- The UT is demonstrably superior to linearization in terms of expected error for all absolutely continuous nonlinear transformations. The UT can be applied with nondifferentiable functions in which linearization is not possible.
- The UT avoids the derivation of Jacobian (and Hessian) matrices for linearizing nonlinear kinematic and observation models. This makes it conducive to the creation of efficient, general-purpose *black box* code libraries.
- Empirical results for several nonlinear transformations that typify those arising in practical applications clearly demonstrate that linearization yields very poor approximations compared to those of the UT.

Beyond analytic claims of unproved accuracy, the UT offers a black box solution to a wide variety of problems arising in both low- and high-level data fusion applications. In particular, it offers a mechanism for seamlessly integrating the benefits of high-level methodologies, such as AI, fuzzy logic, and neural networks, with the low-level workhorses of modern engineering practice, such as covariance intersection and the KF.

Acknowledgments

The authors gratefully acknowledge support from IDAK Industries and the University of Oxford.

References

1. Julier, S.J. and Uhlmann, J.K., Unscented filtering and nonlinear estimation, *Proceedings of the IEEE*, 92(3), 401–422, 2004.
2. Julier, S.J. and Uhlmann, J.K., A nondivergent estimation algorithm in the presence of unknown correlations, *American Control Conference*, 4, 2369–2373, 1997.
3. Kalman, R.E., A new approach to linear filtering and prediction problems, *Transactions of the ASME, Journal of Basic Engineering*, 82, 34–45, 1960.
4. Kushner, H.J., Dynamical equations for optimum nonlinear filtering, *Journal of Differential Equations*, 3, 179–190, 1967.
5. Jazwinski, A.H., *Stochastic Processes and Filtering Theory*, Academic Press, New York, NY, 1970.
6. Maybeck, P.S., *Stochastic Models, Estimation, and Control*, Vol. 2, Academic Press, New York, NY, 1982.

7. Daum, F.E., New exact nonlinear filters, *Bayesian Analysis of Time Series and Dynamic Models*, J.C. Spall, Ed., Marcel Drekker, New York, NY, 199–226, 1988.

8. Gordon, N.J., Salmond, D.J., and Smith, A.F.M., Novel approach to nonlinear/non-Gaussian Bayesian state estimation, *IEEE Proceedings F*, 140(2), 107–113, 1993.

9. Kouritzin, M.A., On exact filters for continuous signals with discrete observations, *IEEE Transaction on Automatic Control*, 43(5), 709–715, 1998.

10. Uhlmann, J.K., Algorithms for multiple target tracking, *American Scientist*, 80(2), 128–141, 1992.

11. Sorenson, H.W., Ed. *Kalman Filtering: Theory and Application*, IEEE Press, New York, NY, 1985.

12. Athans, M., Wishner, R.P., and Bertolini, A., Suboptimal state estimation for continuous time nonlinear systems from discrete noisy measurements, *IEEE Transactions on Automatic Control*, TAC-13(6), 504–518, 1968.

13. Costa, P.J., Adaptive model architecture and extended Kalman–Bucy filters, *IEEE Transactions on Aerospace and Electronic Systems*, AES-30(2), 525–533, 1994.

14. Austin, J.W. and Leondes, C.T., Statistically linearized estimation of reentry trajectories, *IEEE Transactions on Aerospace and Electronic Systems*, AES-17(1), 54–61, 1981.

15. Mehra, R.K. A comparison of several nonlinear filters for reentry vehicle tracking, *IEEE Transactions on Automatic Control*, AC-16(4), 307–319, 1971.

16. Viéville, T. and Sander, P., Using pseudo Kalman filters in the presence of constraints application to sensing behaviours, Technical report, INRIA, April 1992.

17. Dulimov, P.A., Estimation of ground parameters for the control of a wheeled vehicle, Master's thesis, The University of Sydney, 1997.

18. Wolfram, S., *The Mathematica Book*, Wolfram Research, 4th Edition, Cambridge, MA, 1999.

19. Redfern, D., *Maple V Handbook—Release 4*, 2nd Edition, Ann Arbor, MI, 1996.

20. Uhlmann, J.K., Simultaneous map building and localization for real-time applications, Technical Report, Transfer thesis. University of Oxford, 1994.

21. Julier, S.J., Uhlmann, J.K., and Durrant-Whyte, H.F., A new approach for the nonlinear transformation of means and covariances in linear filters, *IEEE Transactions on Automatic Control*, 45(3), 477–482, 2000.

22. Kushner, H.J., Approximations to optimal nonlinear filters, *IEEE Transactions on Automatic Control*, AC-12(5), 546–556, 1967.

23. Holtzmann, J., On using perturbation analysis to do sensitivity analysis: Derivatives vs. differences, *IEEE Conference on Decision and Control*, 37(2), 2018–2023, 1989.

24. Press, W.H., Teukolsky, S.A., Vetterling, W.T., and Flannery, B.P., *Numerical Recipes in C: The Art of Scientific Computing*, 2nd Edition, Cambridge University Press, New York, NY, 1992.

25. Julier, S.J., The spherical simplex unscented transformation, *Proceedings of IEEE American Control Conference*, 3, 2430–2434, 2003.

26. Julier, S.J. and Uhlmann, J.K., A consistent, debiased method for converting between polar and Cartesian coordinate systems, *Proceedings of AeroSense: Acquisition, Tracking and Pointing XI*, SPIE, Vol. 3086, pp. 110–121, 1997.

27. Julier, S.J., A skewed approach to filtering, *Proceedings of AeroSense: The 12th International Symp. Aerospace/Defense Sensing, Simulation, and Controls*, Vol. 3373, pp. 271–282. SPIE, 1998: Signal and Data Processing of Small Targets.

28. Tenne, D. and Singh, T., The higher order unscented filter, *Proceedings of IEEE American Control Conference*, 3, 2441–2446, 2003.

29. Lerro, D. and Bar-Shalom, Y.K., Tracking with debiased consistent converted measurements vs. EKF, *IEEE Transactions on Aerospace and Electonics Systems*, AES-29(3), 1015–1022, 1993.

30. Leonard, J., *Directed Sonar Sensing for Mobile Robot Navigation*, Kluwer Academic Press, Boston, MA, 1991.

31. Schmidt, S.F., Applications of state space methods to navigation problems, *Advanced Control Systems*, Leondes, C.T., Ed., Vol. 3, pp. 293–340, Academic Press, New York, NY, 1966.

32. Chang, C.B., Whiting, R.H., and Athans, M. On the state and parameter estimation for maneuvering reentry vehicles, *IEEE Transactions on Automatic Control*, AC-22(1), 99–105, 1977.

33. Julier, S.J., Comprehensive Process Models for High-Speed Navigation, Ph.D. thesis, University of Oxford, 1997.
34. Ellis, J.R., *Vehicle Handling Dynamics*, Mechanical Engineering Publications, London, UK, 1994.
35. Wong, J.Y., *Theory of Ground Vehicles*, 2nd Edition, Wiley, New York, NY, 1993.
36. Dixon, J.C., *Tyres, Suspension and Handling*, Cambridge University Press, Cambridge, UK, 1991.
37. Julier, S.J. and Durrant-Whyte, H.F., Navigation and parameter estimation of high speed road vehicles, *Robotics and Automation Conference*, 101–105, 1995.
38. Klein, L., *Sensor and Data Fusion Concepts and Applications*, SPIE, 2nd Edition, 1965.

16

Random Set Theory for Multisource-Multitarget Information Fusion

Ronald Mahler

CONTENTS

This chapter deals with finite-set statistics (FISST),[1-3,80] which is also described in practitioner-level detail in the new textbook *Statistical Multisource-Multitarget Information Fusion*.[4] FISST provides a unified, scientifically defensible, probabilistic foundation for the following aspects of multisource, multitarget, and multiplatform data fusion: (1) multisource integration (i.e., detection, identification, and tracking) based on Bayesian filtering and estimation;[5-9] (2) sensor management using control theory;[7,10] (3) performance evaluation using information theory;[11-14] (4) expert systems theory (e.g., fuzzy logic, the

Dempster–Shafer theory of evidence, and rule-based inference);[15–17] (5) distributed fusion;[18] and (6) aspects of situation/threat assessment.[19]

The core of FISST is a multisource-multitarget differential and integral calculus based on the fact that so-called *belief-mass functions* are the multisensor-multitarget counterparts of probability-mass functions. One purpose of this calculus is to enable signal-processing practitioners to directly generalize conventional, engineering-friendly statistical reasoning to multisensor, multitarget, multievidence applications. Another purpose is to extend Bayesian methodologies so that they are capable of dealing with (1) imperfectly characterized data and sensor models and (2) true sensor and target models for multisource-multitarget problems. Therefore, FISST encompasses certain expert-system approaches that are often described as *heuristic*—fuzzy logic, the Dempster–Shafer theory of evidence, and rule-based inference—as special cases of a single Bayesian paradigm.

Following are some examples of applications where FISST techniques have been applied:

- Multisource datalink fusion and target identification[20]
- Naval passive-acoustic antisubmarine fusion and target identification[21]
- Air-to-ground target identification using synthetic aperture radar (SAR)[22]
- Scientific performance evaluation of multisource-multitarget data fusion algorithms[14,23,24]
- Unified detection and tracking using true multitarget likelihood functions and approximate nonlinear filters[25]
- Joint tracking, pose estimation, and target identification using high-range resolution radar (HRRR)[26]

Owing to rapid progress in FISST-related fields since the first edition of this book in 2002, this chapter needed to be rewritten to a large extent.

The chapter is organized as follows. Section 16.1 introduces the basic practical issues underlying Bayes information fusion. Section 16.2 summarizes the basic statistical foundations of single-sensor, single-target tracking and identification. Section 16.3 shows how this familiar paradigm can be naturally extended to *ambiguous* nontraditional information, such as features, attributes, natural-language statements, and inference rules. The mathematical core of FISST—the multisource-multitarget integral and differential calculus—is summarized in Section 16.4. Section 16.5 introduces multisource-multitarget measurement models and their *true* likelihood functions. Section 16.6 describes multitarget motion models and their corresponding *true* multitarget Markov transition densities. The multisource-multitarget Bayes filter and various issues related to it (especially multitarget state estimation) are summarized in Section 16.7. Two new principles of approximate multitarget filtering techniques—the probability hypothesis density (PHD) filter and cardinalized probability hypothesis density (CPHD) filter—are described in Section 16.8. This section includes a brief survey of recent advances involving these filters. Conclusions are presented in Section 16.9.

16.1 Introduction

This section describes the problems that FISST is meant to address, and summarizes the FISST approach for addressing them. The section is organized as follows. Sections 16.1.1 and 16.1.2 describe some basic engineering issues in single-target and multitarget Bayesian inference. The FISST approach is summarized in Section 16.1.3, and Section 16.1.4 shows why this approach is necessary if the *Bayesian Iceberg* is to be confronted successfully.

16.1.1   *Bayesian Iceberg*: Models, Optimality, Computability

Recursive Bayesian nonlinear filtering and estimation have been the most accepted theoretical basis for developing algorithms that are optimal and practical. Our jumping-off point is Bayes' rule:

$$f_{\text{posterior}}(\mathbf{x}|\mathbf{z}) = \frac{f(\mathbf{z}|\mathbf{x})f_{\text{prior}}(\mathbf{x})}{f(\mathbf{z})} \tag{16.1}$$

where $\mathbf{x}$ denotes the unknown quantities of interest (e.g., position, velocity, and target class); $f_{\text{prior}}(\mathbf{x})$, the prior distribution, encapsulates our previous knowledge of $\mathbf{x}$; $\mathbf{z}$ represents a new measurement; $f(\mathbf{z}|\mathbf{x})$, the likelihood function, describes the generation of the measurements; $f_{\text{posterior}}(\mathbf{x}|\mathbf{z})$, the posterior distribution, encapsulates our current knowledge about $\mathbf{x}$; and $f(\mathbf{z})$ is the Bayes normalization factor.

If time has expired before collection of the new information, $\mathbf{z}$, then $f_{\text{prior}}(\mathbf{x})$ cannot be immediately used in Bayes' rule. It must first be extrapolated to a new prior $f_{\text{prior}}^{+}(\mathbf{x})$ that accounts for the uncertainties caused by possible interim target motion. This extrapolation is accomplished using either the Markov time-prediction integral[27]

$$f_{\text{prior}}^{+}(\mathbf{x}|\mathbf{z}) = \int f^{+}(\mathbf{x}|\mathbf{y}) \cdot f_{\text{prior}}(\mathbf{y})d\mathbf{y} \tag{16.2}$$

or solution of the Fokker–Planck equation (FPE).[28] In Equation 16.2, $f^{+}(\mathbf{x}|\mathbf{y}')$ is the Markov transition density. It describes the likelihood that the target will have state $\mathbf{x}$ if it previously had state $\mathbf{y}$.

The density, $f_{\text{posterior}}(\mathbf{x}|\mathbf{z})$, contains all relevant information about the unknown state variables (i.e., position, velocity, and target identity) contained in $\mathbf{x}$.

One could, in principle, plot graphs or contour maps of $f_{\text{posterior}}(\mathbf{x}|\mathbf{z})$ in real time. However, this presupposes the existence of a human operator trained to interpret them. Complete automation is necessary as most data fusion and tracking applications have fewer operators compared to data. To render the information in $f_{\text{posterior}}(\mathbf{x}|\mathbf{z})$ available for completely automated real-time applications, we need a Bayes-optimal state estimator that extracts an estimate of the actual target state from the posterior.

The most familiar Bayes-optimal state estimators are the maximum *a posteriori* (MAP) and expected *a posteriori* (EAP) estimators

$$\hat{\mathbf{x}}^{\text{MAP}} = \arg \max_{\mathbf{x}} f_{\text{posterior}}(\mathbf{x}|\mathbf{z}) \qquad \hat{\mathbf{x}}^{\text{EAP}} = \int \mathbf{x} \cdot f_{\text{posterior}}(\mathbf{x}|\mathbf{z})d\mathbf{x} \tag{16.3}$$

The current popularity of the Bayesian approach is due, in large part, to the fact that it leads to provably optimal algorithms within a relatively simple conceptual framework. Since engineering mathematics is a tool and not an end in itself, this simplicity is a great strength—but also a great weakness. In and of itself, Bayes' rule is nearly content-free—its proof requires barely a single line. Its power derives from the fact that it is merely the visible tip of a conceptual iceberg, the existence of which tends to be forgotten precisely because the rest of the iceberg is taken for granted. In particular, both the optimality and simplicity of the Bayesian framework can be taken for granted only within the confines of standard applications addressed by standard textbooks. When one ventures out of these confines one must exercise proper engineering prudence, which includes verifying that standard textbook assumptions still apply.

16.1.1.1 Bayesian Iceberg: Sensor Models

Bayes' rule exploits, to the best possible advantage, the high-fidelity knowledge about the sensor contained in the likelihood function $f(\mathbf{z}|\mathbf{x})$. If $f(\mathbf{z}|\mathbf{x})$ is too imperfectly understood, then an algorithm will *waste* a certain amount N_{sens} of data trying (and perhaps failing) to overcome the mismatch between model and reality.

Many forms of data (e.g., generated by tracking radars) are so well characterized $f(\mathbf{z}|\mathbf{x})$ can be constructed with sufficient fidelity. Other kinds of data (e.g., SAR images or HRRR range profiles) are proving to be so difficult to simulate that there is no assurance that sufficiently high fidelity will ever be achieved, particularly in real-time operation. In such cases, a potential problem persists—namely, that our algorithm will be Bayes-optimal with respect to an *imaginary sensor*, unless we have the provably true likelihood $f(\mathbf{z}|\mathbf{x})$. That is, we are avoiding the real algorithmic issue: what to do when likelihoods cannot be sufficiently well characterized.

Finally, there are types of data—features extracted from signatures, English-language statements received over datalink, and rules drawn from knowledge bases—that are so ambiguous (i.e., poorly understood from a statistical point of view) that probabilistic approaches are not obviously applicable. This gap in Bayesian inference needs to be filled.

Even if $f(\mathbf{z}|\mathbf{x})$ can be determined with sufficient fidelity, multitarget problems present a new challenge. We will *waste* data—or worse—unless we find the corresponding provably true multitarget likelihood—the specific function $f(\mathbf{z}_1, ..., \mathbf{z}_m|\mathbf{x}_1, ..., \mathbf{x}_n)$ that describes, with the same high fidelity as $f(\mathbf{z}|\mathbf{x})$, the likelihood that the sensor will collect observations $\mathbf{z}_1, ..., \mathbf{z}_m$ (m random) given the presence of targets with states $\mathbf{x}_1, ..., \mathbf{x}_n$ (n random). *Bayes-optimality* is meaningless unless we can construct the provably true $f(\mathbf{z}_1, ..., \mathbf{z}_m|\mathbf{x}_1, ..., \mathbf{x}_n)$.

16.1.1.2 Bayesian Iceberg: Motion Models

Much of what has been said about likelihoods $f(\mathbf{z}|\mathbf{x})$ applies with equal force to Markov densities $f^+(\mathbf{x}|\mathbf{y})$. The more accurately the $f^+(\mathbf{x}|\mathbf{y})$ models target motion, the more effectively the Bayes' rule will apply. Otherwise, a certain amount N_{targ} of data must be expended in overcoming poor motion-model selection.

Once again, however, what does one do in the multitarget situation if $f^+(\mathbf{x}|\mathbf{y})$ is accurate? We must find the provably true multitarget Markov transition density—that is, the specific function $f^+(\mathbf{x}_1, ..., \mathbf{x}_n|\mathbf{y}_1, ..., \mathbf{y}_{n'})$ that describes, with the same high fidelity as $f^+(\mathbf{x}|\mathbf{y})$, how likely it is that a group of targets that previously were in states $\mathbf{y}_1, ..., \mathbf{y}_{n'}$ (n' random) will now be found in states $\mathbf{x}_1, ..., \mathbf{x}_n$ (n random). If we assume that $n = n'$ and that $f^+(\mathbf{x}_1, ..., \mathbf{x}_n|\mathbf{y}_1, ..., \mathbf{y}_n) = f^+(\mathbf{x}_1|\mathbf{y}_1) ... f^+(\mathbf{x}_n|\mathbf{y}_n)$ then we are implicitly assuming that the number of targets is constant and target motions are uncorrelated. However, in real-world scenarios, targets can appear (e.g., multiple independently targetable reentry vehicles [MIRVs] and decoys emerging from a ballistic missile reentry vehicle) or disappear (e.g., aircraft that drop beneath radar coverage) in possibly correlated ways.

Consequently, multitarget filters that assume uncorrelated motion and constant target number may perform poorly against dynamic multitarget environments, and for the same reason, single-target trackers that assume straight-line motion may perform poorly against maneuvering targets. In either case, data are *wasted* in trying to overcome—successfully or otherwise—the effects of motion-model mismatch.

16.1.1.3   Bayesian Iceberg: State Estimation

Care must be taken when we face the problem of extracting an *answer* from the posterior distribution. If the state estimator has unrecognized inefficiencies, then a certain amount, N_{est}, of data will be unnecessarily *wasted* in trying to overcome them, though not necessarily with success. For example, the EAP estimator plays an important role in theory, but often produces erratic and inaccurate solutions when the posterior is multimodal.

In the multitarget case, the dangers of taking state estimation for granted become even grave. For example, we may fail to notice that the multitarget versions of the standard MAP and EAP estimators are not even defined, let alone provably optimal. The source of this failure is, moreover, not some abstruse point in theoretical statistics. Rather, it is due to a familiar part of everyday engineering practice: *units of measurement*.

16.1.1.4   Bayesian Iceberg: Formal Optimality

The failure of the standard Bayes-optimal state estimators in the multitarget case has consequences for *optimality*. A key point that is often overlooked is that many of the classical Bayesian optimality results depend on certain seemingly esoteric mathematical concerns. In perhaps 95% of all engineering applications—which is to say, standard applications covered by the standard textbooks—these concerns can be safely ignored. Danger awaits the unwary in the other 5%; the case with multitarget applications is similar. Because the standard Bayes-optimal state estimators fail in the multitarget case, we must construct new multitarget state estimators and prove that they are well behaved.

16.1.1.5   Bayesian Iceberg: Computability

If the simplicity of Equations 16.1 through 16.3 cannot be taken for granted in so far as modeling and optimality are concerned, then such is even more the case when it comes to computational tractability.[29] The prediction integral and Bayes normalization constant must be computed using numerical integration, and—since an infinite number of parameters are required to characterize $f_{posterior}(\mathbf{x}|\mathbf{y})$ in general—approximation is unavoidable. Naive approximations create the same difficulties as model-mismatch problems. An algorithm must *waste* a certain amount N_{appx} of data overcoming—or failing to overcome—accumulation of approximation error, numerical instability, etc.

16.1.1.6   Bayesian Iceberg: Robustness

The engineering issues addressed in the previous sections (measurement-model mismatch, motion-model mismatch, inaccurate or slowly convergent state estimators, accumulation of approximation error, and numerical instability) can be collectively described as problems of robustness—that is, the *brittleness* that Bayesian approaches can exhibit when reality steeply deviates from assumption. One might be tempted to argue that, in practical application, these difficulties can be overcome by simple *brute force*—that is, by assuming that the data rate is high enough to permit a large number of computational cycles per unit time. In this case—or so the argument goes—the algorithm will override its internal inefficiencies because the total amount, N_{data}, of data that is collected is much larger than the amount, $N_{ineffic\Delta} = N_{sens} + N_{targ} + N_{est} + N_{apps}$, of data required to overcome those inefficiencies.

If this were the case, there would be few tracking and target identification problems left to solve. Current problems are challenging *either because data rates are not sufficiently high or brute force computation cannot be accomplished in real time.*

16.1.2   Why Multisource, Multitarget, Multievidence Problems Are Tricky

One needs systematic and completely probabilistic methodologies for the following:

- Modeling uncertainty in poorly characterized likelihoods
- Modeling *ambiguous* nontraditional data, and the likelihood functions for such data
- Constructing multisource likelihoods for nontraditional data
- Fusing efficiently data from all sources (nontraditional or otherwise)
- Constructing provably true multisource-multitarget likelihood functions from underlying models of the sensors
- Constructing provably true multitarget Markov densities from underlying multitarget motion models, which account for correlated motion and changes in target number
- Constructing provably optimal multitarget state estimators that simultaneously determine target number, target kinematics, and target identity without resorting to operator intervention or optimal report-to-track association
- Developing principled approximate multitarget filters that preserve as much application realism as possible

16.1.3   Finite-Set Statistics

One of the major goals of FISST is to address the *Bayesian Iceberg* issues described in Sections 16.1.1 and 16.1.2. FISST deals with imperfectly characterized data and measurement models by extending Bayesian approaches in such a way that they are robust with respect to these ambiguities. This *robust-Bayes* methodology is described in more detail in Section 16.3. FISST deals with the difficulties associated with multisource and multitarget problems by *directly* extending engineering-friendly, single-sensor, single-target statistical calculus to the multisensor-multitarget realm. This *optimal-Bayes* methodology is described in Sections 16.4 through 16.7. Finally, FISST provides mathematical tools that may help address the formidable computational difficulties associated with multisource-multitarget filtering (whether optimal or robust). Some of these ideas are discussed in Section 16.8.

The basic approach is as follows. A suite of known sensors transmits, to a central data fusion site, the observations they collect regarding targets whose number, positions, velocities, identities, threat states, etc., are unknown. Then it is necessary to carry out the following:

1. Reconceptualize mathematically the sensor suite as a single sensor (a *meta-sensor*)
2. Reconceptualize mathematically the target set as a single target (a *meta-target*) with multitarget state $X = \{\mathbf{x}_1, ..., \mathbf{x}_n\}$ (a *meta-state*)
3. Reconceptualize the current set $Z = \{\mathbf{z}_1, ..., \mathbf{z}_m\}$ of observations, collected by the sensor suite at approximately the same time, as a single measurement (a *meta-measurement*) of the *meta-target* observed by the *meta-sensor*
4. Represent statistically ill-characterized (*ambiguous*) data as random closed subsets Θ of (multisource) observation space

5. Model multitarget multisensor data using a multisensor-multitarget measurement model—a random finite set $\Sigma = T(X) \cup C(X)$, just as single-sensor, single-target data can be modeled using a measurement model $\mathbf{Z} = h(\mathbf{x}, \mathbf{W})$

6. Model the motion of multitarget systems using a multitarget motion model—a randomly varying finite set $\Xi_{k+1} = T_k(X) \cup B_k(X)$, just as single-target motion can be modeled using a motion model $\mathbf{X}_{k+1} = \Phi_k(\mathbf{x}, \mathbf{V}_k)$.

Given this, we can *mathematically reformulate multisensor, multitarget estimation problems as single-sensor, single-target problems*. The basis of this reformulation is the concept of belief-mass. Belief-mass functions are nonadditive generalizations of probability-mass functions. (However, they are not heuristic: they are equivalent to probability-mass functions on certain abstract topological spaces.[1]) That is

- Just as likelihood functions are used to describe the generation of conventional data, $\mathbf{z}$, use generalized likelihood functions to describe the generation of nontraditional data.

- Just as the probability-mass function $p(S|\mathbf{x}) = \Pr(\mathbf{Z} \in S|\mathbf{x})$ of a single-sensor, single-target measurement model is used to describe the statistics of ordinary data, use the belief-mass function $\beta(S|X) = \Pr(\Sigma \subseteq S|X)$ of a multisource-multitarget measurement model Σ to describe the statistics of multisource-multitarget data.

- Just as the probability-mass function $p_{k+1|k}(S|\mathbf{y}) = \Pr(\mathbf{X} \in S|\mathbf{y})$ of a single-target motion model is used to describe the statistics of single-target motion, use the belief-mass function $\beta_{k+1|k}(S|Y) = \Pr(\Xi_{k+1} \subseteq S|Y)$ of a multitarget motion model Ξ_{k+1} to describe multitarget motion.

The FISST multisensor-multitarget differential and integral calculus is the one that transforms these mathematical abstractions into a form that can be used in practice:

- Just as the likelihood function, $f(\mathbf{z}|\mathbf{x})$, can be derived from $p(S|\mathbf{x})$ via differentiation, so the true multitarget likelihood function $f(Z|X)$ can be derived from $\beta(S|X)$ using a generalized differentiation operator called the *set derivative*.

- Just as the Markov transition density $f_{k+1|k}(\mathbf{x}|\mathbf{y})$ can be derived from $p_{k+1|k}(S|\mathbf{y})$ via differentiation, so the true multitarget Markov transition density can be derived from $\beta_{k+1|k}(S|Y)$ via set differentiation.

- Just as $f(\mathbf{z}|\mathbf{x})$ and $p(S|\mathbf{x})$ are related by $p(S|\mathbf{x}) = \int_S f(\mathbf{z}|\mathbf{x})d\mathbf{x}$, so $f(Z|X)$ and $\beta(S|X)$ are related by $\beta(S|X) = \int_S f(Z|X)\delta Z$ where the integral is now a multisource-multitarget *set integral*.

Accordingly, let $Z^{(k)}$: $Z_1, \ldots, Z_k$ be a time sequence of multisource-multitarget observation sets. The multitarget Bayes filter allows *multitarget posterior distributions $f_{k|k}(X|Z^{(k)})$* to be created from the multisource-multitarget likelihood:

$f_{k|k}(\varnothing|Z^{(k)}) = $ posterior likelihood that no targets are present

$f_{k|k}(\{\mathbf{x}_1, \ldots, \mathbf{x}_n\}|Z^{(k)}) = $ posterior likelihood that n targets with states $\mathbf{x}_1, \ldots, \mathbf{x}_n$ are present

From these distributions, simultaneous, provably optimal estimates of target number, kinematics, and identity can be computed without resorting to the optimal report-to-track assignment, characteristic of multihypothesis approaches. Finally, these fundamentals

TABLE 16.1

Parallels between Single-Target and Multitarget Statistics

Random Vector, $\mathbf{Z}$	Finite Random Set, Σ
Sensor	Meta-sensor
Target	Meta-target
Observation-vector, $\mathbf{z}$	Observation-set, Z
State-vector, $\mathbf{x}$	State-set, X
Sensor model, $\mathbf{z} = h(\mathbf{x}, \mathbf{W})$	Multitarget sensor model, $\Sigma = T(X) \cup C(X)$
Motion model, $\mathbf{x}_{k+1} = \Phi_k(\mathbf{x}_k, \mathbf{V}_k)$	Multitarget motion model, $\Xi_{k+1} = \Phi_k(X) \cup B_k(X)$
Differentiation, $dp/d\mathbf{z}$	Set differentiation, $\delta\beta/\delta z$
Integration, $\int_s f(\mathbf{z} \mid \mathbf{x}) d\mathbf{z}$	Set integration, $\int_s f(Z \mid X) \delta Z$
Probability-mass function, $p(S \mid \mathbf{x})$	Belief-mass function, $\beta(S \mid X)$
Likelihood function, $f(\mathbf{z} \mid \mathbf{x})$	Multitarget likelihood function, $f(Z \mid X)$
Posterior density, $f_{k \mid k}(\mathbf{x} \mid Z^k)$	Multitarget posterior, $f_{k \mid k}(X \mid Z^{(k)})$
Markov densities, $f_{k+1 \mid k}(\mathbf{x}_{k+1} \mid \mathbf{x}_k)$	Multitarget Markov densities, $f_{k+1 \mid k}(X_{k+1} \mid X_k)$

enable both optimal-Bayes and robust-Bayes multisensor-multitarget data fusion, detection, tracking, and identification.

Table 16.1 summarizes the direct mathematical parallels between the single-sensor, single-target statistics and the multisensor-multitarget statistics. This parallelism is so close that general statistical methodologies can, with a bit of prudence, be directly translated from the single-sensor, single-target to the multisensor-multitarget case. That is, the table can be thought of as a dictionary that establishes a direct correspondence between the words and grammar in the random vector language and cognate words and grammar of the random set language. Consequently, any *sentence* (any concept or algorithm) phrased in the random vector language can, in principle, be directly *translated* into a corresponding sentence (corresponding concept or algorithm) in the random set language. The correspondence between dictionaries is, of course, not precisely one-to-one. For example, vectors can be added and subtracted, whereas finite sets cannot. Nevertheless, the parallelism is complete enough that, provided some care is taken, 100 years of accumulated knowledge about single-sensor, single-target statistics can be directly brought to bear on multisensor-multitarget problems. This process can be regarded as a general methodology for attacking multisource-multitarget data fusion problems.

16.1.4 Why Random Sets?

Random set theory was systematically formulated by Matheron in the mid-1970s.[30] Its centrality as a unifying foundation for expert systems theory and ill-characterized evidence has become increasingly apparent since the mid-1980s.[1,31,81,82] Its centrality as a unifying foundation for data fusion applications has been promoted since the late-1970s by I.R. Goodman. The basic relationships between random set theory and the Dempster–Shafer theory of evidence were established by Shafer and Logan,[32] Nguyen,[33] and Hestir et al.[34] The basic relationships between random set theory and fuzzy logic can be attributed to Goodman,[35] Orlov,[36] and Hohle.[37] Mahler developed relationships between random set theory and rule-based evidence.[38,39]

FISST, whose fundamental ideas were codified in 1993 and 1994, builds upon this existing body of research by showing that random set theory provides a unified framework for expert systems theory and multisensor-multitarget fusion detection, tracking, identification, sensor management, and performance estimation. FISST is unique in that it provides, under a single probabilistic paradigm, a unified and relatively simple and familiar statistical calculus for addressing all of the *Bayesian Iceberg* problems described earlier.

16.2   Review of Bayes Filtering and Estimation

This section summarizes those aspects of conventional statistics that are most pertinent to tracking, target identification, and information fusion. The foundation of applied tracking and identification—the recursive Bayesian nonlinear filtering equations—are described in Section 16.2.1. The procedure for constructing provably true sensor likelihood functions from sensor models, and provably true Markov transition densities from target motion models, is described in Sections 16.2.2 and 16.2.3, respectively. Bayes-optimal state estimation is reviewed in Section 16.2.4. Data *vectors* have the form $\mathbf{y} = (y_1, ..., y_n, w_1, ..., w_n)$, where $y_1, ..., y_n$ are continuous variables and $w_1, ..., w_n$ are discrete variables.[1] Integrals of functions of such variables involve both summations and continuous integrals.

16.2.1   Bayes Recursive Filtering

Most signal-processing practitioners are familiar with the Kalman filtering equations. Less well known is the fact that the Kalman filter is a special case of the Bayesian discrete-time recursive nonlinear filter.[27,28,40] This more general filter is nothing more than Equations 16.1 through 16.3 applied recursively:

$$f_{k+1|k}(\mathbf{x}|Z^k) = \int f_{k+1|k}(\mathbf{x}|\mathbf{y}) \cdot f_{k|k}(\mathbf{y})d\mathbf{y} \tag{16.4}$$

$$f_{k+1|k+1}(\mathbf{x}|Z^{k+1}) = \frac{f_{k+1}(\mathbf{z}_{k+1}|\mathbf{x}) \cdot f_{k+1|k}(\mathbf{x}|Z^k)}{f_{k+1}(\mathbf{z}_{k+1}|Z^k)} \tag{16.5}$$

$$\hat{\mathbf{x}}_{k|k}^{MAP} = \arg\max_{\mathbf{x}} f_{k|k}(\mathbf{x}|Z^k) \qquad \hat{\mathbf{x}}_{k|k}^{EAP} = \int \mathbf{x} \cdot f_{k|k}(\mathbf{x}|Z^k)d\mathbf{x} \tag{16.6}$$

where

$$f_{k+1}(\mathbf{z}|Z^k) = \int f_{k+1|k}(\mathbf{z}|\mathbf{x}) \cdot f_{k+1|k}(\mathbf{x}|Z^k)d\mathbf{x} \tag{16.7}$$

is the Bayes normalization constant, and where $f_{k+1}(\mathbf{z}|\mathbf{x})$ is the sensor likelihood function; $f_{k+1|k}(\mathbf{x}|\mathbf{y})$ the target Markov transition density; $f_{k|k}(\mathbf{x}|Z^k)$ the posterior distribution conditioned on the observation-stream Z^k: $\mathbf{z}_1, ..., \mathbf{z}_k$; and $f_{k+1|k}(\mathbf{x}|Z^k)$ the time-prediction of $f_{k|k}(\mathbf{x}|Z^k)$ to the time-step of the next observation.

The practical success of Equations 16.4 and 16.5 relies on the ability to construct effectively the likelihood function, $f_{k+1}(\mathbf{z}|\mathbf{x})$, and the Markov transition density, $f_{k+1|k}(\mathbf{x}|\mathbf{y})$. Although likelihood functions sometimes are constructed via direct statistical analysis of data, more typically they are constructed from sensor measurement models. Markov densities typically are constructed from target motion models. In either case, differential and integral calculus must be used, as shown in the following two sections.

16.2.2 Constructing Likelihood Functions from Sensor Models

Suppose that a target with randomly varying state X is interrogated by a sensor that generates observations of the form $Z = h(x) + W$ (where W is a zero-mean random noise vector with density $f_w(w)$) but does not generate missed detections or false alarms. The statistical behavior of Z at time-step k is characterized by its *likelihood function*, $f_{k+1}(z|x)$, which describes the likelihood that the sensor will collect measurement z given that the target has state x. How is this likelihood function computed?

Begin with the *probability-mass function* of the sensor model: $p_{k+1}(S|x) = \Pr(Z \in S|x)$. This is the total probability that the random observation Z will be found in the given region S if the target has state x. The total probability-mass, $p(S|x)$, in region S is the sum of all of the likelihoods in that region:

$$p_{k+1}(S|x) = \int f_{k+1}(y|x)dy$$

So, if E_z is some very small region surrounding the point z with (hyper) volume $V = \lambda(E_z)$, then

$$p_{k+1}(E_z|x) = \int_{E_z} f_{k+1}(y|x)dy \cong f_{k+1}(z|x) \cdot \lambda(E_z)$$

(For example, $E_z = B_{\varepsilon,z}$ could be a hyperball of radius ε centered at z.) In this case,

$$\frac{p_{k+1}(E_z|x)}{\lambda(E_z)} \cong f_{k+1}(z|x)$$

where the smaller the value of $\lambda(E_z)$, the more accurate the approximation. Stated otherwise, the likelihood function, $f_{k+1}(z|x)$, can be constructed from the probability measure, $p_{k+1}(S|x)$, via the limiting process

$$f_{k+1}(z|x) = \frac{\delta p_{k+1}}{\delta z} = \lim_{\varepsilon \to 0} \frac{p_{k+1}(E_z|x)}{\lambda(E_z)} \tag{16.8}$$

The resulting equations

$$p_{k+1}(S|x) = \int_S \frac{\delta p_{k+1}}{\delta z}dz \qquad \frac{\delta}{\delta z}\int_S f_{k+1}(y|x)dy = f_{k+1}(z|x) \tag{16.9}$$

are the relationships that show that $f_{k+1}(z|x)$ is the provably true likelihood function—that is, the density function that faithfully describes the measurement model $Z = h_{k+1}(x, W_{k+1})$. For this particular problem, the *true* likelihood is, therefore

$$f_{k+1}(z|x) = \lim_{\varepsilon \to 0} \frac{\Pr(Z \in B_{\varepsilon,z}|x)}{\lambda(B_{\varepsilon,z})} = \lim_{\varepsilon \to 0} \frac{p_{W_{k+1}}(B_{\varepsilon,z-h_{k+1}(x)})}{\lambda(B_{\varepsilon,z-h_{k+1}(x)})} = f_{W_{k+1}}(z - h_{k+1}(x))$$

16.2.3 Constructing Markov Densities from Motion Models

Suppose that, between the kth and $(k + 1)$st measurement collection times, the motion of the target is best modeled by an equation of the form $\mathbf{X}_{k+1} = \Phi_k(\mathbf{x}) + \mathbf{V}_k$, where $\mathbf{V}_k$ is a zero-mean random vector with density $f\mathbf{v}_k(\mathbf{v})$. That is, if the target had state $\mathbf{x}$ at time-step k, then it will have state $\Phi_k(\mathbf{x})$ at time-step $k + 1$—except possible error in this belief is accounted for by appending the random variation $\mathbf{V}_k$. How would $f_{k+1|k}(\mathbf{y}|\mathbf{x})$ be constructed?

This situation parallels that of Section 16.2.2. The probability-mass function $p_{k+1|k}(S|\mathbf{x}) = \Pr(\mathbf{X}_{k+1} \in S|\mathbf{x})$ is the total probability that the target will be found in region S at time-step $k + 1$, given that it had state $\mathbf{x}$ at time-step k. So,

$$f_{k+1|k}(\mathbf{y}|\mathbf{x}) = \lim_{\varepsilon \to 0} \frac{p_{k+1|k}(B_{\varepsilon,\mathbf{y}}|\mathbf{x})}{\lambda(B_{\varepsilon,\mathbf{y}})} = f_{\mathbf{V}_k}(\mathbf{y} - \Phi_k(\mathbf{x}))$$

is the true Markov density associated with the motion model $\mathbf{X}_{k+1} = \Phi_k(\mathbf{X}_k) + \mathbf{V}_k$. More generally, the equations

$$p_{k+1|k}(S|\mathbf{x}) = \int_S \frac{\delta p_{k+1|k}}{\delta \mathbf{y}} d\mathbf{y} \qquad \frac{\delta}{\delta \mathbf{y}} \int_S f_{k+1|k}(\mathbf{w}|\mathbf{x}) d\mathbf{w} = f_{k+1|k}(\mathbf{y}|\mathbf{x})$$

are the relationships showing that $f_{k+1|k}(\mathbf{y}|\mathbf{x})$ is the provably true Markov density—that is, the density function that faithfully describes the motion model $\mathbf{X}_{k+1} = \Phi_k(\mathbf{x}) + \mathbf{V}_k$.

16.2.4 Optimal State Estimators

An estimator of the state $\mathbf{x}$ is any family $\hat{x}(\mathbf{z}_1, \ldots, \mathbf{z}_m)$ of state-valued functions of the (static) measurements $\mathbf{z}_1, \ldots, \mathbf{z}_m$. *Good* state estimators $\hat{x}$ should be Bayes-optimal in the sense that, in comparison to all other possible estimators, they minimize the Bayes risk

$$R_C(\hat{\mathbf{x}}, m) = \int C(\mathbf{x}, \hat{\mathbf{x}}(\mathbf{z}_1, \ldots, \mathbf{z}_m)) \cdot f(\mathbf{z}_1, \ldots, \mathbf{z}_m \mid \mathbf{x}) \cdot f(\mathbf{x}) d\mathbf{x} d\mathbf{z}_1 \ldots d\mathbf{z}_m$$

for some specified cost (i.e., objective) function $C(\mathbf{x}, \mathbf{y})$ defined by states $\mathbf{x}$, $\mathbf{y}$.[41] Second, they should be statistically consistent in the sense that $\hat{x}(\mathbf{z}_1, \ldots, \mathbf{z}_m)$ converges to the actual target state as $m \to \infty$. Other properties (e.g., asymptotically unbiased, rapidly convergent, stably convergent, etc.) are desirable as well. The most common *good* Bayes state estimators are the MAP and EAP estimators described earlier.

16.3 Extension to Nontraditional Data

One of the most challenging aspects of information fusion has been the highly disparate and ambiguous forms that information can have. Many kinds of data, such as that supplied by tracking radars, can be described in statistical form. However, statistically uncharacterizable real-world variations make the modeling of other kinds of data, such as SAR images, highly problematic.

It has been even more ambiguous how still other forms of data—natural-language statements, features extracted from signatures, rules drawn from knowledge bases—might be mathematically modeled and processed. Numerous expert-system approaches have been proposed to address such problems. But their burgeoning number and variety have led to much confusion and controversy.

This section addresses the question of how to extend the Bayes filter to situations in which likelihood functions and data are imperfectly understood.[4,42] It is in this context that we introduce the two types of data and the four types of measurements that will concern us:

- Two types of data: state-estimates versus measurements
- Four types of measurements: unambiguously generated unambiguous (UGU), unambiguously generated ambiguous (UGA), ambiguously generated ambiguous (AGA), and ambiguously generated unambiguous (AGU)

The four types of measurements are defined as follows:

1. UGU measurements are conventional measurements, as described in Section 16.2.

2. AGU measurements include conventional measurements, such as SAR or HRRR, whose likelihood functions are ambiguously defined because of statistically uncharacterizable real-world variations.

3. UGA measurements resemble conventional measurements in that their relationship to target state is precisely known, but differ in that there is ambiguity regarding what is actually being observed. Examples include attributes or features extracted by humans or digital signal processors from signatures, natural-language statements, rules, etc.

4. AGA measurements are the same as UGA measurements, except that not only the measurements themselves but also their relationship to target state is ambiguous. Examples include attributes, features, natural-language statements, and rules, for which sensor models must be constructed from human-mediated expert knowledge.

The basic approach we shall follow is outlined in the following:

- Represent statistically ill-characterized (*ambiguous*) UGA, AGA, and AGU data as random closed subsets Θ of (multisource) observation space.
- Thus, in general, multisensor-multitarget observations will be randomly varying finite sets of the form $Z = \{\mathbf{z}_1, ..., \mathbf{z}_m, \Theta_1, ..., \Theta_{m'}\}$, where $\mathbf{z}_1, ..., \mathbf{z}_m$ are conventional data and $\Theta_1, ..., \Theta_{m'}$ are *ambiguous* data.
- Just as the probability-mass function $p(S|\mathbf{x}) = \Pr(\mathbf{Z} \in S|\mathbf{x})$ is used to describe the generation of conventional data $\mathbf{z}$, use *generalized likelihood functions* $f(\Theta|\mathbf{x})$ to describe the generation of ambiguous data.
- Construct single-target posteriors $f_{k|k}(\mathbf{x}|Z^k)$ conditioned on all data, whether *ambiguous* or otherwise.
- Proceed essentially as described in Section 16.2.

Section 16.3.1 discusses the issue of modeling data in general: traditional measurements, ambiguous measurements, or state-estimates. Section 16.3.2 introduces the concept of a generalized measurement, and their representation using random sets. Section 16.3.3 introduces the basic random set uncertainty models: imprecise, fuzzy/vague, Dempster–Shafer, and contingent (rules). The next three sections discuss models and generalized likelihood functions for the three types of generalized measurements: UGA (Section 16.3.4), AGA (Section 16.3.5), and AGU (Section 16.3.6). Section 16.3.7 addresses the problem of ambiguous (specifically, Dempster–Shafer) state-estimates. Finally, the extension of the recursive Bayesian nonlinear filtering equations to nontraditional data is discussed in Section 16.3.8.

16.3.1   General Data Modeling

The modeling of observations as vectors $\mathbf{z}$ in some Euclidean space is so ubiquitous that it is commonplace to think of $\mathbf{z}$ as *data* in itself. However, this is not actually the case: $\mathbf{z}$ is actually a mathematical abstraction that serves as a representation of some real-world entity called a *datum*.

The following are examples of actual data that occur in the real world: a voltage, a radio frequency (RF) intensity-signature, an RF I&Q (in-phase and quadrature) signature, a feature extracted from a signature by a digital signal processor, an attribute extracted from an image by a human operator, a natural-language statement supplied by a human observer, a rule drawn from a rule-base, and so on.

All of these measurement types are mathematically meaningless—which is to say, we cannot do anything algorithmic with them—unless we first construct mathematical abstractions that model them. Thus, voltages are commonly modeled as real numbers. Intensity signatures are modeled as real-valued functions or, when discretized into bins, as vectors. I&Q signatures are modeled as complex-valued functions or as complex vectors. Features are commonly modeled using integers, real numbers, feature-vectors, etc.

For these kinds of data, relatively little ambiguity adheres to the representation of a given datum by its associated model. The only uncertainty in such data is that associated with the randomness of the generation of measurements by targets. This uncertainty is modeled by the likelihood function $f_{k+1}(\mathbf{z}|\mathbf{x})$. Thus, it is conventional to think of the $\mathbf{z}$ in $f_{k+1}(\mathbf{z}|\mathbf{x})$ as a *datum* and of $f_{k+1}(\mathbf{z}|\mathbf{x})$ as the full encapsulation of its uncertainty model.

This reasoning will not suffice for data types such as rules, natural-language statements, human-extracted attributes, or more generally, any information involving some kind of human mediation. In reality, $\mathbf{z}$ is a model $\mathbf{z}_D$ of some real-world datum D, and the likelihood actually has the form $f_{k+1}(D|\mathbf{x}) = f_{k+1}(\mathbf{z}_D|\mathbf{x})$.

Consider, for example, a natural-language report supplied by a human observer:

$$D = \text{The target is near sector five.}$$

Besides randomness, two more kinds of ambiguity impede the modeling of this datum. The first kind is due to ignorance. The observer will make random errors because of factors such as fatigue, excessively high data rates, deficiencies in training, and deficiencies in ability. In principle, one could conduct a statistical analysis of the observer to determine a *likelihood function* that models his/her data generation process. In practice, such an analysis is rarely feasible. This fact introduces a nonstatistical component to the uncertainty associated with D, and we must find some way to model that uncertainty to mitigate its contaminating effects.

The second kind of uncertainty is caused by the ambiguities associated with constructing an actionable mathematical model of D. How do we model *fuzzy* and context-dependent concepts such as *near*, for example? Thus, a complete data model for D must have the form $f_{k+1}(\Theta_D|\mathbf{x})$ where Θ_D is a mathematical model of both D and the uncertainties associated with constructing Θ_D; and where $f_{k+1}(\Theta_D|\mathbf{x})$ is a mathematical model of both the process by which Θ_D is generated given $\mathbf{x}$, and the uncertainties associated with constructing $f_{k+1}(\Theta_D|\mathbf{x})$.

To summarize, comprehensive data modeling requires a unified, systematic, and theoretically defensible procedure for accomplishing the following four steps:

1. Creation of the mathematized abstractions that represent individual physical-world observations, including

2. Some approach for modeling any ambiguities that may be inherent in this act of abstraction.

3. Creation of the random variable that, by selecting among the possible mathematized abstractions, models data generation, including

4. Some approach for modeling any ambiguities caused by gaps in our understanding of how data generation occurs.

In conventional applications, steps 1, 2, and 4 are usually taken for granted, so that only the third step remains and ends up being described as the complete data model. If we are to process general and not just conventional information sources, however, we must address the other three steps.

16.3.2 Generalized Measurements

The FISST approach to data that is difficult to statistically characterize is based on the key notion that ambiguous data can be probabilistically represented as random closed subsets of (multisource) measurement space.[1]

Suppose that a data-collection source observes a scene. It does not attempt to arrive at an *a posteriori* determination about the meaning (i.e., the state) of what has been observed. Rather, it attempts only to construct an interpretation of, or opinion about, what it has or has not observed. Any uncertainties due to ignorance are therefore associated with the data-collection process alone and not with the target state.

The simplest instance of nonstatistical ambiguity is an imprecise measurement. That is, the data-collection source cannot determine the value of a measurement, $\mathbf{z}$, precisely but, rather, only to within containment in some measurement-set S. Thus, S is the actual measurement. If one randomizes S, including randomization of position, size, etc., one gets a random imprecise measurement.

Another kind of deterministic measurement is said to be fuzzy or vague. Because any single constraint, S, could be erroneous, the data-collection source specifies a nested sequence $S_1 \subseteq \ldots \subseteq S_e$ of alternative constraints, with the constraint S_i assigned a belief $s_i \geq 0$, which is the correct one, with $s_1 + \cdots + s_e = 1$. If S_e is the entire measurement space then the data-collection source is stipulating that there is some possibility that it may know nothing whatsoever about the value of $\mathbf{z}$. The nested constraint $S_1 \subseteq \ldots \subseteq S_e$, taken together with its associated weights, is the actual *measurement*. It can be represented as a random subset Θ of measurement space by defining $\Pr(\Theta = S_i) = s_i$.

If one randomizes all parameters of a vague measurement (centroid, size, shape, and number of its nested component subsets), one gets a random vague measurement. A (deterministic) vague measurement is just an instantiation of a random vague measurement.

Uncertainty, in the Dempster–Shafer sense, generalizes vagueness in that the component hypotheses no longer need to be nested. In this case Θ is discrete but otherwise arbitrary: $\Pr(\Theta = S_i) = s_i$. By randomizing all parameters, one gets a random uncertain measurement.

All such measurement types have one thing in common: they can be represented mathematically by a single probabilistic concept: a random subset of measurement space. Thus expressed with the greatest mathematical generality, a generalized measurement can be represented as an arbitrary random closed subset Θ of measurement space.

(Caution: The random closed subset Θ is a model of *a single observation collected by a single source*. This subset should not be confused with a multisensor, multitarget observation-set, Σ, whose instantiations $\Sigma = Z$ are finite sets of the form $Z = \{z_1, ..., z_m, \Theta_1, ..., \Theta_{m'}\}$ where $z_1, ..., z_m$ are individual conventional observations and $\Theta_1, ..., \Theta_{m'}$ are random set models of individual ambiguous observations.)

16.3.3 Random Set Uncertainty Models

Recognizing that random sets provide a common probabilistic foundation for various kinds of statistically ill-characterized data is not enough. We must also be able to construct practical random set representations of such data. This section shows how three kinds of ambiguous data can be modeled using random sets: vague, uncertain, and contingent.

16.3.3.1 *Vague Measurements: Fuzzy Logic*

A fuzzy measurement, g, is a fuzzy membership function on measurement space. That is, it is a function that assigns a number $g(z)$ between zero and one to each measurement z in the measurement space. Fuzzy measurements $g(z)$ and $g'(z)$ can be fused using Zadeh's conjunction defined by $(g \wedge g')(z) = \min\{g(z), g'(z)\}$.

The random subset $\Sigma_A(g)$, called the *synchronous random set representation* of the fuzzy subset g, is defined by

$$\Sigma_A(g) = \{z | A \le g(z)\} \tag{16.10}$$

where A is a uniformly distributed random number on $[0, 1]$.

16.3.3.2 *Uncertain Measurements: Dempster–Shafer Evidence*

A Dempster–Shafer measurement, B, is a Dempster–Shafer body of evidence on measurement space. That is, it consists of nonempty focal subsets $B: B_1, ..., B_b$ of measurement space and nonnegative weights $b_1, ..., b_b$ that sum to 1. Dempster–Shafer measurements can be fused using Dempster's combination "*".

Define the random subset Θ of U by $\Pr(\Theta = B_i) = b_i$ for $i = 1, ..., b$. Then Θ is the random set representation of B.[1,33,34,43]

The Dempster–Shafer measurements can be generalized to *fuzzy Dempster–Shafer measurements*. A fuzzy Dempster–Shafer measurement is one in which the focal subsets B_i are *fuzzy membership functions* on measurement space.[44]

Fuzzy Dempster–Shafer measurements can also be represented in random set form, though this representation is too complicated to discuss in the present context.

16.3.3.3 *Contingent Measurements: Rules*

Knowledge-based rules have the form $X \Rightarrow S = if\ X\ then\ S$ where S, X are subsets of measurement space. The rule $X \Rightarrow S$ is said to have been *fired* if its antecedent X is observed, and thus we infer that the consequent S—and thus also $X \cap S$—are also true.

There is at least one way to represent knowledge-based rules in random set form.[38,39] Specifically, let Φ be a uniformly distributed random subset of U—that is, one whose probability distribution is $\Pr(\Phi = S) = 2^{-|U|}$ for all $S \subseteq U$. A random set representation $\Sigma_A(X \Rightarrow S)$ of the rule $X \Rightarrow S$ is

$$\Sigma_\Phi(X \Rightarrow S) = (S \cap X) \cup (X^c \cap \Phi) \tag{16.11}$$

Similar random set representations can be devised for fuzzy rules, composite fuzzy rules, and second-order rules.

16.3.4 Unambiguously Generated Ambiguous Measurements

The simplest kind of generalized measurement is the UGA measurement, which is characterized by the following two points:

1. Modeling the measurement itself involves ambiguity; but
2. The relationship between measurements and target states can be described by a precise state-to-measurement transform model of the familiar form $\mathbf{z} = h(\mathbf{x})$.

Typical applications that require UGA measurement modeling are those that involve human-mediated feature extraction, but in which the features are distinctly associated with target types.

Consider, for example, a feature extracted by a human operator from a SAR image: the number n of hubs/tires. In this case, it is known *a priori* that a target of type ν will have a given number $n = h(\nu)$ of hubs (if a treaded vehicle) or $n = h(\nu)$ of tires (if otherwise). Suppose that the possible values of n are $n = 1, \ldots, 8$. The generalized observation Θ is a random subset of the measurement space $\{1, \ldots, 8\}$.

16.3.4.1 Generalized Likelihood Functions for UGA Measurements

As described in Section 16.2.2, the statistical measurement model typically employed in target detection and tracking has the form $\mathbf{Z} = h(\mathbf{x}) + \mathbf{W}$. Likewise, the corresponding likelihood function has the form $f_{k+1}(\mathbf{z}|\mathbf{x}) = f_\mathbf{W}(\mathbf{z} - h(\mathbf{x}))$. That is, it is the probability (density) that $h(\mathbf{x}) = \mathbf{z} - \mathbf{W}$ or, expressed otherwise, that $h(\mathbf{x}) \in \Theta_\mathbf{z}$ where $\Theta_\mathbf{z}$ is the random singleton subset defined by $\Theta_\mathbf{z} = \{\mathbf{z} - \mathbf{W}\}$.

It can be shown that, for an arbitrary UGA measurement represented as a random subset, Θ, of measurement space, the corresponding measurement model is $h(\mathbf{x}) \in \Theta_\mathbf{z}$ and the corresponding likelihood function is

$$f_{k+1}(\Theta|\mathbf{x}) = \Pr(h(\mathbf{x}) \in \Theta) \tag{16.12}$$

This actually defines a *generalized* likelihood function in the sense that, in general, it integrates to infinity rather than to unity:

$$\int f_{k+1}(\mathbf{z}|\mathbf{x})d\mathbf{z} = \infty$$

Joint generalized likelihood functions can be defined in the obvious manner. Let $\Theta_1, \ldots, \Theta_m$ be generalized measurements. Then their joint likelihood is

$$f_{k+1}(\Theta_1, \ldots, \Theta_m|\mathbf{x}) = \Pr(h(\mathbf{x}) \in \Theta_1, \ldots, h(\mathbf{x}) \in \Theta_m) = \Pr(h(\mathbf{x}) \in \Theta_1 \cap \ldots \cap \Theta_m)$$

For example, consider the hub/tire feature. Then the generalized measurement model $h(v) \in \Theta$ indicates the degree of matching between the data model, Θ, with the known feature, $h(v)$, associated with a target of type v. The generalized likelihood

$$f_{k+1}(\Theta|v) = \Pr(h(v) \in \Theta)$$

is a measure of the degree of this matching.

Because of Equation 16.12, specific formulas can be derived for different kinds of UGA measurements. First, suppose that $\Theta = \Sigma_A(g)$, where $g(z)$ is a fuzzy measurement. That is, it is a fuzzy membership function on measurement space. Then it can be shown that the corresponding likelihood function is

$$f_{k+1}(g|\mathbf{x}) = \Pr(h(\mathbf{x}) \in \Sigma_A(g)) = g(h(\mathbf{x})) \tag{16.13}$$

Next, suppose that B is a Dempster–Shafer measurement. That is, it is a list $B_1, \ldots, B_m$ of subsets of measurement space with corresponding weights $b_1, \ldots, b_m$. If Θ_B is the random set representation of B then it can be shown that the corresponding generalized likelihood function is

$$f_{k+1}(B|\mathbf{x}) = \Pr(h(\mathbf{x}) \in \Theta_B) = \sum_{j=1}^{m} b_j \cdot \mathbf{1}_{B_j}(h(\mathbf{x})) \tag{16.14}$$

Finally, suppose that $\Theta = \Sigma_\Phi(g \Rightarrow g')$ where $g \Rightarrow g'$ is a fuzzy rule on measurement space. That is, the antecedent $g(\mathbf{z})$ and consequent $g'(\mathbf{z})$ are fuzzy membership functions on measurement space. Then it can be shown that the corresponding generalized likelihood function is

$$f_{k+1}(g \Rightarrow g'|\mathbf{x}) = \Pr(h(\mathbf{x}) \in \Sigma_\Phi(g \Rightarrow g')) = (g \wedge g')(h(\mathbf{x})) + \frac{1}{2}(1 - g(h(\mathbf{x}))) \tag{16.15}$$

16.3.4.2 Bayesian Unification of UGA Measurement Fusion

A major feature of UGA measurements is that many familiar expert-system techniques for fusing evidence can be rigorously unified within a Bayesian paradigm. For this unification, we assume that the primary goal of level 1 information fusion is to determine state-estimates and estimates of the uncertainty in those estimates. No matter how complex or disparate various kinds of information might be, ultimately they must be reduced to summary information—state-estimates and, typically, covariances. This reduction itself constitutes a very lossy form of data compression.

Consequently, the unification problem can be restated as follows: what fusion techniques lose no estimation-relevant information? In a Bayesian formalism, this can be restated as: what fusion techniques are Bayes-invariant? That is, what fusion techniques leave posterior distributions unchanged?

Consider fuzzy measurements first. Let $g(\mathbf{z})$ and $g'(\mathbf{z})$ be fuzzy membership functions on measurement space. Then, in general, their joint likelihood is

$$f_{k+1}(g, g'|\mathbf{x}) = \Pr(h(\mathbf{x}) \in \Sigma_A(g') \cap \Sigma_A(g')) = (g \wedge g')(h(\mathbf{x})) = f_{k+1}(g \wedge g'|\mathbf{x}) \tag{16.16}$$

Consequently, the posterior distribution jointly conditioned on g and g' is

$$f_{k+1}(\mathbf{x}|g,g') = f_{k+1}(\mathbf{x}|g \wedge g') \qquad (16.17)$$

That is, fusion of the fuzzy measurements g and g' using Zadeh conjunction yields the same posterior distribution—and thus the same estimates of target state and error—as is obtained using Bayes' rule alone. (This result can be generalized to more general *copula* fuzzy logics.)

Similar results can be obtained for Dempster–Shafer measurements and for fuzzy rules. On the one hand, let o and o' be Dempster–Shafer measurements. Then it can be shown that

$$f_{k+1}(\mathbf{x}|o,o') = f_{k+1}(\mathbf{x}|g * g') \qquad (16.18)$$

where "*" denotes Dempster's combination.

On the other hand, let $g \Rightarrow g'$ be a fuzzy rule on measurement space. Then it can be shown that

$$f_{k+1}(\mathbf{x}|g, g \Rightarrow g') = f_{k+1}(\mathbf{x}|g \wedge g') \qquad (16.19)$$

That is, firing the rule using its antecedent (left-hand side) yields the same result that would be expected from the inference due to rule-firing (right-hand side). Thus rule-firing is equivalent to a form of Bayes' rule.

16.3.4.3 Bayes-Invariant Transformations of UGA Measurements

Much effort has been expended in the expert systems literature on devising *conversions* of one uncertainty representation scheme to another—fuzzy to probabilistic, Dempster–Shafer to probabilistic, to fuzzy, and so on. Such efforts have been hindered by the fact that uncertainty representation formalisms vary considerably in the degree of complexity of information that they encode.

Most obviously, $2^M - 1$ numbers are required to specify a (crisp) basic mass assignment on a finite measurement space with M elements, whereas only $M - 1$ numbers are required to specify a probability distribution $p(\mathbf{z})$. Consequently, any conversion of Dempster–Shafer measurements to probability distributions will result in loss of information.

A second issue is the fact that conversion from one uncertainty representation to another should be consistent with the data fusion methodologies intrinsic to these formalisms. For example, fusion of Dempster–Shafer measurements is commonly accomplished using Dempster's combination $B * B'$. Fusion of fuzzy measurements g, g', however, is usually accomplished using Zadeh's fuzzy conjunction $g \wedge g'$. For Dempster–Shafer measurements to be consistently converted to fuzzy measurements, $B \to g_B$, one should have $g_{B*B'} = g_B \wedge g_{B'}$ in some sense.

As before, the path out of such quandaries is to assume that the primary goal of level 1 information fusion is to determine state-estimates and estimates of the uncertainty in those estimates. In a Bayesian formalism, the conversion problem can be restated as: what conversions between uncertainty representations are Bayes-invariant—that is, leave posterior distributions unchanged?

Given this, it is possible to determine Bayes-invariant transformations between various kinds of measurements. We cannot delve into this issue in any depth in this context. For example, it can be shown that a Bayes-invariant conversion of $B \to g_B$ of Dempster–Shafer measurements to fuzzy measurements is defined by

$$g_B(\mathbf{z}) = \sum_{j=1}^{m} b_j \cdot \mathbf{1}_{B_i}(\mathbf{z}) \tag{16.20}$$

and where, now, fuzzy conjunction must be defined as $(g \wedge g')(\mathbf{z}) = g(\mathbf{z}) \cdot g'(\mathbf{z})$. More details can be found in Ref. 4.

16.3.5 Ambiguously Generated Ambiguous Measurements

An AGA measurement is characterized by two things:

1. Modeling of the measurement itself involves ambiguity (because, for example, of human operator interpretation processes); and
2. The association of measurements with target states cannot be precisely specified by a state-to-measurement transform of the form $\mathbf{z} = h(\mathbf{x})$.

Typical applications that require AGA measurement modeling are those that involve not only human-mediated observations, as with UGA measurements, but also the construction of human-mediated model bases. Consider, for example, the hub/tire feature in the previous section. Suppose that it is not possible to define a function that assigns a hub/tire number $h(\nu)$ to every target type ν. Rather, our understanding of at least some targets may be incomplete, so that we cannot say for sure that type ν has a specific number of hubs/ tires. In this case, it is not only the human-mediated observation that must be modeled; one must rely on the expertise of experts to construct a model of the feature. This introduces a human interpretation process, and the ambiguities associated with it necessitate more complicated models in place of $h(\nu)$. One must also construct a random set model Σ_ν using the expert information that has been supplied.

For example, suppose that target type T1 is believed to have $n = 5$ hubs/tires, but we are not quite sure about this fact. So, on the basis of expert knowledge we construct a random subset Σ_ν of the measurement space $\{1, \ldots, 8\}$ as follows: $\Sigma_\nu = \{5\}$ with probability .6; $\Sigma_\nu = \{4, 5, 6\}$ with probability .3; and $\Sigma_\nu = \{1, 2, 3, 4, 5, 6, 7, 8\}$ with probability .1.

16.3.5.1 Generalized Likelihood Functions for AGA Measurements

The sensor-transform model $h(\mathbf{x})$ for UGA measurements is unambiguous in the sense that the value of $h(\mathbf{x})$ is known precisely—there is no ambiguity in associating the measurement value $h(\mathbf{x})$ with the state value $\mathbf{x}$. However, this assumption is not valid in general.

In mathematical terms, $h(\mathbf{x})$ may be known only up to containment within some constraint $h(\mathbf{x}) \in H$. In this case, $h(\mathbf{x})$ is set valued: $h(\mathbf{x}) \in H_{0,\mathbf{x}}$. Or, we may need to specify a nested sequence of constraints $H_{0,\mathbf{x}} \subseteq H_{1,\mathbf{x}} \subseteq \ldots \subseteq H_{e,\mathbf{x}}$ with associated probabilities $h_{i,\mathbf{x}}$ that $H_{i,\mathbf{x}}$ is the correct constraint. If we define the random subset $\Sigma_\mathbf{x}$ of measurement space by $\Pr(\Sigma_\mathbf{x} = H_{i,\mathbf{x}}) = h_{i,\mathbf{x}}$ then $h(\mathbf{x})$ is random set valued: $h(\mathbf{x}) = \Sigma_\mathbf{x}$. In general, $\Sigma_\mathbf{x}$ can be any random closed subset of measurement space such that $\Pr(\Sigma_\mathbf{x} = \varnothing) = 0$.

The measurement model $h(\mathbf{x}) \in \Theta$ for UGA measurements is generalized to: $\Sigma_\mathbf{x} \cap \Theta \neq \varnothing$. That is, the generalized measurement Θ matches the generalized model $\Sigma_\mathbf{x}$ if it does not directly contradict it. The generalized likelihood function

$$f_{k+1}(\Theta|\mathbf{x}) = \Pr(\Sigma_\mathbf{x} \cap \Theta \neq \varnothing) \tag{16.21}$$

is a measure of the degree of matching between Θ and Σ_x. Specific formulas can be computed for specific situations. Suppose, for example, that both the generalized measurement and the generalized models are fuzzy: $\Theta = \Sigma_A(g)$ and $\Sigma_x = \Sigma_A(h_x)$ where $g(\mathbf{z})$ and $h_x(\mathbf{z})$ are fuzzy membership functions on measurement space. Then it can be shown that the corresponding generalized likelihood function is

$$f_{k+1}(g|\mathbf{x}) = \max_{\mathbf{z}} \min\{g(\mathbf{z}), h_x(\mathbf{z})\} \tag{16.22}$$

Specific formulas can be found for the generalized likelihood functions for other AGA measurement types, for example, Dempster–Shafer measurements.

16.3.6 Ambiguously Generated Unambiguous Measurements

AGU measurements differ from UGA and AGA measurements in that:

1. The measurement $\mathbf{z}$ itself can be precisely specified; but
2. Its corresponding likelihood function $f_{k+1}(\mathbf{z}|\mathbf{x})$ cannot.

Applications that require AGU techniques are those in which statistically uncharacterizable real-world variations make the specification of conventional likelihoods, $f_{k+1}(\mathbf{z}|\mathbf{x})$, difficult or impossible. Typical applications include ground-target identification using SAR intensity-pixel images. SAR images of ground targets, for example, can vary greatly because of the following phenomenologies: dents; wet mud; irregular placement of standard equipment (e.g., grenade launchers); placement of nonstandard equipment; turret articulation for a tank; and so on. One has no choice but to develop techniques that allow one to hedge against unknowable real-world uncertainties.

If the uncertainty in $f_{k+1}(\mathbf{z}|\mathbf{x})$ is due to uncertainty in the sensor state-to-measurement transform model $h(\mathbf{x})$, then AGA techniques could be directly applied. In this case, $h(\mathbf{x})$ is replaced by a random set model Σ_x and we just set $\Theta = \{\mathbf{z}\}$. Applying Equation 16.21 (the definition of an AGA generalized likelihood) we get

$$f_{k+1}(\Theta|\mathbf{x}) = \Pr(\Sigma_x \cap \Theta \neq \varnothing) = \Pr(\Sigma_x \cap \{\mathbf{z}\} \neq \varnothing) = \Pr(\mathbf{z} \in \Sigma_x)$$

Unfortunately, it is often difficult to devise meaningful, practical models of the form Σ_x. In such cases, a different random set technique must be used: the random error bar. We cannot delve into this further in this context. More details can be found in Ref. 4.

16.3.7 Generalized State-Estimates

In the most general sense, measurements are opinions about, or interpretations of, what is being observed. Some data sources do not supply measurements. Instead, they supply *a posteriori* opinions about target state, based on measurements which they do not pass on to us. The most familiar example is a radar which feeds its measurements into a Kalman tracker and then passes on only the time-evolving track data—or, equivalently, the time-evolving posterior distribution.

Such information is far more complex and difficult to process than measurements. We have shown how to deal with state-estimates that have a specific form: they are Dempster–Shafer. The simplest kind of nonstatistical ambiguity in a state-estimate is imprecision. The data source cannot determine the value of a state, $\mathbf{x}$, precisely but, rather, only within containment in some measurement-set S. The state-estimate could also be fuzzy. That is, the data source specifies a nested sequence $S_1 \subseteq \dots \subseteq S_e$ of alternative constraints with associated weights $s_1, \dots, s_e$. Finally, the data source may not supply nested hypotheses—that is, the state-estimate is Dempster–Shafer. Such data can be represented in random set form: $\Pr(\Gamma = S_i) = s_i$.

Using our UGA measurement theory of Section 16.3.4, we have shown how to represent such state-estimates as generalized likelihood functions. We have also shown that the *modified* Dempster's combination of Fixsen and Mahler[45] is equivalent to Bayes' rule, in the sense that using it to fuse Dempster–Shafer state-estimates produces the same posterior distributions as produced by fusing these state-estimates using Bayes' rule alone. More details can be found in Ref. 4.

16.3.8 Unified Single-Target Multisource Integration

We have shown how to model nontraditional forms of data using generalized measurements, generalized measurement models, and generalized likelihood functions. As an immediate consequence, it follows that the recursive Bayes filter of Section 16.2 can be used to fuse multisource measurements and accumulate them over time.

That is, suppose that the measurement stream consists of a time sequence of generalized measurements: Z^k: $\Theta_1, \dots, \Theta_k$. Then Equations 16.4 and 16.5 can be used to fuse and accumulate this data in the usual manner:

$$f_{k+1|k}(\mathbf{x}|Z^k) = \int f_{k+1|k}(\mathbf{x}|\mathbf{y}) \cdot f_{k|k}(\mathbf{y}) d\mathbf{y}$$

(16.23)

$$f_{k+1|k+1}(\mathbf{x}|Z^{k+1}) = \frac{f_{k+1}(\Theta_{k+1}|\mathbf{x}) \cdot f_{k+1|k}(\mathbf{x}|Z^k)}{f_{k+1}(\Theta_{k+1}|Z^k)}$$

(16.24)

where

$$f_{k+1}(\Theta|Z^k) = \int f_{k+1|k}(\Theta|\mathbf{x}) \cdot f_{k+1|k}(\mathbf{x}|Z^k) d\mathbf{x}$$

16.4   Multisource-Multitarget Calculus

This section introduces the mathematical core of FISST—the FISST multitarget integral and differential calculus. In particular, it shows that the belief-mass function of a multitarget sensor model or a multitarget motion model plays the same role in multisensor-multitarget statistics that the probability-mass function plays in single-target statistics.

The integral and derivative are the mathematical basis of conventional single-sensor, single-target statistics. We will show that the basis of multisensor-multitarget statistics is a multitarget integral and a multitarget derivative. We will show that, using the FISST calculus,

1. True multisensor-multitarget likelihood functions can be constructed from the measurement models of the individual sensors, and
2. True multitarget Markov transition densities can be constructed from the motion models of the individual targets.

Section 16.4.1 introduces the concept of a random finite set. The next three sections introduce the fundamental statistical descriptors of a multitarget system: the multitarget probability density function (Section 16.4.2), the belief-mass function (Section 16.4.3), and the probability generating functional (p.g.fl.; Section 16.4.4). The multitarget *set integral* is also introduced in Section 16.4.2. The foundations of multitarget calculus—the functional derivative of a p.g.fl. and its special case, the set derivative of a belief-mass function—are described in Section 16.4.5. Section 16.4.6 lists some of the basic theorems of the multitarget calculus: the fundamental theorem of multitarget calculus; the Radon-Nikodým theorem for multitarget calculus; and the fundamental convolution formula for multitarget probability density functions. Section 16.5.7 lists a few basic rules for the set derivative.

16.4.1   Random Finite Sets

Most readers may already be familiar with the following types of random variables:

- *Random integer, J.* A random variable that draws its instantiations $J = j$ from some subset of integers (all integers, nonnegative integers, and so on).
- *Random number, A.* A random variable that draws its instantiations $A = a$ from some subset of real numbers (all reals, nonnegative reals, the numbers in [0, 1], and so on).
- *Random vector, **Y**.* A random variable that draws its instantiations $\mathbf{Y} = \mathbf{y}$ from some subset of a Euclidean vector space.

It is less likely that the reader is familiar with one of the fundamental statistical concepts of this chapter:

- *Random finite subset, Ψ:* A random variable that draws its instantiations $\Psi = Y$ from the *hyperspace* of all finite subsets of some underlying space (e.g., finite subsets of single-target state space or of single-sensor measurement space).

16.4.2   Multiobject Density Functions and Set Integrals

Let $f(Y)$ be a function of a finite-set variable Y. That is, $f(Y)$ has the form

$f(\varnothing)$ = probability that no targets are present

$f(\{\mathbf{y}_1, ...,\mathbf{y}_n\})$ = probability density that n objects $\mathbf{y}_1, ..., \mathbf{y}_n$ are present

Also, it is assumed that the units of measurement of $f(Y)$ are $u^{-|Y|}$ if u is the unit of measurement of an individual state $\mathbf{y}$. Then the set integral of $f(Y)$ in a region S is[13]

$$\int_S f(Y)\delta Y = f(\varnothing) + \sum_{j=1}^{\infty} \frac{1}{j!} \int_{S \times \cdots \times S} f(\{\mathbf{y}_1, ..., \mathbf{y}_j\})d\mathbf{y}_1 ... d\mathbf{y}_j \tag{16.25}$$

16.4.3   Belief-Mass Functions

Just as the statistical behavior of a random vector $\mathbf{Y}$ is characterized by its probability-mass function $p_{\mathbf{Y}}(S) = \Pr(\mathbf{Y} \in S)$, so the statistical behavior of a random finite set Ψ is characterized by its *belief-mass function:*[1]

$$\beta_{\Psi}(S) = \Pr(\Psi \subseteq S)$$

For example, if Ψ is a random observation-set, then the belief mass $\beta(S|X) = \Pr(\Psi \subseteq S|X)$ is the total probability that all observations in a sensor (or multisensor) scan will be found in any given region S, if targets have multitarget state X.

As a specific example, suppose that $\Psi = \{Y\}$, where Y is a random vector. Then the belief-mass function of Ψ is just the probability-mass function of Y:

$$\beta_{\Psi}(S) = \Pr(\Psi \subseteq S) = \Pr(Y \in S) = p_Y(S)$$

16.4.4 Probability Generating Functionals

Let Ψ be a random finite set with density function $f_{\Psi}(Y)$. For any finite subset Y and any real-valued function $h(y)$ define $h^Y = 1$ if $Y = \varnothing$ and, if $Y = \{y_1, ..., y_n\}$ with $y_1, ..., y_n$ distinct,

$$h^Y = h(y_1) \cdots h(y_n) \tag{16.26}$$

Then the p.g.fl.[4] of Ψ is defined as the set integral:

$$G_{\Psi}\{h\} = \int h^Y \cdot f_{\Psi}(Y)\delta Y \tag{16.27}$$

In particular, if $h(y) = 1_S(y)$ is the indicator function of the set S, then $G_{\Psi}[1_S] = \beta_{\Psi}(S)$.

The intuitive meaning of the p.g.fl. is as follows. If $0 \leq h(y) \leq 1$, then h can be interpreted as the fuzzy membership function of a fuzzy set. The p.g.fl. $G_{\Psi}[h]$ can be shown to be the probability that Ψ is contained in the fuzzy set represented by $h(y)$. That is, $G_{\Psi}[h]$ general-izes the concept of a belief-mass function $\beta_{\Psi}(S)$ from crisp subsets to fuzzy subsets.[4]

16.4.5 Functional Derivatives and Set Derivatives

The gradient derivative (a.k.a. directional or Frechét derivative) of a real-valued function $G(x)$ in the direction of a vector w is

$$\frac{\partial G}{\partial w}(x) = \lim_{\varepsilon \to 0} \frac{G(x + \varepsilon \cdot w) - G(x)}{\varepsilon} \tag{16.28}$$

where for each x the function $w \to (\partial G/\partial w)(x)$ is linear and continuous. Thus,

$$\frac{\partial G}{\partial w}(x) = w_1 \frac{\partial G}{\partial w_1}(x) + \cdots + w_N \frac{\partial G}{\partial w_N}(x)$$

for all $w = (w_1, ..., w_N)$, where the derivatives on the right are ordinary partial derivatives. Likewise, the gradient derivative of a p.g.fl. $G[h]$ in the direction of the function g is

$$\frac{\partial G}{\partial g}[h] = \lim_{\varepsilon \to 0} \frac{G(h + \varepsilon \cdot g) - G[h]}{\varepsilon} \tag{16.29}$$

where for each h the functional $g \to (\partial G/\partial g)[h]$ is linear and continuous. Gradient deriva-tives obey the usual *turn the crank* rules of undergraduate calculus.

In physics, gradient derivatives with $g = \delta_x$ are called functional derivatives.[46] Using the simplified version of this physics notation employed in FISST, we define the functional derivatives of a p.g.fl. $G[h]$ to be[4,47–49]

$$\frac{\delta G}{\delta \varnothing}[h] = \frac{\delta^0 G}{\delta x^0}[h] = G[h] \qquad \frac{\delta G}{\delta x}[h] = \frac{\partial G}{\partial \delta_x}[h] \tag{16.30}$$

$$\frac{\delta G}{\delta X}[h] = \frac{\delta^n G}{\delta x_1 \cdots \delta x_n}[h] = \frac{\partial^n G}{\partial \delta_{x_1} \cdots \partial \delta_{x_n}}[h] \tag{16.31}$$

where $X = \{x_1, \ldots, x_n\}$ with $x_1, \ldots, x_n$ distinct.

The set derivative of a belief-mass function $\beta(S)$ of a finite random set Ξ is a functional derivative of $G[h]$ with $h = 1_S$:

$$\frac{\delta \beta_\Xi}{\delta \varnothing}(S) = \frac{\delta G}{\delta \varnothing}[1_S] = G_\Xi[1_S] \qquad \frac{\delta \beta_\Xi}{\delta x}(S) = \frac{\delta G_\Xi}{\delta x}[1_S] \tag{16.32}$$

$$\frac{\delta \beta_\Xi}{\delta X}(S) = \frac{\delta^n G_\Xi}{\delta x_1 \cdots \delta x_n}[1_S] \tag{16.33}$$

for $X = \{x_1, \ldots, x_n\}$ with $x_1, \ldots, x_n$ distinct.

An alternative way of defining the set derivative is as follows. Let E_x be a very small neighborhood of the point x with (hyper)volume $\varepsilon = \lambda(E_x)$. Then

$$\delta_x(y) \cong \varepsilon^{-1} 1_{E_x}(y)$$

and so,

$$G_\Xi[1_S + \varepsilon \delta_x] \cong G_\Xi[1_S + 1_{E_x}] = G_\Xi[1_{S \cup E_x}] = \beta_\Xi(S \cup E_x)$$

where the second of these equations results from assuming that S and E_x are disjoint. Consequently, it follows that the set derivative can be defined directly from belief-mass functions as follows:

$$\frac{\delta \beta_\Xi}{\delta x}(S) = \lim_{\lambda(E_x) \to 0} \frac{\beta_\Xi(S \cup E_x) - \beta_\Xi(S)}{\lambda(E_x)} \tag{16.34}$$

16.4.6   Key Theorems of Multitarget Calculus

The fundamental theorem of undergraduate calculus states that the integral and derivative are essentially inverse operations:

$$\int_a^b \frac{df}{dx}(y)dy = f(b) - f(a) \qquad \frac{d}{dx}\int_a^x f(y)dy = f(x)$$

Another basic formula of calculus, the Radon-Nikodým theorem, states that the probability-mass function of a random vector $\mathbf{X}$ can be written as an integral

$$\Pr(\mathbf{X} \in S) = \int_S f_{\mathbf{X}}(\mathbf{x})d\mathbf{x}$$

where $f_{\mathbf{X}}(\mathbf{x})$ is the probability density function of $\mathbf{X}$. This section presents analogous theorems for the multitarget calculus.

16.4.6.1 Fundamental Theorem of Multitarget Calculus

The set integral and derivative are inverse to each other:

$$\Pr(\Xi \subseteq S) = \beta_{\Xi}(S) = \int_S \frac{\delta \beta_{\Xi}}{\delta X}(\varnothing)\delta X \qquad f(X) = \left[\frac{\delta}{\delta X}\int_S f(Y)\delta Y\right]_{S=\varnothing}$$

16.4.6.2 Radon-Nikodým Theorem for Multitarget Calculus

The functional derivative and set integral are related by the formula:

$$\frac{\delta G}{\delta X}[h] = \int h^Y \cdot f(X \cup Y)\delta Y$$

16.4.6.3 Fundamental Convolution Formula for Multitarget Calculus

Let $\Xi_1, \ldots, \Xi_n$ be statistically independent random finite subsets and let $\Xi = \Xi_1 \cup \ldots \cup \Xi_n$ be their union. Then the probability density function of Ξ is

$$f_{\Xi}(X) = \sum_{W_1 \cup \ldots \cup W_n = X} f_{\Xi_1}(W_1) \cdots f_{\Xi_n}(W_n)$$

where the summation is taken over all mutually disjoint subsets $W_1, \ldots, W_n$ of X such that $W_1 \cup \ldots \cup W_n = X$.

16.4.7 Basic Differentiation Rules

Practitioners usually find it possible to apply ordinary Newtonian differential and integral calculus by applying the *turn-the-crank* rules they learned as undergraduates. Similar *turn-the-crank* rules exist for the FISST calculus. The simplest of these are

$$\frac{\beta}{\delta Y}[a_1\beta_1(S) + a_2\beta_2(X)] = a_1\frac{\delta\beta_1}{\delta Y}(S) + a_2\frac{\delta\beta_2}{\delta Y}(S) \qquad \text{(sum rule)}$$

$$\frac{\delta}{\delta \mathbf{y}}[\beta_1(S) \cdot \beta_2(X)] = \frac{\delta\beta_1}{\delta \mathbf{y}}(S) \cdot \beta_2(S) + \beta_1(S) \cdot \frac{\delta\beta_2}{\delta \mathbf{y}}(S)$$

$$\text{(product rules)}$$

$$\frac{\delta}{\delta \mathbf{y}}[\beta_1(S) \cdot \beta_2(X)] = \sum_{W \subseteq Z}\frac{\delta\beta_1}{\delta W}(S) \cdot \frac{\delta\beta_2}{\delta(Z-W)}(S)$$

$$\frac{\delta}{\delta \mathbf{y}}f(\beta_1(S), \ldots, \beta_n(S)) = \sum_{i=1}^{n}\frac{\partial f}{\partial x_i}(\beta_1(S), \ldots, \beta_n(S)) \cdot \frac{\delta\beta_i}{\delta \mathbf{y}}(S) \qquad \text{(chain rule)}$$

16.5 Multitarget Likelihood Functions

In the single-target case, probabilistic approaches to tracking and identification (and Bayesian approaches in particular) depend on the ability to construct sensor models together with the likelihood functions for those models. This section shows how to construct multisource-multitarget measurement models and their corresponding true multitarget likelihood functions. The crucial result is as follows:

- The provably true likelihood function $f_{k+1}(Z|X)$ of a multisensor-multitarget problem is a set derivative of the belief-mass function $\beta_{k+1}(S|X) = \Pr(\Sigma \subseteq S|X)$ of the corresponding sensor (or multisensor) model

$$f_{k+1}(Z|X) = \frac{\delta \beta_{k+1}}{\delta Z_{k+1}}(\varnothing|X) \tag{16.35}$$

16.5.1 Multitarget Measurement Models

The following sections illustrate the process of constructing multitarget measurement models for the following successively more realistic situations: (1) no missed detections and no false alarms; (2) missed detections and no false alarms; (3) missed detections and false alarms; and (4) the multisensor case.

16.5.1.1 Case I: No Missed Detections, No False Alarms

Suppose that two targets with states x_1 and x_2 are interrogated by a single sensor that generates observations of the form $Z = h(x) + W$, where W is a random noise vector with density $f_W(w)$. Assume also that there are no missed detections or false alarms, and that observations within a scan are independent. Then, the multitarget measurement is the randomly varying two-element observation-set

$$\Sigma = \{Z_1, Z_2\} = \{h(x_1) + W_1, h(x_2) + W_2\} \tag{16.36}$$

where W_1, W_2 are independent random vectors with density $f_W(w)$.

   Note: We assume that individual targets produce unique observations only for the sake of clarity. Clearly, we could just as easily produce models for other kinds of sensors—for example, sensors that detect only superpositions of the signals produced by multiple targets. One such measurement model is

$$\Sigma = \{Z_1\} = \{h_1(x_1) + h_2(x_2) + W\}$$

16.5.1.2 Case II: Missed Detections

Suppose that the sensor has a probability of detection $p_D < 1$. In this case, observations can have not only the form $Z = \{z_1, z_2\}$ (two detections), but also $Z = \{z\}$ (single detection) or $Z = \varnothing$ (missed detection). The more complex observation model $\Sigma = T_1 \cup T_2$ is needed. It has observation sets T_1, T_2 with the following properties: (a) $T_i = \varnothing$ with probability $1 - p_D$ and (b) T_i is nonempty with probability p_D, in which case, $T_i = \{Z_i\}$. If the sensor has a specific field of view, then p_D will be a function of both the target state and the state of the sensor.

16.5.1.3   Case III: Missed Detections and False Alarms

Suppose the sensor has probability of false alarm $p_{FA} \neq 0$. Then we need an observation model of the form

$$\Sigma = \overset{\text{target}}{T} \cup \overset{\text{clutter}}{C} = T_1 \cup T_2 \cup C$$

where C models false alarms and (possibly state-dependent) point clutter. As a simple example, C could have the form $C = C_1 \cup \ldots \cup C_m$, where each C_j is a clutter genera-tor—which implies that there is a probability, p_{FA}, that C_j will be nonempty (i.e., generator of a clutter-observation). In this case, $C = \{C_i\}$ where C_i is some random noise vector with density $f_{C_i}(\mathbf{z})$. (Notice that $\Sigma = T_1 \cup C$ models the single-target case with false alarms or clutter.)

16.5.1.4   Case IV: Multiple Sensors

In this case, observations will have the form $\mathbf{z}^{[s]} = (\mathbf{z}, s)$ where the integer tag s identifies which sensor originated the measurement. A two-sensor multitarget measurement will have the form $\Sigma = \Sigma^{[1]} \cup \Sigma^{[2]}$ where $\Sigma^{[s]}$ for $s = 1, 2$ is the random multitarget measurement-set collected by the sensor with tag s and can have any of the forms previously described.

16.5.2   Belief-Mass Functions of Multitarget Sensor Models

In this section, we illustrate how multitarget measurement models are transformed into belief-mass functions. The simplest single-sensor measurement model has the form $\Sigma = \{\mathbf{Z}\}$ where $\mathbf{Z}$ is a random measurement-vector. In Section 16.4.3, we have shown that the belief-mass function of Σ is identical to the probability-mass function of $\mathbf{Z}$.

The next most complicated model is that of a single-target with missed detections, $\Sigma = T_1$. Its corresponding belief-mass function is

$$\beta(S|\mathbf{x}) = \Pr(T_1 \subseteq S) = \Pr(T_1 = \varnothing) + \Pr(T_1 \neq \varnothing, \mathbf{Z} \in S)$$

$$= \Pr(T_1 = \varnothing) + \Pr(T_1 \neq \varnothing) \cdot \Pr(\mathbf{Z} \in S|T_1 \neq \varnothing)$$

$$= 1 - p_D + p_D \cdot p_{\mathbf{Z}}(S|\mathbf{x}) \tag{16.37}$$

The two-target missed-detection model has the form $\Sigma = T_1 \cup T_2$. Its belief-mass function is

$$\beta(S|X) = \Pr(T_1 \subseteq S) \cdot \Pr(T_2 \subseteq S)$$

$$= (1 - p_D + p_D \cdot p(S|\mathbf{x}_1)) \cdot (1 - p_D + p_D \cdot p(S|\mathbf{x}_2))$$

where $p(S|\mathbf{x}_i) = \Pr(T_i \subseteq S|T_i \neq \varnothing)$ and $X = \{\mathbf{x}_1, \mathbf{x}_2\}$. Setting $p_D = 1$ yields

$$\beta(S|X) = \Pr(T_1 \subseteq S) \cdot \Pr(T_2 \subseteq S) \tag{16.38}$$

This is the belief-mass function for the model $\Sigma = \{\mathbf{Z}_1, \mathbf{Z}_2\}$.

Finally, suppose that two sensors with identifying tags $s = 1, 2$ collect observation sets $\Sigma = \Sigma^{[1]} \cup \Sigma^{[2]}$. The corresponding belief-mass function has the form

$$\beta(S^{[1]} \cup S^{[2]}|X) = \Pr(\Sigma^{[1]} \subseteq S^{[1]}, \Sigma^{[2]} \subseteq S^{[2]})$$

where $S^{[1]}$, $S^{[2]}$ are subsets of the measurement spaces of the respective sensors. If the two sensors are independent, then the belief-mass function has the form

$$\beta(S^{[1]} \cup S^{[2]}|X) = \Pr(\Sigma^{[1]} \subseteq S^{[1]}) \cdot \Pr(\Sigma^{[2]} \subseteq S^{[2]})$$

16.5.3 Constructing True Multitarget Likelihood Functions

We apply the differentiation formulas of Section 16.4.7 to the belief-mass function $\beta(S|X) = p(S|x_1) \cdot p(S|x_2)$ of the measurement model $\Sigma = \{Z_1, Z_2\}$, where $X = \{x_1, x_2\}$. Then

$$\frac{\delta\beta}{\delta z_1}(S|X) = \frac{\delta}{\delta z_1}\beta(S|X) = \frac{\delta}{\delta z_1}[p(S|x_1) \cdot p(S|x_2)]$$

$$= \frac{\delta p}{\delta z_1}(S|x_1) \cdot p(S|x_2) + p(S|x_1) \cdot \frac{\delta p}{\delta z_1}(S|x_2)$$

$$= f(z_1|x_1) \cdot p(S|x_2) + p(S|x_1) \cdot f(z_1|x_2)$$

$$\frac{\delta^2\beta}{\delta z_2 \delta z_1}(S|X) = \frac{\delta}{\delta z_2}\frac{\delta\beta}{\delta z_1}(S|X) = f(z_1|x_1) \cdot \frac{\delta p}{\delta z_2}(S|x_2) + \frac{\delta p}{\delta z_2}(S|x_1) \cdot f(z_1|x_2)$$

$$= f(z_1|x_1) \cdot f(z_2|x_2) + f(z_2|x_1) \cdot f(z_1|x_2)$$

$$\frac{\delta^3\beta}{\delta z_3 \delta z_2 \delta z_1}(S|X) = \frac{\delta}{\delta z_3}\frac{\delta^2\beta}{\delta z_2 \delta z_1}(S|X) = 0$$

and the higher-order derivatives vanish identically. The multitarget likelihood is, therefore,

$$f(\varnothing|X) = \frac{\delta\beta}{\delta\varnothing}(\varnothing|X) = 0$$

$$f(\{z\}|X) = \frac{\delta\beta}{\delta z}(\varnothing|X) = 0 \tag{16.39}$$

$$f(\{z_1, z_2\}|X) = \frac{\delta^2\beta}{\delta z_2 \delta z_1}(\varnothing|X) = f(z_1|x_1) \cdot f(z_2|x_2) + f(z_2|x_1) \cdot f(z_1|x_2)$$

and where $f(Z|X) = 0$ identically if Z contains more than two elements. More general multitarget likelihoods can be computed similarly.[2,4]

16.6 Multitarget Markov Densities

This section illustrates the process of constructing multitarget motion models and their corresponding true multitarget Markov densities. The construction of multitarget Markov densities strongly resembles that of multisensor-multitarget likelihood functions. The crucial result is as follows:

- The provably true Markov transition density $f_{k+1|k}(Y|X)$ of a multitarget problem is a set derivative of the belief-mass function $\beta_{k+1|k}(S|X)$ of the corresponding multitarget motion model:

$$f_{k+1|k}(Y|X) = \frac{\delta \beta_{k+1|k}}{\delta Y}(\varnothing|X) \qquad (16.40)$$

16.6.1 Multitarget Motion Models

This section considers the following increasingly more realistic situations: (1) multitarget motion models assuming that target number does not change; (2) multitarget motion models assuming that target number can decrease; and (3) multitarget motion models assuming that target number can decrease or increase.

16.6.1.1 Case I: Target Number Is Constant

Assume that the states of individual targets have the form $\mathbf{x} = (\mathbf{y}, c)$, where $\mathbf{y}$ is the kinematic state and c the target type. Assume that each target type has an associated motion model $\mathbf{X}_{c,k+1|k} = \Phi_{c,k}(\mathbf{x}) + \mathbf{W}_{c,k}$. Let

$$\mathbf{X}_{k+1|k} = \Phi_k(\mathbf{x}, \mathbf{W}_k) = (\Phi_{c,k}(\mathbf{x}) + \mathbf{W}_{c,k},\ c)$$

To model a multitarget system in which two targets never enter or leave the scenario, the obvious multitarget extension of the single-target motion model would be $\Xi_{k+1|k} = \Phi_k(X,\mathbf{W}_k)$, where $\Xi_{k+1|k}$ is the randomly varying parameter set at time-step $k + 1$. That is, for the cases $X = \varnothing$, $X = \{\mathbf{x}\}$, or $X = \{\mathbf{x}_1, \mathbf{x}_2\}$, respectively, the multitarget state transitions are

$$\Xi_{k+1|k} = \varnothing$$

$$\Xi_{k+1|k} = \{\mathbf{X}_{k+1|k}(\mathbf{x})\} = \{\Phi_k(\mathbf{x}, \mathbf{W}_k)\}$$

$$\Xi_{k+1|k} = \{\mathbf{X}_{k+1|k}(\mathbf{x}_1), \mathbf{X}_{k+1|k}(\mathbf{x}_2)\} = \{\Phi_k(\mathbf{x}_1, \mathbf{W}_{k,1}), \Phi_k(\mathbf{x}_2, \mathbf{W}_{k,2})\}$$

16.6.1.2 Case II: Target Number Can Decrease

Modeling scenarios in which target number can decrease but not increase is analogous to modeling multitarget observations with missed detections. Suppose that no more than two targets are possible, but that one or more of them can vanish from the scene. One possible motion model would be $\Xi_{k+1|k} = \Phi_k(X)$ where, for the cases $X = \varnothing$, $X = \{\mathbf{x}\}$, or $X = \{\mathbf{x}_1, \mathbf{x}_2\}$, respectively,

$$\Xi_{k+1|k} = \varnothing \qquad \Xi_{k+1|k} = T_k(\mathbf{x}) \qquad \Xi_{k+1|k} = T_k(\mathbf{x}_1) \cup T_k(\mathbf{x}_2)$$

Here $T_k(\mathbf{x})$ is a state-set with the following properties: (a) $T_k(\mathbf{x}) \neq \varnothing$ with probability p_v, in which case $T_k(\mathbf{x}) = \{\mathbf{X}_{k+1|k}(\mathbf{x})\}$, and (b) $T_k(\mathbf{x}) = \varnothing$ (i.e., target disappearance), with probability

$1 - p_v$. In other words, if no targets are present in the scene, this will continue to be the case. If, however, there is one target in the scene, then either this target will persist (with probability p_v) or it will vanish (with probability $1 - p_v$). If there are two targets in the scene, then each will either persist or vanish in the same manner. In general, when n targets are present, one would model $\Xi_{k+1|k} = T_k(\mathbf{x}_1) \cup \ldots \cup T_k(\mathbf{x}_n) \cup B_k$.

16.6.1.3 Case III: Target Number Can Increase and Decrease

Modeling scenarios in which target number can decrease or increase is analogous to modeling multitarget observations with missed detections and clutter. In this case, the general form of the model is

$$\Xi_{k+1|k} = T_k(\mathbf{x}_1) \cup \ldots \cup T_k(\mathbf{x}_n) \cup B_k$$

where B_k is the set of birth targets (i.e., targets that have entered the scene).

16.6.2 Belief-Mass Functions of Multitarget Motion Models

In single-target problems, the statistics of a motion model $\mathbf{X}_{k+1|k} = \Phi_k(\mathbf{x}, \mathbf{W}_k)$ are described by the probability-mass function $p_{k+1|k}(S|\mathbf{x}) = \Pr(\mathbf{X}_{k+1|k} \in S)$, which is the probability that the target state will be found in the region S if it previously had state-vector $\mathbf{x}$. Similarly, suppose that $\Xi_{k+1|k} = \Phi_k(X)$ is a multitarget motion model. The statistics of the finitely varying random state-set $\Xi_{k+1|k}$ can be described by its belief-mass function:

$$\beta_{k+1|k}(S|X) = \Pr(\Xi_{k+1|k} \subseteq S)$$

This is the total probability of finding all targets in region S at time-step $k + 1$ if, in time-step k, they had multitarget state $X = \{\mathbf{x}_1, \ldots, \mathbf{x}_n\}$.

For example, the belief-mass function for independent target motion with no appearing or disappearing targets is, for $n = 2$,

$$\beta_{k+1|k}(S|X) = \Pr(\Phi_k(\mathbf{x}_1) + \mathbf{V}_{k,1}) \cdot \Pr(\Phi_k(\mathbf{x}_2) + \mathbf{V}_{k,2}) = p_{k+1|1}(S|\mathbf{x}_1) \cdot p_{k+1|k}(S|\mathbf{x}_2)$$

16.6.3 Constructing True Multitarget Markov Densities

Multitarget Markov densities[1,4,9,41] are constructed from multitarget motion models in much the same way that multisensor-multitarget likelihood functions are constructed from multisensor-multitarget measurement models. First, construct a multitarget motion model $\Xi_{k+1|k} = \Phi_k(X)$ from the underlying motion models of the individual targets. Second, build the corresponding belief-mass function $\beta_{k+1|k}(S|X) = \Pr(\Xi_{k+1|k} \subseteq S|X)$. Finally, construct the multitarget Markov density $f_{k+1|k}(Y|X)$ from the belief-mass function using the turn-the-crank formulas of the FISST calculus (Section 16.4.7).

For example, consider the case of independent two-target motion with no-target appearance or disappearance. This has the same form as the multitarget measurement model in Section 16.5.3. Consequently, its multitarget Markov density is[30]

$$f_{k+1|k}(Y|X) = f_{k+1|k}(\mathbf{y}_1|\mathbf{x}_1) \cdot f_{k+1|k}(\mathbf{y}_2|\mathbf{x}_2) + f_{k+1|k}(\mathbf{y}_2|\mathbf{x}_1) \cdot f_{k+1|k}(\mathbf{y}_1|\mathbf{x}_2)$$

16.7 Multisource-Multitarget Bayes Filter

Thus far in this chapter we have described the multisensor-multitarget analogs of measurement and motion models, likelihood functions, Markov transition densities, probability-mass functions, and the integral and differential calculus. This section shows how these concepts combine together to produce a direct generalization of ordinary statistics to multitarget statistics, using the multitarget Bayes filter.

The multitarget Bayes filter is introduced in Section 16.7.1, and the issue of its initialization in Section 16.7.2. The failure of the classical Bayes-optimal state estimators in multitarget situations is described in Sections 16.7.3 and 16.7.4. The solution of this problem—the proper definition and verification of Bayes-optimal multitarget state estimators—is described in Section 16.7.5. The remaining two subsections summarize the concept of multitarget miss distance (Section 16.7.6) and of unified multitarget multisource integration.

16.7.1 Multisensor-Multitarget Filter Equations

Bayesian multitarget filtering is inherently nonlinear because multitarget likelihoods $f_{k+1}(Z|X)$ are, in general, highly non-Gaussian even for a Gaussian sensor.[2] Therefore, multitarget nonlinear filtering is unavoidable if the goal is optimal-Bayes detection, tracking, localization, identification, and information fusion. The multitarget Bayes filter equations are

$$f_{k+1|k}(X|Z^{(k)}) = \int f_{k+1|k}(X|Y) \cdot f_{k|k}(Y|Z^{(k)}) \delta Y \tag{16.41}$$

$$f_{k+1|k+1}(X|Z^{(k+1)}) = \frac{f_{k+1}(Z_{k+1}|X) f_{k+1|k}(X|Z^{(k)})}{f_{k+1}(Z_{k+1}|Z^{(k)})} \tag{16.42}$$

where

$$f_{k+1}(Z|Z^{(k)}) = \int f_{k+1|k}(Z|X) \cdot f_{k+1|k}(X|Z^{(k)}) \delta X \tag{16.43}$$

is the Bayes normalization constant, and where $f_{k+1}(Z|X)$ is the multisensor-multitarget likelihood function, $f_{k+1|k}(Y|X)$ the multitarget Markov transition density, $f_{k|k}(X|Z^{(k)})$ the posterior distribution conditioned on the data-stream $Z^{(k)}$: $Z_1, ..., Z_k$, $f_{k+1|k}(X|Z^{(k)})$ the time-prediction of the posterior $f_{k|k}(X|Z^{(k)})$ to time-step $k + 1$, and where the two integrals are set integrals.

16.7.2 Initialization

The initial states of the targets in a multitarget system are specified by a multitarget initial distribution of the form $f_{0|0}(X|Z^{(0)}) = f_0(X)$,[1,6] where $\int f_0(X)\delta X = 1$ and where the integral is a set integral. Suppose that states have the form $x = (y, c)$ where y is the kinematic state variable restricted to some bounded region D of (hyper)volume $\lambda(D)$ and c the discrete state variable(s), drawn from a universe C with N possible members. In conventional statistics, the uniform distribution $u(x) = \lambda(D)^{-1}N^{-1}$ is the most common way of initializing a Bayesian algorithm when nothing is known about the initial state of the target.

The concepts of prior and uniform distributions carry over to multitarget problems, but in this case there is an additional dimension that must be taken into account—target number. For example, suppose that there can be no more than M possible targets in a scene.[1,6] If $X = \{x_1, \ldots, x_n\}$, then the multitarget uniform distribution is defined by

$$u(X) = \begin{cases} n! \cdot N^{-n} \cdot \lambda(D)^{-n} \cdot (M+1)^{-1} & \text{if } X \subseteq D \times C \\ 0 & \text{if otherwise} \end{cases}$$

16.7.3 Multitarget Distributions and Units of Measurement

Multitarget posterior and prior distributions, like multitarget density functions in general, have one peculiarity that sets them apart from conventional density functions: their behavior with respect to units of measurement.[1,2,8] In particular, *the units of measurement of a multitarget prior or posterior f(X) vary with the cardinality of X*. This has important consequences for multitarget state estimation, as described in the next section.

16.7.4 Failure of the Classical State Estimators

In general, the classical Bayes-optimal estimators cannot be extended to the multitarget case. This can be explained using a simple one-dimensional example.[2,4] Let

$$f(X) = \begin{cases} \dfrac{1}{2} & \text{if } X = \varnothing \\ \dfrac{1}{2} \cdot N_{\sigma^2}(x-1) & \text{if } X = \{x\} \\ 0 & \text{if } |X| \geq 2 \end{cases}$$

where the variance σ^2 has units km^2. To compute that classical MAP estimate, find the state $X = \varnothing$ or $X = \{x\}$ that maximizes $f(X)$. Because $f(\varnothing) = 1/2$ is a unitless probability and $f(\{1\}) = 1/2 \cdot (2\pi)^{-1/2} \cdot \sigma^{-1}$ has units of km^{-1}, the classical MAP would compare the values of two quantities that are incommensurable because of mismatch of units. As a result, the numerical value of $f(\{1\})$ can be arbitrarily increased or decreased—thereby getting $X^{\text{MAP}} = \varnothing$ (no target in the scene) or $X^{\text{MAP}} \neq \varnothing$ (target in the scene)—simply by changing units of measurement.

The posterior expectation also fails. If it existed, it would be

$$\int X \cdot f(X)\delta X = \varnothing \cdot f(\varnothing) + \int x \cdot f(x)dx = \frac{1}{2}(\varnothing + 1\,\text{km})$$

Notice that, once again, there is the problem of mismatched units—the unitless quantity $\varnothing$ must be added to the quantity 1 km. Even if the variable x has no units of measurement, the quantity $\varnothing$ still must be added to the quantity 1. If $\varnothing + 1 = \varnothing$, then $1 = 0$, which is impossible. If $\varnothing + 1 = 1$ then $\varnothing = 0$, then the same mathematical symbol represents two different states: the no-target state $\varnothing$ and the single-target state $x = 0$. The same problem occurs if $\varnothing + a = b$ is defined for *any* real numbers a, b since then $\varnothing = b - a$.

16.7.5 Optimal Multitarget State Estimators

We have just seen that the classical Bayes-optimal state estimators do not exist in general multitarget situations. Therefore, new multitarget state estimators must be defined and demonstrated to be statistically well behaved.

In conventional statistics, the maximum likelihood estimator (MLE) is a special case of the MAP estimator, assuming that the prior is uniform. As such, it is optimal and convergent. In the multitarget case, this does not hold true. If $f_{k+1}(Z|X)$ is the multitarget likelihood function, then the units of measurement for $f_{k+1}(Z|X)$ are determined by the observation-set Z (which is fixed) and not by the multitarget state X. Consequently, in multitarget situations the classical MLE is defined[6] even though the classical MAP is not:

$$X^{\text{MLE}} = \arg\max_{n,\mathbf{x}_1,\ldots,\mathbf{x}_n} f(\mathbf{x}_1,\ldots,\mathbf{x}_n), \qquad X^{\text{MLE}} = \arg\max_X f(X)$$

where the second equation is the same as the first, but written in condensed notation. The multitarget MLE will converge to the correct answer if given enough data.[1]

Because the multitarget MLE is not a Bayes estimator, new multitarget Bayes state estimators must be defined and their optimality demonstrated. In 1995, two such estimators were introduced, the *Marginal Multitarget Estimator (MaME)* and the *Joint Multitarget Estimator (JoME)*.[1,2] The JoME is defined as

$$X^{\text{JoME}} = \arg\max_{n,\mathbf{x}_1,\ldots,\mathbf{x}_n} f(\mathbf{x}_1,\ldots,\mathbf{x}_n) \cdot \frac{c^n}{n!}, \qquad X^{\text{JoME}} = \arg\max_X f(X) \cdot \frac{c^{|X|}}{|X|!}$$

where c is a fixed constant, the units of which are the same as those for $\mathbf{x}$. One of the consequences of this is that both the multitarget MLE and the JoME estimate the number $\hat{n}$ and the identities/kinematics $\hat{\mathbf{x}}_1,\ldots,\hat{\mathbf{x}}_{\hat{n}}$ of targets optimally and simultaneously without resorting to optimal report-to-track association. In other words, they optimally resolve the conflicting objectives of detection, tracking, and identification.

16.7.6 Multitarget Miss Distance

FISST provides natural generalizations of the concept of *miss distance* to multitarget situations. Let $X = \{\mathbf{x}_1,\ldots,\mathbf{x}_n\}$ and $Y = \{\mathbf{y}_1,\ldots,\mathbf{y}_m\}$. Then the simplest definition of the multitarget miss distance between X and Y is the Hausdorff distance. This is defined by

$$d_H(X,Y) = \max\{d_0(X,Y), d_0(Y,X)\}, \qquad d_0(X,Y) = \max_{\mathbf{x}}\min_{\mathbf{y}}\|\mathbf{x} - \mathbf{y}\|$$

The Hausdorff distance tends to be relatively insensitive to differences in cardinality between X and Y. Consequently, Hoffman and Mahler[50] have introduced the family of Wasserstein distances. These are defined by

$$d_W^p(X,Y) = \min_C \sqrt[p]{\sum_{i=1}^n \sum_{j=1}^m C_{i,j} \cdot d(\mathbf{x}_i, \mathbf{y}_j)}$$

where the minimum is taken over all so-called *transportation matrices* C. If $n = m$ then this reduces to an objective function for an optimal assignment problem:

$$d_W^p(X,Y) = \min_\sigma \sqrt[p]{\sum_{i=1}^n d(\mathbf{x}_i, \mathbf{y}_{\sigma i})}$$

where the minimum is taken over all permutations σ on the numbers $1,\ldots,n$.

16.7.7 Unified Multitarget Multisource Integration

Suppose that there are a number of independent sources, some of which supply conventional data and others ambiguous data. As in Section 16.5.2, a multisource-multitarget joint generalized likelihood can be constructed of the form

$$f_{k+1}(Z|X) = f(Z^{[1]}, \ldots, Z^{[m]}, \Theta^{[1]}, \ldots, \Theta^{[m']})$$

where

$$Z = Z^{[1]} \cup \ldots \cup Z^{[m]} \cup \Theta^{[1]} \cup \ldots \cup \Theta^{[m']}$$

and where $Z^{[s]}$ denotes a multitarget observation collected by a conventional sensor with identifier $s = 1, \ldots, e$, and where $\Theta^{[s]}$ denotes a multitarget observation supplied by a source with identifier $s = e + 1, \ldots, e + e'$ that collects ambiguous data. Given this, the data can be fused using Bayes' rule. Robust multisource-multitarget detection, tracking, and identification can be accomplished by using the joint generalized likelihood function with the multitarget recursive Bayesian nonlinear filtering equations.

16.8 PHD and CPHD Filters

The single-sensor, single-target Bayesian nonlinear filtering Equations 16.4 and 16.5 are already computationally demanding. Computational difficulties worsen only when attempting to implement the multitarget nonlinear filter (Equations 16.41 and 16.42). This section summarizes two radically new approaches for approximate multitarget nonlinear filtering: the PHD filter[4,48] and its generalization, the cardinalized PHD (CPHD) filter.[4,49]

The PHD and CPHD filters are based on an analogy with single-target tracking. The constant-gain Kalman filter (CGKF)—of which the alpha-beta filter is the most familiar one—is the computationally fastest single-target tracking filter, whereas the single-target Bayes filter representing Equations 16.5 and 16.6 propagate the full Bayes posterior distribution $f_{k|k}(x|Z^k)$; the CGKF propagates only the first-order statistical moment of the posterior—that is, its expected value $x_{k|k}$.

In like fashion, whereas the multitarget Bayes filter of Equations 16.41 and 16.42 propagate the full multitarget Bayes posterior $f_{k|k}(X|Z^{(k)})$, the PHD filter propagates only the first-order multitarget moment of the multitarget posterior—that is, its PHD $D_{k|k}(x|Z^{(k)})$. The CPHD generalizes the PHD filter in that it also propagates the entire probability distribution $p_{k|k}(n|Z^{(k)})$ on target number. As a consequence, the CPHD filter exhibits better performance than the PHD filter, but with increased computational loading.

The derivation of the predictor and corrector equations for both filters require the p.g.fl. and multitarget calculus techniques of Section 16.4.

Subsequently, we introduce the concept of the PHD (Section 16.8.1), the PHD filter (Section 16.8.2), and the CPHD filter (Section 16.8.3). We conclude, in Section 16.8.4, with a summary of recently developed implementations of the PHD and CPHD filters.

16.8.1 Probability Hypothesis Density

The first question that confronts us is: what is the multitarget counterpart of an expected value? Let $f_\Xi(X)$ be the multitarget probability distribution of a random finite state-set Ξ. Then a naïve definition of its expected value would be

$$\bar{\Xi}^{\text{naive}} = \int X \cdot f_\Xi(X) \delta X$$

However, this integral is mathematically undefined since addition $X + X'$ of finite subsets X, X' cannot be usefully defined. Consequently, one must instead resort to a different strategy. Select a transformation $X \to T_X$ that converts finite subsets X into vectors T_X in some vector space. This transformation should preserve basic set-theoretic structure by transforming unions into sums: $T_{X \cup X'} = T_X + T_{X'}$ whenever $X \cap X' = \emptyset$. Given this, one can define an indirect expected value as[47,48]

$$\bar{\Xi}^{\text{indirect}} = \int T_X \cdot f_\Xi(X) \delta X$$

The common practice is to select $T_X = \delta_X$ where $\delta_X(x) = 0$ if $X = \emptyset$ and, otherwise,

$$\delta_X(x) = \sum_{y \in X} \delta_y(x)$$

where $\delta_y(x)$ is the Dirac delta density concentrated at $\mathbf{w}$. Given this,

$$D_\Xi(\mathbf{x}) = \int \delta_X(\mathbf{x}) \cdot f_\Xi(X) \delta X$$

is a multitarget analog of the concept of expected value. It is a density function on single-target state space called the PHD, intensity density, or first-moment density of Ξ or $f_\Xi(X)$.

16.8.2 PHD Filter

We can attempt only a very brief summary of the PHD filter. Additional details can be found in Ref. 48. The PHD filter consists of two equations. The first, or predictor equation, allows the current PHD $D_{k|k}(X|Z^{(k)})$ to be extrapolated to the *predicted PHD*: the PHD $D_{k+1|k}(X|Z^{(k)})$ at the time of the next observation-set Z_{k+1}. The second, or corrector equation, allows the predicted PHD to be data-updated to $D_{k+1|k+1}(X|Z^{(k+1)})$.

The PHD filter predictor equation has the form

$$D_{k+1|k}(\mathbf{x}|Z^{(k)}) = b_{k+1|k}(\mathbf{x}) + \int F_{k+1|k}(\mathbf{x}|\mathbf{y}) \cdot D_{k|k}(\mathbf{y}|Z^{(k)}) \delta Y \qquad (16.44)$$

where the PHD filter *pseudo-Markov transition* is

$$F_{k+1|k}(\mathbf{x}|\mathbf{y}) = p_S(\mathbf{y}) \cdot f_{k+1|k}(\mathbf{x}|\mathbf{y}) + b_{k+1|k}(\mathbf{x}|\mathbf{y}) \qquad (16.45)$$

and where $p_S(\mathbf{y})$ is the probability that a target with state $\mathbf{y}$ at time-step $k + 1$ will survive, $f_{k+1|k}(\mathbf{x}|\mathbf{y})$ the single-target Markov transition density, $b_{k+1|k}(\mathbf{x})$ the PHD of the multitarget distribution $b_{k+1|k}(X)$ of appearing targets, $b_{k+1|k}(\mathbf{x}|\mathbf{y})$ the PHD of distribution $b_{k+1|k}(X|\mathbf{y})$ of targets spawned by targets of state $\mathbf{y}$.

The PHD filter corrector equation, however, has the form

$$D_{k+1|k+1}(\mathbf{x}|Z^{(k+1)}) = L_{Z_{k+1}}(\mathbf{x}) \cdot D_{k+1|k}(\mathbf{x}|Z^{(k)})$$ (16.46)

where the PHD filter *pseudo-likelihood* is

$$L_Z(\mathbf{x}) = 1 - p_D(\mathbf{x}) + \sum_{\mathbf{z} \in Z} \frac{p_D(\mathbf{x}) \cdot L_{\mathbf{z}}(\mathbf{x})}{\lambda c(\mathbf{z}) + D_{k+1|k}[p_D L_{\mathbf{z}}]}$$ (16.47)

and where $p_D(\mathbf{x})$ is the probability of detection of a target with state $\mathbf{x}$ at time-step $k + 1$, $f_{k+1}(\mathbf{z}|\mathbf{x}) = L_{\mathbf{z}}(\mathbf{x})$ the sensor likelihood function, $c(\mathbf{z})$ = physical distribution of Poisson false alarms, λ the expected value of number of false alarms, and where for any function $h(\mathbf{x})$,

$$D_{k+1|k}[h] = \int h(\mathbf{x}) \cdot D_{k+1|k}(\mathbf{x}|Z^{(k)})d\mathbf{x}$$ (16.48)

16.8.3 Cardinalized PHD Filter

The predictor and corrector equations for the CPHD filter are too complex to describe in the present context. Greater details can be found in Refs 4 and 49. Two things should be pointed out, however. First, and unlike the PHD filter, spawning of targets cannot be explicitly modeled in the CPHD predictor equations. Second, and unlike the PHD filter, the false alarm process need not necessarily be Poisson. It can be a more general false alarm process known as an i.i.d. cluster process. Again, details can be found Refs 4 and 49.

As already noted, the CPHD filter generalizes the PHD filter, but at the price of greater computational complexity. Let n be the current number of targets and m observations. Then the PHD filter has computational complexity of order $O(mn)$. The CPHD filter, however, has complexity $O(m^3n)$.

16.8.4 Survey of PHD/CPHD Filter Research

In this section, we briefly summarize current PHD filter research. The PHD filter has usually been implemented using sequential Monte Carlo (a.k.a. particle-system) methods, as proposed by Kjellström (nee Sidenbladh)[51] and by Zajic and Mahler.[52] Instances are provided by Erdinc et al.[53] and Vo et al.[54] Vo, Singh, Doucet, and Clark have established convergence results for the particle-PHD filter.[54,55]

Vo and Ma[56] have devised a closed-form Gaussian-mixture implementation that greatly improves the computational efficiency of the PHD filter. This approach is inherently capable of maintaining track labels.[55,57] Clark and Bell[55] have proved a strong L_2 uniform convergence property for the Gaussian-mixture PHD filter. Erdinc et al.[58] have proposed a purely physical interpretation of both the PHD and the CPHD filters.

Since the *core* PHD filter does not maintain labels for tracks from time-step to time-step, researchers have proposed *peak to track association* techniques for maintaining track labels with particle-PHD filters.[59,60] These authors have demonstrated in 1D and 2D simulations that their track-valued PHD filters can outperform conventional MHT-type techniques (i.e., significantly fewer false and dropped tracks).

Punithakumar et al.[61] have implemented a multiple motion model version of the PHD filter, as have Pasha et al.[62]

Punithakumar et al.[44] have devised and implemented a distributed PHD filter that addresses the problem of communicating and fusing multitarget track information from a distributed network of sensor-carrying platforms.

Balakumar et al.[63] have applied a PHD filter to the problem of tracking an unknown and time-varying number of narrowband, far-field signal sources, using a uniform linear array of passive sensors, in a highly nonstationary sensing environment.

Ahlberg et al.[64] have employed PHD filters for group-target tracking in an ambitious situation assessment simulator system called *IFD03*.

Tobias and Lanterman[65,66] have applied the PHD filter to target detection and tracking using bistatic radio-frequency observations.

Clark et al.[67–70] have applied the PHD filter to 2D and 3D active-sonar problems.

Ikoma et al.[71] have applied a PHD filter to the problem of tracking the trajectories of feature points in time-varying optical images. Wang et al.[72] have employed such methods to tracking groups of humans in digital video.

Zajic et al.[73] report an algorithm in which a PHD filter is integrated with a robust classifier algorithm that identifies airborne targets from HRRR signatures.

El-Fallah et al.[74–76] have demonstrated implementations of a PHD filter–based sensor management approach.[10]

The CPHD filter was first introduced in 2006. As a result, only a few implementations have appeared in the literature. Vo et al.[77] have devised a Gaussian-mixture implementation of the CPHD filter, combined with EKF and unscented Kalman filter (UKF) versions. They have also described detailed performance comparisons of the GM-PHD and GM-CPHD filters.[78,79]

16.9 Summary and Conclusions

FISST was created, in part, to address the issues in probabilistic inference that conventional approaches overlook. These issues include

- Dealing with poorly characterized sensor likelihoods
- Modeling *ambiguous* nontraditional data
- Constructing likelihoods for nontraditional data
- Constructing true likelihoods and Markov densities for multitarget problems
- Developing principled new approximate techniques for multitarget filtering
- Providing a single, fully probabilistic, systematic, and seamlessly unified foundation for multisource-multitarget detection, tracking, identification, data fusion, sensor management, performance estimation, and threat estimation and prediction
- Accomplishing all of these objectives within the framework of a direct and relatively simple and practitioner-friendly generalization of *Statistics 101*

In the past 2 years, FISST has begun to emerge from the realm of basic research and is being applied, with preliminary success, to a range of practical research applications. This chapter has described the difficulties associated with nontraditional data and multitarget problems, as well as summarized how and why FISST resolves them.

For a practitioner-level, textbook-style treatment of the concepts and techniques described in this chapter, see *Statistical Multisource-Multitarget Information Fusion*.[4]

Acknowledgments

The core concepts underlying the work reported in this chapter were developed under internal research and development funding in 1993 and 1994 at the Eagan, MN, division of Lockheed Martin Corporation. This work has been supported at the basic research level since 1994 by the U.S. Army Research Office and the Air Force Office of Scientific Research. Various aspects have been supported at the applied research level by the U.S. Air Force Research Laboratory, SPAWAR Systems Center, DARPA, MDA, and the U.S. Army MRDEC. The content does not necessarily reflect the position or the policy of the government or of Lockheed Martin. No official endorsement should be inferred.

References

1. Goodman, I.R., Mahler, R.P.S., and Nguyen, H.T., *Mathematics of Data Fusion*, Kluwer Academic Publishers, Dordrecht, Holland, 1997.
2. Mahler, R., An introduction to multisource-multitarget statistics and its applications, Lockheed Martin Technical Monograph, March 15, 2000.
3. Mahler, R., 'Statistics 101' for multisensor, multitarget data fusion, *IEEE Aerosp. Electron. Syst. Mag., Part 2: Tutorials*, 19 (1), 53–64, 2004.
4. Mahler, R.P.S., *Statistical Multisource-Multitarget Information Fusion*, Artech House, Norwood, MA, 2007.
5. Mahler, R., Global integrated data fusion, *Proc. 7th Natl. Symp. Sensor Fusion*, I (Unclass), ERIM, Ann Arbor, MI, 187–199, 1994.
6. Mahler, R., A unified approach to data fusion, *Proc. 7th Joint Data Fusion Symp.*, 1994, 154, and *Selected Papers on Sensor and Data Fusion*, Sadjadi, P.A., Ed., SPIE, MS-124, 1996, 325.
7. Mahler, R., Global optimal sensor allocation, *Proc. 1996 Natl. Symp. Sensor Fusion*, I (Unclass.), 347–366, 1996.
8. Mahler, R., Multisource-multitarget filtering: A unified approach, *SPIE Proc.*, 3373, 296, 1998.
9. Mahler, R., Multitarget Markov motion models, *SPIE Proc.*, 3720, 47, 1999.
10. Mahler, R., Multitarget sensor management of dispersed mobile sensors, Grundel, D., Murphey, R., and Paralos, P., Eds., *Theory and Algorithms for Cooperative Systems*, World Scientific, Singapore, 14–21, 2005.
11. Mahler, R., Information theory and data fusion, *Proc. 8th Natl. Symp. Sensor Fusion*, I (Unclass), ERIM, Ann Arbor, MI, 279, 1995.
12. Mahler, R., Unified nonparametric data fusion, *SPIE Proc.*, 2484, 66–74, 1995.
13. Mahler, R., Information for fusion management and performance estimation, *SPIE Proc.*, 3374, 64–74, 1998.
14. Zajic, T. and Mahler, R., Practical information-based data fusion performance estimation, *SPIE Proc.*, 3720, 92, 1999.
15. Mahler, R., Measurement models for ambiguous evidence using conditional random sets, *SPIE Proc.*, 3068, 40–51, 1997.
16. Mahler, R., Unified data fusion: Fuzzy logic, evidence, and rules, *SPIE Proc.*, 2755, 226, 1996.
17. Mahler, R. et al., Nonlinear filtering with really bad data, *SPIE Proc.*, 3720, 59, 1999.
18. Mahler, R., Optimal/robust distributed data fusion: A unified approach, *SPIE Proc.*, 4052, 128–138, 2000.
19. Mahler, R., Decisions and data fusion, *Proc. 1997 IRIS Natl. Symp. Sensor Data Fusion*, I (Unclass), M.I.T. Lincoln Laboratories, 71, 1997.
20. El-Fallah, A. et al., Adaptive data fusion using finite-set statistics, *SPIE Proc.*, 3720, 80–91, 1999.

21. Allen, R. et al., Passive-acoustic classification system (PACS) for ASW, *Proc 1998 IRIS Natl. Symp. Sensor Data Fusion*, 179, 1998.
22. Mahler, R. et al., Application of unified evidence accrual methods to robust SAR ATR, *SPIE Proc.*, 3720, 71, 1999.
23. Zajic, T., Hoffman, J.L., and Mahler, R., Scientific performance metrics for data fusion: New results, *SPIE Proc.*, 4052, 172–182, 2000.
24. El-Fallah, A. et al., Scientific performance evaluation for sensor management, *SPIE Proc.*, 4052, 183–194, 2000.
25. Mahler, R., Multisource-multitarget detection and acquisition: A unified approach, *SPIE Proc.*, 3809, 218, 1999.
26. Mahler, R. et al., Joint tracking, pose estimation, and identification using HRRR data, *SPIE Proc.*, 4052, 195, 2000.
27. Bar-Shalom, Y. and Li, X.-R., *Estimation and Tracking: Principles, Techniques, and Software*, Artech House, Ann Arbor, MI, 1993.
28. Jazwinski, A.H., *Stochastic Processes and Filtering Theory*, Academic Press, New York, NY, 1970.
29. Sorenson, H.W., Recursive estimation for nonlinear dynamic systems, *Bayesian Analysis of Statistical Time Series and Dynamic Models*, Spall, J.C., Ed., Marcel Dekker, New York, NY, 1988.
30. Matheron, G., *Random Sets and Integral Geometry*, Wiley, New York, NY, 1975.
31. Grabisch, M., Nguyen, H.T., and Waler, E.A., *Fundamentals of Uncertainty Calculus with Applications to Fuzzy Inference*, Kluwer Academic Publishers, Dordrecht, Holland, 1995.
32. Shafer, G. and Logan, R., Implementing Dempster's rule for hierarchical evidence, *Artif. Intell.*, 33, 271, 1987.
33. Nguyen, H.T., On random sets and belief functions, *J. Math. Anal. Appt.*, 65, 531–542, 1978.
34. Hestir, K., Nguyen, H.T., and Rogers, G.S. A random set formalism for evidential reasoning, *Conditional Logic in Expert Systems*, Goodman, I.R., Gupta, M.M., Nguyen, H.T., and Rogers, G.S., Eds., North-Holland, 1991, 309.
35. Goodman, I.R., Fuzzy sets as equivalence classes of random sets, *Fuzzy Sets and Possibility Theory*, Yager, R., Ed., Permagon, Oxford, U.K., 1982, 327.
36. Orlov, A.L., Relationships between fuzzy and random sets: Fuzzy tolerances, *Issledovania po Veroyatnostnostatishesk*, Medelironvaniu Realnikh System, Moscow, 1977.
37. Hohle, U., A mathematical theory of uncertainty: Fuzzy experiments and their realizations, *Recent Developments in Fuzzy Set and Possibility Theory*, Yager, R.R., Ed., Permagon Press, Oxford, U.K., 1981, 344.
38. Mahler, R., Representing rules as random sets, I: Statistical correlations between rules, *Inf. Sci.*, 88, 47, 1996.
39. Mahler, R., Representing rules as random sets, II: Iterated rules, *Int. J. Intell. Syst.*, 11, 583, 1996.
40. Ho, Y.C. and Lee, R.C.K., A Bayesian approach to problems in stochastic estimation and control, *IEEE Trans. AC*, AC-9, 333–339, 1964.
41. Van Trees, H.L., *Detection, Estimation, and Modulation Theory, Part I: Detection, Estimation, and Linear Modulation Theory*, Wiley, New York, NY, 1968.
42. Mahler, R., Unified Bayes multitarget fusion of ambiguous data sources, *IEEE Int. Conf. Integration Knowl. Intensive Multi-Agent Syst.* (KIMAS), Cambridge, MA, September 30–October 4, 343–348, 2003.
43. Quinio, P. and Matsuyama, T., Random closed sets: A unified approach to the representation of imprecision and uncertainty, *Symbolic and Quantitative Approaches to Uncertainty*, Kruse, R. and Siegel, P., Eds., Springer-Verlag, New York, NY, 1991, 282.
44. Punithakumar, K., Kirubarajan, T., and Sinha, A., A distributed implementation of a sequential Monte Carlo probability hypothesis density filter for sensor networks, Kadar, I., Ed., *Signal Proc., Sensor Fusion Targ. Recognit. XV*, SPIE, Vol. 6235, Bellingham, WA, 2006.
45. Fixsen, D. and Mahler, R., The modified Dempster-Shafer approach to classification, *IEEE Trans. Syst. Man Cybern.—Part A*, 27, 96, 1997.
46. Ryder, L.H., *Quantum Field Theory*, Second Edition, Cambridge University Press, Cambridge, UK, 1996.

47. Mahler, R., Multitarget moments and their application to multitarget tracking, *Proc. Workshop Estimation, Track. Fusion: A Tribute to Y. Bar-Shalom*, Naval Postgraduate School, Monterey, CA, May 17, 2001, 134.

48. Mahler, R. Multitarget Bayes filtering via first-order multitarget moments, *IEEE Trans. AES*, 39 (4), 1152, 2003.

49. Mahler, R., PHD filters of higher order in target number, *IEEE Trans. Aerosp. Electron. Syst.*, 43 (3), 2005.

50. Hoffman, J. and Mahler, R., Multitarget miss distance via optimal assignment, *IEEE Trans. Syst. Man Cybern. Part A*, 34 (3), 327–336, 2004.

51. Sidenbladh, H., Multi-target particle filtering for the probability hypothesis density, *Proc. 6th Int. Conf. Inf. Fusion*, Cairns, Australia, 2003, International Society of Information Fusion, Sunnyvale, CA, 2003, 800.

52. Zajic, T. and Mahler, R., A Particle-Systems Implementation of the PHD Multitarget Tracking Filter, Kadar, I., Ed., *Signal Proc. Sensor Fusion Targ. Recognit. XII*, SPIE Proc. Vol. 5096, 291, 2003.

53. Erdinc, O., Willett, P., and Bar-Shalom, Y., Probability Hypothesis Density Filter for Multitarget Multisensor Tracking, *Proc. 8th Int. Conf. Inf. Fusion*, Vol. 1, Philadelphia, PA, July 25–28, 2005.

54. Vo, B.-N., Singh, S., and Doucet, A., Sequential Monte Carlo methods for multi-target filtering with random finite sets, *IEEE Trans. AES*, 41 (4), 1224–1245, 2005.

55. Clark, D.E. and Bell, J., Convergence results for the particle PHD filter, *IEEE Trans. Signal Proc.*, 54 (7), 2652–2661, 2006.

56. Vo, B.-N. and Ma, W.-K., A Closed-Form Solution for the Probability Hypothesis Density Filter, *Proc. 8th Int. Conf. Inf. Fusion*, Philadelphia, PA, July 25–29, 2005, 856–863, International Society of. Information Fusion, Sunnyvale, CA, 2005.

57. Clark, D.E., Panta, K., and Vo, B.-N., The GM-PHD filter multiple target tracker, *Proc. 9th Int. Symp. Inf. Fusion*, Florence, Italy, 1–8, July 2006, International Society of Information Fusion, Sunnyvale, CA, 2006.

58. Erdinc, O., Willett, P., and Bar-Shalom, Y., A physical-space approach for the probability hypothesis density and cardinalized probability hypothesis density filters, Drummond, O., Ed., *Signal Processing of Small Targets 2006*, SPIE Proc. Vol. 6236, Bellingham, WA, 2006.

59. Panta, K., Vo, B.-N., Doucet, A., and Singh, S., Probability hypothesis density filter versus multiple hypothesis testing, Kadar, I., Ed., *Signal Proc. Sensor Fusion Targ. Recognit. XIII*, SPIE Vol. 5429, Bellingham WA, 2004.

60. Lin, L., Kirubarajan, T., and Bar-Shalom, Y., Data association combined with the probability hypothesis density filter for multitarget tracking, Drummond, O., Ed., *Signal and Data Proc. of Small Targets 2004*, SPIE Vol. 5428, 464–475, Bellingham, WA, 2004.

61. Punithakumar, K., Kirubarajan, T., and Sinha, A., A multiple model probability hypothesis density filter for tracking maneuvering targets, Drummond, O., Ed., *Signal Data Proc. Small Targets 2004*, SPIE Vol. 5428, 113–121, Bellingham, WA, 2004.

62. Pasha, A., Vo, B.-N., Tuan, H.D., and Ma, W.-K., Closed-form PHD filtering for linear jump Markov models, *Proc. 9th Int. Symp. Inf. Fusion*, Florence, Italy, July 10–13, 2006, International Society of Information Fusion, Sunnyvale, CA, 2006.

63. Balakumar, B., Sinha, A., Kirubarajan, T., and Reilly, J.P., PHD filtering for tracking an unknown number of sources using an array of sensors, *Proc. 13th IEEE Workshop Stat. Signal Proc.*, Bordeaux, France, 43–48, July 17–20, 2005.

64. Ahlberg, A., Hörling, P., Kjellström, H., Jöred, H.K., Mårtenson, C., Neider, C.G., Schubert, J., Svenson, P., Svensson, P., Undén, P.K., and Walter, J., The IFD03 information fusion demonstrator, *Proc. 7th Int.. Conf. on Inf. Fusion*, Stockholm, Sweden, June 28–July 1, 2004, International Society of Information Fusion, Sunnyvale, CA, 2004, 936.

65. Tobias, M. and Lanterman, A.D. Probability hypothesis density-based multitarget tracking with bistatic range and Doppler observations, *IEE Proc. Radar Sonar Navig.*, 152 (3), 195–205, 2005.

66. Tobias, M. and Lanterman, A., Multitarget tracking using multiple bistatic range measurements with probability hypothesis densities, Kadar, I., Ed., *Signal Proc. Sensor Fusion Targ. Recognit. XIII*, SPIE Vol. 5429, 296–305, Bellingham, WA, 2004.
67. Clark, D.E. and Bell, J., Bayesian multiple target tracking in forward scan sonar images using the PHD filter, *IEE Radar Sonar Nav.*, 152 (5), 327–334, 2005.
68. Clark, D.E. and Bell, J., Data association for the PHD filter, *Proc. Conf. Intell. Sensors Sensor Netw. Info. Process.*, Melbourne, Australia, 217–222, December 5–8, 2005.
69. Clark, D.E. and Bell, J., GM-PHD filter multitarget tracking in sonar images, Kadar, I., Ed., *Signal Proc. Sensor Fusion Targ. Recognit. XV*, SPIE Vol. 6235, Bellingham, WA, 2006.
70. Clark, D.E., Bell, J., de Saint-Pern, Y., and Petillot, Y., PHD filter multi-target tracking in 3D sonar, *Proc. IEEE OCEANS05-Europe*, Brest, France, 265–270, June 20–23, 2005.
71. Ikoma, N., Uchino, T., and Maeda, H., Tracking of feature points in image sequence by SMC implementation of PHD filter, *Proc. Soc. Instrum. Contr. Eng. (SICE) Annu. Conf.*, Hokkaido, Japan, Vol. 2, 1696–1701, August 4–6, 2004.
72. Wang, Y.-D., Wu, J.-K., Kassim, A.A., and Huang, W.-M., Tracking a variable number of human groups in video using probability hypothesis density, *Proc. 18th Int. Conf. Pattern Recognit.*, Hong Kong, Vol. 3, 1127–1130, August 20–24, 2006.
73. Zajic, T., Ravichandran, B., Mahler, R., Mehra, R., and Noviskey, M., Joint tracking and identification with robustness against unmodeled targets, Kadar, I., Ed., *Signal Proc. Sensor Fusion Targ. Recognit. XII*, SPIE Vol. 5096, Bellingham, WA, 279–290, 2003.
74. El-Fallah, A., Zatezalo, A., Mahler, R., Mehra, R., and Alford, M., Advancements in situation assessment sensor management, Kadar, I., Ed., *Signal Proc. Sensor Fusion Targ. Recognit. XV*, SPIE Proc. Vol. 6235, 62350M, Bellingham, WA, 2006.
75. El-Fallah, A., Zatezalo, A., Mahler, R., Mehra, R., and Alford, A., Regularized multi-target particle filter for sensor management, Kadar, I., Ed., *Signal Proc. Sensor Fusion Targ. Recognit. XV*, SPIE Proc. Vol. 6235, Bellingham, WA, 2006.
76. El-Fallah, A., Zatezalo, A., Mahler, R., Mehra, K.R., and Alford, M., Unified Bayesian situation assessment sensor management, Kadar, I., Ed., *Signal Proc. Sensor Fusion Targ. Recognit. XIV*, SPIE Proc. Vol. 5809, Bellingham, WA, 253–264, 2005.
77. Vo, B.-T., Vo, B.-N., and Cantoni, A., Analytic implementations of the cardinalized probability hypothesis density filter, *IEEE Trans. Signal Proc.*, 55 (7), 3553–3567, 2006.
78. Vo, B.-N., Vo, B.-T., and Singh, S., Sequential Monte Carlo methods for static parameter estimation in random set models, *Proc. 2nd Int. Conf. Intell. Sensors Sensor Netw. Inf. Process.*, Melbourne, Australia, 313–318, December 14–17, 2004.
79. Vo, B.-T. and Vo, B.-N., Performance of PHD based multi-target filters, *Proc. 9th Int. Conf. Inf. Fusion*, Florence, Italy, 1–8, July 10–13, 2006, International Society of Information Fusion, Sunnyvale, CA.
80. Mahler, R., Random sets: Unification and computation for information fusion—a retrospective assessment, *Proc. 7th Int. Conf. Inf. Fusion*, Stockholm, Vol. 1, 1–20, 2004.
81. Kruse, R., Schwencke, E., and Heinsohn, J., *Uncertainty and Vagueness in Knowledge-Based Systems*, Springer-Verlag, New York, 1991.
82. Goutsias, J., Mahler, R., and Nguyen, H.T. *Random Sets: Theory and Application*, Springer-Verlag, New York, 1997.

17

Distributed Fusion Architectures, Algorithms, and Performance within a Network-Centric Architecture

Martin E. Liggins* and Kuo-Chu Chang

CONTENTS

17.1 Introduction

The concept of network-centric operations is a process that the military is just beginning to struggle with. Concepts such as situation awareness, information fusion, data mining, collaborative agents, domain actors, and fully integrated sensors have permeated the research and the Department of Defense literature.[1–4] Much of this work[1–3] focuses on structural and architectural aspects of network-centric operations such as metadata

---

* The author's affiliation with the MITRE Corporation is provided only for identification purposes and is not intended to convey or imply MITRE's concurrence with, or support for, the positions, opinions, or viewpoints expressed by the author.

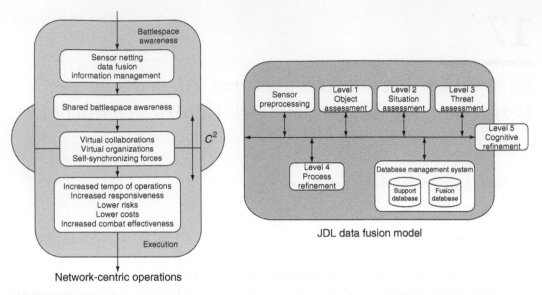

FIGURE 17.1
Network-centric and the Joint Director of Laboratories fusion models.

processes, publishing and subscribing components, or network agents.* For instance, the concept of domain actors or specialized expert agents provides a view of semiautomated networks that balance human–computer interaction. From the fusion side of things, the well-known "levels of fusion" that had its genesis from the Joint Director of Laboratories (JDL[†])[5–7] provides a foundation for processing multiple sensing sources and inferencing conditions (Figure 17.1).

The fusion model begins with data generation and choosing the right types of sensors or sources (level 0),[‡] providing techniques for detecting and identifying targets of interest (level 1), progressing to the development of situational awareness (level 2) and on to higher levels of human awareness and multilevel feedback. Fundamentally however, the JDL model focuses on a centralized framework, for instance, multiple sensors feeding information to a local fusion processing center to provide a set of local target tracks, and then sending the results to decision managers in that region.

The difficulty with developing fusion techniques for the network enterprise is adopting a mathematical framework that squarely integrates fusion processing directly into the network structure. Directly applying centralized fusion concepts to a distributed network is not the answer. As we see, distributed fusion is concerned with the problem of detecting multiple pathways in the network[§] that share common data. This represents a double counting of information (uncertainty), essentially causing the fusion network to become overconfident. Dr. Chee-Yee Chong described this as the *chicken little effect*, where, for example, a commander receives confirmation of battlefield conditions from an assumed

---

* Knowledgeable entities to enable shared information and collaboration, with an ultimate goal of developing situational awareness.
† JDL represents the first attempt to categorize fusion processing into component technologies. Earlier chapters were devoted to discussing the impact and structure of the JDL model and its role in developing fusion technology.
‡ Sensor preprocessing (not explicitly shown in the figure).
§ This is only one of the problems for distributed fusion. Others include allocation of processing to different fusion locations, managing communications, etc.

set of independent sources that in turn used the same upstream source to develop their assessment. Finding and compensating for these common pathways form the primary goal of distributed network fusion.

To show how fusion performance can be affected by a network-centric system, we will build a conceptual network-centric example and develop the corresponding distributed fusion concepts. In doing so, we hope to provide both the fusion community and the network-centric community with an understanding of some of the important concepts, involving distributed fusion.

Figure 17.1 represents one example of how technologies such as fusion are portrayed within the network-centric literature (Ref. 1, p. 89, in part), and fusion as it is often presented in the fusion literature (JDL fusion model). Network-centric system designers have focused on developing the network framework, while assuming that key technologies such as fusion could be easily inserted into the network structure as one of many higher-level processing techniques. Although the JDL model is applicable to a network-centric system, the literature has focused its use on centralized architectures. When dealing with distributed networks, an additional level of complexity is necessary.

We will examine this level of complexity through a sample network-centric system, and in doing so we will provide some insight into how fusion information is passed throughout the network.

The structure of this chapter is as follows. In Section 17.2, we provide a sample network-centric system, involving multiple sensing platforms, each platform concerned with generating locally fused track estimates of targets in their field-of-view and sending the results to be fused globally (e.g., the ground station) with track estimates generated by other local platforms within the network. From this network layout, we develop the corresponding fusion network and use it as an example for the rest of the chapter. Section 17.3 describes the concepts of the information graph. This approach is a convenient method for locating common pathways that affect overall fusion performance. Section 17.4 describes the general fusion concepts used to eliminate redundant information along these common network paths and conditional independence—a concept necessary to understand the fusion techniques that follow. Section 17.5 discusses five common network fusion techniques. The first four techniques rely on the assumption of conditional independence in addressing how each approach treats the two fundamental sources of distributed network error—common process noise and common priors. The fifth technique assumes covariance consistency between track uncertainties, and as such is not based on conditional independence. Finally, Section 17.6 provides an example of fusion performance. We focus on only one aspect of network fusion—the hierarchical fusion architecture. Hierarchical fusion is more common than one would expect and the results are representative of performance issues that occur in many of the network structures.

It is hoped that this chapter will provide both network-centric developers and fusion developers with an understanding of the additional complexities that distributed fusion requires to function within the network-centric environment.

17.2 Distributed Fusion within a Network-Centric Environment

Within the military, complex network-centric operations can be seen at many levels of operation—tactical and strategic. At the tactical level, airborne platforms process measurements from onboard (organic) or nearby sensors to create ground target track estimates (local fusion) of target activity in the local region. These estimates can be fused with similar track estimates from other platforms to form global estimates of overall target activity (global fusion) in the region. In turn, these global estimates may also be sent to higher process or command centers, which may focus on developing an overall estimate of battlefield activity.

These concepts provide the basis for developing an interrelated system of sensors, communication networks, local processing platforms (e.g., aircraft, ground stations), and regional processing centers that integrate and fuse track state estimates and uncertainties from a wide range of sources over a complex information grid used by local battlefield command and higher headquarter centers that could be located halfway around the globe. Network-centric operations provide the means to transform military operations from sets of stovepiped systems that collect data independently into a tightly integrated distribution of information flow. This involves operating conditions that may be hampered by communication bandwidth limits and fixed but differing data structures that, in some cases, limit the quality of information passed. For instance, an unattended ground-based set of acoustic sensors may send target measurements to a local control node within its sensor array, constrained by processing speeds, operating battery life, and communication update rates. These units cannot afford to process and transmit updates at regular intervals or to maintain a database of full track pedigree (track history). Rather, they may send current local track estimates asynchronously to some downstream processing center (e.g., local aircraft) that would take on the role of more elaborate track processing.

Figure 17.2 shows a sample network-centric operation along with the corresponding fusion network. This example will be the basis of our discussions in this chapter. Each of the subsystems—aircraft, acoustic sensor suite, the ground station, and the communication links—will provide the backdrop for decomposing this sample network system into network components that formulate distributed fusion elements. In this case, the illustrated structure initially divides the network into two different substructures, involving hierarchical fusion with and without feedback. More specifically, airborne platforms A and B independently compute (fuse) dynamic track estimates from onboard (organic) sensors. These fused local track state estimates and the corresponding covariance uncertainty errors (F_A and F_B, respectively) are transmitted to a ground station, E, which fuses the results globally (F_E). The ground station is assumed to transmit a refinement of individual track estimates (reduced fused covariance errors) back to each platform. This example

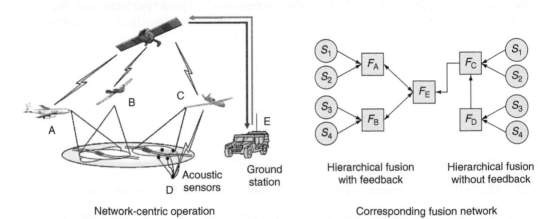

Network-centric operation   Corresponding fusion network

FIGURE 17.2
Comparison of a simplified tactical network-centric system and the corresponding fusion network.

of hierarchical fusion with feedback is shown by the corresponding fusion network on the right. The dual path (double arrows) indicates feedback of the fused global results to each local platform.

The second case involves platforms C and D. In this case, a transmitting node from a set of acoustic sensors (D) feeds local track estimates (F_D) directly to the airborne platform (C), which in turn, fuses these estimates with onboard sensor measurements (F_C). The combined fused tracks are sent to a ground station for further processing (F_E), but the global fused error estimates are not sent back to the local platforms. This case represents hierarchical fusion without feedback.

Other less obvious network examples include, for instance, platform A, where sensors S_1 and S_2 feed F_A. This is an example of centralized fusion—track estimates are formed from the sensors directly on board the platform. Note that hierarchical fusion appears to be similar to centralized fusion but involves multiple stages of track processing. For the fully distributed fusion network, we emphasize sharing fused track estimates directly between each of the three nodes (i.e., airborne platforms A, B, and C). For instance, each of the platforms could share fused track estimates directly with each other by broadcasting locally fused results (fused network of Figure 17.2 linking F_A, F_B, and F_C directly without the ground station).

The fundamental types of fusion architectures are shown in Figure 17.3. Centralized fusion is relatively straightforward, taking in raw measurements directly from the sensors. We usually assume that each sensor generates its own target reports (measurements) independently. Data are collected on a particular target from each sensor; sensor data streams are aligned spatially and temporally, and sensor reports associated with the target are used to formulate a single fused track estimate. The track estimate becomes more refined as more measurements are received.

Within a hierarchical architecture, track results are generated at local fusion processing nodes (centralized architecture), and track estimates are sent in turn to higher nodes (global fusion nodes) to generate a common track estimate. Hierarchical fusion takes advantage of the diversity between local nodes such as the diversity of different sets of sensors to view the target, maintaining local storage of assigned sensor reports and track histories at the originating node, while transmitting refined track estimates. Feedback may be used to take advantage of any fused track improvements developed at the higher nodes by redistributing the reduced error uncertainty estimates back to the local fusion nodes. Hierarchical systems may compensate for limited communication bandwidths that restrict the quantity (and quality) of data from local nodes. Hierarchical formats can ensure considerable communication savings.

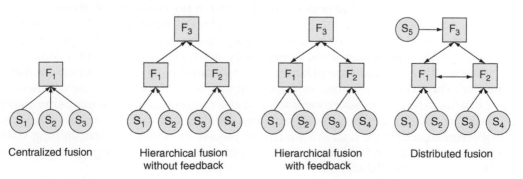

Centralized fusion   Hierarchical fusion without feedback   Hierarchical fusion with feedback   Distributed fusion

FIGURE 17.3
Traditional network fusion architectures.

Distributed architectures do not assume superior or subordinate roles between nodes; we can think of this approach as passing information between equals. Data sharing is based on the needs of each node—requesting information from other platforms to help reduce local track uncertainty error or passing results only at opportune times. In the first case, an aircraft may take advantage of well-placed geometries of other aircrafts to reduce track error or extend track continuity because of local terrain obscuration. In fact, with the advent of agent technology and the knowledge of platform locations, a single platform could improve local estimates by adapting agent techniques to search out and test optimal platform geometries before requesting or even fusing the results from local estimates. In the second case, track results may be transmitted only when track estimates have changed significantly.

Network-centric systems add an additional level of complexity to fusion processing. In the past, as long as we operated within centralized fusion framework, processing *raw* sensor measurements or working with track estimates updated each time new data were received (full-communication rate), the results were easy to deal with and mathematically optimal. However, treatment of track estimates within a network-centric system involves inherently redundant path links between nodes. Regardless of the architecture, these redundancies must be found and the common correlated estimation errors removed before they corrupt subsequent processing. There are two types of common errors to consider: prior estimates and process noise. The complexity of the architecture determines how and where these errors are generated. The fusion process determines how they are removed.

As such, distributed fusion processing should consider three factors: the network architecture, communication links within the network, and applicable fusion algorithms.

Architectures. The network architecture focuses on the node structure between the sensors and the eventual user. The primary architectures considered may involve any combination of centralized, hierarchical (with and without feedback), or distributed fusion. Centralized architectures transmit sensor generated raw measurements directly to a common fusion node. These architectures[2] provide local stovepiped processing centers that limit network-centric development. Any number of fusion texts that discuss this approach can be found.[5-10]

Hierarchical architectures fuse global track estimates from some combination of local fusion processing nodes, forming a subordinate–superior relationship. That is, centralized fusion nodes provide local fused estimates from one or more sensors and send the results to a common command node to develop a global fused track estimate. Hierarchical nodes can involve feedback to reduce local track covariance errors.

Distributed fusion extends the concepts of hierarchical fusion to more general conditions. Under this form, there is no subordinate–superior nodal relationship. Each node can communicate estimates to other nodes as needed and in turn receives estimates.

Communications. Communication in connectivity plays a critical role within a network-centric framework. How each node is connected determines the applicability, performance, and utility of the fusion algorithm. Connectivity pathways and bandwidth constraints determine whether nodes can accept large amounts of raw data or smoothed track estimates are needed. These constraints affect how path links are formed and determine whether full rate or intermediate transmissions are used. Full-rate communication ensures performance optimality by creating in essence supermeasurements but may not be practical under realistic operational settings. Creating hierarchical processing nodes and transmitting track estimates can reduce bandwidth requirements but adds more decision processing layers. Information push or pull dictates the need and complexity of feedback.

Fusion algorithm. Choice of fusion algorithm is based on the network architecture, communication constraints, node processing capabilities and limits, and command and

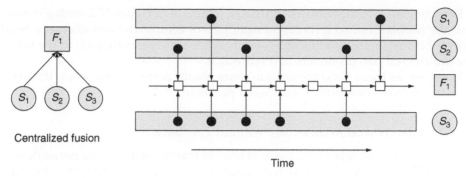

FIGURE 17.4
Information graph for the centralized fusion architecture.

processing protocols. The network structure and operational needs create path link redundancies affected by common track errors: process noise (target dynamics) and prior estimates (fusion with previously communicated track estimates). Choosing the right fusion algorithm is based on the need to minimize these errors. For instance, if target dynamics is considered well-behaved (e.g., tracks generated within predetermined shipping lanes), process noise can probably be ignored. Each fusion algorithm discussed treats these errors differently. Depending on how these assumptions match real-world conditions determines how well the fusion algorithm will perform.

We address how these factors impact network-centric operations. However, first we need to introduce the concepts of the information graph and how it can be used to locate common node links (fusion double counting) within a system network.

17.3 Information Graph

The information graph is a convenient means to understand how fusion process flows impact a network system. In this section we look at the information graph for the fusion architectures in Figure 17.3, based on our network example. Examining these graphs gives us an insight into how fusion is affected by the network system.*

17.3.1 Centralized Architecture

The centralized architecture gives the simplest information graph (Figure 17.4). Each sensor provides an independent set of measurements that is processed at a local fusion node, F_1, to develop and update a common set of track estimates. Updates may be at an asynchronous or synchronous rate. The shaded rectangular boxes (on the right side of the figure) represent measurements generated by each sensor over time. Note that fusion updating can take place even when there are no measurements. Track state estimates are generated directly from the sensor's raw measurements. As such, fused track state estimates are considered optimal.

---

* These concepts are attributed to the work of Chong et al.[11-26] and form the basis for the discussions that follow. Many of the concepts presented are discussed in more detail in these references.

Referring back to the network-centric architecture in Figure 17.2, each airborne platform and the acoustic network represent examples of centralized architectures. Such systems drove the development of fusion research in the early years that led to the formulation of the multitiered JDL fusion model (Figure 17.1).

However, as these systems began to share results across the network, the network complexity began to contribute to additional problems.

17.3.2 Hierarchical Architecture without Feedback

Hierarchical fusion represents the next level of complexity within a network structure. As Figure 17.5 shows, local centralized nodes (F_1 and F_2 representing the aircraft platforms in Figure 17.2) transmit fused target state estimates and track covariance errors to the regional global fusion node, F_3 (e.g., the ground station). The hierarchical architecture on the left side of Figure 17.5 appears to be a natural extension of the centralized approach. However, the information graph (on the right) shows a more complex underlying structure— redundant data pathways feed the global fusion node over time. The figure shows two such pathways by following the two broad-lined paths starting in F_1 at point A and terminating at point E in the fusion center F_3. Track estimates are generated at point A sent to the fusion node F_3 at point B and continue to be processed locally within F_1 over time. At point C, these estimates are updated and once again sent to F_3 at point E. Similarly the fusion center updates its own track estimates from point B at point D. However, when the estimates from F_1 (point C) are fused with the results from point D at point E, this could lead to a smaller fused error due to the common uncertainty originally presented at point A and sent along these two paths. This results in an overconfident global estimate usually much smaller than it should be. Because the information is not mutually exclusive, this redundant information occurs each time a local node sends new updates but continues to process the same error. This could lead to track loss because the redundancy could contribute to an overall track error much smaller than that is reasonable for the target dynamics.

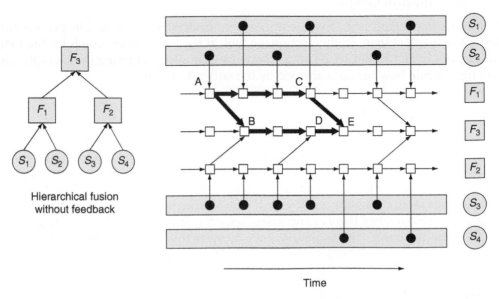

FIGURE 17.5
Information graph for the hierarchical fusion architecture without feedback.

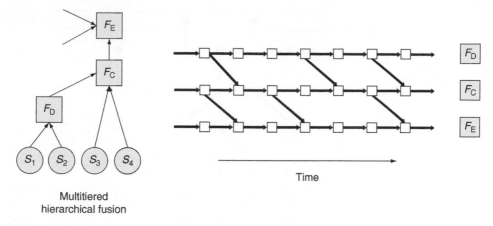

Multitiered
hierarchical fusion

FIGURE 17.6
Multitiered hierarchical fusion flow for the network-centric acoustic sensor flow.

The local nodes (e.g., point A) represent the source of common error that needs to be accounted for and removed each time the fusion center receives a new update. The information graph provides a natural means of determining when and where these redundant pathways are created.

Referring back to Figure 17.5, we can now formulate where these common pathways occur, by finding the common ancestors in the system. For example, the common ancestor originates at the local aircraft nodes, F_A and F_B, when no feedback is used.

Similarly nodes F_C and F_D represent another form of hierarchical fusion without feedback. Tracks are generated at F_D from local acoustic sensors and sent to the aircraft to fuse with track updates processed at F_C. The combined tracks are sent to the ground station. This process will be similar to the multitiered hierarchical flow shown in Figure 17.6, where the sensor inputs have been dropped for simplicity.

17.3.3 Hierarchical Architecture with Feedback

Much of the previous discussion involving hierarchical fusion without feedback is applicable to hierarchical fusion involving feedback. Referring to Figure 17.7, the main difference is that when information is sent back to the local nodes, the common source of error now becomes the higher (global) fusion node, F_3. In this case, the global node generates combined target estimates based on both the local fusion platforms and from any organic sensors. This redundant information, formed at the global fusion center, must be removed before the results are sent back to the individual platforms to reduce local track errors. The advantage of this approach is that the local nodes (F_1 and F_2) will operate with nearly identical uncertainty errors to those at the global fusion node, F_3, essentially improving overall track processing for each local track. The disadvantage is that if the common information in the global estimate is not removed, then it is more difficult for the overall system to recover. That is, the fused global error becomes so small that it is difficult to associate subsequent updates with the same track.

Referring to the information graph in Figure 17.7, note that the fused estimates generated at point A in F_3 are sent back to the lower-level platform, F_1 (at point B), where they are subsequently fused with the current local track state, then updated to the current time at point D. Thus the source of redundancy (common predecessor) for fusion with feedback architectures resides with the fusion center.

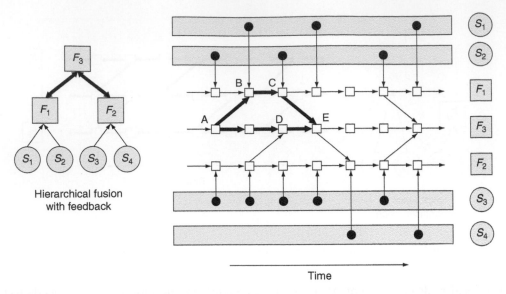

FIGURE 17.7
Information graph for the hierarchical fusion architecture with feedback.

17.3.4 Distributed Architecture

Distributed fusion represents the next level of fusion complexity in a network-centric system. Hierarchical fusion is the simplest form of distributed fusion architectures, and its theoretical development forms the basis for developing more complex distributed fusion systems. As we saw with the two feedback cases of hierarchical fusion, distributed fusion is highly dependent on the architecture and the communication protocol to locate and eliminate the common information sources. Complex architectures, such as cyclic communications, contain multiple common predecessor nodes that form multiple redundant paths, much earlier in the process.[11,21,22]

Broadcast networks are the simplest form of distributed fusion (Figure 17.8) that easily demonstrates network flow redundancy. Assume that after each local fused update, the track estimates are transmitted directly to the other fusion nodes (e.g., F_1 transmits simultaneously to F_2 and F_3). Then path redundancy originating at point A (in this example) must be removed from the fused estimates at points C and D before it is retransmitted back at point E.

To see a distributed network within the network-centric operation in Figure 17.2, it would be as if we removed the ground station and allowed each of the three airborne platforms to broadcast fused track results to each other. Transmissions need not be synchronous. For example, any form of significant change in the track state could be used to initiate a transmission, or a node may generate a request to all other nodes, when it has to reduce large locally generated covariance errors.

As we noted earlier, a complex network-centric operation, such as that represented in Figure 17.2, usually is composed of any combination of the basic fusion network structures in Figure 17.3. Local nodes represent centralized architectures. Hierarchical conditions exist between local fusion centers and a global fusion center. Extending this concept, one could see regional nodes forming further hierarchies with a battlefield command center, or multiple airborne platforms using local agents to seek out other regional airborne systems to help reduce local track uncertainties when their sensors are inadequate (e.g., distributed fusion). However, complex distributed architectures should be constructed with caution

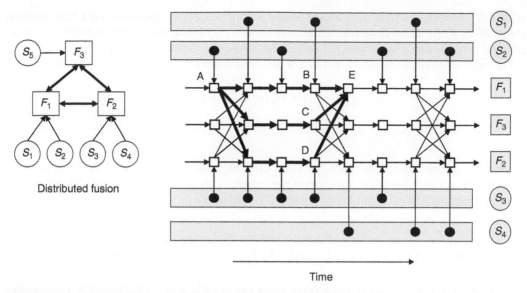

FIGURE 17.8
Information graph for the distributed fusion (broadcast) architecture.

because of the inherent complexity, and the difficulty in finding common sources of error. Only a limited number of simple structured distributed architectures have been studied.* Current research is under way to examine techniques to handle more general networks, such as covariance intersection (CI)[27] and genealogy fusion.[28]

17.4 Fusion Algorithm and Distributed Estimation

Traditional track fusion algorithm development has focused on centralized architectures. These algorithms assume that sensor measurements (reports) are generated independently. Measurement independence is the reason why centralized fusion is theoretically optimal. Distributed fusion researchers have recognized that fusion across a network environment adds an additional level of complexity to standard centralized fusion approaches. This added complexity is driven by the impact of common process noise and common priors. This section lays the foundation for this process.

Revisiting the hierarchical fusion format in Figure 17.5, assume two local nodes are tracking the same target based on individual discrete-time dynamic models of the form:

$$\hat{x}_{F_i}(k+1) = F_{F_i}(k)\hat{x}_{F_i}(k) + G_{F_i}(k)v(k)$$

$$P_{F_i}(k+1) = F_{F_i}(k)P_{F_i}(k)F_{F_i}(k)' + G_{F_i}(k)Q(k)G_{F_i}(k)' \qquad (17.1)$$

$$k = 0, 1, 2, \ldots$$

---

* Other distributed fusion network approaches are discussed in the literature including cyclic, tracklets, and the global restart. Chong et al.[15] discuss these in detail with applicable references given.

where $\hat{x}_{F_i}(k)$ represents the state vector (target location or type) updated over k time updates for node F_i, $P_{F_i}(k)$ is the state transition matrix, and $v(k)$ being a zero-mean white Gaussian process noise with covariance $Q(k)$ and associated covariance error, $P_{F_i}(k+1)$.

For each local node, measurements are received from the sensor mix, modeled as

$$z_{s_i,m}(k+1) = H_{S_i}(k)x_{F_i}(k) + w_{S_i}(k)$$

$$S_i = \text{platform sensor}, k = 0, 1, 2, \ldots$$

(17.2)

where H_{S_i} is the transition from the state and the measurement, and $W_{S_i}(k)$ represents zero-mean white Gaussian measurement noise with covariance $R_{S_i}(k)$. Measurement history from each sensor can be represented by (see Figure 17.5)

$$S_1: Z_{11} = (z_{11}, z_{12}, z_{13}, \ldots)^* \quad \text{and} \quad S_2: Z_{12} = (z_{21}, z_{22}, z_{23}, \ldots) \quad \text{for node } F_1$$

$$S_3: Z_{23} = (z_{31}, z_{32}, z_{33}, \ldots) \quad \text{and} \quad S_4: Z_{24} = (z_{41}, z_{42}, z_{43}, \ldots) \quad \text{for node } F_2$$

Assume that each platform develops a track estimate based on Kalman filter principles. The local track estimates and covariance matrices are transmitted to the fusion center (e.g., ground station, F_3) to develop a global track state estimate of target position and error covariance, $\hat{x}$ and $\hat{P}$. The fusion center maintains these estimates along with previous fused updates (priors) from earlier local track transmissions and extrapolated to the current time. These priors are represented by $\bar{x}$ and $\bar{P}$. Additionally, let x and P represent the true target dynamics.

As each sensor observes the target, a local fused state estimate is formed independently (at nodes F_1 and F_2) and sent to the global fusion center, F_3, to build a common dynamic state estimate of target motion. In other words, a fused global target state estimate is generated on the basis of an assumption of conditional independence between the estimates of F_1 and F_2. To understand this process, let us examine the track state information, generated at each local node. Since the sensor measurements represent the total information present, their collected set can be written as the union of their measurements (e.g., for F_1): $Z_1 = Z_{11} \cup Z_{12} = (z_{11}, z_{12}, z_{13}, \ldots) \cup (z_{21}, z_{22}, z_{23}, \ldots)$. Similarly $Z_2 = Z_{23} \cup Z_{24}$ for fusion node F_2. These measurement sets contain common information that needs to be removed at F_3.

Figure 17.9[21,22] provides a conceptual framework. We use Z_i/Z_j to represent those measurements in Z_i that are not common to Z_j. Then it is easily shown[21] that

$$Z_1 \cup Z_2 = (Z_1/Z_2) \cup (Z_2/Z_1) \cup (Z_1 \cap Z_2)$$

The conditional independence assumption implies the following:

$$p(Z_1 \cup Z_2 | x) = \frac{p(Z_1|x)p(Z_2|x)}{p(Z_1 \cap Z_2|x)}$$

---

* The time index has been removed for simplicity. For instance, S_1 should read $Z_{11}(k) = \{z_{11}(k), z_{12}(k), z_{13}(k),\ldots\}$ indicating measurements from sensor S_1, and the measurement numbers are all received at the same time index k.

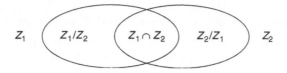

FIGURE 17.9
Conditional relationship between two nodes to be fused.

This leads to a formulation for the common information by the conditional relation:*

$$p(x|Z_1 \cup Z_2) = C^{-1} \frac{p(x|Z_1)p(x|Z_2)}{p(x|Z_1 \cap Z_2)} \tag{17.3}$$

Here, $p(x|Z_1)$ and $p(x|Z_2)$ are the estimates from local nodes F_1 and F_2, respectively. The common information between the nodes is given in the denominator by $p(x|Z_1 \cap Z_2)$, which is equivalent to "subtracting out" the common information. The globally fused (joint) information is represented by $p(x|Z_1 \cup Z_2)$. C is the normalization constant.

For the hierarchical fusion architecture without feedback in Figure 17.5, Equation 17.3 can be adapted to show the common ancestor generated by F_1:

$$p(x|Z_{F_3}(E)) = C^{-1} \frac{p(x|Z_{F_1}(C))p(x|Z_{F_3}(D))}{p(x|Z_{F_1}(A))} \tag{17.4}$$

where the redundant information from $p(x|Z_{F_1}(A))$ is removed.

Similarly, for the hierarchical fusion architecture with feedback in Figure 17.7, we show the common ancestor to be at the global fusion node F_3:

$$p(x|Z_{F_3}(E)) = C^{-1} \frac{p(x|Z_{F_1}(C))p(x|Z_{F_3}(D))}{p(x|Z_{F_3}(A))} \tag{17.5}$$

For the distributed architecture in Figure 17.8, all nodes are able to accept information. Thus for node F_1 receiving information from all three nodes, Equation 17.3 becomes

$$p(x|Z_{F_1}(E)) = C^{-1} \frac{p(x|Z_{F_1}(B))p(x|Z_{F_2}(C))p(x|Z_{F_3}(D))}{p(x|Z_{F_1}(A))^2} \tag{17.6}$$

Note that the common information from A was sent to both nodes B and C and as such is taken out twice (the squared term in the denominator). As discussed previously, two sources of error form the basis for reduced track accuracy at the global fusion center: common process noise and common prior errors. When multiple sources develop track estimates on the basis of the same dynamic model, they develop common estimation errors on the target, although the sensor measurements are independent. Thus, the local dynamic model developed for each local source creates a common estimate of the target motion. This cross-covariance matrix defines the error correlation (common process noise) between the sensor tracks generated between the two local sites and gives rise to a fused

---

* A derivation can be found in Ref. 22 with a proof in Ref. 21.

global covariance matrix, $P = \begin{bmatrix} P_i & P_{ij} \\ P_{ji} & P_j \end{bmatrix}$, where P_i and P_j are the local covariance matrices for F_i and F_j, respectively; P_{ij} and P_{ji} represent the cross-covariance between the local track estimates; that is, P_{ij} is the cross-covariance between $\hat{x}_{F_i}(k)$ and $\hat{x}_{F_j}(k)$, and P_{ji} between $\hat{x}_{F_j}(k)$ and $\hat{x}_{F_i}(k)$. (Henceforth for convenience, we will refer to these state estimates as $\hat{x}_i(k)$ and $\hat{x}_j(k)$, respectively.)

The second source of error between the two local track estimates is on the basis of the common prior information, generated at the global fusion center. As we saw with Figure 17.5, the same covariance error generated at point A was used to update both the local track F_1 and the global fusion center at point B. This common prior has to be removed at the next update at point E.

The next section addresses the development of five different fusion algorithms; each designed to treat these sources of error differently.

17.5   Distributed Fusion Algorithms

There are four types of distributed fusion algorithms that are designed to address the issue of common process noise and common prior errors. These methods assume conditional independence* of the state covariance between the local fusion nodes. They are naïve fusion, cross-covariance fusion, information fusion, and maximum *a posteriori* (MAP) fusion. A fifth fusion algorithm that focuses on covariance consistency is called covariance intersection (CI).† This type of approach does not require the conditional independence assumption. Each algorithm treats correlation errors differently. To compare each approach, we will base our discussion on hierarchical fusion without feedback and assume a linear Gaussian formulation. Equations 17.3 through 17.6, discussed in Section 17.4, are general and apply to either linear or nonlinear conditions. To adapt the following fusion techniques for a particular architecture, we need only to follow a similar approach to develop the equations, as we did with Equations 17.4 through 17.6. A detailed discussion of the algorithms[15–26,29] and a historical perspective[16] of this work can be found in the literature. For instance, variations of this work are presented by Li et al.[30] and Drummond.[31] Recently a new work by Zhu[32] presents a generalized fusion framework to examine statistical dependency between sensors. This material is not within the scope of this paper and will not be discussed.

17.5.1   Naïve Fusion

Naïve fusion assumes that both common process noise (cross-covariance error) and common priors are negligible. This fusion approach represents the simplest of the four types and it is suboptimal. As such, this approach is better suited for simple tracking problems. This involves track motion under relatively benign tracking conditions (single isolated targets with little ambiguity about the dynamic model). Its simplicity involves minimal computational processing.

---

* The material discussed for these algorithms is derived thoroughly from Refs 15–21 and 29.
† This algorithm is a recent addition in the literature and is described in more detail in this handbook.

The global fused state estimation and corresponding covariance error are shown as

$$\hat{x} = P(P_i^{-1}\hat{x}_i + P_j^{-1}\hat{x}_j)$$
$$P = (P_i^{-1} + P_j^{-1})^{-1} \tag{17.7}$$

Note that the fused track covariance is the inverse of the sum of the inverses of the local track covariance matrices. Thus, the fused covariance could have values much smaller than either individual error, which can lead to conditions of overconfidence.*

Because track estimates are generated at each local fusion node separately, a method is needed to link the correct track estimates from both nodes (association). The corresponding performance metric[†] provides a basis for assigning the most appropriate individual track estimates to be fused (between local nodes F_1 and F_2 in Figure 17.5):

$$C_{ij} = \|\hat{x}_i - \hat{x}_j\|^2_{(P_i + P_j)^{-1}} \tag{17.8}$$

This metric represents the Mahalanobis distance metric, commonly used within Kalman filter tracking.

17.5.2   Cross-Covariance Fusion

Cross-covariance fusion takes into account common process noise but assumes all common priors are negligible. The fusion algorithm is represented as follows:

$$\hat{x} = \hat{x}_i + (P_i - P_{ij})(P_i + P_j - P_{ij} - P_{ji})^{-1}(\hat{x}_j - \hat{x}_i)$$
$$P = P_i - (P_i - P_{ij})(P_i + P_j - P_{ij} - P_{ji})^{-1}(P_i - P_{ji}) \tag{17.9}$$

Cross-covariance fusion requires more information to be communicated because of the cross-covariance terms, P_{ij} and P_{ji} (between $\hat{x}_i$ and $\hat{x}_j$).[‡] These results are optimal only within a maximum likelihood sense.[19] Cross-covariance fusion may be applicable under conditions where a target is known to exist (e.g., low background clutter) and tracking parameters need to be determined. Note that when $P_{ij} = P_{ji} = 0$, this form is simply the naïve fusion rule.[§]

The corresponding metric is given by the χ^2 test:

$$C_{ij} = \|\hat{x}_i - \hat{x}_j\|^2_{(P_i + P_j - P_{ij} - P_{ji})^{-1}} \tag{17.10}$$

providing a "polynomial goodness of fit" to the dynamic model.

---

* Fused error should always result in smaller covariances. However, similar to the simple probability of the union between two events, $P(A \cup B) = P(A) + P(B) - P(AB)$, where the common overlap must be subtracted or the result is higher than it should be, the common information between two local track estimates must also be removed.

[†] $\|x\|_p = \sqrt{x^T P x}$, the Euclidian norm for the normalized track state, based on the covariance error. Mori et al.[16] provide these metrics.

[‡] We need only be concerned with one of the cross terms since P_{ij} is the transpose of P_{ji}.

[§] To show this, simply postmultiply the first term in the covariance, P_i, by $(P_i + P_j)^{-1}(P_i + P_j)$ and group common terms.

17.5.3 Information Matrix Fusion

Information fusion assumes that the common prior is dominant, and common process noise is negligible. The fusion algorithm is represented as follows:

$$\hat{x} = P(P_i^{-1}\hat{x}_i + P_j^{-1}\hat{x}_j - \bar{P}^{-1}\bar{x})$$

$$P = (P_i^{-1} + P_j^{-1} - \bar{P}^{-1})^{-1}$$

(17.11)

The common prior (point B in Figure 17.5) represents the cumulative history of the global fused track estimate and covariance error. This includes extrapolating the fused track (F_3), based on the track pedigree and all contributions from the local nodes (F_1 and F_2). The common prior information is noted by $\bar{x}$ and $\bar{P}$.

Information fusion is a natural complement to the information graph. The simplicity of this approach is well suited in a number of ways:

- The information fusion approach is only slightly more complicated than naïve fusion. If the common priors were negligible, the information fusion becomes the naïve fusion rule. This is consistent with the assumption that the local estimates are conditionally independent.

- There is no need to maintain cross-covariance information between the local nodes. This significantly reduces communication bandwidth requirements.

- This approach is optimal when there is no (significant) process noise—the target dynamic model is deterministic, or each local update is transmitted to the fusion center at each update (full-rate communications).

- Information fusion is applicable when target dynamic modeling errors can be quickly established, such as detecting maneuvers quickly and assigning the appropriate target model (e.g., using multiple-filter models).

The corresponding metric is as follows:

$$C_{ij} = \|\hat{x} - \hat{x}_i\|_{(P_i)^{-1}}^2 + \|\hat{x} - \hat{x}_j\|_{(P_j)^{-1}}^2 - \|\hat{x} - \bar{x}\|_{(\bar{P})^{-1}}^2$$

(17.12)

17.5.4 Maximum *A Posteriori* Fusion

MAP fusion represents the most complete but also the most complex form of fusion. MAP fusion incorporates the effects of the common prior and common process noise. The fusion algorithm is represented as follows:

$$\hat{x} = \bar{x} + W_i(\hat{x}_i - \bar{x}) + W_j(\hat{x}_j - \bar{x})$$

$$P = \bar{P} - \Sigma_{x\hat{z}}\Sigma_{\hat{z}\hat{z}}^{-1}\Sigma_{\hat{z}x}'$$

(17.13)

where

$$[W_i \quad W_j] = \Sigma_{x\hat{z}}\Sigma_{\hat{z}\hat{z}}^{-1}$$

$$\Sigma_{x\hat{z}} = [\Sigma_i \quad \Sigma_j] \quad \text{and} \quad \Sigma_{\hat{z}\hat{z}} = \begin{bmatrix} \hat{\Sigma}_{ii} & \hat{\Sigma}_{ij} \\ \hat{\Sigma}_{ji} & \hat{\Sigma}_{jj} \end{bmatrix}$$

(17.14)

$$\hat{z} = [\hat{x}_i' \quad \hat{x}_j']'$$

The coefficients, $W_{k|k} = [W_i \ W_j] = \Sigma_{x\hat{z}} \Sigma_{\hat{z}\hat{z}}^{-1}$, form the gain matrices that determine the effective contribution from each local node. $\Sigma_{x\hat{z}} = [\Sigma_1 \ \Sigma_2]$ represents the cross-covariance between the target state x and the joint local estimates, $\hat{z} = \begin{bmatrix} \hat{x}_i \\ \hat{x}_i \end{bmatrix}$, for nodes F_1 and F_2 (Figure 17.5). Additionally, $\Sigma_{\hat{z}\hat{z}} = \begin{bmatrix} \hat{\Sigma}_{ii} & \hat{\Sigma}_{ij} \\ \hat{\Sigma}_{ji} & \hat{\Sigma}_{jj} \end{bmatrix}$ is the covariance matrix of $\hat{z}$.

This approach deviates from the information fusion algorithm. MAP fusion provides the best linear minimum mean squared error (LMMSE) estimate, when the local state estimates operate under non-Gaussian assumptions, and we assume that the target is static or communication is of full rate (e.g., communicate the updated state estimates after each sensor observation). For less than full-rate communication, MAP may become suboptimal.* Much of the literature focuses on the static condition and therefore the impact of multiple iterations on MAP performance has not been fully studied.[26]

An application of MAP performance would involve conditions where there is a potential for track maneuvering with significant deviation in local track performance. The primary disadvantage of MAP processing is the complexity of the fusion equations. Note that when $\Sigma_{ij} = \Sigma_{ii}\bar{\Sigma}^{-1}\Sigma_{jj}$ (correlation is affected only by the common prior), the MAP fusion process reverts to the information filter.

To summarize, the four fusion rules treat track performance on the basis of their consideration of common process noise and common prior information. Naïve fusion ignores the effects of both sources of error. Cross-covariance fusion is concerned only with common process noise, whereas information fusion is concerned with common priors. Finally MAP fusion accommodates both sources of error.

With the advent of adaptive multiple model fusion algorithms of the four cases, the information fusion process provides the best choice in general.

17.5.5 Covariance Intersection (CI) Fusion

CI deviates from the precepts of the four fusion algorithms, discussed in Sections 17.5.1 through 17.5.4, by not making any claim of conditional independence between the local track estimates.[27,33] The developers of CI recognize that local estimates are correlated. However their basic premise is that conditional independence is "an extremely rare property" and relying on this condition is not a viable means of developing robust track estimate approximations to the centralized case.[27] Each local track covariance must be artificially increased to maintain the assumption of correlation at the fusion node. Researchers continue to debate the merits of CI against that of the other approaches, such as information fusion.

The fundamental nature of CI relies on a geometric interpretation of the Kalman filter. The motivation of this approach gives a conservative estimate whenever correlation between $\hat{x}_1$ and $\hat{x}_2$ is unknown. Under these conditions, the optimal fusion algorithm may be computationally infeasible but can be shown to be bounded by the CI covariance estimate, P_{CI}.[33]

The CI fusion algorithm is given by

$$\hat{x}_{CI} = P_{CI}[\omega P_i^{-1}\hat{x} + (1-\omega)P_j^{-1}\hat{x}_j]$$

$$P_{CI} = [\omega P_i^{-1} + (1-\omega)P_j^{-1}]^{-1}$$

(17.15)

where ω is bounded by a range between $[0, 1]$.

---

* The prior estimates are not the conditional estimates of all the previous estimates. In this case, the priors are based on the LMMSE estimates derived from the previous local estimates.[26]

A close examination of ω leads to some interesting results. The CI fusion equation is similar in form to the simple convex combination algorithm (naïve fusion).[33] That is, the naïve covariance matrices for the two local nodes are P_i and P_j and rewritten for CI as $\omega^{-1}P_i$ and $(1 - \omega)^{-1}P_j$, respectively, where ω is bounded between [0, 1]. The CI covariance matrices provide a conservative bound compared with the naïve fusion approach. Conceptually, the CI approach can be visualized as a weighted assessment of the two overlapping local covariance estimates. Where the traditional naïve fusion approach would result in a fused covariance estimate smaller than the overlapped region between the two local track covariance regions $(P_i^{-1} + P_j^{-1})$, CI subsumes the overlapped area, attributing a higher uncertainty to the local track and having a lower weighted value of ω. For example, if the covariance error for node F_1 has a weighting ω of 0.1, P_{CI} results would be dominated by the uncertainty in F_1, incorporating a fused covariance slightly smaller than P_1 and only a small portion of the error in F_2 (similar in size to the overlapping area). A value of 0.5 for ω would give the same state estimate as in the naïve fusion, but the covariance error would be twice as large:

$$\hat{x}_{CI} = (P_1^{-1} + P_2^{-1})^{-1}[P_1^{-1}\hat{x}_1 + P_2^{-1}\hat{x}_2] = \hat{x}_{NF}$$

$$P_{CI} = 2(P_1^{-1} + P_2^{-1})^{-1} = 2P_{NF}$$

(17.16)

In general, the weight is chosen to minimize either the trace or determinant of P_{CI}.[27,33]

Chong and Mori[33] examine the CI approach from an alternative derivation. They rely on a set-theoretic approach that provides a natural interpretation of the properties of the CI filter. As pointed out, the authors define a tighter bound on the true covariance uncertainty by defining an added weighting factor α^2 that depends on the estimates to be fused. That is,

$$\alpha^2 = (\hat{x}_i - \hat{x}_j)'[\omega^{-1}P_i + (1 - \omega)^{-1}P_j]^{-1}(\hat{x}_i - \hat{x}_j)$$

(17.17)

As long as $\alpha^2 < 1$, which is true as long as the two local estimates overlap, the ellipses are accurate representations of the true track uncertainty. Thus Equation 17.15 can be modified as follows:

$$\hat{x}_{CI} = P_{CI}[\omega P_i^{-1}\hat{x}_i + (1 - \omega)P_j^{-1}\hat{x}_j]$$

$$P_{CI} = (1 - \alpha^2)[\omega P_i^{-1} + (1 - \omega)P_j^{-1}]$$

(17.18)

The key point is that Equation 17.17 can be rewritten into a form similar to the information fusion filter as follows:

$$\alpha^2 = (\omega)\hat{x}_i'P_i^{-1}\hat{x}_i + (1 - \omega)\hat{x}_j'P_j^{-1}\hat{x}_j - \hat{x}_{CI}'P_{CI}^{-1}\hat{x}_{CI}$$

(17.19)

As such, α^2 reassesses a dependency back on the estimates to be fused. The primary difference is that although the estimate lies within the intersection of local covariance errors, there is no probability distribution set to decrease as additional estimates are fused.

We show the equations of each of these approaches in Figure 17.10. Both the original CI and the set-theoretic algorithms are shown. A simple comparison with other fusion approaches can be found in Ref. 33.

Naïve fusion

$$\hat{x} = P(P_i^{-1}\hat{x}_i + P_j^{-1}\hat{x}_j)$$

$$P = (P_i^{-1} + P_j^{-1})^{-1}$$

Cross-covariance fusion

$$\hat{x} = \hat{x}_i + (P_i - P_{ij})(P_i + P_j - P_{ij} - P_{ji})^{-1}(\hat{x}_j - \hat{x}_i)$$

$$P = P_i - (P_i - P_{ij})(P_i + P_j - P_{ij} - P_{ji})^{-1}(P_i - P_{ji})$$

Information fusion

$$\hat{x} = P(P_i^{-1}\hat{x}_i + P_j^{-1}\hat{x}_j - \bar{P}^{-1}\bar{x})$$

$$P = (P_i^{-1} + P_j^{-1} - \bar{P}^{-1})^{-1}$$

Maximum *a posteriori* fusion

$$\hat{x} = \bar{x} + W_i(\hat{x}_i - \bar{x}) + W_j(\hat{x}_j - \bar{x}) \qquad \hat{z} = \begin{bmatrix} \hat{x}_i' & \hat{x}_j' \end{bmatrix}'$$

$$P = \bar{P} - \Sigma_{x\hat{z}}\Sigma_{\hat{z}\hat{z}}^{-1}\Sigma_{\hat{z}x}' $$

$$\begin{bmatrix} W_i & W_j \end{bmatrix} = \Sigma_{x\hat{z}}\Sigma_{\hat{z}\hat{z}}^{-1} \qquad \Sigma_{x\hat{z}} = \begin{bmatrix} \Sigma_i & \Sigma_j \end{bmatrix} \qquad \Sigma_{\hat{z}\hat{z}} = \begin{bmatrix} \hat{\Sigma}_{ii} & \hat{\Sigma}_{ij} \\ \hat{\Sigma}_{ji} & \hat{\Sigma}_{jj} \end{bmatrix}$$

Covariance intersection

Set-theoretic covariance intersection

$$\hat{x}_{CI} = P_{CI}[\omega P_i^{-1}\hat{x}_i + (1-\omega)P_j^{-1}\hat{x}_j] \qquad \hat{x}_{CI} = P_{CI}[\omega P_i^{-1}\hat{x}_i + (1-\omega)P_j^{-1}\hat{x}_j]$$

$$P_{CI} = [\omega P_i^{-1} + (1-\omega)P_j^{-1}]^{-1} \qquad P_{CI} = (1 - \alpha^2)[\omega P_i^{-1} + (1-\omega)P_j^{-1}]$$

FIGURE 17.10
Distributed fusion techniques.

In the next section, we discuss performance comparisons for one of the fusion approaches.* We will restrict ourselves to the hierarchical fusion approach.

17.6 Performance Evaluation between Fusion Techniques

As a core example, we compare performance for the hierarchical fusion architectures. An extended derivation of fusion techniques can be seen in the literature involving closed-form analytic solutions[23–26] and compared analytically.[25,26] The analytic form represents the steady-state fused covariance.

Three forms of the hierarchical fusion approach were used: (1) fusion without feedback (reference Equation 17.4); (2) fusion with feedback (reference Equation 17.5); and (3) fusion with partial feedback (a hybrid of the two).† To ensure consistency between the three approaches, the same track state conditions were used for each fusion technique. The two local nodes are tracking the same target operating under a dynamic model characterized

---

* A complete discussion of the performance comparisons would significantly extend the length of this chapter. An excellent discussion is presented in Refs 23–26.

† Partial feedback involves sending only the state estimate back to the local nodes. In this way, unstable conditions remain at the fusion center and are corrected much faster with local node updates.

by the discrete-time system discussed in Equations 17.1 and 17.2 and using a Kalman filter to create the local tracks. The scenario involves a single target and two sensor nodes. Each node is tracking the target and providing state estimates to a fusion center following the dynamics of Equations 17.1 and 17.2. In this case, the target dynamics[25] are modeled as

$$x_{k+1} = \begin{bmatrix} 1 & T \\ 0 & 1 \end{bmatrix} x_k + \begin{bmatrix} T^2/2 \\ T \end{bmatrix} v_k$$

The measurements for the two sensors are modeled as follows:

$$z_m = [1 \ \ 0]x_k + w_k$$

where the two measurements are assumed to be independent.

For simplicity, each local sensor delays communicating at fixed update intervals, n, with the fusion center. These intervals can be varied, ranging from full rate to a delay of eight updates (where n ranges from one to eight updates).

The curves generated in Figures 17.11 through 17.13 focus on using cross-covariance fusion to provide an upper bound (maximum likelihood—ML) and the information matrix fusion (for n = 1, 2, 4, 8). They were performed using 5000 Monte Carlo simulation

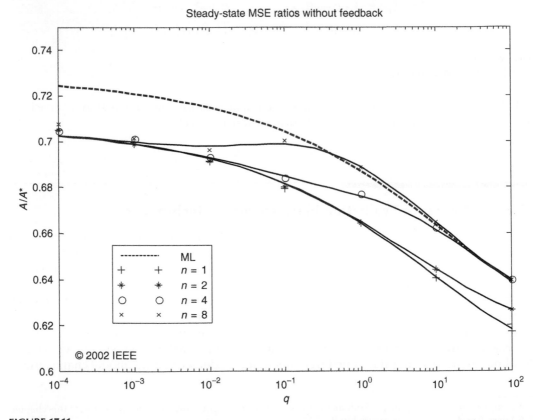

FIGURE 17.11

Hierarchical fusion without feedback: error ellipse area ratios with small process noise. (From Chang, K.C., Tian, Z., and Saha, R., *IEEE Trans. Aerospace Electr. Syst.*, 38, 455, 2002. Copyright IEEE 2002.)

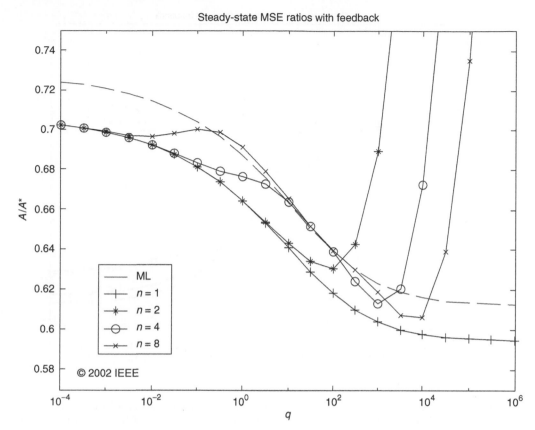

FIGURE 17.12
Hierarchical fusion with complete feedback: error ellipse area ratios with variable process noise. (From Chang, K.C., Tian, Z., and Saha, R., *IEEE Trans. Aerospace Electr. Syst.*, 38, 455, 2002. Copyright IEEE 2002.)

trials. Average root-mean-square (RMS) errors were used to estimate the fused covariance values. Each of the graphs represents area ratios for individual track covariance error matrices (uncertainty region) for the fused covariance, A with A^*, where A^* is the optimal area covered by the steady-state covariance error based on single sensor performance. The comparisons are made over a wide range of track process noise values, q.

The dashed lines represent the analytic results, based on the ML of the cross-covariance (Equation 17.9) and derived in the literature.[24-26] This represents an ideal upper bound to assess overall performance of the hierarchical approach. All remaining performance curves are based on the analytic results[25,26] for various communication rates. These rates show the effect of fused performance when nodes defer communication to the fusion node at fixed intervals. For instance, $n = 1$ represents full communications (optimal performance) and $n = 8$ implies that the local nodes wait eight update cycles to communicate to the global (fused) system.

17.6.1 Fusion without Feedback

Fusion without feedback represents a standard command-based hierarchical network structure. Command nodes or battlefield commanders are interested in obtaining the best information possible to plan effective countermeasures, expand surveillance capability

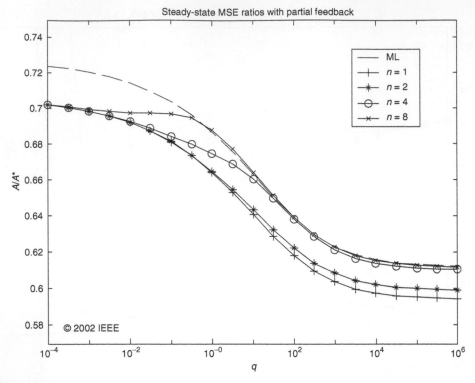

FIGURE 17.13

Hierarchical fusion with partial feedback: error ellipse area ratios with variable process noise. (From Chang, K.C., Tian, Z., and Saha, R., *IEEE Trans. Aerospace Electr. Syst.*, 38, 455, 2002. Copyright IEEE 2002. With permission.)

by adding complementary sensors to improve observability, or just get the best estimate of track activity. Such curves are ideal in designing distributed fusion systems under a range of process noises, q, and determining acceptable limits regarding communication transmission requirements.

Figure 17.11 shows that the analytic results are somewhat sensitive to increasing process noise. For higher noise levels, global fusion performance asymptotically approaches steady state at 50%. Under conditions of increasing process noise, reduced communication updates of the local error to the global fusion center cause the performance to approach the upper bounded limits of ML performance. For conditions where we reduce the local updates by eight time updates ($n = 8$), these errors approach the ML performance bound for relatively small process noise ($q = 0.1$). Under optimal communication updates (full rate, $n = 1$), fusion continues to provide improved overall performance. However, the difference between the bounds is not excessive for fusion without feedback, meaning that reasonable delays would not severely impact system performance.

17.6.2 Fusion with Feedback

Much of the discussion in the previous section is also applicable to fusion with feedback. Fusion with feedback is better suited when each of the local nodes needs to maintain the smallest covariance possible—taking advantage of the fused system.

The primary concern with the full feedback approach is the impact of feedback on the state variables not directly observable (velocity and acceleration). These parameters are severely impacted when the process noise becomes too large. As shown in Figure 17.12, fused performance errors significantly deteriorate with increasing process noise and with communication rates less than full rate. When poorly fused results are sent back to the local nodes, local node performance is directly affected, creating an unstable situation (except when full-rate communication occurs, $n = 1$). In fact, we can see that fusion under these conditions performs worse than the previous "fusion without feedback" case. To compensate for this problem, we examine the effects of partial feedback in the next subsection.

17.6.3 Hierarchical Fusion with Partial Feedback

Partial feedback involves sending only the state estimate without the covariance errors back to the local nodes to contain unstable covariance uncertainties, caused by poorly fused covariance errors. Note that for higher communication intervals, partial feedback performance does degenerate, but the instability problem is reduced and can be controlled sooner. As we will see, this makes partial feedback an effective counter to the full feedback conditions shown earlier.

On the basis of the results in Figure 17.13, we see that the effects of partial feedback in the ratio of the covariance areas provide performance similar to that of hierarchical fusion without feedback.[34]

17.7 Summary

We have presented the details involving distributed multisensor fusion from a network-centric perspective. Both fusion and network-centric operational research has progressed along parallel paths, with very little interaction. In this chapter, we have attempted to provide a fundamental distributed fusion structure that lays the groundwork for incorporating network fusion principles directly into network-centric systems. Fusion algorithms were presented that address the impact of common sources of tracking and target identification error in a network structure—common process noise and common priors. Additionally, we presented CI fusion, an approach that is not concerned with conditional independence assumption. Examples that relate back to the network-centric structures within a battlefield condition were also presented. Finally, we examined performance examples for analytic solutions for hierarchical fusion under conditions of no feedback, full feedback, and partial feedback. Network-centric applications of distributed fusion can be seen in Refs 35–38 involving multiairborne systems, applications to drug interdiction, enhanced radar management, and wireless *ad hoc* networks.

References

1. D. Alberts, J. Garstka, F. Stein, *Network Centric Warfare*, DoD Command and Control Research Program, Library of Congress Cataloging-in-Publication Data, October 2003.
2. Network Centric Warfare: Department of Defense Report to Congress, 31 July 2001.
3. L. Stotts, D. Honey, Gen (R) G. Luck, J. Allen, P. Corpac, Achieving Joint Battle Command in Network Centric Operations, *Proceedings of the SPIE*, Vol. 5441, pp. 1–12, Orlando, FL, July 2004.

4. Y. LaCerte, B. Rickenbach, Net-centric computing technologies for sensor-to-strike operations, *Proceedings of the SPIE,* Vol. 5441, pp. 142–150, Orlando, FL, July 2004.
5. E. Waltz, J. Llinas, *Multisensor Data Fusion,* Artech House, Norwood, MA, 1990.
6. D. Hall, S. McMullen, *Mathematical Techniques in Multisensor Data Fusion,* 2nd Ed., Artech House, Norwood, MA, USA, 2004.
7. D. Hall, J. Llinas (Eds.) *Handbook of Multisensor Data Fusion,* CRC Press, New York, NY, 2001.
8. Y. Bar-Shalom, T. E. Fortmann, *Tracking and Data Association,* Academic Press, Boston, MA, 1988.
9. Y. Bar-Shalom, W. Blair (Eds.) *Multitarget-Multisensor Tracking: Applications and Advances, Vol. III,* Artech House, Norwood, MA, 2000.
10. S. Blackman, R. Popoli, *Design and Analysis of Modern Tracking Systems,* Artech House, Norwood, MA, 1999.
11. C. Y. Chong, S. Mori, K. C. Chang, Distributed multitarget multisensor tracking, in *Multitarget Multisensor Tracking: Advanced Applications,* Y. Bar-Shalom (Ed.) Artech House, Norwood, MA, pp. 247–295, 1990.
12. S. Mori, K. C. Chang, C. Y. Chong, Performance analysis of optimal data association with applications to multiple target tracking, in *Multitarget Multisensor Tracking: Applications and Advances, Vol. II,* Y. Bar-Shalom (Ed.) Artech House, Norwood, MA, pp. 183–236, 1992.
13. S. Mori, C. Y. Chong, E. Tse, R. Wishner, Tracking and classifying targets without a prior identification, *IEEE Transactions on Automatic Control,* 31(10), 401–409, 1986.
14. K. C. Chang, C. Y. Chong, Y. Bar-Shalom, Joint probabilistic data association in distributed sensor networks, *IEEE Transactions on Automatic Control,* 31, 889–897, 1986
15. C. Y. Chong, S. Mori, K. C. Chang, W. Barker, Architectures and algorithms for track association and fusion, *Proceedings Second International Conference on Information Fusion,* Sunnyvale, USA, pp. 239–246, July 1999.
16. S. Mori, W. Barker, C. Y. Chong, K. C. Chang, Track association and track fusion with non-deterministic target dynamics, *IEEE Transactions Aerospace Electronic Systems,* 38(2), 659–668, 2002.
17. C. Y. Chong, Hierarchical estimation, in *Proceedings of 2nd MIT/ONR C3 Workshop,* Monterey, CA, July 1979.
18. C. Y. Chong, Problem characterization in tracking/fusion algorithm evaluation, in *Proceedings of the Third International Society of Information Fusion, Fusion, Vol. 1(10),* pp. MOC 2/26-MOC/ 21, Paris, France, 13 July 2000.
19. K. C. Chang, R. Saha, Y. Bar-Shalom, On optimal track-to-track fusion, *IEEE Transactions on Aerospace and Electronic Systems,* 33(4), 1271–1276, 1997.
20. C. Y. Chong, Distributed architectures for data fusion, in *Proceedings First International Conference on Multisource Multisensor Information Fusion '98,* Las Vegas, pp. 85–91, July 1998.
21. C. Y. Chong, S. Mori, Graphical models for nonlinear distributed estimation, *7th International Conference on Information Fusion,* Stockholm, Sweden, pp. 614–621, July 2004.
22. M. E. Liggins, II, C. Y. Chong, I. Kadar, M. G. Alford, V. Vannicola, S. Thomopoulos, Distributed fusion architectures and algorithms for target tracking, (Invited paper), in *Proceedings of the IEEE,* Vol. 85, Issue 1, pp. 95–107, January 1997.
23. K. C. Chang, R. K. Saha, Y. Bar-Shalom, M. Alford, Performance evaluation of multisensor track-to-track fusion, in *Proceedings of the 1996 IEEE/SICE/RSJ International Conference on Multisensor Fusion and Integration for Intelligent Systems,* pp. 627–631, 1996.
24. K. C. Chang, Z. Tian, R. Saha, Performance evaluation for track fusion with information fusion, in *Proceedings First International Conference on Multisource Multisensor Information Fusion '98,* Las Vegas, pp. 648–654, July 1998.
25. K. C. Chang, Evaluating hierarchical track fusion with information matrix filter, in *Proceedings of the Third International Society of Information Fusion, Fusion, 00,* Paris, France, Vol. 1(10), 13 July 2000.
26. K. C. Chang, S. Mori, Z. Tian, C. Y. Chong, MAP track fusion performance evaluation, in *Proceedings of the Fifth International Society of Information Fusion,* Annapolis, Vol. 1, pp. 512–519, July 2002.

27. S. Julier, J. Uhlmann, General decentralized data fusion with covariance intersection (CI), in *Handbook of Multisensor Data Fusion*, D. Hall and J. Llinas (Eds.) CRC Press, New York, NY, 2001.

28. T. Martin, K. C. Chang, A distributed data fusion approach for mobile ad hoc networks, in *Proceedings of the Eighth International Conference on Information Fusion (Fusion 2005)*, Vol. 2(8), pp. 25–28, July 2005.

29. Y. Bar-Shalom, On the track-to-track correlation problem, *IEEE Transactions on Automatic Control*, 26(2), 571–572, 1981.

30. X. R. Li, Y. Zhu, C. Han, Unified optimal linear estimation fusion—Part I: Unified models and fusion rules, in *Proceedings Third International Conference on Information Fusion 00*, Vol. 1(10), pp. MOC 2/10-MOC 2/17, Paris, France, 2000.

31. O. E. Drummond, Tracklets and a hybrid fusion with process noise, in *Proceedings of SPIE, Signal and Data Processing of Small Targets*, Vol. 3163, pp. 512–524, Orlando, FL, 1997.

32. Y. Zhu, *Multisensor Decision and Estimation Fusion*, Kluwer Academic Publishers, Norwell, MA, 2003.

33. C. Chong, S. Mori, Convex combination and covariance intersection algorithms in distributed fusion, in *ISIF Fusion 2001*, Montreal Quebec, Canada, August 2001.

34. K. C. Chang, Z. Tian, R. Saha, Performance evaluation of track fusion with information matrix filter, *IEEE Transactions on Aerospace Electronic Systems*, 38(2), 455–466, 2002.

35. M. Liggins, C. Y. Chong, Distributed multi-platform fusion for enhanced radar management, in *1997 IEEE National Radar Conference*, pp. 115–119, 1997.

36. C. Y. Chong, M. Liggins, Fusion technologies for drug interdiction, in *Proceedings of the 1994 IEEE Conference on Multisensor Fusion and Integration for Intelligent Systems*, pp. 435–441, Las Vegas, October 1994.

37. C. Y. Chong, S. P. Kumar, Sensor networks: Evolution, opportunities, and challenges, in *Proceedings of the IEEE*, Vol. 91, No. 8, pp. 1247–1256, August 2003.

38. C. Y. Chong, F. Zhao, S. Mori, S. Kumar, Distributed tracking in wireless ad hoc sensor networks, in *Proceedings of the 6th International Conference on Information Fusion*, pp. 431–438, 2003.

18

Foundations of Situation and Threat Assessment

Alan N. Steinberg

CONTENTS

18.1   Scope and Definitions

This chapter aims to explore the underlying principles of situation and threat assessment (STA) and discuss current approaches to automating these processes. As Lambert[1] notes

> Machine based higher-level fusion is acknowledged in the dominant JDL model for data fusion through the inclusion of amorphous "level 2" and "level 3" modules ..., but these higher levels are not yet supported by a standardised theoretical framework.

We hope to take some steps toward developing such a framework.

The past several years have seen much work and, we believe, some progress in these fields. There have been significant developments relating to situation assessment in techniques for characterizing, recognizing, analyzing, and projecting situations. These include developments in mathematical logic, ontology, cognitive science, knowledge discovery and knowledge management.

In contrast, level 3 data fusion—impact or threat assessment—is as yet a formally ill-defined and underdeveloped discipline, but one that is vitally important in today's military and intelligence applications. In this chapter, we examine possible rigorous definitions for level 3 fusion and discuss ways to extend the concepts and techniques of situation and threat assessment.

Issues in developing practical systems for STA derive ultimately from uncertainty in available evidence and in available models of such relationships and situations. Techniques for mitigating these problems are discussed.

18.1.1   Definition of Situation Assessment

Data fusion in general involves the use of multiple data—often from multiple sources—to estimate or predict the state of some aspect of reality. Among data fusion problems are those concerned with estimation/prediction of the state of one or more individuals; i.e., of entities whose states are treated as if they were independent of the states of any other entity. This assumption underlies many formulations of target recognition and tracking problems. These are the province of "level 1" in the well-known Joint Directors of Laboratories (JDL) Data Fusion Model.[2-4]

Other data fusion problems involve the use of context to infer entity states. Such contexts can include relationships among entities of interest and the structure of such relationships. The ability to exploit relational and situational contexts, of course, presupposes an ability to characterize and recognize relationships and situations.

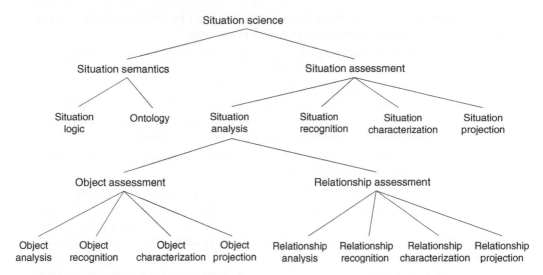

FIGURE 18.1
A taxonomy for situation science.

A situation is a partial world state, in some sense of "partial." Devlin (Ref. 5, p. 31, paraphrased) provides an attractive informal definition for "situation" as "a structured part of reality that is discriminated by some agent." *

Such structure can be characterized in terms of the states of constituent entities and relationships among them.†

As a working definition for *Situation Assessment*, we will use that of the JDL model revision (Chapter 2):

> Situation assessment is the estimation and prediction of structures of parts of reality (i.e. of the aggregation of relationships among entities and their implications for the states of the related entities).[4]

We propose a rough taxonomy of functions related to situation assessment (Figure 18.1). For want of a comprehensive term, we coin *situation science* to span situation assessment and semantics. We define the subordinate functions as follows:

1. *Situation semantics.* Defining situations in the abstract, for example, specifying the characteristics that define a situation as a political situation or a scary situation or, more specifically, as a tennis match or an epidemic. Two interrelated disciplines are *situation ontology* (the determination of the classes of entities involved in situations and their logical dependencies) and *situation logic* (the method for reasoning about such entities). In the terminology of Peircean semiotics, developing situation semantics involves abductive processes that determine general characteristics about types of situations and inductive processes that generalize these characteristics to establish an ontology of situations.[6]

2. *Situation assessment.* Estimating and predicting the structures of parts of reality (i.e., of the aggregation of relationships among entities and their implications for

---

* The provision for a discriminating agent appears to be otiose. To include it in the definition is similar to defining *sound* so as to force an answer to the question of whether a falling tree makes a sound when no one is around to hear. We shall allow that there are situations that no one has noticed.
† We formally define the terms *relation* and *relationship* in Section 18.2.4.1.

the states of the related entities). Situation assessment functions include situation recognition, characterization, analysis, and projection.

3. *Situation recognition.* Classifying situations (whether actual, real-world situations or hypothetical ones) as to situation type. Situation recognition—like signal and target recognition—involves primarily deductive processes.

4. *Situation characterization.* Estimating salient features of situations (i.e., of relations among entities and their implications for the states of the related entities) on the basis of received data. This involves primarily abductive processes.

5. *Situation analysis.* Estimating and predicting relations among entities and their implications for the states of the related entities.[7,8]

6. *Situation projection.* Determining the type and salient features of situations on the basis of contingent data; for example, inferring future situations based on projection of present data or hypothetical data ("what-if?") or inferring past or present situations based on hypothetical or counterfactual data. This involves primarily inductive processes.*

7. *Object assessment.* Estimating the presence and characteristics of individual entities in a situation. The decomposition of object assessment is analogous to that of situation assessment, involving the recognition (e.g., target recognition and combat identification), characterization (e.g., location and activity characterization), analysis (further decomposition), and projection (e.g., tracking) of individual entities.†

8. *Relationship assessment.* Estimating the presence and characteristics of relationships among entities in a situation. The decomposition is analogous to that of situation assessment, involving the recognition, characterization, analysis, and projection of relationships.‡

Situation assessment as a formal discipline has only recently received serious attention in the data fusion community, which has hitherto found it easier to both build and market target identification and tracking technologies.

The newfound interest in situation assessment reflects changes in both the marketplace and the technology base. The change in the former is largely motivated by the transition from a focus on traditional military problems to asymmetrical threat problems. The change in the latter involves developments in diverse aspects of "situation science," to include the fields of situation awareness (SAW),[9,10] belief propagation,[11,12] situation logic,[1,5,13] and machine learning.[14]

18.1.2 Definition of Threat Assessment

In plain terms, threat assessment involves assessing situations to determine whether detrimental events are likely to occur. According to the JDL Data Fusion Model, threat assessment is a level 3 data fusion process. Indeed, the original model[2] used "threat assessment" as the general name for level 3 fusion; indicative of the importance of that topic.

---

* Situation projection is often considered to be within the province of level 3 fusion. We will not agonize inordinately about such boundary disputes in this chapter, which are treated in Chapter 3.

† Object assessment is, of course, the label for level 1 fusion in the JDL model. Such assessment can involve inference both from measurement data, $p(x|Z)$, and from situation context $p(x|s)$.

‡ "Situation awareness" does not appear in this diagram. The correspondence with the three SAW levels in Endsley's model[9,10] is roughly as follows: Endsley's Perception ~ Situation Analysis; Comprehension ~ Situation Recognition and Discovery; Projection ~ Situation Projection.

In subsequent revisions,[3,4,15,16] the concept of level 3 has been broadened to that of impact assessment.*

Level 3 fusion has been the subject of diverse interpretations. Its distinction from level 2 fusion has variously been cast in temporal, ontological, and epistemic terms. Salerno[17] distinguishes situation assessment as being concerned with estimating current states of the world, whereas impact or threat assessment involves predicting future states.

In contrast, Lambert[18,19] sees the distinction not so much a temporal one, but an ontological one: concerning the types of entity being estimated. Situation assessment estimates world states, whereas impact or threat assessment estimates the utility of such states to some assumed agent (e.g., to "us"). In this interpretation, the products of levels 1–3 processing are

- *Level 1*. Representations of objects
- *Level 2*. Representations of relations among objects†
- *Level 3*. Representations of effects of relations among objects

Yet another interpretation, which we will incorporate into our definition, is neither temporal nor ontological, but epistemic: Situation assessment estimates states of situations directly from inference from observations of the situations, whereas impact or threat assessment estimates situational states indirectly from other observations.

That is to say, impact or threat assessment *projects* situational states from information external to these situations. The paradigm case is that of projecting, or predicting future situations from present and past information.‡

Projection from one estimated state to another—whether forward, backward, or lateral in time—is clearly an important topic for system development. So is the estimation of the utility of past, present, or future states.

Projection is, of course, an integral function in level 1 fusion in tracking the kinematic or other dynamic states of individual targets assumed to behave independently from one another. In levels 2 and 3, we are concerned with tracking the dynamic state of multiple targets when such independence cannot be assumed.§

The estimation and exploitation of relationships among multiple targets, the primary topic of situation assessment. It certainly plays a role in situational tracking and prediction as applicable to impact or threat assessment. In particular, impact assessment is generally conceived as a first-person process, that is, as supporting "our" planning by projecting situations to determine consequences as "our" course of action interacts with that of others. This is a common context for threat assessment:

- How will the adversary respond if I continue my present course of action?
- What will be the consequences of the resultant interaction?¶

---

* Threat (impact) assessment is defined in Ref. 2 as the "process of estimation and prediction of effects on situations of planned or estimated/predicted actions by the participants; to include interactions between action plans of multiple players (e.g., assessing susceptibilities and vulnerabilities to estimated/predicted threat actions given one's own planned actions)." Definition has been refined in Chapter 3.

† We would add "… and of structures comprised of such objects and relationships."

‡ This seems to be close to Lambert's position in Section 6 of Ref. 1.

§ The topic of multitarget tracking is, of course, actively being pursued using random set methods (Chapter 16 and Ref. 20). The key distinction is that in situation assessment the dependencies among entity states are the explicit basis for inferring about situational states and, conversely, situational states are used to infer states of component entities and their relationships.

¶ Ceruti et al.[21] note that, "[as] technology improves, the interface between Situation and Threat Assessment may become less distinct due to the larger number of possible interactions between these levels."

Estimation of cost or utility is integral to concerns about consequences. The estimation problem is not appreciably different if we are considering the consequences to us or to somebody else, but most threat assessment discussion is naturally focused on self-protection.

Rather than insisting on a specific definition, we may list the following functions that are relevant to impact assessment:

1. The estimation of *contingent* situations, that is, situations that are contingent upon events that are construed as undetermined. These can include
   a. Projection of future situations conditioned on the evolution of a present situation (as in target or situation tracking)
   b. Projection of past, present, or future situations conditioned on undetermined or hypothetical situations, to include counterfactual assessment, as performed in simulation, "what-if" exercises, historical fiction, or science fiction
2. The estimation of the value of some objective cost function for actual or contingent situations

For breadth of applicability in the present discussion, we apply the term *impact assessment* to all aspects of (1) and (2), while reserving "threat assessment" for some aspects of (1), namely, projection of future situations, as well as factor (2) as it concerns evaluation of the cost of future situations. This rather tedious definition seems to accord generally with the way *threat assessment* is used in practice.

We need to know what is meant by a *threat* ontologically. Let us distinguish the following threat aspects:

1. Threat acts (acts detrimental to "our" interests)
2. Threat agents (agents who may perform threat acts at some time)
3. Threatening agents (agents who may perform threat acts in the future)
4. Threatening situations (situations containing threatening agents and actions)
5. Threatened acts (threat acts that may be performed in the future)
6. Threatened situations (consequences of such acts: "impacts")

Threat assessment involves the prediction of (5) and (6). We perform threat assessment by assessing the capability and intent of potentially threatening agents (3), together with the opportunities for threat acts inherent in the present situation (4).*

18.1.3   Inference in Situation and Threat Assessment

STA—whether implemented by people, automatic processes, or some combination thereof—requires the capability to make inferences of the following types:

1. Inferring the presence and the states of entities on the basis of relationships in which they participate
2. Using entity attributes and relationships to infer additional relationships
3. Recognizing and characterizing extant situations
4. Predicting undetermined (e.g., future) situations

---

* According to the definitions of Section 18.2.4, an *event* is a type of situation. Further discussion of threat ontology is found in Sections 18.3 and 18.4.

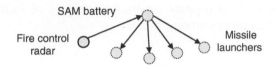

FIGURE 18.2
An example of situation assessment inferences. (From Steinberg, A.N., *Proceedings of the Seventh International Conference on Information Fusion*, Stockholm, 2004. With permission.)

For example, consider the problem of characterizing an air-defense threat (Figure 18.2). Assume that a single radar signal has been detected. On the basis of a stored model of radar signatures, it is recognized that this is a fire control radar of a certain type and it is in a target acquisition mode (level 1 inference). Furthermore, per the recognition model, it is found that radars of this type are strongly associated with a particular type of mobile surface-to-air missile (SAM) battery. Therefore, the presence and activity state of a SAM battery of the given type is inferred (level 2 inference type [1]). Furthermore, the recognition model indicates the expected composition of such a SAM battery to include four missile launchers of a certain type, plus ancillary entities (power generators, command/control vehicle, personnel, and so on). On this basis, these entities in turn are inferred to be present in the vicinity of the observed radar set. Command, control, and communications relationships among these elements can also be inferred (level 2 inference type [2]). Thus, the presence of a SAM battery in target acquisition mode can be construed to be a situation in itself or part of a larger, partly known situation, for example, an enemy air-defense network (level 2 inference type [3]).

All of this can be inferred, with various degrees of certainty, from a single piece of data: the intercept of the radar signal. This can be done, as the example illustrates, on the basis of a *recognition model* of entities, their relationships, the indicators of entity states and relationships, and of expectations for related entities and of additional relationships.

As in level 1 inferencing (i.e., with one-place relations), we can write production rules based on logical, semantic, causal, societal, or material (etc.) relationships among entities and aggregates. Patterns of such inferences are provided in the following subsections.

18.1.3.1 Inferences of Relationships and Entity States

18.1.3.1.1 Patterns of Inference

Estimates of relationship and entity states are used to infer other relationship and entity states. Characteristic inference patterns include the following:
L1 → L1 Deduction:

$$p(R(x)|Q_1(x), s) = \frac{p(Q(x)|R(x), s)p(R(x)|s)}{p(Q(x)|s))} \tag{18.1}$$

for example, estimation of the probability of a single target state $R(x)$ from associated measurements $Q(x)$ or prediction of state $R(x)$ from prior state $Q(x)$ in situation s.* We label this an "L1 → L1 Deduction," as inference is from one level–one (i.e., attributive) state estimate to another. Such inferences are deductive in that the inference is based on the likelihoods and priors shown.

---

* The predicate order superscripts, $R^{(1)}$, etc., have been suppressed for legibility.

Relationships between pairs of entities, or among n-tuples of entities, can similarly be inferred from other relationships among them:

L2 → L2 Deduction:*

$$p(R(x_1,\ldots,x_m)|Q(y_1,\ldots,y_n),s) = \frac{p(Q(y_1,\ldots,x_n)|R(x_1,\ldots,x_m),s)p(R(x_1,\ldots,x_m)|s)}{p(Q(y_1,\ldots,y_n)|s)} \quad (18.2)$$

where $\forall i(x_i \in \{y_1, \ldots, y_n\})$.

Cross-level inference patterns can include the following:

L1 → L2 Deduction:

$$p(R(x_1,\ldots,x_m)|Q_1(x_1),\ldots,Q_m(x_m),s)$$

$$= \frac{p(Q_1(x_1),\ldots,Q_m(x_m)|R(x_1,\ldots,x_m),s)p(R(x_1,\ldots,x_m))|s)}{p(Q_1(x_1),\ldots,Q_m(x_m)|s)} \quad (18.3)$$

L1 → L2 Induction:

$$p(\exists x_2[R(x_1,x_2)]|Q(x_1),s) = \frac{p(Q(x_1)|s)|\exists x_2[R(x_1,x_2)])}{p(Q(x_1),s)} \quad (18.4)$$

L2 → L2 Induction:†

$$p(\exists x_{m+1}[R(x_1,\ldots,x_m,x_{m+1})]|Q(x_1),\ldots,Q(x_m),s)$$

$$= \frac{p(Q(x_1),\ldots,Q(x_m)|s)|\exists x_{m+1}[R(x_1,\ldots,x_m,x_{m+1})])}{p(Q(x_1),\ldots,Q(x_m),s)} \quad (18.5)$$

L2 → L1 Deduction:

$$p(R(x_i)|Q(x_1,\ldots,x_m),s) = \frac{p(Q(x_1,\ldots,x_m),s)|R(x_1))}{p(Q(x_1,\ldots,x_m),s)} \quad (18.6)$$

where $1 \leq i \leq m$.

18.1.3.1.2   Relational States

We find that reasoning about attributes, relations, and situations is facilitated if these concepts are "reified," that is to say, attributes, relations, and situations are admitted as entities in the ontology. Attributes of entities are conveniently treated as one-place relationships.

By explicitly including attributes and relationships into the ontology one is able to reason about such abstractions without the definitional baggage or extensional issues in reductionist formulations (e.g., in which uncertainties in the truth of a proposition involving multiple entities are represented as distributions of multitarget states).‡

---

* This is also the pattern for L2 → L2, L2 → L3, L3 → L2, or L3 → L3 Deduction.

† Also, L2 → L2, L2 → L3, L3 → L2, or L3 → L3 Induction.

‡ A multitarget state of the sort of interest in situation or threat assessment cannot in general be inferred from a set of single-target states $X = \{x_1, \ldots, x_n\}$. For example, from P = "Menelaus is healthy, wealthy and wise," we cannot be expected to infer anything like Q = "Menelaus is married to Helen" (however, in such cases we can sometimes infer $Q \rightarrow \sim P$). It does appear feasible, however, to restrict our ontology of relationships to those of finite and determinate order; that is, any given relation maps from a space of specific finite dimensionality $r{:}X(n) \rightarrow Y$.

We also find it useful to distinguish between an entity X and its state x (say, at a particular time), writing, for example, $X = x$. Here we use capital letters for entity variables, and small letters for corresponding state variables. Thus, we take "$p(x)$" as an abbreviation for "$p(X = x)$" and "$p(x_k)$" for "$p(X = x|k)$".*

A relation is a mapping from n-tuples of entities to a relational state r:

$$R^{(m)} : X_1 \times \cdots \times X_m \to \mathbf{R} \tag{18.7}$$

Dependencies can be given by such expressions as

$$p_{xy}(r) = p(R(x,y) = r) \tag{18.8}$$

$$p_{xy}(r) = \sum_x \sum_y p_{xy}(r)p(x,y) = \sum_x \sum_y p_{xy}(r)p(x|y)p(y) \tag{18.9}$$

Equations 18.1 through 18.6 can be rewritten with this notation in the obvious manner.

18.1.3.2 Inferring Situations

Level 2 inferences have direct analogy to those at level 1. Situation recognition is a problem akin to target recognition. Situation tracking is akin to target tracking.[1]

18.1.3.2.1 Situation Recognition

A situation can imply and can be implied by the states and relationships of constituent entities. The disposition of players and the configuration of the playing field are indicators that the situation is a baseball game, a bullfight, a chess match, an infantry skirmish, an algebra class, or a ballet recital:

$$\exists x_1, \ldots, \exists x_n [R^{(n)}(x_1, \ldots, x_n)] \Rightarrow s \tag{18.10}$$

Often situational inferences can be given in the form of Boolean combinations of quantified expressions:

$$\exists x_1, \ldots, \exists x_n [R_1^{(m_1)}(y_{1_1}, \ldots, y_{m_1}) \ \& \ \ldots \ \& \ R_k^{(m_k)}(y_{1_k}, \ldots, y_{m_k})], $$
$$y_{i_j} \in \{x_1, \ldots x_n\}, 1 \le j \le k, 1 \le i_j \le m_j \tag{18.11}$$

To be sure, we are generally in short supply of situational ontologies that would enable us to write such rules. Issues of ontology science and ontology engineering will be discussed in Section 18.3.

---

* It can be argued that only some entities—called *continuants* in the SNePS ontology discussed in Section 18.3.2.2—persist through state changes.

18.1.3.2.2   Situation Tracking

We may distinguish the following types of dynamic models:

1. *Independent target dynamics.* Each target's motion is assumed not to be affected by that of any other entity; so, multitarget prior probability density functions (pdfs) are simple products of the single target pdfs.

2. *Context-sensitive individual target dynamics.* At each time-step, a particular target x responds to the current situation, that is, to the states of other entities. These may be dynamic but are assumed not to be affected by the state of x. An example is an aircraft flying to avoid a thunderstorm or a mountain.

3. *Interacting multiple targets.* At each time-step, multiple entities respond to the current situation, which may be affected by the possibly dynamic state of other entities. An example is an aerial dogfight.*

Analogous dynamic models are possible for tracking the evolution of situations.

4. *Independent component dynamics.* The situation contains substructures (call them *component situations*) that are assumed to evolve independent from one another such that the situational prior pdfs are simple products of the component situation pdfs. An example might be the independent evolution of wings in insects and birds.

5. *Context-sensitive individual component motion.* At each time-step, a component situation x responds to the current situation, that is, to the states of other entities and component situations. These may be dynamic but are assumed not to be affected by the state of x. An example is an aircraft formation flying to avoid a thunderstorm or a mountain.

6. *Interacting multiple components.* At each time-step, multiple component situations respond to the current situation, which may be affected by the possibly dynamic state of other entities. An example is the interplay of various components of the world economy.

Situation tracking, like target tracking, is far simpler where independent dynamics can be assumed: case (4) or (5). We expect that recent advances in general multitarget tracking will facilitate corresponding advances in general situation tracking.

---

* Tracking targets with type (1) dynamics is clearly a level 1 (i.e., independent-target estimation) fusion problem. Type (2) and (3) dynamics are often encountered, but are generally treated using independent-target trackers; perhaps with a context-dependent selection of motion models, but assuming independence among target tracks. Type (3) cases, at least, suggest the need for trackers that explicitly model multitarget interactions. We may similarly distinguish the following types of measurement models:

1. *Independent measurements.* Measurements of each target are not affected by those of any other entity; so multitarget posterior pdfs are simple products of the single target posterior pdfs.
2. *Context-sensitive multitarget measurements.* Measurements of one target may be affected by the state of other entities as in cases of additive signatures (e.g., multiple targets in a pixel), or occluded or shadowed targets. Other cases involve induced effects (e.g., bistatic illumination, electromagnetic interference, or disruption of the observing medium).

The absent category, interacting multiple measurements, is actually a hybrid of the listed categories, in which entities affect one another's state and thereby affect measurements.

18.1.4 Issues in Situation and Threat Assessment

The relative difficulty of the higher-level situation and impact/threat assessment problems can largely be attributed to the following three factors:

a. *Weak ontological constraints on relevant evidence.* The types of evidence relevant to threat assessment problems can be diverse and can contribute to inferences in unexpected ways. This is why much of intelligence analysis—like detective work—is opportunistic, *ad hoc*, and difficult to codify in a systematic manner.

b. *Weak spatio-temporal constraints on relevant evidence.* Evidence relevant to a level 1 estimation problem (e.g., target recognition or tracking) can be assumed to be contained within a small spatio-temporal volume, generally limited by kinematic or thermodynamic constraints. In contrast, many STA problems can involve evidence that is widespread in space and time, with no easily defined constrains.

c. *Weakly-modeled causality.* Threat assessment often involves inference of human intent and behavior, both as individuals and as social groups. Such inference is basic not only to predicting future events (e.g., attack indications and warning) but also in understanding current or past activity. Needless to say, our models of human intent, planning, and execution are incomplete and fragile to a great extent as compared with the physical models used in target recognition or tracking.[21]

Table 18.1 contrasts these situation/threat assessment factors, which are characteristic of most of today's national security concerns, with the classical "level 1" fusion problems that dominated military concerns of the twentieth century.[22]

Clearly, the problem of recognizing and predicting terrorist attacks is much different from that of recognizing or predicting the disposition and activities of battlefield equipment. Often the key indicators of potential, imminent, or current threat situations are in the relationships among people and equipment that are not in themselves distinguishable from common, nonthreatening entities.

We may summarize the difficulties for STA in such an application as presented in Table 18.2.

TABLE 18.1

Yesterday and Today's Data Fusion Problems

Problem characteristics	Twentieth Century Level 1 fusion: target recognition and tracking	Twenty-First Century Level 2/3 fusion: situation/ threat recognition and tracking
(a) Problem dimensionality	Low: few relevant data types and relations	High: many relevant data types and relations
(b) Problem span	Spatially localized: hypothesis generation via validation gate	Spatially distributed: hypothesis generation via situation/behavior model
(c) Required "target" models	Physical: signature, kinematics models	Human/group behavior: for example, coordination/collaboration, perception, value, influence models

Source: Steinberg, A.N., *Threat Assessment Issues and Methods*, Tutorial presented at Sensor Fusion Conference, Marcus-Evans, Washington, DC, December 2006. With permission.

TABLE 18.2

Data Fusion Issues for Situation Threat Assessment

Issue	System Implications
Unanticipated situations and threats	Reduced dependence on *a priori* knowledge
Consistent management of uncertainty	Comprehensive model of uncertainty
	Generalized belief propagation
Complex and subtle relationships	Efficient search methods
	Managed hypothesis structure
Inferring human and group behavior	Behavior model
Learning and managing models	Integration of abductive and inductive processes (machine learning for ontology management and for situation discovery and projection)
Actionable intelligence	Adaptive to information needs

18.2   Models of Situation Assessment

As noted in the preceding section, there are several approaches to represent and reason about situations. These include

- The JDL Data Fusion Model, as discussed in Refs 2–4, 15, and 16 and in Chapter 3
- Endsley's model of the cognitive aspects of SAW in human beings[9,10]
- The situation theory and logic of Barwise and Perry,[13] and Devlin[5]
- The fusion for SAW initiative of the Australian Defence Science and Technology Organisation (DSTO), led by Lambert[1,18,19,23]

These are discussed in turn in the following subsections.

18.2.1   Situation Assessment in the JDL Data Fusion Model

The definition of the higher levels of data fusion in the JDL model and related excursions has been subject to diverse interpretations.

In early versions of the model,[2] the distinction among levels 1, 2, and 3 is based on *functionality*, that is, types of processes performed:

- *Level 1.* Target *tracking and identification*
- *Level 2. Understanding* situations
- *Level 3. Predicting* and *warning* about threats

In Ref. 3, a revision was suggested in which the levels were distinguished on the basis of *ontology*, that is, the class of targets of estimation:

- *Level 1.* Estimating the states of *individual* objects
- *Level 2.* Estimating the states of *ensembles* of objects
- *Level 3.* Estimating impacts (i.e., costs) of situations

More recently,[4,15,16] we have proposed the split on the basis of *epistemology*, that is, the relationship between an agent (person or system) and the world state that the agent is estimating:

- Level 1 involves estimation of entities *considered as individuals*
- Levels 2 and 3 involve estimation of entities *considered as structures* (termed *situations*)
- Level 3 is distinguished from levels 1 and 2 on the basis of relationship between the estimated states and the information used to estimate that state. Specifically, the distinction is between states that are *observed* and states that are *projected*. By projection, we mean inferences from one set of observed objects and activities to another set that occurs elsewhere in space–time. The paradigm case is in projecting current information to estimate future states

There is no reason to think that there is a single best model for data fusion. As in all scientific disciplines, investigations are made for taxonomies and other abstract representations in data fusion to identify problem space to facilitate understanding of common local problems and of common appropriate solutions within that space.

The fact that functional, ontological, and epistemological criteria all tend to partition the practical problem space in roughly the same way suggests that there may be some ideal categorization of the data fusion problem space. We doubt it.

In the present recommended version of the JDL model (Chapter 3), situation assessment—level 2 in the model—is defined as:

> the estimation and prediction of structures of parts of reality (i.e., of the aggregation of relationships among entities and their implications for the states of the related entities).

This definition follows directly from

1. The definition of *situation* as *a structured part of reality*, due to Devlin[5]
2. The dictionary definition of *structure* as *the aggregate of elements of an entity in their relationships to each other*[24]
3. The JDL model definition of *data fusion* as *combining multiple data to estimate or predict some aspect of the world*[3]

In particular, a given entity—whether a signal, physical object, aggregate or structure—can often be viewed either (1) as an individual whose attributes, characteristics, and behaviors are of interest or (2) as an assemblage of components whose interrelations are of interest. From the former point of view, discrete physical objects (the "targets" of level 1 data fusion) are components of a situation. From the latter point of view, the targets are themselves situations, that is, contexts for the analysis of component elements and their relationships.

According to this definition, then, situation assessment involves the following functions:

- Inferring relationships among (and within) entities
- Using inferred relationships to infer entity attributes

This includes inferences concerning attributes and relationships (a) of entities that are elements of a situation (e.g., refining the estimate of the existence and state variables of an entity on the basis of its inferred relations) and (b) of the entities that are themselves situations (e.g., in recognizing a situation to be of a certain type). According to this definition, situation assessment also includes inferring and exploiting relationships *among* situations, for example, predicting a future situation on the basis of estimates of current and historical situations.

18.2.2   Endsley's Model for Situation Awareness

As defined by Endsley,[10] SAW is defined as *the perception of the elements of the environment, the comprehension of their meaning (understanding), and the projection (prediction) of their status in order to enable decision superiority.*

Endsley's SA model is depicted in Figure 18.3, showing the three levels of *perception, comprehension,* and *projection.* These are construed as aspects or "levels" of mental representation:

- *Level 1. Perception* provides information about the presence, characteristics, and activities of elements in the environment.

- *Level 2. Comprehension* encompasses the combination, interpretation, storage, and retention of information, yielding an organized representation of the current situation by determining the significance of objects and events.

- *Level 3. Projection* involves forecasting future events.

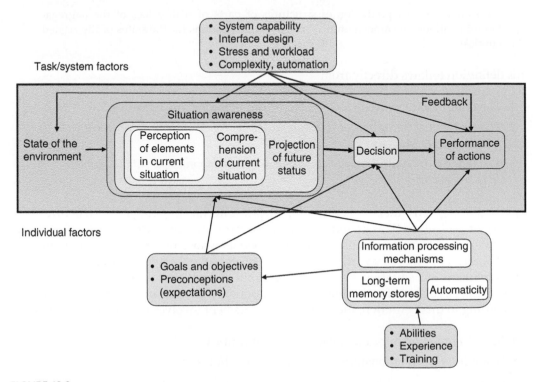

FIGURE 18.3
Endsley's situation awareness model. (From Endsley, M.R., *Situation Awareness Analysis and Measurement,* Lawrence Erlbaum Associates Inc., Mahwah, NJ, 2000. With permission.)

McGuinness and Foy[25] add a fourth level, which they call *resolution*. Resolution provides awareness of the best course of action to follow to achieve a desired outcome to the situation.

The JDL model—including the "situation assessment" level—can be compared and contrasted with this SAW model. First of all, it must be noted that, in these models, assessment is a process; awareness is a product of a process. Specifically, awareness is *a mental state* that can be attained by a process of assessment or analysis.

In an automated situation assessment system for which we are reluctant to ascribe mental states, the product is an *informational state*. It is not an awareness of an actual situation, but a representation of one. The *performance* of a situation assessment system, then, is measured by the fidelity of the representation. The *effectiveness* of a situation assessment system is measured by the marginal utility of responses made on the basis of such representations.

In Table 18.3, we compare in a very notional sense the levels of the JDL Data Fusion Model with those of Endsley's SAW model as augmented by McGuinness and Foy. Both models are subject to various interpretations. We chose interpretations that emphasize the similarities between the models. This is done not only to increase harmony in the world but also to suggest that there is a commonality in concepts and issues that both the fusion and SAW communities are struggling to understand.

We have added a column showing the elements of our taxonomy of situation science. We show JDL levels 0 and 1 as possible beneficiaries of situation analysis processing. Situational analysis can generate or refine estimates of the presence and states of signals/features and objects on the basis of context.

Lambert[23] similarly argues that situation awareness as defined by Endsley is equivalent to levels 1–3 of the JDL model *as performed by people*. This is illustrated in Figure 18.4, which suggests the need for an integrated approach to the assessment of objects, situations,

TABLE 18.3

Comparison Between Joint Directors of Laboratories and Situation Awareness "Levels"

JDL Data Fusion Level[3,4]	Interpretation	Situation Awareness Level[9,10]	Interpretation	Situation Science Taxonomy
0. Signal/feature assessment	Characterization of patterns in received data	—	—	Part of situation analysis
1. Object assessment	Characterization of individual situation components (objects)	1. Perception	Characterization of individual situation components (objects)	Part of situation analysis
2. Situation assessment	Characterization of relations and patterns among these components	2. Comprehension	Characterization of relations and patterns among these components and the implications of such relations and patterns	Situation assessment: Situation analysis, recognition and discovery
3. Impact assessment	Characterization of conditional situations	3. Projection	Characterization of conditional situations	Situation projection
4. Process assessment (and refinement)	Characterization (and improvement) of system's operating state and performance	4. Resolution[25]	Characterization of system's preferred course of action	—

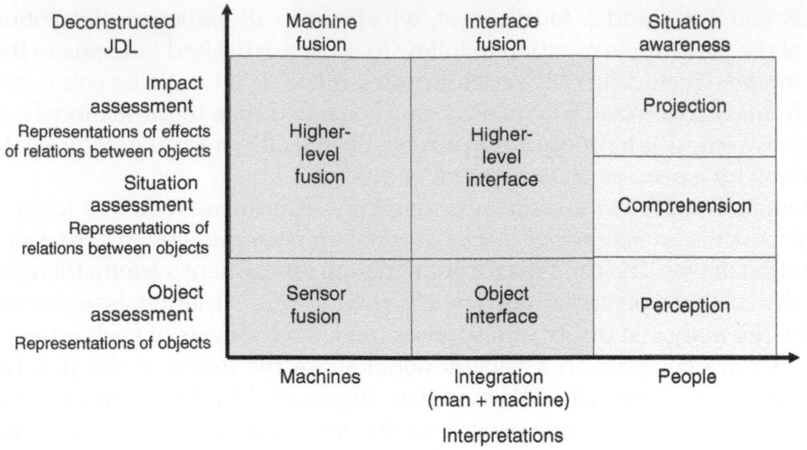

Deconstructed JDL	Machine fusion	Interface fusion	Situation awareness
Impact assessment Representations of effects of relations between objects	Higher-level fusion	Higher-level interface	Projection
Situation assessment Representations of relations between objects			Comprehension
Object assessment Representations of objects	Sensor fusion	Object interface	Perception
	Machines	Integration (man + machine)	People

Interpretations

FIGURE 18.4
Relationship between machine and human fusion processes in achieving situation awareness. (Adapted from Lambert, D.A., *Proceedings of the Tenth International Conference on Information Fusion*, Quebec, 2007. Copyright Commonwealth of Australia. With permission.)

and impacts as performed by machines and people. Responsibility for performing these functions will no doubt evolve as technology improves. In many applications, functional allocation is best performed dynamically on the basis of available resources.

18.2.3 Salerno's Model for Higher-Level Fusion

Salerno et al. at the U.S. Air Force Research Laboratory Information Fusion Branch have developed and refined models for data fusion, with an emphasis on higher-level fusion.[17,26,27] These developments incorporate ideas from the JDL model, from Endsley's SAW model, and from other activities.

We understand Salerno as distinguishing data fusion levels on the basis of *functionality* (the types of assessment performed[4]) rather than on the basis of *ontology* (types of objects estimated). Thus, level 1 can be the province not merely of physical objects, but of concepts, events (e.g., a telephone call), groups, and so on.

Situation assessment (level 2) in this system is the assessment of a situation at a given point in time.

In this scheme, level 1 fusion involves such functions as

- Existence and size analysis (How many?)
- Identity analysis (What/who?)
- Kinematics analysis (Where/when?)

Level 2 involves such functions as

- Behavior analysis (What is the object doing?)
- Activity level analysis (Build up, draw down?)
- Intent analysis (Why?)
- Salience analysis (What makes it important?)
- Capability/capacity analysis (What could they/it do?)*

---

* This list derives from Bossé, Roy and Wark,[28] who however describe them as Situation Analysis products.

Following Endsley, the next step to be performed is projection (which Salerno specifically distinguishes from the JDL level 2/3 processes). Thus, level 3 functions forecast or project the current situation and threat into the future.

In other words, the difference between level 2 and 3 in this scheme is a matter of situation time relative to the assessment time.

18.2.4 Situation Theory and Logic

Situation theory and logic were developed by Barwise and Perry[13] and refined by Devlin.[5] We summarize some of the key concepts.

Situation theory in effect provides a formalism for Lambert's concept that "situation assessment presents an understanding of the world in terms of *situations* expressed as *sets of statements about the world.*"[1] The prime construct of situation logic, the *infon*, is a piece of information, which can be expressed as a statement.*

Note that an infon is not the same as a statement, proposition, or assertion. A statement is a product of human activity, whereas an infon is information that may apply to the world or to some specific situation (including factual and counterfactual situations), independent of people's assertions. Infons that so apply in a given situation are *facts* about that situation. For example, in *Hamlet* it is a fact that Hamlet killed Polonius.

A *situation* is defined as a set of such pieces of information, expressed by a corresponding set of statements. A *real* situation is a subset of reality: a real situation is one in which all pieces of information are *facts* in the real world and the corresponding statements are *true* in the real world.

As in Section 18.1.3.1.2, situation theory "reifies" attributes, relations, and situations. In other words, they are admitted as entities in the ontology. Attributes of entities are treated as one-place relationships.

18.2.4.1 Classical (Deterministic) Situation Logic

Situation logic, as developed by Devlin,[5] employs a second-order predicate calculus, related to the combinatorial logic of Curry and Feys.[29]

The latter represent first-order propositions $R(x_1, ..., x_n)$, involving an m-place predicate "R," $m \geq n$, as second-order propositions $Applies(r, x_1, ..., x_n)$, employing a single second-order predicate "Applies." This abstraction amounts to the reification of relations r corresponding intensionally to predicates R. There being but one second-order predicate, we can often abbreviate "$Applies(r, x_1, ..., x_n)$" as "$(r, x_1, ..., x_n)$."

This is the basis of Devlin's notion of an infon. Under this formulation,[5] an infon has the form

$$\sigma = (r, x_1, ..., x_n, h, k, p) \tag{18.12}$$

for an m-place relation r, entities x_i, location h, and time k, $1 \leq i \leq n \leq m$.†

---

* The term *infon* was coined on analogy with "electron" and "photon," suggesting a discrete "particle"—though not necessarily an elementary, indivisible particle—of information. Infons need not be primitives of situation logic; Devlin (Ref. 5, p. 47) defines an infon to be an equivalence class of a pair ⟨R, C⟩ of a configuration R and a constraint C.

† The provision "$1 \leq i \leq n \leq m$" in the definition allows for infons with free variables as well as for one-place relations (i.e., attributes).

The term p is a polarity, or applicability value, $p \in \{0, 1\}$. We may read "$(r, x_1, ..., x_n, h, k, 1)$" as "relation r applies to the n-tuple of entities $\langle x_1, ..., x_n \rangle$ at location h and time k." Similarly, "$(r, x_1, ..., x_n, h, k, 0)$" can be read "relation r doesn't apply ...."

It will be convenient to make a distinction between relations, construed to be abstract (e.g., marriage), and relationships, which are anchored to sets of referents within a situational context (e.g., Anthony's marriage with Cleopatra or Hamlet's marriage with Ophelia). As the latter indicates, such contexts are not necessarily factual, even within an assumed fictional context.

A relationship is an instantiation of a relation. We formally define a relationship as an n-tuple $\langle r(m), x_1, ..., x_n \rangle$, $n \leq m$, comprising a relation and one or more entities so related. An infon is in effect a relationship concatenated with a polarity and spatio-temporal constraints.

An important feature of this formulation is the fact that infons can be arguments of other infons. This allows the expression of propositional attitudes, fictions, hypothetical and other embedded contexts. For example, consider an infon of the form $\sigma_1 = (believes, x, \sigma_2, h, k, 1)$ to be interpreted as "At place h and time k, x believes that σ_2," where σ_2 is another infon. Similar nested infons can be created with such propositional attitudinal predicates as $R = perceives, hypothesizes, suspects, doubts that, wonders whether$, and so on, or such compounds as "At place h and time k, x_1 asked x_2 whether x_3 reported to x_4 that x_5 believes that σ_2." In this way, the representational scheme of situation logic can be used to characterize and rationalize counterfactual and hypothetical situations, their relations to one another, and their relations to reality.

The second-order formulation also permits inferencing concerning propositions other than assertions. For example, a first-order interrogative "Is it true that σ?" can be restated as the assertion "I ask whether σ" or, in infon notation, "$(ask\ whether, I, \sigma, h_2, k_2, p)$." Similarly, for other modalities, replacing the relation "ask whether" with "demand that," "believe that," "fear that," "pretend that," and so on.[*]

Barwise and Perry[13] and Devlin et al.[4] broaden the expressive power of the second-order predicate calculus by a further admission into the working ontology. The new entities are our friends, *situations*. An operator "$\models$" expresses the notion of contextual applicability, so that "$s \models \sigma$" can be read as "situation s supports σ" or "σ is true in situation s." This extension allows consistent representation of factual, conditional, hypothetical, and estimated information.[†]

Devlin defines situations as sets of infons. As with infons, a situation may be anchored and factual in various larger contexts. For example, the situation of Hamlet's involvement with Ophelia's family is anchored and factual within the context of the situation that is Shakespeare's *Hamlet*, but not necessarily outside that context in the world at large.

Like infons, situations may be nested recursively to characterize, for example, my beliefs about your beliefs or my data fusion system's estimate of the product of your data fusion system. One can use such constructs to reason about such perceptual states as one's adversary's state of knowledge and his belief about present and future world state, for example, our estimate of the enemy's estimate of the outcome of his available courses of action. In this way, the scheme supports an operational net assessment framework allowing reasoning simultaneously from multiple perspectives.[31]

---

[*] In our formulation, such propositional attitudes (as expressed in "x believes that σ" or "x wonders whether σ") are cognitive relationships r and therefore can appear either as the initial argument of an infon or as one of the succeeding n-tuple of arguments.

[†] The operator "$\models$" of situation logic operates on units of information, not units of syntax and is therefore not the same as the Tarski's truth operator.[30] An infonic statement has the form "$s \models \Sigma$," where s is a situation and Σ is a set of infons (Ref. 5., pp. 62f).

This formulation supports our epistemological construal of the JDL Data Fusion Model (Section 18.2.1). Depending on the way information is used, virtually any entity may be treated either as an individual or as a situation. For example, an automobile may be discussed and reasoned about as a single individual or as an assembly of constituent parts. The differentiation of parts is also subject to choices: we may disassemble the automobile into a handful of major assemblies (engine, frame, body, etc.) or into a large number of widgets or into a huge assembly of atoms.

We distinguish, in a natural way, between real situations (e.g., the Battle of Midway) and abstract situations or situation types (e.g., naval battle). A *real situation* is a set of *facts*. An *abstract situation* is equivalent to a set of statements, some or all of which may be false.*

Real-world situations can be classified in the same way as we classify real-world individual entities. Indeed, any entity is—by definition—a kind of situation.

18.2.4.2 Dealing with Uncertainty

Classical situation logic provides no specific mechanism to represent and process uncertainty. STA requires means to deal with uncertainties both in received data (epistemic uncertainty) and in the models we use to interpret such data in relationship and situation recognition (ontological uncertainty).

It is crucial to distinguish among the types of uncertainty that may be involved in ontological uncertainty.†

In particular, a distinction needs to be drawn between

- *Model fidelity.* A predictive model may be uncertain in the sense that the underlying processes to be modeled are not well understood and, therefore, the model might not faithfully represent or predict the intended phenomena. Models of human behavior suffer from this sort of inadequacy. Probabilistic and evidential methods apply to represent such types of uncertainty.[37]

- *Model precision.* The semantics of a model might not be clearly defined, though the subject matter may be well understood. Wittgenstein's well-known problem concerning the difficulty of defining the concept of *game* is a case in point. Fuzzy logic was developed to handle such types of uncertainty.[38]

As suggested, probabilistic, evidential, and fuzzy logic, as well as various *ad hoc* methods, can be and have been employed in representing epistemic and ontological uncertainty. These methods inherit all the virtues and drawbacks of their respective formalisms: the theoretical grounding and verifiability of probability theory, with the attendant difficulty of obtaining the required statistics, and the easily-obtained and robust but *ad hoc* uncertainty assignments of evidential or fuzzy systems.

---

* We distinguish between abstraction and generality. Abstract situations can be more or less general, as can concrete (or fully anchored) situations. Generality is a function of their information content (roughly, the number of independent infons). Consider such abstract situations as conflict, warfare, space warfare, interstellar warfare, the war in Star Wars. These are, once again, to be contrasted with situations that are completely anchored in fact—for example, the Battle of Actium (as it actually occurred)—and hybrids, which are partially anchored in fact, such as the Battle of Actium as depicted by Shakespeare.

† A discussion of the diverse concepts relating to uncertainty is given in Ref. 8, including the models of Bronner,[32] Smithson,[33] Krause and Clark,[34] Bouchon-Meunier and Nguyen,[35] Klir,[36] and Smets.[37]

18.2.4.2.1　Fuzzy Approaches

Some situations can be crisply defined. An example is a chess match, of which the constituent entities and their relevant attributes and relationships are explicitly bounded. Other situations may have fuzzy boundaries. Fuzziness is present in both abstract situation types (e.g., the concepts *economic recession* or *naval battle*) and of concrete situations such as we encounter in the real world (e.g., the 1930s, the Battle of Midway). In such cases, it is impossible to define specific boundary conditions. Both abstract and concrete situations can be characterized via fuzzy membership functions on infon applicability: $\mu_\Sigma(\sigma) \in [0, 1]$, where $\Sigma = \{\sigma | s \models \sigma\}$.

Fuzzy methods have been applied to the problem of semantic mapping, providing a common representational framework for combining information across information sources and user communities.[39]

Research in the use of fuzzy logic techniques for STA has been conducted under the sponsorship of the Fusion Technology Branch of the U.S. Air Force Research Laboratory. The goal was to recognize enemy courses of action and to infer intent and objectives. Both input measurements and production rules are "fuzzified" to capture the attendant uncertainty or semantic variability. The resulting fuzzy relationships are "defuzzified" to support discrete decisions.[40]

18.2.4.2.2　Probabilistic Approaches

Classical situation logic, as described in the preceding section, is a monotonic logic: one in which assertions are not subject to revision as evidence is accrued. It will therefore need to be adapted to nonmonotonic reasoning as required to deal with partial and uncertain evidence. Such "evidence" can take the form of noisy or biased empirical data, whether from electronic sensors or from human observations and inferences. Evidence in the form of *a priori* models—for example, physical or behavior models of targets, or organizational or behavior models of social networks—can also be incomplete, noisy, and biased.

Belief networks (including, but not limited to, Bayesian belief networks) provide mechanisms to reason about relationships and the structure of situations as means for inferring random states of situations.

In Ref. 15 the author proposes modifying the infon of "classical" (i.e., monotonic) situation logic, by replacing the deterministic polarity term p with a probability term. Such a *probabilistic infon* can be used to express uncertainty in sensor measurement reports, track reports, prior models, and fused situation estimates alike. Each such probabilistic infon $\sigma = (r, x_1, ..., x_n, h, k, p)$ is a second-order expression of a *relationship*, stated in terms of the probability p that a relation r applies to an n-tuple of entities $\langle x_1, ..., x_n \rangle$ at some place and time.

The expressive power of the infon (i.e., that of a second-order predicate calculus) can be employed to enhance the expressive capacity of a *probabilistic ontology*.

There has been active development of probabilistic ontologies to address the issues of semantic consistency and mapping.[41–43] Traditional ontological formalisms make no provision for uncertainness in representations.* This limits both the ability to ensure robustness and semantic consistency of developed ontologies. In addition, as noted in Ref. 43, besides the issues of incomplete and uncertain modeling of various aspects of the world, there is a problem of semantic mapping between heterogeneous ontologies that have been developed for specific subdomains of the problem.

A formalism for probabilistic ontology, named PR-OWL has been developed by Costa et al.[41,42] They define probabilistic ontology as follows.

---

* We discuss some major ontology formalisms for STA in Section 18.3.

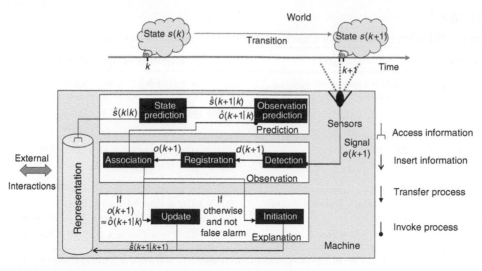

FIGURE 18.5
The STDF model. (From Lambert, D., *Inform. Fusion*, 2007. Copyright Commonwealth of Australia. With permission.)

A probabilistic ontology is an explicit, formal knowledge representation that expresses knowledge about a domain of application. This includes

- Types of entities that exist in the domain
- Properties of those entities
- Relationships among entities
- Processes and events that happen with those entities
- Statistical regularities that characterize the domain
- Inconclusive, ambiguous, incomplete, unreliable, and dissonant knowledge related to entities of the domain
- Uncertainty about all these forms of knowledge

where the term *entity* refers to any concept (real or fictitious, concrete or abstract) that can be described and reasoned about within the domain of application (Ref. 41 quoted in Ref. 43).

18.2.5 State Transition Data Fusion Model

Lambert has developed a model—called the state transition data fusion (STDF) model—intended to comprehend object, situation, and impact assessment.[1,18] At each of these levels, different aspects of the world are predicted, observed (i.e., estimated), and explained as shown in Figure 18.5. The specific representations of state and state transition models for objects, situations, and impacts are given in Table 18.4. As shown,

- In object assessment, the target for estimation is a *state vector* $\underline{u}(k)$, which the fusion system "explains" by $\underline{\hat{u}}(k|k)$.
- In situation assessment, the target for estimation is a *state of affairs* $\Sigma(k)$, which the fusion system "explains" by a set of statements $\hat{\Sigma}(k|k)$.*

---

* In this formulation, like ours of Section 18.1.2, situation assessment involves estimating a current state of affairs (i.e., that most recently observed); impact assessment involves projecting future or past states of affairs.[1]

TABLE 18.4

Lambert's STDF Representation of Object, Situation, and Impact Assessment

Assessment	State $s(k)$	Transition $\{s(t)\|t \in Time\ \&\ t \leq k\}$
Object assessment $\underline{\hat{u}}(k)$	Each $s(k)$ is a *state vector* $\underline{u}(k)$ explained by $\underline{\hat{u}}(k\|k)$	Each $\{s(t)\|t \in Time\ \&\ t \leq k\}$ is an *object* $\underline{u}(k) = \{\underline{u}(t)\|t \in Time\ \&\ t \leq k\}$ explained by $\underline{\hat{u}}(k) = \{\underline{\hat{u}}(t\|t)\|t \in Time\ \&\ t \leq k\}$
Situation assessment $\hat{\Sigma}(k)$	Each $s(k)$ is a *state of affairs* $\Sigma(k)$ explained by $\hat{\Sigma}(k\|k)$	Each $\{s(t)\|t \in Time\ \&\ t \leq k\}$ is a *situation* $\Sigma(k) = \{\Sigma(t)\|t \in Time\ \&\ t \leq k\}$ explained by $\hat{\Sigma}(k) = \{\hat{\Sigma}(t\|t)\|t \in Times\ \&\ t \leq k\}$
Impact assessment $\hat{\underline{S}}(k)$	Each $s(k)$ is a *scenario state* $S(k) = \{\Sigma(n)\|n \in Time\ \&\ n \leq \partial(k)\}$ explained by $\hat{S}(k) = \{\hat{\Sigma}(n\|k)\|t \in Time\ \&\ t \leq \partial(k)\}$	Each $\{s(t)\|t \in Time\ \&\ t \leq k\}$ is a *scenario* $\underline{S}(k) = \{S(t)\|t \in Time\ \&\ t \leq k\} = \{\{\Sigma(n)\| n \in Time\ \&\ n \leq \partial(k)\}\|t \in Time\ \&\ t \leq k\}$ explained by $\hat{\underline{S}}(k) = \{\{\hat{\Sigma}(n\|t)\|n \in Time\ \&\ n \leq \partial(k)\}\|t \in Time\ \&\ t \leq k\}$

Source: Lambert, D., *Inform. Fusion*, 2007. Copyright Commonwealth of Australia. With permission.

- In impact assessment, the target for estimation is a *scenario* $S(\underline{k}) = \{\Sigma(n)\|n \in Time\ \& $ $n \leq \partial(k)\}$ (i.e., a set of situations within some look-ahead time $\partial(k)$), which the fusion system "explains" by $\hat{S}(k)$.

This model has the significant advantages of being formally well grounded and providing unified representation and processing models across JDL levels 1–3.[*]

Incidental concerns include (a) the apparent lack of representation of *relevance* in the models for situation and impact assessment; (b) the somewhat artificial distinction between situations and scenarios; and (c) the characterization of "intent" in terms of intended effects rather than intended actions.

1. This representation of situations (and, thereby, of scenarios) places temporal constraints on constituent facts, but not on any other factors affecting relevance. Therefore, it would seem that situations are configurations of the entire universe.[†]

2. Defining scenarios in terms of sets of situations seems to draw a distinction that weakens the representational power of situation logic. As discussed in Section 18.1.3.2.2, situation tracking is a large, variegated problem that has barely begun to be addressed. We wish to be able to track and predict the evolution of dynamic situations (i.e., of scenarios) of various magnitudes and complexities. An important feature of situation logic is the generality of the definition of "situation," allowing common representation at multiple levels of granularity. Thus, one situation s_1 can be an element of another s_2, such that for all infons σ, $s_2 \models \sigma \Rightarrow s_1 \models \sigma$. Representational power is enhanced by embedding one situation in another, for example, by propositional attitude, as in "$s_2 \models (x$ believes that $\sigma)$" or by subset. A scenario, as commonly understood, is one form of dynamic situation. It may

---

[*] Lambert "deconstructs" the JDL model levels to these three. The extension of the STDF model to JDL levels 0 (signal/feature assessment) and level 4 (system process or performance assessment) is a straightforward exercise; one that we leave for the student.

[†] Lambert does address issues of relevance elsewhere (Section 3 of Ref. 1), so that we are able to reason about subordinate situations.

be decomposed in terms of several types of embedded situations (e.g., temporal stages, facets, and local events). As an example, the enormous, dynamic, multifaceted situation that was World War II included such interrelated component situations as the conquest and liberation of France, the German rail system, the Soviet economy, the disposition of the U.S. fleet before the Pearl Harbor attack, the battle of Alamein, Roosevelt's death, and so on.*

3. Lambert indicates that impact prediction (i.e., the prediction of the evolution of a scenario) involves *intent*, *capability*, and *awareness*, at least where purposeful agents are involved.[†] He characterizes *intent* in terms of an agent's intended effects.[1] We suggest a more precise exegesis in terms of *intended actions* and their *hoped-for effects*. Intended effects are represented by statements about the world at large; intended actions are represented by statements specifically about the agent.[‡] Consider a statement

$$\text{Brutus intended to restore the Republic by killing Caesar} \qquad (18.13)$$

We parse this, ignoring for a moment the times and locations of various events, such as

> *Brutus intends*
> > *(Brutus kills Caesar)*
> *because*
> > *(Brutus hopes*
> > > *((Brutus kills Caesar)*
> > > *will cause*
> > > *(Republic is restored))).*

We can use infon notation to capture the implicit propositional attitude, rendering Equation 18.13 as

$$(Because\,(Intends, Brutus\,\alpha, h_1, k_1, 1), (Hope, Brutus, (Cause, \alpha, \beta, h_{2'}, k_2, p_2), p_1)); \qquad (18.14)$$

where $\alpha = (Kill, Brutus, Caesar, In\ the\ Senate\ House, Ides\ of\ March, 1)$; $\beta = (Restored, Republic, h_2, k_2, 1)$; h_1, k_1 = place and time the intention was held ($k_1 \leq$ Ides of March); h_2, k_2 = place and time the effect was hoped to occur ($k_2 \geq$ Ides of March); p_1 = probability that Brutus maintains this hope; and p_2 = probability that Brutus assigns to the given effect.[§]

---

* The inclusion of the last of these situations emphasizes (a) the admission of brief, localized events as situations and (b) the fuzziness of the boundary conditions for such multifaceted situations as World War II.

† Compare this with Little and Rogova's[44] threat ontology which we have adapted in Section 18.3.2.2 and in previous publications.[21,45,46] That ontology considers *capability*, *opportunity*, and *intent* to be the major factors in predicting intentional actions. We find it useful to distinguish one's *opportunity* to act as a necessary condition for action that can be distinguished from one's *capability* to act. Also, we agree with Lambert that *awareness* is a necessary condition for intentional action, but only as a necessary condition for *intent* to act (as discussed in Section 18.3.3.2). We have previously confused this point.[21,45,46]

‡ Lambert does provide for this distinction in his Mephisto framework[1] in Sections 3.2 and 4.2 and argues that the formulation of intents in terms of intended effects permits a more economical representation.[47]

§ Three points regarding this parsing

1. More precisely, h_2, k_2 should be bound by existential quantifiers. These appear as relational arguments in infon notation; for example, $(\exists, x, \sigma, S, T, p)$, where σ is an infon and "S, T" indicate all places and times

2. It seems preferable to parse (Equation 18.14) in terms of *hope* rather than *belief*, as in "$(Because\,(Intends, Brutus\ \alpha, h_1, k_1, 1), (Believes, Brutus, (Cause, \alpha, \beta, h_2, k_2, p_2), p_1))$"

3. An alternative ontology might parse "intends" in terms of a predicate "*IntendsBecause*"; so that Equation 18.14 is rendered "$(IntendsBecause, Brutus, \alpha, (Hopes, Brutus, (Cause, \alpha, \beta, h_{2'}, k_2, p_2), p_1), h_1, k_1, 1)$". The fact that we have such choices illustrates the slipperiness of ontology engineering.

18.3   Ontology for Situation and Threat Assessment

The type of recognition model that has been generally applied is that of an *ontology*. In its current use in artificial intelligence (AI) and information fusion, an *ontology* is a formal explicit description of concepts in a domain of discourse.[48] In other words, an ontology is a representation of semantic structure. In AI, ontologies have been used for knowledge representation ("knowledge engineering") by expressing concepts and their relationships formally, for example, by means of mathematical logic.[49]

Genesereth and Nilsson[50] define an *ontology* as an *explicit specification of a conceptualization: the objects, concepts, and other entities that are assumed to exist in some area of interest and the relationships that hold among them*. Similarly, to Kokar,[51] an *ontology* is composed of sets of concepts of object classes, their features, attributes, and interrelations.

It will be noted that this definition of an *ontology* is essentially the same as that given in Section 18.1.1 for an *abstract situation*. Indeed, an ontology is a sort of abstract situation. The problem is in clearly defining what sort of abstract situation it is. In classical metaphysics, Ontology (with a capital "O" and *sans* article) represented the essential order of things.[52] "Accidental" facts of the world—those not inferable from the concepts involved—are not captured in the classical ontology. As we shall see, many such accidental facts are essential to most practical inferencing and, therefore, need to be captured in the knowledge structures that we now call ontologies.

Many recent examples of ontologies rely heavily on such relations as "is_a" (as in "a tree is a plant") and "part_of" ("a leaf is part of a tree").

However, such a concept of ontology is too narrow to meet the needs of situation or threat assessment in the real world. We wish to be able to recognize and predict relationships well beyond those that are implicit in the semantics of a language, that is, those that are true by definition. Relationships to be inferred and exploited in situation assessment can include

- Logical/semantic relationships, for example, definitional, analytic, taxonomic, and mereologic ("part_of") relationships
- Physical relationships, to include causal relationships ("electrolysis of water yields hydrogen and oxygen" or "the short circuit caused the fire") and spatio-temporal relationships (e.g., "Lake Titicaca is in Bolivia" or "the moon is in Aquarius")
- Functional relationships, to include structural or organizational roles ("the treasurer is responsible for receiving and disbursing funds for the organization" or "Max is responsible for the company's failure")
- Conventional relationships, such as ownerships, legal and other societal conventions ("in the United Kingdom, one drives on the left side of the road")
- Cognitive relationships, such as sensing, perceiving, believing, fearing ("w speculates that x believes that y fears an attack by z")

Therefore, we will want an expanded concept of *ontology* to encompass all manners of expectations concerning a universe of discourse. Of course, the certainty that we can ascribe to expectations varies enormously. We expect no uncertainty in mathematical relations, but considerable uncertainty in inferring conventional relationships. Depending on the source of the uncertainty—for example, lack of model fidelity or precision—the class

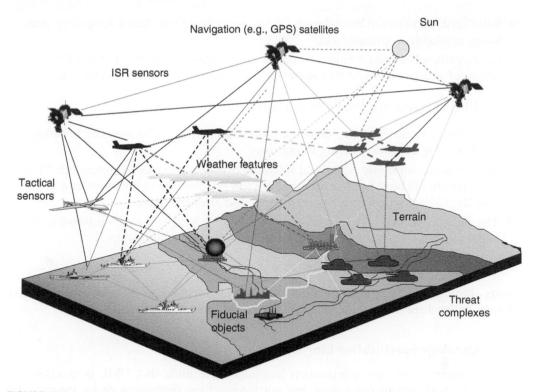

FIGURE 18.6
Representative relations of interest in a tactical military situation. (From Steinberg, A.N., *Proceedings of the Eighth International Conference on Information Fusion*, Phildelphia, 2005. With permission.)

and relation definitions in our ontology will need to be tagged with the appropriate confidence metrics, such as probability or fuzzy membership functions.*

It is evident that relationships of many of the types discussed are generally not directly observable, but rather must be inferred from the observed attributes of entities and their situational context. Indirect inference of this sort is by no means unique to situation assessment. Inference of relationships and entity states in given relationships is also essential to model-based target recognition, in which the spatial and spectral relationships among components are inferred and these, in turn, are used to infer the state of the constituted entity.

The realm of relations and relationships is vast. For example (illustrated in Figure 18.6), relationships of interest in tactical military applications can include

- Relationships among objects in an adversary's force structure (deployment, kinetic interaction, organization role/subordination, communication/coordination, type similarity, etc.)
- Relationships among friendly sensor and response elements (spatio-temporal alignment, measurement calibration, confidence, communication/coordination, etc.)
- Relationships between sensors and sensed entities (intervisibility, assignment/cueing, sensing, data association, countermeasures, etc.)

---

* The preceding is a list of representative expected relationships; expected because of assumed models of semantics, nature, society, or psychology. There are, of course, "accidental" or contingent physical or societal relationships: this cat has no tail, the moon is in Aquarius, George is older than Bill, etc.

- Relationships between components of opposing force structures (targeting, jamming, engaging, capturing, etc.)
- Relationships between entities of interest and contextual entities (terrain features, cultural features, solar and atmospheric effects, weapon launch, impact points, etc.)

Some additional points to be made about relations:

1. A simple relation can have complex hidden structure: "The monster is eating" implies "The monster is eating something."
2. One relation can imply additional relations between situation components: "The monster ate the bus" implies things about the relative size of the monster and the bus, their relative positions before, during, and after the meal.
3. The order of a relation is not necessarily fixed: "x bought a car"—ostensibly the expression of a 2-place relationship—implies a 5-place relationship.

$$\exists w \exists y \exists z \exists t [Car(w) \, \& \, x \text{ buys } w \text{ from } y \text{ for amount of money } z \text{ at time } t \, \& \, t < Now]^*$$

Various ontology specification languages are discussed in Section 18.3.1. Some ontologies that have defined specifically for STA are introduced in Section 18.3.2.

18.3.1 Ontology Specification Languages

Kokar[51] compares ontology specification languages. He finds that UML is excellent for visualization and human processing, but not for machine processing or for information exchange among machines or agents. Just the converse is true of XML, which includes no semantics. Some semantics are provided by RDF (resource description framework) and more so by RDFS (RDF schema). Significantly greater expressiveness is provided by OWL.

OWL (web ontology language) allows representation of data, objects and their relationships as annotations in terms of an ontology. Its description logic (OWL DL) was initially defined to be decidable for a semantic web.

Object query language (OQL) is a specification defined by the object data management group (ODMG) to support an object-oriented, SQL-like query language.[52] Kokar[51] defines a general-purpose OWL reasoner to answer queries formulated in OQL. The trace of the reasoner provides explanations of findings to users. Multiple ontologies and annotations can be combined using the colimit of category theory.

As discussed in Section 18.2.4.2.2, an OWL variant, PR-OWL, has developed an upper-ontology as a framework for probabilistic ontologies.[41,42,53] PR-OWL employs a logic, MEBN, that integrates classical first-order logic with probability theory.[53] It is claimed[43] that PR-OWL is expressive enough to represent even the most complex probabilistic models and flexible enough to be used by diverse Bayesian probabilistic tools (e.g., Netica, Hugin, Quiddity*Suite, JavaBayes, etc.) based on different probabilistic methods (e.g., probabilistic relational models, Bayesian networks, etc.).

A significant limitation of OWL and its variants is in allowing only first-order binary relations within the formal models. This has been found to be a significant constraint in representing the complex relational structures encountered in STA.[44,51,54]

Protégé 2000 is an open-source tool developed at Stanford University, which provides "domain-specific knowledge acquisition systems that application experts can use to enter

---

* This is the point of the situation logic construct discussed in footnote * in Section 18.2.4.1.

and browse the content knowledge of electronic knowledge bases."[55] Protégé 2000 provides basic ontology construction and ontology editing tools.[44]

Protégé 2000, however, lacks the expressive capacity needed to represent complex relationships and higher-level relationships such as the ones occurring in STA. Protégé 2000 is based on a rudimentary description logic so that it is capable of modeling only hierarchical relations that are unary or binary ("is-a," "part_of," and the like).

CASL is the Common Algebraic Specification Language approved by IFIP WG1.3.[56,57] CASL allows the expression of full first-order logic, plus predicates, partial and total functions, and subsorting. There are additional features for structuring, or "specification in the large." A sublanguage, CASL-DL, has been defined to correspond to OWL DL, allowing mappings and embeddings between the two representations.[58]

SNePS (Semantic Networks Processing System) is a propositional semantic network, developed by the Computer Science and Engineering Department of the State University of New York, Buffalo, NY, under the leadership of Stuart Shapiro. The original goal in building SNePS was to produce a network that would be able to represent virtually any concept that is expressible in natural language. SNePS involves both a logic-based and a network- (graph-) based representation. Any associative-network representation can be duplicated or emulated in SNePS networks. This dual nature of SNePS allows logical inference, graph matching, and path-matching to be combined into a powerful and flexible representation formalism and reasoning system. A great diversity of path-based inference rules are defined.[59,60]

IDEF5 (Integrated Definition 5) is a sophisticated standard for building and maintaining ontologies. It was developed to meet the increasing needs for greater expressive power and to serve the use of model-driven methods and of more sophisticated AI methods in business and engineering.

The schematic language of IDEF5 is similar to that of IDEF1 and IDEF1X, which were designed for modeling business activities. IDEF5 adds to this schematic language the expressive power found in IDEF3 to include representation of temporal dynamics, full first-order logic, and sufficient higher-order capabilities to enable it to represent ontologies.

IDEF5's expressive power allows representation of *n*-place first-order relations and 2-place second-order relations. As such, it is able to capture the fundamental constructs which situation logic expresses via infons: higher-order and nested relations, relational similarity, type/token distinctions, propositional attitudes and modalities, and embedded contexts.[61]*

18.3.2 Ontologies for Situation Threat Assessment

There have been several efforts to define taxonomies or ontologies of relations, situations, and threats.

18.3.2.1 Core Situation Awareness Ontology (Matheus et al.)

Matheus et al.[63] have developed a formal ontology of SAW, drawing on concepts from Endsley, the JDL model, and the situation theory of Barwise et al. Their core SAW ontology is depicted in Figure 18.7 as a UML diagram. As they describe it,

---

* IDEF5 does lack some of the specialized representations of IDEF1/1X, so that it would be a cumbersome tool for such purposes as designing a relational database.[62]

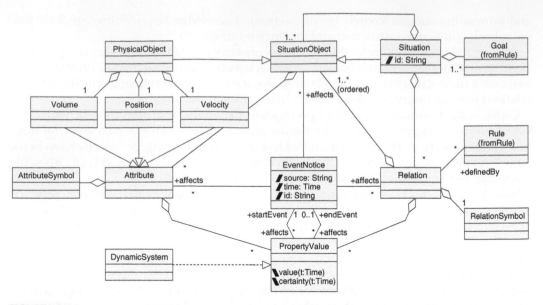

FIGURE 18.7
Core SA ontology. (From Matheus, C.J., Kokar, M., and Baclawski, K., *Proceedings of the Ninth International Conference on Information Fusion*, Florence, Italy, 2003. With permission.)

The *Situation* class … defines a situation to be a collection of *Goals*, *SituationObjects*, and *Relations*. *SituationObjects* are entities in a situation—both physical and abstract—that can have characteristics (i.e., *Attributes*) and can participate in relationships. *Attributes* define values of specific object characteristics, such as weight or color. A *PhysicalObject* is a special type of *SituationObject* that necessarily has the attributes of *Volume*, *Position*, and *Velocity*. *Relations* define the relationships between ordered sets of *SituationObjects*. For example, inRangeOf(X,Y) might be a *Relation* representing the circumstance when one *PhysicalObject*, X, is within firing range of a second *PhysicalObject*, Y.[63]

The Core SA Ontology is intended to define the *theory of the world* that describes the classes of objects and their relationships that are relevant to SAW.

18.3.2.2 Ontology of Threat and Vulnerability (Little and Rogova)

Little and Rogova[44] have developed an ontology for threat assessment. This has been applied to threat assessment problems in such domains as natural disaster response[64,65] and urban/asymmetric warfare.[66]

This ontology builds on the basic formal ontology developed by Smith and Grenon for SNePS.[59,67–69]

A key SNePS distinction that influences Little and Rogova's threat ontology is that between continuants and determinants.

Continuants are described as "spatial" items: entities that occupy space and endure through time, though subject to change. Examples are Napoleon's army, Heraclitus' river, Aristotle's boat, and (possibly) Schrödinger's cat.

Determinants are described as "temporal" entities. Examples are a military exercise, the firing of a weapon or a thought process. The SNePS ontology distinguishes relation-types that can occur between spatial (SNAP) and temporal (SPAN) items.[67–69]

This threat ontology considers *capability, opportunity,* and intent to be the principal factors in predicting (intentional) actions.

- *Capability* involves an agent's physical means to perform an act.
- *Opportunity* involves spatio-temporal relationships between the agent and the situation elements to be acted upon.*
- *Intent* involves the will to perform an act.

Consider the following uses:

1. x is able to jump 4 m (Capability)
2. x is standing near the ditch and the ditch is 4 meters wide (Opportunity)
3. = (1) + (2): x is prepared to jump over the ditch (Capability + Opportunity)
4. x intends to jump over the ditch (Intent)

Note that *intent* is defined in terms of the will to act, not the desire for a particular outcome. Therefore, if

5. x wants to get to the other side of the ditch (Desire, not Intent)

   then x may have the intent:

6. x intends to jump over the ditch to get to the other side of the ditch

Also, as mentioned, *awareness* in such an example is a condition of *intent*, rather than of *capability* or *opportunity*: x must be aware of the ditch if he intends to jump over it.

More precisely, it is *hope for* one's capability and opportunity, rather than one's awareness, which is a condition for intent. So

7. x hopes that (3) is a necessary condition for (4)

Regarding (7), it is, of course, not necessary that x believes or hopes that the ditch is 4 m wide or that he is able to jump 4 m; only that he hopes that he is able to jump the width of the ditch.

Furthermore, one may intend to perform a series of acts with dependencies among them. Thus, (6) might be amenable to parsing as (8) or (9):

8. x intends to jump over the ditch once he improves his jumping skills (acquires the capability) and walks to the ditch (acquires the opportunity)
9. x intends to get to the other side of the ditch (by some means, whether identified or not, for example, by jumping)

Little and Rogova make a further key distinction between *potential* and *viable* threats. A potential threat is one for which either capability, opportunity, or intent is lacking. One

---

* Note that *opportunity* is defined only in spatio-temporal terms. In previous writings, we distinguished *capability* from *opportunity* as factors that are, respective, internal to and external to the agent.[21,45,46] So doing, however, allows for such confusing applications as "a nearby latter provides an *opportunity* to scale a wall whereas the possession of a ladder provides a *capability* to scale a wall." We do have some residual uneasiness about defining opportunity solely in spatio-temporal terms. It would seem that a target's vulnerability is a matter of opportunity; as in "x can enter the fort because the gate was left unlocked (or because its walls are paper-thin)." Therefore, we prefer to characterize *opportunity* as involving *the relationships between an agent and the situation elements to be acted upon.*

could argue that a viable threat—defined as one for which all three conditions are met—is not only viable but also inevitable.

18.4   A Model for Threat Assessment

Impact assessment—and, therefore, threat assessment—involves all the functions of situation assessment, but applied to *projected* situations. That is to say, impact assessment concerns situations inferred on the basis of observations other than of the situations themselves, such as

- Inferences concerning potential future situations
- Inferences concerning counterfactual present or past situations (e.g., what might have happened if Napoleon had won at Waterloo?)
- Inferences concerning historical situations on the basis of evidence extrinsic to the situation (e.g., surmising about characteristics of the original Indo-European language on the basis of evidence from derivative languages)

Threat assessment—a class of impact assessment—involves assessing projected *future* situations to determine the likelihood and cost of detrimental events.

Recall our threat taxonomy: threatening agents and situations; threatened agents, events and situations. Given our focus on *intentional* detrimental events, we will use the term "attack" broadly to refer to such events.*

Threat assessment includes the following functions:

- *Threat event prediction.* Determining likely threat events ("threatened events" or "attacks"): who, what, where, when, why, how
- *Indications and warning.* Recognition that an attack is imminent or under way
- *Threat entity detection and characterization.* Determining the identities, attributes, composition, location/track, activity, capability, intent of agents, and other entities involved in a threatened or current attack
- *Attack (or threat event) assessment.*
  - Responsible parties (country, organization, individuals, etc.) and attack roles
  - Intended target(s) of attack
  - Intended effect (e.g., physical, political, economic, psychological effects)
  - Threat capability (e.g., weapon and delivery system characteristics)
  - Force composition, coordination, and tactics (goal and plan decomposition)
- *Consequence assessment.* Estimation and prediction of event outcome states (threatened situations) and their cost/utility to the responsible parties, to affected parties, or to the system user, which can include both intended and unintended consequences

---

* It would be naïve to assume that the concern in threat assessment is only with attacks *per se*. Rather, as discussed below, we are concerned with any event or change in the world situation that changes the Bayesian cost to us. An adversary's withdrawal, resupply, or leadership change is a concern in threat assessment to the extent that it affects the adversary's capability or opportunity or intent to conduct detrimental activity.

To support all of these processes, it will be necessary to address the following issues, as noted in Ref. 70:

1. Representation of domain knowledge in a way that facilitates retrieval and processing of that knowledge.
2. Means of obtaining the required knowledge.
3. Representation and propagation of uncertainty.

Issues (1)–(3) are, of course, pervasive across STA problems. Threat assessment involves two aspects that raise additional issues. These are the aspects of human intention and event prediction.

4. In the many threats of concern that are the results of human intent, threat event prediction involves the recognition and projection of adversarial plans. This can involve methods for spatial and temporal reasoning; it certainly involves methods for recognizing human goals, priorities, and perceptions.
5. Whether the concern is for human or other threats, threat assessment involves prediction of events and of situations, requiring models of causal factors and of constraints in projecting from known to unknown situational states.[70]

18.4.1 Threat Model

Threats are modeled in terms of potential and actualized relationships between threatening entities (which may or may not be people or human agencies) and threatened entities, or "targets" (which often include people or their possessions).

We define a *threatening situation* to be a situation in which *threat events* are likely. A threat event is defined as an event that has adverse consequences for an agent of concern; paradigmatically, for "us." Threat events may be intentional or unintentional (e.g., potential natural disasters or human error). We call an intentional threat event an *attack*.

In the present discussion, we are not directly concerned with unintentional threats such as from natural disasters. Rather, we are interested in characterizing, predicting, and recognizing situations in which a willful agent intends to do harm to something or somebody. This, of course, does not assume that intended threat actions necessarily occur as intended or at all, or that they yield the intended consequences. Furthermore, we need to recognize threat assessment as an aspect of the more general problems of assessing intentionality and of characterizing, recognizing, and predicting intentional acts in general.

Looney and Liang[71] formulate threat assessment as a process that applies the situation assessment result to

1. Predict intentions of the enemy
2. Determine all possible actions of the enemy
3. Recognize all opportunities for hostile action by the enemy
4. Identify all major threats to friendly force resources
5. Predict the most likely actions of the enemy
6. Determine the vulnerabilities of friendly forces to all likely enemy actions using match-up of forces, weapon types, and preparedness

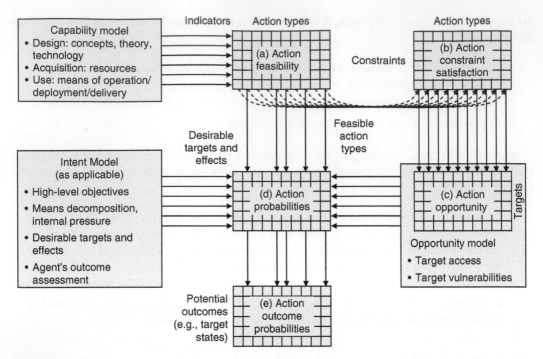

FIGURE 18.8
Elements of a threat situation hypothesis. (From Steinberg, A.N., Llinas, J., Bisantz, A., Stoneking, C., and Morizio, N., *Proceedings of the MSS National Symposium on Sensor and Data Fusion*, McLean, VA, 2007. With permission.)

7. Determine favorable offensive actions of friendly forces with favorable match-ups to thwart all likely actions of the enemy with least cost

8. Determine optimal targeting for all offensive actions*

We base our threat model on the ontology of Little and Rogova,[44] with intent redefined as per footnote * in Section 18.3.2.2. Indicators of threat situations relate to the *capability*, *opportunity*, and—where purposeful agents are involved—*intent* of agents to carry out such actions against various targets.

Figure 18.8 shows the elements of a representative threat(ening) situation hypothesis and the principal inference paths.

The three principal elements of the threat situation—an entity's capability, opportunity, and intent to affect one or more entities—are each decomposed into subelements. The Threat assessment process (a) generates, evaluates, and selects hypotheses concerning entities' capability, intent, and opportunity to carry out various kinds of attack and (b) provides indications, warnings, and characterizations of attacks that occur.

An entity's *capability* to carry out a specific type of threat activity depends on its ability to design, acquire, and deploy or deliver the resources (e.g., weapon system) used in that activity. The hypothesis generation process searches for indicators of such capability and generates a list of feasible threat types.

---

* We would consider functions (7) and (8)—like McGuinness and Foy's "resolution"[25]—to be process assessment/refinement functions that fall outside of threat assessment *per se*.

Intent is inferred by means of goal decomposition, based on decision models for individual agents and for groups of agents.

The postulated threat type constrains a threat entity's *opportunity* to carry out an attack against particular targets (e.g., to deploy or deliver a type of weapon as needed to be effective against the target). Other constraints are determined by the target's accessibility and vulnerability to various attack types, and by the threat entity's assessment of opportunities and outcomes.

Capability, opportunity, and intent all figure in inferring attack–target pairing. Some authors have pursued threat assessment on the basis of capability and intent alone.[72] These factors, of course, are generally more observable and verifiable than is intent. However, capability and opportunity alone are not sufficient conditions for an intentional action.

The threat assessment process evaluates threat situation hypotheses on the basis of likelihood, in terms of the threat entity's *expected perceived net pay-off*. It should be noted that this is an estimate, not of the actual outcome of a postulated attack, but an estimate of the *threat entity's estimate* of the outcome.

The system's ontology provides a basis for inferencing—capturing useful relationships among entities that can be recognized and refined by the threat assessment process. An ontology can capture a diversity of inference bases, to include logical, semantic, physical, and conventional contingencies (e.g., societal, legal, or customary contingencies). The represented relationships can be conditional or probabilistic.

This ontology will permit an inferencing engine to generate, evaluate, and select hypotheses at various fusion levels. In data fusion level 0, 1, and 2, these are hypotheses concerning signals (or features); individuals; and relationships and situations, respectively. A level 3 (threat or impact assessment) hypothesis concerns potential situations, projecting interactions among entities to assess their outcomes. In Lambert's[1] term, these are scenario hypotheses. In intentional threat assessment we are, of course, concerned with interactions in which detrimental outcomes are intended.

18.4.2 Models of Human Response

As noted in Ref. 18, tools are needed to assess multiple behavior models for hostile forces during threat assessment. For example, game theoretic analysis has been suggested for application to level 3 fusion. Combinatorial game theory can offer a technique to select combinations in polynomial time out of a large number of choices.

Problems in anticipating and influencing adversarial behaviors result from, at a minimum, the following three factors:

1. *Observability.* Human psychological states are not directly observed but must be inferred from physical indicators, often on the basis of inferred physical and informational states.

2. *Complexity.* The causal factors that determine psychological states are numerous, diverse, and interrelated in ways that are themselves numerous and diverse.

3. *Model fidelity.* These causal factors are not well understood, certainly in comparison with the mature physical models that allow us to recognize and predict target types and kinematics.

Nonetheless, however complex and hidden human intentions and behaviors may be, they are not random.

In Refs 46 and 73, we propose methods for representing, recognizing, and predicting such intentions and behaviors. We describe an agent response model, developed as a

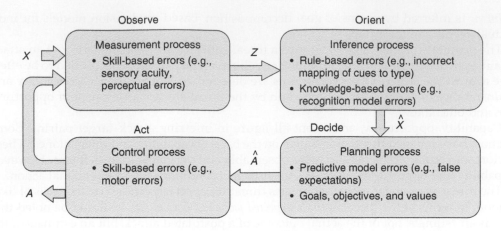

Red = Possibly intentional factors

FIGURE 18.9
Components of an action process and constituent error factor.

predictive model of human actions. This model is based on a well-known methodology in cognitive psychology for characterizing and predicting human behavior: Rasmussen's skill-rule-knowledge (SRK) model.[74] We have formulated the SRK model in terms of a formal probabilistic model for characterizing such behavior.[73]

Actions of a responsive agent—to include actions of information reporting—are decomposed by means of four process models:

1. *Measurement model.* $p(Z_{M_w}^k | X, w)$; probabilities that agent w will generate measurement sets Z in world states X

2. *Inference model.* $p(\hat{X}_{I_w}^k | Z_{M_w}^k)$; probabilities that agent w will generate world state estimates $\hat{X}$, given measurement sets Z

3. *Planning model.* $p(\hat{A}_{P_w}^k | \hat{X}_{I_w}^k)$; probabilities that agent w will generate action plans $\hat{A}$, given world state estimate $\hat{X}$

4. *Control model.* $p(A_{C_w}'^k | \hat{A}_{P_w}^k)$; probabilities that agent w will generate actions A', given action plans $\hat{A}$

These process models can generally be assumed to be conditionally independent, arising from mutually independent system components.* Thus, they can be modeled serially as in Figure 18.9, which shows representative SRK performance factors for each component. As shown, this Measurement-Inference-Planning-Control (MIPC) model can be viewed as a formal interpretation of Boyd's well-known "OODA-Loop."

Among the four constituent models, measurement and control are familiar territory in estimation and control theory. Inference and planning model processes are, respectively, the provinces of information fusion and automatic planning, respectively. Inference corresponds to Endsley's perception, comprehension, and projection. Planning involves the generation, evaluation, and selections of action plans (as discussed in Chapter 3 and

---

* It is well recognized that human perception, comprehension, and projection (inference) are often driven by expectations and desires; that is, by prior plans and inferences concerning their expected outcomes. These factors are modeled as feedback in the measurement/inference/planning/control loop (an action component A in Figure 18.9).

Refs 4 and 16). Such actions can include external activity (e.g., control of movement, of sensors and effectors) and control of internal processes (e.g., of the processing flow), (see Ref. 17).

Inference and planning are "higher-level" functions in the old dualistic sense that human perception, comprehension, and projection and planning are considered "mental" processes, whereas "measurement" (sensation) and motor control are "physical" processes. The former are certainly less directly observable, less well understood and, therefore, less amenable to predictive modeling.

By decomposing actions into constituent elements, and these into random and systematic components, we can isolate those components that are difficult to model. Having done that, we are able—at the very least—to assess the sensitivity of our behavior model to such factors.

We often have reliable predictive models for measurement and control in automated systems. Our ability to predict inference and planning performance in automated systems is relatively primitive, to say nothing of our ability to predictively model human cognition or behavior. As shown, each of the four models can involve both random and bias error components. However, note that only planning can involve *intentional* error components.

On the assumption of conditional independence, we can write expressions for biases in inference, planning and control analogous to the familiar one for measurement:

$$Z_M^k = h_s^k(\beta_M, X_t^k) + v_M^k \tag{18.15}$$

$$\hat{X}_I^k = h_I^k(\beta_I, Z_M^k) + v_I^k \tag{18.16}$$

$$\hat{A}_P^k = h_P^k(\beta_P, \hat{X}_I^k) + v_P^k \tag{18.17}$$

$$A_C^k = h_C^k(\beta_C, \hat{A}_P^k) + v_C^k \tag{18.18}$$

Terms are defined as follows (the agent index has been suppressed): Z_M^k is the measurement set as of time k; $\hat{X}_I^k$ is the estimated world state; $\beta_M, \beta_I, \beta_P, \beta_C$ are systematic bias terms for measurements, inference, planning, and control, respectively; $v_M^k, v_I^k, v_P^k, v_C^k$ are noise components of measurement, inference, planning, and control, respectively (generally non-Gaussian).

Intentionality occurs specifically as factors in the bias term β_P of Equation 18.17. This term, then, is the focus of the assessment of an agent's intentionally directed patterns of behavior.

The transforms h, particularly those for inference and planning, can be expected to be highly nonlinear. Equation 18.18 is the control model dual of Equation 18.15, with explicit decomposition into random and systematic error components. Equation 18.17 is a planning model, which—as argued by Bowman[75]—has its dual in data association:

$$\hat{Y}_A^k = h_A^k(\beta_A, Z_M^k) + v_A^k \tag{18.19}$$

for association hypotheses $\hat{Y}_A^k \in 2^{Z_M^k}$ (i.e., hypotheses within the power set of measurements).

18.5   System Engineering for Situation and Threat Assessment

We now consider issues in the design of techniques for STA. In Section 18.5.1, we discuss the implications for data fusion to perform STA—first in terms of functions within a processing node, then in terms of an architecture that includes such nodes. The very important topics of semantic registration and confidence normalization are treated in Section 18.5.2, whereas Sections 18.5.3 and 18.5.4 discuss issues and techniques for data association and state estimation, respectively. Implications for efficiency in processing and data management are addressed in Sections 18.5.5 and 18.5.6, respectively.

18.5.1   Data Fusion for Situation and Threat Assessment

18.5.1.1   Data Fusion Node for Situation and Threat Assessment

Figure 18.10 depicts the data fusion node model that we have been using to represent all data fusion processes, regardless of fusion level (e.g., in Chapters 3 and 22). The figure is annotated to indicate the specific issues involved in data fusion for situation and threat assessment.

As seen in the figure, the basic data fusion functions are data alignment (or common referencing), data association, and state estimation.

- *Data alignment for situation assessment* involves preparing available data for data fusion, that is, for associating and combining multiple data.
- *Data association for situation assessment* involves deciding how data relate to one another: generating, evaluating, and selecting hypotheses concerning relationships and relational structures.
- *State estimation for situation assessment* involves the functions defined in Section 18.1.3; viz.,

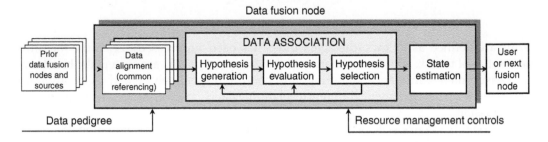

Data alignment

- How to reason across disparate data types and sources (numeric and symbolic)?
  - Common formatting
  - Spatio-temporal registration
  - Semantic registration
  - Confidence normalization

Data association

- How does evidence fit together?
  - Relevance
  - Relationships
- How to evaluate and compare complex data structures?

State estimation/prediction

- How to discover and characterize relationships?
- How to recognize situations?
- How to discover situations?
- How to predict situations?

FIGURE 18.10
Generic data fusion node for situation and threat assessment.

- Inferring relationships among (and within) entities, to include (a) entities that are elements of a situation and (b) entities that are themselves treated as situations (i.e., to static and dynamic "structures," to include everything we commonly refer to as situations, scenarios, aggregates, events, and complex objects)
- Using inferred relationships to infer attributes of entities in situations and of situations themselves (situation recognition or characterization)

Data fusion for impact assessment (and, therefore, for threat assessment) involves all of these functions; state estimation concerns *projected*, rather than observed, situations. Threat assessment involves projecting and estimating the cost of future detrimental situations.

18.5.1.2 Architecture Implications for Adaptive Situation Threat Assessment

As noted, the STA process will need to generate, evaluate, and select hypotheses concerning present and projected situations.

We anticipate that processes for situation assessment—and *a forteriori*, for threat assessment—will need to be adaptive and opportunistic. Thus, we expect that a processing architecture that could be constructed on the basis of modules per Figure 18.5 will require tight coupling between inferencing and data acquisition (collection management and data mining) to build, evaluate, and refine situation and threat hypotheses. The simplest form of information exploitation involves open-loop data collection, processing, and inferencing, illustrated in Figure 18.11a. Here, "Blue" collects and exploits whatever information his sensors/sources provide. Naturally, entities in the sensed environment may react to their environment, as illustrated by "Red's" OODA-loop.

Of course, the only aspects of Red's behavior that are potentially observable by Blue are the consequences of Red's physical acts. Red's observation, orientation, and decision

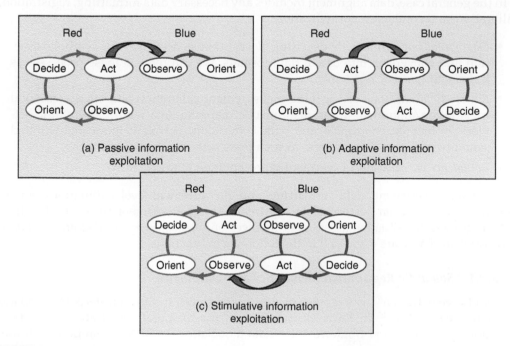

FIGURE 18.11

Passive, adaptive, and stimulative information exploitation. (From Steinberg, A.N., *Proceedings of the MSS National Symposium on Sensor and Data Fusion*, McLean, VA, 2006. With permission.)

processes* are hidden, except insofar as they involve physical activity (e.g., electromagnetic emissions of active sensors). Estimation and prediction activity within these hidden processes must be inferred indirectly, as discussed in Section 18.4.2.

In such an open-loop approach—typical of most data fusion systems today—Blue observes and interprets such activities, but response decisions and actions are explicitly outside the purview of the data fusion process.

Information exploitation is often conceived as a closed-loop process: TPED, TPPU, and so on, as illustrated in Figure 18.11b. Here data collectors and the fusion process are adaptively managed to improve Blue's fusion product. This can include prediction of future and contingent states of entities and situations (e.g., Red's actions and their impact). They can include predicting the information that would be needed to resolve among possible future states and the likelihood that candidate sensing or analytic actions will provide such information. That is, this closed-loop information exploitation process projects the effects of the environment on one's own processes, but does not consider the effects of one's own actions on the environment.

In contrast, *closed-loop stimulative intelligence*, Figure 18.11c, involves predicting, exploiting, and guiding the response of the system environment (and, in particular, of purposeful agents in that environment) to one's own actions. The use of such stimulative techniques for information acquisition is discussed in Ref. 76.

18.5.2 Data Alignment in Situation and Threat Assessment

Data alignment (often referred to as *common referencing*) involves normalizing incoming data as necessary for association and inferencing. Data alignment functions operate on both reported data and associated metadata to establish data pedigree and render data useful for the receiving data fusion node.[16]

In the general case, data alignment includes any necessary data formatting, registration/calibration, and confidence normalization.

- *Data formatting* can include decoding/decryption as necessary for symbol extraction and conversion into useful data formats. We can refer to these processes as *lexical* and *syntactic registration*.
- *Registration/calibration* can include measurement calibration and spatio-temporal registration/alignment. It can also include mapping data within a semantic network, to include, coordinate, and scale conversion, as well as more sophisticated semantic mapping. We can refer to such processes as *semantic registration*.
- *Confidence normalization.* Consistent representation of uncertainty.

In situation assessment, data formatting and measurement-level calibration/registration usually have been performed by upstream object assessment fusion nodes. If so, data alignment for situation assessment involves only *semantic registration* and *confidence normalization*. These are discussed in the following subsections.

18.5.2.1 Semantic Registration: Semantics and Ontologies

"Semantic registration" was coined as the higher-level fusion counterpart to spatio-temporal registration.[18] In lower-level fusion, spatio-temporal registration provides a common spatio-temporal framework for data association. In STA, semantic registration

* Or, equivalently, his measurement, inference and planning processes, as per the MIPC model of Section 18.4.2.

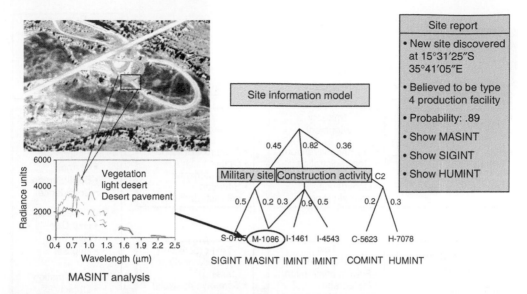

FIGURE 18.12
An example of multiple intelligence deduction. (From Steinberg, A.N. and Waltz, E.L., *Perceptions on Imagery Fusion*, Presented at NASA Data Fusion/Data Mining Workshop, Sunnyvale, CA, 1999. With permission.)

involves the translation of sensor fusion object assessments and higher-level assessments into a common semantic framework for concept association.

Semantic registration permits semantic interoperability among information processing nodes and information users. As discussed in Section 18.3, an ontology provides the means for establishing a semantic structure. Processing may be required to map between separate ontologies as used by upstream processes.[77]

Because information relevant to STA is often obtained from a diversity of source types, it will be necessary to extract information into a format that is common both syntactically and semantically. In the field of content-based image retrieval (CBIR), this is called the problem of the semantic gap.

Figure 18.12 depicts a concept for combining information extracted from human and communications intelligence (HUMINT and COMINT) natural language sources, as well as from various sensor modalities, such as imagery intelligence (IMINT) and measurement and signals intelligence (MASINT).[78] The tree at the bottom of the figure represents the accumulation of confidence indicators (e.g., probabilities) from diverse sources supporting a particular situation hypothesis. Such fusion across heterogeneous sources requires lexical, syntactic, and semantic registration, as well as spatio-temporal registration and confidence normalization.

Figure 18.13 illustrates a process for bridging this gap, so that information from such diverse sources as imagery and natural language text can be extracted as necessary for fusion.

Techniques and developments for performing these functions are discussed in Ref. 79.

18.5.2.2 Confidence Normalization

Confidence normalization involves assigning consistent and (at least in principle) verifiable confidence measures to incoming data.

The concept of confidence normalization is tied to that of *data pedigree*. Data pedigree can include any information that a data fusion or resource management node requires to maintain its formal and mathematical processing integrity.[16]

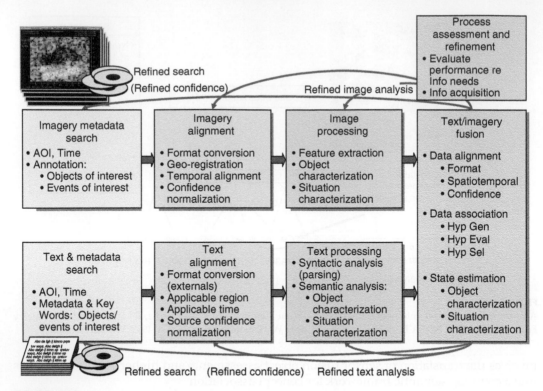

FIGURE 18.13
Fusion of imagery and text information.

Pedigree metadata involves two components:

1. Content description: data structure, elements, ontologies, constraints, etc.
2. Origin of the data and processing history[77]

The latter requires processes to estimate the error characteristics of data sources and of the processes that transform, combine, and derive inferences from such data before arrival at a given user node (i.e., to a node that performs further processing or to a human user of the received data). One way of maintaining consistency across fusion nodes and fusion levels is by exchanging pedigree metadata along with the data that they describe.[16,77,80]

The problem of characterizing data sources and data processing is particularly acute in STA, because of the heavy involvement in source data generation, transformation, combination, and inferencing.

We may conceive the flow of information to be exploited by an STA process as an *open communications network*, that is, as an unconstrained network of agents that may interact on the basis of diverse capabilities, motives, and allegiances.[81] Agent interactions within such a network may be

- Intentional (e.g., by point-to-point or broadcast communications or publication)
- Unintentional (e.g., by presenting active or passive signatures, reflective cross sections, electromagnetic or thermal emissions, or other detectable physical interactions with the environment)

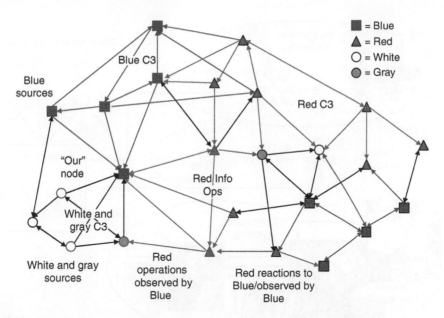

Blue C3

Blue sources

"Our" node

Red C3

Red Info Ops

White and gray C3

White and gray sources

Red operations observed by Blue

Red reactions to Blue/observed by Blue

■ = Blue
▲ = Red
○ = White
● = Gray

FIGURE 18.14

An example of open communications network. (From Steinberg, A.N., *Proceedings of the Ninth International Conference on Information Fusion*, Florence, Italy, 2006. With permission.)

Agents can interact with one another and with nonpurposeful elements of their environment in complex ways. As illustrated in Figure 18.14, if we consider ourselves a node in a network of interacting agents, our understanding of received data requires the characterization of proximate sources and, recursively, of their respective sources. That is, we need to characterize the *pedigree* of our received data.*

Such data, coming from a variety of sources with different levels of reliability, is often incomplete, noisy, and statistically biased. The concern for bias is especially acute when sources of information are people, who have the potential for intentional as well as inadvertent distortion of reported data.

HUMINT and COMINT, as well as open-source document intelligence (OSINT), are key sources of information in threat assessment. These sources involve information generated by human sources, and typically interpreted and evaluated by human analysts. Although intelligence analysts may be presumed to report honestly, there is always the possibility of inadvertent bias. In the case of external subjects—authors of print or electronic documents, interrogation subjects, and so on—it is not often clear how much credence to ascribe.

Consider the situation of a HUMINT reporting agent. As shown in Figure 18.15, there are three modes by which such an agent may receive information:

1. Passive observation (e.g., as a forward observer)

2. Direct interaction (e.g., by interrogation of civilians, or enemy combatants)

3. Receiving information from third-party sources (e.g., from civilian informants, from other friendly force members, from print or electronic media such as radio, TV, bloggers, e-mail, telephone, text messaging, and so on)

---

* In Ref. 46, we broaden the notion of *open communications network* by expanding the notion of communications to include *any* interaction among agents: one may communicate with another either by an e-mail message, a dozen roses, or a punch in the nose. This expanded notion allows us to characterize the interactions of concern to threat assessment in Section 18.4.1.

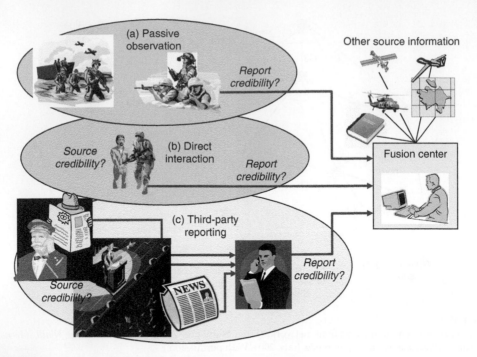

FIGURE 18.15
Modes of collecting human-generated information. (From Steinberg, A.N., Llinas, J., Bisantz, A., Stoneking, C., and Morizio, N., *Proceedings of the MSS National Symposium on Sensor and Data Fusion*, McLean, VA, 2007. With permission.)

Human reporting agents and human subjects can have various degrees—and possibly time-varying degrees—of allegiance, common purpose, cooperativeness, information fidelity, and controllability.

In mode (1) there is a concern for errors caused by faults in the reporting agent's observation, inference, decision, and reporting processes. In modes (2) and (3), the concern is for such faults in his sources, compounded with those of the reporting agent.

Estimating the credibility and possible biases in information from human sources is critical to exploiting such information and combining it with other information products (e.g., sensor data). Consider an analyst's dilemma when one source with historically good performance provides a report with low certainty, whereas a second with historically less-reliable source provides a conflicting report with greater certainty.[77]

Systematic methods are needed to recognize and correct distorted information from external subjects of the aforementioned types as reported in HUMINT, COMINT, and OSINT. These methods will also need to recognize and correct distortions—whether intentional or unintentional—contributed by the observers, interrogators, interpreters, analysts, and others in the information exploitation chain. This is, in effect, a problem of data alignment similar to those of spatio-temporal and semantic registration discussed in Section 18.5.2.1. The result may be expressed as likelihood values appended to individual report elements; as needed for inferring entity states, behaviors, and relationships as part of STA.[46]

Data reliability should be distinguished from data certainty in that reliability is a measure of the confidence in a measure of certainty (e.g., confidence in an assigned probability). Reliability has been formalized in the statistical community as second-order uncertainty.[82,83]

Rogova and Nimier[84] have assessed available approaches to assessing source reliability. They differentiate cases where

- It is possible to assign a numerical degree of reliability (e.g., second-order uncertainty) to each source (in this case values of reliability may be "relative" or "absolute" and they may or may not be linked by an equation such as $\Sigma R i = 1$).
- Only a relative ordering of the reliabilities of the sources is known but no specific values can be assigned.
- A subset of sources is reliable but the specific identity of these is not known.

Rogova and Nimier argue that there are basically two strategies for downstream fusion processes to deal with these cases:

1. Strategies explicitly utilizing known reliabilities of the sources.
2. Strategies for identifying the quality of data input to fusion processes and elimination of data of poor reliability (this still requires an ability to process the "reasonably reliable" data).[84]

The agent response model described in Section 18.4.2 can be used for characterizing the performance of human information sources.

Actions of a responsive agent—to include actions of information reporting—are decomposed by means of the following four process models, as discussed in Section 18.4.2.

- *Measurement model.* $p(Z_{M_w}^k \mid X, w)$; probabilities that agent w will generate measurement sets Z in world states X. In the examples of Figure 18.15, these would include visual observations or received verbal or textual information.
- *Inference model.* $p(\hat{X}_{I_w}^k \mid Z_{M_w}^k)$; probabilities that agent w will generate world state estimates $\hat{X}$, given measurement sets Z.
- *Planning model.* $p(\hat{A}_{P_w}^k \mid \hat{X}_{I_w}^k)$; probabilities that agent w will generate action plans $\hat{A}$, given world state estimate $\hat{X}$. In the source characterization application, this can be planned reporting.
- *Control model.* $p(A'_{C_w}^k \mid A'_{P_w}^k)$; probabilities that agent w will generate actions A', given action plans $\hat{A}$; for example, actual reporting (subject to manual or verbal errors).

It is evident that an agent's reporting bias is distinct from, but includes measurement bias. Characterizing reporting biases and errors involves the fusion of four types of information:

1. Information within the reports themselves (e.g., explicit confidence declarations within a source document or an analyst's assessment of the competence discernable in the report)
2. Information concerning the reporting agent (prior performance or known training and skills)

3. Information about the observing and reporting conditions (e.g., adverse observing conditions or physically or emotionally stressful conditions that could cause deviations from historical reporting statistics)

4. Information from other sources about the reported situation that could corroborate or refute the reported data (e.g., other human or sensor-derived information or static knowledge bases)

Information of these types may be fused to develop reporting error models in the form of probability density functions (pdfs). The techniques used derive from standard techniques developed for sensor calibration and registration:[81,85]

- *Direct filtering (absolute alignment) methods* are used to estimate and remove biases in confidence-tagged commensurate data in multiple reports, using standard data association methods to generate, evaluate, and select hypotheses that the same entities are being report on.

- *Correspondence (relative alignment) methods* such as cross-correlation and polynomial warping methods are used to align commensurate data when confidence tags are missing or are themselves suspect.

- *Model-based methods*—using ontologies or causal models relating to the reported subject matter—are used to characterize biases and errors across noncommensurate data types (e.g., when one source reports a force structure in terms of unit type and designation, another reports in terms of numbers of personnel and types of equipment).

The source characterization function can serve as a front-end to the STA processing architecture. Alternatively, it can be integrated into the entity state estimation process, so that states of source entities—including states of their measurement, inference, planning, and control processes—are estimated in the same way as are states of "target" entities. In this way, "our" node in the open network of Figure 18.14 performs level 1 assessment of other nodes, across allegiance and threat categories. Pedigree data are traced and accumulated via level 2 processes.

18.5.3 Data Association in Situation and Threat Assessment

Data association involves generating, evaluating, and selecting hypotheses regarding the association of incoming data:

- A signal/feature (level 0) hypothesis contains a set of measurements from one or more sensors, hypothesized to be the maximal set of measurements of a perceived signal or feature.

- An object (level 1) hypothesis contains a set of observation reports (possibly products of level 0 fusion), hypothesized to be the maximal set of observation reports of a perceived entity.

- A situation (level 2) hypothesis is a set of entity state estimates (possibly products of level 1 fusion), hypothesized to constitute a relationship or situation.

- A threat (level 3) hypothesis is a situation hypothesis, projected on the basis of level 1, 2, and 3 hypotheses.

These hypothesis classes are defined in terms of an upward flow of information between data fusion levels, as shown in Figure 18.16.

However, inference can flow downward as well, as noted in the inductive examples of Section 18.1.3. Therefore, a signal, individual or situation can be inferred partially or

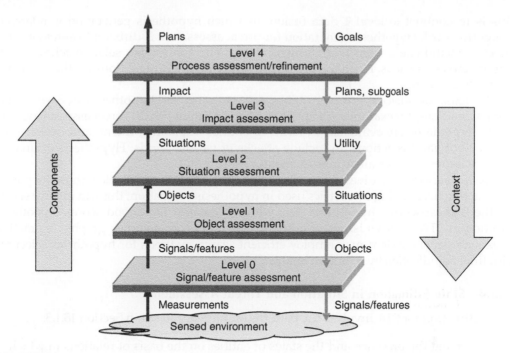

FIGURE 18.16
Characteristic process flow across the fusion "levels." (From Steinberg, A.N. and Bowman, C.L., *Proceedings of the MSS National Symposium on Sensor and Data Fusion*, June 2004. With permission.)

completely on the basis of a contextual individual, situation or impact. For example, a level 2 (situation) hypothesis may contain level 1 hypotheses that contain no level 0 products. Such would be the case if an entity was not directly observed, but inferred on the basis of its situational context (as exemplified in Figure 18.2).

As noted in Ref. 77, a system is needed that can generate, evaluate, and present multiple hypotheses regarding intent and hostile COAs. Hostile value functions need to be inferred, using for example, techniques for inferring Markov representations of the deterministic parts of enemy behavior. Multiple hypotheses as expressed in the probability distribution of enemy intent are not usually well known. Perhaps it is possible to select most likely case when pressed to do so. Probable as well as possible actions need to be considered. Dependences can be revealed with alternative generation. A related technique is reinforcement learning.

Hypothesis generation involves

- Determining relevant data (data batching)
- Generating/selecting candidate scenario hypotheses (abduction)

The open world nature of many situation assessment problems poses serious challenges for hypothesis generation. This is particularly so in the case of threat assessment applications because of the factors listed in Section 18.1.4; viz.,

- Weak spatio-temporal constraints on relevant evidence
- Weak ontological constraints on relevant evidence
- Dominance of relatively poorly modeled processes (specifically, human, group, and societal behavior, *vice* simple physical processes)

This is in contrast to level 1 data fusion, in which hypothesis generation is relatively straightforward. Hypothesis generation for threat assessment is difficult to automate and, therefore, will benefit from human involvement. The analyst can select batches of data that he finds suspicious. He may suggest possible situation explanations for the automatic system to evaluate.

Hypothesis evaluation assigns plausibility scores to generated hypotheses. In Bayesian net implementations, for example, scoring is in terms of a joint pdf. In threat assessment, generated hypotheses are evaluated in terms of entities' capability, intent, and opportunity to carry out various actions, to include attacks of various kinds. Hypothesis evaluation involves issues of uncertainty management.

Hypothesis selection involves comparing competing hypotheses to select those for subsequent action. Depending on metrics used in hypothesis evaluation, this can involve selecting the highest-scoring hypothesis (e.g., with the greatest likelihood score). Hypothesis selection in STA involves issues of efficient search of large, complex graphs. A practical threat assessment system may employ efficient search methods for hypothesis selection, with final adjudication by a human analyst.[86]

18.5.4   State Estimation in Situation and Threat Assessment

State estimation for STA involves the patterns of inference listed in Section 18.1.3:

1. Inferring the presence and the states of entities on the basis of relations in which they participate
2. Inferring relationships on the basis of entity states and other relationships
3. Recognizing and characterizing extant situations
4. Projecting undetermined (e.g., future) situations

Considerable effort has been devoted in the last few years to apply to STA the rich body of science and technology in such fields as pattern recognition, information discovery and extraction, model-based vision, and machine learning.

Techniques that are employed in STA generally fall into three categories: data-driven, model-driven, and compositional methods.

- *Data-driven methods* are those that discover patterns in the data with minimal reliance on *a priori* models. Prominent among such techniques that have been used for STA are link-analysis methods. Data-driven methods can employ *abductive* inference techniques to explain discovered patterns.

- *Model-driven methods* recognize salient patterns in the data by matching with *a priori* models. Such techniques for STA include graph matching and case-based reasoning. Model-driven methods typically involve *deductive* inference processes.

- *Compositional methods* are hybrids of data- and model-driven methods. They adaptively build and refine explanatory models of the observed situation by discovering salient characteristics and composing hypotheses to explain the specific characteristics and interrelations of situation components. In general, these involve *inductive* (generalization) and *abductive* (explanatory) methods integrated with *deductive* (recognition) inference methods to build and refine partial hypotheses concerning situation components.

Figure 18.17 shows the ways that such techniques could operate in a hybrid, compositional STA system. Data-driven techniques search available data, seeking patterns. Model-driven

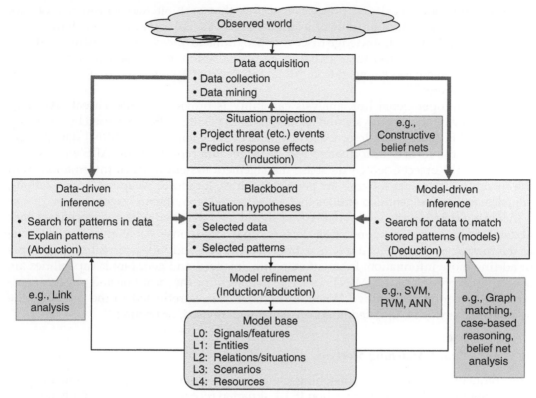

FIGURE 18.17
Notional hybrid situation and threat assessment architecture.

techniques search the model set, seeking matching data. Discovered structures—hypothesized situation fragments—are maintained on a blackboard. Inductive techniques project situations, for example, to predict threat situations and threat events. They also predict the effectiveness of candidate system actions in the projected future situations, including the effectiveness of candidate data acquisition actions to resolve, refine, and complete the situation hypotheses.

Together with this major operational flow (thick arrows in the figure) is a maintenance flow, which evaluates and refines the model base.

We can touch upon only a representative selection of these techniques.

18.5.4.1 Link Analysis Methods

Link analysis is an important technique in situation analysis (as defined in Section 18.1.1). Link analysis methods discover associations of one sort or another among reported entities. They have been used extensively in law enforcement and counter-terrorism to link people, organizations, and activities. This is a sophisticated technology to search for linkages in data. However, it tends to generate unacceptable false alarms. While classic link analysis methods are effective in detecting linkages between people and organizations, their means for recognizing operationally significant relationships or behavior patterns are generally weak. In addition, finding anomalous, possibly suspicious patterns of activity requires assumptions concerning joint probabilities of individual activities, for which statistics are generally lacking.

There are serious legal and policy constraints that prohibit widespread use of link discovery methods. As a prominent example, DARPA's TIA program was chartered to develop data mining or knowledge discovery programs to uncover relationships to detect terrorist organizations and their activities. In response to widespread privacy concerns, the program's name was changed from "Total Information Awareness" to "Terrorist Information Awareness."[87]

A currently operational link analysis capability is provided in the Threat HUMINT Reporting, Evaluation, Analysis, and Display System (THREADS) developed by Northrop Grumman. THREADS was designed to deal with heavy volumes of HUMINT data that are primarily in the form of unformatted text messages and reports. THREADS employs the Objectivity/DB object-oriented database management system to track information from the incoming messages looking for people, facilities, locations, weapons, organizations, vehicles, or similar identified entities that are extracted from the messages.

Incoming data reports are monitored, analyzed, and matched with the stored network of detected links, possibly generating alerts for analyst response. An analyst can select one of the messages and review a tree of extracted attributes and links, correcting any errors or adding more information. Inherent mapping, imagery, and geographic capabilities are used in interpreting the HUMINT data; e.g., to determine the locations and physical attributes of threat-related entities. When an updated case is returned to the database, the stored link network is updated: links are generated, revised, or pruned.[88]

18.5.4.2 Graph Matching Methods

STA frequently employs graph matching techniques for situation recognition and analysis. According to our definitions in Section 18.1.1, situation recognition fundamentally involves the matching of relational structures found in received data to stored structures defining situation-classes. A survey of the many powerful graph matching methods that have been developed is found in Ref. 89.

Graph matching involves (a) processing the received data to extract a relational structure in the form of a semantic net and (b) comparing that structure with a set of stored graphs to find the best match. Matching is performed using the topologies of the data-derived and model graph (generally called the *data graph* and the *template graph*, respectively). To be of use in threat assessment applications, the graph matching process must incorporate techniques to (a) resolve uncertain and incomplete data; (b) efficiently search and match large graphs; and (c) extract, generate and validate template graphs.

In the simplest case, the search is for 1:1 isomorphism between data and template graphs. In most practical applications, however, the data graph will be considerably larger than the template graph, incorporating instantiation specifics and contextual information that will not have been modeled. In such cases, the template graph will at best be isomorphic with a subgraph of the data graph. This is typically the case in automatic target recognition, in which a target of interest is viewed within a scene, features of which may be captured in the data graph but will not have been anticipated in the target template graph.

In addition, parts of the template graph may not be present in the data graph, so that matching is between a subgraph of the former with that of the latter (i.e., partial isomorphism). In the target recognition example, the target may be partially occluded or obscured. A three-dimensional target viewed from one perspective will occlude part of its own structure.

The search time for isomorphic subgraphs is generally considerably longer than that for isomorphic graphs.[90]

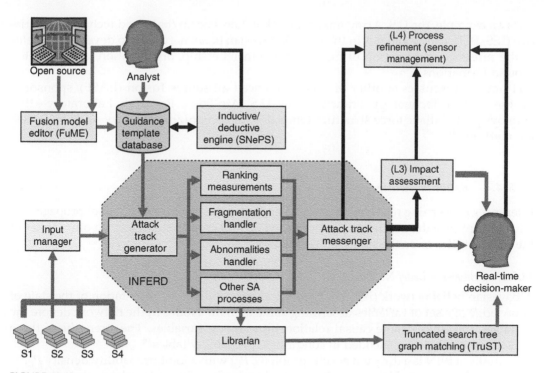

FIGURE 18.18
INFERD high-level fusion: situation awareness, impact assessment, and process refinement. (From Sudit, M., *CMIF Information Fusion Technologies*, CMIF Internal Presentation, 2005. With permission.)

Similar problems occur in STA domains. Often, the template does not capture the real-world variability in the modeled target or situation class. This, of course, is the expectation in threat assessment, given the variability inherent in intelligent, opportunistic adversaries as well as in complex attack contexts. To operate robustly in such cases, isomorphism matching criteria must be relaxed. A formulation of inexact graph matching is presented in Ref. 91.

Researchers at the State University of New York at Buffalo have developed a graph matching system, information fusion engine for real-time decision-making (INFERD), which they have applied to threat assessment problems in such domains as natural disaster response,[64,65] cyber attacks,[92] and urban/asymmetric warfare.[93]

INFERD is an information fusion engine that adaptively builds knowledge hierarchically from data fusion levels 1 through 4 (Figure 18.18).[94] The most recent version of INFERD incorporates advanced features to minimize the dependence on perfect or complete *a priori* knowledge, while allowing dynamic generation of hypotheses of interest. The most important among these features is abstraction of *a priori* knowledge into meta-hypotheses called Guidance Templates. In level 2 fusion, template nodes (generated from feature trees) are composed into acyclic directed graphs called "Attack Tracks". The latter are used by INFERD to represent the current situation. Attack Tracks are instantiated over a Guidance Template, a meta-graph containing every "known" possible node and edge.

18.5.4.3 Template Methods

Graph matching is a class of model-driven methods. Other methods using a variety of templating schemes have been developed.

As an example, the U.S. Army imagery exploitation system/balanced technology initiative (IES/BTI), uses templates in the form of Bayesian belief networks to develop estimates of the presence and location of military force membership, organization, and expected ground formations.[72,95]

Hinman[96] discusses another project—enhanced all-source fusion (EASF), sponsored by the Fusion Technology Branch of the U.S. Air Force Research Laboratory—that employs probabilistic force structure templates for recognizing military units by Bayesian methods.[97]

18.5.4.4 Belief Networks

Inferences concerning relationships are amenable to the machinery of belief propagation, whereby entity states are inferred conditioned on other entity states (i.e., on other states of the same entities or on the states of other entities).

18.5.4.4.1 Bayesian Belief Networks

A Bayesian belief network (BBN) is a graphical, probabilistic representation of the state of knowledge of a set of variables describing some domain. Nodes in the network denote the variables and links denote causal relationships among variables. The strengths of these causal relationships are encoded in conditional probability tables.[98]

Formally, a BBN is a directed acyclic graph (G, P_G) with a joint probability density P_G.

Each node is a variable in a multivariate distribution. The network topology encodes the assumed causal dependencies within the domain: lack of links among certain variables represents a lack of direct causal influence. BBNs are, by definition, causal. That is to say, influence flows from cause to effect or from effect to cause without loops: there is no directed path from any node back to itself that can be represented by a directed graph. For example, graphs such as given in Figures 18.19a and 18.19b are feasible BBNs, but not ones like Figure 18.19c.

The joint distribution for a BBN is uniquely defined by the product of the individual distributions for each random variable.

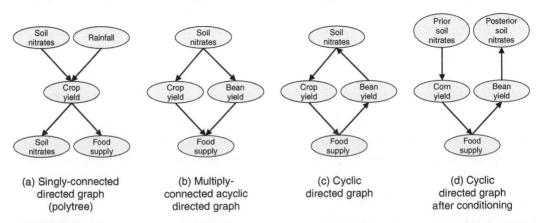

(a) Singly-connected directed graph (polytree)

(b) Multiply-connected acyclic directed graph

(c) Cyclic directed graph

(d) Cyclic directed graph after conditioning

FIGURE 18.19
Sample network topologies.

Belief in the node X of a belief network is the probability distribution $p(X|e)$ given all the evidence received. The posterior probability of a node state $X = x$ after receiving evidence e_X is

$$p(x|e_X) = \frac{(e_X|x)p(x)}{p(e_X)} = \alpha\lambda(x)p(x) \tag{18.20}$$

Belief revision is accomplished by message passing between the nodes. Figures 18.19a and 18.19b show examples of causal networks, such as BBNs. Causal networks strictly distinguish evidence as causes or effects:

- *Causal evidence.* Evidence that passes to a variable node X from nodes representing variables that causally determine the state of the variable represented by X, designated $\pi(X) = p(e_X^+)$
- *Diagnostic evidence.* Concerning effects of X, $\lambda(X) = p(e_X^-)$

The plus and minus signs reflect the conventional graphical depiction, as seen in Figure 18.20, in which causal evidence is propagated as downward-flowing and diagnostic evidence as upward-flowing messages.

The *a posteriori* belief in the value of a state of X is

$$
\begin{aligned}
Bel(X) &= p(X|e_X^+, e_X^-) \\
&= \frac{p(e_X^+, e_X^-, X)}{p(e_X^+, e_X^-)} \\
&= \frac{p(e_X^-|e_X^+, X)p(X|e_X^+)p(e_X^+)}{p(e_X^+, e_X^-)} \\
&= \alpha p(e_X^-|e_X^+, X)p(X|e_X^+) \\
&= \alpha\pi(X)\lambda(X)
\end{aligned}
\tag{18.21}
$$

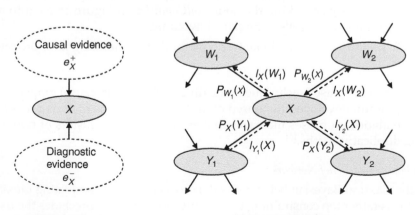

(a) Belief revision in a Bayesian net

(b) Message passing example

FIGURE 18.20
Belief revision in causal Bayesian networks.

$\lambda(X) = p(e_x^-)$ is the joint likelihood function on X:

$$\lambda(x) = \prod_i p(y_i|x)$$

(18.22)

for the respective observed states y_i of all causal children Y_i of X. This is illustrated in Figure 18.2b.

Evidence from a parent W to a child node X is passed as conditionals on the probability distribution of that evidence:

$$\pi_W(x) = p(x|W) = \sum_j p(x|w_j)p(w_j)$$

(18.23)

$$\pi(x) = \sum_{w_1,...,w_k} p(x|w_1,...,w_k)\prod_{i=1}^k \pi_{W_i}(x)$$

(18.24)

As is evident from the summation notation, Bayesian belief networks deal in dependencies among discrete-valued variables. In dealing with a continuous variable, it is necessary to "discretize" the data (i.e., subdivide the range of the variables into a mutually exclusive set of intervals). Das[98] discusses optimal maximum entropy solutions to discretization.

Evaluating large graphs can be daunting, and various pruning strategies can be employed to eliminate weak dependencies between nodes: trading optimality for speed.[98]

There is a concern for maintaining information integrity in multiply-connected graphs. For example, in Figure 18.19b, the state of *SoilNitrates* affects the states of both *CornYield* and *BeanYield*, which in turn affect the state of *FoodSupply*. The problem arises in double-counting the effects of the common variable, *SoilNitrates*, on *FoodSupply*.

Pearl[11] discusses three techniques that solve this problem with various degrees of success:

- *Clustering.* Forming compound variables to eliminate multiple connectivity (e.g., combining *CornYield* and *BeanYield* in Figure 18.19b to a common node)
- *Conditioning.* Decomposing networks to correspond to multiple instantiations of common nodes (e.g., breaking the node *SoilCondition* in Figure 18.19c into a prior and a posterior node to reflect the causal dynamics)
- *Stochastic simulation.* Using Monte Carlo methods to estimating joint probabilities at a node

One of the most widely used technique for dealing with loops in causal graphs is the Junction Tree algorithm, developed by Jensen et al.[99,100] Like Pearl's clustering algorithm, the Junction Tree algorithm passes messages that are functions of clusters of nodes, thereby eliminating duplicative paths.[101]

18.5.4.4.2 Generalized Belief Networks

As noted, the classical Bayesian belief network is concerned with evaluating *causal influence networks*. This is rather too constraining for our purposes in that it precludes the use of complex random graph topologies of the sorts encountered in the data graphs and template graphs needed for many STA problems, for example, graphs with loops as shown in Figure 18.19c.

Yedida et al.[101] provide a generalization that mitigates this restriction. Their Generalized Belief Propagation formulation—which they show to be equivalent to Markov Random

Field and to Bethe's free energy approximation formulations—models belief in a state x_j of a node X

$$b_X(x_j) = k\phi_X(x_j) \prod_{W \in N(X)} m_{W,X}(x_j) \tag{18.25}$$

in terms of "local" evidence $\phi_X(x_j)$ and evidence passed as messages from other nodes W in the immediate neighborhood $N(X)$ in the graph that represents the set of relationships in the relevant situation:

$$m_{W,X}(x_j) = \sum_{w_i} \phi_W(w_i)\psi_{W,X}(w_i, x_j) \prod_{Y \in N(W) \setminus X} m_{Y,W}(w_i) \tag{18.26}$$

Joint belief, then, is given as

$$b_{W,X}(w_i, x_j) = k\psi_{W,X}(w_i, x_j)\phi_W(w_i)\phi_X(x_j) \prod_{Y \in N(W) \setminus X} m_{Y,W}(w_i) \prod_{Z \in N(X) \setminus W} m_{Z,X}(x) \tag{18.27}$$

As before, k is normalizing constant ensuring that beliefs sum to 1.

18.5.4.4.3 Learning of Belief Networks

Learning belief networks entails learning both the dependency structure and conditional probabilities. Das[98] evaluates applicability of various techniques as a function of prior knowledge of the network structure and data observability (Table 18.5).

18.5.4.4.4 Belief Decision Trees

Additional induction methods have been developed based, not on joint probability methods, but on the Dempster–Shafer theory of evidence.[102–104] These methods are designed to reduce the sensitivity of the decision process to small perturbations in the learning set, which has been a significant problem with existing decision tree methods. Because of the capability to express various degrees of knowledge—from total ignorance to full knowledge—Dempster–Shafer theory is an attractive way of building recognition systems that are forgiving in the absence of well-modeled domains.

TABLE 18.5

Techniques for Learning of Belief Networks

Structure Observability	Known Network Structure	Unknown Network Structure
Fully observable variables	Maximum likelihood (ML) estimation Bayesian Dirichlet (BD)	Entropy, Bayesian or MDL score based (e.g., K2) Dependency constraint based
Partially observable variables	Expectation maximization (EM) Gibb's sampling Gradient descent based (e.g., APN)	Expectation maximization (EM) (e.g., structured EM)

Source: Das, S., *Proceedings of the Eighth International Conference on Information Fusion*, Philadelphia, 2005. With permission.

Vannoorenberghe[102] augments the Belief Decision Tree formulation with several machine-learning techniques—including bagging and randomization—and claims strikingly improved performance.

18.5.4.5 *Compositional Methods*

In assessing well-modeled situations (e.g., a chess match or a baseball game), it should be possible to define a static architecture for estimating and projecting situations.

This, however, is not the case in the many situations where human behavior is relatively unconstrained (as in most threat assessment problems of interest).

In particular, adapting to unanticipated behavior of interacting agents requires an ability to create predictive models of behaviors that a purposeful agent might exhibit and determine the distinctive observable indicators of those behaviors.

Often, however, the specific behaviors that must be predicted are unprecedented and unexpected, making them unsuitable for template-driven recognition methods. This drives a need to shift from the *model recognition paradigms* familiar in automatic target recognition to a *model discovery paradigm*.

Under such a paradigm, the system will be required to *compose* models that explain the available data, rather than simply to *recognize* preexisting models in the data. In such a system, the assessment process will need to be adaptive and opportunistic. That is to say, it cannot assume that it has a predefined complete set of threat scenarios available for simple pattern-matching recognition (e.g., using graph matching). Rather, situation and event hypotheses must be generated, evaluated, and refined as the understanding of the situation evolves.

A process of this sort has been developed under the DARPA evidence extraction and link detection (EELD) program. The system, continuous analysis and detection of relational evidence (CADRE) was developed by BAE Systems/Alphatech. CADRE discovers and interprets subtle relations in data.

CADRE is a link detection system that takes in a threat pattern and partial evidence about threat cases and outputs threat hypotheses with associated inferred agents and events. CADRE uses a Prolog-based frame system to represent threat patterns and enforce temporal and equality constraints among pattern slots. Based on rules involving uniquely identifying slots in the pattern, CADRE triggers an initial set of threat hypotheses, and then refines these hypotheses by generating queries for unknown slots from constraints involving known slots. To evaluate hypotheses, CADRE scores each local hypothesis using a probabilistic model to create a consistent, high-value global hypothesis by pruning conflicting lower-scoring hypotheses.[105]

Another approach is to represent the deployed situation by means of a constructive Belief Network as illustrated in Figure 18.21.

An individual object, such as the mobile missile launcher at the bottom of the figure, can be represented either as a single node in the graph, or as a subgraph of components and their interrelationships. In this way the problems of Automatic Target Recognition (ATR), scene understanding and situation assessment are together solved via a common belief network at various levels of granularity.

Such an approach was implemented under the DARPA multi-source intelligence correlator (MICOR) program.[106] The system composed graphs representing adversary force structures, including the class and activity state of military units, represented as nodes in the graph.

To capture the uncertainty in relational states as well as in entity attributive states, the Bayesian graph method was extended by the use of labeled directed graphs.

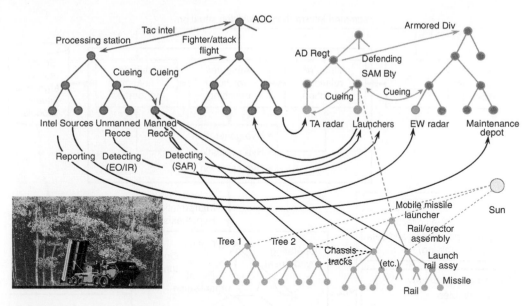

FIGURE 18.21
A representative graph of a tactical military situation. (From Steinberg, A.N., *Threat Assessment Issues and Methods*, Tutorial presented at Sensor Fusion Conference, Marcus-Evans, Washington, DC, December 2006. With permission.)

Applying this to the generalized belief propagation formulation of Section 18.5.4.4.2, we expand Equation 18.26 by marginalizing over relations:

$$m_{W,X}(x_j) = \sum_{w_i} \phi_W(w_i)\psi_{W,X}(w_i, x_j) \prod_{Y \in N(W) \backslash X} m_{Y,W}(w_i)$$

$$= \sum_{w_i} \phi_W(w_i) \sum_r p(w_i, x_j \mid r(W, X)) f(r(W, X)) \prod_{Y \in N(W) \backslash X} m_{Y,W}(w_i) \tag{18.28}$$

The effect is to build belief networks in which entity state variables and relational variables are nodes.

A reference processing architecture for adaptively building and refining situation estimate models is shown in Figure 18.22. The architecture elaborates the notional blackboard concept of Figure 18.17. The specific design extends one developed for model-based scene understanding and target recognition.

This adaptive process for model discovery iteratively builds and validates *interpretation hypotheses*, which attempt to explain the available evidence. A feature extraction process searches available data for indicators and differentiators (i.e., supporting and discriminating evidence) of situation types of interest (e.g., movements of weapons, forces, or other resources related to threat organizations).

Hypothesis generation develops one or more interpretation hypotheses, which have the form of labeled directed graphs (illustrated at the bottom of the figure). In a threat assessment application, interpretation hypotheses concern the capability, opportunity, and intent of agents to carry out various actions. These conditions for action are decomposed into mutually consistent sets and evaluated against the available evidence.

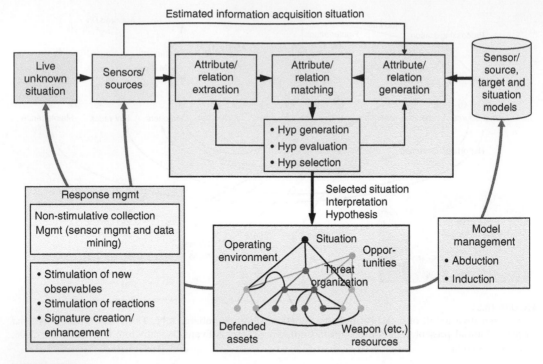

FIGURE 18.22
Reference architecture for SA/TA. (From Steinberg, A.N., *Threat Assessment Issues and Methods*, Tutorial presented at Sensor Fusion Conference, Marcus-Evans, Washington, DC, December 2006. With permission.)

Threat hypotheses include

- Inference of threatening situations: the capability, opportunities, and intent of one entity x to (adversely) affect an entity y
- Prediction of threat situations and events (e.g., attacks): interactions whereby entities (adversely) affect entities of interest

Threat situations and threat events are inferred on the basis of the attributes and relationships of the entities involved.

Estimates of physical, informational, and psychological states of such entities are used to infer both *actual* and *potential relationships* among entities. Specifically, threat situations are inferred in terms of one entity's capability, opportunity, and intent to affect (other) entities. A predicted or occurring threat event is characterized in terms of its direction (i.e., planning, command, and control), execution, and outcome.

The process closes the loop by means of a resource management function that nominates information acquisition actions. A utility/probability/cost model is used to predict the cost-effectiveness of such actions in terms of (a) the predicted utility of particular information, (b) the probability of attaining such information given various actions, and (c) the cost of such actions.[107] Utility is calculated in terms of the expected effectiveness of specific information to refine hypotheses or to resolve among competing hypotheses as needed to support the current decision needs (i.e., to map from possible world space to decision space).[21,45]

Information acquisition actions can include the intentional stimulation of the information environment to elicit information, as shown in Figure 18.11c. Stimulation to induce information sources to reveal their biases requires an inventory of models for various classes of behavior.[76] A key research goal is the development and validation of methods for the systematic understanding and generalization of such models—the abductive and inductive elements of Figure 18.22.

18.5.4.6 Algorithmic Techniques for Situation and Threat Assessment

Various techniques applicable to situation recognition and discussed in Section 18.5 have been applied to problems of impact and threat assessment.

Hinman[96] discusses a range of techniques that have been assessed for these purposes under the sponsorship of the Fusion Technology Branch of the U.S. Air Force Research Laboratory. These include

- Bayesian techniques, using force structure templates for recognizing military units.[97]

- Knowledge-based approaches for behavior analysis, specifically for vehicle recognition but with distinct applicability to the recognition of organizations and their activities. Hidden Markov Models are used to capture doctrinal information and uncertainty.[108]

- Neural networks for situation assessment, using back propagation training in a multilayer neural network to aggregate analysts' pairwise preferences in situation recognition. Dempster–Shafer methods are used to resolve potential conflicts in these outputs.[109]

- Neural networks for predicting one of a small number of threat actions (attack, retreat, feint, hold). A troop deployment analysis module performs a hierarchical constrained clustering on the battle map. The resulting battlefield cluster maps are processed by a tactical situation analysis module, using both rule-based and neural network methods to predict enemy intent.[110]

- Fuzzy logic techniques were evaluated to recognize enemy courses of action and to infer intent and objectives (discussed in Section 18.2.4).[111]

- Genetic algorithms are used in a planning-support tool called FOX, which rapidly generates and assesses battlefield courses of action (COAs). Wargaming techniques enable a rapid search for desirable solutions.[112] As such, the tool can be applied to impact assessment by generating potential friendly and adversary COAs and predicting their interactions and consequences.

18.5.5 Data Management

Data management is an important issue in developing and operating practical systems for STA. In particular, there is the need to manage and manipulate very large graphical structures. It is not uncommon for data graphs used in link analysis to have thousands of nodes. Network representations of situation hypotheses can have several hundred nodes. Furthermore, systems must have means to represent the sorts of uncertainty typical of military or national security threat situations. Often, this is performed by entertaining multiple hypotheses.

This leads to issues of data structure, search, and retrieval. Solutions have been developed at both the application and infrastructure layers.

18.5.5.1 Hypothesis Structure Issues

McMichael and coworkers at CSIRO have developed what they call *grammatical methods* for representing and reasoning about situations.[113–115]

Their approach has numerous attractive features. Expressive power is obtained by a compositional grammar. Much like a generative grammar for natural language, this compositional grammar obtains expressive power by allowing the recursive generation of a very large number of very complex structures from a small number of components. The method extends Steedman's Combinatory Categorical Grammar[116] to include grammatical methods for both set and sequence combinations. The resulting Sequence Set Combinatory Categorical Grammar generates parse trees from track data. These trees are parsed to generate situation trees.[115]

Dynamic situations are readily handled. Such a situation is partitioned into set and sequence components, rather analogous to the SPAN and SNAP constructs of SNePS.[59,60] *Set components* are entities that persist over time despite changes: physical objects, organizational structures, and the like. *Sequence components* are dynamic states of the set components. These can be *stages* of individual dynamic entities. They can also be *episodes* of dynamic situations. Both sets and sequences are amenable to hierarchical decomposition: a military force may have a hierarchy of subordinate units; a complex dynamic situation can involve a hierarchy of subordinate situations. Therefore, a situation can be represented as a tree structure—called a *sequence set tree*—as illustrated in Figure 18.23.

This hierarchical graphical formulation has value not only in its representational richness, but in allowing efficient processing and manipulation. *The aim of the representational scheme is to maximize the state complexity (expressive power) while minimizing the complexity of what is expressed.*[117]

Dependency links among tree nodes—to include continuity links between stages of a dynamically changing component and interaction links between components—are represented within the tree structure. So doing transforms situation representations for highly connected graphs such as in Figures 18.14 and 18.21, which cannot be searched in polynomial time, to trees that can.

A pivot table algorithm is used for fast situation extraction.[114]

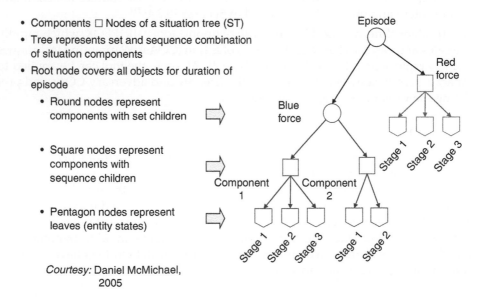

Courtesy: Daniel McMichael,
2005

FIGURE 18.23
An example of McMichael's situation tree. (From Steinberg, A.N., *Threat Assessment Issues and Methods*, Tutorial presented at Sensor Fusion Conference, Marcus-Evans, Washington, DC, December 2006. With permission.)

18.5.5.2 Data Repository Structure Issues

Let us examine the implications for data management of the STA issues listed in Table 18.2. Table 18.6 shows features of a database management system that help to address these issues.

Many current systems maintain large databases using a relational data base management system (RDBMS), such as Oracle. An RDBMS performs well in the following cases:

- *Ad hoc* queries, for which relationships do not need to be remembered
- Static change policy for the database and schema
- Centralized database server
- Storage of data that is logically amenable to representation in two-dimensional tables

However, as data become more dynamic and more complex, in terms of greater interconnectivity and greater number of data types, the overhead in an RDBMS becomes excessively cumbersome.

Furthermore, the demand for larger, distributed enterprises in intelligence analysis increases this burden.

Therefore, in many of the applications of interest—those driven by issues such as listed in Table 18.6—it is found that an object-oriented data base management system (OODBMS) should be preferred.

In contrast to an RDBMS, an OODBMS such as Objectivity/DB has the following desired features:

- Data of any complexity can be directly represented in the database.
- Relationships are persisted, so that there is no need to rediscover the relationship upon each use (e.g., by matching a local key to a foreign key).

TABLE 18.6

Data Management Issues for Situation Threat Assessment

Issue	Distributed	Scalable	Responsive	Flexible	Representation of Complex Data Structures	Assured Availability
Unanticipated situations and threats	√	√	√	√	—	√
Consistent management of uncertainty	√	√	—	√	√	√
Complex and subtle relationships	—	—	—	—	√	—
Inferring human and group behavior	√	—	√	—	√	—
Learning and managing models	—	√	—	—	√	—
Actionable intelligence	—	—	√	√	—	√

- Direct storage of objects and relationships provides efficient storage and searching of complex trees and networks of objects.
- Applications with data that have many relationships or many objects with many interrelationships benefit greatly.
- Support for complex data and inheritance enables creation of a comprehensive schema that models the union of large numbers of disparate data sources.
- A comprehensive global view to data can open up many data mining opportunities.
- Once objects are stored in the database, any number of arbitrary structures can be created on top of the data to rapidly find and access those objects again, such as maps, lists, sets, trees, and specialized indices. Multidimensional indices can be created over the data objects.[118]

18.6 Summary

Every technological discipline undergoes a formative phase in which the underlying concepts and principles are only vaguely perceived. During that phase, much unfocused energy is inefficiently expended, as researchers experiment with a multiplicity of fragmentary methods, with little sense of concerted progress.

The disciplines of situation and, especially, threat assessment have been in that nascent phase since people began thinking systematically about them, perhaps in the 1970s. Eventually, a rigorous conceptual foundation will be established, enabling profound progress to be made rapidly and efficiently.

We are seeing very encouraging signs that we are nearing that point. Recent developments in inferencing methods, knowledge representation, and knowledge management are beginning to bear fruit in practical STA systems.

References

1. D. Lambert, A blueprint for higher-level information fusion systems, *Information Fusion*, 2008 (forth coming).
2. F.E. White, A model for data fusion, *Proceedings of the First National Symposium on Sensor Fusion*, GACIAC, IIT Research Institute, Chicago, IL, vol. 2, 1988.
3. A.N. Steinberg, C.L. Bowman and F.E. White, Revisions to the JDL Model, *Joint NATO/IRIS Conference Proceedings*, Quebec, October 1998; reprinted in *Sensor Fusion: Architectures, Algorithms and Applications*, Proceedings of the SPIE, vol. 3719, 1999.
4. A.N. Steinberg and C.L. Bowman, Rethinking the JDL data fusion model, *Proceedings of the MSS National Symposium on Sensor and Data Fusion*, June 2004.
5. K. Devlin, *Logic and Information*, Press Syndicate of the University of Cambridge, Cambridge, MA, 1991.
6. E.T. Nozawa, Peircean semeiotic: A new engineering paradigm for automatic and adaptive intelligent systems, *Proceedings of the Third International Conference on Information Fusion*, vol. 2, pp. WEC4/3–WEC410, July 10–13, 2000.
7. J. Roy, From data fusion to situation analysis, *Proceedings of the Fourth International Conference on Information Fusion*, vol. II, pp. ThC2-3–ThC2-10, Montreal, 2001.
8. A.-L. Jousselme, P. Maupin and É. Bossé, Uncertainty in a situation analysis perspective, *Proceedings of the Sixth International Conference on Information Fusion*, vol. 2, pp. 1207–1214, Cairns, Australia, 2003.

9. M.R. Endsley, Toward a theory of situation awareness in dynamic systems, *Human Factors*, 37(1), 1995.
10. M.R. Endsley, Theoretical underpinnings of situation awareness: A critical review, in M.R. Endsley and D.J. Garland (Eds.), *Situation Awareness Analysis and Measurement*, Lawrence Erlbaum Associates Inc., Mahwah, NJ, 2000.
11. J. Pearl, *Probabilistic Reasoning in Intelligent Systems: Networks of Plausible Inference*, Morgan Kaufmann Publishers, San Mateo, CA, 1988.
12. J.S. Yedidia, W.T. Freeman and Y. Weiss, Understanding belief propagation and its generalizations, in G. Lakemeyer and B. Nebel (Eds.), *Exploring Artificial Intelligence in the New Millennium*, Morgan Kaufmann Publishers, New York, NY, 2002.
13. J. Barwise and J. Perry, *Situations and Attitudes*, Bradford Books, MIT Press, Cambridge, MA, 1983.
14. A. Khalil, *Computational Learning and Data-Driven Modeling for Water Resource Management and Hydrology*, PhD dissertation, Utah State University, 2005.
15. A.N. Steinberg, Unifying data fusion levels 1 and 2, *Proceedings of the Seventh International Conference on Information Fusion*, Stockholm, 2004.
16. J. Llinas, C. Bowman, G. Rogova, A. Steinberg, E. Waltz and F. White, Revisiting the JDL data fusion model II, in P. Svensson and J. Schubert (Eds.), *Proceedings of the Seventh International Conference on Information Fusion*, Stockholm, 2004.
17. J.J. Salerno, Where's level 2/3 fusion: A look back over the past 10 years, *Proceedings of the Tenth International Conference on Information Fusion*, Quebec, 2007.
18. D.A. Lambert, A unification of sensor and higher-level fusion, *Proceedings of the Ninth International Conference on Information Fusion*, pp. 1–8, Florence, Italy, 2006.
19. D.A. Lambert, Tradeoffs in the design of higher-level fusion systems, *Proceedings of the Tenth International Conference on Information Fusion*, Quebec, 2007.
20. R. Mahler, Random set theory for target tracking and identification, in D.L. Hall and J. Llinas (Eds.), *Handbook of Multisensor Data Fusion*, CRC Press, London, 2001.
21. A.N. Steinberg, Threat assessment technology development, *Proceedings of the Fifth International and Interdisciplinary Conference on Modeling and Using Context (CONTEXT'05)*, pp. 490–500, Paris, July 2005.
22. Steinberg, A.N., *Threat Assessment Issues and Methods*, Tutorial presented at Sensor Fusion Conference, Marcus-Evans, Washington, DC, December 2006
23. D.A. Lambert, Situations for situation awareness, *Proceedings of the Fourth International Conference on International Fusion*, Montreal, 2001.
24. *Merriam-Webster's Collegiate Dictionary*, Tenth edition, Merriam-Webster, Inc., Springfield, MA, 1993.
25. B. McGuinness and L. Foy, A subjective measure of SA: The Crew Awareness Rating Scale (CARS), *Proceedings of the First Human Performance: Situation Awareness and Automation Conference*, Savannah, GA, October 2000.
26. J. Salerno, M. Hinman and D. Boulware, Building a framework for situation awareness, *Proceedings of the Seventh International Conference on Information Fusion*, Stockholm, July 2004.
27. J. Salerno, M. Hinman and D. Boulware, The many faces of situation awareness, *Proceedings of the MSS National Symposium on Sensor and Data Fusion*, Monterey, CA, June 2004.
28. É. Bossé, J. Roy and S. Wark, *Concepts, Models and Tools for Information Fusion*, Artech House, Boston, MA, 2007.
29. H. Curry and R. Feys, *Combinatory Logic, Volume 1*, North-Holland Publishing Company, Amsterdam, 1974.
30. A. Tarski, The semantic conception of truth and the foundations of semantics, *Philosophy and Phenomenological Research* IV(3), 341–376, 1944 (Symposium on Meaning and Truth).
31. *Doctrinal Implications of Operational Net Assessment (ONA)*, United States Joint Forces Command, Joint Warfare Center Joint Doctrine Series, Pamphlet 4, February 24, 2004.
32. G. Bronner, *L'incertitude, Volume 3187 of Que sais-je?* Presses Universitaires de France, Paris, 1997.
33. M. Smithson, *Ignorance and Uncertainty: Emerging Paradigms*, Springer-Verlag, Berlin, 1989.
34. P. Krause and D. Clark, *Representing Uncertain Knowledge: An Artificial Intelligence Approach*, Kluwer Academic Publishers, New York, NY, 1993.

35. B. Bouchon-Meunier and H.T. Nguyen, *Les incertitudes dans les syst`emes intelligents, Volume 3110 of Que sais-je?* Press Universitaires de France, Paris, 1996.
36. G.J. Klir and M.J. Wierman, *Uncertainty-Based Information: Elements of Generalized Information Theory, Volume 15 of Studies in Fuzziness and Soft Computing,* 2nd edition. Physica-Verlag, Heidelberg, New York, NY, 1999.
37. P. Smets, Imperfect information: Imprecision—uncertainty, in A. Motro and P. Smets (Eds.), *Uncertainty Management in Information Systems: From Needs to Solutions,* Kluwer Academic Publishers, Boston, MA, 1977.
38. L. Wittgenstein, *Philosophical Investigations,* Blackwell Publishers, Oxford, 2001.
39. K. Perusich and M.D. McNeese, Using fuzzy cognitive maps for knowledge management in a conflict environment, *IEEE Transactions on Systems, Man and Cybernetics—Part C,* 26(6), 810–821, 2006.
40. P. Gonsalves, G. Rinkus, S. Das and N. Ton, A hybrid artificial intelligence architecture for battlefield information fusion, *Proceedings of the Second International Conference on Information Fusion,* Sunnyvale, CA, 1999.
41. P.C.G. Costa, *Bayesian Semantics for the Semantic Web,* Doctoral dissertation, Department of Systems Engineering and Operations Research, George Mason University, Fairfax, VA, 2005.
42. P.C.G. Costa, K.B. Laskey and K.J. Laskey, Probabilistic ontologies for efficient resource sharing in semantic web services, *Proceedings of the Second Workshop on Uncertainty Reasoning for the Semantic Web (URSW 2006),* held at the Fifth International Semantic Web Conference (ISWC 2006), Athens, GA, 2006.
43. K.B. Laskey, P.C.G. da Costa, E.J. Wright and K.J. Laskey, Probabilistic ontology for net-centric fusion, *Proceedings of the Tenth International Conference on Information Fusion,* Quebec, 2007.
44. E.G. Little and G.L. Rogova, An ontological analysis of threat and vulnerability, *Proceedings of the Ninth International Conference on Information Fusion,* pp. 1–8, Florence, Italy, 2006.
45. A.N. Steinberg, An approach to threat assessment, *Proceedings of the Eighth International Conference on Information Fusion,* vol. 2, p. 8, Philadelphia, 2005.
46. A.N. Steinberg, Predictive modeling of interacting agents, *Proceedings of the Tenth International Conference on Information Fusion,* Quebec, 2007.
47. D.A. Lambert, Personal communication, August 2007.
48. M. Uschold and M. Grüninger, Ontologies: Principles, methods and applications, *Knowledge Engineering Review,* 11(2), 93–155, 1996.
49. B. Krieg-Brückner, U. Frese, K. Lüttich, C. Mandel, T. Mossakowski and R.J. Ross, Specification of an ontology for route graphs, in Freksa et al. (Eds.), *Spatial Cognition IV,* LNAI 3343, Springer-Verlag, Berlin, Heidelberg, 2005.
50. M.R. Genesereth and N.J. Nilsson, *Logical Foundations of Artificial Intelligence,* Morgan Kaufmann, Los Altos, CA, 1987.
51. M.M. Kokar, *Ontologies and Level 2 Fusion: Theory and Application,* Tutorial presented at International Conference on Information Fusion, 2004.
52. Plato, *Parmenides,* First Edition (fragment), Collection of the Great Library of Alexandria, Egypt, 370 BC.
53. K.B. Laskey, *MEBN: A Logic for Open-World Probabilistic Reasoning* (Research Paper), C41-06-01, http://ite.gmu.edu/klaskey/index.html, 2005.
54. M. Kokar, Choices in ontological languages and implications for inferencing, *Presentation at Center for Multisource Information Fusion Workshop III, Critical Issues in Information Fusion,* Buffalo, NY, September, 2004.
55. *Protégé Ontology Editor and Knowledge Acquisition System, Guideline Modeling Methods and Technologies,* Open Clinical Knowledge Management for Medical Care, http://www.openclinical.org/gmm_protege.html.
56. E. Astesiano, M. Bidoit, B. Krieg-Brückner, H. Kirchner, P.D. Mosses, D. Sannella and A. Tarlecki, CASL: The common algebraic specification language, *Theoretical Computer Science,* 286, 2002.
57. P.D. Mosses (Ed.), *CASL Reference Manual, Volume 2960 of Lecture Notes in Computer Science,* Springer-Verlag, Berlin, 2004.

58. K. Lüttich, B. Krieg-Brückner and T. Mossakowski, *Ontologies for the semantic web in CASL, W3C Recommendation*, http://www.w3.org/TR/owl-ref/, 2004.

59. S.C. Shapiro and W.J. Rapaport, The SNePS Family, *Computers & Mathematics with Applications*, 23(2–5), 243–275, 1992. Reprinted in F. Lehmann (Ed.), *Semantic Networks in Artificial Intelligence*, pp. 243–275, Pergamon Press, Oxford, 1992.

60. S.C. Shapiro and the SNePS Implementation Group, *SNePS 2.6.1 User's Manual*, Department of Computer Science and Engineering, University at Buffalo, The State University of New York, Buffalo, NY, October 6, 2004.

61. *IDEF5 Ontology Description Capture Overview*, Knowledge Based Systems, Inc., http://www.idef.com/IDEF5.html.

62. *IDEF5 Method Report, Information Integration for Concurrent Engineering (IICE)*, Knowledge Based Systems, Inc., http://www.idef.com/pdf/Idef5.pdf.

63. C.J. Matheus, M. Kokar and K. Baclawski, A core ontology for situation awareness, *Proceedings of the Ninth International Conference on Information Fusion*, vol. 1, pp. 545–552, Florence, Italy, 2003.

64. E. Little and G. Rogova, Ontology meta-modeling for building a situational picture of catastrophic events, *Proceedings of the Eighth International Conference on Information Fusion*, Philadelphia, 2005.

65. P.D. Scott and G.L. Rogova, Crisis management in a data fusion synthetic task environment, *Proceedings of the Seventh International Conference on Information Fusion*, Stockholm, July 2004.

66. R. Nagi, M. Sudit and J. Llinas, An approach for level 2/3 fusion technology development in urban/asymmetric scenarios, *Proceedings of the Ninth International Conference on Information Fusion*, pp. 1–5, Florence, Italy, 2006.

67. P. Grenon and B. Smith, SNAP and SPAN: Towards dynamic spatial ontology, *Spatial Cognition and Computation*, 4(1), 69–104, 2004.

68. P. Grenon, Spatiotemporality in basic formal ontology: SNAP and SPAN, upper-level ontology and framework for formalization, *IFOMIS Technical Report Series*, (http://www.ifomis.unisaarland.de/Research/IFOMISReports/IFOMIS%20Report%2005_2003.pdf), 2003.

69. B. Smith and P. Grenon, The cornucopia of formal-ontological relations, *Dialectica* 58(3), 279–296, 2004.

70. K.B. Laskey, S. Stanford and B. Stibo, *Probabilistic Reasoning for Assessment of Enemy Intentions*, Publications #94–25, George Mason University, 1994.

71. C.G. Looney and L.R. Liang, Cognitive situation and threat assessments of ground battlespaces, *Information Fusion*, 4(4), 297–308, 2003.

72. T.S. Levitt, C.L. Winter, C.J. Turner, R.A. Chestek, G.J. Ettinger and A.M. Sayre, Bayesian inference-based fusion of radar imagery, military forces and tactical terrain models in the image exploitation system/balanced technology initiative, *IEEE International Journal on Human-Computer Studies*, 42, 667–686, 1995.

73. A.N. Steinberg, J. Llinas, A. Bisantz, C. Stoneking and N. Morizio, Error characterization in human-generated reporting, *Proceedings of the MSS National Symposium on Sensor and Data Fusion*, McLean, VA, 2007.

74. J. Rasmussen, Skills, rules, and knowledge: Signals, signs, and symbols, and other distractions in human performance models, *IEEE Transactions on Systems, Man, and Cybernetics* SMC-13(3), 257–266, 1983.

75. C.L. Bowman, The data fusion tree paradigm and its dual, *Proceedings of the National Symposium on Sensor Fusion*, 1994.

76. A.N. Steinberg, Stimulative intelligence, *Proceedings of the MSS National Symposium on Sensor and Data Fusion*, McLean, VA, 2006.

77. M.G. Ceruti, A. Ashenfelter, R. Brooks, G. Chen, S. Das, G. Raven, M. Sudit and E. Wright, Pedigree information for enhanced situation and threat assessment, *Proceedings of the Ninth International Conference on Information Fusion*, pp. 1–8, Florence, Italy, 2006.

78. Steinberg, A.N. and Waltz, E.L., *Perceptions on Imagery Fusion*, Presented at NASA Data Fusion/Data Mining Workshop, Sunnyvale, CA, 1999

79. J.D. Erdley, Bridging the semantic gap, *Proceedings of the MSS National Symposium on Sensor and Data Fusion*, McLean, VA, 2007.

80. C.J. Matheus, D. Tribble, M.M. Kokar, M.G. Ceruti and S.C. McGirr, Towards a formal pedigree ontology for level-one sensor fusion, *The 10th International Command and Control Research and Technology Symposium (ICCRTS 2005)*, June 2005.

81. A.N. Steinberg, Open networks: Generalized multi-sensor characterization, *Proceedings of the Ninth International Conference on Information Fusion*, pp. 1–7, Florence, Italy, 2006.

82. P. Walley, *Statistical Reasoning with Imprecise Probabilities*, Chapman & Hall, London, 1991.

83. J. Llinas, New challenges for defining information fusion requirements, *Plenary Address, Tenth International Conference on Information Fusion*, International Society for Information Fusion, Quebec, 2007.

84. G. Rogova and V. Nimier, Reliability in information fusion: literature survey, *Proceedings of the Seventh International Conference on Information Fusion*, pp. 1158–1165, Stockholm, 2004.

85. E.R. Keydel, Multi-INT registration issues, *IDGA Image Fusion Conference*, Washington, DC, 1999.

86. B.V. Dasarathy, Information fusion in the context of human-machine interface, *Information Fusion*, 6(2), 2005.

87. P. Rosenzeig and M. Scardaville, The need to protect civil liberties while combating terrorism: legal principles and the Total Information Awareness program, *Legal Memorandum #6*, http://www.heritage.org/Research/HomelandSecurity/lm6.cfm, February 6, 2003.

88. *Objectivity Platform in THREADS—Northrop Grumman Mission Systems Application Assists U.S. Counter Terrorist and Force Protection Analysts in the Global War on Terrorism*, Objectivity, Inc. Application Note, 2007.

89. D. Conte, P. Foggia, C. Sansone and M. Vento, Thirty years of graph matching in pattern recognition, *International Journal of Pattern Recognition and Artificial Intelligence*, 18(3) 265–298, 2004.

90. D.F. Gillies, *Computer Vision Lecture Course*, http://www.homes.doc.ic.ac.uk/~dfg/vision/v18.html.

91. A. Hlaoui and W. Shengrui, A new algorithm for inexact graph matching, *Proceedings of the Sixteenth International Conference on Pattern Recognition*, vol. 4, pp. 180–183, 2002.

92. M. Sudit, A. Stotz and M. Holender, Situational awareness of a coordinated cyber attack, *Proceedings of the SPIE Defense & Security Symposium*, vol. 5812, pp. 114–129, Orlando, FL, March 2005.

93. M. Sudit, R. Nagi, A. Stoltz and J. Delvecchio, Dynamic hypothesis generation and tracking for facilitated decision making and forensic analysis, *Proceedings of the Ninth International Conference on Information Fusion*, Florence, Italy, 2006.

94. Sudit, M., *CMIF Information Fusion Technologies*, CMIF Internal Presentation, 2005.

95. M.C. Stein and C.L. Winter, Recursive Bayesian fusion for force estimation, *Proceedings of the Eighth National Symposium on Sensor Fusion*, 1995.

96. M.L. Hinman, Some computational approaches for situation assessment and impact assessment, *Proceedings of the Fifth International Conference on Information Fusion*, vol. 1, pp. 687–693, Annapolis, MD, 2002.

97. M. Hinman and J. Marcinkowski, Final results on enhanced all source fusion, *Proceedings of the SPIE, Sensor Fusion: Architectures, Algorithms and Applications IV*, vol. 4051, pp. 389–396, 2000.

98. S. Das, Tutorial AM3: An integrated approach to data fusion and decision support, part I: Situation assessment, *Proceedings of the Eighth International Conference on Information Fusion*, Philadelphia, 2005.

99. F.V. Jensen and F. Jensen, Optimal Junction Trees. Uncertainty in Artificial Intelligence, 1994.

100. A.L. Madsen and F.V. Jensen, LAZY propagation: A junction tree inference algorithm based on lazy evaluation, *Journal of Artificial Intelligence*, 113(1), 1999.

101. Y.S. Yedida, W.T. Freeman and Y. Weiss, Understanding belief propagation and ts generalization, in G. Lakemeyer and B. Nevel (Eds.), *Exploring AI in the New Millennium*, pp. 239–269, 2002.

102. P. Vannoorenberghe, On aggregating belief decision trees, *Information Fusion*, 5(3), 179–188, 2004.

103. T. Fenoeux and M.S. Bjanger, Induction of decision trees from partially classified data using belief functions, *Proceedings of SMC2000*, Nashville, TN, IEEE, 2000.

104. Z. Elouedi, K. Mellouli and P. Smets, Belief decision trees: Theoretical foundations, *International Journal of Approximate Reasoning*, 18(2–3), 91–124, 2001.

105. N. Pioch, D. Hunter, C. Fournelle, B. Washburn, K. Moore, K.E. Jones, D. Bostwick, A. Kao, S. Graham, T. Allen and M. Dunn, CADRE: Continuous analysis and discovery from relational evidence, *Integration of International Conference on Knowledge Intensive Multi-Agent Systems*, pp. 555–561, 2003.

106. A.N. Steinberg and R.B. Washburn, Multi-level fusion for Warbreaker intelligence correlation, *Proceedings of the National Symposium on Sensor Fusion*, 1995.

107. R.C. Whitehair, *A Framework for the Analysis of Sophisticated Control*, PhD dissertation, University of Massachusetts CMPSCI Technical Report 95, February 1996.

108. C. Burns, A knowledge based approach to information fusion, *Presentation to the Air Force Scientific Advisory Board*, 2000.

109. G. Rogova, P. Losiewicz and J Choi, Connectionist approach to multi-attribute decision making under uncertainty, *AFRL-IF-RS-TR-1999-12*, 2000.

110. W. Wright, Artificial neural systems (ANS) fusion prototype, *AFRL-IF-RS-TR-1998-126*, 1998.

111. P. Gonsalves, G. Rinkus, S. Das and N. Ton, A Hybrid Artificial Intelligence Architecture for Battlefield Information Fusion, *Proceedings of the Second International Conference on Information Fusion*, Sunnyvale, CA, 1999.

112. J.S. Schlabach, C. Hayes and D.E. Goldberg, SHAKA-GA: A genetic algorithm for generating and analyzing battlefield courses of action (white paper), 1997, Cited in *Evolutionary Computation*, 7(1), 45–68, 1999.

113. G. Jarrad, S. Williams and D. McMichael. *A framework for total parsing, Technical Report 03/10*, CSIRO Mathematical and Information Sciences, Adelaide, Australia, January 2003.

114. D. McMichael, G. Jarrad, S. Williams and M. Kennett, Modelling, simulation and estimation of situation histories, *Proceedings of the Seventh International Conference on Information Fusion*, pp. 928–935, Stockholm, Sweden, June 2004.

115. D. McMichael and G. Jarrad, Grammatical methods for situation and threat analysis, *Proceedings of the Eighth International Conference on Information Fusion*, vol. 1, p. 8, Philadelphia, June 2005.

116. M. Steedman, *The Syntactic Process*, MIT Press, Cambridge, MA, 2000.

117. D. McMichael and G. Jarrad, *Grammatical Methods for Situation and Threat Analysis*, Tutorial Presented at the Eighth International Conference on Information Fusion, vol. 1, p. 8, Philadelphia, July 2005.

118. OODBMS vs. ORDMS vs. RDBMS: the pros and cons of different database technologies (white paper), Objectivity, Inc., 2006, http://www.objectivity.com/pages/downloads/whitepaper.asp#.

19

Introduction to Level 5 Fusion: The Role of the User

Erik Blasch

CONTENTS

19.1 Introduction

Before automated information fusion (IF) can be qualified as a technical solution to a problem, we must consider a defined need, function, and capability. The need typically includes data gathering and exploitation to augment the current workload such as a physician desiring assistance in detecting a tumor; an economist looking for inflationary trends; or a military commander trying to find, locate, and track targets. "Information needs" can only be designated by these users who will utilize the information fusion system (IFS) developments.[1] Once the needs have been defined, decomposing the needs into actionable functions requires data availability and observability. Finally, transforming the data into information to satisfy a need has to become a realizable capability. All too often, IF engineers gather user information needs and start building a system without developing in the interaction, the guidance, or the planned role of the users.

Level 5 fusion, preliminarily labeled "user refinement,"[2] places emphasis on developing the IF process model by utilizing the cognitive fusion capability of the user (1) to assist in the design and analysis of emerging concepts; (2) to play a dedicated role in control and management; and (3) to understand the necessary supporting activities to plan, organize, and coordinate sensors, platforms, and computing resources for the IFS. We will explore three cases—the prototype user, the operational user, and the business user. "Prototype users" must be involved throughout the entire design and development process to ensure that the system meets specifications. "Operational users" focus on behavioral aspects including *neglect, consult, rely,* and *interact.*[3] Many times, operational users lack trust and credibility in the fusion system because they do not understand the reasoning behind the fusion-system. To facilitate design acceptance, the user must be actively involved in the control and management of the data fusion system (DFS). "Business users" focus on refinement functions for *planning, organizing, coordinating, directing,* and *controlling* aspects in terms of resources, approval processes, and deployment requirements.

This chapter emphasizes the roles that users play in the IF design. To adequately design systems that will be acceptable, the prototype, operational, and business users must work hand-in-hand with the engineers to fully understand what IFSs can do (i.e., augment the users task functions) and can not do (i.e., make decisions without supporting data).[4] Together, engineers and users can design effective and efficient systems that have a defined function. As an example of a poorly defined system, one can only look at marketing brochures that infer that a *full spectrum of situational awareness will be provided of a city.* Engineers look at the statement and fully understand that to achieve full situational awareness (SA) (such as tracking and identifying all objects in the city), the entire city must be observable at all times, salient features must be defined for distinguishing all target types, and redundant sensors and platforms must be available and working reliably. However, users reading the brochure expect that any city location will be available to them at their fingertips, information will always be timely, and the system will require no maintenance (e.g., *a priori* values). To minimize the gap between system capability and user expectation that facilitates acceptance of fusion systems, level 5 fusion is meant to focus specifically on the process refinement opportunities for users.

This chapter is organized as follows. Section 19.2 focuses on issues surrounding the complexity of IF designs and the opportunities for users to aid in providing constraints and controls to minimize the complexity. Section 19.3 overviews the data fusion information group (DFIG) model, an extension to the Joint Directors of Laboratories (JDL) model. Section 19.4 details user refinement (UR) whereas Section 19.5 discusses supporting topics. Section 19.6 presents an example that develops a system prototype where operational users are involved in a design. Section 19.7 concludes with a summary.

19.2 User Refinement in Information Fusion Design

19.2.1 User Roles

IFSs are initiated, designed, and fielded based on a need: such as data compression,[5] spatial coverage, and task assistance. Inherently behind every fusion system design is a user in question. Users naturally understand the benefits of fusion such as multisensor *communication* with their eyes and ears,[6] multisensor *transportation control* of vehicles that augment movement,[7] and multisensor *economic management* of daily living through computers, transactions, and supply of goods and services.[8] Although opportunities for successful fusion abound, some of the assumptions are not well understood. For efficient communication, a point-to-point connection is needed. For transportation control, societal protocols are understood such as roadways, vehicle maintenance, and approved operators. For economic management, organization and business operations are contractually agreed on. On the basis of these examples, IFS designs also require individual, mechanical, and economic support dependent on operational, prototype, and business users.

Fusion designs can be separated into two modes: micro and macro levels. The *micro level* involves using a set of defined sensors within a constrained setting. Such examples include a physician requiring a fused result of a PET–MRI image, a driver operating a car, or a storeowner looking at the financial report. In each of these cases, the settings are well defined, the role of the fusion systems is constrained, and the decisions are immediate. The more difficult case is the *macro level* that deals with military decision support systems,[9] political representatives managing a society, or a corporate board managing a trans-continental business. The complexity of the macro level generally involves many users and many sensors operating over a variety of data and information sources with broad availability. An example of an aggregation of micro- and macro-level fusion concepts can be seen by the multilevel display in Figure 19.1. As shown, individual users sought to spatially minimize the information by placing the displays together. Although this is not the ideal "information-fusion" display, it does demonstrate that "business users" afforded the resources, "operational users" trusted the system, and that "prototype users" were able to provide the technology. To manage large, complex systems and hope that IF can be a useful tool to augment the macro-level systems, the appropriate constraints, assumptions, and defined user roles must be subscribed up front. One goal is to reduce the complexity of fusion at the macro level by creating a common operational picture[10] across sensors, targets, and environments.

The user (level 5) can take on many roles: such as a prioritizer, sensor, operator, controller, and planner, which can design, plan, deploy, and execute an IF design. Although there are a host of roles that could be determined, one taxonomy is presented in Table 19.1. This table is based on trying to field a typical IFS such as a target tracker and an identifier. The first role is that of a prioritizer, which determines what objects are of interest, where to look, and what supporting database information is needed. Most fusion algorithms require some initializing information such as *a priori* probabilities, objective functions, processing weights, and performance parameters. The second role is that of a sensor, which includes reporting of contextual and message reports, making decisions on data queries, and drafting information requests. The third role is that of an operator, which

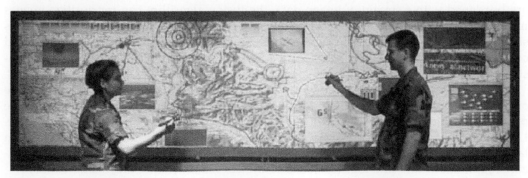

FIGURE 19.1
User information fusion of colocated displays.

TABLE 19.1

Users and their Refinements in the Stage of an Information Fusion Design

User	Conceptualizing	Designing	Deployment	Execution	Reporting
Prioritizer	Systems, database	Database size	Access	Maintenance of the database	Intelligence reports
Sensor	Visual observations	Training	People on station	Reports	HUMINT reports
Operator	Interface design to augment assigned task	Which actions in the interface	Monitor sensor images, signals	Monitor and report information	Screen selection
Controller	Sensor planning	Sensor suite	Turn on sensor platform routing	Steer the sensor to spots	Spot looks
Planner	Architecture	Budgets	Platform availability	Platform resources (e.g., gas)	Task orders

includes data verification such as highlighting images, filtering text reports, and facilitating communication and reporting. Many times the operator is trained for a specific task and has a defined role (which distinguishes the role vs. that of a sensor). A controller is a "business" operator who routes sensors and platforms, determines the mission of the day, and utilizes the IF output for larger goals. Finally, the planner is not looking at today's operations, but tomorrow's (e.g., how can the data gathering be better facilitated in the future to meet undefined threats and situations forthcoming).

The most fundamentally overlooked opportunities for successful fusion design is in the *conceptual phase*. Engineers take advantage of the inherent benefits of data fusion within the initial design process: developing, for instance, Bayesian rules that assume that exemplar data, *a prior* probabilities, and sensors will be available. Without adequate "user refinement" of the problem through constraints, operational planning, and user-augmented algorithms; the engineering design has little chance of being deployed. By working with a team of users, as in Table 19.1, we can work within operational limits in terms of available databases, observation exploitation capabilities, task allocations, platforms, and system-level opportunities. The interplay between the engineers and the users can build trust and efficacy in the system design.

In the *design phase*, users can aid in characterizing fusion techniques that are realizable. Two examples come to mind—one is object detection and the other is sensor planning. Because no automatic target recognition (ATR) fusion algorithm can reduce the false alarm rate (FAR) to zero, although simultaneously providing near-perfect probability of detection, individual users must somehow determine acceptable thresholds. Because operators are burdened with a stream of image data, the fusion algorithm can reprioritize images important to the user for evaluation (much like a prescreener). Another example includes sensor planning. IF requires the right data to integrate, and thus, having operators and controllers present in the design phase can aid in developing concept-of-operations constraints based on estimated data collection possibilities.

For macro-level systems, deployment, execution, and reporting within IF designs are currently notional since complexity limits successes. One example is in training,[5] where prototype and operational users work with the system designers to better the interface. Lessons from systems engineering can aid in IF designs. Systems engineering focuses on large, complex, and multiple levels of uncertainty systems. Many concepts such as optimization, human factors, and logistics focus on system's maintenance and sustainability.

Figure 19.2 highlights the concept that systems engineering costs escalate if users at all levels are not employed up front in the IF design. At the information problem definition

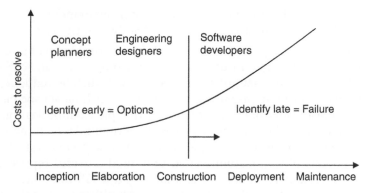

FIGURE 19.2
Design cycle—maybe test and evaluation.

stage, the research team has to determine what types of information is observable and available, and which information can be correlated with sensor design, data collection, and algorithm data exploitation. Thus, information needs and user tasks drive the concept design. During the development phase, which includes elaboration and construction, design studies, trade-offs, and prototypes need to be tested and refined (which includes the expected user roles). Llinas[11]* indicates that many of the design developments are prototypes; however, one of the main concerns in the development phase is to ensure that the user be a part of the fusion process to set priorities and constraints. One emerging issue is that IF designs are dependent on software developers. To control costs of deployment and maintenance, software developers need to work with users to be able to include operator controls such as data filtering, algorithm reinitialization, and process initiation.

19.2.2 Prioritization of Needs

To reduce information to a dimensionally minimal set requires the user to develop a hierarchy of needs to allow one to perform active reasoning over collected and synthesized data. Some needs that can be included in such a hierarchy are:

Things. Objects, number and types; threat, whether harmful, passive, or helpful; location, close or far; basic primitives, features, edges, and peaks; existence known or unknown, that is, new objects; dynamics, moving or stationary.

Processes. Measurement system reliability (uncertainty); ability to collect more data; delays in the measurement process; number of users and interactions.

The user must prioritize information needs for designers. The information priority is related to the information desired. The user must have the ability to choose or select the objects of interest and the processes from which the raw data is converted into the fused information. One of the issues in the processing of fused information is related to the ability to understand the information origin or pedigree. To determine an area of interest, for instance, a user must seek answers to basic questions:

Who—is controlling the objects of interest (i.e., threat information)
What—are the objects and activities of interest (i.e., diagnosis, products, and targets)
Where—are the objects located (i.e., regions of interest)
When—should we measure the information (i.e., time of day)

---

* See Chapter 25 of this handbook.

How—should we collect the information (i.e., sensor measurements)

Which—sensors to use and the algorithms to exploit the information

To reason actively over a complex prioritized collection, a usable set of information must be available along with collection uncertainties on which the user can act.[12] Ontology, *the seeking of knowledge*,[13] indicates that goals must be defined, such as reducing uncertainty. The user has to deal with many aspects of uncertainty, such as sensor bias, communication delays, and noise. Heisenberg uncertainty principle exemplifies the challenge of observation and accuracy.[14] Uncertainty is a measure of doubt,[15] which can prevent a user from making a decision. The user can supplement a machine-fusion system to account for deficiencies and uncertainties. Jousselme et al. presented a useful representation of the different types of uncertainty.[16] A state of ignorance can result in an error. An error can occur between expected and measured results caused by distortion of the data (mis-registered) or incomplete data (not enough coverage). With insufficient sensing capabilities, lack of context for situational assessment, and undefined priorities, IFSs will be plagued by uncertainty. Some of the categories of uncertainty are

Vague—not having a precise meaning or clear sensing[17]

Probability—belief in element of the truth

Ambiguity—cannot decide between two results

Priority of information needs and uncertainty reduction is based on *context*, which necessitates the user to augment IF designs with changing priorities, informal data gathering, and real-time mission planning.

19.2.3 Contextual Information

Context information, as supplied by a user, can augment object, situational, and threat understanding. There are certain object features the user desires to observe. By selecting objects of interest, target models can be called to assess feature matches.

Modeled target features can be used to predict target profiles and movement characteristics of the targets for any set of aspect angles. Object context includes a pose match between kinematic and feature relationships. Targeting includes selecting priority weights for targets of interest, and information modifications based on location (i.e., avenues of approach and mobility corridors). By properly exploiting the sensor, algorithm, and target, the user can select the ideal perspective from which to identify a target. Examples of contextual information aids include tracking,[18] robotics,[19] and medicine.

SA usually means that the user's team has some *a priori* world information. Aspects of context in situational assessment could include the sociopolitical environment of interest, *a priori* historical data, topology, and mission goals. Contextual information such as road networks, geographical information, and weather is useful.[20] In a military context, targets can be linked by observing selected areas, segmented by boundaries as shown in Figure 19.3. If there is a natural boundary of control, then the targets are more likely to be affiliated with the group defending that area. Types of situation contexts include

- Sociopolitical information
- Battlefield topography and weather
- Adjacent hostile and friendly force structures
- Tactical or strategic operational doctrine
- Tactical or strategic force deployments
- Supporting intelligence data

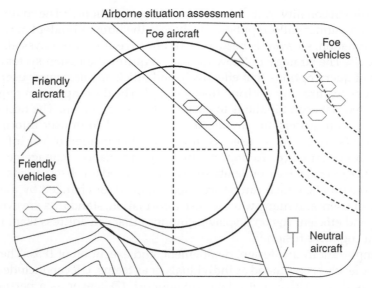

FIGURE 19.3
Situation assessment display.

From Figure 19.3, we see that if targets are grouped near other targets of the same allegiance, we could expect the classification of a target to the same as that of other spatially adjacent targets. Likewise, we could subscribe to the fact that trucks are on roads and only tanks can be off-road. Thus, information context plays a role in classifying the objects, developing situation understanding, and assessing threat priorities. The method by which a user integrates contextual information to manage the fusion process is *cognitive*.

19.2.4 Cognitive Fusion

Formally, IF integrates the machine and cognitive fusion functions:

IF is a formal mathematics, techniques, and processes to combine, integrate, and align data originating from different sources through association and correlation for decision making. The construct entails obtaining greater quality, decreased dimensionality, and reduced uncertainty for the user's application of interest.

We can subdivide the field of fusion technology into a number of categories:

Data fusion. Combining and organizing numerical entities for analysis
Sensor fusion. Combining devices responding to a stimulus
IF. Combining data to create knowledge
Display fusion. Presenting data simultaneously to support human operations
Cognitive fusion. Integrating information in user's mental model.

Display fusion is an example of integrating cognitive and machine fusion. Users are required to display machine-correlated data from individual systems. Figure 19.1 shows a team of people coordinating decisions derived from displays. Since we desire to utilize IF where appropriate, we assume that combining different sets of data from multiple sources would improve performance gain (i.e., increased quality and reduced uncertainty). Uncertainty about world information can be reduced through systematic and appropriate data combinations, but how to display uncertainty is not well understood. Because the world is complex (based on lots of information, users, and machines) we desire timely responses and extended dimensionality over different operating conditions.[21] The computer or fusion system is adept at numerical computations whereas the human is adept at reasoning over displayed information, so what is needed is an effective fusion of physical, process, and cognitive models. The strategy of "how to fuse" requires the fusion system to display a pragmatic set of query options that effectively and efficiently *decompose* user information needs into signal, feature, and decision fusion tasks.[12] The decomposition requires that we prioritize selected data combinations to ensure a knowledge gain. The decomposition of tasks requires fusion models and architectures developed from user information needs. Information needs are based on user requirements that include object information and expected fusion performance. To improve fusion performance, management techniques are necessary to direct machine or database resources based on observations, assessments, and mission constraints, which can be accepted, adapted, or changed by the user. Today, with a variety of users and many ways to fuse information, standards are evolving to provide effective and efficient fusion display techniques that can address team requirements for distributed cognitive fusion.[22]

Users are important in the design, deployment, and execution of IFSs. They play active roles at various levels of command as individuals or as teams. UR can include a wide range of roles such as (1) prioritizer with database support, (2) sensor as a person facilitating communication and reporting, (3) an operator who sits at an human–computer interface (HCI) terminal monitoring fusion outputs, (4) controller who routes sensors and platforms, and (5) planner who looks at the system-level operations. Recognizing the importance of the roles of the user, the DIFG sought to update the JDL model.

19.3   Data Fusion Information Group Model

Applications for multisensor IF require insightful analysis of how these systems will be deployed and utilized. Increasingly complex, dynamically changing scenarios arise, requiring more intelligent and efficient reasoning strategies. Decision making (DM) is integral to information reasoning, which requires a pragmatic knowledge representation for user interaction.[23] Many IF strategies are embedded within systems; however, the user-IFS must be rigorously evaluated by a standardized method that can operate over a variety of locations, including changing targets, differing sensor modalities, IF algorithms, and users.[24] A useful model is one that represents critical aspects of the real world.

The IF community has rallied behind the JDL process model with its revisions and developments.[25-27] Steinberg indicates that "the JDL model was designed to be a *functional model*—a set of definitions of the functions that could comprise any data fusion system." The JDL model, like Boyd's user-derived observe, orient, decide, and act (OODA) *loop* and the MSTAR predict, extract, match, and search (PEMS) *loop*, has been interpreted as a *process model*—a control-flow diagram specifying interactions among functions. It has been also categorized as a *formal mode* involving a set of axioms and rules for manipulating entities such as probabilistic, possibilistic, and evidential reasoning.

Figure 19.4 shows a user-fusion model that builds a *user model*[28] that emphasizes user interaction within a functional, process, and formal model framework. For example, SA includes the user's *mental model*.[29] The mental model provides a representation of the world as aggregated through the data gathering; IF design; and the user's perception of the social, political, and military situations. The interplay between the fusion results (formal), protocol (functional), control (process), and perceptual (mental) models form the basis of UR.

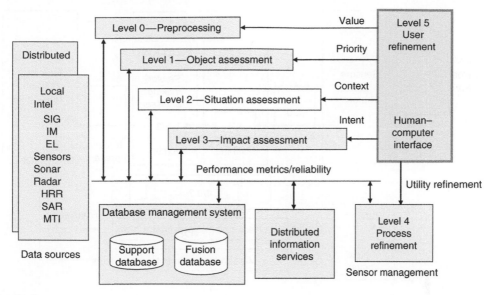

FIGURE 19.4
User-fusion model.

The current team* (now called the *data fusion information group*) assessed this model, shown in Figure 19.5, to advocate the user as a part of management (control) versus estimation (fusion). In the DFIG model,[†] the goal was to separate the IF and management functions. Management functions are divided into sensor control, platform placement, and user selection to meet mission objectives. level 2 (SA) includes tacit functions that are inferred from level 1 explicit representations of object assessment. Because algorithms cannot process data not observed, data fusion must evolve to include user knowledge and reasoning to fill these gaps. The current definitions, based on the DFIG recommendations, include

Level 0: *Data assessment.* Estimation and prediction of signal/object observable states on the basis of pixel/signal level data association (e.g., information systems collections)

Level 1: *Object assessment.* Estimation and prediction of entity states on the basis of data association, continuous state estimation, and discrete state estimation (e.g., data processing)

Level 2: *Situation assessment.* Estimation and prediction of relations among entities, to include force structure and force relations, communications, etc. (e.g., information processing)

Level 3: *Impact assessment.* Estimation and prediction of effects on situations of planned or estimated actions by the participants, to include interactions between

---

* Frank White, Otto Kessler, Chris Bowman, James Llinas, Erik Blasch, Gerald Powell, Mike Hinman, Ed Waltz, Dale Walsh, John Salerno, Alan Steinberg, David Hall, Ron Mahler, Mitch Kokar, Joe Karalowski, and Richard Antony.

† The views expressed in the chapter are those of the authors and do not reflect the official position of the DFIG.

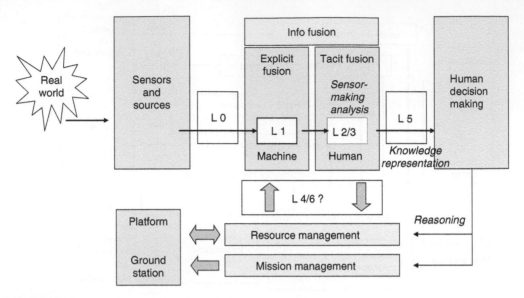

FIGURE 19.5
DFIG 2004 model.

action plans of multiple players (e.g., assessing threat actions to planned actions and mission requirements, performance evaluation)

Level 4: *Process refinement (an element of resource management).* Adaptive data acquisition and processing to support sensing objectives (e.g., sensor management and information systems dissemination, command/control)

Level 5: *UR (an element of knowledge management).* Adaptive determination of who queries information and who has access to information (e.g., information operations), adaptive data retrieved and displayed to support cognitive DM and actions (e.g., HCI)

Level 6: *Mission management (an element of platform management).* Adaptive determination of spatial-temporal control of assets (e.g., airspace operations) and route planning and goal determination to support team DM and actions (e.g., theater operations) over social, economic, and political constraints.

19.3.1 Sensor Management

The DFIG model updates include acknowledging the resource management of (1) sensors, (2) platforms that may carry one or multiple sensors, and (3) user control; each with differing user needs based on levels of command. The user not only plans the platform location but can also reason over situations and threats to alter sensing needs. For a timeline assessment, (1) the user plans days in advance for sensing needs, (2) platforms are routed hours in advance, and (3) sensors are slewed in seconds, to capture the relevant information.

Level 4, *process refinement*, includes sensor management and control of sensors and information. To utilize sensors effectively, the IFS must explore service priorities, search methods (breadth or depth), and determine the scheduling and monitoring of tasks. Scheduling is a control function that relies on the aggregated state-position, state-identity, and uncertainty management for knowledge reasoning. Typical methods used are (1) objective cost function for optimization, (2) dynamic programming (such as neural-network [NN] methods and reinforcement-learning based on a goal), (3) greedy information-theoretic approaches for optimal task assignment,[30] or (4) Bayesian networks that aggregate probabilities to accommodate rule constraints. Whichever method is used, the main idea is to reduce the uncertainty.

Level 4 control should entail hierarchical control functions as well. The system is designed to accommodate five classes of sensing (static, event, trigger, trend, and drift), providing a rich source of information about the user's information needs. The most primitive sensor class, *state*, allows the current value of an object to be obtained, whereas the *event* class can provide notification that a change of value has occurred. A higher-level sensor class, *trigger*, enables the measurement of the current value of a sensor when another sensor changes its value, that is, a temporal combination of *state* and *event* sensing. The design supports the logging of sensor data over time, and the production of simple statistics, known as *trend* sensing. A fifth sensor class, *drift*, is used for detecting when trend statistics cross a threshold. This can be used to indicate that a situation trends toward an undesirable condition, perhaps a threat. There is also a need for checking the integrity, reliability, and quality of the sensing system.

Issues must also involve *absolute and relative assessment* as well as local and global referencing. Many IF strategies are embedded within systems; however the user-IFS must be rigorously evaluated by a standardized approach that assesses a wide range of locations, changing targets, differing sensor modalities, and IF algorithms.

19.3.2 User Interaction with Design

The IF community has seen numerous papers that assumed a data-level, bottom-up, level 1 object refinement design, including target tracking and ID. Some efforts explored level 1 aggregation for SA.[31,32] However, if the fusion design approach was top-down, then the community would benefit to start the IF designs by asking the customer what they need. The customer is the IFS user, whether the business person, analyst, or commander. If we ask the customers what they want, they would most likely request something that affords reasoning and DM, given their perceptual limitations.[33] Different situations will drive different needs; however, the situation is never fully known *a priori*. There are general constructs that form an initial set of situation hypotheses, but in reality not all operating conditions will be known. To minimize the difficulty of trying to assess all known/ unknown *a priori* situational conditions, the user helps to develop these *a priori* conditions that help to define the more relevant models. In the end, IFS designs would be based on the user, not strictly the machine or the situation.

Management includes business, social, and economic effects on the fusion design. Business includes managing people, processes, and products. Likewise IF is concerned with managing people, sensors, and data. The ability to design an IFS with regard to physical, social, economic, military, and political environments would entail user reasoning about the data to infer information. User control should be prioritized as user, mission, and sensor management. For example, if sensors are on platforms, then the highest ranking official coordinating a mission determines who is in charge of the assets (not under automatic control). Once treaties, air space, insurance policies, and other documentation are in place, only then would we want to turn over control to an automatic controller (e.g., sensor).

UR, level 5 fusion could form the basis to (1) advocate standard fusion metrics, (2) recommend a rigorous fusion evaluation and training methodology, (3) incorporate dynamic decision planning, and (4) detail knowledge presentation for fusion interface standards. These issues are shown in Figure 19.6.

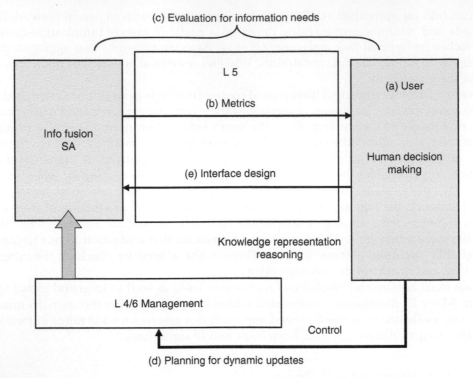

FIGURE 19.6
User refinement of fusion system needs.

19.4 User Refinement

Level 5, UR,[2] functions include (1) selecting models, techniques, and data; (2) determining the metrics for DM; and (3) performing higher-level reasoning over the information based on the user's needs as shown in Figure 19.7. The user's goal is to perform a task or a mission. The user has preconceived expectations and utilizes the levels 0–4 capabilities of a machine to aggregate data for DM. To perform a mission, a user employs engineers to design a machine to augment his/her workload. Each engineer, as a vicarious user, imparts decision assumptions into the algorithm design. However, since the machine is designed by engineers, system controls might end up operating the fusion system in ways different from what the user needs. For example, the user plans ahead (forward reasoning) whereas the machine reasons over collected data (backward reasoning). If a delay exists in the IFS, a user might deem it useless for planning (e.g., sensor management—where to place sensor for future observations).

A user (or IF designer) is forced to address situational constraints. We define UR operations as a function of responsibilities. Once an IF design is ready, the user can act in a variety of ways: *monitoring* a situation in an active or passive role or *planning* by either reacting to new data or providing proactive control over a future course of actions (COAs). When a user interacts with an IFS, it is important that they supply knowledge reasoning. The user has the abilities to quickly grasp where key target activity is occurring and use this information to reduce the search space for the fusion system algorithm and hence, guide the fusion system process.

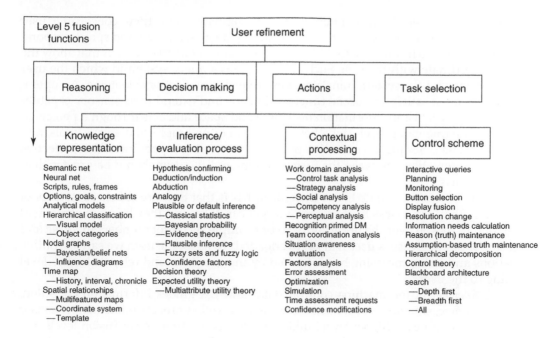

FIGURE 19.7
Level 5: User-refinement functions.

19.4.1   User Action

User actions have many meanings that could be conveyed by SA and assessment needs. One of the key issues of SA (defined in Section 19.5) is that the IF must map to the user's perceptual needs: spatial awareness,[34] neurophysiological,[35] perceptual,[36] and psychological,[37] and that those needs combine perceptually and cognitively.[6,38,39] If the display/delivery of information is not consistent with the user expectations, all is lost. The machine cannot reason, as Brooks stated in "Elephant's don't play chess."[40] A machine cannot deal with unexpected situations.[41] For this reason, there are numerous implications for (1) incorporating the user in the design process, (2) gathering user needs, and (3) providing the user with available control actions.

Users are not Gaussian processors, yet they do hypothesis confirmation. In hypothesis reasoning, we can never prove the null condition, but we can disprove the alternative hypothesis. Level 5 is intended to address the cognitive SA that includes knowledge representation and reasoning methods. The user defines a fusion system, for without a user, there is no need to provide fusion of multisensory data. The user has a defined role with objectives and missions. Typically, the IF community has designed systems that did not require user participation. However, through years of continuing debate, we find that the user plays a key role in the design process by

Level 0: determining what and how much data *value* to collect
Level 1: determining the target *priority* and where to look
Level 2: understanding scenario *context* and user role
Level 3: defining what is a threat and adversarial *intent*
Level 4: determining which sensors to deploy and activate by assessing the *utility* of information
Level 5: designing user interface controls (Figure 19.1)

Kokar and coworkers[42] stress ontological and linguistic[43] questions concerning user interaction with a fusion system such as semantics, syntactics, efficacy, and spatiotemporal queries. Developing a framework for UR requires *semantics* or *interface actions* that allow the system to coordinate with the user. Such an example is a query system in which the user seeks questions and the system translates these requests into actionable items. An operational system must satisfy the users' functional needs and extend their sensory capabilities. A user fuses data and information over time and space and acts through a perceived world reference (mental) model—whether in the head or on graphical displays, tools, and techniques.[44] The IFS is just an extension of the user's sensing capabilities. Thus, effective and efficient interactions between the fusion system and the user should be greater than the separate parts.

The reason why users will do better than machines is that they are able to reason about the situation, they can assess likely routes a target can take, and they bring in contextual information to reason over the uncertainty.[16] For example, as target numbers increase and processing times decrease, a user will get overloaded. To assist the user, we can let routine calculations and data processing be performed by the computer and relegate higher-level reasoning to the user.

Process refinement of the machine controls data flow. UR guides information collection, region of coverage, situation and context assessment as well as process refinement of sensor selections. Although each higher level builds on information from lower fusion levels, the refinement of fused information is a function of the user. Two important issues are *control* of the information and the *planning* of actions. There are many social issues related to centralized and distributed control such as team management and individual survivability. For example, the display shown in Figure 19.1 might be globally distributed, but centralized for a single commander for local operations. Thus, the information displayed on an HCI should not only reflect the information received from lower levels, but also afford analysis and distribution of commander actions, directives, and missions. Once actions are taken, new observable data should be presented as operational feedback. Likewise, the analysis should be based on local and global operational changes and the confidence of new information. Finally, execution should include updates including orders, plans, and actions for people carrying out mission directives. Thus, process refinement is not the fusing of signature, image, track, and situational data, but that of *decision information fusion* (DEC-IF) for functionality.

DEC-IF is a refined assessment of fused observational information, globally and locally. DEC-IF is a result on which a user can plan and act. Additionally, data should be gathered by humans to aid in deciding what information to collect, analyze, and distribute to others. One way to facilitate the receipt, analysis, and distribution of actions is that of the OODA loop. The OODA loop requires an interactive display for functional cognitive DM. In this case, the display of information should orient a user to assess a situation based on the observed information (i.e., SA). These actions should be assessed in terms of the user's ability to deploy assets against the resources, within constraints, and provide opportunities that capitalize on the environment.

Extended information gathering can be labeled as *action information fusion* (ACT-IF) because an assessment of possible/plausible actions can be considered. The goal of any fusion system is to provide the user with a functionally actionable set of refined information. Taken together, the user-fusion system is actually a *functional sensor*, whereas the traditional fusion model just represents sensor observations. The UR process not only determines *who* wants the data and whether they can refine the information, but also *how* they process the information for knowledge. Bottom-up processing of information can take the form of physically sensed quantities that support higher-level functions. To include machine automation as well as IF, it is important to afford an interface that allows the user to interact with the system (i.e., sensors). Specifically, the top-down user can query necessary information to plan and focus on specific situations while the automated system can work in the background to update on the changing environment. Thus, the functional fusion system is concerned with processing the entire situation while a user can typically focus on key activities. The computer must process an exhaustive set of combinations to arrive at a decision, but the human is adept at cognitive reasoning and is sensitive to the changes in situational context.

The desired result of data processing is to identity entities or locally critical information spatially and temporally. The difficulty in developing a complete understanding is the sheer amount of data available. For example, a person monitoring many regions of interest (ROIs) waits for some event or *threat* to come into the sensor domain. Under continuously collecting sensors, most processed data are entirely useless. Useful data occurs when a sensed *event* interrupts the monitoring system and initiates a threat update. Decision makers only need a subset of the entire data to perform successful SA and impact assessment to respond to a threat. Specifically, the user needs a time-event update on an object of interest. Thus, data reduction must be performed. A fusion strategy can reduce this data over time as well as space, by creating entities. Entity data can be fused into a single common operating picture for enhanced cognitive fusion for action; however, user control options need to be defined.

19.4.2 User Control

Table 19.2 shows different user management modes that range from passive to active control.[45] In the table, we see that the best alternative for fusion systems design is "management by interaction." Management by interaction can include both compensatory and noncompensatory strategies. The *compensatory strategies* include *linear additive* (weighting inputs), *additive difference* (compares alternative strategies), or *ideal point* (compare alternative against desired strategies) methods. The *noncompensatory* alternative selective strategies include *dominance* (strategy alternatives equal in all dimensions except one), *conjunctive* (passes threshold on all dimensions), *lexicographic* (best alternative strategy in one dimension, if equal strategies are based on second dimension), *disjunctive* (passes threshold on one dimension), *single feature difference* (best alternative strategy over alternative dimensions of greatest difference), *elimination by aspects* (choose one strategy based on probabilistic elimination of less important dimensions), and *compatibility test* (compare alternative strategies based on weighted threshold dimensions) strategies. In the strategies presented, both the user and the DFS require rules for DM. Because the computer can only do computations, the user must select the best set of rules to employ.

One method to assess the performance of a user or fusion process for target identification (ID) (for instance) is through the use of receiver operator curves (ROC). Management by interaction includes *supervisory* functions such as monitoring, data reduction and mining, and control. An ROC plots the hit rate (HR) versus the FAR. The ROC could represent any modality such as hearing, imaging, or fusion outputs. As an example, we can assess the performance of a user-fusion system by comparing the individual ROC to the performance of the integrated user-fusion system. The user may act in three ways (1) *management-by-ignorance*—turn off fusion display, do manual manipulation (cognitive-based); (2) *management-by-acceptance*—agree with the computer (i.e., automatic); or (3) *management-by-interaction*—adapt with output, reasoning (i.e., associative). What is needed is the reduction of data that the user needs to process.

For instance, consider that the human might desire *acceptance*—such as the case when the HR is high and independent of the FAR; or the user might desire *ignorance*—when we want FAR to be low, but track length high. However an appropriate combination of the two

TABLE 19.2

Continuum of System Control and Management for Operators

Management Mode	Automation Functions	User Functions	Level of Automation	Level of Involvement
Autonomous operation	Complete autonomous operation; system may or may not be capable of being bypassed	Controller has no active role; monitors fault detection; controller not usually informed	Very high	Very low
Management by exception	Essentially autonomous operation; automatic decision selection; system informs controller and monitors responses	Controller is informed of system intent; may intervene by reverting to lower level	High	Low
Management by consent	Decisions are made by automation; controller must consent to decisions before implementation	Controller consent to decisions; controller may select alternatives and decision options	Medium	Some
Management by delegation	Automation takes action only as directed by controller; level of assistance is selectable	Controller specifies strategy and may specify level of computer authority	Some	Medium
Management by interaction	Automation cues operator to key attributes and fusion of incoming data	Directs, controls, plans, and coordinates fusion for action; proactive and predictive control	High	High
Assisted control	Control automation is not available; database information is available	Direct authority over all decisions; voice control and coordination	Low	High
Unassisted control	Complete computer failure; no assistance is available	Procedural control, unaided decision making	Very low	Very high

Source: Adapted from Billings, C.E., *Aviation Automation: The Search for the Human-Centered Approach*, Lawrence Erlbaum Associates, Mahwah, NJ, 1997.

approaches would lead to a case of *interaction*—adjust decisions (hit [H]/false alarm [FA]) based on fusion ROC performance for unknown environments, weak fusion results, or high levels of uncertainty. From Figure 19.8 we see there is the possibility for fusion gain (moving the curve to the upper left) is associated with being able to process more information. One obvious benefit of the user to the fusion system is that the user can quickly rule out obvious false alarms (thereby moving the curve left). Likewise, the fusion system can process more data, thereby cueing more detections (moving the curve up). Operating on the upper left of the 2D curve would allow the user to increase confidence in DM over an acceptable FAR.

We have outlined user control actions; now let us return to estimation.

19.4.3 User Interaction with Estimation

It is important to understand which activities are best suited for user engagement with a DFS. We need to understand the impact of the user on fusion operations, communication delays, as well as target density. For example, UR includes monitoring to ensure that the computer is operating properly. Many UR functions are similar to business management where the user is not just operating independently with a single machine but operating with many people to gather facts and estimate states. Typically, one key metric is reaction time[44] that determines which user strategy can be carried out efficiently.

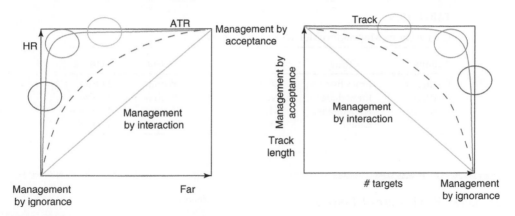

FIGURE 19.8
2D ROC analysis combined user actions and fusion system performance gain.

UR can impact near- and far-term analyses over which the human must make decisions. For the case of a *short-term analysis,* the human *monitors* (active vs. passive) the system by being engaged with the system or just watching the displays. For the *far-term* analysis, the human acts as a *planner*[46] (reactive vs. proactive) in which the user estimates long-term effects and consequences of information. The information presented includes not only the time span associated with collecting the data, but also the time delay needed to display the data (latent information), as well as the time between the presentation of information on the screen to the user's perception and reaction time. User reaction time includes physical responses, working associations, and cognitive reasoning. User focus and attention for action engagement can be partitioned into immediate and future information needs. Engagement actions can be defined as

Predictive actions. Projective, analytical approach to declare instructions in advance
Reactive actions. Immediate, unthinking response to a stimulus
Proactive actions. Active approach requiring an evaluation of the impact of the data on future situations
Passive actions. Waiting, nonthought condition allowing current state to proceed without interaction

The user's role is determined by both the type of DFS action and user goals. In Table 19.3, we see that the action the user chooses determines the time horizon for analysis and a mapping of the user actions.

Other user data metrics needed in the analysis and presentation of data and fusion information include integrity—formal correctness (i.e., uncertainty analysis); consistency—enhancing redundancy, reducing ambiguity, or double counting (i.e., measurement origination); and traceability—backward inference chain (i.e., inferences assessed as data is translated into information).

Integrity, consistency, and traceability map into trust, confidence, and assurance that the fused data display is designed to meet user objectives. To be able to relate the user needs for the appropriate design action, we use the automatic, associative, and cognitive levels of IF. To perform the actions of planning, prediction, and proactive control for fusion assessment, the user must be able to trust the current information and be assured that the fusion system is adaptable and flexible to respond to different requests and queries. Typical HCI results discussed in the literature offer a number of opportunities for the fusion community: user models, metrics, evaluation, decision analysis, work domain assessment, and display design, as detailed in Section 19.5.

TABLE 19.3

User Actions Mapped to Metrics of Performance

Actions	Time	Thought	Need	Future
Predictive	Projective	Some	Within	Future
Reactive	Immediate	None	Within	Present
Proactive	Anticipatory	Much—active	Across	Future +
Passive	Latent	Delayed/none	Across	Present

19.5 User-Refinement Issues

Designing complex and often-distributed decision support systems—which process data into information, information into decisions, decisions into plans, and plans into actions—requires an understanding of both fusion and DM processes. Important fusion issues include timeliness, mitigation of uncertainty, and output quality. DM contexts, requirements, and constraints add to the overall system constraints. Standardized metrics for evaluating the success of deployed and proposed systems must map to these constraints. For example, issues of *trust, workload*, and *attention*[47] of the user can be assessed as an added quality for an IFS to balance control management with perceptual needs.

19.5.1 User Models in Situational Awareness

Situation assessment is an important concept of how people become aware of things happening in their environment. SA can be defined as *keeping track of prioritized, significant events, and the conditions in one's environment*. Level 2 SA is the estimation and prediction of relations among entities, to include force structure and relations, communications, etc., which require adequate user inputs to define these entities. The human-in-the-loop (HIL) of a semiautomated system must be given adequate SA. According to Endsley, *SA is the perception of the elements in the environment within a volume of time and space, the comprehension of their meaning, and the projection of their status in the near future*.[48–50] This now-classic model, shown in Figure 19.9, translates into three levels:

- Level 1 SA: *Perception* of environment elements
- Level 2 SA: *Comprehension* of the current situation
- Level 3 SA: *Projection* of future states

Operators of dynamic systems use their SA to determine actions. To optimize DM, the SA provided by an IFS should be as precise as possible with respect to the objects in the environment (level 1 SA). An SA approach should present a fused representation of the data (level 2 SA) and provide support for the operator's projection needs (level 3 SA) to facilitate operator's goals. From the SA model presented in Figure 19.9, *workload*[51] is a key component of the model that affects not only SA, but also user decision and reaction time.

To understand how the human uses the situation context to refine the SA, we use the *recognition primed decision-making* (RPD) model,[52] shown in Figure 19.10. The RPD model develops user DM capability based on the current situation and past experience. The RPD model shows the goals of the user and the cues that are important. From the IFS, both the user and the IFS algorithm can cue each other to determine data needs. The user can cue the IFS by either selecting the data of interest or choosing the sensor collection information.

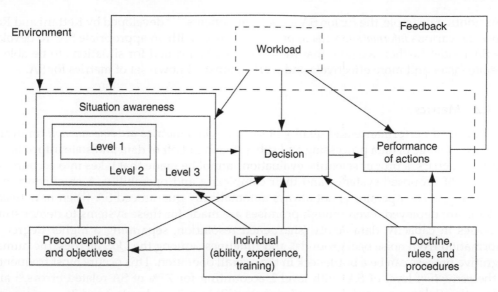

FIGURE 19.9
Endsley's situation awareness model.

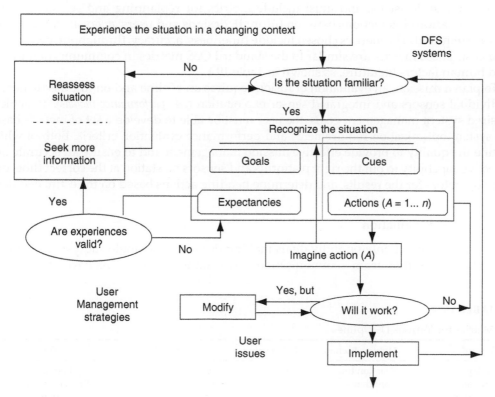

FIGURE 19.10
Recognition primed decision-making model for situational awareness.

As another example, the fusion SA model components[53,54] developed by Kettani and Roy show the various *information needs* to provide the user with an appropriate SA. To develop the SA model further, we note that the user must be primed for situations to be able to operate faster and more effectively and work against a known set of metrics for SA.

19.5.2 Metrics

Standardized performance and quality of service (QoS) metrics are *sine qua non* for evaluating every stage of data processing and subsystem hand off of data and state information. Without metrics, a proper scientific evaluation cannot be made that takes into account the diversity of proposed systems, and their complexity and operability. A chief concern is adequate SA. The operator is not always automatically guaranteed SA when new fusion systems are deployed. Even though promises are made for these systems to demonstrate increases in capacity, data acuity (sharpened resolution, separating a higher degree of information from noise backgrounds), and timeliness among the QoS metrics, the human cognitive process can be a bottleneck in the overall operation. This occurs most frequently at the *perception* level of SA, with level 1 accounting for 77% of SA-related errors,[51] and indirectly affecting *comprehension* and *projection* DM acuity in levels 2 and 3.

Dynamic DM requires: (1) SA/image analysts (IA), (2) dynamic responsiveness to changing conditions, and (3) continual evaluation to meet throughput and latency requirements. These factors usually impact the IFS directly, require an interactive display to allow the user to make decisions, and must include metrics for replanning and sensor management.[55] To afford interactions between future IF designs and users' *information needs*, metrics are required. The metrics chosen include timeliness, accuracy, throughput, confidence, and cost. These metrics are similar to the standard QoS metrics in communication theory and human factors literature, as shown in Table 19.4.

To plan a mission, an evaluation of fusion requires an off-line and on-line assessment of individual sensors and integrated sensor exploitation (i.e., *performance models*). To achieve desired system management goals, the user must be able to develop a set of metrics based on system optimization requirements and performance evaluation criteria. Both of which define the quality to provide effective mission management and to ensure an accurate and effective capability to put the right platforms and sensors on station at the correct time, prepare operators for the results, and determine pending actions based on the estimated data.

19.5.3 User Evaluation

IF does not depend only on individuals making decisions, or single subsystems generating analyses. Rather, team communication and team-based decisions have always been

TABLE 19.4

Metrics for Various Disciplines

Comm	Human Factors	Info Fusion	ATR/ID	Track
Delay	Reaction time	Timeliness	Acquisition/run time	Update rate
Probability of error	Confidence	Confidence	Probability of hit, probability of FA	Probability of detection
Delay variation	Attention	Accuracy	Positional accuracy	Covariance
Throughput	Workload	Throughput	Number of images	Number of targets
Security	Trust	Reliability	Authorize	Cooperative
Cost	Cost	Cost	Collection platforms	Number of assets

integral to complex operations, whether civilian or military, tactical or strategic, municipal, federal, or intergovernmental. Team communication takes many forms across operational environments and changes dynamically with new situations. Communication and DM can be joint, allocated, shared, etc., and this requires that the maintenance of any *team-generated SA* be rigorously evaluated.

As computer processing loads increase, both correct and incorrect ways are available to design IFSs. The *incorrect way* is to fully automate the system, providing no role for the user. The *correct way* is to build the role of the operator into the design at inception and test the effectiveness of the operator to meet system requirements. Fusionists are aware of many approaches, such as level 1 target ID and tracking methods that only support the user, but do not replace the user. In the case of target ID, a user typically does better with a small number of images. However, as the number of collected images and throughput increases, a user does not have the time to process every image. One key issue for fusion is the timeliness of information.

Because every stage of IF DM adds to overall delays—in receiving sensor information, in presenting fused information to the user, and in the information user's processing capacity—overall system operation must be evaluated in its entirety. The bottom-level component in the human-system operation involves data from the sensors. Traditionally, the term *data fusion* has referred to bottom-level, data-driven fusion processing, whereas the term *information fusion* refers to the conversion of already-fused data into meaningful and preferably relevant information useful to higher cognitive systems, which may or may not include humans.

Cognitive processes—human or machine equivalent "smart" algorithms—require non-trivial processing *time* to reach a decision, at both individual and team levels. This applies, analogously, to individual components and intercomponent/subsystem levels of machine-side DM. DM durations that *run shorter* than the interarrival time of sensor data lead to starvation within the DM process; DM durations that *run longer* than the interval between new data arrivals lead to dropped data. Both cases create situations prone to error, whether the DM operator is human or machine. The former case has consequences of starving the human's cognitive flow, causing wandering attention spans, and loss of key pieces of information held in short-term memory, which could have been relevant to later-arriving data. In the latter case, data arrives while the information is being processed, or the operator is occupied, so this data often passes out of the system without being used by the operator. This occurs when systems are not adequately designed to store excess inputs to an already overburdened system. These excess inputs are dropped if there is no buffer in which to queue the data, or if the buffer is so small as to overwrite data. The operator and team must have the means to access quality information and sufficiently process the data from different perspectives within a timely manner. In addition to the metrics that establish core quality (reliability/integrity) for information, there are issues surrounding information security. Precision performance measures may be reflected by *content validity* (whether the measure adequately measures the various facets of the target construct), *construct validity* (how well the measured construct relates to other well-established measures of the same underlying construct), or *criterion-related validity* (whether measures correlate with performance on some task). The validity of these IF performance measures needs to be verified in an operational setting where operators are making cognitive decisions.

19.5.4 Cognitive Processing in Dynamic Decision Making

There are many ways to address the users' roles (reactive, proactive, and preventive) in system design. Likewise, it is also important to develop metrics for the interactions to enhance the users' abilities in a defined role or task. One way to understand how a user operates is to address the *automatic* actions performed by the user (which includes eye–arm movements). Additionally, the user *associates* past experiences to current expectations, such as identifying targets in an image. Intelligent actions include the *cognitive* functions of reasoning and understanding. Rasmussen[37] developed these ideas by mapping behaviors into decision actions including skills, rules, and knowledge through a means-end decomposition.

When tasked with an SA analysis, a user can respond broadly by one of three manners: reactive, proactive, or preventive, as shown in Figure 19.11. In the example that follows, we use a target tracking and ID exemplar to detail the user cognitive actions. In a *reactive mode*, the user makes a rapid detection to minimize damage or repeat offense. An IFS would gather information from a sensor grid detection of *in situ* threats and is ready to act. In this model, the system interprets and alerts users to *immediate threats*. The individual user selects the immediate appropriate response (in seconds) with aid of sensor warnings of nonlethal or lethal threats.

In the *proactive mode*, the user utilizes sensor data to anticipate, detect, and capture needed information before an event. In this case, a sensor grid provides surveillance based on prior intelligence and predicted target movements. A multi-INT sensor system could detect and interpret anomalous behavior and alert an operator to *anticipated threats* in minutes. Additionally, the directed sensor mesh tracks individuals back to dwellings and meeting places, where troops respond quickly and capture the insurgents, weapons, or useful intelligence.

The mode that captures the entire force over a period of time (i.e., an hour) is the *preventive mode*. To prevent *potential threats* or actions, we would (1) increase insurgent risk (i.e., arrested after being detected), (2) increase effort (i.e., make it difficult to act), and (3) lower payoff of action (i.e., reduce the explosive damage). The preventive mode includes an intelligence database to track events before they reach deployment.

The transition and integration of decision modes to perceptual views is shown in Figure 19.12. The differing displays for the foot soldier or command center would correlate with the DM mode of interest. For the reactive mode, the user would want an actual location of the *immediate threats* on a physical map. For the *anticipated threats*, the user would want the predicted locations of the adversary and the range of possible actions.

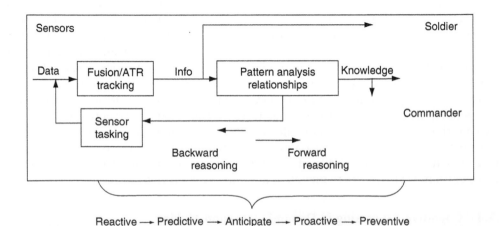

FIGURE 19.11
Reasoning in proactive strategies.

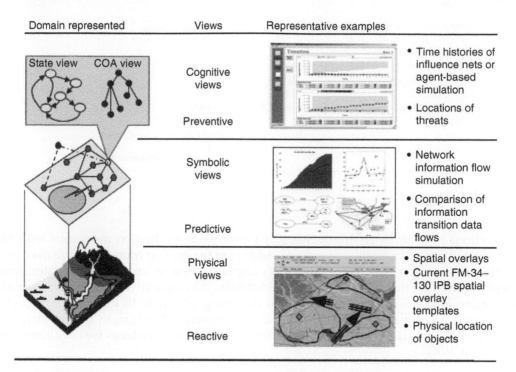

FIGURE 19.12
Categories of analytic views. (Extracted from Waltz, E., Data Fusion in Offensive and Defensive Information Operations, NSSDF, June 2000.)

Finally, for the *potential threats*, the user could utilize behavior analysis displays that piece together aggregated information of group affiliations, equipment stores, and previous events to predict actions over time. These domain representations were postulated by Waltz[39] as cognitive, symbolic, and physical views that capture differing perceptual needs for intelligent DM.

Intelligent DM employs many knowledge-based information fusion (KBIF) strategies such as neural networks, fuzzy logic, Bayesian networks, evolutionary computing, and expert systems.[56] Each KBIF strategy has different processing durations. Furthermore, each strategy differs in the extent to which it is constrained by the facility and its use within the user-fusion system. OODA loop helps to model a DM user's planned, estimated, or predicted actions. Assessing susceptibilities and vulnerabilities to detect/estimate/predict threat actions, in the context of planned actions, requires a concurrent timeliness assessment. Such assessment is required for adequate DM, yet is not easily attained.

This is similarly posed in the Endsley model of SA: the *"projection"* level 3 of SA maps to this assessment processing activity. DM is most successful in the presence of high levels of projection SA, such as high accuracy of vulnerability or adversarial action assessments. DM is enhanced by correctly anticipating effects and state changes that will result from potential actions. The nested nature of effect-upon-effect creates difficulty in making estimations within an OODA cycle. For instance, effects of own-force intended COA are affected by the DM cycle time of the threat instigator and the ability to detect and recognize them. Three COA processes[57] to reduce the IFS search space are as follows: (1) manage similar COA, (2) plan related COA, and (3) select novel COA, shown in Figure 19.13.

Given that proactive and preventive actions have a similar foundational set of requirements for communicating actions, employing new sensor technologies, and managing

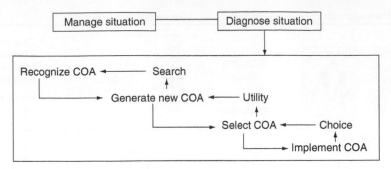

FIGURE 19.13
Course of action model.

interactions in complex environments, any performance evaluation must start with the construct of generally applicable metrics. Adding to the general framework, we must also create an evaluation analysis that measures responsiveness to unanticipated states, since these are often the first signals of relevant SA threats. A comparative performance analysis for future applications requires intelligent reasoning, adequate IF knowledge representation, and process control within realistic environments. The human factors literature appropriately discusses the effects of realistic environments as a basis for cognitive work analysis (CWA).

19.5.5 Cognitive Work Analysis/Task Analysis

The key to supporting user reasoning is in designing a display that supports the users task and work roles. A task analysis (TA) is the modeling of user actions in the context of achieving goals. In general, actions are analyzed to move from a current state to a goal state. Different approaches based on actions include[58]

- *Sequential TA.* It includes organizing activities and describing them in terms of the perceptual, cognitive, and manual user behavior to show the man-machine activities.

- *Timeline TA.* It is an approach that assesses temporal demands of the tasks or sequences of actions and compared time availability for task execution.

- *Hierarchical TA.* It represents the relationships between which tasks and subtasks need to be done to meet operating goals.

- *Cognitive Task Analysis (CTA).* It describes the cognitive skills and abilities needed to perform a task proficiently, rather than the specific physical activities carried out in performing the task. CTA is used to analyze and understand task performance in complex real-world situations, especially those involving change, uncertainty, and time pressure.

Level 5 issues require an understanding of the workload, time, and information availability. Workload and time can be addressed through a TA; however, user performance is a function of the task within the mission. A CWA includes (1) defining what task the user is doing, (2) structuring the order of operations, (3) ensuring that the interface supports SA, (4) defining the decomposition of required tasks, and (5) characterizing what the human actually perceives.[59] The CWA looks at the tasks the human is performing and the understanding of how the tasks correlate with the information needs for intelligent actions.

The five stages of a CWA are (1) work domain, (2) control task, (3) strategy, (4) social-organizational, and (5) worker competency analysis. Vicente[59] shows that a CWA requires an understanding of the environment and the user capabilities. An evaluation CWA requires starting from the work domain and progressing through the fusion interface to the cognitive domain. Starting from the work domain, the user defines control tasks and strategies (i.e., utilizing the fusion information) to conduct the action. The user strategies, in the context of the situation, require an understanding of the social and cognitive capabilities of the user. Social issues would include no-fly zones for target ID and tracking tasks, whereas cognitive abilities include the reaction time of the user to a new situation. Elements of the CWA process include reading the fusion display, interpreting errors, and assessing sensory information and mental workloads. We are interested in the typical work-analysis concepts of fusion/sensor operators in the detection of events and state changes, and pragmatic ways to display the information.

19.5.6 Display/Interface Design

A display interface is a key to allowing the user to have control over data collection and fusion processing.[60,61] Without designing a display that matches the cognitive perception, it is difficult for the user to reason over a fused result. Although many papers address interface issues (i.e., multimodal interfaces), it is of concern to the fusion community to address cognitive issues such as ontological relations and associations.[62,63] Fusion models are based on processing information for DM. A user-fusion model would highlight the fact that fusion systems are designed for human consumption. In effect, the top-down approach would explore an information need—pull through queries to the IFS (*cognitive fusion*), whereas a bottom-up approach would push data combinations to the user from the fusion system (*display fusion*). The main issues for user-fusion interaction include ontology development for initiating queries, constructing metrics for conveying successful fusion opportunities, and developing techniques to reduce uncertainty[1] and dimensionality. In Section 19.6, we highlight the main mathematical techniques for fusion based on the user-fusion model for DM.

An inherent difficulty resides in the fusion system that processes only two forms of data. If one sensor perceives the correct entity and the other does not, there is a *conflict*. Sensor conflicts increase the cognitive workload and delay action. However, if time, space, and spectral events from different sensors are fused, conflicts can be resolved with an emergent process to allow for better decisions over time. The strategy is to employ multiple sensors from multiple perspectives with UR. Additionally, the HCI can be used to give the user a global and local SA/IA to guide attention, reduce workload, increase trust, and afford action.

Usability evaluation is a *user-centered, user-driven* method that defines the system as the connection between the application software and hardware. It focuses on whether the system delivers the right information in the appropriate way for users to complete their tasks. Once user requirements have been identified, good interface designs transform these requirements into display elements, and organize them in a way that is compatible with how users perceive and utilize information, and how they are used to support work context and to use conditions.[33] The following items are useful principles for interface design:

- *Consistency.* Interfaces should be consistent with experiences, conform to work conventions, and facilitate reasoning to minimize user errors, information retrieval, and action execution.
- *Visually pleasing composition.* Interface organization includes balance, symmetry, regularity, predictability, economy, unity, proportion, simplicity, and groupings.

- *Grouping.* Gestalt principles provide six general ways to group screen elements spatially: principles of proximity, similarity, common region, connectedness, continuity, and closure.
- *Amount of information.* Too much information could result in confusion. Using Miller's principle that people cannot retain more than 7 ± 2 items in short-term memory, chunking data using this heuristic could provide the ability to add more relevant information on the screen as needed.
- *Meaningful ordering.* The ordering of elements and their organization should have meaning with respect to the task and information processing activities.
- *Distinctiveness.* Objects should be distinguishable from other objects and the background.
- *Focus and emphasis.* Salience of objects should reflect the relative importance of focus.

Typically, nine usability areas are considered when evaluating interface designs with users:

- *Terminology.* Labels, acronyms, and terms used
- *Workflow.* Sequence of tasks in using the application
- *Navigation.* Methods used to find application parts
- *Symbols.* Icons used to convey information and status
- *Access.* Availability and ease of access of information
- *Content.* Style and quality of information available.
- *Format.* Style of information conveyed to the user
- *Functionality.* Application capabilities
- *Organization.* Layout of the application screens

The usability criteria typically used for evaluation of these usability areas are noted below:[33]

- *Visual clarity.* This displayed information should be clear, well-organized, unambiguous, and easy-to-read to enable users to find required information, draw the user's attention to important information, and allow the user to see where information should be entered quickly and easily.
- *Consistency.* This dimension conveys that the way the system looks and works should be consistent at all times. Consistency reinforces user expectations by maintaining predictability across the interface.
- *Compatibility.* This dimension corresponds to whether the interface conforms to existing user conventions and expectations. If the interface is familiar to users, it will be easier for them to navigate, understand, and interpret what they are looking at and what the system is doing.
- *Informative feedback.* Users should be given clear informative feedback on where they are, and what actions were taken, whether successful, and what should be taken.
- *Explicitness.* The way the system works and is structured should be clear to the user.

TABLE 19.5

Interface Design Activities

	Work Domain Activities
Info requirements	Cognitive work analysis, SA analysis, contextual inquiry—task analysis, scenario-based design, participatory design
Interface design	User interface design principles, participatory design
Evaluation	Situation awareness analysis, usability evaluation—scenario-based design, participatory design, usability evaluation

- *Appropriate functionality.* The system should meet the user requirements and needs when carrying out tasks.
- *Flexibility and control.* The interface should be sufficiently flexible in structure—the way information is presented and in terms of what the user can do, to suit the needs and requirements of all users, and to allow them to feel in control.
- *Error prevention and correction.* A system should be designed to minimize the possibility of user error, with built-in functions to detect when these errors occur. Users should be able to check their inputs and to correct potential error before inputs are processed.
- *User guidance, usability, and support.* Informative, easy-to-use, and relevant guidance should be provided to help the users understand and use the system.

Work domain issues should be considered in the environment in which processes will take place, such as the natural setting of a command post. *Ecological interface design* is the methodology concerned with evaluating user designs as related to stress, vigilance, and the physical conditions such as user controls (Table 19.5).

19.6 Example: Assisted Target Identification through User-Algorithm Fusion

Complex fusion designs are built from a wide range of experiences. A large set of business, operational, and prototype users are involved in deploying a fusion system. One such example, the automated fusion technology (AFT) system, is the assisted target recognition for time-critical targeting program that seeks to enhance the operational capability of the digital command ground station.[64] The AFT system consists of the data fusion processor, change detection scheduler, image prioritizer, automatic target cuer (shown in Figure 19.14), and a reporting system. Some of the user needs requested from IA include the following:

1. Reduce noninterference background processing (areas without targets and opportunity cost)
2. Cue them to only high probability images with high confidence (accuracy)
3. Reduce required workload and minimize additional workload (minimize error)
4. Reduce time to exploit targets (timeliness)
5. Alert IA to possible high-value targets in images extracted from a larger set of images on the image deck (throughput)

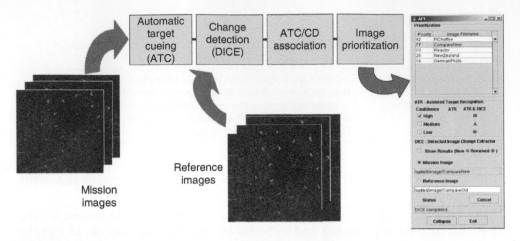

FIGURE 19.14
AFT system.

There are many reports on the BAE real-time moving and stationary automatic target recognition (RT-MSTAR) and Raytheon detected imagery change extractor (DICE) algorithms. Developments for the RT-MSTAR came from enhancements of the MSTAR DARPA program.[65] Additionally, the ATR algorithms from MSTAR have undergone a rigorous testing and evaluation over various operating conditions.[66] The key for AFT is to integrate these techniques for the operationally viable VITec™ display. Figure 19.15 shows a sample image of the AFT system.

Although extensive research has been conducted on object-level fusion, change detection, and platform scheduling, these developments are limited by the exhaustive amount of information to completely automate the system. Users have been actively involved in design reconfiguration, user control, testing,[67] and deployment. As the first phase of deployment, the automatic fusion capabilities were used to reorder the "image deck" so that user verification proceeded pragmatically. The key lesson learned was that complete IF solutions were still not reliable enough, and tactful UR controls are necessary for acceptance. For the AFT system, the worthy elements of object-level fusion include support for sensor observations, cognitive assistance in detecting targets, and providing an effective means for target reporting. These capabilities combine to increase fusion system efficacy for time savings and extended spatial coverage, as detailed in Figure 19.16.

19.7 Summary

If a machine presents results that conflict with the user's expectations, the user would experience cognitive dissonance. The user needs reliable information that is accurate, timely, and confident. Figure 19.17 shows that for DM there are technological operational conditions that affect the user's ability to make informed decisions.[68] The IFS must produce quality results for effective and efficient DM by increasing fused output quality. An IF design should be robust to object, data, and environment model variations.[69]

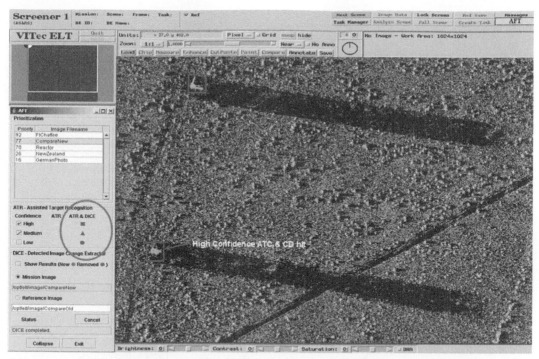

(a)

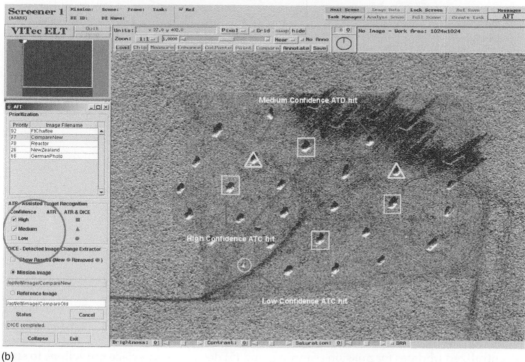

(b)

FIGURE 19.15
Output embedded in VITec.

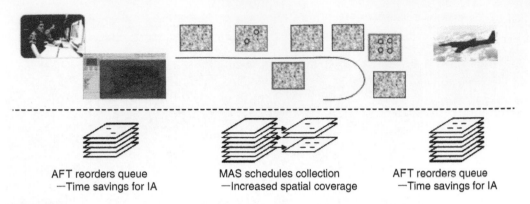

AFT reorders queue   MAS schedules collection   AFT reorders queue
—Time savings for IA   —Increased spatial coverage   —Time savings for IA

FIGURE 19.16
AFT "image deck" reprioritization with "bonus" images.

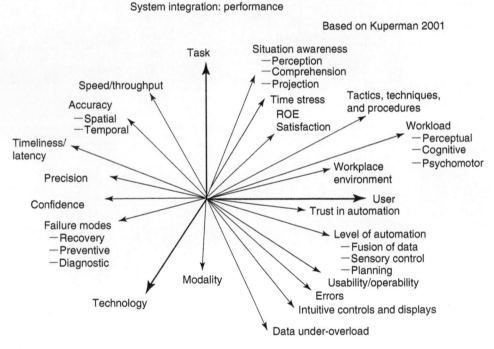

FIGURE.19.17
Machine-user interaction performance.

This chapter highlights important IF issues, complementing Hall's,[4]* for UR (sensor, user, and mission [SUM] management) including (1) designing for users, (2) establishing a standard set of metrics for cost function optimization, (3) advocating rigorous fusion evaluation criteria to support effective DM, (4) planning for dynamic DM and requiring decentralized updates for mission planning, and (5) designing interface guidelines to support user's control actions and user efficacy of IFSs.

---

* Also see Chapter 1 of this handbook.

References

1. Hall, D. Knowledge-Based Approaches, in *Mathematical Techniques in Multisensor Data Fusion*, Chapter 7, pp. 239, Artech House, Boston, MA, 1992.
2. Blasch, E. and P. Hanselman, Information Fusion for Information Superiority, *NAECON* (*National Aerospace and Electronics Conference*), pp. 290–297, October 2, 2000.
3. Blasch, E. and S. Plano, Level 5: User Refinement to Aid the Fusion Process, *Proceedings of SPIE*, Vol. 5099, 2003.
4. Hall, D. Dirty Little Secrets of Data Fusion, *NATO Conference*, 2000.
5. Hall, J.M., S.A. Hall and T. Tate, Removing the HCI Bottleneck: How the Human–Computer Interface (HCI) Affects the Performance of Data Fusion Systems, in D. Hall and J. Llinas (Eds.), *Handbook of Multisensor Data Fusion*, Chapter 19, pp. 1–12, CRC Press, Boca Raton, FL, 2001.
6. Cantoni, V., V. Di Gesu and A. Setti (Eds.), *Human and Machine Perception: Information Fusion*, Plenum Press/Kluwer, New York, 1997.
7. Abidi, M. and R. Gonzalez (Eds.), *Data Fusion in Robotics and Machine Intelligence*, pp. 560, Academic Press, Orlando, FL, 1992.
8. Blasch, E. Decision Making for Multi-Fiscal and Multi-Monetary Policy Measurements, *Fusion98*, Las Vegas, NV, pp. 285–292, July 6–9, 1998.
9. Waltz, E. and J. Llinas, Data Fusion and Decision Support for Command and Control, *Multisensor Data Fusion*, Chapter 23, Artech House, Boston, MA, 1990.
10. Looney, C.G. Exploring Fusion Architecture for a Common Operational Picture, in *Information Fusion*, Vol. 2, pp. 251–260, December 2001.
11. Llinas, J. Assessing the Performance of Multisensor Fusion Processes, in D. Hall and J. Llinas (Eds.), *Handbook of Multisensor Data Fusion*, Chapter 20, CRC Press, Boca Raton, FL, 2001.
12. Fabian, W. Jr., and E. Blasch, Information Architecture for Actionable Information Production, (for Bridging Fusion and ISR Management), *NSSDF 02*, August 2002.
13. Blasch, E. Ontological Issues in Higher Levels of Information Fusion: User Refinement of the Fusion Process, *Fusion03*, July 2003.
14. Blasch, E. and J. Schmitz, Uncertainty Issues in Data Fusion, *NSSDF 02*, August 2002.
15. Tversky, K. and D. Kahneman, Utility, Probability & Decision Making, in D. Wendt and C.A.J. Vlek (Eds.), *Judgment Under Uncertainty*, D. Reidel Publishing Co., Boston, MA, 1975.
16. Jousselme, A.-L., P. Maupin and E. Bosse, Formalization of Uncertainty in Situation Analysis, DSTO Conference, 2003.
17. *Webster's Dictionary*, 1998.
18. Anthony, R.T. Data Management Support to Tactical Data Fusion, in D. Hall and J. Llinas (Eds.), *Handbook of Multisensor Data Fusion*, Chapter 18, CRC Press, Boca Raton, FL, 2001.
19. Wu, H. Sensor Data Fusion for Context-Aware Computing Using Dempster–Shafer Theory, PhD Thesis, Carnegie Mellon University, 2003.
20. Antony, R.T. *Principles of Data Fusion Automation*, Artech House, Boston, MA, 1995.
21. Torrez, W. Information Assurance Considerations for a Fully Netted Force for Strategic Intrusion Assessment and Cyber Command and Control, *Fusion01*, 2001.
22. Nilsson, M. and T. Ziemke, Rethinking Level 5: Distributed Cognition and Information Fusion, *Fusion06*, 2006.
23. Blasch, E. Situation, Impact, and User Refinement, *Proceedings of SPIE*, Vol. 5096, April 2003.
24. E. Blasch, M. Pribilski, B. Daughtery, B. Roscoe and J. Gunsett, Fusion Metrics for Dynamic Situation Analysis, *Proceedings of SPIE*, Vol. 5429, April 2004.
25. [JDL] US Department of Defense, Data Fusion Sub-panel of the Joint Directors of the Laboratories, Technical Panel for C3, "Data Fusion Lexicon," 1991.
26. Steinberg, A.N., C. Bowman and F. White, Revisions to the JDL Data Fusion Model, *NATO/IRIS Conference*, October 1998.
27. Llinas, J., C. Bowman, G. Rogova, A. Steinberg, E. Waltz and F. White, Revisiting the JDL Data Fusion Model II, *Fusion04*, 2004.

28. Kaupp, T., A. Makarenko, F. Ramos and H. Durrant-Whyte, Human Sensor Model for Range Observations, *IJCAI Workshop Reasoning with Uncertainty in Robotics*, Edinburgh, Scotland, July 2005.
29. Endsley, M.R. Design and Evaluation for Situation Awareness Enhancement, *Proceedings of Human Factors Society*, pp. 97–101, 1988.
30. Musick, S. and R. Malhotra, Chasing the Elusive Sensor Manager, *IEEE NAECON*, 1994.
31. Schubert, J. Robust Report Level Cluster-to-Track Fusion, *Fusion02*, 2002.
32. Schubert, J. Evidential Force Aggregation, *Fusion03*, 2003.
33. Blasch, E. Assembling an Information-Fused Human–Computer Cognitive Decision Making Tool, *IEEE AESS*, pp. 11–17, June 2000.
34. Welsh, R. and B. Blasch, *Foundations of Orientation and Mobility*, American Foundation for the Blind, NY, 1980.
35. Blasch, E. and J. Gainey, Jr. Physio-Associative Temporal Sensor Integration, *Proceedings of SPIE*, Orlando, FL, pp. 440–450, April 1998.
36. Gibson, J.J. *The Senses Considered as Perceptual Systems*, Waveland Press, Inc., Prospect Heights, IL, 1966.
37. Rasmussen, J. *Information Processing and Human–Machine Interaction: An Approach to Cognitive Engineering*, North-Holland, New York, 1986.
38. Kadar, I. Data Fusion by Perceptual Reasoning and Prediction, *Proceedings of Tri-Service Data Fusion Symposium*, Johns Hopkins University, June 1987.
39. Waltz, E. Data Fusion in Offensive and Defensive Information Operations, *NSSDF*, June 2000.
40. Brooks, R. Elephants Don't Play Chess, *Robotics and Autonomous Systems*, Vol. 6, pp. 3–15, 1990.
41. Lambert, D.A. Situations for Situation Awareness, *Proceedings of Fusion 2001*, Montreal, Quebec, Canada, 7–10 August, 2001.
42. Matheus, C.J., M.M. Kokar and K. Baclawski, A Core Ontology for Situational Awareness, *Proceedings of the Sixth International Conference on Information Fusion (Fusion03)*, Cairns, Australia, July 8–11, 2003, pp. 545–552.
43. McMichael, D. and G. Jarrad, Grammatical Methods for Situation and Threat Analysis, *Fusion05*, 2005.
44. Wickens, C.D. *Engineering Psychology and Human Performance*, Scott, Foresman, & Company, Glenview, IL, 1984.
45. Billings, C.E. *Aviation Automation: The Search for the Human-Centered Approach*, Lawrence Erlbaum Associates, Mahwah, NJ, 1997.
46. Suchman, L.A. *Plans and Situated Actions: The Problem of Human–Machine Communication*. Cambridge University Press, Cambridge, England, 1987.
47. Blasch, E.P. and S. Plano, JDL Level 5 Fusion Model: User Refinement Issues and Applications in Group Tracking, *Proceedings of SPIE*, Vol. 4729, Aerosense, 2002.
48. Gilson, D., D. Garland and J. Koonce, *Situational Awareness in Complex Systems*, Aviation Human Factors Series, Defense Technical Information Center, Dayton, OH, 1994.
49. Endsley, M.R. Toward a Theory of Situational Awareness in Dynamic Systems, *Human Factors J.*, Vol. 37, pp. 32–64, 1996.
50. Endsley, M.R. and D.J. Garland (Eds.), *Situation Awareness Analysis and Measurement*, Lawrence Erlbaum, Mahwah, NJ, 2000.
51. Endsley, M.R., B. Bolte and D.G. Jones, *Designing for Situation Awareness: Approach to User-Centered Design*, Taylor and Francis, London, 2003.
52. Klein, G.A. Recognition-Primed Decisions, in W.B. Rouse (Ed.), *Advances in Man–Machine Systems Research*, Vol. 5. JAI Press, Greenwich, CT, pp. 47–92, 1989.
53. Kettani, D. and J. Roy, A Qualitative Spatial Model for Information Fusion and Situation Analysis, *Proceedings of the Third International Conference on Information Fusion*, Paris, France, July 2000.
54. Roy, J., S. Paradis and M. Allouche, Threat Evaluation for Impact Assessment in Situation Analysis Systems, *Proceedings of SPIE*, Vol. 4729, 2002.

55. Xiong, N. and P. Svensson, Multisensor Management for Information Fusion: Issues and Approaches, *Information Fusion*, Vol. 3, pp. 163–186, 2002.
56. Hall, D.L. and S.A. McMullen, *Mathematical Techniques in Multisensor Data Fusion*, Artech House Inc., Norwood, MA, 2004.
57. Zsambok, C.E., G.A. Klein, L.R. Beach, G.L. Kaempf, D.W. Klinger, M.L. Thordsen, and S.P. Wolf, *Decision-Making Strategies in the AEGIS Combat Information Center*. Klein Associates, Fairborn, OH, 1992.
58. Burns, C.M. and J.R. Hajdukiewicz, *Ecological Interface Design*, CRC Press, New York, NY, 2004.
59. Vicente, K.J. *Cognitive Work Analysis, Toward Safe, Productive and Healthy Computer-Based Work*, Lawrence Erlbaum, Mahwah, NJ, 1999.
60. Salerno, J., M. Hinman and D. Boulware, Building a Framework for Situational Awareness, *Proceedings of Fusion 2004*, Stockholm, Sweden, 2004.
61. Matheus, C., M. Kokar, K. Baclawski, J.A. Letkowski, C. Call, M. Hinman, J. Salerno, and D. Boulware, SAWA: An Assistant for Higher-Level Fusion and Situation Awareness, *Proceedings of SPIE*, Vol. 5813, 2005.
62. Kokar, M., C.J. Matheus, J. Letkowski, and K. Baclawski, Association in Level 2 Fusion, *Proceedings of SPIE: International Society for Optical Engineering*, Vol. 5434, pp. 228–237, 2004.
63. Matheus, C.J., K.P. Baclawski and M. Kokar, Derivation of Ontological Relations Using Formal Methods in a Situation Awareness Scenario, *Proceedings of SPIE: International Society for Optical Engineering*, Vol. 5099, 2003.
64. Blasch, E. Assisted Target Recognition through User–Algorithm Fusion, *NSSDF 06*, 2006.
65. Burns, T., R. Douglass, R. Hummel and K. Kasprick, *DARPA MSTAR Handbook*, Washington, DC, 2006.
66. Ross, T.D. and J.C. Mossing, The MSTAR Evaluation Methodology, *Proceedings of SPIE*, Vol. 3721, *Algorithms for Synthetic Aperture Radar Imagery VI*, p. 705, April 1999.
67. Baumann, J.M., J.L. Jackson III, G.D. Sterling and E. Blasch, RT-ROC: A Near-Real-Time Performance Evaluation Tool, *Proceedings of SPIE*, April 2005.
68. Kuperman, G. Human Systems Interface (HIS) Issues in Assisted Target Recognition (ASTR), *NAECON97*, pp. 37–48, 1997.
69. Ross, T. and M. Minardi, Discrimination and Confidence Error in Detector Reported Scores, *Proceedings of SPIE*, Vol. 5427, pp. 342–353, *Algorithms for Synthetic Aperture Radar Imagery XI*, 2004.

20

Perspectives on the Human Side of Data Fusion: Prospects for Improved Effectiveness Using Advanced Human–Computer Interfaces

David L. Hall, Cristin M. Hall, and Sonya A. H. McMullen

CONTENTS

20.1 Introduction

Even before the 9/11 attack, the U.S. military and intelligence communities had expended huge amounts of resources to develop new types of sensors and surveillance methods. Advanced sensors range from the development of nanoscale, smart sensors in distributed sensor networks (Smart Dust)[1] to national level sensors collecting image and signal data. This trend has enabled an ever-increasing ability to collect data in a vast "vacuum cleaner" approach. The concept is illustrated in Figure 20.1.

It is tempting to view the fusion process as primarily a data- or energy-driven process, that is, energy is collected by sensors (either actively or passively) and is transformed into signals, images, scalar, or vector quantities. These quantities, in turn, are translated into state vectors, labels, and interpretive hypotheses by automated estimation and reasoning processes. Ultimately, one or more humans observe the results to develop an understanding of an event or situation. Indeed, this is a common (implicit) view of the information fusion process. In Figure 20.1, the "level" (e.g., level 0, level 1, etc.) refers to the levels of functions identified in the Joint Directors of Laboratories (JDL) data fusion process model.[2]*

---

* Also see Chapter 3 of this handbook.

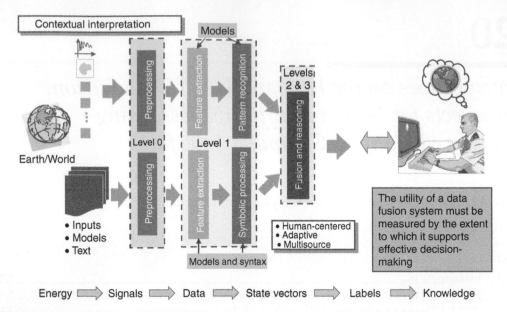

FIGURE 20.1
Transformation of energy into knowledge.

Extensive research in the data fusion community has been conducted to develop techniques for level 1 fusion (the data or sensing side of the process). The bulk of the literature on multisensor data fusion focuses on the automation of target tracking and automatic target recognition.[3]* Although such research is needed, current problems involve complexities such as identifying and tracking individual people and groups of people, monitoring global activities and recognition of events that may be a precursor to terrorist activities. The requisite data for this analysis involves sensor data (including signals, images, and vector quantities), textual information (from web sites and human reports), and utilization of models. This process of analysis is very human intensive, requiring teams of analysts to search for data, interpret the results, develop alternative hypotheses, and assess the consequences of such hypotheses.

Our perception is that, by and large, researchers have started at "the wrong end" of the data fusion process. That is, researchers have started at the input side and sought to address methods for processing sensor data to automatically develop a situation database and display (e.g., target tracks, icons representing target identification, etc.). We argue that research on the user side of the data fusion process has been relatively neglected. It should be noted that the level 5 fusion concept was introduced in 2000 by Hall et al.,[4] in part, to call attention to the need to focus on the human side of the fusion process. Level 5 processing recognizes that the data fusion system needs to actively interact with a user to guide the focus of attention, assist via cognitive aids, and improve the human-in-the-loop processing of multisensor data.

Analysts face a situation in which they are immersed in a sea of data (drowning in data), but thirst for knowledge about the meaning of the data. Although this is beginning to be addressed by a number of researchers,[5–8] there is a continued need to provide cognitive aids to support the process of analysis. We have previously suggested a combination of new

---

* And Chapter 1 of this handbook.

directions to improve the human side of fusion, including utilization of three-dimensional (3D) visualization, intelligent advisory agents, and interactive gaming techniques.[9] In this chapter, we explore recent advances in human–computer interaction technologies that provide a basis for significant improvements in increasing the human side of data fusion and make recommendations for research directions.

We suggest that the current situation is analogous to a pilot attempting to fly an aircraft by providing quantitative directions directly to the aircraft control structures (e.g., move the left aileron down 3.7°, advance the engine throttles by 23%). It would be literally impossible to fly an aircraft under such conditions. Instead, pilots fly modern aircraft by "electronic wire" in which computer interfaces map human physical motions into commands for controlling physical motions such as moving the flaps. By analogy, we still require that human users interact with databases and search engines by creating Boolean queries with specific terms (e.g., find textual information that contains the terms, *Fallujah, weapon systems*, and *terrorist*). It is, therefore, not surprising that interaction with huge databases is frustrating to analysts, and that searches involve extensive work that focuses the analyst's attention on the data interaction rather than the inference process. New methods are required to improve the interaction of humans with data fusion systems and to increase the level of effectiveness of the "human-in-the-loop" in a fusion process.

20.2 Enabling Human–Computer Interface Technologies

The rapid evolution in human–computer interface (HCI) technologies provides opportunities for new concepts on how analysts/users could interact with a data fusion system.[10] Key technological advances include 3D immersive visualization environments, sonification, and haptic interfaces. A brief overview of these and their application to data fusion is provided in Sections 20.2.1–20.2.3.

20.2.1 Three-Dimensional Visualization Techniques/Environments

Perhaps the most spectacular visual display concept is that of the full-immersion display or virtual reality. *PC Magazine*'s online encyclopedia[11] defines *virtual reality* as *an artificial reality that projects you [user] into a 3D space generated by a computer* (p. 1). *PC Magazine* further states, "virtual reality (VR) can be used to create an illusion of reality or imagined reality and is used both for amusement as well as serious training" (p. 1). In 1991, Thomas DeFanti and Dan Sandin conceived the concept of the CAVE virtual reality system.[12] The prototype CAVE was developed and tested at the Electronic Visualization Laboratory at the University of Illinois in Chicago the following year. The CAVE is essentially a small room,

> Approximately ten feet-wide on each side, with three walls and a floor. Projectors, situated behind the walls, projected computer-generated imagery onto the walls and floor. Two perspective-corrected images were drawn for each frame, one for the right eye and one for the left. Special glasses were worn that ensured that each eye saw only the image drawn for it. This created a stereoscopic effect where the depth information encoded in the virtual scene restored and conveyed to the eyes of those using it.[12 (paragraph 2)]

Figure 20.2 shows the three-sided CAVE.

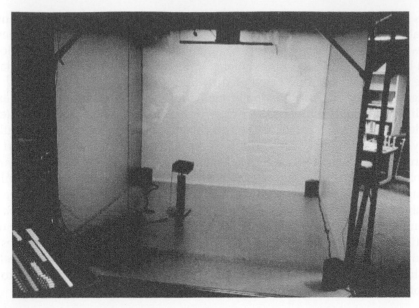

FIGURE 20.2
CAVE at the Electronic Visualization Laboratory (http://cave.ncsa.uiuc.edu/about.html, 2000–2004).

Leetaru reports that,

> The CAVE works by reproducing many of the visual cues that your brain uses to deci-pher the world around you. Information such as the differing perspectives presented by your eyes, depth occlusion, and parallax (to name a few) are all combined into the single composite image that you are conscious of while the rest is encoded by your brain to provide you with depth clues. The CAVE must reproduce all of this complex informa-tion in real-time, as you move about in the CAVE.[12] (paragraph 5)

The user(s) immersed in the CAVE can visualize complex images and data in three dimensions similar in nature to *Star Trek: The Next Generation's* "Holodeck."[13] Only the CAVE user must wear special shutter glasses that project a separate image for each eye. The user must also control the visualization using a control device that can be either a wand, glove, or full-body suit, which is tracked by the computer system through electromagnetic sensors allowing for the presentation of the proper viewpoint of the visualization scene. A key advantage of the CAVE technology over head-mounted 3D displays is the ability for multiple observers to simultaneously see the 3D display and each other. This provides an opportunity for multiple analysts to collaborate on an assessment of data, directly compar-ing notes and literally pointing out observations to each other.

The CAVE is only one of a number of virtual reality systems available. For example, Infinity Wall™ provides a one-wall display that gives a 3D visual illusion whereas Immersa-Desk™ developed by the Electronic Vision Laboratory Company allows a user sitting in front of a terminal to see a 3D illusion.[14] The Elumen Company has developed a desktop system that provides a full-color, immersive display, with 360° projection and a 180° field of view. The hemispherical screen is positioned vertically so as to fill the field of view of the user, creating a sense of immersion. The observer loses the normal depth cues, such as edges, and perceives 3D objects beyond the surface of the screen.

There are clear applications of 3D visualization techniques to multisensor data fusion. The most obvious example is simply creating 3D situation displays of various kinds.

FIGURE 20.3
Example of three-dimensional display using height to represent time.

Researchers at Penn State University[7,9,10] have developed 3D displays that use the third dimension (height above the floor—on which a situation map is displayed) to represent time of day or time of year. In this example (see Figure 20.3) multisource reports related to the same geographic area are shown in a column above the situation map. Thus, data concerning a specific geolocation are shown in a column above the map whereas data occurring at the same time of day or time of year are shown in a plane parallel to the situation map. In these displays, data occurring in a sphere or ellipsoid are data in a con-strained geo-spatial-temporal volume of interest. Other experiments we have conducted use height above the floor (or map) to represent different levels of abstraction or process-ing. For example, different layers of processing/pattern recognition are shown in different hierarchical layers.

Other examples of related research by researchers at Penn State University include using 3D representations of data to allow a user to "surf through" large data sets. An example is shown in Figure 20.4. In this figure, provided by Wade Shumaker,[15] thousands of data points can be observed simultaneously by stacking the data in a translucent data cube and allowing key data (e.g., reports that meet specified criteria) to be turned into different colors and different degrees of transparency.

Finally, Seif El-Nasr and coworkers[9] have explored techniques to map a strategy-based game interface within the analyst domain and measure its utility in enhancing analyst productivity, speed, and quality of the hypotheses generated. Several explored visualiza-tion techniques for analysts have assumed a removed point of view with a geographic information system (GIS).[16–19] These are similar to the visualization methods used in strategy-based games. The interfaces have evolved to adopt a very minimal context-sensitive interface to visualize dynamic and static information and events. The research is focusing on developing several mappings between analyst's data and the current types of displays or visualizations used by strategy-based games. The goal is to use gaming tech-niques as well as cinematic techniques to visually depict analyst-related data in a way that allows analysts to quickly grasp the situation and the event.

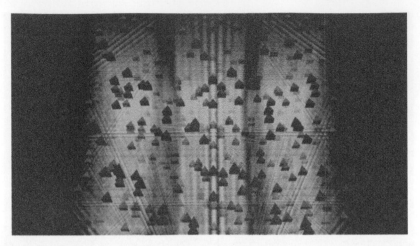

FIGURE 20.4
Example of three-dimensional display of large data set. (From Shumaker, W., Private communication to D.L. Hall, September, 2007.)

20.2.2 Sonification

Another area for interaction with computers involves the concept of sonification—converting data into sound effects[20] for an auditory display. A number of software packages exist to support such transformations. These "audio displays" can be used to improve the ability of users to perform pattern recognition in complex data. For example, Ballora et al.[21] have developed a method using digital music software to transform the sequence of intervals between successive human heartbeats into an electroacoustic sound track. They were able to demonstrate the ability of nonspecialists (with limited training) to readily distinguish among the following conditions of heart: (1) healthy heart behavior, (2) congestive heart failure, (3) atrial fibrillation, and (4) obstructive sleep apnea.

In the defense community, sonar operators have long used a combination of sound (listening to transformed acoustic signatures) and visual displays of acoustic spectral patterns to detect, classify, and characterize underwater phenomena such as submarines, whale sounds, and so on.[22] In a recent work at Penn State University, we have conducted experiments to use sonification for characterizing uncertainty. A Geiger counter analogy has been used in a 3D visualization environment to represent the uncertain location of a target using 3D sound. Geiger counter sounds are used to allow a user to understand that the location of a target lies within an uncertainty region rather than at a single point. This tends to mitigate the effect of a visual display in which an error ellipsoid is viewed as a "bulls eye"—namely, the uncertainty region tends to be ignored and the target location is viewed as being in the center of the error ellipsoid rather than somewhere within the error region. Finally, we have also experimented with sonification to provide an aural representation of correlation. In a 3D visualization environment, our application allows users to retrieve multisource data reports about a situation, event, or activity. After the user specifies a set of data, reputed to be correlated, we compute a general measure of correlation (MOC) on the basis of common report elements such as location, time, and identity characteristics.[23] The MOC is translated into a harmonic sound with the degree of harmony based on the level of correlation. Highly correlated reports produce a pleasing/harmonious sound whereas uncorrelated reports produce a discordant sound. Thus, a user can quickly determine whether his/her hypothesis regarding the relationship of reports is well-founded or not.

20.2.3 Haptic Interfaces

Haptic interfaces are devices that use controlled feedback to provide a user the sensation that he or she is actually touching a figure in a computer. New research is being conducted on the mechanics, ergonomics, and utility of haptic interfaces (see http://www.hapticssymposium.org/). Devices range from single "pen-like" devices to gloves and other methods. Using a pen configuration, for example, a user can observe a figure on a computer screen, obtain a sense of touch of the object, and manipulate the object by carving, shaving, or gouging the surface of the figure. The apparent density and surface hardness of the figure can be readily changed via a slider scale. There has been extensive research on haptic interfaces for medical applications such as training surgeons. An example of the use of such an interface for data fusion might be to allow a user to "touch" the surface of an error ellipsoid to provide a sense of the second-order uncertainty (viz., the uncertainty of the uncertainty). A soft or "squishy" error surface would indicate that the uncertainty was not well known whereas a relatively hard surface for the ellipsoid would indicate a well-known error boundary.

20.3 The Way Ahead: Recommendations for New Research Directions

To make significant progress to improve the effectiveness of human-centered information fusion, we recommend several approaches (to be conducted in parallel).

20.3.1 Innovative Human–Computer Interface Designs

Innovative designs should be sought to support data fusion and understanding. Although this appears obvious, there is a tendency to simply reformat the same situation displays using the "display technology du jour." What is needed is true innovation, analogous to Richard Feynman's invention of Feynman diagrams to represent the interaction of fundamental particles in physics (see http://scienceworld.wolfram.com/physics/Feynman Diagram.html). Feynman's diagrams transformed the way physicists understood particle interactions and have created a new analysis approach. Ironically, Feynman[24] has reported that he originally developed the diagrams to help himself understand new developments in particle physics. We need the equivalent of Feynman diagrams for situation awareness and threat assessment. Similarly, Tufte[25] has made a career of showing how creative ways of displaying information provide new insights. He has provided multiple examples of how new graphic representations lead to new analysis results. It is uncertain how to elicit or motivate such innovative designs. Possible techniques might include conducting contests analogous to creative writing or creative art contests to solicit radical new ways of envisioning data.

20.3.2 Cross-Fertilization

Another potential area for improved design is cross-fertilization—applying data analysis and visualization methods from one domain into another. Could a standard tool such as the Hertzsprung-Russell diagram (see, for example, http://aspire.cosmic-ray.org/labs/star_life/hr_diagram.html) used routinely in astronomy to analyze stellar structure be applied to aspects of data fusion? Could techniques used to understand gene phenomena in biology be useful for situation assessment? A systematic attempt at cross-fertilization might be instructive and a source of new visualization and analysis techniques. An example of numerous types of graphic displays is provided by the "periodic table of visualization" (see http://www.visual-literacy.org/periodic_table/periodic_table.html).

20.3.3 Mining the Abnormal

Inferences about human cognition and perception have often been accurately made from observations about the abnormal or pathological cases of functioning. *The Diagnostic and Statistical Manual of Mental Disorders*, Fourth Edition Text Revision (DSM-IV TR, 2000)[26] defines many mental health disorders that include changes in perception and thought process. Psychotic disorders such as schizophrenia include reports of delusions, hallucinations, and heightened sensitivity to stimuli. These disorders are also referred to as thought disorders in the field of clinical psychology because of the tendency for delusions (ideas that are believed but have no basis in reality) that affect behavior. Hallucinations, or perceptions that occur without stimulation, have been reported most commonly for the senses of hearing and sound, although olfactory (smell) and tactile (touch) hallucinations have also been reported. According to the DSM-IV there are no laboratory findings that can diagnose schizophrenia; however, there have been associated anatomic differences between those with the disorder and those without. Although psychotic disorders are usually associated with perceptual and thought processes, there are other disorders that have significant associated sequelae. Anxiety disorders such as posttraumatic stress disorder (PTSD) include symptoms of heightened arousal and over-reaction to certain stimuli and have been associated with problems with concentration and attention. Depressive mood disorders have also been shown to have associated problems with concentration. Perhaps, another point of insight might be an examination of those with autism spectrum disorders (ASD). Formerly thought of as "idiot savants" or "autistic savants," some individuals with a higher-functioning form of autism known as Asperger's disorder have social interaction problems, difficulty in interpreting social cues, and associated repetitive and stereotyped behaviors. These behaviors are thought to be self-stimulating and include rocking, hand-flapping, preoccupation with parts of objects, and inflexible adherence to routines. Sometimes, these individuals may have highly developed behaviors such as decoding (reading but not necessarily comprehension), mathematics, and music. This is not a typical presentation of ASD; however, these behaviors and associated strengths through neurological and medical inquiry may help us understand normal behaviors. Similarly, case studies in the areas of stroke, brain injury, and seizure disorders have provided invaluable insight into brain functioning, language processing, personality, and impulse control. These kinds of studies and literature review may help us understand more fully the capabilities and limitations of human reason and behavior. Ramachandran[27] has provided examples of how the study of abnormal behavior and cognitive anomalies can lead to insights about perception in ordinary cognition.

20.3.4 Quantitative Experiments

The areas of cognitive psychology and even consumer marketing research have examined important aspects of the limitations of human decision making. Level 5 data fusion implies a generalized problem with a need for empirical research that bridges the gap between cognitive science and human–computer interfaces. The following heuristics are just a small sample of potential problems with human judgment that need to be systematically studied to optimize data visualization (see, for example, Ref. 28):

1. Humans judge the probability of an uncertain event according to the representativeness heuristic. Humans judge based on how obviously the uncertain event is similar to the population in which it was derived rather than considering actual base rates and the true likelihood of an occurrence (e.g., airplane crashes being considered more likely than they actually are).

2. Humans also judge based on how easily one can recall relevant instances of a phenomenon and have particular trouble when it confirms a belief about the self and the world. For example, the vast media coverage of airplane accidents and minimal coverage of automobile accidents contributes to fear of flying even though statistically it is much safer. Media coverage, however, gives people dramatic incidences of airplane crashes to recall more than automobile accidents.

3. Even the order and context in which options are presented to people influences the selection of an option. People tend to engage in risk aversion behavior when faced with an option involving potential losses and risk-seeking behavior when faced with potential gains.

4. People tend to have an inflated view of their actual skills, knowledge, and reasoning abilities. Most individuals see themselves as "above average" in many traits, including sense of humor, driving abilities, and other skills even though not everyone who reports this way could be.

5. Humans have a tendency to engage in confirmation-biased thinking, in which analysts tend to believe a conclusion if it seems to be true (based on the context in which it is presented) even when the logic is actually flawed.[29]

One challenge in visualization and understanding of human–computer interaction as it applies to data fusion is a tendency to develop prototype displays or tools and use simple expressions of user interest to determine their utility. Hall et al.[29] have argued for a systematic approach to conduct experiments that can quantitatively assess the utility and effect of displays and cognitive tools to support data fusion. An example of such experiments is the work of McNeese et al. who have developed a "living laboratory" approach (http://minds.ist.psu.edu/) to evaluate cognitive aids and collaboration tools. McNeese has developed a tool called NeoCITIES that simulates an ongoing sequence of events or activities. The simulator provides an environment in which individuals or teams can be provided with an evolving situation (with incoming data reports), and are required to make decisions regarding the situation and response. Thus, tools such as intelligent agents, collaboration tools, and visualization aids can be introduced and their effect can be observed on the decision-making efficacy of team.

20.3.5 Multisensory Experiments

Finally, we suggest that experiments be conducted in which the multisensory capabilities of the human user/analyst is deliberately exploited to improve data understanding. That is, experiments should be conducted in which a combination of 3D visualization, sonification, and haptic interfaces is used to improve data understanding and analysis. In addition, research should seek to simultaneously exploit human natural language ability along with sensory-based pattern recognition (viz., by automated generation of semantic metadata to augment traditional image and signal data). The techniques developed by

Wang and coworkers[30,31] are especially relevant for information fusion at the semantic level.[32] Thus, multisource data and information could be interpreted using multisensory capability and multibrain functions (language processing and pattern recognition). This approach should reduce the "impedance" mismatch between a human and data, and assist in conserving the ultimate limited resource—human attention units.

References

1. http://www-bsac.eecs.berkeley.edu/archive/users/warneke-brett/SmartDust/.
2. Steinberg, A. and C. L. Bowman, Revisions to the JDL data fusion model. In *Handbook of Multisensor Data Fusion*, D. Hall and J. Llinas (Eds.), CRC Press, Boca Raton, FL, 2001.
3. Hall, D. and A. Steinberg, Dirty secrets in multisensor data fusion. In *Handbook of Multisensor Data Fusion*, D. Hall and J. Llinas (Eds.), CRC Press, Boca Raton, FL, 2001.
4. Hall, M. J., S. A. Hall, and T. Tate, Removing the HCI bottleneck: How the human-computer interface (HCI) affects the performance of data fusion systems. In *Handbook of Multisensor Data Fusion*, D. Hall and J. Llinas (Eds.), CRC Press, Boca Raton, FL, 2001.
5. Yen, J., X. Fan, S. Sun, M. McNeese and D. Hall, Supporting antiterrorist analysis teams using agents with shared RPD Process. In *Proceedings of the IEEE International Conference on Computational Intelligence for Homeland Security and Personal Safety*, Venice, Italy, July 21–22, 2004.
6. Connors, E. S., P. Craven, M. McNeese, T. Jefferson, Jr., P. Bains, and D. L. Hall, An application of the AKADAM approach to intelligence analyst work. In *Proceedings of the 48th Annual Meeting of the Human Factors and Ergonomics Society*, Human Factors and Ergonomics Society, New Orleans, LA, 2004.
7. McNeese, M. D. and D. L. Hall, User-centric, multi-INT fusion for homeland defense. In *Proceedings of the 47th Annual Meeting of the Human Factors and Ergonomics Society*, Human Factors and Ergonomics Society, Santa Monica, CA, pp. 523–527, Oct 13–17, 2003.
8. Patterson, E. S., D. D. Woods, and D. Tinapple, Using cognitive task analysis (CTA) to seed design concepts for intelligence analysis under data overload. In *Proceedings of the Human Factors and Ergonomics Society 43rd Annual Meeting*, Human Factors and Ergonomics Society, Minneapolis, MN, 2001.
9. Hall, D., M. Seif El-Nasr, and J. Yen, A three pronged approach for improved data understanding: 3-D visualization, use of gaming techniques, and intelligent advisory agents. In *Proceedings of the NATO N/X Conference on Visualization*, Copenhagen, Denmark, October 17–20, 2006.
10. D. L. Hall, Increasing operator capabilities through advances in visualization. In *3rd Annual Sensors Fusion Conference: Improving the Automation, Integration, Analysis and Distribution of Information to the Warfighter*, Washington, DC, Nov. 29–Dec. 1, 2006.
11. PC Magazine. (n.d.). Definition of: Virtual reality, Retrieved on 24 August 2007 from http://www.pcmag.com/encyclopedia_term/0,2542,t=virtual+reality&i=53945,00.asp.
12. Leetaru, K., The CAVE at NCSA: About the CAVE, Retrieved on August 4, 2007 from http://cave.ncsa.uiuc.edu/about.html, 2000–2004.
13. Roddenberry, G. (Creator) *Star Trek: The Next Generation*, Paramount Pictures, 1987.
14. Hall, S. A., An investigation of cognitive factors that impact the efficacy of human interaction with computer-based complex systems. A Human Factors Research Project Submitted to the Extended Campus in Partial Fulfillment of the Requirements of the Degree of Master of Science in Aeronautical Science, Embry-Riddle Aeronautical University, Extended Campus, 2001.
15. Shumaker, W., Private communication to D. L. Hall, September, 2007.
16. Risch, J. S., D. S. Rex, S. T. Dowson, T. B. Walters, R. A. May and B. D. Moon, The STARLIGHT information visualization system. In *Proceedings of IEEE Conference on Information Visualization*, 42, 1997.

17. Rex, B., *Starlight Approach to Text Spatialization for Visualization*, Cartography Special Group Sponsored Workshop, 2002.
18. Chen, H., H. Atabakhsh, C. Tseng, D. Marshall, S. Kaza, S. Eggers, H. Gowda, A. Shah, T. Peterson and C. Violette, Visualization in law enforcement. *CHI 2005 Extended Abstracts on Human Factors in Computing Systems*, Portland, OR, 2005.
19. Corporation, K. C. COPLINK, White Paper, Tucson, AZ, 2005.
20. Madhyastha, T. M. and D. A. Reed, Data sonification: Do you see what I hear? *IEEE Software*, 12(2): 45–56, 1995.
21. Ballora, M., B. Pennycook, P. Ch. Ivanov, A. L. Goldberger, and L. Glass, Detection of obstructive sleep apnea through auditory display of heart rate variability. In *Proceedings of Computers in Cardiology 2000*, IEEE Engineering in Medicine and Biology Society, 2000.
22. Baker, L., Sub hunters: Detecting the enemy beneath the sea, *Pakistan Daily Times*, August 30, 2007.
23. Hall, D. and S. McMullen, *Mathematical Techniques in Multisensor Data Fusion*, ARTECH House, Boston, MA, 2004.
24. Feynman, R., *Surely You're Joking Mr. Feynman: Adventures of a Curious Character*, Norton & Co., New York, NY, 1997.
25. Tufte, E., *The Visual Display of Quantitative Information*, 2nd edn., Graphics Press, Cheshire, CT, 2001.
26. *Diagnostic and Statistical Manual of Mental Disorders*, 4th edn., Text Revision (DSM-IV-TR), American Psychiatric Association, Washington, DC, 2000.
27. Ramachandran, V. S., *A Brief Tour of Human Consciousness: From Impostor Poodles to Purple Numbers*, PI Press, Essex, UK, 2004.
28. Heuer, R., *The Psychology of Intelligence Analysis*, Norinka Books, CIA Center for the Study of Intelligence, Washington, DC, 1999.
29. Hall, C., S. A. H. McMullen, and D. L. Hall, Cognitive engineering research methodology: A proposed study of visualization analysis techniques. In *Proceedings of the NATO Workshop on Visualizing Network Information*, Copenhagen, Denmark, October 17–20, 2006.
30. Li, J. and J. Z. Wang, Automatic linguistic indexing of pictures by a statistical modeling approach, *IEEE Trans. Pattern Anal. Machine Intell.*, 25(9): 1075–1088, 2003.
31. Wang, J. Z., J. Li, and G. Wiederhold, SIMPLIcity: Semantics-sensitive integrated matching for picture libraries, *IEEE Trans. Pattern Anal. Machine Intell.*, 23(9): 947–963, 2001.
32. Hall, D. L., Beyond level "N" fusion: Performing fusion at the information level. In *Proceedings of the MSS National Symposium on Sensor and Data Fusion*, San Diego, CA, August 13–15, 2002.

21

Requirements Derivation for Data Fusion Systems

Ed Waltz and David L. Hall

CONTENTS

21.1 Introduction

The design of practical systems requires the translation of data fusion theoretic principles, practical constraints, and operational requirements into a physical, functional, and operational architecture that can be implemented, operated, and maintained. This translation of principles to practice demands a discipline that enables the system engineer or architect to perform the following basic functions:

- Define user requirements in terms of functionality (qualitative description) and performance (quantitative description).
- Synthesize alternative design models and analyze/compare the alternatives in terms of requirements and risk.
- Select optimum design against some optimization criteria.
- Allocate requirements to functional system subelements for selected design candidates.
- Monitor the as-designed system to measure projected technical performance, risk, and other factors (e.g., projected life cycle cost) throughout the design and test cycle.
- Verify performance of the implemented system against top- and intermediate-level requirements to ensure that requirements are met, and to validate the system performance model.

The discipline of system engineering, pioneered by the aerospace community to implement complex systems over the past four decades, has been successfully used to implement both research and development and large-scale data fusion systems. This approach is characterized by formal methods of requirement definition at a high level of abstraction, followed by decomposition to custom components that can then be implemented. More recently, as information technology has matured, the discipline of enterprise architecture design has also developed formal methods for designing large-scale enterprises using commercially available custom software as well as hardware components. Both of these disciplines contribute sound methodologies for implementing data fusion systems.

This chapter introduces each approach before comparing the two to illustrate their complementary nature and the utility of each. The approaches are not mutually exclusive, and methods from both may be applied to translate data fusion principles to practice.

21.2   Requirements Analysis Process

Derivation of requirements for a multisensor data fusion system must begin with the recognition of a fundamental principle: *there is no such thing as a data fusion system*. Instead, there are applications to which data fusion techniques can be applied. This implies that generating requirements for a generic data fusion system is not particularly useful (although one can identify some basic component functions). Instead, the particular application or mission to which the data fusion is addressed drives the requirements. This concept is illustrated in Figures 21.1a and 21.1b.[1]

Figure 21.1a indicates that the requirements analysis process begins with an understanding of the overall mission requirements. What decisions or inferences are sought by the overall system? What decisions or inferences do the human users want to make? The analysis and documentation of this is illustrated at the top of the figure. An understanding of the anticipated targets supports this analysis, the types of threats anticipated, the environment in which the observations and decisions are to be made, and the operational doctrine. For Department of Defense (DoD) applications—such as automated target recognition—this would entail specifying the types of targets to be identified (e.g., army tanks and launch vehicles) and other types of entities that could be confused for targets (e.g., automobiles and school buses). The analysis must specify the environment in which the observations are made, the conditions of the observation process, and sample missions or engagement scenarios. This initial analysis should clearly specify the military or mission needs and how these would benefit from a data fusion system.

From this initial analysis, system functions can be identified and performance requirements associated with each function. The Joint Directors of Laboratories (JDL) data fusion process model can assist with this step. For example, the functions related to communications/message processing could be specified. What are the external interfaces to the system? What are the data rates from each communications link or sensor? What are the system transactions to be performed?[2] These types of questions assist in the formulation of the functional performance requirements. For each requirement, one must also specify how the requirement can be verified or tested (e.g., via simulations, inspection, and effectiveness analysis). A requirement is vague (and not really a requirement) unless it can be verified via a test or inspection.

Ideally, the system designer has the luxury of analyzing and selecting a sensor suite. This is shown in the middle of the diagram that appears in Figure 21.1a. The designer performs a survey of current sensor technology and analyzes the observational phenomenology (i.e., how the inferences to be made by the fusion system could be mapped to

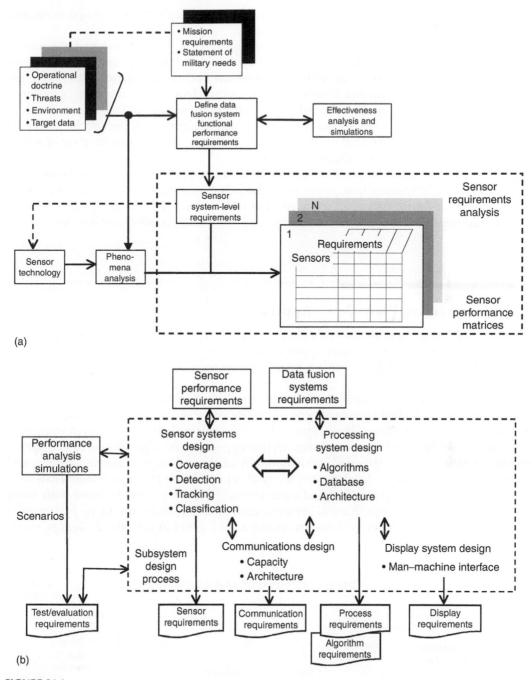

FIGURE 21.1
Requirements flow-down process for data fusion.

observable phenomena such as infrared spectra and radio frequency measurements). The result of this process is a set of sensor performance measures that link sensors to functional requirements, and an understanding of how the sensors could perform under anticipated conditions. In many cases, of course, the sensors have already been selected (e.g., when designing a fusion system for an existing platform such as a tactical aircraft). Even in such cases, the designer should perform the sensor analysis to understand the operation and contributions of each sensor in the sensor suite.

The flow-down process continues as shown in Figure 21.1b. The subsystem design/analysis process is shown within the dashed frame. At this step, the designer explicitly begins to allocate requirements and functions to subsystems such as the sensor subsystem, the processing subsystem, and the communications subsystem. These must be considered together because the design of each subsystem affects the design of the others. The processing subsystem design entails the further selection of algorithms, the specific elements of the database required, and the overall fusion architecture (i.e., the specification of where in the process flow the fusion actually occurs).

This requirement analysis process results in well-defined and documented requirements for the sensors, communications, processing, algorithms, displays, and test and evaluation requirements. If performed in a systematic and careful manner, this analysis provides a basis for an implemented fusion system that supports the application and mission.

21.3   Engineering Flow-Down Approach

Formal systems engineering methods are articulated by the U.S. DoD in numerous classical military standards[3] and defense systems engineering guides. A standard approach for development of complex hardware and software systems is the *waterfall approach* shown in Figure 21.2.

This approach uses a sequence of design, implementation, and test and evaluation phases or steps controlled by formal reviews and delivery of documentation. The waterfall approach begins at the left side of Figure 21.2 with system definition, subsystem design, preliminary design, and detailed design. In this approach, the high-level system requirements are defined and partitioned into a hierarchy of increasingly smaller subsystems and components. For software development, the goal is to partition the requirements to a level of detail so that they map to individual software modules comprising no more than about 100 executable lines of code. Formal reviews, such as a requirement review, preliminary design review (PDR), and critical design review (CDR), are held with the designers, users,

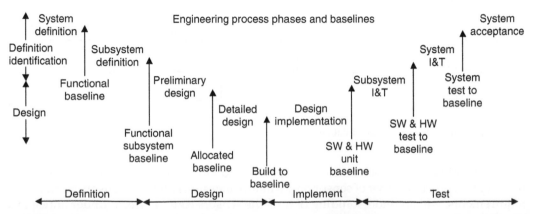

FIGURE 21.2
System engineering methodology.

and sponsors to obtain agreement at each step in the process. A baseline control process is used, so that requirements and design details developed at one phase cannot be changed in a subsequent phase without a formal change/modification process.

After the low-level software and hardware components are defined, the implementation begins. (This is shown in the middle of Figure 21.2.) Small hardware and software units are built and aggregated into larger components and subsystems. The system development proceeds to build small units, integrate these into larger entities, and test and evaluate evolving subsystems. The test and integration continues until a complete system is built and tested (as shown on the right side of Figure 21.2). Often a series of *builds* and tests are planned and executed.

Over the past 40 years, numerous successful systems have been built in this manner. Advantages of this approach include

- The ability to systematically build large systems by decomposing them into small, manageable, testable units
- The ability to work with multiple designers, builders, vendors, users, and sponsoring organizations
- The capability to perform the development over an extended period of time with resilience to changes in development personnel
- The ability to define and manage risks by identifying the source of potential problems
- The ability to formally control and monitor the system development process with well-documented standards and procedures

This systems engineering approach is certainly not suitable for all system developments. The approach is most applicable for large-scale hardware and software system. Basic assumptions include the following:

- The system to be developed is of sufficient size and complexity that it is not feasible to develop it using less formal methods.
- The requirements are relatively stable.
- The requirements can be articulated via formal documentation.
- The underlying technology for the system development changes relatively slowly compared to the length of the system development effort.
- Large teams of people are required for the development effort.
- Much of the system must be built from scratch rather than purchased commercially.

Over the past 40 years, the formalism of systems engineering has been very useful for developing large-scale DoD systems. However, recent advances in information technology have motivated the use of another general approach.

21.4 Enterprise Architecture Approach

The rapid growth in information technology has enabled the construction of complex computing networks that integrate large teams of humans and computers to accept, process, and analyze volumes of data in an environment referred to as the *enterprise*. The development of enterprise architectures requires the consideration of functional operations and the allocation of these functions in a network of human (cognitive), hardware (physical), or software components.

The enterprise includes the collection of people, knowledge (tacit and explicit), and information processes that deliver critical knowledge (often called *intelligence*) to analysts and decision-makers, enabling them to make accurate, timely, and wise decisions. This definition describes the enterprise as a process that is devoted to achieving an objective for its stakeholders and users. The enterprise process includes the production, buying, selling, exchange, and promotion of an item, substance, service, and system. The definition is similar to that adopted by DaimlerChrysler's extended virtual enterprise, which encompasses its suppliers:

A DaimlerChrysler coordinated, goal-driven process that unifies and extends the business relationships of suppliers and supplier tiers to reduce cycle time, minimize systems cost, and achieve perfect quality.[4]

This all-encompassing definition brings the challenge of describing the full enterprise, its operations, and its component parts. Zachman has articulated many perspective views of an enterprise information architecture and has developed a comprehensive framework of descriptions to thoroughly describe an entire enterprise.[5,6] The following section describes a subset of architecture views that can represent the functions in most data fusion enterprises.

21.4.1　The Three Views of the Enterprise Architecture

The enterprise architecture is described in three views (as shown in Figure 21.3), each with different describing products. These three interrelated perspectives or architecture views

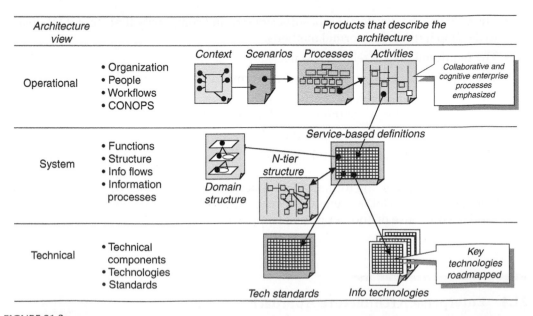

FIGURE 21.3
Three architecture views are described in a variety of products.

are outlined by the DoD in their description of the Command, Control, Communication, Computation, Intelligence, Surveillance, and Reconnaissance (C4ISR) framework.[7] They include the following:

1. Operational architecture (OA) is a description (often graphical) of the operational elements, business processes, assigned tasks, workflows, and information flows required to accomplish or support the C4ISR function. It defines the type of information, the frequency of exchange, and tasks supported by these information exchanges. This view uniquely describes the human role in the enterprise and the interface of human activities to automated (machine) processes.

2. Systems architecture (SA) is a description, including graphics, of the systems and interconnections providing for or supporting functions. The SA defines the physical connection, location, and identification of the key nodes, circuits, networks, and war-fighting platforms, and it specifies system and component performance parameters. It is constructed to satisfy OA requirements per standards defined in the technical architecture. The SA shows how multiple systems within a subject area link and interoperate and may describe the internal construction or operations of particular systems within the architecture.

3. Technical architecture (TA) is a minimal set of rules governing the arrangement, interaction, and interdependence of the parts or elements whose purpose is to ensure that a conformant system satisfies a specified set of requirements. TA identifies the services, interfaces, standards, and their relationships. It provides the technical guidelines for implementation of systems upon which engineering specifications are based, common building blocks are built, and product lines are developed.

The primary products that describe the three architecture views (Figure 21.3) include the following:

1. Context diagram. The intelligence community context that relates shareholders (owners, users, and producers)

2. Scenarios. Selected descriptions of problems representing the wide range of situations expected to be confronted and solved by the enterprise

3. Process hierarchy. Tree diagrams that relate the intelligence community business processes and describe the functional processes that are implemented as basic services

4. Activity diagrams. Sequential relationships between business processes that are described in activity sequence diagrams

5. Domain operation. The structure of collaborative domains of human virtual teams (user community)

6. Service descriptions. Define core and special software services required for the enterprise—most commercial and some custom

7. *n*-Tier structure diagram. The *n*-tier structure of the information system architecture is provided at a top level (e.g., two-tier systems are well-known client-server tiers; three-tier systems are partitioned into data warehouse, business logic, and presentation layers)

8. Technical standards. Critical technical standards that have particular importance to data fusion business processes, such as data format standards and data service standards (e.g., SQL and XML)

9. Information technology roadmaps. Projected technology needs and drivers that influence system growth and adoption of emerging technologies are critical components of an enterprise; they recognize the highly dynamic nature of both the enterprise and information technology as a whole.

21.5 Comparison of Approaches

The two design methods are complementary in nature and both provide helpful approaches for decomposing problems into component parts and developing data fusion solutions. Comparison of the major distinguishing characteristics of the approaches (Table 21.1) illustrates the strengths of each approach for data fusion system implementation:

- Perspective. System engineering seeks specific (often custom) solutions to meet all specific functional requirements; system architecting begins with components and seeks to perform the set of use cases (accepting requirements' flexibility) with the optimum use of components (minimizing custom-designed components).

TABLE 21.1

Comparison of System-Level Design Approaches

	System Engineering	System Architecting
Perspective	Design from first principles	Design from best standard components
	Optimize design solution to meet all functional requirements; quantify risk to life cycle cost and to implementation cost and schedule	Optimize design solution to implement use cases; quantify risk to operational performance and enterprise future growth (scalability, component upgrade)
	Provide traceability from requirements to implementation	Provide traceability between multiple architecture views
Starting assumptions	Top-level problem requirements exist and are quantified	Functional components (e.g., software components) exist
	Requirements are specific, technical, and quantified	Requirements tend to be general, user-oriented applications, and subjective
	Emphasis on functional models	Emphasis on use-case models
Methodology	Structured problem decomposition	Use case modeling
	Requirements flow-down (deriving and allocating requirements)	Data and functional modeling
	Design, integrate custom components	Multiple architecture views construction (multiple functional perspectives)
		Design structure, integrate standard components
Basis of risk analysis	Implementation (cost, schedule) and performance risk as a function of alternative design approaches	Operational utility (over life cycle) risk as a function of implementation cost
Design variable perspective	Requirements are fixed; cost is variable	Cost is fixed; requirements are variable
Applicable data fusion systems	One-of-a-kind intelligence and military surveillance systems	General business data analysis systems
	Mission-critical, high-reliability systems	Highly networked component-based, service-oriented systems employing commercial software components

- Starting assumptions. System engineering assumes that top-level system requirements exist and are quantifiable. The requirements are specific and can be documented along with performance specifications. System engineering emphasizes functional models. By contrast, system architecting assumes that functional components exist and that the requirements are general and user oriented. The emphasis is on use-case models.

- Methodology. The methodology of system engineering involves structured problem decomposition and requirements flow down (as described in Section 21.2). The design and implementation proceed in accordance with a waterfall approach. System architecting involves use-case modeling and data and functional modeling. Multiple functional perspectives may be adopted, and the focus is on integration of standard components.

- Risk analysis. Systems engineering addresses risks in system implementation by partitioning the risks to subsystems or components. Risks are identified and addressed by breaking the risks into manageable smaller units. Alternative approaches are identified to address the risks. In the system architecting approach, risk is viewed in terms of operational utility over a life cycle. Alternative components and architectures address risks.

- Design variables. System engineering assumes that the system requirements are fixed but that development cost and schedule may be varied to meet the requirements. By contrast, system architecting assumes that the cost is fixed and the requirements may be varied or traded off to meet cost constraints.

- Application. The system engineering approach provides a formal means of deriving, tracking, and allocating requirements to permit detailed performance analysis and legal contract administration. This is often applied on one-of-a-kind systems, critical systems, or unique applications. The architectural approach is appropriate for the broader class of systems, where many general approaches can meet the requirements (e.g., many software products may provide candidate solutions).

The two basic approaches described here are complimentary. Both hardware and software developments will tend to evolve toward a hybrid utilization of systems engineering and architecting engineering. The rapid evolution of information technology and the appearance of numerous commercial-off-the-shelf tools provide the basis for the use of methods such as architecting engineering. New data fusion systems will likely involve combinations of traditional systems engineering and architecting engineering approaches, which will provide benefits to both the implementation and the user communities.

21.6 Requirements for Data Fusion Services

Defense systems have adopted more distributed, network-centric architectures that loosely couple sensors, sources, data, processes, and presentation (displays); these architectures impose new requirements on data fusion process that allow them to be used as a general application or service by many network users. Network-centric approaches employ a service-oriented architecture (SOA) framework, where services are published, discovered, and dynamically applied as part of an adaptive information ecosystem. These network-centric enterprise services (NCES) are aware of other services, are adaptive to the threat environment, and are reusable.

In such a framework, a data fusion service is one of many published resources for all participants in the network (humans and other services) to use, and the fusion service is required to meet the following general capabilities so that all participants can access it in a standardized way:

- The data fusion functionality must be designed to be as widely accessible as possible; to be a generally available fusion service, it should minimize restrictions on data types, conditions, and pedigrees of source data.

- The SOA requires the tagging of all data (e.g., intelligence, nonintelligence, raw, and processed) with metadata to enable discovery of data. The data fusion service can discover needed data (e.g., sensor data, geospatial or map data, calibration data, etc.), and display services can readily locate the fused products posted by the fusion service.

- The fusion service must advertise its capabilities within standard service registries for access by other services, expressing the service interfaces using standard metadata descriptions.

- The fusion service must communicate with other services using standard protocols.

- The fusion service must post or publish data to shared spaces to provide access to all users, except when limited by security, policy, or regulations.

These general provisions allow a fusion service to perform the *many-to-many* exchanges typical of a network environment, rather than the more limited interoperability provided by traditional point-to-point interfaces.

The U.S. DoD has published a detailed checklist of requirements that guide program managers and developers to meet these service requirements.[8] The sharing of data among sensor, fusion, and other services is also guided by policies and instructions that assure reliable exchange and use of data.[9] The SOA approach imposes stringent simultaneous quality of service (QoS) demands that also require data fusion services to be real time, scalable (to service many users simultaneously, with assured response and accuracy), and secure. Legg has enumerated the key factors that must be considered to specify a distributed multisensor fusion surveillance system, including sensor data processing and distribution, tracking, sensor control, computing resources, and performance to operate in an SOA network.[10]

Network-centric distributed computing environments leverage open standards and use the open architecture to dynamically interconnect services to respond to operational needs. Consider, for example, the following operation of a hypothetical tactical network that employs fusion services to illustrate a network-centric concept of operations.

A tactical military unit is ambushed and comes under intense fire in a desolate tribal area; the commander requests immediate targeting and available firepower support information via a CRITICAL_SUPPORT service.

1. The CRITICAL_SUPPORT service issues a network search for all available sensor services and information source services on the ambush force:

   a. The area search discovers (calls) a Tactical Fusion (TACFUSION) service that correlates area sensor data over the past 12 h, filtering for unusual movements.

   b. The TACFUSION service also filters recent Situational Reports (SITREPS) and Human Intelligence (HUMINT) reports (tagged and available in repositories) about neighboring tribal areas for similar ambush patterns and reports of recent movements.

c. The TACFUSION service discovers (calls) a SENSOR_SEARCH service to identify immediately available sensors that can provide surveillance of the ambush forces, tracking, and targeting. The SENSOR_SEARCH reports back that a Predator unmanned air vehicle returning to base is within 5 min time over target (if redirected) and has sufficient fuel to provide 25 min of Electro-Optical/Synthetic Aperture Radar (EO/SAR) coverage of the area and targeting. Unfortunately, the aircraft has expended its weapons and cannot support direct attack. The SENSOR_SEARCH service also identifies another national asset that may provide targeting quality information, but it will not be available for almost an hour.

d. The TACFUSION service also searches for all Blue Force tracking data published by the BLUE_TRACK to identify potential supporting sensors within the range of the ambush force; none are found.

e. The TACFUSION service also checks for unattended ground sensors in the area and identifies a small emplaced sensor net within range of the ambush force capable of detecting and tracking vehicles used by the force.

f. The TACFUSION service reports to the CRITICAL_SUPPORT service its need of the Predator sensors and follow-on national asset sensor data; it also begins to publish a stream of emerging data on the ambush force using prior data and the ground sensors, and awaits Predator data to become available.

2. The CRITICAL_SUPPORT service also requests a search for available supporting fire control, and identifies two distant tactical aircraft (25 min out) with direct attack munitions, and an artillery unit that is marginally in-range, but capable of supporting area fires.

3. The CRITICAL_SUPPORT service requests use of the Predator sensors (as another available net service) from the TACTICAL_C2 service, and also requests available fire control support from the TACTICAL_C2 service.

4. The TACTICAL_C2 service then issues a request to the FIRE_FUSION service to develop targeting solutions for immediately suppressing area fires, and when the Predator data come online to develop coarse target tracking for handoff to the tactical aircraft when they arrive on-station bringing attack munitions.

The DoD has developed a body of architectural and engineering knowledge to provide guidance in the design, implementation, maintenance, and use of the such net-centric solutions for military applications, called the Net-Centric Enterprise Solutions for Interoperability (NESI).[11] NESI provides specific technical guidance to assure that service developers are compliant with the Net-centric Checklist and other data sharing directives.

References

1. Waltz, E. and Llinas, J., *Multisensor Data Fusion*, Artech House, Norwood, MA, 1990.
2. Finkel, D., Hall, D.L., and Beneke, J., Computer performance evaluation: the use of a time-line queuing method throughout a project life cycle, *Model. Simul.*, 13, 729–734, 1982.
3. Classical DoD military standards for systems engineering include MIL-STD-499B, Draft Military Standard: Systems Engineering, HQ/AFSC/EN, Department of Defense, "For Coordination Review" draft, May 6, 1992, and NSA/CSS Software Product Standards Manual, NSAM

81-3, National Security Agency. More recent documents that organize the principles of systems engineering include: EIA/IS 632, Interim Standard: Systems Engineering, Electronic Industries Alliance, December 1994; Systems Engineering Capability Assessment Model SECAM (version 1.50), INCOSE, June 1996; and the ISO standard for system life cycle processes, ISO 15288.

4. DaimlerChrysler Extended Enterprise, see http://supplier.chrysler.com/purchasing/extent/index.html

5. Zachman, J.A., A framework for information systems architecture, *IBM Syst. J.*, 26(3), 276–292, 1987.

6. Sowa, J.F. and Zachman, J.A., Extending and formalizing the framework for information systems architecture, *IBM Syst. J.*, 31(3), 590–616, 1992.

7. Joint Technical Architecture, Version 2.0, Department of Defense, October 31, 1997 (see paragraph 1.1.5 for definitions of architecture and the three architecture views).

8. Net-Centric Checklist, Office of the Assistant Secretary of Defense for Networks and Information Integration/Department of Defense Chief Information Officer, Version 2.1.3, May 12, 2004.

9. See *DoD Net-Centric Data Strategy*, May 9, 2003 and Instruction DoD Directive 8320.2 *Data Sharing in a Net-Centric DoD*, December 2, 2004.

10. Legg, J. A., *Distributed Multisensor Fusion System Specification and Evaluation Issues*, Australian Defence Science and Technology Organisation (DSTO), DSTO–TN–0663, October 2005.

11. Net-Centric Enterprise Solutions for Interoperability (NESI), Version 1.3, June 16, 2006; this six-part document is a collaborative activity of the USN PEO for C4I and Space, the USAF Electronic Systems Center, and the Defense Information Systems Agency.

22

Systems Engineering Approach for Implementing Data Fusion Systems

Christopher L. Bowman and Alan N. Steinberg

CONTENTS

22.1 Scope

This chapter defines a systematic method for characterizing and developing data fusion and resource management (DF&RM) systems. This is defined in the form of a technical architecture at the applications layer.

Although DF&RM can and has been performed by people with various degrees of auto-mated tools, there is increasing demand for more automation of data fusion and response management functions. In particular, there is a need for such processes operating in open, distributed environments. The exponential increase in DF&RM software development costs has increased the demand to cease building one-of-a-kind DF&RM software systems. Software technical architectures help to achieve this by defining the standard software components, interfaces, and engineering design methodology. This provides a *toolbox* to guide software designers and to organize the software patterns that they develop (thus easing reusability).

The intent of this chapter is to provide a common, effective foundation for the design and development of DF&RM systems. The DF&RM technical architecture provides guide-lines for selecting among design alternatives for specific DF&RM system developments.

This system's engineering approach has been developed to provide

- A standard model for representing the requirements, design components, their interfaces, and performance assessment of DF&RM systems
- An engineering methodology for developing DF&RM systems to include the selection of the DF&RM system architecture and the analysis of technique alternatives for cost-effective satisfaction of system requirements

This DF&RM technical architecture[1-3] builds on a set of Data Fusion Engineering Guide-lines, which was developed in 1995–1996 as part of the U.S. Air Force Space Command's Project Correlation.* The present work extends these guidelines by proposing a technical architecture as defined by the Department of Defense (DoD) Architecture Framework for system engineering (www.defenselink.mil/nii), thereby establishing the basis for rigorous problem decomposition, system architecture design, and technique application.

Integral to the guidelines is the use of a functional model for characterizing diverse DF&RM system architectures. This architecture functional paradigm has been found to successfully capture the salient operating characteristics of the diverse automatic and manual approaches that have been employed across great diversity of data fusion applica-tions. This DF&RM system paradigm can be implemented in human, as well as automated processes, and in hybrids involving both.

The recommended architecture concept represents DF&RM systems as interlaced net-works—earlier termed DF&RM *trees*—of data fusion processing nodes and resource management processing nodes that are amenable to standard *dual* representations. The interlacing of data fusion nodes with resource management nodes enables fast response loops. The close relationship between fusion and management, both in interactive opera-tion and in their underlying design principles, plays an important part in effective system design and in comparing alternative methods for achieving DF&RM system designs.

The largely heuristic methods presented in the Project Correlation Data Fusion Engi-neering Guidelines are amenable to more rigorous treatment. System engineering, and specifically data fusion system engineering, can be viewed as a planning (i.e., resource management) process. Available techniques and design resources can be applied to meet an objective criterion: cost-effective system performance. Therefore, the techniques of optimal planning can be applied to building optimal systems. Furthermore, the formal duality between data fusion and resource management first propounded by Bowman, enables data fusion design principles developed over the past 30+ years to be applied to

---

* A closely related set of guidelines[3] for selecting among data correlation and association techniques, devel-oped as part of the same project, is discussed in Chapter 23 of this handbook.

the *dual* problems of resource management. As system engineering is itself a resource management problem, the resource management architecture principles described herein can be used to organize the building of any software system.*

In summary, this chapter provides a description of how the DF&RM dual node network (DNN) architecture organizes and unifies applications-layer software approaches to DF&RM. Refinements of the Joint Directors of Laboratories (JDL) fusion model levels and dual resource management process model levels are proposed to take advantage of their duality to bootstrap the less mature resource management field. The levels are used to divide and conquer DF&RM software developments. This enables the batching of data belief functions for combination within fusion and the batching of resource tasking within management. Comparisons to other fusion models and examples of the application of the DNN for affordable fusion and management processing using these levels are discussed.

22.2 Architecture for Data Fusion

22.2.1 Role of Data Fusion in Information Processing Systems

Data fusion is the process of combining data/information to estimate or predict the state of some aspect of the world.[1] Fusion provides fused data from the sources to the user and to resource management, which is the process of planning/controlling response capabilities to meet mission objectives (Figure 22.1).

In this section, refinements of the JDL data fusion model and *dual* extensions for resource management are described and applied in the DF&RM DNN software technical architecture. The motivation for this and earlier versions of the JDL data fusion model has been to facilitate understanding of the types of problems for which data fusion is applicable and to define a useful partitioning of DF&RM solutions.

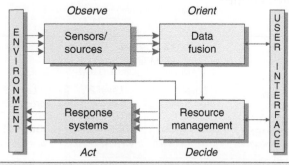

- *Data fusion* is the process of combining data/information to estimate or predict the state of some aspect of the world.

- *Resource management* is the process of planning/ controlling response capabilities to meet mission objectives.

FIGURE 22.1
Fusion and management lie between *observe* and *act*.

---

* The *dual* relationship between Data Fusion and Resource Management is discussed in Section 22.2.3.

The JDL data fusion model is a *functional* model. It was never conceived as a process model or as a technical architecture. Rather, it has been motivated by the desire to reduce the community's perceived confusion regarding many elements of fusion processes. The model was developed to provide a common frame of reference for fusion discussions and to facilitate the understanding and recognition of the types of problems for which data fusion is applicable and also as an aid to recognizing commonality among problems and the relevance of candidate solutions.

As stated by Frank White, the original JDL panel chairman, *much of its (JDL model) value derives from the fact that identified fusion functions have been recognizable to human beings as a model of functions they were performing in their own minds when organizing and fusing data and information. It is important to keep this human centric sense of fusion functionality since it allows the model to bridge between the operational fusion community, the theoreticians and the system developers.*

The framework of the model has been useful in categorizing investment in automation and highlighting the difficulty of building automated processes that provide functionality in support of human decision processes particularly at the higher levels where reasoning and inference are required functions. The dual resource management levels described herein are offered as a starting place from which to evolve the same benefits for the resource management community.

The suggested revised data fusion function at the functional levels[2] are based on the entities of interest to information users (Figure 22.2) with revised fusion levels defined as follows:

- Level 0 (L0). Signal/feature assessment—estimation and prediction of signal or feature states

- Level 1 (L1). Entity assessment—estimation and prediction of entity parametric and attributive states (i.e., of entities considered as individuals)

- Level 2 (L2). Situation assessment—estimation and prediction of the structures of parts of reality (i.e., of relationships among entities and their implications for the states of the related entities)

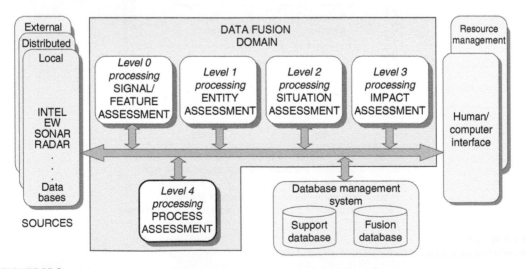

FIGURE 22.2
Recommended revised data fusion model. (From Steinberg, A.N. and Bowman, C.L., *Rethinking the JDL Fusion Levels*, NSSDF JHAPL June, 2004.)

- Level 3 (L3). Impact assessment—estimation and prediction of the utility/cost of signal, entity, or situation states; including predicted impacts given a system's alternative courses of action (COAs)
- Level 4 (L4). Process assessment—estimation and prediction of a system's performance as compared to given desired states and measures of effectiveness

The JDL fusion levels provide a historically useful and natural batching of fusion nodes (i.e., into signals, entities, relationships, and impacts) that should be considered in a fusion system design to achieve the knee-of-the-curve in performance versus cost. The level 1 entities of interest are defined by the user as individual objects. Level 0 characterizes patterns in received data to estimate types and characteristics of features and signals. This may be done without necessarily posting a specific causative level 1 entity. Their relationships of interest (e.g., as defined by an ontology of their aggregates, attacks, cues, etc.) are the focus of level 2 processing. This level 2 processing can be represented with multiple predicates as described in Ref. 2, with single predicates (e.g., as defined by a taxonomy of entity classes) used for the entity attributes estimated in level 1. The use of the DNN architecture for level 2 and level 3 fusion provides the system designer with techniques to interlace the levels 2 and 3 processes and the functional partitioning for each DF&RM node. The processing at these DF&RM levels are not necessarily performed in sequential order and two or more levels may need to be integrated into common nodes to achieve increased performance albeit at higher complexity/cost.

The most significant change, from earlier versions[1,2] is the removal of process management from fusion level 4. Process management is now found within the new resource management levels, as suggested in Ref. 1. Process assessment is retained in the fourth level of JDL model for fusion (i.e., process assessment has always been included in the JDL level 4 fusion). The consideration of fusion process management as a resource management function provides a rich set of tools for segmenting and developing process management software using DNN resource management nodes much as they are applied for resource management process refinement (see Section 22.4). The restriction of level 4 fusion to process assessment provides the benefit for developing performance evaluation (PE) systems using the DNN data fusion architecture (see Section 22.3). Examples of levels 2 and 3 fusion processes are given in Sections 22.1 and 22.2, respectively. A discussion of how the DF&RM DNN architecture unifies numerous DF&RM models is given in Section 22.5. Before presenting these the role and significant components of the applications-layer DNN architecture are reviewed in Sections 22.2 and 22.3, respectively.

22.2.2 The Role for the DNN Architecture

Information technology and concepts are evolving rapidly in the global web. Paraphrasing Frank White, the evolution toward client services including web services, agent-based computing, semantic web with its ontology development offers the potential for a widely distributed information environment. Network services are receiving the bulk of the attention at present but the importance of the information services is becoming recognized. Information services center on content including data pedigree, contextual metadata, and contribution to the situational awareness. Data and information fusion is cited as a central, enabling technology to satisfy the inherent informational needs. From the fusion community perspective the ability to achieve automated fusion (e.g., belief function combination) is dependent on the data pedigree, metadata, context services, and the architecture standards that evolve to prescribe them. Thus, the user identifiable functional components of the fusion process as embodied in a fusion architecture are critical to providing affordable distributed information services.

The increasing demand for more automation of data fusion and response management functions operating in open distributed environments makes it imperative to halt the exponential increase in DF&RM software development costs and building of one-of-a-kind DF&RM software systems.

Architectures are frequently used as mechanisms to address a broad range of common requirements to achieve interoperability and affordability objectives. An architecture (IEEE definition from ANSI/IEEE Standard 1471-2000) is a structure of components, their relationships, and the principles and guidelines governing their design and evolution over time. An architecture should

- Identify a focused purpose with sufficient breadth to achieve affordability objectives
- Facilitate user understanding/communication
- Permit comparison, integration, and interoperability
- Promote expandability, modularity, and reusability
- Achieve most useful results with least cost of development

As such, an architecture provides a *functional toolbox* to guide software designers and organize the software patterns that they develop (thus easing reusability).

The DoD framework for architectures (Department of Defense Architecture Framework [DoDAF]) provides a common basis for developing architectures and defines standard presentation products for each role: operational, systems, and technical architectures. The DF&RM DNN architecture is a technical architecture as shown in Figure 22.3. The DoD framework promotes effective communications of operational analysis and systems engineering from design to operations. It is used to visualize and define operational and technical concepts, plan process improvement, guide systems development, and improve interoperability. The DoD architecture framework is complementary to the Defense Information Systems Agency (DISA) Technical Architecture Framework for Information Management (TAFIM), the Common Operating Environment (Figure 22.4), and the layered network enterprise (Figure 22.5). The DF&RM DNN architecture provides a unification

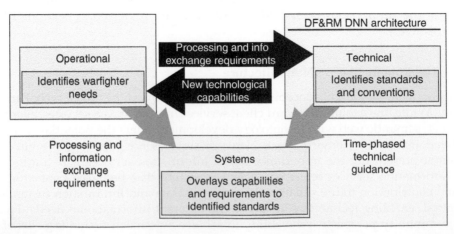

FIGURE 22.3
Role for DF&RM DNN architecture within the DoD architecture framework.

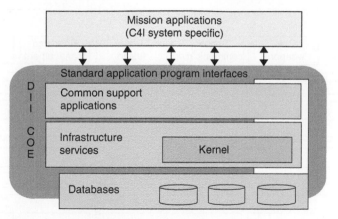

FIGURE 22.4
The simplified COE/C4I architecture.

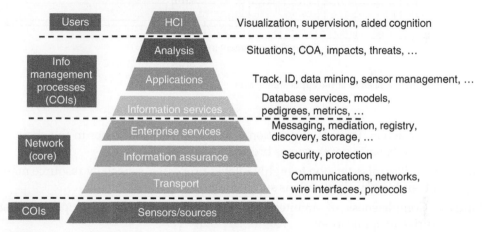

FIGURE 22.5
Layered view of functionality within a networked enterprise.

and organization of the DF&RM algorithmic approaches at the applications layer as depicted in Figure 22.6.

The criteria for the selection of the DNN architecture for DF&RM at the applications layer includes the following:

Breadth. Permits representation and comparison of all aspects of fusion

Depth. Sufficient level of detail to represent and compare all approaches to fusion (e.g., individual fusion patterns)

Affordability. Provides engineering guidelines for the application of the architecture to fusion system development to achieve useful results while minimizing development costs

Understandability. Familiarity in community and facilitation of user understanding and communication of all fusion design alternatives

Hierarchical. Sufficient module structure to enable incorporation of legacy codes at varying levels of granularity (e.g., whole fusion node, data association, hypothesis scoring, estimation, tracking filters, attribute/ID confidence updating)

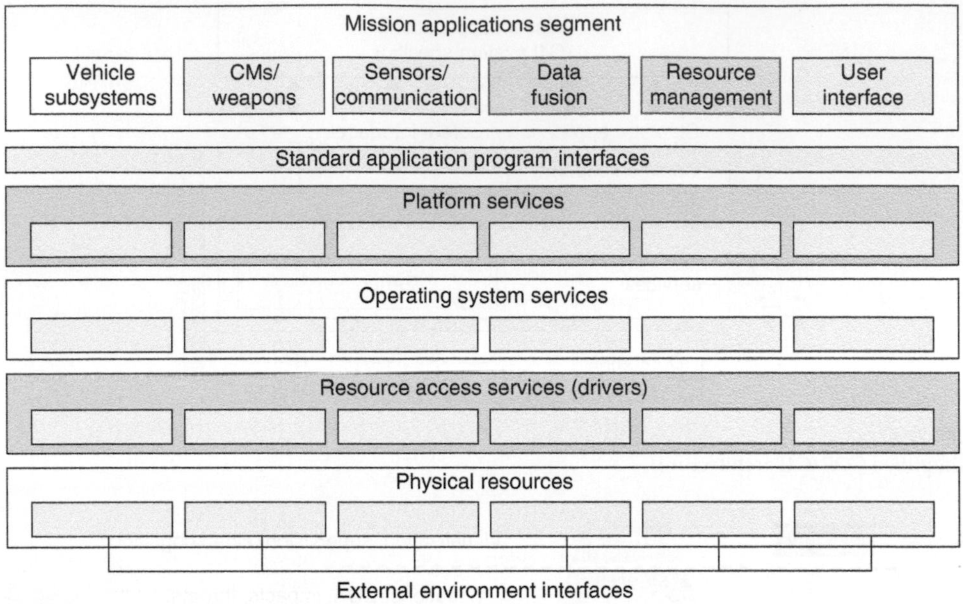

FIGURE 22.6
A DF&RM architecture is needed at the applications layer.

Levels of abstraction. Permits abstraction of fusion processes higher in the chain of command to balance depth versus breath as needed

Extendibility. Ease of extension to highly related processes, especially resource management, and to new applications supporting code reuse

Maturity. Completeness of architecture components and relationships yielding reduced risk in applications

22.2.3 Components of the DNN Technical Architecture for DF&RM System Development

As data fusion is the process of combining data/information to estimate or predict the state of some aspect of the world[1] and resource management is the process for managing/controlling response capability to meet mission objectives, it is important to remember the definitions of optimal estimation and control. Namely

> The *estimation problem* consists of determining an approximation to the time history of a system's response variables from erroneous measurements.[5]
>
> An *optimal estimator* is a computational algorithm that processes measurements to deduce a minimum error estimate of the state of a system by utilizing: knowledge of system and measurements dynamics, assumed statistics of system noises and measurements errors, and initial condition information.[6]
>
> The *control problem* is that of specifying a manner in which the control input should be manipulated to force the system to behave in some desired fashion. If a performance measure is introduced to evaluate the quality of the system's behavior, and the control input is to be specified to minimize or maximize this measure, the problem is one of *optimal control*.[5]

The solution of the Riccati equation provided by estimation theory provides the solution of the quadratic optimal control problem as described in Figure 22.7. Data fusion contains the estimation process that is defined by the data preparation, and association function such as resource management contains the control process that is defined by the task preparation and planning functions. As such the estimation and control duality can be extended to DF&RM (Figure 22.8). Figure 22.9 depicts this for the original data fusion tree architecture developed over 20 years ago.[7,8]

DF&RM systems can be implemented using a network of interacting fusion and management nodes. The dual DF&RM levels provide insight into techniques to support the designs for each. The key components in the DF&RM DNN architecture are depicted

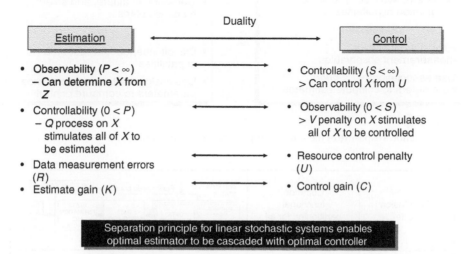

FIGURE 22.7
The estimation and control duality and the separation principle.

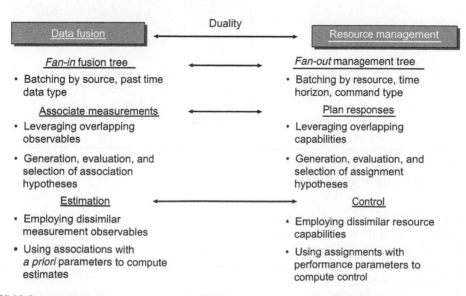

FIGURE 22.8
Duality extensions yield similar methodology for data fusion and resource management.

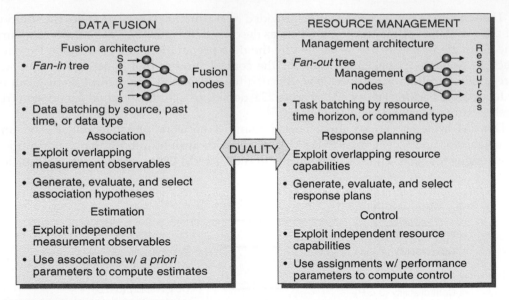

FIGURE 22.9
DF&RM duality allows exposure of the solution approaches and enables software reuse.

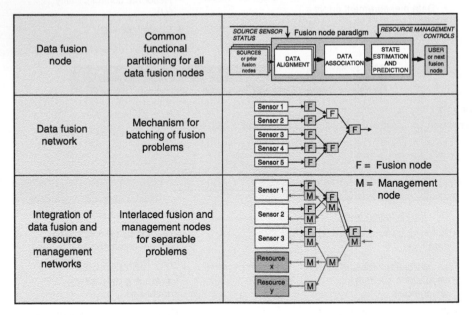

FIGURE 22.10
Dual node network architecture components.

in Figure 22.10. The fusion and management node networks are interlaced to provide faster local feedback as well as data sharing and response coordination. A sample interlaced DF&RM node network is shown in Figure 22.11. The dual fan-in fusion and fan-out management networks are composed of DF&RM nodes whose *dual* processing components are described in Figure 22.12.

The Kalman estimate is the expected value of the state, *x*, given the measurements. This estimate is optimal for any symmetric and convex performance measure. The separation

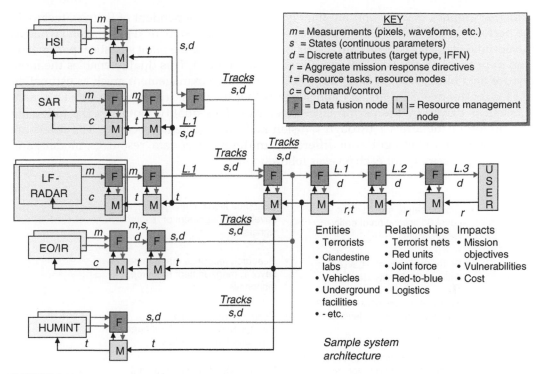

FIGURE 22.11
Sample interlaced tree of DF&RM nodes.

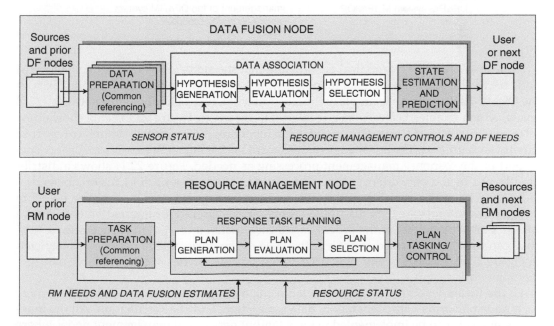

FIGURE 22.12
DF&RM node duality facilitates understanding of alternatives and reuse.

principle[9] implies that the optimal filter gain matrix is independent of the parameters in the performance loss function (and dependent on the plant statistical parameters) whereas the optimal feedback control gain matrix is independent of the plant statistical parameters (and dependent on the performance objective parameters). This duality enables the more mature field of data fusion to rapidly advance resource management much as estimation did for control over 35 years ago, and suggests that fusion is driven by the data and management is driven by the performance objectives. This duality motivates the following description of the levels 1 through 4 fusion and management processes. The levels are partitioned due to the significant differences in the types of data, resources, models, and inferencing necessary for each level as follows:

- Signal/feature assessment—Level 0: estimation of entity feature states

- Entity assessment—Level 1: estimation of entity states

- Situation assessment—Level 2: estimation of entity relationship states

- Impact assessment—Level 3: estimation of the mission impact of fused states

- Process assessment—Level 4: estimation of the DF&RM system MOE/MOP states

- Resource signal management—Level 0: management of resource signals

- Resource response management—Level 1: management resource responses

- Resource relationship Management—Level 2: management of resource relationships

- Mission objective management—Level 3: management of mission objectives

- Process/knowledge management—Level 4: management of the DF&RM system engineering and products

These levels provide a canonical partitioning that should be considered during DF&RM node network design. Though not mandatory, such partitioning reduces the complexity of the DF&RM software. An example of the clustering of DF&RM nodes by the above 0/1/2/3 levels, which enables fast response feedback, is shown in Figure 22.13. This shows how interlaced fusion and management nodes can be organized by level into similar response timeliness segments. The knowledge management for the interactions with the user occurs as part of the response management processing as needed. Figure 22.14 shows a distributed level 1 fusion network over time that has individual sensor level 1 fusion nodes providing tracks to the ownship multiple sensor track-to-track fusion nodes. The ownship fusion track outputs are then distributed across all the aircraft for internetted cooperative fusion. Further batching of the processing of each fusion level (i.e., to reduce complexity/ cost) can then be performed to achieve the knee-of-the-curve fusion network, as depicted in Figure 22.15. A dual knee-of-the-curve design process can be used for resource management network design.

All the fusion levels can be implemented using a fan-in network of fusion nodes where each node performs data preparation, data association, and state estimation. All the management levels can be implemented using a fan-out network of management nodes where each node performs response task preparation, response task planning, and resource state control. These levels are not necessarily processed in numerical order or in concert with the corresponding dual fusion level. Also, any one can be processed on their own given

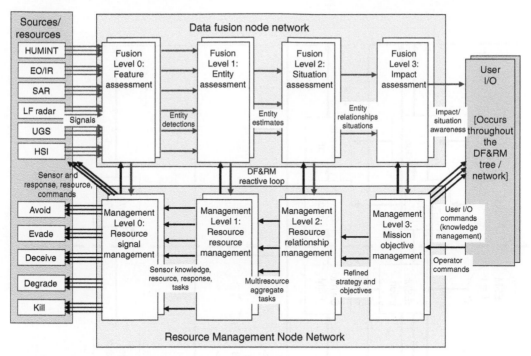

FIGURE 22.13
Sample interlaced sequential level DF&RM system network.

their corresponding inputs. The DF&RM function-processing levels are based on what the user defines as his entities and resources of interest. The five DF&RM dual processing levels are described in more detail below in their dual pairs per level:

- Resource signal management level 0. Management to task/control specific resource response actions (e.g., signals, pulses, waveforms, etc.)

- Signal/feature assessment level 0. Fusion to detect/estimate/perceive specific source entity signals and features

- Resource response management level 1. Management to task/control continuous and discrete resource responses (e.g., radar modes, countermeasures, maneuvering, communications)

- Entity assessment level 1. Fusion to detect/estimate/perceive continuous parametric (e.g., kinematics, signature) and discrete attributes (e.g., IFF, class, type, ID, mode) of entity states

- Resource relationship management level 2. Management to task/control relationships (e.g., aggregation, coordination, conflict) among resource responses

- Situation assessment level 2. Fusion to detect/estimate/comprehend relationships (e.g., aggregation, causal, command/control, coordination, adversarial relationships) among entity states

- Mission objective management level 3. Management to establish/modify the objective of levels 0, 1, 2 action, response, or relationship states

- Impact assessment level 3. Fusion to predict/estimate the impact of levels 0, 1, or 2 signal, entity, or relationship states

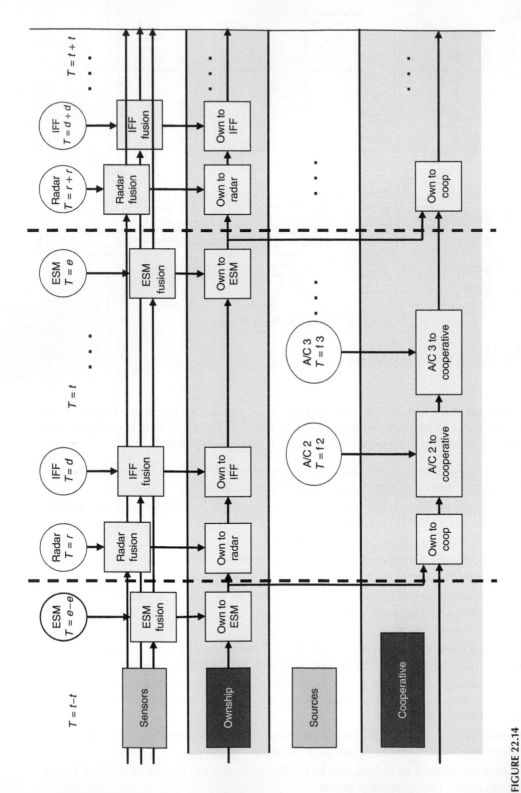

FIGURE 22.14
Sequential distributed fusion network over time, sensors, and platforms.

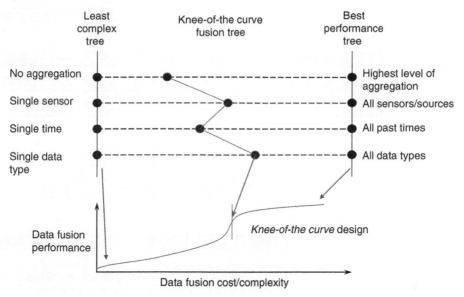

Least complex tree | Knee-of-the curve fusion tree | Best performance tree

No aggregation — Highest level of aggregation

Single sensor — All sensors/sources

Single time — All past times

Single data type — All data types

Data fusion performance

Knee-of-the curve design

Data fusion cost/complexity

FIGURE 22.15
Fusion network selected to balance performance and complexity.

- Process/knowledge management level 4: Design, adjudication, and concurrency management to control the system engineering, distributed consistent tactical picture (CTP), and product relevant context concurrency
- Process assessment level 4: Performance, consistency, and context assessment to estimate the fusion system measures of performance/effectiveness (MOP/MOE)

22.2.4   DF&RM Node Processing

22.2.4.1   *Data Fusion Node Processing*

Data fusion is the process of combining data/information to estimate or predict the state of some aspect of the world. An example of the fusion node processes is shown in Figure 22.16.

The fusion node paradigm for all levels of fusion involves the three basic functions of data preparation, data association, and entity-state estimation, as shown in Figure 22.17. The means for implementing these functions and the data and control flow among them will vary from node to node and from system to system. Nonetheless, the given paradigm has been a useful model for characterizing, developing, and evaluating automatic and human-aided fusion systems.

Data preparation (sometimes termed *data alignment* or *common referencing*) preprocesses data received by a node to permit comparisons and associations among the data. Alignment involves functions for

- Common formatting and data mediation
- Spatiotemporal alignment
- Confidence normalization

Data association assigns data observations—received in the form of sensor measurements, reports, or tracks—to hypothesized entities (e.g., vehicles, ships, aircraft, satellites,

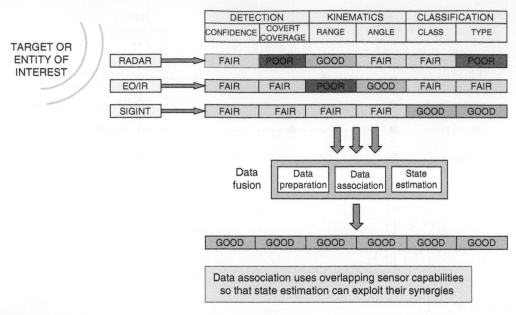

FIGURE 22.16
Sensor fusion exploits sensor commonalities and differences.

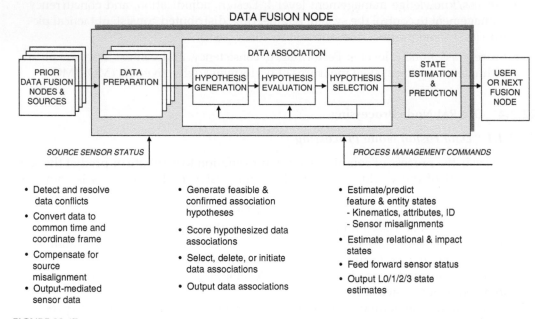

FIGURE 22.17
The DF&RM DNN technical architecture fusion node functional paradigm.

and abnormal events). We will refer to the resulting entity estimates based on this association as tracks. The general level 1 data association problem is a labeled set-covering problem where the labels on the associated data sets are for false or valid entity tracks. In many applications the set-covering problem can be simplified to a set-partitioning problem. An example of set partitioning for abnormal event track association is to limit an abnormality report to associating to only one event track. Before a sensor observation and its associated information are fused with an existing track, the hypothesis that the observation is of the same entity as that represented by the track information can be generated, evaluated, and one or many selected.* The association process is accomplished via the following three functions:

1. Hypothesis generation. Identifying sets of sensor reports and existing tracks (report associations) that are feasible candidates for association

2. Hypothesis evaluation. Determining the confidence (i.e., quality) of the report to track association or nonassociation as defined by a particular metric

3. Hypothesis selection. Determining which of the report-to-track associations, track propagations or deletions, and report track initiations or false alarm declarations the fusion system should retain and use for state estimation

State estimation and prediction use the given fusion node batch of data to refine the estimates reported in the data and infer entity attributes and for level 2 fusion their relations. For example, kinematics measurements or tracks can be filtered to refine kinematics states. Kinematics and other measurements can be compared with entity and situation models to infer features, identity, and relationships of entities in the observed environment. This can include inferring states of entities other than those directly observed. This also applies to mission impact prediction in level 3 fusion based on course action hypothesis generation, evaluation, and selection.

22.2.4.2 Resource Management Node Processing

Resource management is the process of planning/controlling response capabilities to meet mission objectives. An example of the dual resource management node processes for sensor management is shown in Figure 22.18.

Given this duality, the maturity of data fusion can be used to better understand the less mature resource management field, much like the duality of estimation did for control over 30 years ago. In doing so, the fusion system development process becomes intertwined with the resource management system development within each design phase. As multinodal data fusion networks are useful in partitioning the data association and state estimation problems, so are resource management networks useful in partitioning planning and control problems. The data fusion node performs an association/estimation process, and the resource management node performs a response planning and execution process. Both of these DF&RM node networks—one synthetic (i.e., constructive) and the other analytic (i.e., decompositional)—are characteristically recursive and hierarchical.

As depicted in Figure 22.19, the resource management node contains functions that directly correspond to those of the *dual* data fusion node. The resource management node is described as follows:

1. Task preparation—converts prioritized needs to prioritized candidate tasks for the current resource management node

---

* There has been significant progress in developing multisensor and multitarget data fusion systems that do not depend on explicit association of observations to tracks (see Chapter 16). Once again, the data fusion node paradigm is meant to be comprehensive; every system should be describable in terms of the functions and structures of the DF&RM DNN architecture. This does not imply that every node or every system need include every function.

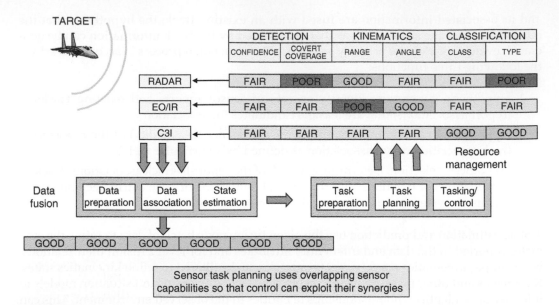

FIGURE 22.18
Sensor management exploits sensor commonalities and differences.

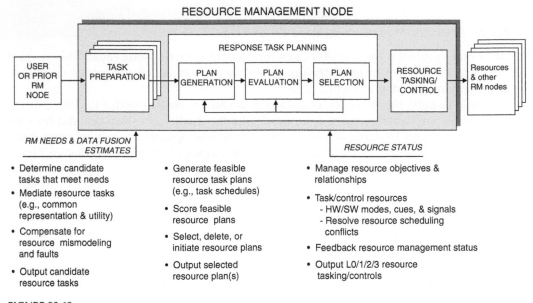

FIGURE 22.19
Duality of DF&RM nodes facilitates understanding and reuse.

2. Response task planning:

   a. Plan generation—generate feasible resource task plans (e.g., candidate sequences of tasks that compose a feasible response plan)

   b. Plan evaluation—evaluate alternative feasible resource response task plans (e.g., with Bayesian cost, possibilistic, expert system rules, *ad hoc*, etc.)

   c. Plan selection—select, delete, or initiate response task plans (e.g., deterministic, probabilistic, multiple hypothesis testing (MHT), n-dimensional partitioning, or set covering)

3. Control (i.e., resource tasking plan execution)—generates the resource control commands to implement the selected resource allocation plan

Planning is a process that is analogous to *data association* in data fusion. Planning functions that correspond to association hypothesis generation, evaluation, and selection are involved: (a) plan generation involves searching over a number of possible actions for assembly into candidate multiple or single resource plan segments, which are passed to (b) plan evaluation and (c) plan selection. *Plan generation, plan evaluation,* and *plan selection* offer challenges similar to those encountered in designing the corresponding *dual* data fusion functions. As with *hypothesis generation* in data fusion, *plan generation* involves potentially massive searches, which must be constrained in practical systems. The objective is to reduce the number of feasible plans for which a detailed evaluation is required. Challenges for *plan evaluation* lie in the derivation of efficient scoring schemes capable of reflecting the expected utility of alternative resource response plans. Candidate plans, that is, schedules of tasking for system resources, can be assembled recursively. The level of planning is adapted on the basis of

a. The assessed utility relative to the mission goals of the given plan segment as currently developed
b. The time available for further planning

For case (a), near-term plan segments tend to be constructed in greater detail than far-term ones, for which the added expenditure in planning resources may outweigh the confidence that the plan will still be appropriate at execution time.

Planning can be accomplished by recursively partitioning a goal into candidate sets of subgoals (*plan generation*) and combining them into a composite higher-level plan (*plan selection*). At each level, candidate plans are evaluated with regard to their effectiveness in achieving assigned goals, the global value of each respective goal, and the cost of implementing each candidate plan (*plan evaluation*). By evaluating higher-level cost/payoff impacts, the need for deeper planning or for selecting alternate plans is determined. In many applications, *plan selection* can be an NP-hard (i.e., not proven to be a linearly hard) problem—a search of multiple resource allocations over n future time intervals. Also, *dual* MHT and probabilistic planning/control approaches are exposed for consideration. In addition, *dual* fusion and management separation principles can be derived.

Contentions for assigning available resources can be resolved by prioritized rescheduling on the basis of time sensitivity and predicted utility of contending allocations. Each resource management node can receive commands from level 2 resource management nodes that are determining resource coordination and conflict plans. For example, in countermeasures management in a combat aircraft, evaluation of self-protection actions must consider detectable signature changes, flight path changes, or emissions that could interfere with another sensor. level 2 resource management nodes respond by estimating the impact of such effects on the mission objectives that are maintained by level 3 resource management nodes. In this way, response plans responsive to situation-dependent mission goals are assembled in a hierarchical fashion.

22.2.4.3   Comparison of the Dual Data Association and Response Planning Functions at Each DF&RM Level

Figure 22.20 describes the data association processes when the fusion processing is segmented by fusion level. Figure 22.21 does the same for response planning functions when the resource management processing is segmented by resource management level.

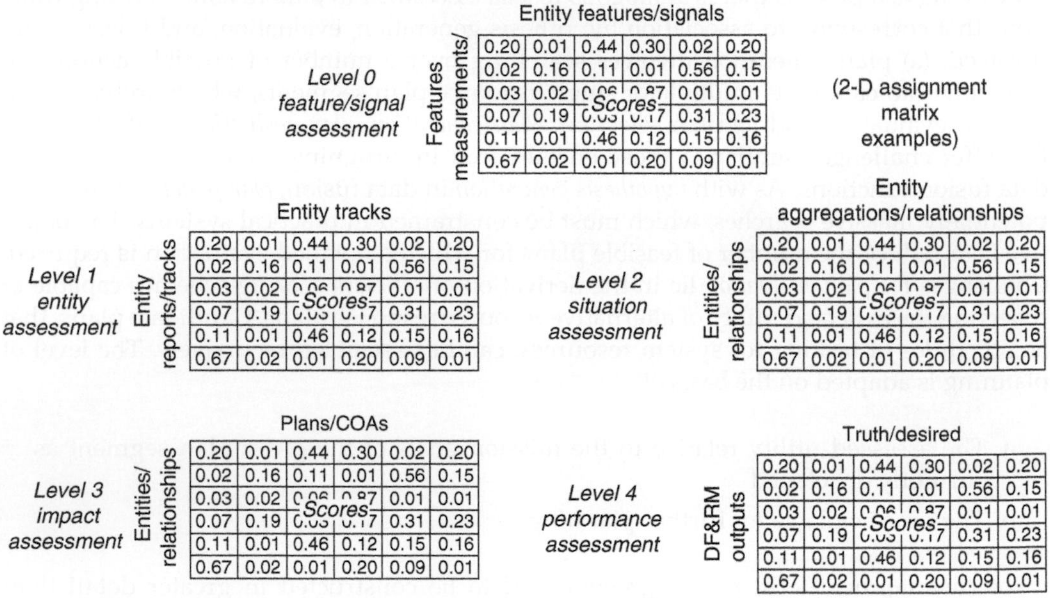

Entity features/signals

Level 0 feature/signal assessment — Features measurements/

0.20	0.01	0.44	0.30	0.02	0.20
0.02	0.16	0.11	0.01	0.56	0.15
0.03	0.02	*Scores*		0.01	0.01
0.07	0.19			0.31	0.23
0.11	0.01	0.46	0.12	0.15	0.16
0.67	0.02	0.01	0.20	0.09	0.01

(2-D assignment matrix examples)

Entity tracks

Level 1 entity assessment — Entity reports/tracks

0.20	0.01	0.44	0.30	0.02	0.20
0.02	0.16	0.11	0.01	0.56	0.15
0.03	0.02	*Scores*		0.01	0.01
0.07	0.19			0.31	0.23
0.11	0.01	0.46	0.12	0.15	0.16
0.67	0.02	0.01	0.20	0.09	0.01

Entity aggregations/relationships

Level 2 situation assessment — Entities/relationships

0.20	0.01	0.44	0.30	0.02	0.20
0.02	0.16	0.11	0.01	0.56	0.15
0.03	0.02	*Scores*		0.01	0.01
0.07	0.19			0.31	0.23
0.11	0.01	0.46	0.12	0.15	0.16
0.67	0.02	0.01	0.20	0.09	0.01

Plans/COAs

Level 3 impact assessment — Entities/relationships

0.20	0.01	0.44	0.30	0.02	0.20
0.02	0.16	0.11	0.01	0.56	0.15
0.03	0.02	*Scores*		0.01	0.01
0.07	0.19			0.31	0.23
0.11	0.01	0.46	0.12	0.15	0.16
0.67	0.02	0.01	0.20	0.09	0.01

Truth/desired

Level 4 performance assessment — DF&RM outputs

0.20	0.01	0.44	0.30	0.02	0.20
0.02	0.16	0.11	0.01	0.56	0.15
0.03	0.02	*Scores*		0.01	0.01
0.07	0.19			0.31	0.23
0.11	0.01	0.46	0.12	0.15	0.16
0.67	0.02	0.01	0.20	0.09	0.01

FIGURE 22.20
Data association problems occur at each fusion level.

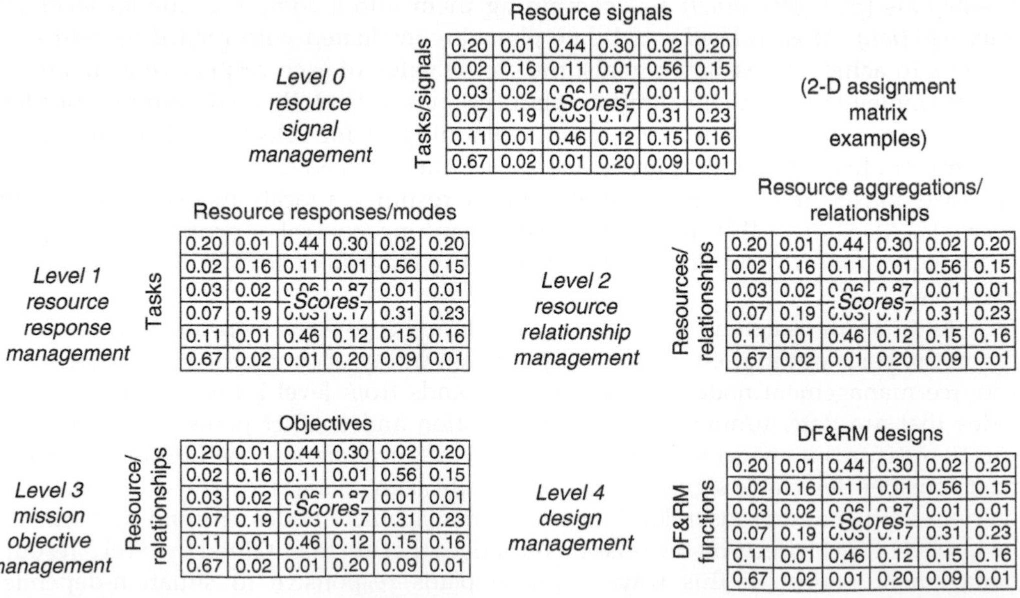

Resource signals

Level 0 resource signal management — Tasks/signals

0.20	0.01	0.44	0.30	0.02	0.20
0.02	0.16	0.11	0.01	0.56	0.15
0.03	0.02	*Scores*		0.01	0.01
0.07	0.19			0.31	0.23
0.11	0.01	0.46	0.12	0.15	0.16
0.67	0.02	0.01	0.20	0.09	0.01

(2-D assignment matrix examples)

Resource responses/modes

Level 1 resource response management — Tasks

0.20	0.01	0.44	0.30	0.02	0.20
0.02	0.16	0.11	0.01	0.56	0.15
0.03	0.02	*Scores*		0.01	0.01
0.07	0.19			0.31	0.23
0.11	0.01	0.46	0.12	0.15	0.16
0.67	0.02	0.01	0.20	0.09	0.01

Resource aggregations/relationships

Level 2 resource relationship management — Resources/relationships

0.20	0.01	0.44	0.30	0.02	0.20
0.02	0.16	0.11	0.01	0.56	0.15
0.03	0.02	*Scores*		0.01	0.01
0.07	0.19			0.31	0.23
0.11	0.01	0.46	0.12	0.15	0.16
0.67	0.02	0.01	0.20	0.09	0.01

Objectives

Level 3 mission objective management — Resource/relationships

0.20	0.01	0.44	0.30	0.02	0.20
0.02	0.16	0.11	0.01	0.56	0.15
0.03	0.02	*Scores*		0.01	0.01
0.07	0.19			0.31	0.23
0.11	0.01	0.46	0.12	0.15	0.16
0.67	0.02	0.01	0.20	0.09	0.01

DF&RM designs

Level 4 design management — DF&RM functions

0.20	0.01	0.44	0.30	0.02	0.20
0.02	0.16	0.11	0.01	0.56	0.15
0.03	0.02	*Scores*		0.01	0.01
0.07	0.19			0.31	0.23
0.11	0.01	0.46	0.12	0.15	0.16
0.67	0.02	0.01	0.20	0.09	0.01

FIGURE 22.21
Response planning problems occur at each management level.

22.3 Data Fusion System Engineering Process

The DF&RM DDN architecture components and their implied relationships have been discussed in Sections 22.2.3 and 22.2.4. The engineering process discussed in this chapter involves a hierarchical decomposition of system-level requirements and constraints. Goals and constraints are refined by feedback of performance assessments. This is an example of a *goal-driven* problem-solving approach, which is the typical approach in system engineering. More general approaches to problem-solving are presented by Bowman et al. and by Steinberg.[25] As depicted in Figure 22.22, goal-driven approaches are but one of four approaches applicable to various sorts of problems. In cases where it is difficult to state system goals—as in very experimental systems or in those expected to operate in poorly understood environments—the goal-driven system engineering approach will integrate elements of data-, technique-, or model-driven methods.[25]

This section describes DF&RM DNN architecture principles and guidelines governing DF&RM software design and evolution over time as shown in Figure 22.23. A key insight, first presented in Ref. 10 is in formulating the system engineering process as a resource management problem; allowing the application of Bowman's model[4] of the duality between data fusion and resource management. The DF&RM DNN technical architecture provides standardized representation for

- Coordinating the design of data fusion processes
- Comparing and contrasting alternative designs
- Determining the availability of suitable prior designs

The four phases of the DNN architecture engineering methodology are defined as follows:

1. Fusion system role design. Analyze system requirements to determine the relationship between a proposed data fusion system and the other systems with which it interacts.
2. Fusion tree design. Define how the data is batched to partition the fusion problem.

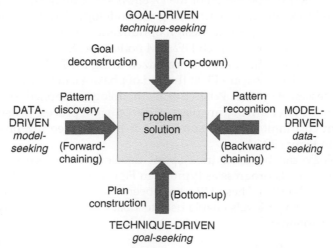

FIGURE 22.22
Types of problem-solving approaches.

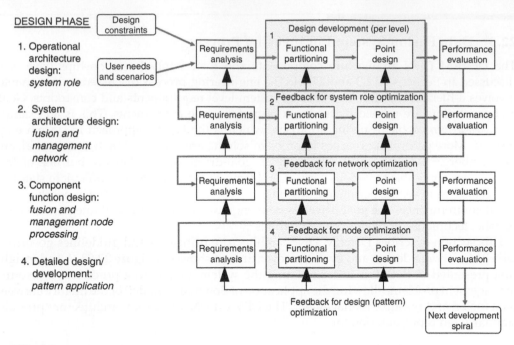

FIGURE 22.23
The DNN architecture DF&RM system engineering process.

3. Fusion node design. Define the data and control flow within the nodes of a selected fusion tree.

4. Fusion algorithm design. Define processing methods for the functions to be performed within each fusion node (e.g., data association, data alignment, and state estimation/prediction).

More specifically, the CONOPS and blackbox DF&RM system role optimization is performed in phase 1. The DF&RM levels described in Section 22.2.3 typically provide the basis for the highest segmentation of the DF&RM software development process during phase 2 of the DNN software engineering methodology. The requirements and MOP refinement are tailored to support the design optimization for each sequential phase. The optimization of the algorithms for each DF&RM node is performed in phase 3. The selection or development of software patterns organized using the DNN architecture *toolbox* is performed in phase 4. The software PE at the end of phase 4 can be accomplished with the level 4 process assessment system. When this methodology is applied to the level 4 system development then the level 4 MOPs are used. This recurrent process is usually cut-off here due to PE system cost limitations. A summary of this engineering processing for data fusion is shown in Figure 22.24.

This methodology can be used in a spiral development process to support rapid prototyping. These spirals progress as typified in Figure 22.25 to achieve increasing technology readiness levels (TRL). These TRLs have been adapted from NASA to describe technology maturity. The lower levels enable user feedback to reduce risk. Their abbreviated descriptions are as follows:

- TRL 1. Basic principles observed and reported
- TRL 2. Technology concept and application formulated

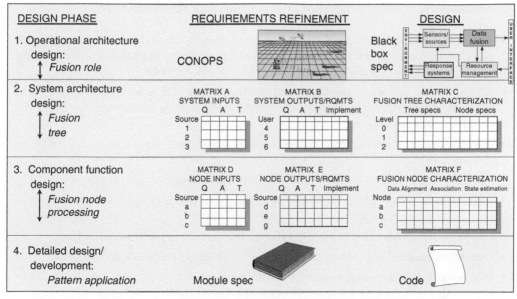

Note: Q A T = Data quality, availability, timeliness

FIGURE 22.24
Data fusion system engineering phases and products.

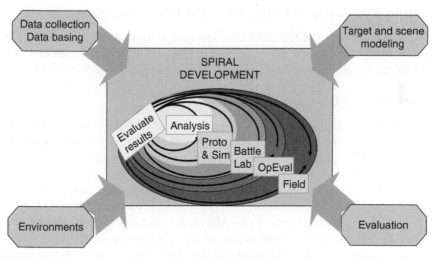

FIGURE 22.25
Spiral development cycles lead to increased technology readiness levels.

- TRL 3. Proof of concept
- TRL 4. Component validated in laboratory environment
- TRL 5. Component validated in relevant environment
- TRL 6. Prototype demonstration in relevant environment
- TRL 7. Prototype demonstration in operations environment
- TRL 8. System completed and qualified through test and demonstration
- TRL 9. System proven through successful mission operations

In summary the benefits of DF&RM DNN architecture are summarized as follows:

- The extension of the estimation and control duality to fusion and management, which enables the more mature fusion field to bootstrap management technology and the unification and coordination of their developments.

- A top-down software development methodology for specifying the system role, interlacing DF&RM components and relationships, and designing reusable DF&RM node solutions.

- Hierarchical framework for understanding alternative solutions to DF&RM, as well as its PE.

- Specification of a DNN, which enables divide and conquer partitioning of F&M.

- Algorithmic forum (paradigm) for balancing system operation, sensor responsibilities, and system allocations.

- Curse of dimensionality mitigation due to the duality separation principal.

- Quick insertion of algorithm patterns due to modular software architecture.

- Framework for assessing algorithms and understanding relative utility for algorithms that are directly compatible with open-layered architectures.

- Provision of a common representation and language for trade-offs.

22.4 Sample Applications of the DF&RM DNN Architecture

This section describes how the DF&RM DNN technical architecture can be applied to the higher fusion levels and resource management levels. It then describes how other data fusion models can be represented and compared with the DNN architecture.

22.4.1 Level 2 Fusion DNN Application Example

An example of the correspondence of the fusion node functions across fusion levels is described next for the hypothesis selection function within data association for level 2 fusion. Across the five fusion levels, the declarations (e.g., detections) of entity features, entities, their relationships, their predicted activities, and their correspondence to truth are accomplished in the hypothesis selection function within data association.

For level 2 fusion the data association function can apply deterministic, MHT, or probabilistic approaches to pick the highest confidence relationships, multiple alternative relationships, or the probabilistic combination of the alternative relationships to estimate the relationship states (e.g., the estimation of the size, shape, count, etc., of a cluster, the strength and type of jamming signal, or other relationship state updates).

More specifically, if the relationship of interest is aggregation, then each entity association confidence to its feasible aggregates can be entered into the aggregate-to-entity feasible association matrix. This matrix can be searched to select the best aggregate association for each entity or an MHT set of aggregations can be selected to create alternative scenes from which alternative aggregate states (e.g., size, mean separation, count, etc.) are estimated. Alternatively a probabilistic weighting of the alternative aggregations can be used to estimate the aggregate states similar to what is done in level 1 entity-state estimation using the probabilistic data association filter (PDAF) approaches. In summary, for each level 2 node the labeling of entities with relationships (i.e., discrete association in the data association function) and the estimation of the relationship states are tailored for that node.

24.4.2 Level 3 Fusion DNN Application Example

Impact prediction is based on alternative projected states. This typically requires additional information on the planned actions and projected reactions for the entities (e.g., in the battlespace). The utility function is usually based on mission success criteria. Both of these (i.e., projected actions and utility function) are not necessary for levels 0–2 fusion (e.g., because levels 0–2 typically only *predicts the fused state forward* to the current time tag for the next data to be fused). Utility assessment in the estimation portion of level 3 processing is based on the associated projected reactions.

The *data association* function in each fusion node at all levels refers to a decision process that is not an estimation process. The role for this decision process for level 3 can be more easily understood by first describing its role in level 1 fusion processing and its extension to all fusion levels. The association process for level 1 is a *labeled set covering* function that provides selected *associations* (e.g., scored clusters or assignments of 1 or more with false or true labels) to be used for each entity-state estimation (e.g., deterministic or probabilistic). The existence of this function in the DNN fusion architecture enables the fusion designer to be enlightened about the alternative approaches that take advantage of a data association function at each fusion level. In addition, the architecture provides the organization for the data association software patterns that have been implemented in other applications.

Specifically for level 3 fusion, the DNN architecture exposes the software designer to consider a segmentation of the level 3 processing (e.g., by data source, entity type, location, etc.) and association decision processes that provide selected *associations* usually between alternative COAs for the host platform, other single entities, and groups of entities (e.g., just as in level 1). A sample of the level 3 data association functions are as follows:

- Hypothesis generation to determine the feasible COAs of the entities or aggregates of interest for impact prediction
- Hypothesis evaluation techniques such as probabilistic (e.g., max *a posteriori*, likelihood, etc.), possibilistic (e.g., fuzzy, evidential), symbolic, unified, and so on, besides hypothesis evaluation scoring algorithms
- Hypothesis selection techniques such as deterministic, MHT, etc. and search algorithms (e.g., Munkres, Jonker-Volgenant-Castanon [JVC], Lagrangian relaxation, etc.) for the COAs association to be used as the basis of impact prediction

In level 3 utility state estimation, the selected COAs can be used to generate alternative utility assessments (e.g., using an MHT approach) or averaged using probabilistic data association such as done in level 1 fusion entity-state estimation. Thus, the designer can extend his knowledge of level 1 fusion approaches to the corresponding types of data association approaches to be used as the basis for the utility (e.g., threat) state estimation. Note that level 3 fusion provides the impact assessment for the user and for automated resource management nodes. Specific resource management nodes may need to predict the fused state in alternative fashions with respect to alternative resource plans.

22.4.3 Level 4 DF&RM DNN Application Example

The difference between mission impact (utility) assessment (of levels 0–2 states) and the level 4 fusion system process assessment is that the former predicts the *situation* utility functions based on the observed situation and the latter estimates the *DF&RM system* MOEs/MOPs based on a comparison of the DF&RM system output to external sources of reality and the desired responses. Both can be done on- or off-line. For example, fusion performance assessment can provide the user an estimate of the trust in online fusion products. By considering PE as a data fusion function the designer receives the benefits of the DNN architecture that provides an engineering methodology for the PE system development. This includes an understanding of the alternative batching of the DF&RM outputs that defines the development of a PE node network (e.g., batching of the data to be evaluated over DF&RM level, time, nodes, and output data types) and the functional partitioning of the PE nodes. For example, the DNN architecture exposes alternative PE node functional alternatives such as consideration of track-to-truth association hypothesis generation, evaluation, and selection that can be deterministic, MHT, probabilistic, symbolic, or unified, and the organization of existing and the new reusable data fusion SW patterns. As another example, the PE node network design phase of the DNN process exposes the design trade where the level 2 fusion PE nodes can be based solely on the level 1 track-to-truth association or can be integrated with the level 1 PE node to determine the track-to-truth association based on both the entity-fused states and their relationships.

As another example, the DNN architecture enables a comparison of Wasserstein metric approach to PE with other approaches. The Wasserstein metric as proposed is used for association and for the PE MOPs estimation. The distance metric, $d(x_i,y_j)$, used in Wasserstein can be tailored to one application MOP (e.g., with increased error costs for targets of interest). Alternatively, the $d(x_i,y_j)$ can be the track-to-truth association score, and then the Wasserstein $C_{i,j}$ used to compute the PE MOPs within the DNN architecture. The approach, as given in Hoffman and Mahler,[11] computes a deterministic association of track-to-truth with no false alarms or missing tracks when the number of tracks and truths are equal. However, unless the transportation matrix search is changed (to include the case when the number of tracks differ with the number of truths), the track-to-truth associations are not assignments with false tracks and missed truths; and when the number of tracks matches the number of truths, associations are forced for distant track-to-truth best matches. Solutions to these problems can be formulated within the DNN architecture. For example, the transportation matrix search can be changed to first use a different association distance metric to declare nonassociations and to incorporate track and truth existence over multiple time points. In addition, the DNN architecture exposes the issue that separating the association metric from the MOP can provide flexibility to the PE designer to enable an association metric that is akin to how an operator would interpret the fused results (e.g., for countermeasure response) with the MOP metric tailored to the mission requirements.

Finally, the DNN architecture enables PE node designs to be interlaced with system design resource management node networks, especially for automated system design methodologies. PE nodes tend to have significant interactions with their dual level 4 processes in resource management in that they provide the performance estimates for a DF&RM solution that are used to propose a better DF&RM solution. The new level 4 resource management function for optimizing the mapping from problem-to-solution space is usually referred to as system engineering. This dual resource management level 4 provides a representation of the system engineering problem that partitions its solution into the resource management node processes: design preparation (resolve design needs, conflicts, and mediation), design planning (design generation, evaluation, and selection), and design implementation. As with the use of the DNN for all the other DF&RM levels, the payoff is primarily for software implementations (e.g., automated system design), although some insight into user fusion and management processes is provided.

22.4.4   Dual RM DNN Application Example

It is expected that the dual management levels will serve to improve understanding of the management alternatives and enable better capitalization of the significant differences in the resource modes, capabilities, and types as well as mission objectives. Thus the less mature resource management field can be bootstrapped with the more mature data fusion field technology much as the duality between estimation and control enabled the solution of the Riccati equation to be applied to the formulation and solution of the quadratic optimal control problem in the 1960s. As with the data fusion *levels*, the resource management levels are also not necessarily processed in order and may need to be integrated with other resource management levels or data fusion levels such as when the locally linear separation principle does not apply.

The process management portion of the JDL fusion model level 4 can be implemented with a set of resource management node within various levels of a Resource Management node network. Process management of a DF&RM system involves as the adaptive data acquisition and processing to support mission objectives (e.g., DF&RM network management, DF&RM node process management, data management to drive adaptive DF&RM networks, data/information/response dissemination, and sensor replanning management). This enables the developer to integrate fusion process management software nodes into appropriate levels within the resource management node network that is interlaced with the fusion node network segmentation of the DF&RM problem. Each data fusion or resource management process management node can then be designed using the DNN resource management node components when it is appropriate to apply the separation principle.

Process management also includes the DF&RM model management function. Typically data mining and ontology development functions generate the models used for DF&RM. For fusion these models store prior knowledge of entity parametric and attribute signatures, entity relationships, and the entity-state evolution sequences (e.g., COAs). The association of data to models determines the levels 0–3 states where previous events instantiate alternative entity behavior and relationship patterns and arriving events further substantiate, cause deletions, or instantiate new hypothesized patterns.

Data mining abductive (hypothesis creation), deductive (hypothesis selection), and inductive (hypothesis validation) reasoning can determine and adapt DF&RM levels 0–3 model parameters on- or off-line. Data mining can be interlaced with fusion level processes to provide such modeling. The model management is performed in the interlaced resource management nodes.

Historically there has been an overlap of the data mining and fusion communities. As Ed Waltz et al.[12] states, *The role for abduction is to reason about a specific target, conjecturing and hypothesizing to discover an explanation of relationships to describe a target (i.e., hypothesis creation). The role for induction is to generalize the fundamental characteristics of the target in a descriptive model. The next step is to test and validate the characteristics on multiple cases (i.e., hypothesis validation).* The more deductive traditional data fusion processes (JDL levels 0–3) are based on these data mining models. A primary purpose of mining is to discover and model some as aspect of the world rather than to estimate and predict the world state based on combining multiple signals.

The model information on the feasible level 2 fusion relationships and level 3 impact pattern recognition can be captured using an ontology. The applicable ontology definitions according to Webster's are as follows:

- *Ontology.* (1) The branch of metaphysics dealing with the nature of being, reality, or ultimate substance (cf. phenomenology); (2) particular theory about being or reality.
- *Phenomenology.* (1) The philosophical study of phenomena, as distinguished from ontology; (2) the branch of a science that classifies and describes its phenomena without any attempt at metaphysical explanation.
- *Metaphysics.* (1) The branch of philosophy that deals with first principles and seeks to explain the nature of being or reality (ontology); it is closely associated with the study of nature of knowledge (epistemology).

As stated by Mieczyslaw Kokar, *An ontology is an explicit specification of a conceptualization: the objects, concepts, and other entities that are assumed to exist in some area of interest and the relationships that hold among them. Definitions associate the names of entities in the universe of discourse (e.g., classes, relations, functions, or other objects) with human-readable text describing what the names mean, and formal axioms that constrain the interpretation and well-formed use of these terms. (An ontology is) a statement of a logical theory. An agent commits to an ontology if its observable actions are consistent with the definitions in the ontology (knowledge level). A common ontology defines the vocabulary with which queries and assertions are exchanged among agents.*[13] Ontologies are used to capture entity relationships in level 2 fusion whereas level 1 fusion typically uses taxonomies for entity attributes. Taxonomy is a type of ontology that is a complete and disjoint representation in a set-theoretical format. Taxonomies have been used extensively to represent the entity attributes (e.g., entity classification trees) within level 1 fusion. A formal ontology provides

1. A shared lexicon of relevant terms
2. A formal structure capable of capturing (i.e., representing) all relations between terms within the lexicon
3. A methodology for providing a consistent and comprehensive representation of physical and nonphysical items within a given domain

As held by Eric Little,[14] the ontological construction is ideally derived from a synergistic relation with user information needs. The user/task information needs inform and bound the ontological structure. The user needs can be developed from cognitive work analysis techniques, empirical, or *ad hoc* top-down or bottom-up approaches. The ontological development structures and validates one's common-sense abstractions of the world. The development is usually top-down, rationally driven, and reflects the epistemologically independent structure of the existing world. When the fusion and management separation principle is being applied, process management may dynamically manage which ontological or data mining processes that allowed to be run on the fusion processor based on the current mission state.

By considering process management (i.e., for either fusion or management processes) as a resource management function the designer receives the benefits of the resource management architecture in his design decisions. For example, fusion process management (i.e., refinement) exposes the following design alternatives:

- Process management is part of the overall management of the processor that fusion resides on and may need to consider or coordinate with the demands on the processor by other external users of the processing.
- The fusion process management may need to take direction from or coordinate with the needs of the other resources to provide tailored and timely fused results.
- The fusion process management may need to be interlaced with a fan-out fusion process management network batched over time, data sources, or data types so as to divide and conquer the fusion process management function even if there is no coordination with any resource management function.
- Each fusion process management node may want to utilize software patterns (e.g., fusion process plan generation, evaluation, and selection followed by tasking/ control) organized within the DNN architecture.

Similar advantages are afforded to the other resource management functions. This also includes the proposed level 5, user refinement which is an element of knowledge management. As stated by Eric Blasch at Air Force Research Laboratory (AFRL), user refinement provides an adaptive determination of who queries information, who has access to information (e.g., information operations), and adaptive data retrieved and displayed to support cognitive decision making and actions (e.g., altering the sensor display). The DNN provides insight into the development of the knowledge management and the user interface (e.g., displays, audio, etc.). Management functions within the overall resource management system and their integration with the data fusion network.

22.4.5 DF&RM System Engineering as a Level 4 Resource Management Problem

System engineering is a process for determining a mapping from a problem space to a solution space in which the problem is to build a system to meet a set of requirements. Fundamental to the system engineering process (as in all resource management processes) is a method for representing the structure of a problem in a way that exposes alternatives and is amenable to a patterned solution. The DNN architecture for resource management permits a powerful general method for system engineering (e.g., for DF&RM system engineering). It does so by providing a standardized formal representation that allows formal resource allocation theory and methods to be applied. As a resource management process, system engineering can be implemented as a hierarchical, recursive planning and execution process (e.g., as DNN level 4 resource management).

The Data Fusion Engineering Guidelines can be thought of as the design specification for a resource management process, the *phases* being levels in a hierarchical resource management tree. A network-structured resource management process is used to build, validate, and refine a system concept (which, in this application, may be a network-structured resource management or data fusion process).

To illustrate how design management and process assessment (e.g., PE) can be implemented as dual processes much like the lower levels of fusion and management. Figures 22.26 and 22.27 show examples of dual levels 1–3 fusion and management and corresponding dual level 4 fusion and management node network designs, respectively. The system engineering process shown in Figure 22.27 is a fan-out (top-down) resource

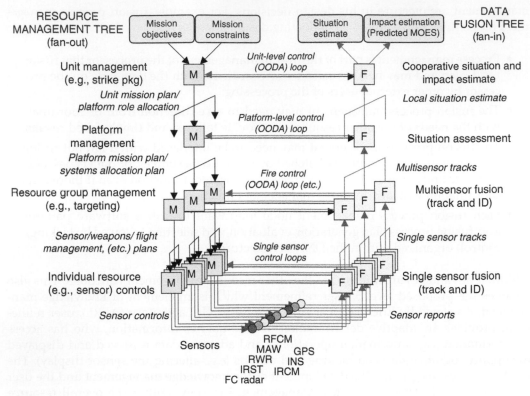

FIGURE 22.26
Dual DF&RM tree architecture for fighter aircraft.

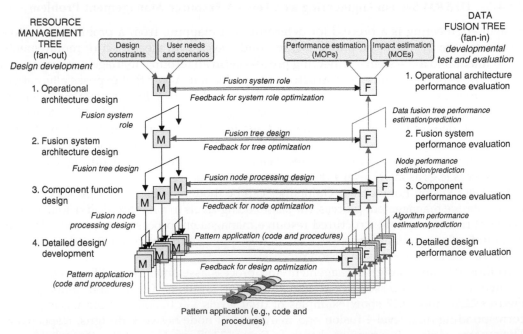

FIGURE 22.27
Dual DF&RM tree architecture for DF system engineering.

management tree interlaced with a fan-in (bottom-up) data fusion tree, similar to that shown in Figure 22.26. Design goals and constraints flow down from the system level to allocations over successively finer problem partitioning.

22.5 The DF&RM Model Unification Provided by the DNN Architecture

The DF&RM DNN architecture provides a unification of numerous other proposed fusion models when tailored to support affordable software development. Section 22.2.1 described how the Boyd observe, orient, decide, act (OODA) loop model and the JDL data fusion model have been incorporated into the DNN. This section describes the unification provided by the DNN architecture for the following models:

1. Dasarathy model[15]
2. Bedworth and O'Brien's omnibus model[16]
3. Kovacich taxonomy[17]
4. Endsley's model[18]. The DNN architecture also unifies the fusion approaches described in Refs. [19–46] under a single functional framework.

22.5.1 Dasarathy Fusion Model

The Dasarathy fusion model partitions the levels 0–1 fusion problem into categories that are based on the data, feature, and object types of input/output (I/O) for each. These are shown boxed in bold in Figure 22.28 with the extensions to level 2 and 3 fusion and response functions.

INPUT	OUTPUT					
	Data	Features	Objects	Relations	Impacts	Responses
Data	Signal detection *DAI-DAO*	Feature extraction *DAI-FEO*	Gestalt-based object extract *DAI-DEO*	Gestalt-based situation assessment *DAI-RLO*	Gestalt-based impact assessment *DAI-IMO*	Reflexive response *DAI-RSO*
Features	Model-based detection/feature extraction *FEI-DAO*	Feature refinement *FEI-FEO*	Object characterization *FEI-DEO*	Feature-based situation assessment *FEI*	Feature-based impact assessment *FEI-IMO*	Feature-based response *FEI-RSO*
Objects	Model-based detection/estimation *DEI-DAO*	Model-based feature extraction *DEI-FEO*	Object refinement *DEI-DEO*	Entity-relational situation assessment *DEI-RLO*	Entity-based impact assessment *DEL-IMO*	Entity-relational based response *DEI-RSO*
Relations	Context-sensitive detection/estimation *RLI-DAO*	Context-sensitive feature extraction *RLI-FEO*	Context-sensitive object refinement *RLI-DEO*	Micro/macro situation assessment *REI-RLO*	Context-sensitive impact assessment *RLI-IMO*	Context-sensitive response *RLI-RSO*
Impacts	Cost-sensitive detection/estimation *IMI-DAO*	Cost-sensitive feature extraction *IMI-FEO*	Cost-sensitive object refinement *IMI-DEO*	Cost-sensitive situation assessment *IMI-RLO*	Cost-sensitive impact assessment *IMI-RLO*	Cost-sensitive response *IMI-RSO*
Responses	Reaction-sensitive detection/estimation *RSI-DAO*	Reaction-sensitive feature extraction *RSI-FEO*	Reaction-sensitive object refinement *RSI-DEO*	Reaction-sensitive situation assessment *RSI-RLO*	Reaction-sensitive impact assessment *RSI-RLO*	Reaction-sensitive response *RSI-RSO*
	Level 0		Level 1	Level 2	Level 3	Level 4

FIGURE 22.28
Extension of the Dasarathy model using the data fusion levels and response management processes.

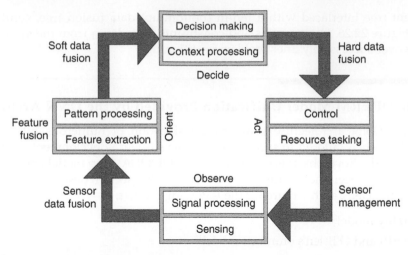

FIGURE 22.29
The Bedworth and O'Brien's omnibus model.

22.5.2 Bedworth and O'Brien's Omnibus Model

The Bedworth and O'Brien omnibus model, as depicted in Figure 22.29, applies the OODA model in a feedback process. The omnibus model can replace a selected portion of itself (i.e., they show all four OODA loop parts inserted into the *Orient* box).

The DNN can be viewed as extending this using its tree of interacting fusion (i.e., orient) or management (i.e., decide) nodes (e.g., segmenting the sensors and effectors from interlaced fan-in fusion and fan-out management processes). The distinctions of *sensor, soft, and hard data fusion* and *sensor management* (i.e., the labels of the arrows in the Figure 22.29) are not made in the DNN model. The DNN provides structure for the DF&RM functions to be implemented in each of the OODA boxes so as to provide a hierarchical modular architecture to achieve affordability objectives (e.g., enabling software reuse and *pattern* tool development through standard hierarchical functional components).

22.5.3 The Kovacich Fusion Taxonomy

The Kovacich fusion taxonomy can be viewed as an extension of one proposed by Drummond. The fusion local nodes are allowed to be sensor nodes or track nodes. The sensor nodes are fusion nodes that generate unfused sensor data and the track nodes are fusion nodes that fuse one or more sensor nodes or other track nodes. The matrix in Figure 22.30 specifies the four basic fusion network structures that are allowed.

Each fusion node consists of the following subfunctions:

- Alignment. Aligns input data to a common frame
- Association. Associates input data to existing entities
- Estimation. Estimates, predicts entity state
- Management. Decides entity existence
- Distribution. Distributes entities

This is similar to the DNN fusion node except the management and distribution functions are inserted into their fusion node as compared with the DNN that segments

Local nodes	Centralized	Distributed
Sensor nodes	Centralized sensor to global	Distributed sensor to global
Track nodes	Centralized local track to global	Distributed local track to global

FIGURE 22.30
The four types of fusion network structures.

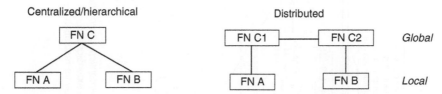

FIGURE 22.31
The Kovacich fusion network structure types.

management from fusion and includes communications management within the resource management node network. Their general N node fusion network is built using a combination of centralized/hierarchical and distributed fusion node networks. In the centralized/hierarchical the local output of fusion nodes A and B are fused into a global product at fusion node C. In the distributed, the local output of fusion node A is fused locally into a global product at fusion node C1, which exchanges information with fusion node C2 that, in turn, fuses the local output of fusion node B, as shown in Figure 22.31. The DNN can be seen as extending the types of DF&RM node networks allowed.

22.5.4   The Endsley Model

Endsley's model can be summarized as containing the following functions:

1. Perception
   a. Explicit objects, events, states, and values
2. Comprehension (what is)
   b. Implicit meanings, situation types
3. Projection (what if)
   a. Future scenarios, possible outcomes
   b. Intentions, COAs

The Endsley model has rough correspondence to JDL and DNN fusion levels 1, 2, and 3, respectively. The focus is on user situation awareness techniques and less on providing a methodology (e.g., functional partitioning) for computer automation. The Endsley model is summarized and compared with the 1998 version[1] of the JDL model in Figure 22.32. Note that this version shows process refinement as the level 4 fusion process. The DNN architecture segments this into process assessment as a fusion process and implements the process management portion of it as part of the resource management network of nodes.

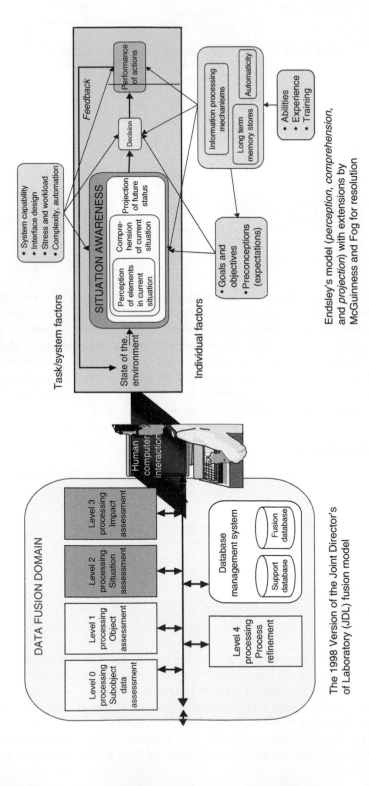

FIGURE 22.32
The Endsley model compared to the JDL model.

References

1. Steinberg, A.N., Bowman, C.L., and White, F.E., *Revisions to the JDL Model*, Joint NATO/IRIS Conference Proceedings, Quebec, Montreal, Canada, October 1998; and in Sensor Fusion: Architectures, Algorithms, and Applications, *Proceedings of the SPIE*, Vol. 3719, 1999.
2. Steinberg, A.N. and Bowman, C.L., *Rethinking the JDL Fusion Levels*, NSSDF JHAPL June, 2004.
3. Bowman, C.L., "The Data Fusion Tree Paradigm and its Dual", *Proceedings of 7th National Symposium on Sensor Fusion*, 1994.
4. Steinberg, A.N. and Bowman, C.L., "Development and Application of Data Fusion Engineering Guidelines", *Proceedings of 10th National Symposium on Sensor Fusion*, 1997.
5. Meditch, J.S., *Stochastic Optimal Linear Estimation and Control*, McGraw Hill, New York, 1969.
6. The Analytical Sciences Corporation, *Applied Optimal Estimation*, Ed. A. Gelb, MIT Press, Cambridge, MA, 1974.
7. Bowman, C.L., Morefield, C.L., and Murphy, M.S., *Multi-Sensor Multi-Target Recognition and Tracking*, 13th Asilomar Conference on Circuits, Systems, & Computers, pp. 329–333, August 1980.
8. Bowman, C.L. and Morefield, C.L., *Multi-Sensor Fusion of Target Attributes and Kinematics*, 19th IEEE Conference on Decision and Control, Vol. 19, pp. 837–839, December 1980.
9. Wonham, W.M., On the Separation Theorem of Stochastic Control, PM-38, NASA Electronics Research Center, Cambridge, MA, January 22, 1968; and *SIAM J. Control*, Vol. 6, p. 312, 1968.
10. Bowman, C.L., "The Role of Process Management in a Defensive Avionics Hierarchical Management Tree", *Tri-Service Data Fusion Symposium Proceedings*, Johns Hopkins University, June 1993.
11. Hoffman, J. and Mahler, R., "Multi-Target Miss Distance via Optimal Assignment", *IEEE Trans. Syst. Man Cybern. A*, Vol. 34(3), pp. 327–336, 2004.
12. Llinas, J., Bowman, C.L., Rogova, G., Steinberg, A., Waltz, E., and White, F., "Revisiting the JDL Data Fusion Model II", *Proceedings of the 7th International Conference on Information Fusion*, July, 2004.
13. Kokar, M., Private communication with Christopher Bowman.
14. Llinas, J., *On the Scientific Foundations of Level 2 Fusion*, University of Prague Fusion Conference, 2003.
15. Dasarathy, B., *Decision Fusion*, IEEE Computer Society Press, New York, NY, 1994.
16. Bedworth, M. and O'Brien, J.O., "The Omnibus Model: A New Model of Data Fusion?" *Proceedings of the 2nd International Conference on Information Fusion*, 1999.
17. Kovacich, G.L. and Jones, A., "What InfoSec Professionals Should Know About Information Warfare Tactics by Terrorists", *Comput. Secur.*, Vol. 21(2), pp. 113–119, 2002.
18. Endsley, M.R., Bolte, B., and Jones, D.G., *Designing for Situation Awareness: An Approach to User-Centered Design*, Taylor and Francis, Inc., New York, 2003.
19. Llinas, J. et al., *Data Fusion System Engineering Guidelines*, Technical Report 96-11/4, USAF Space Warfare Center (SWC) Talon-Command project report, Vol. 2, 1997.
20. Dasarathy, B., "Sensor Fusion Potential Exploitation-Innovative Architectures and Illustrative Applications", *IEEE Proceedings*, Vol. 85(1), 1997.
21. Bowman, C.L., "Affordable Information Fusion via an Open, Layered, Paradigm-Based Architecture", *Proceedings of 9th National Symposium on Sensor Fusion*, Monterey, CA, March 1996.
22. White, F.E., "A Model for Data Fusion", *Proceedings of 1st National Symposium on Sensor Fusion*, Vol. 2, 1988.
23. Llinas, J., Bowman, C.L., and Hall, D.L., *Project Correlation Engineering Guidelines for Data Correlation Algorithm Characterization*, TENCAP SEDI Contractor Report, SEDI-96-00233, 1997; and SWC Talon-Command Operations Support Technical Report 96-11/4, 1997.

24. Steinberg, A. and Bowman, C.L., "Development and Application of Data Fusion Engineering Guidelines", *Proceedings of 10th National Symposium on Sensor Fusion*, 1997.
25. Steinberg, A., Problem-Solving Approaches to Data Fusion, *Proceedings, Fourth International Conference on Information Fusion*, Montreal, 2001.
26. *C4ISR Architecture Framework, Version 1.0*, C4ISR ITF Integrated Architecture Panel, CISA-0000-104-96, 7 June 1996.
27. Steinberg, A., "Data Fusion System Engineering", *Proceedings of 3rd International Symposium for Information Fusion, Fusion2000*, Paris, Vol. 1, pp. MOD5/3–MOD5/10, 10–13 July 2000.
28. Moore, J.C. and Whinston, A.B., "A Model of Decision-Making with Sequential Information Acquisition", *Decis. Support Syst.*, Vol. 2(4), pp. 285–307, 1986; part 2, Vol. 3(1), pp. 47–72, 1987.
29. Steinberg, A., "Adaptive Data Acquisition and Fusion", *Proceedings of 6th Joint Service Data Fusion Symposium*, Vol. 1, pp. 699–710, 1993.
30. Steinberg, A., "Sensor and data fusion", *The Infrared and Electro-Optical Systems Handbook*, Vol. 8, Chapter 3, 1993, pp. 239–341.
31. Llinas, J., Johnson, D., and Lome, L., "Developing Robust and Formal Automated Approaches to Situation Assessment", presented at *Situation Awareness Workshop*, Naval Research Laboratory, September 1996.
32. Steinberg, A., "Sensitivities to Reference System Performance in Multiple-Aircraft Sensor Fusion", *Proceedings of 9th National Symposium on Sensor Fusion*, 1996.
33. Blackman, S.S., *Multiple Target Tracking with Radar Applications*, Artech House, Norwood, MA, 1986.
34. Bar-Shalom, Y. and X.-R. Li, *Estimation and Tracking: Principles, Techniques, and Software*, Artech House, Norwood, MA, 1993.
35. Gelfand, A., Colony, M., Smith, C., Bowman, C., Pei, R., Huynh, T., and Brown, C., Advanced Algorithms for Distributed Fusion, *SPIE*, Orlando, March 2008.
36. Waltz, E. and Llinas, J., *Multisensor Data Fusion*, Artech House, Norwood, MA, 1990.
37. Jordan, J.B. and Howan, C., "A Comparative Analysis of Statistical, Fuzzy and Artificial Neural Pattern Recognition Techniques", *Proceedings of SPIE Signal Processing, Sensor Fusion, and Entity Recognition*, Vol. 1699, pp. 166–176, 1992.
38. Hall, D.L., *Mathematical Techniques in Multisensor Data Fusion*, Artech House, Boston, MA, 1992.
39. Goodman, I.R., Nguyen, H.T., and Mahler, R., *Mathematics of Data Fusion (Theory and Decision Library)*. Series B, Mathematical and Statistical Methods, Vol. 37, Kluwer Academic Press, Dordrecht, The Netherlands, 1997.
40. Steinberg, A.N. and Washburn, R.B., Multi-Level Fusion for War Breaker Intelligence Correlation, *Proceedings of 8th National Symposium on Sensor Fusion*, 1995, pp. 137–156.
41. Kastella, K., "Joint Multi-Target Probabilities for Detection and Tracking", *SPIE*, Vol. 3086, pp. 122–128, 1997.
42. Stone, L.D., Barlow, C.A., and Corwin, T.L., *Bayesian Multiple Target Tracking*, Artech House, Boston, MA, 1999.
43. Bar-Shalom, Y. and Fortmann, T.E., *Tracking and Data Association*, Academic Press, San Diego, CA, 1988.
44. Mahler, R., "The Random Set Approach to Data Fusion", *Proceedings of SPIE*, Vol. 2234, 1994.
45. Bowman, C.L., "Possibilistic Versus Probabilistic Trade-Off for Data Association", *Proceedings of SPIE*, Vol. 1954, pp. 341–351, April 1993.
46. Pearl, J., *Probabilistic Reasoning in Intelligent Systems: Networks of Plausible Interference*, Morgan Kaufman Series in Representation and Reasoning, Los Altos, CA, 1988.

23

Studies and Analyses within Project Correlation: An In-Depth Assessment of Correlation Problems and Solution Techniques[*]

James Llinas, Capt. Lori McConnell, Christopher L. Bowman, David L. Hall, and Paul Applegate

CONTENTS

---

[*] This chapter is based on a paper by James Llinas et al., Studies and analyses within project correlation: an in-depth assessment of correlation problems and solution techniques, in *Proceedings of the 9th National Symposium on Sensor Fusion*, March 12–14, 1996, pp. 171–188.

23.1   Introduction

The *correlation* problem is one in which both measurements from multiple sensors and additional inputs from multiple nonsensor sources must be optimally allocated to estimation processes that produce (through data/information fusion techniques) fused parameter estimates associated with hypothetical targets and events of interest. In the most general sense, this problem is one of combinatorial optimization, and the solution strategies involve application and extension of existent methods of this type.

This chapter describes a study effort, *Project Correlation*, which involved stepping back from the many application-specific and system-specific solutions and the extensively described theoretical approaches to conduct an assessment and develop guidelines for moving from *problem space* to *solution space*. In other words, the project's purpose was to gain some understanding of the engineering design approaches for solution development and assess the scalability and reusability of solution methods according to the nature of the problem.

Project Correlation was a project within the U.S. Air Force Tactical Exploitation of National Capabilities (AFTENCAP) program. The charter of AFTENCAP was to *exploit all space and national system capabilities for warfighter support*. It was not surprising therefore that the issue of how to cost-effectively correlate such multiple sources of data/information is of considerable importance. Another AFTENCAP charter tenet was to *influence new national system design and operations*; it was in the context of this tenet that Project Correlation sought to obtain the generic/reusable engineering guidelines for effective correlation problem solution.

23.1.1   Background and Perspectives on This Study Effort

The functions and processes of correlation are part of the functions and processes of data fusion. (See Refs 1 and 2, for reviews of data fusion concepts and mathematics.) As a component of data fusion processing, correlation suffers from some of the same problems as other parts of the overall data fusion process (which has been maturing for approximately 20 years): a lack of an adequate scientifically based foundation of knowledge to serve as the basis for engineering guidelines with which to approach and effectively solve problems. In part, the lack of this knowledge is the result of relatively few comparative studies that assess and contrast multiple solution methodologies on an equitable basis. A search of modern literature on correlation and related subjects, for example, revealed a small number of such comparative studies and many singular efforts for specialized algorithms. In part, the goal of the effort described in this chapter was to attempt to overlay or map onto these prior works an equitable basis for comparing and assessing the problem spaces in which these (individually described) algorithms work reasonably well. The lack of an adequate literature base of quantitative comparative studies forced such judgments to become subjective, at least to some degree. As a result, an experienced team was assembled to cooperatively form these judgments in the most objective way possible; none of the evaluators has a stake in, or has been in the business of, correlation algorithm development. Moreover, as an augmentation of this overall study, peer reviews of the findings were conducted via a conference and open session in January 1996 and a workshop and presentation at the National Symposium on Sensor Fusion in April 1996.

Others have attempted such characterizations, at least to some degree. For example, Blackman describes the Tracker-Correlator problem space with two parameters: sampling

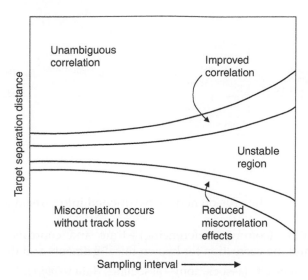

FIGURE 23.1
Interpretation of MTT correlation in a closely spaced target environment.

interval and intertarget spacing.[3] This example is, as Blackman remarks, *simplified but instructive.* Figure 23.1 shows three regions in this space:[3]

1. The upper region of *unambiguous correlation*—characterized by widely spaced targets and sufficiently short sampling intervals.

2. An *unstable region*—in which targets are relatively close (in relation to sensor resolution) and miscorrelation occurs regardless of sampling rate.

3. A region where miscorrelation occurs without track loss—consisting of very closely spaced targets and where miscorrelation occurs, however, measurements are assigned to some track, resulting in no track loss but degradations in accuracy.

As noted in Figure 23.1, the boundaries separating these regions are a function of the two parameters and are also affected by other aspects of the processing. For example, detection probability (Pd) is known to strongly affect correlation performance, so that alterations in Pd can alter the shape of these regions. For the unstable region, Blackman cites some related studies that show that this region may occur for target angular separations of about two to five times the angular measurement standard deviation. Other studies quoted by Blackman show that unambiguous tracking occurs for target separations of about five times the measurement error standard deviation. These boundaries are also affected by the specifics of correlation algorithms, all of which have several components.

23.2   A Description of the Data Correlation Problem

One way to effectively architect a data fusion process is to visualize the process as a tree-type structure with each node of the fusion tree process having a configuration such as that shown in Figure 23.2. The partitioning strategy for such a data fusion tree is beyond the scope of this chapter and is discussed by Bowman[4] and in detail in Chapter 22 of this handbook.

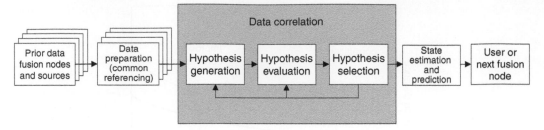

FIGURE 23.2
Data fusion tree node.

The processing in each data fusion node is partitioned into three distinguishable tasks:

1. Data preparation (common referencing)—time and coordinate conversion of source data, and the correction for known misalignments and data conflicts
2. Data correlation or DC (association)—associates data to *objects*
3. State estimation and prediction—updates and predicts the *object* state (e.g., kinematics, attributes, and identity [ID])

This study focused on the processes in the shaded region in Figure 23.2 labeled *Data correlation*. Note that DC is segmented into three parts, each involved with the association hypotheses that cluster reports together to relate them to the *objects*:

- In *hypothesis generation* (HG), the current data and prior selected hypotheses are used to generate the current correlation hypotheses via feasibility gating, pruning, combining, clustering, and object aggregation. That is, alternate hypotheses are defined that represent feasible associations of the input data, including, for example, existing information (e.g., tracks, previous reports, or *a priori* data). Feasibility is defined in a manner that effectively reduces the number of candidates for evaluation and selection (e.g., by a region centered on a time-normalized track hypothesis where measurements that fall within that region are accepted as being possibly associated with that track).

- In *hypothesis evaluation* (HE), each of these feasible correlation hypotheses are evaluated using kinematic, attribute, ID, and *a priori* data as needed to rank the hypotheses (with a score reflecting *closeness* to a candidate object or hypothesis) for more efficient selection searching.[5] Evaluation techniques include numerical (Bayesian, possibilistic), logical (symbolic, nonmonotonic), and neural (nonlinear pattern recognition) methods, as well as hybrids of these methods.[6]

- *Hypothesis selection* (HS) involves a (hopefully efficient) search algorithm that selects one or more association hypotheses to support an improved state estimation process (e.g., to resolve *overlaps* in the measurement/hypothesis matrix). This algorithm may also provide feedback to aid in the generation and evaluation of new hypotheses to initiate the next search. The selection functions include elimination, splitting, combining, retaining, and confirming correlation hypotheses to maintain tracks and aggregated objects.

Most simply put, HG nominates a set of hypotheses to which observations (based on domain/problem insight) can be associated. The HE step develops and computes a metric, which reflects the degree to which any observation is associable (accounting for various errors and other domain effects) to that hypothesis. In spite of the use of such metrics, ambiguities can remain in deciding on how best to allocate the observations. As a result, an HS function (typically an *assignment* problem solution algorithm) is used to achieve an optimal or near-optimal allocation of the observations (e.g., maximum likelihood (ML) based).

Note that the *objects* discussed here are, first of all, hypothetical objects—on the basis of some signal threshold being exceeded, an object, or perhaps more correctly, some *causal factor* is believed to have produced the signal. Typically, the notion of an object's existence is instantiated by starting, in software, an estimation algorithm that attempts to compute (and predict) parameters of interest regarding the hypothetical *object*. In the end, the goal is to correlate the *best* (according to some optimization criteria) ensemble of measurements from multiple input sources to the estimation processes, so that, by using this larger quantity of information, improved estimates result.

23.3 Hypothesis Generation

23.3.1 Characteristics of Hypothesis Generation Problem Space

The characterization of the HG problem involves many of the issues related to HE discussed in Section 23.3.2. These issues include the nature of the input data available, knowledge of target characteristics and behavior, target density, characteristics of the sensors and our knowledge about their performance, the processing time frame, and characteristics of the mission. These are summarized in Table 23.1.

23.3.2 Solution Techniques for Hypothesis Generation

The approach to developing a solution for HG can be separated into two aspects: (1) hypothesis enumeration and (2) identification of feasible hypotheses. A summary of HG solution techniques is provided in Table 23.2. Hypothesis enumeration involves developing a global set of possible hypotheses based on physical, statistical, or explicit knowledge about the observed environment. Hypothesis feasibility assessment provides an initial screening of the total possible hypotheses to define the feasible hypotheses of subsequent processing (i.e., for HE and HS processing). These functions are described in Sections 23.3.2.1 and 23.3.2.2.

23.3.2.1 Hypothesis Enumeration

1. *Physical models*—Physical models can be used to assist in the definition of potential hypotheses. Examples include intervisibility models to determine the possibility of a given sensor observing a specified target (with specified target-sensor interrelationships, environmental conditions, etc.), models for target motion (i.e., to *move* a target from one location in time to the time of a received observation via dynamic equations of motion, terrain models, etc.), and models of predicted target signature for specific types of sensors.

2. *Syntactic models*—Models can be developed to describe how targets or complex entities are constructed. This is analogous to the manner in which syntactic rules are used to describe how sentences can be correctly constructed in English. Syntactical models may be developed to identify the necessary components (e.g., emitters, platforms, sensors, etc.) that comprise more complex entities, such as a surface-to-air missile (SAM) battalion.

TABLE 23.1

Aspects of the Hypothesis Generation Problem Space

Problem Characteristics	Impact on HG	Comments
Input data available	Characteristics of the input data (e.g., availability of locational information, observed entity attributes, entity ID information, etc.) affect factors that can be used to distinguish among entities and, hence, define alternate hypotheses	Reliability and availability of the input data impacts the hypothesis generation function
Knowledge of target characteristics/behavior	The extents to which the target characteristics are known to affect HG. In particular, if the target's kinematic behavior is predictable, then the predicted positions of the target may be used to establish kinematic gates for eliminating unlikely observation-entity pairings. Similarly, ID and attribute data can be used, if known, to reduce the combinatorics	The definition of what constitutes a target (or entity), clearly affects the HG problem. Definition of complex targets, such as a military unit (e.g., a SAM entity), may entail observation of target components (e.g., an emitter) that must be linked hierarchically to the defined complex entity. Hence, hierarchical syntactic reasoning may be needed to generate a potential hypothesis
Target density	The target density (i.e., intertarget spacings) relative to sensor accuracy affects the level of ambiguity about potential observation-entity assignments	If targets are widely spaced relative to sensor accuracy, identifying multiple hypotheses may not be necessary
Sensor knowledge/characteristics	The characteristics and knowledge of the sensor characteristics affect HG. Knowledge of sensor uncertainty may improve the predictability of the observation residuals (i.e., the difference between a predicted observation and actual observation, based on the hypothesis that a particular known object is the *cause* of an observation). The number and type of sensors affect viability of HG approaches	The more accurately the sensor characteristics are known, the more accurately feasible hypotheses can be identified
Processing time frame	The time available for hypothesis generation and evaluation affects HG. If the data can be processed in a batch mode (i.e., after all data are available), then an exhaustive technique can be used for HG/HE. Alternatively, hypotheses can be generated after sets of data are available. In extremely time-constrained situations, HG may be based on sequential evaluation of individual hypotheses	The processing time frame also affects the allowable sophistication of HG processing (e.g., multiple versus single hypotheses) and the complexity of the HE metric (i.e., probability models, etc.)
Mission characteristics	Mission requirements and constraints affect HG. Factors such as the effect (or penalties) for miscorrelations and required tracking accuracy affect which techniques may be used for HG	Algorithm requirements for HG (and for any other data fusion technique) are driven and motivated by mission constraints and characteristics

TABLE 23.2

Solution Techniques for Hypothesis Generation

Processing Function	Solution Techniques	Description	References
Hypothesis enumeration	Physical models	Models of sensor performance, signal propagation, target motion, intervisibility, etc., to identify possible hypotheses	2
	Syntax-based models	Use of syntactical representations to describe the makeup (component entities, interrelationships, etc.) of complex entities such as military units	2
	Doctrine-based models	Definition of tactical scenarios, enemy doctrine, anticipated targets, sensors, engagements, etc. to identify possible hypotheses	1
	Probabilistic models	Probabilistic models of track initiation, track length, birth/death probabilities, etc. to describe possible hypotheses	3, 7
	Ad hoc models	*Ad hoc* descriptions of possible hypotheses to explain available data, may be based on exhaustive enumeration of hypotheses (e.g., in a batch processing approach)	
Identification of feasible hypotheses	Pattern recognition	Use of pattern recognition techniques, such as cluster analysis, neural networks, or gestalt methods to identify *natural* groupings in input data	8, 9
	Gating techniques	Use of *a priori* parametric boundaries to identify feasible observation pairings and eliminate unlikely pairs; techniques include kinematic gating, probabilistic gating, and parametric range gates	10, 11
	Logical templating	Use of prespecified logical conditions, temporal conditions, causal relations, entity aggregations, etc. for feasible hypotheses	12, 13
	Knowledge-based methods	Establishment of explicit knowledge via rules, scripts, frames, fuzzy relationships, Bayesian belief networks, and neural networks	14, 15

3. *Doctrine-based scenarios*—Models or scenarios can also be developed to describe anticipated conditions and actions for a tactical battlefield or battle space. Thus, anticipated targets, emitters, relationships among entities, entity behavior (e.g., target motion, emitter operating anodes, sequences of actions, etc.), engagement scenarios, and other information can be represented.

4. *Probabilistic models*—Probabilistic models can be developed to describe possible hypotheses. These models can be developed on the basis of a number of factors, such as *a priori* probability of the existence of a target or entity, expected number of false correlations or false alarms, and the probability of a track having a specified length.

5. *Ad hoc models*—In the absence of other knowledge, *ad hoc* methods may be used to enumerate potential hypotheses, including (if all else fails) an exhaustive enumeration of all possible report-to-report and report-to-track pairs.

The result of hypothesis enumeration is the definition or identification of possible hypotheses for subsequent processing. This step is critical to all subsequent processing.

Failure to identify realistic possible causes (or *interpretations*) for received data (e.g., such as countermeasures and signal propagation phenomena) cannot be recovered in subsequent processing (i.e., the subsequent processing is aimed at reducing the number of hypotheses and ultimately selecting the most likely or feasible hypotheses from the superset produced at this step), at least in a deductive, or model-based approach. It may be possible, in association processes involving learning-based methods, to adaptively create association hypotheses in real time.

23.3.2.2 Identification of Feasible Hypotheses

The second function required for HG involves reducing the set of possible hypotheses to a set of feasible hypotheses. This involves eliminating unlikely report-to-report or report-to-track pairs (hypotheses) based on physical, statistical, explicit knowledge, or *ad hoc* factors. The challenge is to reduce the number of possible hypotheses to a limited set of feasible hypotheses, without eliminating any viable alternatives that may be useful in subsequent HE and HS processing. A number of automated techniques are used for performing this initial *pruning*. These are listed in Table 23.2; space limitations prevent further elaboration.

23.3.3 HG Problem Space to Solution Space Map

A mapping between the HG problem space and solution space is summarized in Table 23.3.* The matrix shows a relationship between characteristics of the HG problem (e.g., input data and output data) and the classes of solutions. Note that this matrix is not especially prescriptive in the sense of allowing a clear selection of solution techniques based on the character of the HG problem. Instead, an engineering design process,[16] must be used to select the specific HG approach applicable to a given problem.

23.4 Hypothesis Evaluation

23.4.1 Characterization of the HE Problem Space

The HE problem space is described for each batch of data (i.e., fusion node) by the characteristics of the data inputs, the type of score outputs, and the measures of desired performance. The selection of HE techniques is based on these characteristics. This section gives a further description of each element of the HE problem space.

23.4.1.1 Input Data Characteristics

The inputs to HE are the feasible associations with pointers to the corresponding input data parameters. The input data are categorized according to the available data type, the level of its certainty, and its commonality with the other data being associated, as shown in Table 23.4. Input data includes both recently sensed data and *a priori* source data. All data types have a measure of certainty, albeit possibly highly ambiguous, corresponding to each data type.

---

* The solution space abbreviations PM, SM, DM, etc., refer to the solution techniques in Table 23.2.

TABLE 23.3

Mapping Between HG Problem Space and Solution Space

Problem Space	Solution Space								
	Hypothesis Enumeration					ID of Feasible Hypothesis			
	PM	SM	DM	PrM	*Ad Hoc*	PR	GT	LT	KB
Input data categories									
ID attributes	N	Y		Y	Y	Y	Y		Y
Kinematic data	Y			Y	Y	Y	Y		Y
Parameter attributes	Y	Y		Y	Y	Y	Y		Y
A priori sensor/scenario			Y	Y	Y				Y
Linguistic data		Y	Y	N	Y				Y
Space-time patterns		Y	Y	Y	Y	Y			Y
High uncertainty		Y	Y		Y		N		Y
Unknown structures			N	N	Y	Y	N		Y
Error PDF	Y			Y		Y	Y		Y
Target characteristics	Y	Y	Y	Y			Y		Y
Signal propagation models	Y			Y					Y
Output data									
Report-to-report	Y		N	Y	Y	Y	Y	N	N
Report-to-track	Y		N	Y	Y	Y	Y	Y	Y
Track-to-track	Y			Y	Y	Y	Y	Y	Y
Spatiotemporal	Y			Y	Y	Y	Y	Y	Y
Multispectral	Y			Y	Y	Y	Y	N	N
Cross-level	Y	Y	Y	Y	Y	N	Y	Y	Y
Multisite sources	Y			Y	Y		Y	Y	Y
Multiscenes	Y		Y	Y	Y		Y	Y	Y
2D set partitioning	Y		Y	Y	Y		Y	Y	Y
ND set partitioning	Y		Y	Y	Y		Y	Y	Y
Requirements/constraints									
Single scan (*N*=1)	Y	N	N	Y	Y	Y	Y	N	N
Multiple scans (*N*=*n*)	Y	Y	Y	Y	Y	Y	Y	Y	Y
Batch (*N*=~)	Y	Y	Y	Y	Y	Y	Y	Y	Y
Limited processing	Y	Y	N	Y	Y	Y	Y	N	N
Short decision time	Y	Y	N	Y	Y	Y	Y	N	N

TABLE 23.4

Input Data Characteristics

Input Data Categories	Description	Examples of Inputs
Identity (ID)	Discrete/integer valued	IFF, class, type of platform/emitter
Kinematics	Continuous-geographical	Position, velocity, angle, range
Attributes/features	Continuous-nongeographical	RF, PRI, PW, size, intensity, signature
A priori sensor/scenario data	Association hypothesis stats	P_D, P_{FA}, birth/death statistics, coverage
Linguistic	Syntactic/semantic	Language, HUMINT, message content
Object aggregation in space-time	Scripts/frames/rules	Observable sequences, object aggregations
High uncertainty-in-uncertainty	Possibilistic (fuzzy, evidential)	Free text, *a priori* and measured object ID
Unknown structure/patterns	Analog/discrete signals	Pixel intensities, RF signatures
Partially known error statistics	Nonparametric data	$P(R \mid H)$ only partially known
Partial and conflicting data	Missing, incompatible, incorrect	Wild points, closed versus open world, stale
Differing dimensions	Multidimensional discrete/continuous	3D and 2D evaluated against ND track
Differing resolutions	Coarseness of discernability	Sensor resolution differences: radar + IR
Differing data types	Hybrid types and uncertainties	Probabilistic, possibilistic, symbolic

TABLE 23.5

Output Data Characteristics

Score Output Categories	Description	Examples of Outputs
Yes/no association	0/1 logic (no scoring)	High confidence only association
Discrete association levels	Integer score	Low/medium/high confidence levels
Numerical association score	Continuous real-valued score	Association probability/confidence
Multiscores per association	ND (integer or real per dim)	Separate score for each data group
Confidence function with score	Score uncertainty functional	Fuzzy membership or density function
No explicit association	State estimates directly on data	No association representation

23.4.1.2 Output Data Characteristics

The characteristics of the HE outputs needed by HS also drive the selection of the HE techniques. Table 23.5 describes these HE output categories, which are partitioned according to the output variable type: logical, integer, real, N-dimensional (ND), functional, or none. Most HE outputs are real-valued scores reflecting the confidence in the hypothesized association. However, some output a discrete confidence level (e.g., low, medium, or high), whereas others output multiple scores (e.g., one per data category) or scores with higher-order statistics (e.g., fuzzy, evidential, or random set). For some batches of data, no HE scoring is performed, and only a yes/no decision on the feasibilities is output for HS. *No explicit association* refers to those rare cases where the data association function is not performed (i.e., the data is only implicitly associated in performing state estimation directly on all of the data). An example in image processing is the estimation of object centroid or other features, based on intensity patterns without first clustering the pixel intensity data.

23.4.2 Mapping of the HE Problem Space to HE Solution Techniques

This section describes the types of problems for which the solution techniques are most applicable (i.e., mapping problem space to solution space). A preliminary mapping of this type is shown in Table 23.6; final guidelines were developed by Llinas et al.[16] The *ad hoc* techniques are used when the problem is easy (i.e., performance requirements are easy to meet) or the input data errors are ill defined. Probabilistic techniques are selected according to the error statistics of the input data. Namely, ML techniques are applied when there is no useful *a priori* data, otherwise maximum *a posteriori* (MAP) are considered. Chi-square (CHI) techniques are applied for data with Gaussian statistics (e.g., without useful ID data), especially when there is data of differing dimensions where ML and MAP would have to use expected values to maintain constant dimensionality. Neyman-Pearson (NP) techniques are statistically powerful and are used as the basis for nonparametric techniques (e.g., sign test and Wilcoxon test). Conditional algebra event (CAE) techniques are useful when the input data is given, conditioned on different events (e.g., linguistic data). Rough sets (Rgh) are used to combine/score data of differing resolutions. Information/entropy (Inf) techniques are used to select the density functions and score data whose error statistics are not known. Further discussion of the various implications of problem-to-solution mappings is provided by Llinas et al.[16]

TABLE 23.6

Mapping HE Problem Space to HE Solution Techniques

Problem Space	Probabilistic							Possibilistic				Logic Symbolic					Neural		Hybrid
Solution Space	Ad Hoc	ML	MAP	NP	CHI	CAE	Rgh	INF	DS	Fuzzy	S/F	NM	ES	C-B	PR	PD	HC	Super	RS
Input data categories																			
Identity (ID)	Y	Y	Y		N														Y
Kinematics		Y	Y		Y														Y
Attributes/features		Y	Y		Y														Y
A priori sensor data		N	Y		N														Y
Linguistic						Y				Y	Y	Y							Y
Object aggregation											Y								
High uncertainty									Y	Y									
Unknown structure								Y									N	Y	
Nonparametric data	Y		Y					Y											
Partial data															Y	Y			
Differing dimension					Y														
Differing resolution							Y												
Differing data types																			Y
Score output categories																			
Yes/no association	Y						N	N											
Discrete scores	Y						N	N											
Numerical scores		Y	Y	Y	Y	Y	Y	Y											
Multiscores per						Y	Y	Y	Y										
Confidence functional										Y									
No explicit scores																	Y	Y	
Performance measures																			
Low-cost software	Y	Y	Y	Y	Y	N	N	N	N	N				N	N	N	N	Y	N
Compute efficiency	Y	Y	Y	Y	Y	N									N	N	Y	Y	
Score accuracy	N	Y	Y	N	Y	Y											N	N	
User adaptability											Y		Y	Y	Y		N	Y	
Self-trainable														Y			N	Y	
Self-coding														Y			N	Y	
Robustness to error											Y	Y	Y	Y	Y	Y	N	Y	
Result explanation											Y	Y	Y	Y	Y	Y	N	N	

Note: ML—maximum likelihood; MAP—maximum *a posteriori*; NP—Neyman-Pearson; CHI—chi-square; CAE—conditional algebra event; Rgh—rough sets; INF—information/entropy; DS—Dempster-Shafer; Fuzzy—fuzzy set; S/F—scripts/frames/rules; NM—nonmonotonic; ES—expert systems; C-B—case-based; PR—partitioned representations; PD—power domains; HC—hard-coded; Super—supervised; RS—random set.

23.5 Hypothesis Selection

When the initial clustering, gating, distance/closeness metric selection, and fundamental approach to HE are completed, the overall correlation process reaches a point where the *most feasible* set of both multisensor measurements and multisource inputs exist. The inputs are *filtered*, in essence, by the preprocessing operations and the remaining inputs allocated or *assigned* to the appropriate estimation processes that can exploit them for improved computation and prediction of the states of interest. This process is HS, in which the hypothesis set comprises all of the possible/feasible assignment *patterns* (set permutations) of the inputs to the estimation processes; thus, any single hypothesis is one of the set of feasible assignment patterns. This chapter focuses on position and ID estimation from such assigned inputs as the states of interest. However, the hypothesis generation–evaluation–selection process is also relevant to the estimation processes at higher levels of abstraction (e.g., wherein the states are *situational states* or *threat states*), and the state estimation processes, unlike the highly numeric methods used for level 1 estimates, are reasoning processes embodied in symbolic computer-based operations.

So, what exists as *input to the hypothesis selection* process in effect is, at this point, a matrix (or matrices) where the typical dimensions are the indexed input data/information/measurement set on one hand, and the indexed state estimation systems or processes, along with the allowed ambiguity states, on the other hand (i.e., the *other* states or conditions, beyond those state estimates being maintained, to which the inputs may be assigned). Simply put, for the problems of interest described here, the two dimensions are the indexed measurements and the indexed position or ID state estimation processes. (Note, however, as discussed later in this chapter, that assignment problems can involve more than two dimensions.)

In any case, the matrix/matrices are populated with the closeness measures, which could be considered *costs* of assigning any single measurement to any single estimator (resulting from the HE solution). Despite the effort devoted to optimizing the HG and HE solutions, considerable ambiguity (many feasible hypotheses) can still result. The costs in these matrices may directly be the values of the *distance* or scoring metrics selected for a particular approach to correlation, or a newly developed cost function specifically defined for the HS step. The usual strategy for defining the optimal assignment (i.e., selecting the optimal hypothesis) is to find that hypothesis with the lowest total cost of assignment. Recall, however, that there are generally two conditions wherein such matrices develop: (1) when the input systems (e.g., sensors) are initiated (turned on) and (2) when the dynamic state estimation processes of interest is being maintained in a recursive or iterative mode.

As noted earlier, these assignment or association matrices, despite the careful preprocessing of the HG and HE steps, may still involve ambiguities in how to best assign the inputs to the state estimators. That is, the cost of assigning any given input to any of a few or several estimators may be reasonable or allowable within the definition of the cost function and its associated thresholds of acceptable costs. If this condition exists across many of the inputs, identifying the total-lowest-cost assignments of the inputs becomes a complex problem. The *central problem* to be solved in HS is that of defining a way to select the hypothesis with minimum total cost from all feasible/permissible hypotheses for any given case; often, this involves large combinations of possibilities and leads to a problem in *combinatorial optimization*. In particular, this problem—called the *assignment problem* in the domain of combinatorial optimization—is applicable to many cases other than the measurement assignment problem presented in this chapter and has been well studied by the mathematical and operations research community, as well as by the data fusion community.

23.5.1 The Assignment Problem

The goal of the assignment problem is to obtain an optimal way of assigning various available N resources (in this case, typically, measurements) to various N or M (M can be less than, greater than, or equal to N) *processes* that require them (in our case estimation processes, typically). Each such feasible assignment of the $N \times N$ problem (a permutation of the set N) has a cost associated with it, and the usual notion of optimality equates to minimum cost. Although special cases allow for multiassignment (in which resources are shared), for many problems, a typical constraint allows only one-to-one assignments; these problems are sometimes called *bipartite matching* problems.

Solutions for the typical and special variations of these problems are provided by mathematical programming and optimization techniques. (Historically, some of the earliest applications were to multiworker/multitask problems and many operations research and mathematical programming texts motivate assignment problem discussions in this context.) This problem is also characterized in the related literature according to the nature of the mathematical programming or optimization techniques used as *solutions* applied for each special case. Not surprisingly, because the underlying problem model has broad applicability to many specific and real problems, the literature describes certain variations of the assignment problem and its solution in different (and sometimes confusing) ways. For example, assignment-type problems also arise in analyzing *flows in networks*. Ahuja et al., in describing network flows, divide their discussion into six topics:[17]

1. Applications
2. Basic properties of network flows
3. Shortest path problems
4. Maximum flow problems
5. Minimum cost flow problems
6. Assignment problems

In their presentation, they characterize the assignment problem as a *minimum cost flow problem on a network*.[17] This characterization, however, is exactly the same as asserted in other applications. In network parlance, however, the assignment problem is now called a *variant of the shortest path problem* (which involves determining directed paths of smallest cost from any node X to all other nodes). Thus, *successive shortest path* algorithms solve the assignment problem as a sequence of N shortest path problems (where N = number of resources = number of processes in the (square) assignment problems).[17] In essence, this is the bipartite matching problem restated in a different way.

23.5.2 Comparisons of Hypothesis Selection Techniques

Many technical, mathematical aspects comprise the assignment problem that, given space limitations, are not described in this chapter. For example, there is the crucial issue of solution complexity in the formal sense (i.e., in the sense of *big O* analyses), and there is the dilemma of choosing between multidimensional solutions and two-dimensional (2D) solutions and all that is involved in such choices; in addition, many other topics remain for the process designer. The solution space of assignment problems at level 1 can be thought of as comprising (a) *linear and nonlinear mathematical programming* (with some emphasis on integer programming), (b) *dynamic programming and branch-and-bound methods* as part of the family of methods employing implicit enumeration strategies, and (c) *approximations and*

TABLE 23.7

Frequently Cited Level 1 Assignment Algorithms (Generally: Deterministic, 2D, Set Partitioned)

	Applicability		
Algorithms	**Problem Space**	**Processing Characteristics**	**Runtime Performance**[a]
Hungarian[18]	Square matrices; optimal pairwise algorithm	Primal-dual; steepest descent	$O(nS(n,m,C))$; or O ([# trks + # msmts]**3)
Munkres[19]	Square matrices; optimal pairwise algorithm		
Bourgeois-Lassale;[20] see also Refs 21 and 22	Rectangular matrices; optimal pairwise algorithm	B-L faster than squaring-off method of Kaufmann	
Stephans-Krupa[22,23]	Sparse matrices; optimal pairwise algorithm		
JVC[24,25]	Sparse matrices; optimal pairwise algorithm	Appears to be the fastest of the traditional methods; S-K second fastest to NC; sparse Munkres third; JV is augmenting cycle approach	
Auction types[26]	Optimal pairwise algorithm	Primal-dual, coordinate descent; among the fastest algorithms; parallelizable versions developed; appears much faster than N algorithm (as does JVC)	$O(n^{**}2mC)$; scaled version $O(nmlognC)$; others show $O(n^{**}1/2mlogC)$
Primal simplex/ alternating basis[27]	Applied to relatively large matrices (1000–4500 nodes); optimal pairwise algorithm	Moderate speed for large problems	
Signature methods[28]	Optional pairwise algorithm	Dual simplex approach	$O(n^{**}3)$

[a] O = worst case, n = # of nodes, m = # of arcs, C = upper bound on costs, S = successive shortest path solution time for given parameters.

heuristics. Although this generalization is reasonable, note that the assignment problems of the type experienced for level 1 data fusion problems arise in many different application areas; as a result, many specific solution types have been developed over the years, making broad generalizations difficult.

Table 23.7 summarizes the conventional methods used for the most frequently structured versions of assignment problems for data fusion level 1 (i.e., deterministic, 2D, set-partitioned problem formulations). Furthermore, these are solutions for the linear case (i.e., linear objective or cost functions) and linear constraints. Without doubt, this is the most frequently discussed case in the literature and the solutions and characteristics cited in Table 23.7 represent a reasonable benchmark in the sense of applied solutions (but not in the sense of improved optimality).

Note that certain formulations of the assignment problem can lead to nonlinear formulations, such as the quadratic assignment. Additionally, the problem can be formulated as a *multiassignment* problem as for a set-covering approach, although these structures usually result in linear multiassignment formulations if the cost/objective function can be treated as separable. Otherwise, it, too, will typically be a nonlinear, quadratic-type formulation.[29] No explicitly related nonlinear formulation for level 1 type applications was observed in the literature; however, Ref. 30 is a useful source for solutions to the quadratic problem. The key work in multiassignment structures for relevant fusion-type problems is by Tsaknakis et al., who form a solution analogous to the auction-type approach.[29]

This chapter introduces additional material focused on comparative assessments of assignment algorithms. These materials were drawn, in part, from works completed during the strategic defense initiative, or SDI (Star Wars), era and reported at the *SDI Tracking Panels* proceedings assembled by the Institute for Defense Analysis, and, in part, from a variety of other sources. Generally, the material dates to about 1990 and could benefit from updating.

One of the better sources, in the sense of its focus on the same issues/problems of interest to this study, is the study by Washburn.[31] In one segment of a larger report, Washburn surveys and summarizes various aspects of what he calls *data partitioning, gating,* and *pairwise; multiple hypothesis;* and *specialized* object correlation algorithms. *Data partitioning and gating* equate to the terms *hypothesis generation and evaluation* used here, and the term *correlation* equates to the term *hypothesis selection* (input data set assignment) used here. Washburn's assessments of multiple hypothesis class of correlation algorithms that he reviewed are presented in Table 23.8. Washburn repeats a critical remark related to comparisons of these algorithms: for those cases where a multiscan or multiple data set approach is required to achieve necessary performance levels (this was perhaps typical on SDI, especially for midcourse and terminal engagements), the (exact) optimal solution is unable to be computed, and so comparisons to the true solution are infeasible, and a *relaxed* approach to comparison and evaluation must be taken. This is, of course, the issue of developing an optimal solution to the ND case, which is an NP-hard problem. This raises again the key question of what to do if an exact solution is not feasible.

23.5.2.1 2D Versus ND Performance

One of the only comparisons of 2D versus ND for a common problem that was captured in this study was a reference in the Washburn, 1992 report to Allen et al. (1988) that examined comparative performance for a 10-target/3-passive sensor case.[31,41] These results are shown in Figure 23.3.[31] Essentially, *pairwise*—2D—solutions performed considerably worse than the three-dimensional (3D) approaches; branch-and-bound, RELAX, and backtrack all achieved average tracking of eight of ten targets—far below perfection.

This effort involved examining several other comparison and survey studies. To summarize, the research found that, among other factors, performance is affected by angular accuracy-to-angular target separation, number of targets, and average target density in the HG process gates, coding language, degree of parallelization, and available reaction time.

23.5.3 Engineering an HS Solution

As mentioned at the outset, this project sought to develop a set of engineering guidelines to guide data fusion process engineers in correlation process design. Such guidelines have, in fact, been developed,[16] and an example for the HS process is provided in this section. In the general case, a fairly complex feasible hypothesis matrix exists. The decision tree guideline for engineering these cases is shown in Figure 23.4. Although the order of the questions or criteria shown in the figure is believed to be correct in the sense of maximal ordered partitioning of the problem/solution spaces, the decisions could possibly be determined by some other sequence. Because the approaches are so basically different and involve such different philosophies, the first question is whether the selected methodology is *deterministic* or *probabilistic*. If it is deterministic, then the engineering choices follow the *top* path, and vice versa. Some of the trade-off factors affecting this choice are shown in Table 23.9, which concentrate on the deterministic family of methods.

TABLE 23.8

Multiple Hypothesis Object Correlation Algorithms Summary

Algorithm	Functional Performance	Processing Requirements
Multidimensional row-column (MRC)[32,33]	Worst functional performance of multiple hypothesis algorithms	$O([n_s - 1] \cdot m_G)$ operations. Selection process is sequential, but nonsequential approaches could be applied for parallel processing
Multidimensional maximum marginal return[34,35]	Performs better than pairwise correlation and better than MRC. Worse than MST or backtracking M³R approach, but may have acceptable performance in sparse scenarios	$O(n_s - 1] \cdot m_G \log m_G)$ operations. Parallel algorithm has been developed for binary tree processor
Backtracking and look ahead M³R[35]	Performed significantly better than M³R by using backtracking heuristic to improve solutions	$O([n_s - 1] \cdot m_G \log m_G)$ operations. More complicated logic was found difficult to parallelize
Monte Carlo M³R[35]	Uses randomization to change effect of branch ordering on M³R and improve solution. Annealing approach converges to optimal solution (as computed by Branch-and-Bound)	Has $O(n_{MC}[n_s - 1] \cdot m_G \log m_G)$ processing requirements where n_{MC} Monte Carlo iterations are required. This number may be very large and evaluations are needed to determine how small for adequate performance. Parallelizes completely over n_{MC} Monte Carlo iterations
Minimal spanning tree (MST)[34] Tsaknakis	Performance depends on bipartite assignment used. With optimal, MST obtained best performance in Ref. 34 evaluations against alternative correlation algorithms (including Branch-and-Bound) for small scenarios	Processing depends on assignment algorithm used. Optimal auction algorithm (which can be parallelized), processing is $O(n_s^2[n_T + n_M] \cdot m_G \log[n_T + n_M] \cdot C)$. Parallelizes over n_s^2 factor, as well as over bipartite assignment algorithm
Viterbl correlation[33,36]	Special cost structure gives limited applicability or poor performance if approximations are used. No evaluations for SDI tracking problems	Processing requirement is unacceptably large for large scenarios
Branch-and-Bound[34,35,37–39]	Optimal multiple hypothesis algorithm. Obtained best performance in Ref. 34 evaluations against alternative correlation algorithms with same scan depth n_s	Uses various backtracking heuristic to speed up search, but potentially searches all possible solutions. Has very large requirements for even small dense scenarios. May be feasible if data partitioning produces small groups of data. Parallel algorithm developed in Ref. 38
Relaxation[40]	Iterative algorithm that generates feasible solution at each iteration and approaches optimal solution. Little evaluation in realistic scenarios to determine convergence rate	Processing depends on bipartite assignment algorithm used in each iteration and on the number of iterations required. $O(n_R [n_s - 1] \cdot [n_T + n_M] \cdot m_G \log[n_T + n_M] \cdot C)$ operations using auction with n_R relaxation iterations

23.5.3.1 Deterministic Approaches

If the approach is deterministic, the next crucial decision relates to the quantity of input data that will be dealt with at any one time. In the tracking world, this usually reflects the *number of scans* or *batches of data* that will be considered within a processing interval. These issues reflect the more complex case of correlating and assigning *ensembles* of data

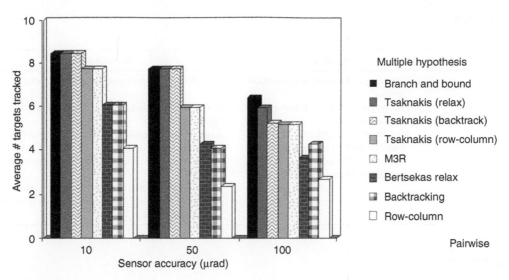

FIGURE 23.3
Functional comparison of correlation algorithms.

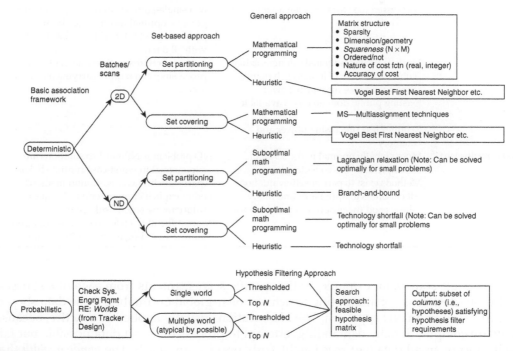

FIGURE 23.4
Representative HS decision flow diagram.

at a time rather than single contact at a time. In the typical cases where data ensembles are processed, they are frequently delineated by source type—often implying the type of sensor—or they are grouped by scan time. In a multisensor or multisource (or multitime) architecture, the (near-globally optimal) *natural* formulation for assignment processing would be *n-dimensional*, with *n* designating either the number of batch segments

TABLE 23.9

Trade-off 1

Design Decision	Positives	Negatives
Deterministic	Wide array of formal mathematical methods	Potentially incorrect specific assignments
Probabilistic	Proven useful for few tracks plus high clutter	Not widely applied to other than special cases as noted
	Finesses specific assignments by probabilistically using all measurements in a gate (an *all-neighbors* approach)	Assignment processing not based on proven formalized methods; removes assignment from conventional correlation processing, embeds it in tracker
		For many problems, not conceptually correct; measurements truly only come from one source in most (but not all) cases

TABLE 23.10

Trade-off 2

Design Decision	Positives	Negatives
2D (~single scan)	Each 2D solution is optimal for the data processed (i.e., for the given assignment matrix)	Primarily that optimality is local in the sense of a single-scan worth of data, and that more globally optimal result comes from processing WEI data (ideal is batch solution with all data)
	Most, if not all, traditional solutions exhibit cubic (in number of inputs) polynomial-time, worst-case runtime behavior	Not easy to retrospectively adjust for errors in processing; can require carrying alternative scenes
	Existing public-domain codes available; easier to code	
	Requires less processing power and memory	
ND (~multiscan)	More globally optimal in the sense of multiple data sets considered	ND problem is NP-hard so solutions are exponential in worst-case runtime behavior
	Methods exist to permit self-evaluation of closeness to optimality, and many show good (near-optimal) performance	Apart from branch-and-bound methods or other explicit enumeration techniques, solutions are strictly suboptimal
		More difficult to code; some codes nonpublic; more demanding of computing resources

from a single source, for example, or the specific-scan segments from each of n sources. In the strictest sense of attempting to achieve near-global optimality, these data should be processed in an ND approach. The ND solutions usually applied in practice, because they are computationally demanding, attempt to process the most data possible but fall well short of an all data (i.e., true batch), truly optimal approach. The *window width* that such techniques can handle is yet another specific design choice to be made in the ND case. Because of the computational aspects, in part, there are applications where the data are segmented into source-specific sets, and a sequence of 2D assignment solutions are applied to obtain a satisfying (*good enough*, knee-of-the-curve performance), but strictly suboptimal result. The second choice is depicted in Table 23.10. Obviously this design decision separates the assignment solutions into quite different categories; the basic trade-off is in the nature of the optimality sought versus the solution complexity and runtime behavior.

For either the 2D or ND cases, the next decision has to do with whether the problem formulation is of the set-partitioning or set-covering type. This decision, as will be seen, also separates the potential solution types into considerably different categories, because these formulations are entirely different views of the problem. Recall that the set-partitioning scheme is one that divides the notional allocations of inputs into mutually exclusive sets with single inputs eventually being assigned, exclusively, to a particular state. The set-covering approach also divides the input data ensemble, but into nonexclusive sets such that single inputs can be assigned to more than one state. Only in the sense of permitting multiple assignments does the set-covering construct has some similarity to the notion of probabilistic assignment. However, the general set-covering approach makes assignments deterministically, as opposed to probabilistically, and also takes a different approach to defining allocable sets. The probabilistic approach permits multiassignment as determined largely by the gating process defined by HG; all measurements in a gate are probabilistically assigned to the track, and to the degree that gates overlap, multiassignment of specific measurements exists. Generalized set-covering would ideally consider a number of factors on which to define its overlapping sets and is conceptually more robust than the probabilistic approach. So what is at issue here is one's *view of the world*, in the sense of how the data should be grouped, which, in turn, affects the eventual structure of the assignment matrix or other construct that provides the input to the assignment problem and its solution.

As noted in Table 23.11 and shown in Figure 23.4, the set-partitioning view of the problem leads to a situation where there are many more known and researched solutions, so the fusion node designer will have wider solution choices if the problem is set up this way.

The next design choice is whether to employ mathematical programming-type solutions or some type of heuristic approach. The mathematical programming solution set includes linear programming (LP) methods, integer and dynamic programming, and a wide variety of variations of these types. Although mathematically elegant and formal, these methods can become computationally demanding for problems of a large scale, whether they are 2D or ND. So, in many practical situations, an approximation-based or heuristic approach—which will develop a reasonably accurate solution and require considerably less computing resources—is preferable. Often heuristics can be embedded as subroutines within otherwise formal methods.

TABLE 23.11

Trade-off 3

Design Decision	Positives	Negatives
Set partitioning	For many, if not most, cases, this is the true view of the world in that singular data truly associate to only one causal factor or object (see trade-off 1, Deterministic). This assertion is conditioned on several factors, including sensor resolution and target spacing, but represents a frequent case	Lacks flexibility in defining feasible data sets
	Much larger set of researched and evaluated solution types	Does not solve multispectral or differing resolution or crossing-object-type problems
Set covering	More flexible in formulating feasible data groupings; often the true representation of real-world data sets	More complex formulation and solution approach

23.6 Summary

DC can be viewed as a three-step process of HG, HE, and HS. The first step limits the number of possible associations by permitting or *gating through* only those that are feasible. These potential associations are evaluated and quantitatively scored in the second step and assigned in the third.

Each of these stages can be performed in a variety of ways, depending on the specific *nature* of the DC problem. This *nature* has less to do with the specific military application than its required characteristics, such as processing efficiency and types of input data and output data, knowledge of how the characteristics relate to the statistics of the input data and the contents of the supporting database. The data fusion systems engineer needs to consider the system's user-supplied performance-level requirements, implementation restrictions (if any), and characteristics of the system data. In this way, the application can be considered as *encoded* into an engineering-level *problem space*. This problem space is then mapped into a solution space via the guidelines of the type discussed in this chapter.

Beginning with HG, the designer needs to consider how the input data gets batched (e.g., single-input, single-sensor scan, multiple-sensor scan, or seconds between update *frames*) and what metric(s) to use. The goal should be a process that is less computationally intensive than the HE step and coincides with available input data (position, attribute, ID), and the list of possible hypotheses data (e.g., old track, new track, false alarm, ECM/deception, or anomalous propagation ghosting). Section 23.3 discusses these design factors and provides a preliminary mapping of gating techniques to problem space.

The next function, HE, considers much of the same material as HG; however, where the previous function evaluated candidate hypotheses on a pass/fail basis, this function must grade the goodness of unrejected hypotheses. Again, the type of input data must be considered together with its uncertainty characteristics. Probabilistic, possibilistic, and script-based techniques can be applied to generate scores that are, for example, real, integer, or discrete. These outputs are then used by the selection stage to perform the final assignment.

The selection stage accepts the scores to determine an *optimum* allocation of the gated-through subset of a batch of data to targets, other measurements, or other possible explanations for the data (e.g., false alarms). Optimum, in this case, means that the cost (value) of the set of associations (measurement-track, measurement-measurement), when taken as a whole, is minimized (maximized). The tools that handle ambiguous measurements are solutions to what is called the *assignment problem*. Solutions to this problem can be computationally intensive; therefore, efficient solutions that cover both the 2D and ND matrices of varying geometries are required. Section 23.5 provides a brief tutorial to the assignment problem and lists the numerical characteristics of popular techniques.

As mentioned previously in this chapter, the evolving guidelines and mappings presented in Sections 23.3 through 23.5 were presented to an open audience at the government-sponsored Combat Information/Intelligence Symposium in January 1996 and at the (U.S.) National Symposium on Sensor Fusion in April 1996; however, further peer review is encouraged. The basis of such reviews should be the version of the engineering guidelines as presented in Ref. 16. Further, this chapter has treated these three individual processes as isolated parts, rather than part of a larger, integrated notion of correlation processing. Thus, additional work remains to be performed to assess the interprocess sensitivities in design choices, so that effective, integrated solutions result from the application of these guidelines.

References

1. Waltz, E. and Llinas, J., *Multi-sensor Data Fusion*, Artech House, Norwood, MA, 1990.
2. Hall, D.L., *Mathematical Techniques in Multisensor Data Fusion*, Artech House, Norwood, MA, 1992.
3. Blackman, S., *Multiple Target Tracking with Radar Applications*, Artech House, Norwood, MA, 1986.
4. Bowman, C.L., The data fusion tree paradigm and its dual, in *Proc. 7th Nat'l Fusion Symp.*, invited paper, Sandia Labs, NM, March 1994.
5. Bowman, C., Max likelihood track correlation for multisensor integration, in *18th IEEE Conference on Decision and Control*, December 1979.
6. Bowman, C.L., Possibilistic versus probabilistic trade-off for data association, in *Proc. Signal and Data Processing of Small Targets*, Vol. 1954, pp. 341–351, SPIE, April 1993.
7. Bar-Shalom, Y. and Tse, E., Tracking in a cluttered environment with probabilistic data association, *Automatica*, 11, 451–460, 1975.
8. Aldenderfer, M.S. and Blashfield, R.K., Cluster analysis, Sage University Paper Series on *Quantitative Applications in the Social Sciences*, Paper No. 07-044, London, UK, 1984.
9. Fukunaga, K., *Introduction to Statistical Pattern Recognition*, 2nd edn., Academic Press, New York, 1990.
10. Blackman, S., *Multiple Target Tracking with Radar Applications*, Artech House, Norwood, CT, 1985.
11. Uhlmann, J.K. and Zuniga, M.R., Results of an efficient gating algorithm for large-scale tracking scenarios, *Naval Research Reviews*, 43(1), 24–29, 1991.
12. Hall, D.L. and Linn, R.J., Comments on the use of templating for multisensor data fusion, in *Proc. 1989 Tri-Service Data Fusion Symp.*, Vol. 1, p. 345, May 1989.
13. Noble, D.F., Template-based data fusion for situation assessment, in *Proc. 1987 Tri-Service Data Fusion Symp.*, Vol. 1, pp. 152–162, June 1987.
14. Jackson, P., *Introduction to Expert Systems*, Addison-Wesley, Reading, MA, 1999.
15. Pearl, J., *Probabilistic Reasoning in Intelligent Systems: Networks of Plausible Inference*, Morgan Kaufmann Series on Representation and Reasoning, San Mateo, CA, 1988.
16. Llinas, J. et al., Engineering Guidelines for Data Correlation Algorithm Characterization, Vol. 3 of Final Reports on Project Correlation, June 1996.
17. Ahuja, R.K. et al., Network flows, in *Handbook in OR and MS*, Vol. I, GL Nemhauser et al., Eds., Elsevier Science Publishers, North-Holland, 1989, Chapter IV.
18. Kuhn, H.W., The Hungarian method for the assignment problem, *Naval Research Logistics Quarterly*, 2, 83–97, 1955.
19. Munkres, J., Algorithms for the assignment and transportation problems, *J. SIAM*, 5(1), 32–38, 1957.
20. Bourgeois, F. and Lassale, J., An extension of the Munkres algorithm for the assignment problem to rectangular matrices, *Comm ACM*, 14(12), 802–804, 1971.
21. Silver, R., An algorithm for the assignment problem, *Comm ACM*, 3, 605–606, 1960.
22. Kaufmann, A., *Introduction a la Combinatorique en Veu de ses Applications*, Dunod Publishers, Paris, 1968.
23. Salazar, D.L., Application of Optimization Techniques to the Multi-Target Tracking Problem, Master's Thesis, University of Alabama at Huntsville, AL, 1980.
24. Drummond, D.E., Castanon, D.A. and Bellovin, M.S. Comparison of 2D assignment algorithms for sparse, rectangular, floating point cost matrices, *J. SDI Panels Tracking*, 4, 81–87, 1990.
25. Jonker, R. and Volgenant, A., A shortest augmenting path algorithm for dense and sparse linear assignment problems, *Computing*, 38(4), 325–340, 1987.
26. Bersakas, D.P. and Eckstein, J., Dual coordinate step methods for linear network flow problems, *Math Prog., Series B*, 42, 203–243, 1988.

27. Barr, R.S. et al., The alternating basis algorithms for assignment problems, *Math Prog.*, 13, 1–13, 1977.

28. Balinski, M.L., Signature methods for the assignment problem, *Oper. Res.*, 33(3), 427–536, 1985.

29. Tsaknakis, H., Doyle, B.M. and Washburn, R.B., Tracking closely spaced objects using multiassignment algorithms, in *Proc. DFS 91*, 1991.

30. Finke, G. et al., Quadratic assignment problems, in *Annals of Discrete Mathematics*, 31, Elsevier Science Pub., Amsterdam, 1987.

31. Washburn, R.B., Tracking algorithms, Section 2, in a report on the SDI Algorithm Architecture program, provided to TRW as prime, September 1992.

32. Liggins, M. and Kurien, T., Report-to-target assignment in multitarget tracking, *Proceedings of the 1988 IEEE Conference on Decision and Control*, Austin, TX, December 1988.

33. Wolf, J.K., Viterbi, A.M. and Dixon, G.S., Finding the best set of K paths through a trellis with application to multitarget tracking, *IEEE Trans. Aerospace Electr. Systems*, AES-25, 287–296, 1988.

34. Allen, T.G., Feinberg, L.B., LaMaire, R.O., Pattipati, K.R., Tsaknakis, H., Washburn, R.B., Wren, W., Dobbins, T. and Patterson, P., Multiple information set tracking correlator (MISTC), TR-406, Final Report, Alphatech, Inc., Burlington, MA, and Honeywell, Inc., Space and Strategic Avionics Division, September 1988, USASDC Contract No. DASG60-87-C-0015.

35. Parallel processing of battle management algorithms, Final Report, CDRL Item A003, AT&T Bell Labs, Whippany, NJ, December 1989.

36. Castanon, D.A., Efficient algorithms for finding the K best paths through a trellis, *IEEE Trans. Aerospace Electr. Systems*, 26(2), 405–410, 1989.

37. Erbacher, J.A., Todd, J. and Hopkins, C.H., SDI tracker/correlator algorithm implementation document and algorithm design document, R-041-87, Contract No. N00014-86-C-2129, VERAC, San Diego, CA, April 1987.

38. Allen, T.G., Cybenko, G., Angelli, C. and Polito, J., Parallel processing for multitarget surveillance and tracking, Technical Report TR-360, Alphatech, Burlington, MA, January 1988.

39. Nagarajan, V., Chidambara, M.R. and Sharma, R.N., Combinatorial problems in multitarget tracking—a comprehensive solution, *IEEE Proc.*, 134, 113–118, 1987.

40. Pattipati, K.R., Deb, S., Bar-Shalom, Y. and Washburn, R.B., Passive multisensor data association using a new relaxation algorithm, *IEEE Trans. Automatic Control*, February 24, 1989.

41. Allen, T.G. et al., Multiple Information Set Tracking Correlator (MISTC), Technical Report TR-406, Alphatech Inc., Burlington, MA, September 1988.

24

Data Management Support to Tactical Data Fusion

Richard Antony

CONTENTS

24.1   Introduction

Historically, database management systems (DBMS) associated with data fusion applications provided storage and retrieval of sensor-derived parametric and text-based data, fusion products, and algorithm components such as templates and exemplar sets. The role of DBMS rapidly expanded as multimedia data sources, such as imagery, video, and graphic overlays, became important components of fusion algorithms. Today, DBMS are widely recognized as a critical component of the overall system design.

In seeking the proficiency and robustness of human analysts, fusion algorithms increasingly employ context-sensitive problem domain knowledge. In tactical applications, relevant context includes existing weather conditions, local natural domain features (e.g., terrain/elevation, surface materials, vegetation, rivers, and drainage regions), and manmade features (e.g., roads, airfields, and mobility barriers), all of which have a spatial component that must be stored, searched, and manipulated to support a spectrum of both real-time fusion and forensic-like data mining applications. In addition to crisp (Boolean) spatial representations, the database must support semantic (fuzzy) spatially organized information.

The objective of this chapter is to provide a brief description of the data management challenges and suggest approaches for overcoming current limitations. Section 24.2 introduces DBMS requirements. Section 24.3 discusses spatial, temporal, and hierarchical reasoning that are key underlying requirements for advanced data fusion automation. Section 24.4 discusses critical database design criteria. Section 24.5 presents the concept of an object-oriented representation of space, showing that it is fully analogous to the traditional object representation of entities that exist within a domain. Section 24.6 briefly describes a composite database system consisting of an integrated representation of both spatial and nonspatial objects. Section 24.7 discusses reasoning approaches and presents a comprehensive example to demonstrate the application of the proposed architecture. Section 24.8 introduces the requirements for combined Boolean and fuzzy database operations and offers a number of motivational examples. Section 24.9 offers a brief summary.

24.2   Database Management Systems

In general, data fusion applications require access to data sets that are maintained in a variety of forms, including

- Text
- Tables (e.g., track files, equipment characteristics, and logistical databases)
- Entity-relationship graphs (e.g., organizational charts, functional flow diagrams, explicit transactions, and deduced relationships)
- Maps (e.g., natural and cultural features)
- Images (e.g., optical, forward-looking infrared radar, and synthetic aperture radar)
- Three-dimensional (3D) physical models (e.g., terrain, buildings, and vehicles)

Perhaps the simplest data representation form is the *flat file*. Because it lacks organizational structure, access efficiency tends to be low. Database *indexing* seeks to overcome the inefficiency of the exhaustive search. A database index is analogous to a book index in the sense that it affords direct access to information. Just as a book might use multiple index dimensions, such as a subject index organized alphabetically and a figure index organized numerically, a DBMS can provide multiple, distinct indexes for data sets. Numerous data representation schemes exist, including hierarchical, network, and relational data models. Each of these models support some form of indexing.

Following the development of the *relational data model* in 1970, relational database management system (RDBMS) development experienced explosive growth for more than two decades. The relational model maintains data sets in tables. Each row of a table stores one occurrence of an entity, and columns maintain the values of an entity's attributes. To facilitate rapid search, tables can be indexed with respect to either a particular attribute or a linear combination of attributes. Multiple tables that share a primary *key* (a unique attribute) can be viewed together as a composite table (e.g., linking personnel data and corporate records through an employee's social security number). Because the relational model fundamentally supports only linear search dimensions, it affords inefficient access to data that exhibit significant dependencies across multiple dimensions. As a consequence, a traditional RDBMS tends to be suboptimal for managing 2D or 3D spatially organized information.

To overcome this limitation, commercial geographic information systems (GIS) were developed that offered direct support to the management and manipulation of spatially organized data. A GIS typically employs vector- and grid-based 2D representations of points, lines, and regions, as well as 3D representations of surfaces stored in triangulated irregular networks (TIN).* Some GIS support 3D spatial data structures such as octrees. As the utility of GIS systems became more evident, hybrid data management systems were built by combining a GIS and a RDBMS. Although well intentioned, such approaches to data management tended to be inefficient and difficult to maintain.

During the early 1990s, object-oriented reasoning became the dominant reasoning paradigm for large-scale software development programs. In this paradigm, objects

- Contain data, knowledge bases, and procedures
- Inherit properties from parent objects based on an explicitly represented object hierarchy
- Communicate with and control other objects

As a natural response to the widespread nature of the object-oriented development environment, numerous commercial object-oriented database management systems (OODBMS) were developed to offer enhanced support to object-oriented reasoning approaches. In general, OODBMS permit users to define new data types as needed (extensibility), while hiding implementation details from the user (encapsulation). In addition, such databases allow explicit relationships to be defined between objects. As a result, the OODBMS offers more flexible data structures and more *semantic expressiveness* than the strictly table-based relational data model.

To accommodate some of the desirable attributes of the object-oriented data model, numerous extensions to the relational data model were developed. The *object-relational* model eventually evolved into databases that support tables, large binary objects (such

---

* Although the discussions throughout this chapter focus on 2D spatial representations, the concepts also apply to 3D.

as images), vector-based spatial data, and entity-relationship-entity data (Resource Description Framework [RDF] standardized by the World Wide Web Consortium).

24.3 Spatial, Temporal, and Hierarchical Reasoning

To initiate the database requirements discussion, consider the following spatially oriented query:

Find all roads west of River 1, south of Road 10, and not in Forest 11.

Given the data representation depicted in Figure 24.1, humans can readily identify all roads that lie within the specified query window. However, if the data set were presented in a vector (i.e., tuple-based) form, considerable analysis would be required to answer the query. Thus, though the two representations might be technically *equivalent*, the representation in Figure 24.1 permits a human to readily perceive all relevant spatial relationships about the various features, whereas a vector representation does not. In addition, because humans can perceive a boundary-only representation of a region as representing the boundary *plus* the associated enclosed area, humans can perform a 2D set operation virtually *by inspection*. With a vector-based representation form, however, all spatial relationships among features must be discovered by computational means. In addition to internal database operations, data representation tends to dramatically impact the efficiency of machine-based reasoning.

To capitalize on a human's facility for spatial reasoning, military analysts historically plotted sensor reports and analysis products on clear acetates overlaid on appropriately scaled maps and map products. Soft-copy versions of this paradigm are still in widespread use. The highly intuitive nature of such data representations supports spatial *focus of attention*, effectively limiting the size of the search space. Only those acetates containing features relevant to a particular stage of analysis need be considered, leading to a potential reduction in the size of the search space. All analysis products generated are in a standard, fully registered form and, thus, are directly usable in subsequent analysis. With its many virtues, the acetate overlay-reasoning paradigm provides a convenient metaphor for studying machine-based approaches to spatial reasoning.

In the tactical situation awareness problem domain, spatially organized information can exhibit a dynamic, time-varying character. Thus, data fusion applications often involve

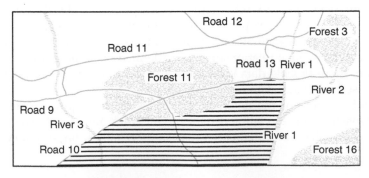

FIGURE 24.1
Two-dimensional map-based representation supporting search for all roads that meet a set of spatial constraints.

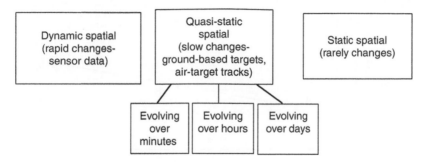

FIGURE 24.2
Three general classes of spatiotemporal data.

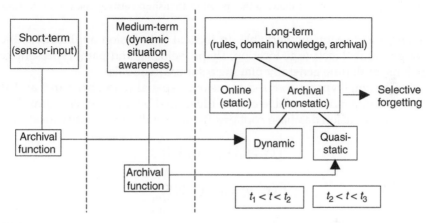

FIGURE 24.3
Possible implementation strategy for temporally sensitive spatial data.

combined spatial and temporal (spatial–temporal) representations. Thus, database searches must be capable of retrieving data that is close in both space and time to new sensor-derived information, for instance. If domain entity locations are maintained as discrete 3-tuples (x_i, y_i, t_i), indexing or sorting along any *individual* dimension is straightforward.* Although searching across multiple dependent search dimensions tends to be inefficient, modern GIS and DBMS provide indexed support to spatial–temporal data. Such support, however, represents somewhat myopic support to reasoning.

Rather than reviewing the large body of literature associated with spatial database design and spatiotemporal reasoning, this chapter focuses on a conceptual level view of the problem. Figure 24.2 offers a temporal sensitivity taxonomy for spatial data, and Figure 24.3 maps the three classes of temporal sensitivity to a human memory metaphor. Later discussions will focus on a potential implementation strategy based on Figure 24.3.

On the surface, treating time as an independent variable (or dimension) seems quite reasonable. However, treating time as an independent variable requires the spatial dimensions of a spatial–temporal representation to become the dependent dimensions. Because we seek a representation that supports object-oriented reasoning, and objects are typically associated with physical or aggregate entities that exist within a 2D or 3D *world*,

---

* To simplify the discussion, only 2D spatial data structures are addressed throughout this chapter. Extensions to 3D are straightforward.

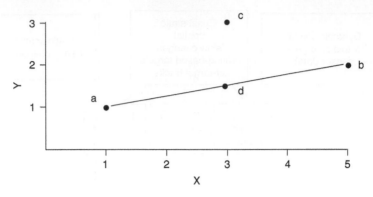

FIGURE 24.4
Two-dimensional depiction of data set showing the spatial relationships between the various features.

such an arrangement seems unworkable. In tactical data fusion applications, the spatial attributes of physical objects (e.g., location, shape, velocity, and path) represent key problem dimensions, with time generally playing a subordinate role.

The benefits of employing a *true n*-dimensional spatial data structure* and therefore treating time as a dependent variable can be illustrated by the following simple example. Suppose an existing fusion database contains the *temporally sorted* entity locations:

$$(x_a, y_a, t_a) = (1, 1, 1)$$

$$(x_c, y_c, t_c) = (3, 3, 5)$$

$$(x_b, y_b, t_b) = (5, 2, 10)$$

Assume a new sensor report, indicating a target detection at $(x_d, y_d, t_d) = (3, 1.5, 5)$, must be fused with the existing data set. Direct comparison of this new report with the existing database suggests that query point d is *near* point c (i.e., $t_c = t_d$) and *not near* either point a or b. If the database was sorted by all three dimensions, with the y-coordinate selected as the primary index dimension, point d would appear to be *near* point a and *not near* either point b or c. Suppose that the detections occurring at points a and b represent successive detections of the same entity (i.e., the endpoints of a track segment). As illustrated in Figure 24.4, which depicts the existing data set and all new detections in a *true* 2D spatial representation (rather than a list-oriented, tuple-based representation), query point d is readily discovered to be *spatially distant* from all three points, but *spatially close* to the line segment a–b.

To accommodate the temporal dimension, the distance between (x_d, y_d) and the *interpolated* target position along the line segment a–b (at time t_d) can be readily computed. If the target under track is known to be a wheeled vehicle moving along a road network, data associations can be based on a *road-constrained* trajectory, rather than a *linear* trajectory.

Although target reports typically occur at discrete times (e.g., at radar revisit rates), the trajectory of physical objects is a continuous function in both space and time. Despite the fact that the closest line segment to an arbitrary point in space can be derived from

---

* A structure that preserves both the natural spatial search dimensions and the true character of the data (e.g., area for image-like data and volume for 3D data).

tuple-based data, much more efficient data retrieval is possible if all spatial features are explicitly represented within the database. Thus, by employing a *time-coded*, *explicit* spatial representation (and not just preserving line segment endpoints), a database can effectively support a search along continuous dimensions in both space and time. With such a representation, candidate track segments can be found through highly localized (true 2D) spatial searches about point **d**.

Representing moving targets by continuous, time-referenced spatially organized trajectories not only matches the characteristics of physical phenomenon, but also supports highly effective database search dimensions. Equally important, such representations preserve spatial relationships between dynamic objects and other static and dynamic spatially organized database entities. Thus, there exist multiple justifications for subordinating temporal indexing to spatial indexing.

Both semantic and spatial reasoning are intrinsically hierarchical. Semantic object reasoning relies heavily on explicitly represented object hierarchies; similarly, spatial reasoning can benefit from multiple-resolution spatial representations. Time is a monotonically increasing function; therefore, temporal phenomena possess a natural sequential ordering, which can be treated as a specialized hierarchical representation. These considerations attest to the appropriateness of subordinating hierarchical reasoning to both semantic and spatial reasoning and, in turn, subordinating temporal reasoning to hierarchical reasoning.

Figure 24.5 depicts the hierarchical relationship that exists between the three reasoning classes. At the highest level of abstraction, reasoning can be treated as either *spatial* or *semantic* (nonspatial). Each of these reasoning classes can be handled using either *hierarchical* or *nonhierarchical* approaches. Hierarchical spatial reasoning employs *multiple-resolution spatial* representations, and hierarchical nonspatial reasoning uses *tree-structured semantic* representations. Each of these classes, in turn, may or may not be temporally sensitive.

If we refer to *temporal reasoning* as *dynamic reasoning* and *nontemporal reasoning* as *static reasoning*, there exist four classes of *spatial reasoning* and four classes of *nonspatial reasoning*: (1) dynamic, hierarchical; (2) static, hierarchical; (3) dynamic, nonhierarchical; and (4) static, nonhierarchical. To effectively support data fusion automation, a database must accommodate each of these reasoning classes. Because nonhierarchical reasoning is a special case of hierarchical reasoning, and static reasoning is a special case of dynamic reasoning, data fusion applications are adequately served by a DBMS that provides *dynamic, hierarchical spatial representations* and *dynamic, hierarchical semantic representations*. Thus, supporting these two key reasoning classes represents the primary database design criterion.

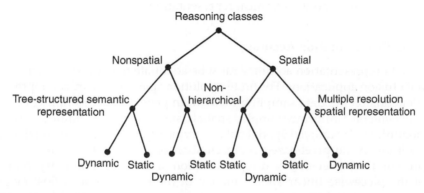

FIGURE 24.5
Taxonomy of the three principal reasoning classes.

24.4 Database Design Criteria

This section addresses key database design criteria to support advanced algorithm development. Data representation can have a profound impact on algorithm development. The preceding discussion regarding spatial, temporal, and hierarchical reasoning highlighted the benefit of representations that provide natural search dimensions and preserve significant relationships among domain objects. In addition to supporting key database operations, such as search efficiency, intuitive data representations facilitate the development of sophisticated algorithms that emulate the top-down reasoning process of human analysts.

24.4.1 Intuitive Algorithm Development

In analyzing personnel data, for example, table-based representations are much more natural (and useful) than tree-based data structures. Hierarchical data structures are generally more appropriate for representing the organization of a large company than a spatially organized representation. Raster representations of imagery tend to support more intuitive image processing algorithms than do vector-based representations of point, line, and region boundary-based features.

The development of sophisticated machine-based reasoning benefits from full integration of both the semantic and the spatial data representations, as well as integration among multiple representations of a given data type. For example, a river possesses both semantic attributes (class, nominal width, and flow-rate) and spatial attributes (beginning location, ending location, shape description, and depth profile). At a reasonably low resolution, a river's shape might be best represented as a *lineal* feature; at a higher resolution, a *region-based* representation is likely more appropriate.

24.4.2 Efficient Algorithm Performance

For algorithms that require large supporting databases, data representation form can significantly affect algorithm performance efficiency. For business applications that demand access to massive table-based data sets, the relational data model supports more efficient algorithm performance than a semantic network model. Two-dimensional template matching algorithm efficiency tends to be higher if true 2D (map-like) spatial representations are used rather than vector-based data structures. Multiple level-of-abstraction semantic representations and multiple-resolution spatial representations tend to support more efficient problem solving than do nonhierarchical representations.

24.4.3 Data Representation Accuracy

In general, data representation accuracy must be adequate to support the widest possible range of data fusion applications. For finite resolution spatially organized representations, accuracy depends on the data sampling method. In general, data sampling can be either *uniform* or *nonuniform*. Uniformly sampled spatial data are typically maintained as integers, whereas nonuniformly sampled spatial data are typically represented using floating-point numbers. Although the pixel-based representation of a region boundary is an integer-based representation, a vector-based representation of the same boundary maintains the vertices of the piecewise linear approximation of the boundary as a floating-point list. For a given memory size, nonuniformly sampled representations tend to provide higher accuracy than uniformly sampled representations.

24.4.4   Database Performance Efficiency

Algorithm performance efficiency relies on the six key database efficiency classes described in the following paragraphs.

24.4.4.1   *Storage Efficiency*

Storage efficiency refers to the relative storage requirements among alternative data representations. Figure 24.6 depicts two similar polygons and their associated vector, raster, pixel boundary, and quadtree representations. With a raster representation (Figures 24.6c and 24.6d), A/Δ^2 nodes are required to store the region, where A is the area of the region, and Δ is the spatial width of a (square) resolution cell (or pixel). Accurate replication of the region shown in Figure 24.6a requires a resolution cell size four times smaller than that required to replicate the region in Figure 24.6b. Because the pixel boundary representation

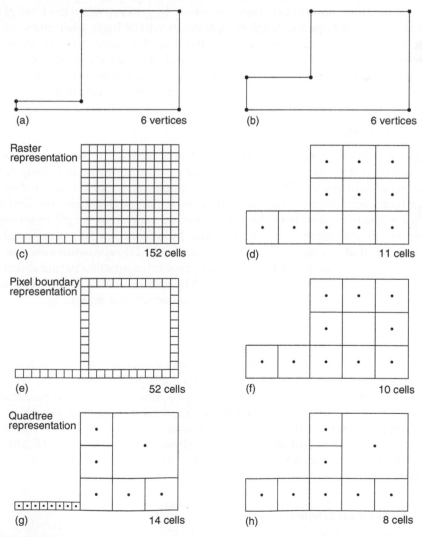

FIGURE 24.6
Storage requirements for four spatial data representations of two similarly shaped regions.

(Figures 24.6e and 24.6f) maintains only the boundary nodes of the region, the required node count is proportional to P/Δ, where P is the perimeter of the region. For fixed Δ, the ratio of the node count for the raster representation relative to the pixel boundary representation is $A/(\Delta P)$. Although a quadtree representation (Figures 24.6g and 24.6h) stores both the region boundary and its interior, the required node count, for large regions, is proportional to the region perimeter rather than its area.[1] Thus, for grid-based spatial decompositions, the region node count depends on the maximum resolution cell size and the overall size of the region (either its area or perimeter).

Whereas storage requirements for grid-based representations tend to be dependent on the size of the region, storage requirements for nonuniform decompositions are sensitive to the small-scale feature details. Consequently, for nonuniform representations, no simple relationship exists between a region's size and its storage requirements. For example, two regions of the same physical size possessing nearly identical uniform decomposition storage requirements might have dramatically different storage requirements under a nonuniform decomposition representation. Vector-based polygon representations are perhaps the most common nonuniform decomposition. In general, when the level of detail in the region boundary is high, the polygon tuple count will be high; when the level of detail is low, few vertices are required to represent the boundary accurately. Because the level of detail of the boundaries in both Figures 24.6a and 24.6b are identical, both require the same number of vertices. In general, for a given accuracy, nonuniform decompositions have significantly lower storage requirements than uniform decompositions.

24.4.4.2 Search Efficiency

Achieving search efficiency requires effective control of the search space size for the range of typical database queries. In general, search efficiency can be improved by storing or indexing data sets along effective search dimensions. For example, for vector-represented spatial data sets, associating a bounding box (i.e., vertices of the rectangle that forms the smallest outer boundary of a line or region) with the representation provides a simple indexing scheme that affords significant search reduction potential. Many kinds of data possess natural representation forms that, if preserved, facilitate the search. 2D representations that preserve the essential character of maps, images, and topographic information permit direct access to data along very natural 2D search dimensions. Multiple-resolution spatial representations potentially support highly efficient top-down spatial search and reasoning.

24.4.4.3 Overhead Efficiency

Database overhead efficiency includes both *indexing* efficiency and *data maintenance* efficiency. Indexing efficiency refers to the cost of creating data set indices, whereas data maintenance efficiency refers to the efficiency of re-indexing and reorganization operations, including tree-balancing required following data insertions or deletions. Because natural data representations do not require separate indexing structures, such representations tend to support overhead efficiency. Although relatively insignificant for static databases, database maintenance efficiency can become a significant factor in highly dynamic data sets.

24.4.4.4 Association Efficiency

Association efficiency refers to the efficient determination of relationships among data sets (e.g., inclusion, proximity). *Natural* data representations tend to significantly enhance association efficiency over vector-based spatial representations because they tend to pre-serve the inherent organizational characteristics of data. Although relational database tables can be joined (via a static, single-dimension explicit key), the basic relational model does not support efficient data set association for data that possess correlated attributes (e.g., combined spatial and temporal proximity).

24.4.4.5 Complex Query Efficiency

Complex query efficiency includes both set operation efficiency and complex clause evaluation efficiency. *Set operation efficiency* demands efficient Boolean and fuzzy set operations among point, line, and region features. *Complex clause evaluation efficiency* requires query optimization for compound query clauses, including those with mixed spatial and semantic constraints.

24.4.4.6 Implementation Efficiency

Implementation efficiency is enhanced by a database architecture and associated data structures that support the effective distribution of data, processing, and control. Figure 24.7 summarizes these 12 key design considerations.

24.4.5 Spatial Data Representation Characteristics

Many spatial data structures and numerous variants exist. The taxonomy depicted in Figure 24.8 provides an organizational structure that is useful for comparing and contrast-ing spatial data structures. At the highest level of abstraction, sampled representations of 2D space can employ either uniform (regular) or nonuniform (nonregular) decompositions. Uniform decompositions generate *data-independent* representations, whereas nonuniform decompositions produce *data-dependent* representations.

With certain exceptions (e.g., fractal-like data), low-resolution representations of spatial features tend to be uniformly distributed and high-resolution representations of spatial data tend to be nonuniformly distributed. Consequently, uniform decompositions tend to

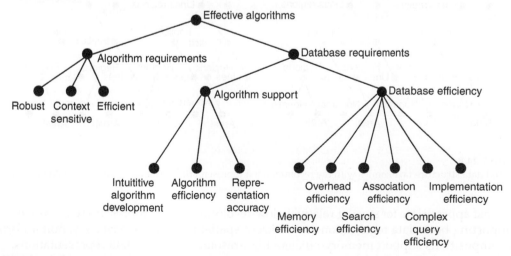

FIGURE 24.7
High-level summary of the database design criteria.

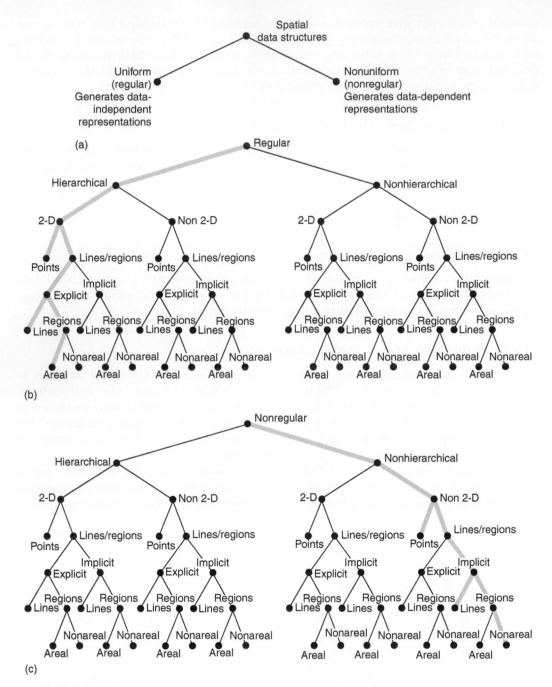

FIGURE 24.8
Spatial data structure taxonomies with the recommended representation branches outlined in bold gray lines.

be most appropriate for storing relatively low-resolution spatial representations, as well as supporting both data registration and efficient spatial indexing. Conversely, nonuniform decompositions support memory-efficient high-resolution spatial data representations.

Hierarchical decompositions of 2D spatial data support the continuum from highly global to highly local spatial reasoning. Global analysis typically begins with relatively low-resolution data representations and relatively global constraints. The resulting analysis products are then progressively refined through higher-resolution analysis and the application of successively more local constraints. For example, a U.S. Interstate road map provides a low-resolution representation that supports the first stage of a route-planning strategy. Once a coarse resolution plan is established, various state, county, and city maps can be used to refine appropriate portions of the plan to support excursions from the Interstate highway to visit tourist attractions or to find lodging and restaurants. Humans routinely employ such top-down reasoning, permitting relatively global (often near-optimal) solutions to a wide range of complex tasks.

Data decompositions that preserve the inherent 2D character of spatial data are defined to be *2D data structures*. 2D data structures provide natural spatial search dimensions, eliminating the need to maintain separate search indices. In addition, such representations preserve Euclidean distance metrics and other spatial relationships. Thus, a raster representation is considered a *2D data structure*, whereas tuple-based representations of lines and region boundaries are *non-2D data structures*.

Explicit data representations literally depict a set of data, whereas *implicit* data representations maintain information that permits reconstruction of the data set. Thus, a raster representation is an explicit data representation because it explicitly depicts line and region features. A vector representation, which maintains only the endpoint pairs of piecewise continuous lines and region boundaries, is considered an implicit representation. In general, implicit representations tend to be more memory-efficient than explicit representations.

Specific feature spatial representations maintain individual point, line, and region features, and *composite feature* spatial representations store the presence or absence of multiple features. Specific spatial representations are most effective for performing spatial operations with respect to individual features; composite representations are most effective for representing classes of spatial features.

Areal-based representations maintain a region's boundary and interior. *Nonareal-based* representations explicitly store only a region's boundary. Boolean set operations among 2D spatially organized data sets inherently involve both boundaries and region interiors. As a result, areal-based representations tend to support more efficient set operation generation than nonareal-based representations among all classes of spatial features.

Data that are both regular and hierarchical support efficient tree-oriented spatial search. Regular decompositions utilize a fixed grid size at each decomposition level and provide a fixed resolution relationship between levels, enabling registered data sets to be readily associated at all resolution levels. Registered data representations that are both regular and areal-based support efficient set operations among all classes of 2D spatial features. Because raster representations are both regular and areal-based, set intersection generation requires only simple Boolean AND operations between respective raster cells. Similarly, set union can be computed as the Boolean OR between the respective cells.

Spatial representations that are both regular and 2D preserve spatial adjacency among all classes of spatial features. They employ a data-independent, regular decomposition and, therefore, do not require extensive database rebuilding operations following insertions and deletions. In addition, because no separate spatial index is required, re-indexing operations are unnecessary. Spatially local changes tend to have highly localized effects on the representation; as a result, spatial decompositions that are both regular and 2D are relatively dynamic data structures. Consequently, database maintenance efficiency is generally enhanced by regular 2D spatial representations.

Data fusion involves the composition of both dynamic (sensor data, existing fusion products) and static (tables of equipment, natural and cultural features) data sets; consequently, data fusion algorithm efficiency is enhanced by employing compatible data representation forms for both dynamic and static data. For example, suppose the (static) road network was maintained in a nonregular vector-based representation, whereas the (dynamic) target track file database was maintained in a regular 2D data structure. The disparate nature of the representations would make the fusion of these data sets both cumbersome and inefficient. Thus, maintaining static and dynamic information with identical spatial data structures potentially enhances both database *association efficiency* and *maintenance efficiency*.

Spatial representations that are *explicit, regular, areal,* and 2D support efficient set operation generation, offer natural search dimensions for fully registered spatial data, and represent relatively dynamic data structures. Because they possess key characteristics of analog 2D spatial representations (e.g., paper maps and images), the representations are defined to be *true 2D* spatial representations. Spatial representations possessing one or more of the following properties violate the requirements of true 2D spatial representations:

- Spatial data is not stored along 2D search dimensions (i.e., spatial data stored as list or table-based representations).
- Region representations are nonareal-based (i.e., bounding polygons).
- Nonregular decompositions are employed (e.g., vector representations and R-trees).

True 2D spatial representations tend to support both intuitive algorithm development and efficient spatial reasoning. Such representations preserve key 2D spatial relationships, thereby supporting highly intuitive spatial search operations (e.g., northwest of point A, beyond the city limits, or inside a staging area). Because grid-based representations store both the boundary and the interior of a region, intersection and union operations among such data can be generated using straightforward cell-by-cell operations. The associated computational requirements scale linearly with data set size independent of the complexity of the region data sets (e.g., embedded holes or multiple disjoint regions). Conversely, set operation generation among regions based on non-true-2D spatial representations tend to require combinatorial computational requirements.

Computational geometry approaches to set intersection generation require the determination of all line segment intersection points; therefore, computational complexity is of the order $m \times n$, where m and n are the number of vertices associated with the two polygons. When one or both regions contain embedded holes, dramatic increases in computational complexity can result. For analysis and fusion algorithms that rely on set intersections among large complex regions, algorithm performance can be adversely impacted. Although the use of bounding rectangles (bounding boxes) can significantly enhance search and problem-solving efficiency for vector-represented spatial features, they provide limited benefit for multiple-connected regions, extended lineal features (e.g., roads, rivers, and topographic contour lines), and directional and proximity-based searches.

Specific spatial representations that are explicit, regular, areal, 2D, and hierarchical (or hierarchical true 2D) support highly efficient top-down spatial search and top-down areal-based set operations among specific spatial features. With a quadtree-based region representation, for example, set intersection can be performed using an efficient top-down, multiple-resolution process that capitalizes on two key characteristics of Boolean set operations. First, the intersection product is a proper subset of the smallest region being intersected. Second, the intersection product is both *commutative* and *associative* and, therefore, independent of the order in which the regions are composed. Thus, rather than exhaustively ANDing all nodes in all regions, intersection generation can be reformulated as a search problem where the smallest region is selected as a variable resolution spatial search *window* for interrogating the representation of the next larger region.[1] Nodes in the second smallest region that are determined to be within this search window are known to be in the intersection product. Consequently, the balance of the nodes in the larger region need not be tested.

If three or more regions are intersected, the product that results from the previous intersection product becomes the search window for interrogating the next larger region. Consequently, whereas the computational complexity of polygon intersection is a function of the product of the number of vertices in each polygon, the computational complexity for region intersection using a regular grid-based representation is related to the number of tuples that occur within cells of the physically smallest region. Composite spatial representations that are explicit, regular, areal, 2D, and hierarchical support efficient top-down spatial search among classes of spatial features.

Based on the relationships between spatial data decomposition classes and key database efficiency classes, a number of generalizations can be formulated:

- Hierarchical, true 2D spatial representations support the development of highly intuitive algorithms.
- For tasks that require relatively low-resolution spatial reasoning, algorithm efficiency is enhanced by hierarchical, true 2D spatial representations.
- Implicit, nonregular spatial representations support representation accuracy.
- Specific spatial representations are most appropriate for reasoning about specific spatial features; conversely, composite spatial representations are preferable for reasoning about classes of spatial features. For example, a true 2D composite road network representation would support more efficient determination of the closest road to a given point in space than a representation maintaining individual named roads. With the latter representation, a sizable portion of the road database would have to be interrogated.
- Finite resolution data decompositions that are implicit, nonregular, nonhierarchical, nonareal, and non-2D tend to be highly storage-efficient.
- Spatial indexing efficiency is enhanced by hierarchical, true 2D representations that support natural, top-down search dimensions.
- Spatially local changes require only relatively local changes to the underlying representation; therefore, database maintenance efficiency is enhanced by regular, dynamic, 2D spatial representations.
- For individual spatial features, database search efficiency is enhanced by spatial representations that are specific, hierarchical, and true 2D, and that support distributed search.
- Search efficiency for classes of spatial features is enhanced by composite hierarchical spatial feature representations.
- True 2D spatial representations preserve spatial relationships among data sets; therefore, for individual spatial features, database association efficiency is enhanced by hierarchical true 2D spatial representations of specific spatial features. For classes of spatial features, association efficiency is enhanced by hierarchical, true 2D composite feature representations.

- For specific spatial features, complex query efficiency is enhanced by hierarchical, true 2D representations of individual features; for classes of spatial features, complex query efficiency is enhanced by hierarchical, composite, true 2D spatial representations.
- Finally, database implementation efficiency is enhanced by data structures that support the distribution of data, processing, and control.

These general observations are summarized in Table 24.1.

24.4.6 Database Design Tradeoffs

Object-oriented reasoning potentially supports the construction of robust, context-sensitive fusion algorithms, enabling data fusion automation to benefit from the development of an effective OODBMS. This section explores design tradeoffs for an object-oriented database that seek to achieve an effective compromise among the algorithm support and database efficiency issues listed in Table 24.1. As previously discussed, the principal database design requirement is support for dynamic, hierarchical spatial reasoning and dynamic, hierarchical semantic reasoning. Consequently, an optimal database must provide data structures that facilitate storage and access to both temporally and hierarchically organized spatial and nonspatial information.

TABLE 24.1

Spatial Representation Attributes Supporting Nine Spatial Database Requirements

Spatial Database Requirements	Regular	Nonregular	Hierarchical	Nonhierarchical	Areal	Nonareal	2D	Non-2D	Static Structures	Dynamic Data Structures	Distributed	Nondistributed	Explicit Feature Representation	Implicit Feature Representation	Specific-Feature Based	Composite-Feature Based
Intuitive algorithm development	•		•		•		•						•		•	•
Spatial reasoning algorithm efficiency																
Individual features	•		•		•		•						•		•	
Feature classes	•		•		•		•						•			•
Representation accuracy		•						•						•		
DB storage efficiency		•	•			•	•							•		
DB overhead efficiency																
Indexing	•		•		•		•						•		•	
Maintenance	•						•			•			•		•	
DB search efficiency																
Individual features	•	•	•	•						•			•		•	
Feature classes	•	•	•	•						•			•			•
DB association efficiency																
Individual features	•	•	•	•						•			•		•	
Feature classes	•	•	•	•						•			•			•
DB query efficiency																
Individual features	•	•	•	•							•		•		•	
Feature classes	•	•	•	•							•		•			•
DB implementation efficiency											•					

Nonspatial (or semantic) declarative knowledge can be represented as n-tuples, arrays, tables, transfer functions, frames, trees, and graphs. Modern semantic object databases provide effective support to complex, multiple level-of-abstraction problems that (1) possess extensive parent–child relationships, (2) benefit from problem decomposition, or (3) demand global solutions. Because object-oriented representations permit the use of very general internal data structures, and the associated reasoning paradigm fully embraces hierarchical representations at the semantic object level, conventional object databases intrinsically support hierarchical semantic reasoning. Semantic objects can be considered relatively dynamic data structures because temporal changes associated with a specific object tend to affect only that object or closely related objects.

The character and capabilities of a *spatial object* database are analogous to those of a semantic object database. A spatial object database must support top-down, multiple level-of-abstraction (i.e., multiple resolution) reasoning with respect to classes of spatial objects, as well as permit efficient reasoning with specific spatial objects. Just as the semantic object paradigm requires an explicitly represented semantic object hierarchy, a spatial object database requires an equivalent spatial object hierarchy. Finally, just as specific entities in conventional semantic object databases possess individual semantic object representations, specific spatial objects require individual spatial object representations.

24.5 Object Representation of Space

Consider the query:

Determine the class-1 road closest to query point (x_1, y_1).

In a database that maintains only implicit representations of individual spatial features, this *feature-class query* could potentially require the interrogation of all class-1 road representations. Just as a hierarchical representation of semantic objects permits efficient class-oriented queries, a hierarchical representation of space supports efficient queries with respect to classes of spatial objects. At the highest level of abstraction, an *object representation of space* consists of a single object that characterizes the entire area of interest (Asia, a single map sheet, a division's area of interest). At each successively lower level of the spatial object hierarchy, space is decomposed into progressively smaller regions that identify spatial objects (specific entities or entity classes) associated with that region. Just as higher-order semantic objects possess more general properties than their offspring, higher-order object representations of space characterize the general properties of 2D space.

Based on the principles outlined in the last section, an object representation of 2D space must satisfy the properties summarized in Table 24.2. As previously mentioned, a true 2D spatial representation possesses the first five characteristics listed in Table 24.2. With the addition of the sixth property, the spatial object hierarchy provides a uniform hierarchical spatial representation for classes of point, line, and region features. The *pyramid data structure* fully satisfies these first six properties. Because multiple classes of spatial objects (e.g., roads, waterways, and soil type) can be maintained within each cell, a pyramid representation is well suited to maintain a composite feature representation. In a pyramid representation, all the following are true:

1. Spatially local changes require only relatively local changes to the data structure.
2. Limited re-indexing is required following updates.
3. Extensive tree-balancing operations are not required following insertions or deletions.

TABLE 24.2

Key Representation Characteristics Required by the Three Distinct Spatial Data Representation Classes

	Key Spatial Data Types		
Representation Characteristics	Object Representation of 2D Space	Low-Resolution Spatial Representation	High-Resolution Spatial Resolution
1 Finite resolution	•	•	•
2 Regular/nonregular	Regular	Regular	Nonregular
3 Areal/nonareal	Areal	Areal	Nonareal
4 2D/non-2D	2D	2D	Non-2D
5 Explicit/implicit	Explicit	Explicit	Implicit
6 Hierarchical/nonhierarchical	Hierarchical	Hierarchical	Nonhierarchical
7 Specific features/composite features	Composite features	Specific features	Specific features
8 Relatively dynamic data structure	•	•	•
9 Distributed representation potential	•	•	•
10 Low/high precision	Low	Low	High

As a result, the pyramid is a relatively *dynamic data structure*. In addition, the hierarchical and grid-based character of the pyramid data structure enables it to readily accommodate data distribution. Therefore, a pyramid data structure fully satisfies all nine requirements for an object representation of 2D space.

24.5.1   Low-Resolution Spatial Representation

As summarized in Table 24.2, an effective low-resolution spatial representation must possess 10 key properties. With the exception of the composite feature representation property, the low-resolution spatial representation requirements are identical to those of the object representation of 2D space. Whereas a composite-feature-based representation supports efficient spatial search with respect to classes of spatial features, a specific feature-based representation supports effective search and manipulation of specific point, line, and region features. A regular region quadtree possesses all of the properties presented in Table 24.2, column 3.

24.5.2   High-Resolution Spatial Representation

An effective high-resolution spatial representation must possess the 10 properties indicated in the last column of Table 24.2. Vector-based spatial representations clearly meet the first four criteria. Because they use nonhierarchical representations of specific features and employ implicit piecewise linear representations of lines and region boundaries, vector representations also satisfy properties 5–7. Changes to a feature require a modification of the property list of a single feature; therefore, vector representations tend to be relatively dynamic data structures. A representation of individual features is self-contained, so vector representations can be processed in a highly distributed manner. Finally, vector-based representations inherently provide high precision. Thus, vector-based representations satisfy all of the requirements of the high-resolution spatial representation.

Feature	Hybrid representation	Example
Point	Node location and within node offset	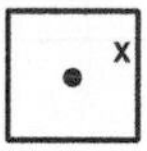
Line	Node location of each associated node and list of vertices (node offset form) of piecewise line segments within each node	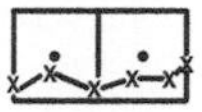
Region	List of minimal region quadtree nodes; line feature representation within all boundary leaf nodes	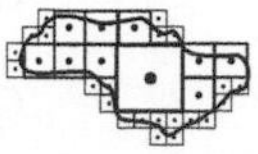

FIGURE 24.9
Quadtree-indexed vector spatial representation for points, lines, and regions.

24.5.3 Hybrid Spatial Feature Representation

Traditionally, spatial database design has involved the selection of either a single spatial data structure or two or more alternative, but substantially independent, data structures (e.g., vector and quadtree). Because of the additional degrees of freedom it adds to the design process, the use of a hybrid spatial representation offers the potential for achieving a near-optimal compromise across the spectrum of design requirements. Perhaps the most straightforward design approach for such a hybrid data structure is to directly integrate a multiple-resolution, low-resolution spatial representation and a memory-efficient, high-resolution spatial representation.

As indicated in Figure 24.9, the quadtree data structure can form the basis of an effective hybrid data structure, serving the role of both a low-resolution spatial data representation and an efficient spatial index into high-accuracy vector representations of point, line, and region boundaries. Table 24.3 offers a coarse-grain evaluation of the effectiveness of the vector, raster, pixel boundary, region quadtree, and the recommended hybrid spatial representation based on the database design criteria. Table 24.4 summarizes the key characteristics of the recommended spatial object representation and demonstrates that it addresses the full spectrum of spatial data design issues.

24.6 Integrated Spatial/Nonspatial Data Representation

To effectively support data fusion applications, spatial and nonspatial data classes must be fully integrated. Figure 24.10 depicts a high-level view of the resultant semantic/spatial database kernel depicting both explicit and implicit links between the various data structures. Because a pyramid data structure can be viewed as a *complete quadtree*, the pyramid and the low-resolution spatial representation offer a unified structure, with the latter

TABLE 24.3

Comparison of Spatial Data Representations Based on Their Ability to Support Database Design Criteria

Design Criteria	Representations				
	Vector-Based[a]	Raster-Based	Pixel Boundary	Region Quadtree	Hybrid Spatial Representation
Intuitive algorithm development	Poor	Good	Moderate	Good	Good
Computationally efficient algorithms	Poor	Moderate	Poor	Good	Good
Representation accuracy	Good	Poor	Poor	Poor	Good
Data storage efficiency	Good	Poor	Moderate	Moderate	Moderate
DB overhead efficiency	Good	Good	Good	Moderate	Moderate
Spatial search efficiency					
Specific features	Moderate	Good	Moderate	Good	Good
Feature classes	Poor	Good	Poor	Moderate	Good
Complex Boolean query efficiency					
Specific features	Moderate	Good	Moderate	Good	Good
Feature classes	Poor	Moderate	Poor	Moderate	Good
DB implementation efficiency	Good	Good	Good	Good	Good

[a] Assumes the use of bounding bo xes.

TABLE 24.4

Summary Characteristics of the Three Spatial Representation Classes

Database Element	Spatial Representation Attributes											
	Regular	Nonregular	Hierarchical	Nonhierarchical	Areal	Nonareal	2D	Non-2D	Explicit	Implicit	Specific	Composite
Object representation of 2D space (pyramid)	•		•		•		•		•			•
Hybrid representation of specific object												
Low-resolution index	•		•		•		•			•	•	
High-resolution representation		•		•		•		•	•	•	•	
Composite representation: pyramid and hybrid spatial representation	•	•	•	•	•	•	•	•	•	•	•	•

effectively an extension to the former. Therefore, the quadtree data structure serves as the link between the pyramid and the individual spatial feature representations and provides a hierarchical spatial index to high-resolution vector-represented point, line, and region features.

The integrated spatial and semantic object representation permits efficient top-down search for domain objects based on a combination of semantic and spatial constraints. The

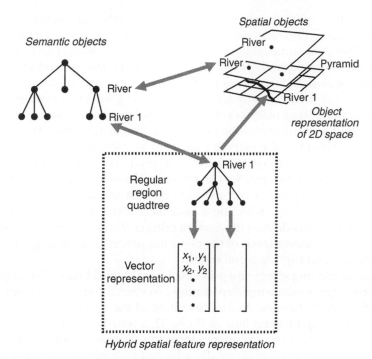

FIGURE 24.10
High-level view of the integrated semantic/spatial object database.

fully integrated data structures support efficient and effective manipulation of spatial and nonspatial relations, and the computation of spatial and semantic distance metrics among domain objects. The quadtree-indexed data structure preserves the precision of the highest-fidelity spatial data without compromising the memory efficiency of the overall system.

24.7 Sample Application

This section illustrates key benefits of the proposed database system design by applying the concepts to a simple, easily visualized path-planning application. Humans routinely plan simple tasks in a nearly subconscious manner (e.g., walking from one room to another and picking up a glass of water); however, other tasks require considerable deliberation. Consider, for instance, planning an automobile trip from Washington, District of Columbia to Portland, Oregon. Adding constraints to the problem, such as stops at nearby national parks and other attractions along the way, can significantly complicate the task of generating an optimal plan. One approach to planning such a trip would be to accumulate all the county road maps through which such a route might pass; however, the amount of data would be overwhelming. The task would be much less daunting if the planner first acquired a map of the U.S. Interstate highway system used this map to select a coarse resolution plan. Then, a small collection of state maps would probably suffice for planning routes to specific attractions located some distance from an interstate highway. Finally, for locating a friend's residence, a particular historical landmark, or some other point of interest, detailed county or even city maps could be used to refine the details of the overall plan.

Clearly, the top-down path development approach is both more natural and much more efficient than a single level-of-abstraction approach, exploiting the continuum from global to highly local objectives and evaluation metrics. Given a time- and resource-constrained vacation, the relative importance of competing objectives might need to be weighted to generate something close to an optimal plan.

In path planning, as with virtually all decision making, effective evaluation criteria are needed to select among large numbers of candidate solutions. Path selection metrics could be as simple as seeking to minimize the overall path length, minimize travel time, or maximize travel speed. However, the planner may want to minimize or maximize a weighted combination of several individual metrics or to apply fuzzier (more subjective) measures of performance. Whereas crisp metrics can be treated as *absolute* metrics (e.g., the shortest path length or the fastest speed), fuzzy metrics are most appropriate for representing *relative* metrics or constraints (e.g., selecting a *moderately* scenic route that is *not much longer* than the shortest route). In addition to selection criteria, various problem constraints could be appropriate at the different levels of the planning process. These might include regular stops for fuel, meals, and nightly motel stays.

In addition to developing effective *a priori* plans, real-world tasks often require the ability to perform real-time replanning. Replanning can span the spectrum from global modifications to highly local changes. As an illustration of the latter, suppose a detailed plan called for travel in the right-hand lane of Route 32, a four-lane divided highway. During plan execution, an accident is encountered blocking further movement in the rightmost lane requiring the initial plan to be locally adjusted to avoid this impasse. More global modifications would be required if, before reaching Route 32, the driver learned of a recent water main break that was expected to keep the road closed to traffic the entire day.

In general, top-down planning begins by establishing a coarse-grain skeletal solution, which is then recursively refined to the appropriate level of detail. Multiple level-of-abstraction planning tends to minimize the size of both the search and decision spaces, making it highly effective. Based on both its efficacy and efficiency, top-down planning readily accommodates replanning, a critical component in complex, dynamic problem domains.

24.7.1 Problem-Solving Approach

This section outlines a hierarchical path-planning algorithm that seeks to develop one or more routes between two selected locations for a wheeled military vehicle capable of both *on-road* and *off-road* movement. Rather than seeking a mathematically optimal solution, we strive for a near-optimal global solution by emulating a human-oriented approach to path selection and evaluation.

In general, path selection is potentially sensitive to a wide range of geographical features (i.e., terrain, elevation changes, soil type, surface roughness, and vegetation), environmental features (i.e., temperature, time of day, precipitation rate, visibility, snow depth, and wind speed), and cultural features (i.e., roads, bridges, the location of supply caches, and cities). Ordering the applicable domain constraints from the *most* to the *least* significant effectively treats path development as a top-down, constraint-satisfaction task.

In general, achieving *high quality* global solutions requires that global constraints be applied before more local constraints. Thus, in stage 1 (the highest level of abstraction), path development focuses on route development, considering only *extended barriers* to ground travel. Extended barriers are those cultural and man-made features that cannot be readily circumnavigated. As a natural consequence, they tend to have a more profound impact on route selection than smaller-scale features. Examples of extended features include rivers, canyons, neutral zones, ridges, and the Great Wall in China. Thus, in terms of the supporting database, first-stage analysis involves searching for extended barriers that lie between the path's *start* and *goal* state.

If an extended barrier is discovered, candidate barrier-crossing locations (e.g., bridges or fording locations for a river or passes for a mountain range) must be sought. One or more of these barrier-crossing sites must be selected as candidate high-level path subgoals. As with all stages of the analysis, an evaluation metric is required to either *prune* or rank candidate subgoals. For example, the selected metric might retain the n closest bridges possessing adequate weight-carrying capacity. For each subgoal that satisfies the selection metric, the high-level barrier-crossing strategy is reapplied, searching for extended barriers and locating candidate barrier-crossing options until the final goal state is reached. The product of stage 1 analysis is a set of coarse resolution candidate paths represented as a set of subgoals that satisfy the high-level path evaluation metrics and global domain constraints.

Because the route is planned for a wheeled vehicle, the existing road network has the next most significant impact on route selection. Thus, in stage 2, road connectivity is established between all subgoal pairs identified in stage 1. For example, the shortest road path or the shortest m road paths are generated for those subgoals discovered to be *on* or *near* a road. When the subgoal is not near a road or when path evaluation indicates that all candidate paths provide only low-confidence solutions, overland travel between the subgoals would be indicated. Upon completion of stage 2 analysis, the coarse resolution path sets that have been developed during stage 1 will be refined with the appropriate road-following segments.

In stage 3, overland paths are developed between all subgoal pairs not already linked by high-confidence, road-following paths. Whereas *extended barriers* were associated with the most global constraints on mobility, *nonextended barriers* (e.g., hills, small lakes, drainage ditches, fenced fields, or forests) represent the most significant mobility constraints for overland travel. Ordering relevant nonextended barriers from stronger constraints (i.e., larger barriers and no-go regions) to weaker constraints (i.e., smaller barriers and slow-go regions) will extend the top-down analysis process. At each successive level of refinement, a selection metric, sensitive to progressively more local path evaluation constraints, is applied to the candidate path sets. The path refinement process terminates when one or more candidate paths have been generated that satisfy all path evaluation constraints. Individual paths are then rank-ordered against selected evaluation metrics.

Whereas traditional path development algorithms generate plans based on brute force optimization by minimizing *path resistance* or other similar metric, the hierarchical constraint-satisfaction-based approach just outlined emulates a more human-like approach to path development. Rather than using simple, single level-of-abstraction evaluation metrics (path resistance minimization), the proposed approach supports more powerful reasoning, including concatenated metrics (e.g., *maximal concealment from one or more vantage points* plus *minimal travel time to a goal state*). A path that meets both of these requirements might consist of a set of road segments not visible from specified vantage points, as well as high-mobility off-road path segments for those sections of the roadway that are visible from those vantage points. Hierarchical constraint-based reasoning captures the character of human problem-solving approaches, achieving the spectrum from global to more local subgoals, producing intuitively satisfying solutions. In addition, top-down, recursive refinement tends to be more efficient than approaches that attempt to directly generate high-resolution solutions.

24.7.2   Detailed Example

This section uses a detailed example of the top-down path-planning process to illustrate the potential benefits of the integrated semantic and spatial database discussed in Section 24.6. Because the database provides both natural and efficient access to both hierarchical semantic information and multiple-resolution spatial data, it is well suited to problems that are best treated at multiple levels of abstraction. The tight integration between semantic and spatial representation allows effective control of both the search space and the solution set size.

The posed problem is to determine one or more *good* routes for a wheeled vehicle from the start to the goal state depicted in Figure 24.11. Stage 1 begins by performing a spatially anchored search (i.e., anchored by both the start and goal states) for extended mobility barriers associated with both the cultural and the geographic feature database. As shown in Figure 24.12, the highest level-of-abstraction representation of the object representation of space (i.e., the top-level of the pyramid) indicates that a river, which represents an extended ground-mobility barrier, exists in the vicinity of both the start and the goal states. At this level of abstraction, it cannot be determined whether the extended barrier lies between the two points.

The pyramid data structure supports highly focused, top-down searching to determine whether ground travel between the start and the goal states is blocked by a river. At the next higher-resolution level, however, ambiguity remains. Finally, at the third level of the pyramid, it can be confirmed that a river lies between the start and the goal states. Therefore, an efficient, global path strategy can be pursued that requires breaching the identified barrier. Consequently, bridges, suitable fording locations, or bridging operations become candidate subgoals.

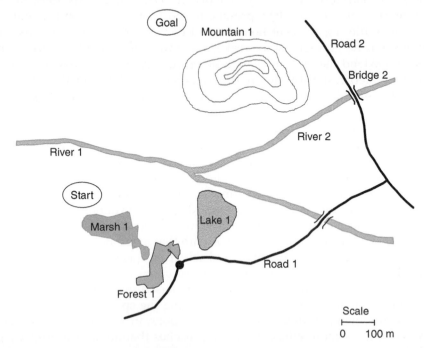

FIGURE 24.11
Domain mobility map for path development algorithm.

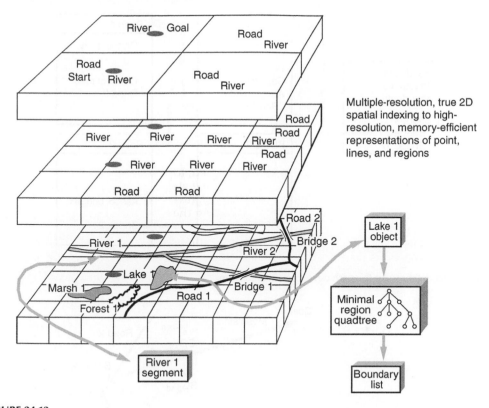

Multiple-resolution, true 2D spatial indexing to high-resolution, memory-efficient representations of point, lines, and regions

FIGURE 24.12
Top-down multiple-resolution spatial search, from the start toward the goal node, reveals the existence of a river barrier.

If, however, no extended barrier had been discovered in the cells shared by the start and the goal states (or in any intervening cells) at the outset of stage 1 analysis, processing would terminate without generating any intervening subgoals. In this case, stage 1 analysis would indicate that a direct path to the goal is feasible.

Whereas conventional path-planning algorithms operate strictly in the spatial domain, a flexible top-down path-planning algorithm supported by an effectively integrated semantic and spatial database can operate across both the semantic and the spatial domains. For example, suppose nearby bridges are selected as the primary subgoals. Rather than perform spatial search, direct search of the semantic object (River 1) could determine nearby bridges. Figure 24.13 depicts attributes associated with that semantic object, including the location of a number of bridges that cross the river. To simplify the example, only the closest bridge (Bridge 1) will be selected as a candidate subgoal (denoted $SG_{1,1}$). Although this bridge could have been located via spatial search in both directions along the river (from the point at which a line from the start to the goal state intersects River 1), a semantic-based search is more efficient.

To determine if one or more extended barriers lie between $SG_{1,1}$ and the goal state, a spatial search is reinitiated from the subgoal in the direction of the goal state. High-level spatial search within the pyramid data structure reveals another *potential* river barrier. Top-down spatial search once again verifies the existence of a second extended barrier (River 2). Just as before, the closest bridging location, denoted as $SG_{1,2}$, is identified by evaluating the bridge locations maintained by the semantic object (River 2). Spatial search

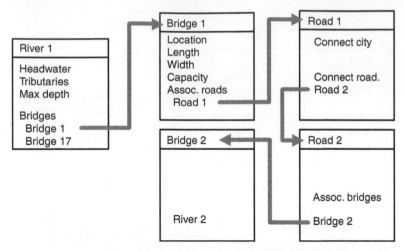

FIGURE 24.13
Semantic object database for the path development algorithm.

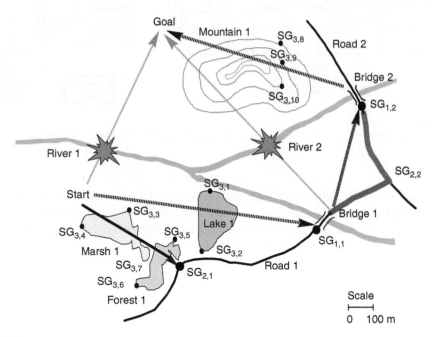

FIGURE 24.14
Subgoals associated with all three stages of the path development algorithm.

from Bridge 2 toward the goal state reveals no extended barriers that would interfere with ground travel between the $SG_{1,2}$ and the goal state.

As depicted in Figure 24.14, the first stage of the path development algorithm generates a single path consisting of the three subgoal pairs (start, $SG_{1,1}$), ($SG_{1,1}$, $SG_{1,2}$), and ($SG_{1,2}$, goal) satisfying the global objective of reaching the goal state by breaching all extended barriers. Thus, at the conclusion of stage 1, the primary alternatives to path flow have been identified.

In stage 2, road segments connecting adjacent subgoals that are on or near the road network must be identified. The semantic object representation of the bridges identified as subgoals during the stage 1 analysis also identify their road association; therefore, a road network solution potentially exists for the subgoal pair ($SG_{1,1}$, $SG_{1,2}$). Algorithms are widely available for efficiently generating minimum distance paths within a road network. As a result of this analysis, the appropriate segments of Road 1 and Road 2 are identified as members of the candidate solution set (shown in bold lines in Figure 24.14).

Next, the paths between the start state and $SG_{1,1}$ are investigated. $SG_{1,1}$ is known to be on a road and the start state is not; therefore, determining whether the start state is near a road is the next objective. Suppose the assessment is based on the fuzzy qualifier *near* shown in as Figure 24.15. Because the detailed spatial relations between features cannot be economically maintained with a semantic representation, spatial search must be used. Based on the top-down, multiple-resolution object representation of space, a road is determined to exist within the vicinity of the start node. A top-down spatially localized search within the pyramid efficiently reveals the closest road segment to the start node. Computing the Euclidean distance from that segment to the start node, the start node is determined to be near a road with degree of membership 0.8.

Because the start node has been determined to be near a road, in addition to direct overland travel toward Bridge 1 (start, $SG_{1,1}$), an alternative route exists based on overland travel to the nearest road (subgoal $SG_{2,1}$) followed by road travel to Bridge 1 ($SG_{2,1}$, $SG_{1,1}$). Although a spectrum of variants exists between direct travel to the bridge and direct travel to the closest road segment, at this level of abstraction only the primary alternatives must be identified. Repeating the analysis for the path segment ($SG_{1,2}$, goal), the goal node is determined to be not near any road. Consequently, overland route travel is required for the final leg of the route.

In stage 3, all existing nonroad path segments are refined based on more local evaluation criteria and mobility constraints. First, large barriers, such as lakes, marshes, and forests, are considered. Straight-line search from the start node to $SG_{1,1}$ reveals the existence of a large lake. Because circumnavigation of the lake is required, two subgoals are generated ($SG_{3,1}$ and $SG_{3,2}$) as shown in Figure 24.14, one representing clockwise travel and the other counter-clockwise travel around the barrier. In a similar manner, spatial search from the start state toward $SG_{2,1}$ reveals a large marsh, generating, in turn, two additional subgoals ($SG_{3,3}$ and $SG_{3,4}$).

Spatial search from both $SG_{3,3}$ toward $SG_{2,1}$ reveals a forest obstacle (Forest 1). Assuming that the forest density precludes wheeled vehicle travel, two more subgoals are generated representing a northern route ($SG_{3,5}$) and a southern route ($SG_{3,6}$) around the forest.

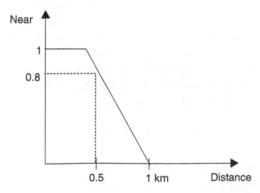

FIGURE 24.15
Membership function for fuzzy metric *near*.

Because a road might pass through the forest, a third strategy must be explored (road travel through the forest). The possibility of a road through the forest can be investigated by testing containment or generating the intersection between Forest 1 and the road database.

The integrated spatial/semantic database discussed in Section 24.6 provides direct support to containment testing and intersection operations. With a strictly vector-based representation of roads and regions, intersection generation might require interrogation of a significant portion of the road database; however, the quadtree-indexed vector spatial representation presented permits direct spatial search of that portion of the road database that is within Forest 1.[1] Suppose a dirt road is discovered to intersect the forest. As no objective criterion exists for evaluating the *best* subpath(s) at this level of analysis, an additional subgoal ($SG_{3,7}$) is established. To illustrate the benefits of deferring decision making, consider the fact that although the length of the road through the forest could be shorter than the travel distance around the forest, the road may not enter and exit the forest at locations that satisfy the overall path selection criteria.

Continuing with the last leg of the path, spatial search from $SG_{1,2}$ to the goal state identifies a mountain obstacle. Because of the inherent flexibility of a constraint-satisfaction-based problem-solving paradigm, a wide range of local path development strategies can be considered. For example, the path could be constrained to employ one or more of the following strategies:

1. Circumnavigate the obstacle ($SG_{3,8}$)
2. Remain below a specified elevation ($SG_{3,9}$)
3. Follow a minimum terrain gradient ($SG_{3,10}$)

Figure 24.16 shows the path-plan subgoal graph following stages 1–3. Proceeding in a top-down fashion, detailed paths between all sets of subgoals can be recursively refined based on the evaluation of progressively more local evaluation criteria and domain constraints. Path evaluation criteria at this level of abstraction might include (1) the minimum mobility resistance, (2) minimum terrain gradient, or (3) maximal speed paths.

Traditional path-planning algorithms generate global solutions by using highly local nearest-neighbor path extension strategies (e.g., gradient descent), requiring the generation of a combinatorial number of paths. Global optimization is typically achieved by rank ordering all generated paths against an evaluation metric (e.g., shortest distance or maximum speed). Supported by the semantic/spatial database kernel, the top-down path-planning algorithm just outlined requires significantly smaller search spaces when compared to traditional, single-resolution algorithms. Applying a single high-level constraint that eliminates the interrogation of a single 1 km × 1 km resolution cell, for example, could potentially eliminate search-and-test of as many as 10,000, 10 m × 10 m resolution cells. In addition to efficiency gains, due to its reliance on a hierarchy of constraints, a top-down approach potentially supports the generation of more robust solutions. Finally, because it emulates the problem-solving character of humans, the approach lends itself to the development of sophisticated algorithms capable of generating intuitively appealing solutions.

In summary, the hierarchical path development algorithm

1. Employs a reasoning approach that effectively emulates manual approaches
2. Can be highly robust because constraint sets are tailored to a specific vehicle class

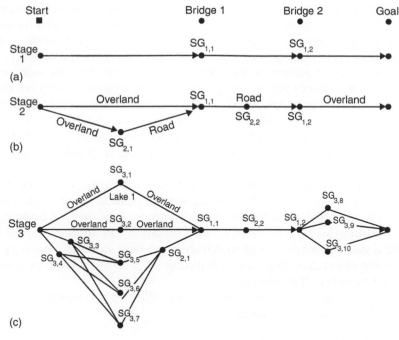

FIGURE 24.16
Path development graph following (a) stage 1, (b) stage 2, and (c) stage 3.

3. Is dynamically sensitive to the current domain context
4. Generates efficient global solutions

The example outlined in this section demonstrates the utility of the database kernel presented in Section 24.6. By facilitating the efficient, top-down, spatially anchored search and fully integrated semantic and the spatial object search, the spatial/semantic database provides direct support to a wide range of demanding, real-world problems.

24.8 Mixed Boolean and Fuzzy Reasoning

24.8.1 Introduction

Conventional sensors typically generate systematic (if imperfect) location descriptions, be they range cells, error ellipses, or angle- or range-only measurements. Humans, however, tend to describe locations in a nonsystematic fashion. Although semantic location descriptions tend to be somewhat ambiguous, they often prove entirely satisfactory and even preferred for communicating concepts to other humans. When a human says that he met a friend *at the intersection of Riverdale Rd and Sycamore St, in front of First Virginia Bank,* or *along the road that runs past the American Embassy,* another person familiar with the area would have little trouble *placing* these locations. Ambiguous natural language descriptions, however, tend to be problematic when the objective is support to an automated multisource fusion system.

Because natural language effectively encourages vague descriptions of locations, capturing the essence of this vagueness in a *rigorous* fashion is essential. From its inception in the 1960s, fuzzy set theory[2] has sought to support language understanding and semantic reasoning. Quite apart from fuzzy set theory and its formalisms, the notion of assigning *mathematical* meanings to adjectives and adverbs such as *tall, fast,* and *near* seems quite natural. Although arguments persist that probability theory can credibly handle the concept of vagueness, for many people, it takes a significant intuitive leap to go from *semantic vagueness* (mostly red, fairly tall, reasonably handsome) to a *probability of occurrence space*. We sidestep both the theoretical and the philosophical arguments associated with fuzzy set theory and focus instead on a practical implementation of mixed crisp and fuzzy spatial reasoning.

By their very nature, crisp spatial modifiers such as *inside, outside, to the west of,* and *higher are absolute* qualifiers. Although their meaning may depend to some extent on the referenced features, it is not tied to problem context. The concept of *outside* a building (in other words, NOT building) leaves little room for interpretation. The same cannot be said about *near* a building. *Near* is clearly a context-sensitive modifier/qualifier that may be highly dependent on the object class (e.g., near a building vs. farm vs. a town vs. a country), as well as on other aspects of the problem.

To manage and manipulate mixed crisp and fuzzy features, it is convenient to transform one or both representations into a common *space*.

- Crisp descriptions can be fuzzified.
- Fuzzy descriptions can be transformed into crisp (Boolean-like) descriptions.
- Both crisp and fuzzy representations can be transformed into a new space that better supports mixed representation analysis.

The first two approaches alter the precision of the data sets, whereas the last approach need not.

An example of the latter approach that shows promise indexes *crisp* descriptions of points, lines, and regions using a hierarchical regular decomposition referred to as the *quadtree-indexed vector* representation.[1] In this representation, a point feature is indexed by one quadtree cell. A line feature is indexed by a collection of adjacent quadtree cells, whereas a region feature is indexed by the set of quadtree cells that cover both the region boundary and its interior. *Fuzzy* features are indexed identically and then modified to accommodate generalizations such as *near* and *not near*.

Figure 24.17a shows the indexing cells for all buildings (crisp features) in a sample data set. Figure 24.17b depicts the selected fuzzy set *near* function, and Figure 24.17c shows the indexing cells that represent 100% *near* all buildings. Figure 24.18 depicts indexing cells and the road cells plus the 100% *near* cells for a simple road network. Figure 24.19 shows the combined road and building data set, as well as results of selected mixed Boolean and fuzzy set operations. Because all analysis relies on indexing space operations, the accuracy of the underlying data is not altered in any way. Because of the *regular* and *area*-based nature of the quadtree decomposition, both indexing space search and set operation generation tend to be efficient.

An additional complication of using natural language input is that automated approaches must be able to translate from vague semantic location descriptions to formal indexed spatial representations. Examples include *near* a park, *in front of* a named building, *between* two buildings, *across the street* from a named location, *away* from road intersections, *not far* from the center of town. To emphasize the need to translate from semantic location descriptions to more traditional formal spatial representations, consider the

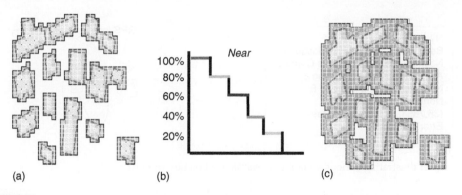

FIGURE 24.17
(a) Indexing cells of 15 buildings, (b) fuzzy set near function, and (c) 100% near buildings.

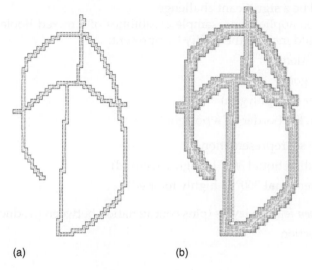

FIGURE 24.18
(a) Indexing cells of road network and (b) 100% *near* road network.

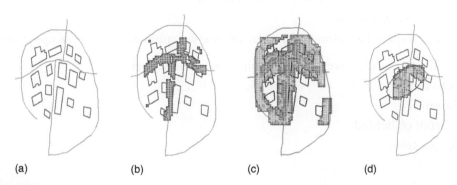

FIGURE 24.19
(a) Combined road and building data set, (b) indexing cells associated with 100% near all roads and 100% near all buildings, (c) indexing cells 60–100% near all roads and all buildings, and (d) indexing cells from (c) that intersect error ellipse.

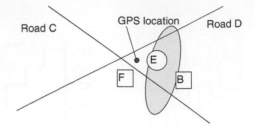

Near Building B
Between Building B and Mosque E
Near intersection of Roads C and D
Along Road D near Mosque E
Near Road C and Mosque E
Northeast of Building F
(36.7654, 24.12345) degrees
Error ellipse

FIGURE 24.20
Sample data set illustrating various semantically *similar* location descriptions.

example shown in Figure 24.20. All eight locations described in the left-hand box refer to roughly the same region of space. Although the reader can readily verify this assertion, inferring the similarity of this set of descriptions using strictly semantic-based reasoning approaches would be a significant challenge.

In terms of fusion applications, sample capabilities of a mixed Boolean and fuzzy reasoning system would include the ability to represent:

Crisp representations
   Inside (a polygon)
   On the border (of a polygon)
   Inside and on the border (of a polygon)

Semantic/fuzzy set representations
   Near/along (directional 360° to highly focused)
   Not near (directional 360° to highly focused)

Operators *between* feature layers (plus concatenation between products)
   AND/intersection
   OR/union

Operations within a single layer
   AND (e.g., find intersections of specified (all) roads)
Search selected data layers (find feature classes)
   Near a selected point
   Within a spatial window (defined by arbitrary combinations of set of operations)

Derived operators (derived from combinations of base operations)
   Between
   Around
   In front of, behind
   Across from

24.8.2   Extended Operations

Extended operations are either derivable from the base operations or require specialized functions. *Between* and *across the street* are two examples. Figure 24.21a depicts the result

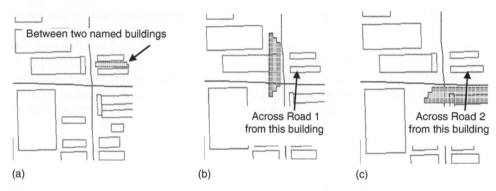

FIGURE 24.21
Examples of extended set operations. Fuzzy near gray scale is the same as in Figure 24.17b.

of ANDing two *directional nears* that effectively creates the concept *between. Across the street from the Bank* requires multiple layers of reasoning. The first step involves determining the closest road to the named building (*near* Bank AND all roads). If the building happened to be near multiple roads, the query result would be ambiguous, generating two or more possible road instantiations. (It is worth noting that this kind of ambiguity would exist whether the query was handled by a human or by a machine.) Assuming just one road is closest, the road cells directly in front of the bank are determined (*near* Bank AND Road 1). Across Road 1 from the Bank is then generated by region growing from these select road cells, but only on the opposite side of the road from the Bank as depicted in Figure 24.21b. Choosing Road 2 rather than Road 1 results in the solution shown in Figure 24.21c.

Relevant context can fall across a spectrum from highly local to highly global. As an example of more global context support, suppose a message refers to a grocery store and a specified road (Road 9) but gives no indication to either the address or the name of the establishment. As long as the supporting GIS contains the named road (Road 9) and identifies all building classes, candidate building(s) can potentially be identified.

Search building layer (grocery store) *near* Road 9.

If more than one grocery store is discovered, it might be helpful to apply other heuristics to narrow the selection, for example, *weight selections closest to where an individual being tracked lives* or weight based on candidate's ranking in the *key location* database.

Even when weaker constraints are provided, it might still be possible to provide useful products. Suppose a suspect is reported to have entered the large building at the intersection of Jennifer and Vine Streets. The desired road intersection can be found by ANDing the two named roads. A *near* filter (spatial window) center on the discovered road intersection would then be applied to the building database. Depending on the relative size of all discovered buildings, the query might produce one or more candidates. In the worst case, if all the buildings happened to be roughly the same size, four alternative hypotheses would result. Once again, other context constraints can be applied to rank-order the candidates. If the two largest buildings near the intersection, for example, happened to be a movie theater and a lady's dress shop, the movie theater might be the more likely candidate. If the observation contained the additional qualification that the building was north of the intersection, only a single building might satisfy the constraint set.

Even when location descriptions are highly ambiguous, some level of automated support might still be possible. Suppose an observation stated that a man ran out of the bank and disappeared in the nearby city park. Intersecting *near city parks* with *near commercial banks* is likely to generate only a small number of bank matches. If the bank had been further described as being along a street lined by restaurants, intersecting the resultant regions with *near all restaurant districts* is likely to further narrow the candidate list.

Exploitation of ambiguous information possibly resident in different databases assembled for different purposes has a wide range of possible applications. The development of a generic context support capability that handles natural language location translation and flexible manipulation of mixed Boolean and fuzzy spatially organized constraints appears to be a realistic objective. Beyond serving a wide range of applications, such a system could offer a systematic approach for managing spatial uncertainty.

24.9 Summary and Conclusions

All-source situation awareness development relies on the effective combination of a wide range of data and knowledge sources, including the maximal use of relevant sensor-derived (e.g., imagery, overlays, and video), human-generated information, and nonsensor-derived information (e.g., topographic features; cultural features; and past, present, and future weather conditions). Sensor and human-supplied data provides dynamic information that feeds the analysis process; however, relatively static domain-context knowledge provides equally valuable information that constrains the interpretation of sensor-derived information. Owing to the potentially large volume of both sensor and nonsensor-derived databases, the character and capability of the supporting DBMS can significantly impact both the effectiveness and the efficiency of machine-based reasoning.

This chapter outlined a number of top-down design considerations for designing an object database kernel that supports the development of both effective and efficient data fusion algorithms. At the highest level of abstraction, the near-optimal database kernel consists of two classes of objects: semantic and spatial. Because conventional OODBMS provide adequate support to semantic object representations, the chapter focused on the design for the spatial object representation.

A spatial object realization consisting of an object representation of 2D space integrated with a hybrid spatial representation of individual point, line, and region features was shown to achieve an effective compromise across all design criteria. An object representation of 2D space provides a spatial object hierarchy metaphorically similar to a conventional semantic object hierarchy. Just as a semantic object hierarchy supports top-down semantic reasoning, a spatial object hierarchy supports top-down spatial reasoning. A hybrid spatial representation, the quadtree-indexed vector representation, supports an efficient top-down search and analysis and high-precision refined analysis of individual spatial features. Both the object representation of 2D space and the multiple-resolution representation of individual spatial features employ the identical quadtree decomposition. Therefore, the quadtree-indexed vector representation is a natural extension of the object representation of 2D space.

Acknowledgments

Preparation and revision of this chapter was funded by U.S. Army CERDEC I2WD, Fort Monmouth, New Jersey.

All figures shown in Section 24.8 were generated by the Context Fusion Support (CFS) tool currently under development in support of the U.S. Army Soft Target Exploitation and Fusion (STEF) ATO program.

References

1. R. Antony, *Principles of Data Fusion Automation*, Artech House Inc., Boston, MA, 1995.
2. L. Zadeh, Fuzzy sets, *Information and Control*, 8, 338–353, 1965.

25

Assessing the Performance of Multisensor Fusion Processes

James Llinas

CONTENTS

25.1 Introduction

In recent years, numerous prototypical systems have been developed for multisensor data fusion (DF). A paper by Hall and co-workers[1] describes more than 50 such systems developed for Department of Defense (DoD) applications some 15 years ago. Such systems have become ever more sophisticated. Indeed, many of the prototypical systems summarized by Hall and co-workers[1] use advanced identification techniques such as knowledge-based or expert systems, Dempster–Shafer interface techniques, adaptive neural networks, and sophisticated tracking algorithms.

Although much research has been performed to develop and apply new algorithms and techniques, much less work has been performed to formalize the techniques for determining how well such methods work or to compare alternative methods against a common problem. The issues of system performance and system effectiveness are keys to establishing, first, how well an algorithm, technique, or collection of techniques performs in a technical sense and, second, the extent to which these techniques, as part of a system, contribute to the probability of success when that system is employed on an operational mission. An important point to remember in considering the evaluation of DF processes is that those processes are either a component of a system (if they were designed-in at the beginning) or they are enhancements to a system (if they have been incorporated with the intention of performance enhancement). In other words, it is not usual that the DF processes are *the* system under test; DF processes are said to be designed *into* systems rather than being systems in their own right. What is important to understand in this sense is that the DF processes contribute some marginal or piecewise improvement to the overall system, and if the contribution of the DF process *per se* needs to be calculated, it must be done while holding other factors fixed. If the DF processes under examination are enhancements, it is important that such performance must be evaluated in comparison to an agreed-to baseline (e.g., without DF capability, or presumably a *lesser* DF capability). These points will be discussed in more detail later.

It is also important to mention that our discussion in this chapter is largely about *automated* DF processing (although we will make some comments about human-in-the-loop aspects later), and by and large such processes are enabled through software. Thus, it should not be surprising that remarks made herein draw on or are similar to concerns for test and evaluation (T&E) of complex software processes.

System performance at level 1, for example, focuses on establishing how well a system of sensors and DF algorithms may be utilized to achieve estimates of or inferences about location, attributes, and identity of platforms or emitters. Particular measures of performance (MOPs) may characterize a fusion system by computing one or more of the following:

- Detection probability—probability of detecting entities as a function of range, signal-to-noise ratio, and so on
- False alarm rate—rate at which noisy or spurious signals are incorrectly identified as valid targets
- Location estimate accuracy—the accuracy with which the position of an entity is determined
- Identification probability—probability of correctly identifying an entity as a target
- Identification range—the range between a sensing system and target at which the probability of correct identification exceeds an established threshold
- Time from transmission to detect—time delay between a signal emitted by a target (or by an active sensor) and the detection by a fusion system
- Target classification accuracy—ability of a sensor suite and fusion system to correctly identify a target as a member of a general (or particular) class or category

These MOPs measure the ability of the fusion process as an information process to transform signal energy either emitted by or reflected from a target, to infer the location, attributes, or identity of the target. MOPs are often functions of several dimensional parameters used to quantify, in a single variable, a measure of operational performance.

Conversely, measures of effectiveness (MOEs) seek to provide a measure of the ability of a fusion system to assist in completion of an operational mission. MOEs may include

- Target nomination rate—the rate at which the system identifies and nominates targets for consideration by weapon systems
- Timeliness of information—timeline of availability of information to support command decisions
- Warning time—time provided to warn a user of impending danger or enemy activity

- Target leakage—percentage of enemy units or targets that evade detection
- Countermeasure immunity—ability of a fusion system to avoid degradation by enemy countermeasures

At an even higher level, measures of force effectiveness (MOFEs) quantify the ability of the total military force (including the systems having DF capabilities) to complete its mission. Typical MOFEs include rates and ratios of attrition, outcomes of engagement, and functions of these variables. In the overall mission definition other factors such as cost, size of force, force composition, and so on may also be included in the MOFEs.

This chapter presents top-down, conceptual, and methodological ideas on the T&E of DF processes and systems, describes some of the tools available that are needed to support such evaluations, and discusses the range of measures of merit useful for quantification of evaluation results.

25.2 Test and Evaluation of the Data Fusion Process

Although, as has been mentioned in the preceding section, the DF process is frequently part of a larger system process (i.e., DF is often a *subsystem* or *infrastructure* process to a larger whole) and thereby would be subjected to an organized set of system-level test procedures, this section develops a stand-alone, top-level model of the T&E activity for a general DF process. This characterization is considered the proper starting point for the subsequent detailed discussions on metrics and evaluation, because it establishes a viewpoint or framework (a context) for those discussions, and also because it challenges the DF process architects to formulate a global and defendable approach to T&E.

In this discussion, it is important to understand the difference between the terms *test* and *evaluation*. One distinction (according to *Webster's Dictionary*) is that testing forms a basis for evaluation. Alternately, testing is a process of conducting trials to prove or disprove a hypothesis—here, a hypothesis regarding the characteristics of a procedure within the DF process. Testing is essentially laboratory experimentation regarding the active functionality of DF procedures and, ultimately, the overall process (*active* meaning during their execution—not statically analyzed).

However, evaluation takes its definition from its root word: value. Evaluation is thus a process by which the value of DF procedures is determined. Value is something measured in context; it is because of this that a context must be established.

The view taken here is that the T&E activities will both be characterized as having the following components:

- A *philosophy* that establishes or emphasizes a particular point of view for the tests and evaluations that follow. The simplest example of this notion is reflected in the so-called *black box* or *white box* viewpoints for T&E, from which either external (input/output behaviors) or internal (procedure execution behaviors) are examined (a similar concern for software processes in general, as has been noted). Another point of view revolves around the research or development goals established for the program. The philosophy establishes the high-level statement regarding context and is closely intertwined with the program goals and objectives, as discussed in Section 25.2.1.
- A set of *criteria* according to which the quality and correctness of the T&E results or inferences will be judged.

- A set of *measures* through which judgments on criteria can be made, and a set of *metrics* on which the measures depend and, importantly, *which can be measured during T&E experiments.*
- An *approach* through which tests and analyses can be defined and conducted that
  - Are consistent with the philosophy
  - Produce results (measures and metrics) that can be effectively judged against the criteria

25.2.1 Establishing the Context for Evaluation

Assessments of delivered value for defense systems must be judged in light of system or program goals and objectives. In the design and development of such systems, many translations of the stated goals and objectives occur as a result of the systems engineering process, which both analyzes (decomposes) the goals into functional and performance requirements and synthesizes (reassembles) system components intended to perform in accordance with these requirements. Throughout this process, however, the program goals and objectives must be kept in view because they establish the context in which value will be judged.

Context, therefore, reflects what the program and the DF process (or system within it) are trying to achieve—that is, what the research or developmental goals (the purposes of building the system at hand) are. Such goals are typically reflected in the program name, such as a *Proof of Concept* program or *Production Prototype* program. Many recent programs involve *demonstrations* or *experiments* of some type or other, with these words reflecting in part the nature of such program goals or objectives.

Several translations must occur for the T&E activities themselves. The first of these is the translation of goals and objectives into T&E philosophies; that is, philosophies follow from statements about goals and objectives. Philosophies primarily establish points of view or perspectives for T&E that are consistent with, and can be traced to, the goals and objectives: they establish the *purpose* of investing in the T&E process. Philosophies also provide guidelines for the development of T&E criteria, for the definition of meaningful T&E cases and conditions, and, importantly, a sense of a *satisfaction scale* for test results and value judgments that guides the overall investment of precious resources in the T&E process. That is, T&E philosophies, while generally stated in nonfinancial terms, do in fact establish economic philosophies for the commitment of funds and resources to the T&E process. In today's environment (it makes sense categorically in any case), notions of affordability must be considered for any part of the overall systems engineering approach and for system development, to include certainly the degree of investment to be made in T&E functions.

25.2.2 T&E Philosophies

Establishing a philosophy for T&E of a DF process is also tightly coupled to the establishment of what the DF process boundaries are. In general, it can be argued that the T&E of any process within a system should attempt the longest extrapolation possible in relating process behavior to program goals; that is, the evaluation should endeavor to relate process test results to program goals *to the extent possible.* This entails first understanding the DF process boundary, and then assessing the degree to which DF process results can be related to superordinate processes; for defense systems, this means assessing the degree to which DF results can be related to mission goals. Philosophies aside, certain *acid tests* should always be conducted:

- Results with and without fusion (e.g., multisensor vs. single sensor or some *best* sensor)
- Results as a function of the number of sensors or sources involved (e.g., single sensor, 2, 3, ..., N sensor results for a common problem)

The last two points are associated with defining some type of *baseline* against which the candidate fusion process is being evaluated. In other words, these points address the question, *Fusion as compared to what*? If it is agreed that DF processing provides some marginal benefit, then that gain must be evaluated in comparison to the *unenhanced* or baseline system. That comparison also provides the basis for the cost-effectiveness trade-off in that the relative costs of the baseline and fusion-enhanced systems can be compared to the relative performance of each.

Other philosophies could be established, however, such as

- *Organizational*: A philosophy that examines the benefits of DF products accruing to the system-owning organization and, in turn, subsequent superordinate organizations in the context of organizational purposes, goals, and objectives (no *platform* or *mission* may be involved; the benefits may accrue to an organization).
- *Economic*: A philosophy that is explicitly focused on some sense of economic value of the DF results (weight, power, volume, and so on) or cost in a larger sense, such as the cost of weapons expended, and so on.
- *Informal*: The class of philosophies in which DF results are measured against some human results or expectations.
- *Formal*: The class of philosophies in which the evaluation is carried out according to appropriate formal techniques that prove or otherwise rigorously validate the program results or internal behaviors (e.g., proofs of correctness, formal logic tests, and formal evaluations of complexity).

The list is not presented as complete but as representative; further consideration would no doubt uncover many other perspectives.

25.2.3 T&E Criteria

Once having espoused one or another of the philosophies, there exists a perspective from which to select various criteria, which will collectively provide a basis for evaluation. It is important at this step to realize the exact meaning and subsequent relationships impacted by the selection of such criteria.

There should be a functionally complete hierarchy that emanates from each criterion as follows:

- **Criterion**—a standard, rule, or test upon which a judgment or decision can be made (this is a formal dictionary definition),

which leads to the definition of

- **Measures**—the *dimensions* of a criterion, that is, the factors into which a criterion can be divided and, finally,

- **Metrics**—those attributes of the DF process or its parameters or processing results
  which are considered easily and straightforwardly quantifiable or able to be defined
  categorically, which are relatable to the measures, and which are *observable*

Thus, there is, in the most general case, a functional relationship as follows:

Criterion = fct [(Measure$_i$ = fct (Metric$_i$, Metric$_j$···), Measure$_j$ = fct (Metric$_k$, Metric$_l$···), etc.]

Each metric, measure, and criterion also has a scale that must be considered. Moreover,
the scales are often incongruent so that some type of normalized *figure of merit* approach
may be necessary to integrate metrics on disparate scales and construct a unified, quantitative parameter for making judgments.

One reason to establish these relationships is to provide for *traceability* of the logic
applied in the T&E process. Another rationale, which argues for the establishment of these
relationships, is in part derived from the requirement or desire to estimate, even roughly,
predicted system behaviors against which to compare actual results. Such prediction must
occur at the metric level; predicted and actual metrics subsequently form the basis for
comparison and evaluation. The prediction process must be functionally consistent with
this hierarchy. For level 1 numeric processes, prediction of performance expectations can
often be done, to a degree, on an analytical basis. (It is assumed here that in many T&E
frameworks the *truth* state is known; this is certainly true for simulation-based experimentation but may not be true during operational tests, in which case comparisons are often
done against consensus opinions of experts.) For levels 2 and 3 processes, which generally
employ heuristics and relatively complex lines of reasoning, the ability to predict the metrics with acceptable accuracy must usually be developed from a sequence of exploratory
experiments. Failure to do so may in fact invalidate the overall approach to the T&E process because the fundamental traceability requirement that are described in this context
would be confounded.

Representative criteria focused on the DF process *per se* are listed below for the numerically dominated level 1 processes, and the symbolic-oriented level 2 and 3 processes.

Level 1 Criteria	Level 2, 3 Criteria
Accuracy	Correctness in reasoning
Repeatability/consistency	Quality or relevance of decisions/advice/recommendations
Robustness	Intelligent behavior
Computational complexity	Adaptability in reasoning (robustness)

Criteria such as computational efficiency, time-critical performance, and adaptability
are applicable to all levels whereas certain criteria reflect either the largely numeric or the
symbolic processes that distinguish these fusion-processing levels.

Additional conceptual and philosophical issues regarding what constitutes *goodness* for
software of any type can, more or less, alter the complexity of the T&E issue. For example,
there is the issue of reliability versus trustworthiness. Testing oriented toward measuring reliability is often *classless*, that is, it occurs without distinction of the type of failure
encountered. Thus, reliability testing often derives an unweighted likelihood of failure,
without defining the class or, perhaps more importantly, a measure of the severity of the
failure. This perspective derives from a philosophy oriented to the unweighted confor-
mance of the software with the software specifications, a common practice within the DoD
and its contractors.

It can be asserted, based on the argument that exhaustive path testing is infeasible for complex software, that trustworthiness of software is a more desirable goal to achieve via the T&E process. Trustworthiness can be defined as a measure of the software's likelihood of failing catastrophically. Thus, the characteristic of trustworthiness can be described by a function that yields the probability of occurrence for all significant levels of severe failures. This probabilistic function provides the basis for the estimation of a confidence interval for trustworthiness. The system designer/developer (or customer) can thus have a basis for assuring that the level of failures will not, within specified probabilistic limits, exceed certain levels of severity.

25.2.4 Approach to T&E

The final element of this framework is called the *approach* element of the T&E process. In this sense, approach means a set of activities, which are both procedural and analytical, that generates the *measure* results of interest (via analytical operations on the observed metrics) as well as provides the mechanics by which decisions are made based on those measures and in relation to the criteria. The approach consists of two components:

- A *procedure*, which is a metric-gathering paradigm (it is an experimental procedure)
- An *experimental design*, which defines (1) the test cases, (2) the standards for evaluation, and (3) the analytical framework for assessing the results.

Aspects of experimental design include the formal methods of classical, statistical experimental, design.[2] Few, if any, DF research efforts in the literature have applied this type of formal strategy, presumably as a result of cost limitations. Nevertheless, there are the serious questions of sample size and confidence intervals for estimates, among others, to deal with in the formulation of any T&E program, since simple comparisons of mean values, and so on under unstructured test conditions may not have very much statistical significance in comparison to the formal requirements of a rigorous experimental design. Such DF efforts should at least recognize the risks associated with such analyses.

This latter point relates to a fundamental viewpoint taken here about the T&E of DF processes: the DF process can be considered a function that operates on random variables (the noise-corrupted measurements or other uncertain inputs, that is, those which have a statistical uncertainty) to produce estimates that are themselves random variables and therefore have a distribution. Most would agree that the inputs to the DF process are stochastic in nature (sensor observation models are almost always based on statistical models); if this is agreed, then any operation on those random variables produces random variables. It could be argued that the DF processes, separated from the sensor systems (and their noise effects), are deterministic *probability calculators*; in other words, processes that given the same input (the same random variable) produce the same output (the same output random variable).[3] In this context, we would certainly want and expect a DF algorithm, if no other internal stochastic aspects are involved to generate the same output when given a fixed input. It could therefore be argued that some portion of the T&E process should examine such repeatability. But DeWitt[3] also agrees that the proper approach for a *probabilistic predictor* involves stochastic methods such as those that examine the closeness of distributions. (DeWitt raises some interesting epistemological views about evaluating such processes, but, as in his report, we also do not wish to *plow new ground in that area*, although recognizing its importance.) Thus, we argue here for T&E techniques that somehow account for and consider the stochastic nature of the DF results when exposed to *appropriately representative* input, such as by employment of

Monte Carlo–based experiments, analysis of variance methods, distributional closeness, and statistically designed experiments.

25.2.5   The T&E Process: A Summary

This section suggests a framework for the definition and discussion of the T&E process for the DF process and DF-enhanced systems; this framework is summarized in Figure 25.1. Much of the rationale and many of the issues raised are derived from good systems engineering concepts but are intended to sensitize DF researchers to the need for formalized T&E methods to quantify or otherwise evaluate the marginal contributions of the DF process to program/system goals. This formal framework is consistent with the formal and structured methods for the T&E of C3 systems in general—see, for example, Refs 4 and 5. In addition, since fusion processes at levels 2 and 3 typically involve the application of knowledge-based systems, further difficulties involving the T&E of such systems or processes can also complicate the approach to evaluation since, in effect, human reasoning strategies (implemented in software), not mathematical algorithms, are the subject of the tests. Improved formality of the T&E process for knowledge-based systems, using a framework similar to that proposed here, is described in Ref. 6. Little, if any, formal T&E work of this type, with statistically qualified results, appears in the DF literature. As DF procedures, algorithms, and technology mature, the issues raised here will have to be dealt with, and the development of guidelines and standards for DF process T&E undertaken. The starting point for such efforts is an integrated view of the T&E domain—the proposed process is one such view, providing a framework for discussion among DF researchers.

25.3   Tools for Evaluation: Testbeds, Simulations, and Standard Data Sets

Part of the overall T&E process just described involves the decision regarding the means for conducting the evaluation of the DF process at hand. Generally, there is a cost versus quality/fidelity trade-off in making this choice, as is depicted in Figure 25.2.[7] Another characterization of the overall range of possible tools is shown in Table 25.1. Over the past several years, the defense community has built up a degree of testbed capability for studying various components of the DF process. In general, these testbeds have been associated with a particular program and its range of problems, and—except in one or two instances—the testbeds have permitted *parametric-level* experimentation but not *algorithm-level* one. That is, these testbeds, as software systems, were built from *point* designs for a given application wherein normal control parameters could be altered to study attendant effects, but these testbeds could not (at least easily) permit replacement of such components as a tracking algorithm. Recently, some new testbed designs are moving in this direction. One important consequence of building testbeds that permit algorithm-level test and replacement is of course that such testbeds provide a consistent basis for system evolution over time, and in principle such testbeds, in certain cases, could be shared by a community of researcher-developers. In an era of tight defense research budgets, algorithm-level shareable testbeds, it is suspected and hoped, will become the norm for the DF community. A snapshot of some representative testbeds and experimental capabilities is shown in Table 25.2. An inherent difficulty (or at least an issue) in testing DF algorithms warrants discussion because it fundamentally results from the inherent complexity of the DF process: the complexity of

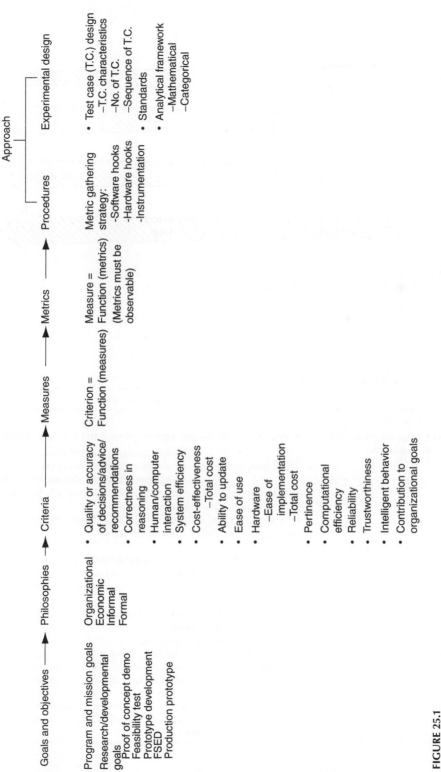

FIGURE 25.1
Test and evaluation activities for the data fusion process.

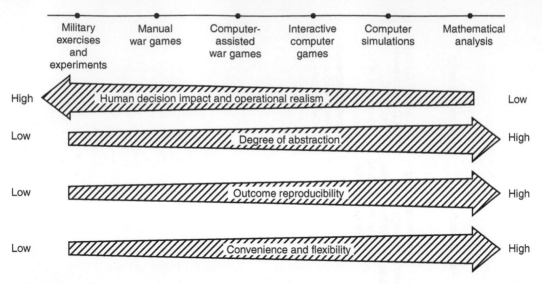

FIGURE 25.2
Applicability of modeling technique. (From Przemieniecki, J.S., *Am. Inst. Aero. Astro. Education Series*, Washington, DC, 1990.)

TABLE 25.1

Generic Spectrum of Evaluation Tools

Toolset	Characteristics
I. Digital simulations	
• Level 1: engineering models	• Relatively high fidelity; explore physics and 1 versus 1 problem
• Level 2: 1 versus N	⎰ Explore engagement effects
• Level 3: M versus N	} Engagement models ⎨ Fidelity decreases with Level
• Level 4: organizational level	⎱ Engagement complexity increases with Level
• Level 5: theater level	• Individualized, *ad hoc* simulations for tracking, ID, detection
• Numerous DF process models	Statistical qualification usually feasible
II. Hybrid simulations Man-in-the-loop or equipment-in-the-loop	Important effects of real humans and equipment; more costly; statistical qualification often unaffordable
III. Specialized field data collection/calibration	Real-world physics, phenomenology; relatively costly; often used to verify/validate digital simulations; good for phenomenological modeling but not for behavior modeling; statistically controlled in most cases
IV. Test range data collection	Real-world physics, humans, equipment; relatively costly; can do limited engagement effects studies; some behavioral effected modeled; statistically uncontrolled
V. Military exercises	Real-world physics, humans, equipment, and tactics/doctrine; costly; data difficult to collect/analyze; extended engagement effects studies at least feasible; extended behavioral effects modeled; statistically uncontrolled
VI. Combat operations	For example, desert storm; actual combat effects with real adversaries; data difficult to collect/analyze; high-fidelity enemy behavioral data; statistically uncontrolled

TABLE 25.2

Representative Multisensor Data Fusion Testbeds

Testbed	General Purpose	Characteristics
(1) Multisensor, multitarget data fusion testbed (Rome Lab, Air Force)[8]	Compare the performance of various *level 1* fusion algorithms (association, tracking, and identification)	Three major components: scenario generator, platform simulator, data analysis and display. Monte Carlo capability
(2) Advanced sensor exploitation testbed (ASET) (Rome Lab, Air Force)[9]	Large testbed to study *level 2* fusion algorithms primarily for order of battle estimation	Major components: scenario generator, sensor simulators, C^3I simulator, fusion element, timing and control, evaluation, nominal man-in-the-loop capability, parametric flexibility
(3) National testbed (SDIO)[10]	Evaluate candidate fusion concepts for Strategic Defense Initiative (SDI)/ *level 1* tracking and discrimination functions. Study effects of interelement communication links	Large-scale, multi-Cray (plus other large computers) type environment to simulate broad range of functions in SDI problem. Purposeful man-in-the-loop capability
(4) Surveillance testbed (SDIO)[11]	High-fidelity background and sensor (radar, optical) simulator for broad range of SDI *level 1* fusion algorithm testing	Algorithm-level test and replacement design. *Framework* and *Driver* concept that separates simulation and analysis activities
(5) NATO data fusion demonstrator (NATO)[12]	Initial configuration is to study *level 1-2-3* fusion processes for airland battle (initially army) applications	*Client-Server* design concept to permit algorithm-level test and replacement
(6) All-source systems evaluation testbed (ASSET) (University of Virginia)[13]	To evaluate data association and correlation algorithms for ASAS-type airland battle applications	Algorithm-level test and replacement design; standardized metrics; connected to Army C^2SW
(7) AWACS fusion evaluation testbed (FET) (Mitre, Ref. 14)	Provide analysis and evaluation support of multisensor integration (MSI) algorithms for AWACS level 1 applications	Algorithm-level test and replacement design; permits live and simulate data as driver; standardized MOEs; part of Air Defense C^2 (ADC^2) Laboratory

the DF process may make it infeasible or unaffordable to evolve, through experimentation, DF processing strategies that are optimal for other than level 1 applications. This issue depends on the philosophy with which one approaches testbed design. Consider that even in algorithmic-replaceable testbeds, the *test article* (a term for the algorithm under test) will be tested *in the framework of the surrounding algorithms available from the testbed library*. Hence, a tracking algorithm will be tested while using a separate detection algorithm, a particular strategy for track initiation, and so on. Table 25.3 shows some of the testable (replaceable) DF functions for the SDI surveillance testbed developed during the SDI program. Deciding on the granularity of the test articles (i.e., the plug-replaceable level of fidelity of algorithms to be tested) is a significant design decision for a DF testbed designer. Even if the testbed has, for example, multiple detection algorithms in its inventory, cost constraints will probably not permit combinatorial testing for optimality. The importance therefore of clearly thinking about and expressing the T&E philosophy, goals, and objectives, as described in Section 25.2, becomes evident relative to this issue. In many real-world situations, it is likely, therefore, that T&E of DF processes will espouse a *satisficing* philosophy, that is, be based on developing solutions that are good enough because cost and other practical constraints will likely prohibit extensive testing of wide varieties of algorithmic combinations.

Standardized challenge problems and associated data sets are another goal that the DF community should be seeking. To this date, very little calibrated and simultaneously collected data on targets of interest with known truth conditions exist. Some contractors have

TABLE 25.3

SDI Surveillance Testbed: Testable Level 1 Fusion Functions

Test Article	Function
Bulk filter	Reject measurement data from nonlethal objects
Track initialization	Data association and filters for starting tracks from measurement data only
Track continuation	Data association and filters to improve existing track information using new measurement data
Differential initialization	Data association and filters for starting new tracks using measurement data and existing track information
Cluster track	Data association and filters that maintain group information for multiple objects
Track fusion	Data association and filters for combining track information from multiple sources
Track file editing	Reject improbable measurements from a sequence of measurements
Feature calculation	Calculate discrimination features from measurement data
Classification	Provide an estimate of an object's lethality
Discrimination of data fusion	Combine discrimination information from multiple sources
Radar search and acquisition	Search of a handover volume
Sensor tasking/radar scheduling	Scheduling of sensor resources

invested in the collection of such data, but those data often become proprietary. Alternatives to this situation include artificially synthesizing multisensor data from individual sensor data collected under nonstandard conditions (not easy to do in a convincing manner), or employing high-fidelity sensor and phenomenological simulators.

Some attempts have been made to collect such data for community-wide application. One of the earliest of such activities was the 1987 DARPA HI-CAMP experiments that collected pulse Doppler radar and long-wavelength infrared data on fairly large U.S. aircraft platforms under a limited set of observation conditions. The Army (Night Vision Laboratory) has also collected data (ground-based sensors, ground targets) for community use under a 1989 program called Multisensor Fusion Demonstration, which collected carefully ground-truthed and calibrated multisensor data for DF community use. More recently, DARPA, in combination with the Air Force Research Laboratory Sensors Directorate, made available a broad set of synthetic aperture radar (SAR) data for ground targets under a program known as MSTAR.[8] However, the availability of such data to support algorithm development and DF system prototypes is extremely limited and represents a serious detriment and cost driver to the DF community.

However, similar to the DF process and its algorithms, the tool sets and data sets for supporting DF research and development are just beginning to mature. Modern designs of true testbeds permitting flexible algorithm-level test-and-replace capability for scientific experimentation are beginning to appear and are at least usable within certain subsets of the DF community; it would be encouraging to at least see *plans* to share such facilities on a broader basis as short-term, prioritized program needs are satisfied—that is, in the long term, these facilities should enter a national inventory. The need for data sets from real sensors and targets, even though such sensor-target pairs may be representative for only a variety of applications, is a more urgent need of the community. Programs whose focus is on algorithm development must incur redundant costs of data collection for algorithm demonstrations with real data when, in many cases, representative real data would suffice. Importantly, the availability of such data sets provides a natural framework for comparative analyses when various techniques are applied against a common or baseline problem as represented by the data. Comparative analyses set the foundation for developing deeper understanding of what methods work where, for what reasons, and for what cost.

25.4   Relating Fusion Performance to Military
Effectiveness: Measures of Merit

Because sensors and fusion processes are contributors to improved information accuracy, timeliness, and content, a major objective of many fusion analyses is to determine the effect of these contributions to military effectiveness. This effectiveness must be quantified, and numerous quantifiable measures of merit can be envisioned; for conventional warfare such measures might be engagement outcomes, exchange ratios (the ratio of blue-red targets killed), total targets serviced, and so on as previously mentioned. The ability to relate DF performance to military effectiveness is difficult because of the many factors that relate improved information to improved combat effectiveness and the uncertainty in modeling them. These factors include

- Cumulative effects of measurement errors that result in targeting errors
- Relations between marginal improvements in data and improvements in human decision making
- Effects of improved threat assessment on survivability of own forces

These factors and the hierarchy of relationships between DF performance and military effectiveness must be properly understood in order for researchers to develop measures and models that relate them. In other words, there is a large conceptual distance between the value of improved information quality, as provided by DF techniques, and its effects on military effectiveness; this large distance is what makes such evaluations difficult. The Military Operations Research Society[9] has recommended a hierarchy of measures that relate performance characteristics of C3 systems (including fusion) to military effectiveness (see Table 25.4). *Dimensional parameters* are the typical properties or characteristics that directly define the elements of the DF system elements, such as sensors, processors, communication channels, and

TABLE 25.4

Four Categories of Measures of Merit

Measure	Definition	Typical Examples
Measures of force effectiveness (MOFEs)	Measure of how a C3 system and the force (sensors, weapons, C3 system) of which it is a part perform military missions	Outcome of battle; cost of system; survivability; attrition rate; exchange ratio; weapons on targets
Measures of effectiveness (MOEs)	Measure of how a C3 system performs its functions within an operational environment	Target nomination rate; timeliness of information; accuracy of information; warning time; target leakage; countermeasure immunity; communications survivability
Measures of performance (MOPs)	Measures closely related to dimensional parameters (both physical and structural) but measure attributes of behavior	Detection probability; false alarm rate; location estimate accuracy; identification probability; identification range; time from detect to transmission; communication time delay; sensor spatial coverage; target classification accuracy
Dimensional parameters	The properties or characteristics inherent in the physical entities whose values determine system behavior and the structure under question, even when not operating	Signal-to-noise ratio; operations per second; aperture dimensions; bit-error rates; resolution; sample rates; antijamming margins; cost

Source: Sweet, R., Command and control evaluation workshop, MORS C² MOE Workshop, Military Operations Research Society, January 1985.

so on. (These are equivalent to the *metrics* defined in Section 25.2.) They directly describe the behavior or structure of the system and should be considered to be typical measurable specification values (bandwidth, bit-error rates, physical dimensions, and so on).

MOPs are measures that describe the important behavioral attributes of the system. MOPs are often functions of several dimensional parameters to quantify, in a single variable, a significant measure of operational performance. Intercept and detection probabilities, for example, are important MOPs that are functions of several dimensional parameters of both the sensors and the detailed signal processing operations, DF processes, and the characteristics of the targets being detected.

MOEs gauge the degree to which a system or militarily significant function was successfully performed. Typical examples, as shown in Table 25.4, are target leakage and target nomination rate.

MOFEs are the highest-level measures that quantify the ability of the total military force (including the DF system) to complete its mission. Typical MOFEs include rates and ratios of attrition, outcome of engagements, and functions of these variables. To evaluate the overall mission, factors other than outcome of the conflict (e.g., cost, size of force, composition of force) may also be included in the MOFEs.

Figure 25.3 depicts the relationship between a set of *surveillance* measures for a two-sensor system, showing the typical functions that relate lower-level dimensional parameters upward to higher-level measures. In this example, sensor coverage (spatial and frequency), received signal-to-noise ratios, and detection thresholds define sensor-specific detection and false alarm rate MOPs labeled *measures of detection processing performance (MODPP)* on the figure.

Alternately, the highest-level measures are those that relate P_D (or other detection-specific parameters) to mission effectiveness. Some representative metrics are shown at the top of Figure 25.3, such as P_{Kill}, cost/kill, miss distance, and so on. These metrics could be developed using various computer models to simulate endgame activities, while driving the detection processes with actual sensor data. As mentioned earlier, there is a large *conceptual distance* between the lowest-level and highest-level measures. Forming the computations linking one to the other requires extensive analyses, data and parameters, simulation tools, and so on, collectively requiring possibly significant investments.

The next level down the hierarchy represents the viewpoint of studying surveillance site effectiveness; these measures are labeled *measures of surveillance site effectiveness (MOSSE)* in Figure 25.3. Note, too, that at this level there is a human role that can enter the evaluation; of frequent concern is the workload level to which an operator is subjected. That is, human factor-related measures enter most analyses that range over this evaluation space, adding yet other metrics and measures into the evaluation process; these are not elaborated on here, but they are recognized as important and possibly critical.

Lower levels measure the effectiveness of employing multiple sensors in generating target information (labeled *measures of multisensor effectiveness [MOMSE]* in Figure 25.3), the effectiveness of data combining or fusing *per se* (labeled *measures of data fusion system performance [MODFSP]*), and the effectiveness of sensor-specific detection processes (labeled *MODPP*), as already mentioned.

In studying the literature on surveillance and detection processes, it is interesting to see analogous varying perspectives for evaluation. Table 25.5 shows a summary of some of the works examined, where a hierarchical progression can be seen, and compares favorably with the hierarchy of Figure 25.3; the correspondence is shown in the *Level in Hierarchy* column in Table 25.5.

Metrics (a) and (b) in Table 25.5 reflect a surveillance system-level viewpoint; these two metrics are clearly dependent on detection process performance, and such interactive

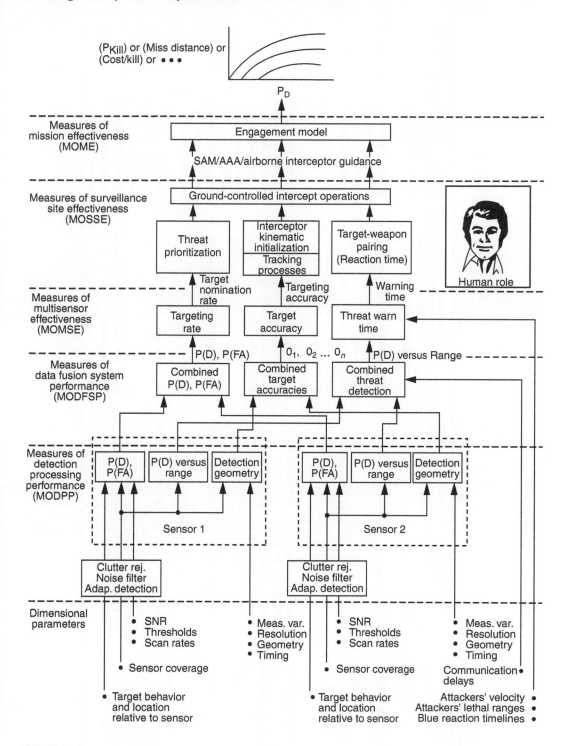

FIGURE 25.3
Hierarchical relationship among fusion measures.

TABLE 25.5

Alternative Concepts and Metrics for Evaluation

Evaluation Point of View	Metric Name	Level in Hierarchy	Calculation	Remarks	Reference
(a) Tracking accuracy/reliability/survivability	Average track (target) *exposure*	MOSSE	Average time during which N surv. system signals are on target	Simulation-based, scenario-dependent analysis	10
(b) Site and system computational workload	Signal loading	MOSSE	Compute sampling distribution for received signal density at (site, system)	Simulation-based scenario-dependent analysis	10
(c) Track continuity	Track purity	MOSSE/MOMSE	Percent correctly associated measurements in a given track; fct (avg. innov. std. dev., target density)	Deterministic target dynamics	11
(d) Composite (M/N) track initiation logic	System operating characteristic (SOC)	MOSSE/MOMSE	Compute $\tilde{P}_{D}$, target track detection probability over several scans	Markov chain model, includes effects of FA	12
(e) Steady-state RMS position error	Tracker operating characteristic (TOC)	MOSSE/MOMSE	Compute numerically steady-state $\tilde{P}(P_D, P_F)$	Probability Data Association (PDA) tracker (Poisson clutter)	13
(f) "Energy management" for optimum target tracking (optimal pulse sharing, sampling/integration periods)	Steady-state MSE	MOSSE/MODFSP	Compute MSE as fct (parameters)	Empirical formulation, square-law detector, Swerling II (single sensor)	14
(g) Optimum threshold management (track initiation, delete, continuation)	*Nagarajan*	MOSSE/MODFSP	Various metrics	Constant false alarm rate (CFAR) assumed (single sensor)	15
(h) Detection error control (threshold control)	Min prob. (error) (global)	MODPP	Relate Min POE to interhypothesis distance metric	Distributed (binary) decision fusion; Blackwell theorem	16

effects could be studied.[10] Track purity, metric (c), a concept coined by Mori et al.,[11] assesses the percentage of correctly associated measurements in a given track, and so it evaluates the association/tracking boundary (MOSSE/MOMSE of Figure 25.3). As commented in the table, this metric is not explicitly dependent on detection performance but the setting of association gates (and thus the average innovations standard deviation, which depends on P_D), so a track purity-to-detection process connection is clear.

Metrics (d) and (e), the system operating characteristic (SOC) and tracker operating characteristic (TOC), developed by Bar-Shalom and co-workers,[12,13] form quantitative bases for connecting track initiation, SOC, and tracker performance, TOC, with P_D, and thus detection threshold strategy performance (the MOSSE/MOMSE boundary in Figure 25.3). SOC evaluates a composite track initiation logic, whereas TOC evaluates the state error covariance, each as connected to, or a function of, single-look P_D.

Metric (f) is presented by Kurniawan et al.[14] as a means to formulate optimum energy management or pulse management strategies for radar sensors. The work develops a semiempirical expression for the mean square error (MSE) as a function of controllable parameters (including the detection threshold), thereby establishing a framework for optimum control in the sense of MSE. For metric (g), Nagarajan et al.[15] formulate a similar but more formally developed basis for sensor parameter control, employing several metrics. Relationships between the metrics and P_D/threshold levels are established in both cases, so the performance of the detection strategy can be related to, among other things, MSE for the tracker process in a fashion not unlike the TOC approach. Note that these metrics evaluate the interrelationships across two hierarchical levels, relating MOSSE to MODFSP.

Metric (h), developed by Hashlamoon and Varshney[16] for a distributed binary decision fusion system, is based on developing expressions for the Min (probability of error, POE) at the global (fusion) level. Employing the Blackwell theorem, this work then formulates expressions for optimum decision making (detection threshold setting) by relating Min (POE) to various statistical distance metrics (distance between H_0, H_1 conditional densities), which directly affect the detection process.

The lowest levels of metrics, as mentioned in the preceding paragraph, are those intimately related to the detection process. These are the standard probabilistic measures P_D and P_{fa} and, for problems involving clutter backgrounds, the metrics that are composed of the set known as *clutter filter performance measures*. The latter has a standard set of IEEE definitions and has been the subject of study of the Surface Radar Group of the AES Radar Panel.[17] The set is composed of

- Multitarget Identification (MTI) improvement factor
- Signal-to-clutter ratio improvement
- Subclutter visibility
- Interclutter visibility
- Filter mismatch loss
- Clutter visibility factor

In the context of the DF process, researchers working at level 1 have been most active in the definition and nomination of measures and metrics for evaluation. In particular, the tracking research community has offered numerous measures for evaluation of tracking, association, and assignment functions.[18–22] In the United States, the Automatic Target Recognizer Working Group (ATRWG), involved with level 1 classification processing, has also been active in recommending standards for various measures.[23]

The challenges in defining measures to have a reasonably complete set across the DF process clearly lie in the level 2–level 3 areas. To assess exactly the *goodness* of a situation or threat assessment is admittedly difficult but certainly not intractable; for example, analogous concepts employed for assessing the quality of images come to mind as possible candidates. A complicating factor for evaluation at these levels is that the *final* determinations of situations or threats typically involve *human* DF—that is, final interpretation of the automated DF products by a human analyst. The human is therefore the final interpreter and effecter/decision-maker in many DF systems, and understanding the interrelationship of MOEs will require understanding a group of *transfer functions* that characterize the translation of information about situation and threat elements into eventual engagement outcomes; one depiction of these interrelationships is shown in Figure 25.4. This figure

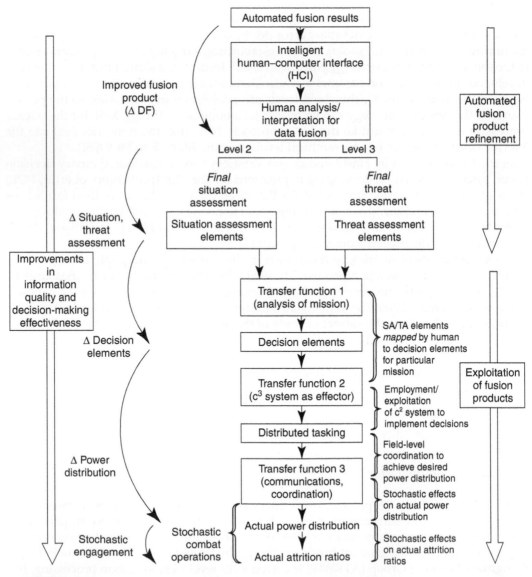

FIGURE 25.4
Interrelationships between human data fusion and system and mission effectiveness.

begins, at the top, with the final product of the automated DF process; all algorithmic and symbolic processing associated with fusion has occurred by this point. That product is communicated to a human through an HCI for both levels 2 and 3 as shown. Given this cognitive interpretation (of the displayed automated results), the human must *transfer* this interpretation, via considerations of

- Decision elements associated with decisions for the given mission (transfer function 1)
- Considerations of the effectiveness of the C3 system, and coordination of communications aspects, to influence the power distribution of his forces on the battlefield (transfer functions 2 and 3)

The implemented decisions (via the C3 and communications systems) result in revisions to force deployment, which then (stochastically) yield an engagement outcome, and the consequent force-effectiveness measures and results. Thus, if all the aspects of human decision making are to be formally accounted for, the conceptual distance mentioned earlier is extended even further, as described in Figure 25.4.

There is yet one more, and important, aspect to think about in all this: ideally, the DF process is implemented as a dynamic (i.e., runtime dynamic) adaptive feedback process— that is, we have level 4, process refinement, at work during runtime, in some way. Such adaptation could involve adaptive sensor management, enabling some intelligent logic to dynamically improve the quality of data input, or it could involve adaptive algorithm management, enabling a logic that switches algorithms in some optimal or near-optimal way. Hence, the overall fusion process is not fixed during runtime execution, so that temporal effects need to be considered. Control theorists talk about evaluating alternative control trajectories while discussing evaluation of specific control laws that enable some type of adaptive logic or equations. DF process analysts will therefore have to think in a similar way and evaluate performance and effectiveness as a function of time. This hints at an evaluation approach that focuses on different phases of a mission and the prioritized objectives at each time-phase point or region. Combined with the stochastic aspects of fusion process evaluation, this temporal dependency only further complicates the formulation of a reasonable approach.

25.5 Summary

Developing an understanding of the relationships among these measures is a difficult problem, labeled here the *interconnectivity of MOEs* problem. The toolkit/testbed needed to generate, compare, and evaluate such measures can be quite broad in scope and itself represent a development challenge as described in Section 25.3.

Motivated in part by the need for continuing maturation of the DF process and the amalgam of techniques employed, and in part by expected reductions in defense research budgets, the DF community must consider strategies for the sharing of resources for research and development. A part of such resources includes standardized (i.e., with approved models) testbed environments, which will offer an economical basis not only for testing of DF techniques and algorithms, but also, importantly, a means to achieve optimality or at least properly satisfying performance of candidate methods under test. However, an important adjunct to shareable testbeds is the standardization of both the overall approach to evaluation and the family of measures involved. This chapter has attempted to offer some ideas for discussion on several of these very important matters.

References

1. Linn, R.J., Hall, D.L., and Llinas, J., A survey of multisensor data fusion systems, presented at the SPIE, Sensor Fusion Conference, In: *Data Structures and Target Classification*, Orlando, FL, pp. 13–29, April, 1991.
2. Villars, D.S., *Statistical Design and Analysis of Experiments for Development Research*, W.M.C. Brown Co., Dubuque, IA, 1951.
3. DeWitt, R.N., Principles for testing a data fusion system, PSR (Pacific-Sierra Research, Inc.) Internal Report, March 1998.
4. Anon, A., Methodology for the test and evaluation of C3I systems, Vol. 1—Methodology Description, Report 1ST-89-R-003A, prepared by Illgen Simulation Technologies, Santa Barbara, CA, May 1990.
5. *International Test and Evaluation Association*, Special Journal Issue on Test and Evaluation of C3IV, Vol. VIII, 2, 1987.
6. Llinas, J. et al., The test and evaluation process for knowledge-based systems, Rome Air Development Center Report RADC-TR-87-181, November 1987.
7. Przemieniecki, J.S., Introduction to mathematical methods in defense analysis, *Am. Inst. Aero. Astro.* Education Series, Washington, DC, 1990.
8. Bryant, M. and Garber, F., SVM classifier applied to the MSTAR public data set, *Proc. SPIE*, Vol. 3721, pp. 355–360, 1999.
9. Sweet, R., Command and control evaluation workshop, MORS C² MOE Workshop, Military Operations Research Society, January 1985.
10. Summers, M.S. et al., Requirements on a fusion/tracker process of low observable vehicles, Final Report, Rome Laboratory/OCTM Contract No. F33657-82-C-0118, Order 0008, January 1986.
11. Mori, S. et al., Tracking performance evaluation: Prediction of track purity, Signal and Data Processing of Small Targets, *SPIE*, Vol. 1096, p. 215, 1989.
12. Bar-Shalom, Y. et al., From receiver operating characteristic to system operating characteristic: Evaluation of a large-scale surveillance system, *Proc. EASCON 87*, Washington, DC, October 1987.
13. Fortmann, T.E., Bar-Shalom, Y., Scheffe, M., and Gelfand, S., Detection thresholds for tracking in clutter-A connection between estimation and signal processing, *IEEE Transactions on Automatic Control*, Vol. 30, pp. 221–229, March 1985.
14. Kurniawan, Z. et al., Parametric optimization for an integrated radar detection and tracking system, *IEEE Proc. (F)*, 132, 1, February 1985.
15. Nagarajan, V. et al., New approach to improved detection and tracking performance in track-while-scan-radars; Part 3: Performance prediction, optimization, and testing, *IEEE Proc. (F), Communications, Radar and Signal Processing*, 134, 1, pp. 89–112, February 1987.
16. Hashlamoon, W.A. and Varshney, P.K., Performance aspects of decentralized detection structures, *Proc. 1989 Int. Conf. SMC*, Cambridge, MA, Vol. 2, pp. 693–698, November 1989.
17. Ward, H.R., Clutter filter performance measures, *Proc. 1980 Int. Radar Conf.*, Arlington, VA, pp. 231–239, April 1980.
18. Wiener, H.L., Willman, J.H., Kullback, J.H., and Goodman, I.R., *Naval Ocean-Surveillance Correlation Handbook 1978*, Naval Research Laboratory, 1978.
19. Schweiter, G.A., MOPs for the SDI tracker-correlator, Daniel H. Wagner Associates Interim Memorandum to Naval Research Laboratory, 1978.

20. Mifflin, T., Ocean-surveillance tracker-correlator measures of performance, Naval Research Laboratory Internal Memorandum, 1986.

21. Askin, K., SDI-BM tracker/correlator measures of performance, Integrated Warfare Branch Information Technology Division, April 4, 1989.

22. Belkin, B., Proposed operational measures of performance, Daniel H. Wagner Associates Internal Memorandum, August 8, 1988.

23. Anon., Target recognizer definitions and performance measures, prepared under the auspices of the Automatic Target Recognizer Working Group Data Collection Guidelines Subcommittee of the Data Base Committee, ATRWG No. 86-001, approved February 1986.

26

Survey of COTS Software for Multisensor Data Fusion*

Sonya A. H. McMullen, Richard R. Sherry, and Shikha Miglani

CONTENTS

26.1 Introduction

In 1993, D. L. Hall and R. J. Linn[1] conducted a survey of commercial off the shelf (COTS) software to support development of data fusion systems for applications such as automatic target recognition, identification-friend-foe-neutral processing, and battlefield surveillance. In the survey, they described a number of emerging packages containing basic algorithms for signal processing, image processing, statistical estimation, and prototyping of expert systems. Since the publication of that paper, extensive progress has been made in data fusion for both Department of Defense (DoD) applications as well as non-DoD applications. The basic algorithms and techniques for data fusion have evolved[2] and engineering standards are beginning to emerge for system design[3] and requirement derivation and analysis.[4] Numerous data fusion systems have been developed for DoD applications[5,6] and systems are beginning to be developed for non-DoD applications such as the condition-based monitoring of complex mechanical systems.[7†] In addition, non-DoD applications such as data mining, pattern recognition, and knowledge discovery have spurred the development of commercial software tools[8] and general packages such as MATLAB[9] and Mathematica.[10] Because of these rapid developments along with the emergence of COTS packages for prototyping data fusion applications, it was deemed worthwhile to update

---

* This chapter is an update of original paper: Sherry, R.R. and Hall, S.A. (2002). "A survey of COTS software for multi-sensor data fusion; what's new since Hall and Linn?" *Proceedings of the MSS National Symposium on Sensor and Data Fusion*, San Diego, CA.
† Reference Chapters 21, 22, 27, and 28 of this Handbook for updated information on the above references.

the survey conducted by Hall and Linn. S. A. Hall and R. R. Sherry updated the original survey by Hall and Linn and presented that survey at the 2002 MSS National Symposium on Sensor and Data Fusion. This chapter provides an update of that paper and presents a summary of the updated survey, identifies some applicable COTS software, and provides a *survey of surveys* related to data fusion systems and software.

26.2   Taxonomy for Multisensor Data Fusion

Numerous experts have described the concept of the multisensor data fusion process in-depth. Hall, Hall and Oue.[11] have previously provided the following description. Traditionally, data fusion systems have been developed to ingest information from multiple sensors and sources to provide information for automated situation assessment, or to assist a human in development of situation assessments.[12–14] Extensive research has focused on the transformation from sensor data to target locations, target identification, and (in limited cases) contextual interpretation of the target information. Numerous systems have been developed to perform this processing to achieve or support situation assessment.[5,6] The data fusion process is often represented conceptually by the Joint Directors of Laboratories (JDL) data fusion process model illustrated in Figure 26.1. The JDL augmented model shown in Figure 26.1 shows the level 5 process recommended by Hall et al.[15] (and introduced at the same conference by Blasch) to explicitly account for functions associated with human–computer interactions (HCI).

26.3   Survey of COTS Software and Software Environments

Three significant advances have shaped the evolution of COTS data fusion applications software since the publication of Hall and Linn's survey in 1993. First is the rapid

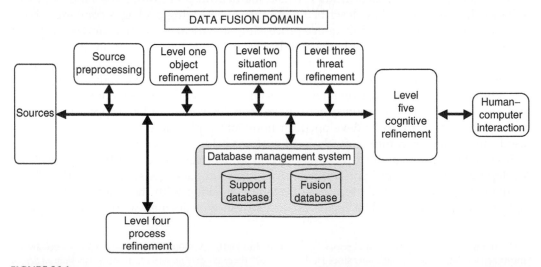

FIGURE 26.1
Revised Joint Directors of Laboratories (JDL) data fusion process model. (From Hall, M.J., Hall, S.A., and Tate, T., *Proceedings of the 2000 MSS National Symposium on Sensor and Data Fusion*, October 2000.)

development of specialized software algorithm packages used within technical computation environments such as MATLAB,[16] Mathematica,[17] and Mathcad.[18] These environments provide unified platforms for mathematical computation, analysis, visualization, algorithm development, and application. The second development is the evolution of general-purpose data fusion software packages with adaptive capabilities to a diverse set of data fusion applications. The third development (described in Section 26.5) involves the development of tools such as the Adobe Flex 2.0 cross-platform development framework for prototyping web-based applications.

26.3.1 Special Purpose COTS Software

Table 26.1 shows a summary of some of the specialized software packages. These are all compatible with at least one of the three mathematical environments previously described. There are literally hundreds of software packages available. Hence, Table 26.1 is by no means a complete listing of the available commercial products. However, it provides a basis for a more in-depth search for such products. Use of these types of packages in conjunction with MATLAB, Mathematica, or Mathcad algorithms, visualization features, and user-interface capabilities can significantly reduce development time for prototype application simulations. Projects initially prototyped and implemented in MATLAB or Mathematica environment can be imported to C and C++ to generate parallel C code from high-level MATLAB programs, or MathCode C++[19] for generation of C++ code from Mathematica code.

An example of experience utilizing MATLAB as a prototype environment is described by West et al.[20] They indicate that MATLAB-based tracking is feasible and relatively efficient. However, they noted that when the MATLAB *compiler* translated the MATLAB (*.m) files into C coding, there were many calls to MATLAB.dll functions and the resulting runtimes were similar to standard (*.m) file utilization. West et al.[20] observed a dramatic runtime improvement (speed-up) by recoding functions in the C language.

26.3.2 General Purpose Data Fusion Software

The second class of COTS software involves packages specifically designed to perform data fusion. A summary of these is provided in Table 26.2 and a brief summary of example packages is provided below.

The Lockheed Martin data fusion workstation[21] is directed at integrated multisensor, multitarget classification based on fusing 14 types of sensor inputs including passive-acoustic, ESM/ELINT, magnetic field, electric field, inverse synthetic aperture radar, and geolocation. The data fusion workstation consists of three principal components: (1) the embedded data fusion system, (2) a graphical user interface, and (3) a data fusion input simulator. The fusion engine is fuzzy logic based on Dempster–Shafer reasoning. The software is objected-oriented C++ code, which runs on Silicon graphics and Sub SPARCstation platforms. The Lockheed Martin data fusion system software is utilized as an embedded function in the Rotorcraft Pilot's Associate.

Hall and Kasmala[22] at the Pennsylvania State University Applied Research Laboratory have developed a visual programming tool kit for level 1 data fusion. The tool kit is modeled after the tool *Khoros* developed for collaborative development of image-processing algorithms. The top-level interface allows a user to define a level 1 processing flow by a *point and click* manipulation of graphic symbols representing types of fusion algorithms. The user can define multiple processing streams to analyze single sensor data or multiple processing streams to fuse multisensor data. A library of routines is available and can be

TABLE 26.1

Examples of COTS Software for Use in Mathematical Software Environments

Category of Algorithms/Techniques	JDL Model Level	Program Name	Source
Image Processing and Mapping	0	DIPimage	Delft University of Technology, http://www.diplib.org/home2
		Mapping Toolbox	Mathworks, http://www.mathworks.com/products/mapping/tryit.html
		Image Processing Extension Pack	Adept Scientific International, http://www.adeptscience.com/products/mathsim/mathcad/add-ons/image.html
		Digital Image Processing for Mathematica	HalloGram Publishing, http://hallogram.com/mathematica/digitalimage/index.html
		ICALAB	Brain Science Institute, http://www.bsp.brain.riken.go.jp/ICALAB/
		PyrTools	Laboratory for Computational Vision, http://www.cns.nyu.edu/~eero/software.html
Pattern Recognition and Classification	1, 2, 3	PRT	Innovative Dynamics, Inc., http://www.idiny.com/prt.html
		Statistical Pattern Recognition Toolbox	Czech Technical University, http://cmp.felk.cvut.cz/cmp/software/stprtool/index.html
		PRTools	Delft University of Technology, http://www.prtools.org/
		DADiSP	Adept Scientific, http://www.adeptscience.co.uk/products/dataanal/dadisp/modules/nn.html
		TEMPLAR	Clay Scott, DSP Group, Rice University, http://www.dsp.rice.edu/software/templar.shtml
Neural Networks	1, 2, 3	Netlab	Aston University, http://www.ncrg.aston.ac.uk/netlab/
		Netpack	Radboud University of Nijmegen, The Netherlands, http://www.snn.ru.nl/nijmegen/netpack.php3
		NNSYSID	Technical University of Denmark, http://www.iau.dtu.dk/research/control/nnsysid.html
		NNCNTL	Technical University of Denmark, http://www.iau.dtu.dk/research/control/nnctrl.html
		Back Propagated Neural Network	Steve Hunka, University of Alberta, http://library.wolfram.com/infocenter/MathSource/475/
		Simulating Neural Networks	James A. Freeman, http://library.wolfram.com/infocenter/Books/3485/
		LMAM-OLMAM	Demokritos, http://iit.demokritos.gr/~abazis/toolbox/
Fuzzy Logic/Control	2, 3, 4	FISMAT	http://www.mathtools.net/files/net/fismat.zip
		Fuzzy Logic	Wolfram Research, http://www.wolfram.com/products/applications/fuzzylogic/
Bayesian Networks	1, 2, 3	Bayes Net	Kevin Murphy, http://bnt.sourceforge.net/
		Hidden Markov Model	Kevin Murphy, University of British Colombia, http://www.cs.ubc.ca/~murphyk/Software/HMM/hmm.html
		Kalman	University of Oxford, http://www.robots.ox.ac.uk/~sjrob/Outgoing/software.html
		VBHMM; VBMFA	SUNY, Buffalo, http://www.cse.buffalo.edu/faculty/mbeal/software.html
Digital Signal Processing and Filters	0, 1	DSP-KIT	Meodat Gmbh, http://www.meodat.de/PCSys/dsp-kit.html

(continued)

TABLE 26.1   (Continued)

Category of Algorithms/Techniques	JDL Model Level	Program Name	Source
		DirectDSP	Signalogic, http://www.signalogic.com/matlab.shtml
		DSP Builder	Altera, http://www.altera.com/products/software/system/products/dsp/dsp-builder.html
		FIR and IIR Filter Design	Rice University, http://www.dsp.rice.edu/software/rufilter.shtml
		DSP Developer	SDL, http://www.sdltd.com/
		Kalman Filter	Kevin Murphy, University of British Colombia, http://www.cs.ubc.ca/~murphyk/Software/Kalman/kalman.html
		KALMTOOL	Technical University of Denmark, http://www.iau.dtu.dk/research/control/kalmtool.html
		WarpTB	Helsinki University of Technology, http://www.acoustics.hut.fi/software/warp/
		ReBEL	Oregon Health & Science University, http://choosh.csee.ogi.edu/rebel/
		Signals Processing Packages	Brian Evans, http://www.ece.utexas.edu/~bevans/projects/symbolic/spp.html
		Signal Processing Extension Pack	Adept Scientific plc, http://www.adeptscience.co.uk/products/mathsim/mathcad/add-ons/signal.html
Tracking and Positioning	1	AuTrakMatlab	AuSIM, http://www.ausim3d.com/products/autrak.html
		DynaEst	Carnegie Mellon University, http://vasc.ri.cmu.edu/old_help/Software/Dynaest/dynaest.html
Wavelet Analysis	0, 1, 2	Wavelab	Stanford University, http://www.stat.stanford.edu/~wavelab/
		Rice Wavelet Toolbox	Rice University, http://www.dsp.rice.edu/software/rwt.shtml
		MWM	Rice University, http://www.dsp.rice.edu/software/mwm.shtml
		Wavelet Explorer	Wolfram Research, http://www.wolfram.com/products/applications/wavelet/
		ForWaRD	Rice University, http://www.dsp.rice.edu/software/ward.shtml
Self-Organizing Maps	1, 2	SOM Toolbox	Helsinki University of Technology, http://www.cis.hut.fi/projects/somtoolbox/
Graph Theory and Search	1	Graph Theory Toolbox	University of California at Berkeley, http://www.cs.ubc.ca/~murphyk/Software/graph.zip
		Anneal	Jeff Stern, http://library.wolfram.com/infocenter/MathSource/445/
		Combinatorica	Stony Brook University, http://www.cs.sunysb.edu/~skiena/combinatorica/combinatorica.html
		Heuristic Search Techniques	Raza Ahmed, Brian L. Evans, http://library.wolfram.com/infocenter/MathSource/607/
Speech Processing	5	VOICEBOX	Imperial College of Science, Technology and Medicine, http://www.ee.ic.ac.uk/hp/staff/dmb/voicebox/voicebox.html
	5	COLEA	University of Texas, Dallas, http://www.utdallas.edu/%7Eloizou/speech/colea.html
Database	N/A	DBTool	http://energy.51.net/dbtool/
		Database Toolbox	Mathworks, http://www.mathworks.com/products/database/

TABLE 26.2

General Purpose Data Fusion Software

Software/Product Name	Source/Developer	Description
Lockheed Martin Data Fusion Workstation[21]	Lockheed Martin	• Integrated multisensor, multitarget tool • Fuses data from 14 sensor inputs including: passive-acoustic, ESM/ELINT, magnetic field, electric field, inverse synthetic aperture radar and geolocation • Components: graphical user interface, embedded data fusion system, data fusion input simulator • Fusion engine based on Dempster–Shafer reasoning • Platform: Silicon Graphics, Sub SPARCstation
PSU Applied Research Laboratory Data Fusion Toolkit[22]	Penn State Applied Research Laboratory	• Provides a visual interface for utilizing and developing Level 1 fusion algorithms • Platform: PC with Windows
Brooks and Iyengar[23]	R.R. Brooks & S.S. Iyengar	• C language routines as companion software to their data fusion book • General techniques including neural networks, Kalman filter, etc.
HydroFACT[24]	Fusion and Control Technology, Inc.	• Department of Energy funded tool to combine data associated with geophysical, geologic, and hydrologic data
KnowledgeBoard[25]	SAIC	• Information visualization and fusion for multisource data including video, XML-based documents, and images
ImageLock Data Fusion[26]	PCI Geomatics	• Semiautomatic geometric correction and fusion of image data
ECognition[27]	Definiens Imaging	• Fusion and preprocessing of image data
Boeing NDE S/W[28]	Boeing	• Data fusion workstation for nondestructive evaluation (NDE) and process monitoring
CASE ATTI[29]	Defense R&D Canada	• Test bed for evaluating multisensor data fusion systems against requirements
Pharos[30]	Sonardyne	• Acoustic navigation software for subsea acoustic geopositioning
Objectivity/DB[31]	Objectivity	• Object-Oriented Database Management System • Data fusion applications in government and complex manufacturing applications
DNBi Supply Management Solution[32]	Open Ratings, Inc.	• Analyzes and links essential supplier information—including key performance metrics, financial indexes, and legal and government compliance requirements—with an organization's own global supply base information • Supply risk assessment for businesses in aerospace, automotive, and industrial equipment manufacturing
Black Coral LIVE[33] and Black Coral MOBILE[33]	Black Coral, Inc.	• Geospatial data fusion and collaboration that permit teams of users to share various types of information, track assets and to communicate in real time using digital mapping software • Target areas: police, fire and rescue, emergency health and medical services, defense, intelligence, homeland security, emergency management, environmental agencies, forest services, transportation facilities, oil and gas companies, pipeline companies, chemical/petrochemical companies, hydro electric utilities, telephone companies

selected for processes such as signal conditioning, feature extraction, and pattern recognition or decision-level fusion. The tool kit was originally developed for applications involving condition monitoring of mechanical systems. However, the algorithms are general purpose and can be used on any signal or vector data.

Brooks and Iyengar[23] provide a suite of C language routines as companion software to their book on multisensor fusion. These routines collectively have been labeled the sensor fusion toolkit and contain implementations of machine learning, learned meta-heuristics, neural networks, simulated annealing, genetic algorithms, tabu search, a Kalman filter, and their own distributed dynamic sensor algorithm.

The Department of Energy (DoE) has funded research directed at developing models for groundwater flow and contaminant transport. This work has resulted in development of the hydrogeologic data fusion computer model.[24] This model can be used to combine various types of geophysical, geologic, and hydrologic data from diverse sensor types to estimate geologic and hydrogeologic properties such as ground water flow. A commercial version of this software is available under the name Hydro-FACT from Fusion and Control Technology, Inc.

KnowledgeBoard[25] is a software framework for information collection, data fusion, and information visualization. KnowledgeBoard is directed at the fusion of distributed high-level multisource data including XML-based documents, video, web-based content, relational and object databases, flat files, etc. Heterogeneous data sources are displayed within a contextual framework to support user-level fusion of the information and to support improved knowledge generation.

PCI Geomatics has developed the ImageLock Data Fusion software[26] that is designed to perform semiautomatic geometric correction, fuse the corrected multisensor imagery, and perform noise and artifact removal. The ImageLock Data Fusion software is a component in PCI's EASI/PACE remote sensing image processing software package.

eCognition[27] by Definiens Imaging is designed to extract features from high-resolution satellite and aerial imagery. eCognition provides for multisource fusion from a large variety of formats, sensors, and platforms at any spatial and spectral resolution.

Boeing has also developed a data fusion workstation for fusing data from nondestructive examination (NDE) testing and from process monitoring.[28] NDE and process monitoring is applicable throughout the life cycle (manufacturing, repair, in-service inspection) of numerous components.[28] The benefit of the data fusion system is that it aids in cross correlation of data, which results in the reduction in ambiguity from NDE inspections.[28] Overall, this tool has served to reduce technical labor requirements as well as evaluation variability particularly for aerospace components.[28]

Defense R&D Canada—Valcartier has developed a test bed for evaluating multisensor data fusion systems against system requirements.[29] This test bed is called CASE ATTI (Concept Analysis and Simulation Environment for Automatic Target Tracking and Identification). One of the main objectives of the development of CASE ATTI was to allow the testing and comparison of different sensor fusion algorithms and techniques without extensive recoding. CASE ATTI allows use of either externally generated sensor data or data generated by a high fidelity simulator that emulates the behavior of targets, sensor systems, and the environment. CASE ATTI runs on UNIX and Windows platforms and can be used with multiple computers across a local area network.

Objectivity/DB[31] by objectivity is an object-oriented database management system dealing with large volumes of high dimensional interrelated data. Objectivity/DB is a database engineer with a database library linked to application programs rather than a separate data server process. The developers of Objectivity/DB claim that the package integrates with applications software through standard language interfaces such as C#/.NET, C++, java, SQL, XML, and others. They also cite applications to areas such as bioinformatics, manufacturing, and military situation assessment.

Open Ratings's DNBi Supply Management Solution[32] incorporates machine learning and data mining technologies that identify pattern from large sets of noisy data sets and maintains a supplier database of more than 100 million records. In addition, there is a collaborative workspace for collecting and organizing supplier information when an issue arises—including historical supplier performance intelligence, user notes, internet links, supplier-completed assessments, external files, and contingency plans.

Black Coral LIVE[33] and Black Coral MOBILE[33] are integrated products that facilitate situational awareness and interoperability allowing multiple organizations in emergency response and military environments to communicate, collaborate, and coordinate for distributed personnel and tactical teams. The system utilizes the ESRI ArcGIS software for displaying and manipulating geographical information (and associated overlay data), and provides functions to support distributed users collaborating via annotation of maps and collective development of an evolving situation display.

26.4 A Survey of Surveys

To assist in identifying additional tools, Table 26.3 identifies sources (papers, reports, and websites) that provide additional surveys of relevant software. In addition to these resources, annual software surveys are available via the ACM computing surveys. The website for the ACM special interest group SIGGRAPH (http://www.siggraph.org/) provides general information on graphics and visualization software, whereas the commercial company Zoom Info surveys and identifies companies who offer specialized services or products. Hence, Zoom Info can be used to find data fusion products and related companies.

26.5 Discussion

This brief survey shows the acceleration of the development and application of data fusion software methods across a growing number of application domains. Since Hall and Linn's original survey, several new trends can be observed:

1. There is an explosion of commercial software tools for a wide variety of component techniques such as signal processing, image processing, pattern recognition, data mining, and automated reasoning.
2. New mathematical environments such as MATLAB, Mathcad, and Mathematica provide extensive tool kits (including methods for signal processing, neural nets, image processing, etc.) that are useful for rapid prototyping and evaluation of data fusion methods.
3. A limited number of more general purpose tool kits are available. These are quasi-COTS tools (they are available for use or purchase but do not have the general utilization or support of software such as MATLAB or Mathematica).

TABLE 26.3

Software Surveys

Developer/Source	Focus	Reference
D. L. Hall and R. J. Linn	Survey of COTS S/W to support data fusion	Ref. 1
D. L. Hall, R. J. Linn, and J. Llinas	Survey of DoD data fusion systems	Ref. 6
M. L. Nichols	Survey of DoD data fusion systems	Ref. 5 and Chapter 27 of this handbook.
Aerospace Corp.	Survey and analysis of 34 tools for multisource data fusion at the NAIC	"An investigation of tools/systems for multisource data fusion at the national air intelligence center" by C.N. Mutchler; technical report prepared for the NAIC; Aerospace Report TOR-2001(1758)-0396e; 2001
M. Goebel and L. Gruenwald	Survey of data mining and knowledge discovery S/W	Ref. 8
B. Duin	Web-based links to tools associated with pattern recognition, neural networks, machine learning, image processing and vision, and OCR and handwriting recognition	http://www.ph.tn.tudelft.nl/PRInfo/software.html
EvoWeb	List of software tools for pattern recognition and classification maintained by the European Commission's IST program	http://evonet.lri.fr/evoweb/resources/software/bbsubcatlist.php?id=6.02&desc=Pattern%20recognition%20and%20classification
Pattern recognition on the web	Large set of links to training materials, tutorials, and software tools for pattern recognition, statistics, computer vision, classifiers, filters, and other S/W	http://cgm.cs.mcgill.ca/~godfried/teaching/pr-web.html
Carnegie Mellon University (CMU)	StatLib—data, software, and information for the statistics community maintained by researchers at CMU; extensive list of software and resources	http://lib.stat.cmu.edu/
Data and Analytic Center for Software (DACS)	Extensive list and comments on software for data mining, data warehousing, and knowledge discovery tools	http://www.thedacs.com/databases/url/key.php?keycode=222:225
M. Wexler	Open directory of software, databases, and reference materials on data mining; list of 87 tool vendors	http://dmoz.org/Computers/Software/Databases/Data_Mining/
Open Channel Foundation	Open Channel Foundation publishes software from academic and research institutions; software available for development of expert systems and pattern recognition	http://www.openchannelfoundation.org/discipline/Artificial_Intelligence_and_Expert_Systems/

(continued)

TABLE 26.3   (Continued)

Developer/Source	Focus	Reference
Compinfo—The Computer Information Center	Link to sources of information about computers software and information technology	http://www.compinfo-center.com/
OR/MS Today	Survey of 125 vendors of statistical software	http://www.lionhrtpub.com/orms/surveys/sa/sa-surveymain.html
Zoom Info	Directory of data fusion software vendors	http://www.zoominfo.com/Industries/software-mfg/software-development-design/data-fusion.html
U.S. Army Topographic Engineering Center (TEC)	Survey of Terrain Visualization Software	http://www.tec.army.mil/TD/tvd/survey/
KDnuggets	Software suites for data mining and knowledge discovery	http://www.kdnuggets.com/software/suites.html
Scientific Web	Scientific web—modules and add-ons for Mathematica for database, digital image processing, dynamic visualization, Fuzzy logic and finance	http://www.scientificweb.com/software/wolfram/mathmode.html
Scientific Web	Scientific web—modules and add-ons for Matlab with toolboxes for finance, database, communication, adaptive optics, bioinformatics, data acquisition, data structures and control system	http://www.scientificweb.com/software/mathworks/matlmode.html
Hearne Scientific Software	Mathcad add-on collection for data analysis, financial and economic analysis, graphing and contouring, market research, statistics, risk management	http://www.hearne.com.au/products/mathcad/add-on/collection/
SKB's Virtual Cave	Matlab toolboxes for data fusion algorithms	http://stommel.tamu.edu/~baum/toolboxes.html
Wolfram	Mathematica software solutions and third party applications for data fusion applications	Ref. 10
SIGGRAPH2007	List of companies which provides software/hardware for data fusion algorithms	http://esub.siggraph.org/cgi-bin/cgi/idECatList.html&CategoryID=1
Northern Illinois University	Web resource for graph theory text books and software	Northern Illinois University, http://www.math.niu.edu/~rusin/known-math/index/05CXX.html

4. The software development environment (including new languages such as Visual Basic, Visual C++, JAVA Script, and XML) provides an improved basis for development of object-oriented, interoperable software. In this breed of new software is Adobe Flex 2.0[35] cross-platform development framework. The purpose of Flex 2.0 is to create software tools that quickly build and deploy rich internet applications (RIA) within the enterprise or across the web. As a result, Flex 2.0 creates intelligent desktop applications that augment overall user experience and assists users in analyzing data and business processes for productive corporate decisions.

Flex applications are developed with MXML and object-oriented programming language Actionscript using the Flex builder integrated development environment or a standard text editor. The Flex compiler then compiles the source code into bytecode as binary SWF file. This file is executed by flash player at runtime on all major browsers and operating systems. Developers can use Flex's prebuilt visual, service, and behavior components from the class library of components and containers, and customize them by extending these components or their base classes to achieve the desired look and feel. Flex clients can be used in conjunction with any server environment such as JSP, ASP, ASP.NET, ColdFusion, and PHP. The data request over the Internet can be made via standard HTTP calls or web-services. In addition, Java server based Flex data services enable data transfer, data synchronization and conflict management, and real-time data messaging. Furthermore, to develop data dashboard for interactive analysis, Flex charting provides a library of extensible charting component.[35]

Recently, Flex has been used in analytical applications as a user-interface technology to analyze trends and monitor applications for better business decisions.[34]

In general, these developments provide an increasingly efficient environment to develop and use data fusion algorithms. However, there is still a need for the widespread availability of a standard package for multisensor data fusion.

References

1. D. L. Hall and R. J. Linn, "Survey of commercial software for multisensor data fusion," *Proceedings of the SPIE Conference on Sensor Fusion and Aerospace Applications*, Orlando, FL, 1993.

2. D. L. Hall and J. Llinas, eds. *Handbook of Multisensor Data Fusion*, CRC Press Inc., Boca Raton, FL, 2001.

3. C. L. Bowman and A. N. Steinberg, "A systems engineering approach for implementing data fusion systems," chapter 16 in *Handbook of Multisensor Data Fusion*, D. L. Hall and J. Llinas, eds., CRC Press Inc., Boca Raton, FL, 2001.

4. E. Waltz and D. L. Hall, "Requirements derivation for data fusion systems," chapter 15 in *Handbook of Multisensor Data Fusion*, D. L. Hall and J. Llinas, eds., CRC Press Inc., Boca Raton, FL, 2001.

5. M. L. Nichols, "A survey of multisensor data fusion systems," chapter 22 in *Handbook of Multisensor Data Fusion*, D. L. Hall and J. Llinas, eds., CRC Press Inc., Boca Raton, FL, 2001.

6. D. L. Hall, R. J. Linn, and J. Llinas, "A survey of data fusion systems," *Proceedings of the SPIE Conference on Data Structures and Target Classification*, pp. 13–36, 1991.

7. C. S. Byington and A. K. Garga, "Data fusion for developing predictive diagnostics for electro-mechanical systems," chapter 23 in *Multisensor Data Fusion*, D. L. Hall and J. Llinas, eds., CRC Press Inc., Boca Raton, FL, 2001.

8. M. Goebel and L. Gruenwald, "A survey of data mining and knowledge discovery software tools," *SIGKDD Explorations*, Vol. I, issue 1, pp. 20–33, 1999.

9. http://www.mathworks.com/.

10. http://www.wolfram.com/products/mathematica/index.html.

11. M. J. Hall, S. A. Hall, and C. Oue, "A word (may) be worth a thousand pictures: on the use of language representation to improve situation assessment," *Proceedings of the 2001 MSS National Symposium on Sensor and Data Fusion*, Vol. II, June 2001.

12. E. Waltz and J. Llinas, *Multisensor Data Fusion*, Artech House, Inc., Norwood, MA, 1990.

13. D. L. Hall, *Mathematical Techniques for Multisensor Data Fusion*, Artech House, Inc., Norwood, MA, 1992.

14. D. L. Hall and J. Llinas, "An introduction to multisensor data fusion," *Proceedings of the IEEE*, Vol. 85, No. 1, January 1997.

15. M. J. Hall, S. A. Hall, and T. Tate, "Removing the HCI bottleneck: how the human computer interface (HCI) affects the performance of data fusion systems," *Proceedings of the 2000 MSS National Symposium on Sensor and Data Fusion*, Vol. II, San Antonio, TX, pp. 89–104, October 2000.

16. *Developers of MATLAB and Simulink for Technical Computing*. The MathWorks. Retrieved August 13, 2007, from the World Wide Web: http://www.mathworks.com/.

17. *Mathematica*. Wolfram Research. Retrieved August 13, 2007, from the World Wide Web: http://www.wolfram.com/products/mathematica/index.html.

18. *Mathcad*. MathSoft. Retrieved August 13, 2007, from the World Wide Web: http://www.ptc.com/products/mathcad/mathcad14/promo.htm.

19. *MathCode C++*. MathCore AB. Retrieved August 13, 2007, from the World Wide Web: http://www.mathcore.com/products/mathcode/mathcodec++.php.

20. P. West, D. Blair, N. Jablonski, B. Swanson, and T. Armentrout, "Development and real-time testing of target tracking algorithms with AN/SPY-1 radar using Matlab," presented at the *Fourth ONR/GTRI Workshop on Target Tracking and Sensor Fusion*, Monterey, CA, 2001.

21. A. Pawlowski and P. Gerken, "Simulator, workstation and data fusion components for onboard/off-board multi-targeted multisensor data fusion," *IEEE/AIAA Digital Avionics Systems*, 1998.

22. D. L. Hall and G. A. Kasmala, "A visual programming tool kit for multisensor data fusion," *Proceedings of the SPIE AeroSense 1996 Symposium*, Orlando, FL, Vol. 2764, pp. 181–187, April 1996.

23. R. R. Brooks and S. S. Iyengar, *Multi-Sensor Fusion: Fundamentals and Applications with Software*, Prentice Hall, Upper Saddle River, NJ, 1998.

24. *Hydrogeologic Data Fusion*. Innovative Technology, 1999. Retrieved August 13, 2007, from the World Wide Web: http://apps.em.doe.gov/ost/pubs/itsrs/itsr2944.pdf.

25. *KnowledgeBoard*. SAIC. Retrieved August 13, 2007, from the World Wide Web: http://www.saic.com/products/software/knowledgeboard/.

26. *ImageLock Data Fusion Package*. PCI Geomatics. Retrieved August 13, 2007, from the World Wide Web: http://www.pcigeomatics.com/cgi-bin/pcihlp/IMAGELOCK.

27. *eCognition—object-oriented image analysis*. Definiens Imaging. Retrieved August 13, 2007, from the World Wide Web: http://www.definiens.com/article.php?id=6.

28. R. Bossi and J. Nelson. *NDE Data Fusion*. Boeing. Retrieved August 13, 2007, from the World Wide Web: http://www.ndt.net/abstract/asntf97/053.htm.

29. *CASE ATTI: A Test Bed for Sensor Data Fusion*. Defence R&D Canada—Valcartier. Retrieved August 13, 2007, from the World Wide Web: www.valcartier.drdc-rddc.gc.ca/poolpdf/e/137_e.pdf.

30. http://www.sonardyne.co.uk/Products/PositioningNavigation/systems/pharos.html.

31. http://www.objectivity.com/data-fusion.shtml.
32. http://www.openratings.com/solutions/solutions_overview.html.
33. http://www.blackcoral.net.
34. A. Lorenz and G. Schoppe, *Developing SAP Applications with Adobe Flex*, Galileo Press, Bonn, Germany, 2007.
35. *Adobe Flex 2.0*. Adobe. Retrieved August 13, 2007, from the World Wide Web: http://www.adobe.com/products/flex/whitepapers/pdfs/flex2wp_technicaloverview.pdf.

27

Survey of Multisensor Data Fusion Systems

Mary L. Nichols

CONTENTS

27.1   Introduction

During the past two decades, extensive research and development (R&D) on multisensor data fusion has been performed for the Department of Defense (DoD). By the early 1990s, an extensive set of fusion systems had been reported for a variety of applications ranging from automated target recognition (ATR) and identification-friend-foe-neutral (IFFN) systems to systems for battlefield surveillance. Hall et al.[1] provided a description of 54 such systems and an analysis of the types of fusion processing, the applications, the algorithms, and the level of maturity of the reported systems. Subsequent to that survey, Llinas and Antony[2] described 13 data fusion systems that performed automated reasoning (e.g., for situation assessment) using the blackboard reasoning architecture. By the mid-1990s, extensive commercial off-the-shelf (COTS) software was becoming available for different data fusion techniques and decision support. Hall and Linn[3] described a survey of COTS software for data fusion (an update of which is provided in this handbook) and Buede[4,5] performed surveys and analyses of COTS software for decision support.

This chapter presents a new survey of data fusion systems for DoD applications. The survey was part of an extensive effort to identify and assess DoD fusion systems and activities. This chapter summarizes 79 systems and provides an assessment of the types of fusion processing performed and their operational status.

27.2   Recent Survey of Data Fusion Activities

A survey of DoD operational, prototype, and planned data fusion activities was performed in 1999–2000. The data fusion activities that were surveyed had disparate missions and provided a broad range of fusion capabilities. They represented all military services. The survey emphasized the level of fusion provided (according to the JDL model described in many chapters of this book, such as Chapter 3) and the capability to fuse different types of intelligence data. A summary of the survey results is provided here.

In the survey, a data fusion system was considered to be more than a mathematical algorithm used to automatically achieve the levels of data fusion described in, for example, Chapter 2. In military applications, data fusion is frequently accomplished by a combination of the mathematical algorithms (or *fusion engines*) and display capabilities with which a human interacts. Hence, the activities range from relatively small-scale algorithms to large-scale command, control, intelligence, surveillance, and reconnaissance (C4I) systems, which use specific algorithms—such as trackers—in conjunction with a sophisticated display of data from multiple intelligence (multi-INT) data types.

The objective in identifying the unique data fusion activities was to isolate the unique capabilities, both mathematical and display-related, of the activity. A master list was initiated, and the researcher applied expert judgment in eliminating activities for any of the following reasons: (1) obsolete systems, (2) systems that were subsumed by other systems, (3) systems that did not provide unique fusion capabilities, (4) systems that were only data fusion enablers, and (5) systems that emphasized visualization.

Table 27.1 lists the resulting 79 unique DoD activities with their primary sponsoring service or organization. The list is intended to be a representative, rather than exhaustive, survey of all extant DoD fusion activities. The R&D activities, as well as the prototypical systems, are shown in bold type.

27.3 Assessment of System Capabilities

A primary goal of the survey was to understand the JDL fusion capabilities of the current operational fusion activities, as well as to emphasize the R&D activities. Recognizing that the activities often provided more than one level of fusion, all of the activities were assigned one or more levels. For instance, a typical operational fusion activity, such as the global command and control system (GCCS), provides specialized algorithms for tracking to achieve level 1 (object refinement) in addition to specific display capabilities aimed at providing the necessary information from which the analyst can draw level 2 (situation refinement) inferences.

The capabilities were then counted for both the operational and nonoperational activities, and a histogram was generated (Figure 27.1). Note that level 2 fusion was divided into two components, given that many operational military data fusion systems are said to facilitate level 2 fusion through the display fusion of various INTs from which an analyst can draw level 2 inferences.

The majority of operational data fusion activities provide level 1 fusion. These activities include weapon systems, such as the advanced tomahawk weapons control system (ATWCS), and trackers, such as the processor independent correlation and exploitation system (PICES). A less common operational capability is algorithmic level 2 fusion, which is provided by some operational systems such as the all-source analysis system (ASAS).

The majority of the operational systems that are geared toward intelligence analysis have emerged from a basic algorithm to track entities of interest. In most cases, the trackers have operated from signals intelligence (SIGINT). Gradually, these systems evolved by adding a display not only to resolve ambiguities in the tracking, but also to bring in additional INTs. As a result, most of the systems provide an underlying algorithmic fusion, in addition to a display that accommodates multi-INT data.

Algorithmic fusion to achieve level 3 fusion is uncommon among the operational systems, and none of the operational systems provide level 4 fusion. These capabilities,

TABLE 27.1

Recent Survey of DoD Data Fusion Activities

Activity Acronym	Data Fusion Activity	Primary Service
ABI	AWACS broadcast intelligence	USAF
ADSI	Air defense systems integrator	Joint
AEGIS	AEGIS weapon system	USN
AEPDS	Advanced electronic processing and dissemination system	USA
AMSTE	Affordable moving surface target engagement	DARPA
ANSFP	Artificial neural system fusion prototype	USAF
ARTDF	Automated real-time data fusion	USMC
ASAS	All-source analysis system	USA
ATW	Advanced tactical workstation	USN
ATWCS	Automated tomahawk weapons control system USN	USN
CAMDMUU	Connectionist approach to multiattribute decision making under uncertainty	USAF
CEC	Cooperative engagement capability	USN
CEE	Conditional events and entropy	USN
CV	Constant vision	USAF
DADFA	Dynamic adaptation of data fusion algorithms	USAF
DADS	Deployable autonomous distributed system	USN
DDB	Dynamic database	DARPA
E2C MCU	E2C mission computer upgrade	USN
EASF	Enhanced all source fusion	USAF
EAT	Enhanced analytical tools	USAF
ECS	Shield engagement coordination system	USAF
ENT	Enhancements to NEAT templates	USAF
ESAI	Expanded situation assessment and insertion	USAF
FAST	Forward area support terminal	USA
GALE-Lite	Generic area limitation environment lite	Joint
GCCS	Global command and control system	Joint
GCCS A	Global command and control system army	USA
GCCS I3	Global command and control system integrated imagery and intelligence	Joint
GCCS M	Global command and control system maritime	USN
GDFS	Graphical display fusion system	USN
GISRC	Global intelligence, surveillance, and reconnaissance capability	USN
Hercules		Joint
IAS	Intelligence analysis system	USMC
IDAS	Interactive defense avionics system	USAF
IFAMP	Intelligent fusion and asset management processor	USAF
ISA	Intelligence situation analyst	USAF
ISAT	Integrated situation awareness and targeting	USA
IT	Information trustworthiness	USA
ITAP	Intelligent threat assessment processor	USAF
JIVA	Joint intelligence virtual architecture	Joint
JSTARS CGS	Joint surveillance target attack radar subsystem common ground station	Joint
JTAGS	Joint tactical ground station	Joint
KBS4TCT	Knowledge-based support for time critical targets	USAF
LOCE	Linked operational intelligence centers Europe	Coalition
LSS	Littoral surveillance system	USN
MDBI&U	Multiple database integration and update	USAF

(continued)

TABLE 27.1   (Continued)

Activity Acronym	Data Fusion Activity	Primary Service
MITT	Mobile integrated tactical terminal	USA
Moonlight	Moonlight	Coalition
MSCS	Multiple source correlation system	USAF
MSFE	Multisource fusion engine	USAF
MSI	[E2C] Multisensor integration	USN
MSTS	Multisource tactical system	USAF
NCIF	Network-centric information fusion	USAF
NEAT	Nodal exploitation and analysis tool	USAF
OBATS	Off-board augmented theater surveillance	USAF
OED	Ocean surveillance information system evolutionary development	USN
Patriot	Patriot weapon system	Joint
PICES	Processor independent correlation exploitation system	USN
QIFS	Quarterback information fusion system	USAF
SAFETI	Situation awareness from enhanced threat information	USAF
SCTT	SAR contextual target tracking	USA
SMF	SIGINT/MTI fusion	USA
Story Teller	EP3 story teller	USN
TADMUS	Tactical decision making under stress	USN
TAS	Timeline analysis system	USAF
TBMCS	Theater battle management core systems	USAF
TCAC	Technical control and analysis center	USMC
TCR	Terrain contextual reasoning	USA
TDPS	Tactical data processing suites	USAF
TEAMS	Tactical EA-6B mission support	USN
TERPES	Tactical electronic reconnaissance processing evaluation system	USMC
TES	Tactical exploitation system	USA
TIPOFF	TIBS integrated processor and online fusion function	Joint
TMBR	Truth maintenance belief revision	USAF
TRAIT	Tactical registration of airborne imagery for targeting	USAF
TSA	Theater situation awareness	USAF
UGBADFT	Unified generalized Bayesian adaptive data fusion techniques	USAF
VF	Visual fusion	USA
WECM	Warfighter electronic collection and mapping	USA

however, are being developed within the R&D community and do exist in prototypical systems. In addition, R&D efforts are also focusing on level 1 fusion, but generally with new intelligence data types or new combinations of intelligence data types.

The activities were also analyzed for their capability to fuse data algorithmically from multi-INT types. The use of more than one intelligence data type is becoming increasingly critical to solving difficult military problems. Multi-INT data types collected on a single entity can increase the dimensionality of that entity. Common intelligence data types that are in use today include SIGINT, infrared intelligence (IR), imagery intelligence (IMINT), moving target indicator (MTI), measurement and signatures intelligence (MASINT), and a combination of two or more of the above or multi-INT data types.

The pie charts in Figure 27.2 illustrate the capabilities to fuse data algorithmically from multi-INT data types for the surveyed activities. There are two pie charts (operational and

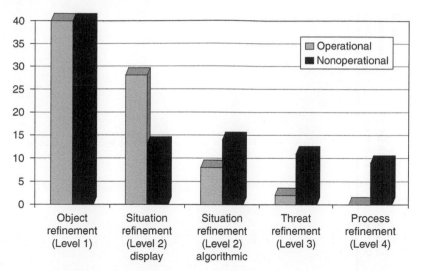

FIGURE 27.1
Comparison of fusion capabilities.

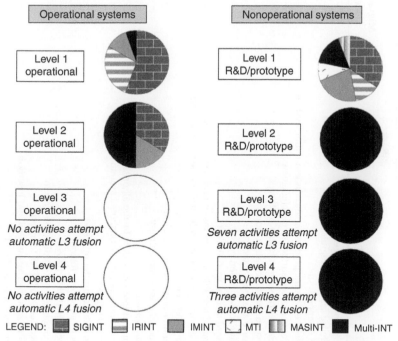

FIGURE 27.2
Algorithmic fusion by intelligence data type.

nonoperational) for each of four JDL levels of fusion. The pie chart in the upper left corner of the figure is interpreted as follows. All systems that provide algorithmic fusion to achieve level 1 fusion (object refinement) were tallied according to the intelligence data type(s) used to achieve the fusion. The chart shows that the majority of operational systems use only SIGINT to achieve level 1 fusion. By contrast, the pie chart in the upper right corner shows, for nonoperational systems, a wider variety in the usage of other intelligence data types.

TABLE 27.2

Status of Data Fusion Activities

Activity Acronym	Data Fusion Activity	Migration System	Legacy System	R&D/ Prototype	Specialized Functions and Others
ABI	AWACS broadcast intelligence	—	—	—	X
ADSI	Air defense systems integrator	—	—	—	X
AEGIS	AEGIS weapon system	X	—	—	—
AEPDS	Advanced electronic processing and dissemination system	—	X	—	—
AMSTE	Affordable moving surface target engagement	—	—	X	—
ANSFP	Artificial neural system fusion prototype	—	—	X	—
ARTDF	Automated real-time data fusion	X	—	—	—
ASAS	All-source analysis system	X	—	—	—
ASF	Adaptive sensor fusion	—	—	X	X
ATW	Advanced tactical workstation	—	—	—	—
ATWCS	Automated tomahawk weapons control system USN	X	—	—	—
CAMDMUU	Connectionist approach to multiattribute decision making under uncertainty	—	—	X	—
CEC	Cooperative engagement capability	—	—	X	—
CEE	Conditional events and entropy	—	—	X	X
CV	Constant vision	—	—	—	—
DADFA	Dynamic adaptation of data fusion algorithms	—	—	X	—
DADS	Deployable autonomous distributed system	—	—	X	—
DDB	Dynamic database	—	—	X	X
E2C MCU	E2C mission computer upgrade	—	—	X	—
EASF	Enhanced all-source fusion	—	—	X	—
EAT	Enhanced analytical tools	—	—	X	—
ECS	Shield engagement coordination system	—	—	X	—
ENT	Enhancements to NEAT templates	—	—	X	—
ESAI	Expanded situation assessment and insertion	—	—	—	—
FAST	Forward area support terminal	—	X	—	—
GALE-Lite	Generic area limitation environment lite	X	—	—	—
GCCS	Global command and control system	X	—	—	—
GCCS A	Global command and control system army	X	—	—	—
GCCS I3	Global command and control system integrated imagery and intelligence	X	—	—	—

Abbreviation	System				
GCCS M	Global command and control system maritime				X
GDFS	Graphical display fusion system	X			
GISRC	Global intelligence, surveillance, and reconnaissance capability		X		
Hercules			X		
IAS	Intelligence analysis system				X
IDAS	Interactive defense avionics system	X			
IFAMP	Intelligent fusion and asset management processor		X		
ISA	Intelligence situation analyst	X			
ISAT	Integrated situation awareness and targeting		X		
IT	Information trustworthiness		X		
ITAP	Intelligent threat assessment processor		X		
JIVA	Joint intelligence virtual architecture				X
JSTARS CGS	Joint surveillance target attack radar subsystem common ground station	X			
JTAGS	Joint tactical ground station	X			
KBS4TCT	Knowledge-based support for time critical targets		X		X
LOCE	Linked operational intelligence centers Europe				X
LSS	Littoral surveillance system				X
MDBI&U	Multiple database integration and update		X		
MITT	Mobile integrated tactical terminal			X	
Moonlight					
MSCS	Multiple source correlation system	X			
MSFE	Multisource fusion engine	X			
MSI	[E2C] Multisensor integration	X	X		
MSTS	Multisource tactical system				
NCIF	Network-centric information fusion	X	X		
NEAT	Nodal exploitation and analysis tool		X		
OBATS	Off-board augmented theater surveillance		X		
OED	Ocean surveillance information system evolutionary development				
Patriot	Patriot weapon system	X			
PICES	Processor independent correlation exploitation system	X			
QIFS	Quarterback information fusion system	X			
SAFETI	Situation awareness from enhanced threat information	X	X		
SCTT	SAR contextual target tracking		X		
SMF	SIGINT/MTI fusion		X		

(continued)

TABLE 27.2 (Continued)

Activity Acronym	Data Fusion Activity	Migration System	Legacy System	R&D/Prototype	Specialized Functions and Others
Story Teller	EP3 Story teller	—	—	—	X
TADMUS	Tactical decision making under stress	—	—	X	—
TAS	Timeline analysis system	—	—	—	—
TBMCS	Theater battle management core systems	X	—	—	—
TCAC	Technical control and analysis center	X	—	—	—
TCR	Terrain contextual reasoning	—	X	X	—
TDPS	Tactical data processing suites	—	X	—	—
TEAMS	Tactical EA-6B mission support	—	—	—	X
TERPES	Tactical electronic reconnaissance processing evaluation system	—	X	—	—
TES	Tactical exploitation system	X	—	—	—
TIPOFF	TIBS integrated processor and online fusion function	—	—	X	X
TMBR	Truth maintenance belief revision	—	—	X	X
TRAIT	Tactical registration of airborne imagery for targeting	—	—	X	X
TSA	Theater situation awareness	—	—	—	X
UGBADFT	Unified generalized Bayesian adaptive data fusion techniques	—	—	X	—
VF	Visual fusion	—	—	X	—
WECM	Warfighter electronic collection and mapping	—	—	X	—

Other conclusions from this figure are that level 2 fusion, which is achieved by operational systems, is primarily achieved using multi-INT data according to the pie chart in the lower left corner, second row. All nonoperational activities use multi-INT data to achieve level 2 fusion. Few operational systems automatically integrate multi-INT data. Most data fusion systems display multi-INT data—sometimes on a common screen. Selected R&D systems are tackling the algorithmic integration of multi-INT data.

A common belief is that the realm of military data fusion is marked by numerous duplicative activities, which seems to imply a lack of organization and coordination. In reality, the 79 activities in this study reflect various relationships and migration plans. A close examination of the pedigree, current relationships, and progeny demonstrates some commonality and a greater degree of coordination than is often apparent. In addition, the DoD has established several migration systems to which existing fusion systems must evolve. Examples of these are the GCCS and the distributed common ground station (DCGS).

Specific operational intelligence systems have been identified as migration systems; others are viewed as legacy systems that will eventually be subsumed by a migration system. For nonintelligence systems, the fusion capabilities are frequently highly specialized and tailored to the overall mission of the system, such as weapon cueing. In addition, the nonoperational systems are generally highly specialized in that they are developing specific technologies, but they also frequently leverage other operational and nonoperational activities.

Table 27.2 shows the division of these activities into several categories:

1. Migration systems that are converging to a common baseline to facilitate interoperability with other systems
2. Legacy systems that will be subsumed by a migration system
3. Government-sponsored R&D and prototypes
4. Highly specialized capabilities that are not duplicated by any other system

The R&D activities, as well as the prototypical systems, are shown in bold type.

In conclusion, the recent survey of DoD data fusion activities provides a snapshot of current and emerging fusion capabilities in terms of their level of fusion and their usage of various types of intelligence data. The survey also reflects a greater degree of coordination among military data fusion activities than was recognized by previous surveys.

References

1. Hall, D. L., Linn, R. J., and Llinas, J., A survey of data fusion systems, *Proc. SPIE Conf. Data Struct. Target Classif.*, Orlando, FL, April 1991, 147B, 13–29.
2. Llinas, J. and Antony, R. T., Blackboard concepts for data fusion applications: Blackboard systems, *Int. J. Pattern Recognit. Artif. Intell.*, 7(2), 285–308, 1993.
3. Hall, D. L. and Linn, R. J., Survey of commercial software for multi-sensor data fusion, *Proc. SPIE Conf. Sensor Fusion: Aerosp. Appl.*, Orlando, FL, 1991.
4. Buede, D., Software review: Overview of the MCDA market, *J. Multi-Criteria Decis Anal.*, 1, 59–61, 1992.
5. Buede, D., Superior design features of decision analytic software, *Comput. Oper. Res.*, 19(1), 43–57, 1992.

28

Data Fusion for Developing Predictive Diagnostics for Electromechanical Systems

Carl S. Byington and Amulya K. Garga

CONTENTS

28.1   Introduction

Condition-based maintenance (CBM) is a philosophy of performing maintenance on a machine or system only when there is objective evidence of need or impending failure. By contrast, time-based or use-based maintenance involves performing periodic maintenance after specified periods of time or hours of operation. CBM has the potential to decrease life-cycle maintenance costs (by reducing unnecessary maintenance actions), increase operational readiness, and improve safety.

Implementation of CBM involves *predictive diagnostics* (i.e., diagnosing the current state or health of a machine and predicting time to failure based on an assumed model of anticipated use). CBM and predictive diagnostics depend on multisensor data, such as vibration, temperature, pressure, and presence of oil debris, which must be effectively fused to determine machinery health. Indeed, Hansen et al. suggested that predictive diagnostics involve many of the same functions and challenges demonstrated in more traditional Department of Defense (DoD) applications of data fusion (e.g., signal processing, pattern recognition, estimation, and automated reasoning).[1] This chapter demonstrates the potential for technology transfer from the study of CBM to DoD fusion applications.

28.1.1   Condition-Based Maintenance Motivation

CBM is an emerging concept enabled by the evolution of key technologies, including improvements in sensors, microprocessors, digital signal processing, simulation modeling, multisensor data fusion, and automated reasoning. CBM involves monitoring the health or status of a component or system and performing maintenance based on that observed health and some predicted remaining useful life (RUL).[2-5] This *predictive maintenance* philosophy contrasts with earlier ideologies, such as *corrective maintenance*—in which action is taken after a component or system fails—and *preventive maintenance*—which is based on event or time milestones. Each involves a cost trade-off. Corrective maintenance incurs low maintenance cost (minimal preventive actions), but high performance costs caused by operational failures. Conversely, preventive maintenance produces low operational costs, but greater maintenance department costs. Moreover, the application of statistical safe-life methods (which are common with preventive maintenance) usually leads to very conservative estimates of the probability of failure. The result is the additional hidden cost associated with the disposal of components that still retain significant RUL.

Another important consideration in most applications is the operational availability (a metric that is popular in military applications) or equipment effectiveness (more popular in industrial applications). Figure 28.1 illustrates regions of high total cost when overly corrective or overly preventive maintenance dominate. These regions also provide a lower total availability of the equipment. On the corrective side, equipment neglect typically leads to more operational failures during which time the equipment is unavailable. On the preventive side, the equipment is typically unavailable because it is being maintained much of the time. An additional concern that affects availability and cost in this region is the greater likelihood of maintenance-induced failures.

The development of better maintenance practices is driven by the desire to reduce the risk of catastrophic failures, minimize maintenance costs, maximize system availability, and increase platform reliability. These goals are desirable from the application arenas of aircraft, ships, and tanks to industrial manufacturing of all types. Moreover, given that maintenance is a key cost driver in military and commercial applications, it is an

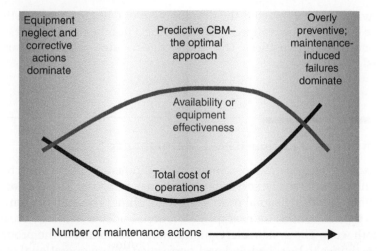

FIGURE 28.1
Condition-based maintenance provides the best range of operational availability or equipment effectiveness.

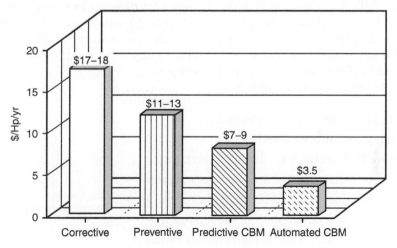

FIGURE 28.2
Moving toward condition monitoring and the optimal level of maintenance provided dramatic cost savings in the electric industry.

important area in which to focus research and development efforts. At nuclear power plants, for example, the operations and maintenance portion of the direct operating costs (DOC) grew by more than 120% between 1981 and 1991, a level more than twice as great as the fuel cost component.[6]

A more explicit cost savings can be seen in Figure 28.2 derived from an Electric Power Research Institute study to estimate the costs associated with different maintenance practices in the utility industry. The first three columns were taken directly from the study, and the fourth is estimated from unpublished cost studies. Clearly, predictive practices provide cost benefit. The estimated 50% additional cost savings derived from predictive condition monitoring to automated CBM is manifested by the manpower cost focused on data collection/analysis efforts and cost avoidances associated with continuous monitoring and fault prediction.[7]

Such cost savings have motivated the development of CBM systems; furthermore, substantially more benefit can be realized by automating a number of the functions to achieve improved screening and robustness. This knowledge has driven CBM research and development efforts.

28.2   Aspects of a Condition-Based Maintenance System

CBM uses sensor systems to diagnose emerging equipment problems and to predict how long equipment can effectively serve its operational purpose. The sensors collect and evaluate real-time data using signal processing algorithms. These algorithms correlate the unique signals to their causes, for example, vibrational sideband energy created by developing gear-tooth wear. The system alerts maintenance personnel to the problem, enabling maintenance activities to be scheduled and performed before operational effectiveness is compromised.

The key to effectively implementing CBM is the ability to detect, classify, and predict the evolution of a failure mechanism with sufficient robustness, and at a low enough cost, to use that information as a basis to plan maintenance for mission- or safety-critical systems. *Mission critical* refers to those activities that, if interrupted, would prohibit the organization from meeting its primary objectives (e.g., completion and delivery of 2500 control panels to meet an OEM's assembly schedule). Safety-critical functions must remain operational to ensure the safety of humans (e.g., airline passengers).

Thus, a CBM system must be capable of

- Detecting the start of a failure evolution
- Classifying the failure evolution
- Predicting RUL with a high degree of certainty
- Recommending a remedial action to the operator
- Taking the indicated action through the control system
- Aiding the technician in making the repair
- Providing feedback for the design process

These activities represent a closed-loop process with several levels of feedback, which differentiates CBM from preventive or time-directed maintenance. In a preventive maintenance system, time between overhaul (TBO) is set at design, based on failure mode effects and criticality analyses (FMECA), and experience with like machines' mortality statistics. The general concept of a CBM system is shown in Figure 28.3.[8]

28.3   The Diagnosis Problem

Multisensor data fusion has been recognized as an enabling technology for both military and nonmilitary applications. However, improved diagnosis and increased performance do not result automatically from increased data collection. The data must be contextually filtered to extract information that is relevant to the task at hand. Another key requirement

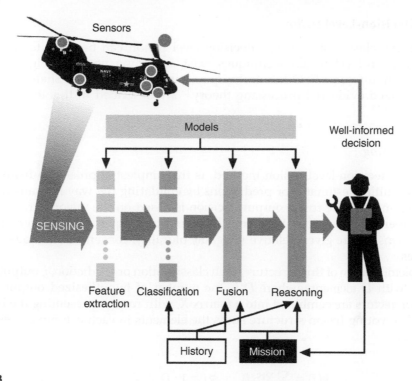

FIGURE 28.3
The success of condition-based maintenance systems depends on (1) the ability to design or use robust sensors for measuring relevant phenomena; (2) real-time processing of the sensor data to extract useful information (e.g., features or data characteristics) in a noisy environment, and to detect parametric changes that could indicate impending failure conditions; (3) fusion of multisensor data to obtain improved information (beyond what is available from a single sensor); (4) micro- and macro-level models that predict the temporal evolution of failure phenomena; and (5) automated approximate reasoning capable of interpreting the results of the sensor measurements, processed data, and model prediction in the context of an operational environment. (From Hall, D.L. et al., *Proceedings of the International Gas Turbine Institute Conference*, ASME, Birmingham, England, June 1997.)

that justifies the use of data fusion is low false alarms. In general, there is a trade-off between missed detections and false alarms, which is greatly influenced by the mission or operation profile. If a diagnostic system produces excessive false alarms, personnel will likely ignore it, resulting in an unacceptably high number of missed detections. However, presently data fusion is rarely employed in monitoring systems, and, when it is used, it is usually an afterthought. Data fusion can most readily be employed at the feature or decision levels.

28.3.1 Feature-Level Fusion

Diagnosis is most commonly performed as classification using feature-based techniques.[9] Machinery data are processed to extract features that can be used to identify specific failure modes. Discriminant transformations are often utilized to map the data characteristic of different failure mode effects into distinct regions in the feature subspace. Multisensor systems frequently use this approach because each sensor may contribute a unique set of features with varying degrees of correlation with the failure to be diagnosed. These features, when combined, provide a better estimate of the object's identity. Examples of this approach are illustrated in the applications section.

28.3.2 Decision-Level Fusion

Following the classification stage, decision-level fusion can be used to fuse identity. Several decision-level fusion techniques exist, including voting, weighted decision, and Bayesian inference.[10] Other techniques, such as Dempster–Shafer's method[11,12] and generalized evidential processing theory,[13] are described in this text and in other publications.[14,15]

28.3.2.1 Voting

Voting, as a decision-level fusion method, is the simplest approach to fusing the outputs from multiple estimates or predictions by emulating the way humans reach some group agreement.[10] The fused output decision is based on the majority rule (i.e., maximum number of votes wins). Variations of voting techniques include weighted voting (in which sensors are given relative weights), plurality, consensus methods, and other techniques.

For implementation of this structure, each classification or prediction, i, outputs a binary vector, x_i, with D elements, where D is the number of hypothesized output decisions. The binary vectors are combined into a matrix X, with row i representing the input from sensor i. The voting fusion structure sums the elements in each column as described by Equation 28.1.

$$y(j) = \sum_{i=1}^{N} X(i,j) \quad \forall j = 1 : D \tag{28.1}$$

The output, $y(j)$, is a vector of length D, where each element indicates the total number of votes for output class j. At time k, the decision rule selects the output, $d(k)$, as the class that carries the majority vote, according to

$$d(k) = \arg\max_j y(j) \tag{28.2}$$

28.3.2.2 Weighted Decision Fusion

A weighted decision method for data fusion generates the fused decision by weighting and combining the outputs from multiple sensors. *A priori* assumptions of sensor reliability and confidence in the classifier performance contribute to determining the weights used in a given scenario. Expert knowledge or models regarding the sensor reliability can be used to implement this method. In the absence of such knowledge, an assumption of equal reliability for each sensor can be made. This assumption reduces the weighted decision method to voting. Note that at the other extreme, a weighted decision process could selectively weight sensors so that, at a particular time, only one sensor is deemed to be credible (i.e., weight = 1), whereas all other sensors are ignored (i.e., weight = 0).

Several methods can be used for implementing a weighted decision fusion structure. Essentially, each sensor, i, outputs a binary vector, x_i, with D elements, where D is the number of hypothesized output decisions. A binary *one*, in position j, indicates that the data was identified by the classifier as belonging to class j. The classification vector from sensor i becomes the ith row of an array, X, that is passed to the weighted decision fusion structure. Each row is weighted, using the *a priori* assumption of the sensor reliability.

Subsequently, the elements of the array are summed along each column. The following equation describes this process mathematically:

$$y(j) = \sum_{i=1}^{N} w_i X(i, j) \qquad \forall j = 1 : D \tag{28.3}$$

The output, $y(j)$, is a row vector of length D, where each element indicates the confidence that the input data from the multiple sensor set has membership in a particular class. At time k, the output decision, $d(k)$, is the class that satisfies the maximum confidence criteria of

$$\sum_{i=1}^{N} w_i = 1 \tag{28.4}$$

This implementation of weighted decision fusion permits future extension in two ways. First, it provides a path to the use of confidence as an input from each sensor. This would allow the fusion process to utilize fuzzy logic (FL) within the structure. Second, it enables an adaptive mechanism to be incorporated that can modify the sensor weights as data are processed through the system.

28.3.2.3 Bayesian Inference

Bayes' theorem[16–18] serves as the basis for the Bayesian inference technique for identity fusion. This technique provides a method for computing the *a posteriori* probability of a particular outcome, based on previous estimates of the likelihood and additional evidence. Bayesian inference assumes that a set of D mutually exclusive (and exhaustive) hypotheses or outcomes exists to explain a given situation.

In the decision level, multisensor fusion problem, Bayesian inference is implemented as follows. A system exists with N sensors that provide decisions on membership to one of D possible classes. The Bayesian fusion structure uses *a priori* information on the probability that a particular hypothesis exists and the likelihood that a particular sensor is able to classify the data to the correct hypothesis. The inputs to the structure are (1) $P(O_j)$, the *a priori* probabilities that object j exists (or equivalently that a fault condition exists), (2) $P(D_{k,i}|O_j)$, the likelihood that each sensor, k, will classify the data as belonging to any one of the D hypotheses, and (3) D_k, the input decisions from the K sensors. The following equation describes the Bayesian combination rule.

$$P(O_j|D_1, \ldots, D_K) = \frac{P(O_j)\prod_{k=1}^{K} P(D_k|O_j)}{\sum_{i=1}^{N} P(O_j)\prod_{k=1}^{K} P(D_k|O_j)} \tag{28.5}$$

The output is a vector with element j representing the *a posteriori* probability that the data belong to hypothesis j. The fused decision is made based on the maximum *a posteriori* probability criteria given in

$$d(k) = \arg\max_j [P(O_j|D_1, \ldots, D_K)] \tag{28.6}$$

A basic issue with the use of Bayesian inference techniques involves the selection of the *a priori* probabilities and the likelihood values. The choice of this information has a significant impact on the performance. Expert knowledge can be used to determine these inputs. In the case where the *a priori* probabilities are unknown, the user can resort to the principle of indifference, where the prior probabilities are set to be equal, as in Equation 28.7.

$$P(O_j) = \frac{1}{N} \qquad (28.7)$$

The *a priori* probabilities are updated in the recursive implementation as described by Equation 28.8. This update sets the value for the *a priori* probability in iteration t equal to the value of the *a posteriori* probability from iteration $(t - 1)$.

$$P_t(O_j) = P_{t-1}(O_j \mid D_1, \ldots, D_K) \qquad (28.8)$$

28.3.3 Model-Based Development

Diagnostics model development can proceed down a purely data-driven, empirical path or model-based path that uses physical and causal models to drive the diagnosis. Model-based diagnostics can provide optimal damage detection and condition assessment because empirically verified mathematical models at many state conditions are the most appropriate knowledge bases.[19] The Pennsylvania State University Applied Research Laboratory (Penn State ARL) has taken a model-based diagnostic approach toward achieving CBM that has proven appropriate for fault detection, failure mode diagnosis, and, ultimately, prognosis.[20]

The key modeling area for CBM is to develop models that can capture the salient effects of faults and relate them to virtual or external observables. Some fundamental questions arise from this desired modeling. How can mathematical models of physical systems be adapted or augmented with separate damage models to capture symptoms? Moreover, how can model-based diagnostics approaches be used for design in CBM requirements such as sensor type, location, and processing requirements?

In the model-based approach, the physical system is captured mathematically in the form of empirically validated computational or functional models. The models possess or are augmented with damage association models that can simulate a failure mode of given severity to produce a symptom that can be compared to measured features. The failure mode symptoms are used to construct the appropriate classification algorithms for diagnosis. The sensitivity of the failure modes to specific sensor processing can be compared for various failure modes and evaluated over the entire system to aid in the determination of the most effective CBM approach.

28.3.3.1 Model-Based Identification and Damage Estimation

Figure 28.4 illustrates a conceptual method for identifying the type and amount of degradation using a validated system model. The actual system output response (event and performance variables) is the result of nominal system response plus fault effects and uncertainty. The model-based analysis and identification of faults can be viewed as an optimization problem that produces the minimum residual between the predicted and the actual response.

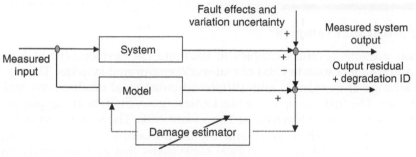

FIGURE 28.4
Adaptive concept for deterministic damage estimation.

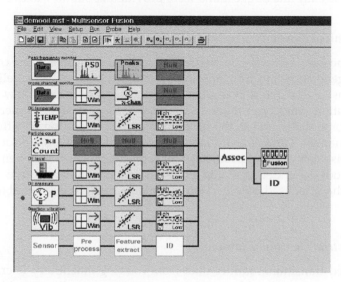

FIGURE 28.5
The Penn State ARL multisensor fusion toolkit is used to combine data from multiple sensors, improving the ability to characterize the current state of a system.

28.4 Multisensor Fusion Toolkit

A multisensor data fusion toolkit was developed at the Penn State ARL to provide the user with a standardized visual programming environment for data fusion (Figure 28.5).[21] With this toolkit, the user can develop and compare techniques that combine data from actual and virtual sensors. The detection performance and the number of false alarms are two of the metrics that can be used for such a comparison.

The outputs of one or more state/damage estimates can be combined with available usage information, based on feature vector classification. This type of a tool is an asset because it utilizes key information from multiple sensors for robustness and presents the results of the fusion assessment, rather than just a data stream. Furthermore, the tool is very useful for rapid prototyping and evaluation of data analysis and data fusion algorithms. The toolkit was written in Visual C++ using an object-oriented design approach.

28.5 Application Examples

This section presents several examples to illustrate the development of a data fusion approach and its application to CBM of real-world engineered systems. The topics chosen represent a range of machinery with different fundamental mechanisms and potential CBM strategies. The first example is a mechanical (gear/shaft/bearing) power transmission that has been tested extensively at Penn State ARL. The second example uses fluid systems (fuel/lubrication/hydraulic), and the third example focuses on energy storage devices (battery/fuel cells). All are critical subsystems that address fundamental CBM needs in the DoD and industry. Developers of data fusion solutions must carefully select among the options that are applicable to the problem. Several pitfalls in using data fusion were identified recently and suggestions were provided about how to avoid them.[22]

28.5.1 Mechanical Power Transmission

Individual components and systems, where a few critical components are coupled together in rotor power generation and transmission machinery, are relatively well understood as a result of extensive research that has been conducted over the past few decades. Many notable contributions have been made in the analysis and design, in increasing the performance of rotor systems, and in the fundamental understanding of different aspects of rotor system dynamics. More recently, many commercial and defense efforts have focused on vibration/noise analysis and prediction for fault diagnostics. Many employ improved modeling methods to understand the transmission phenomena more thoroughly,[23-26] whereas others have focused on detection techniques and experimental analysis of fault conditions.[27-33]

28.5.1.1 *Industrial Gearbox Example*

Well-documented transitional failure data from rotating machinery is critical for developing machinery prognostics. However, such data is not readily or widely available to researchers and developers. Consequently, a mechanical diagnostics test bed (MDTB) was constructed at the Penn State ARL for detecting faults and tracking damage on an industrial gearbox (Figure 28.6).

28.5.1.1.1 *System and Data Description*

The MDTB is a motor-drivetrain-generator test stand. (A complete description of the MDTB can be found in Ref. 34.) The gearbox is driven at a set input speed using a 30 Horse power (HP), 1750 rpm AC (drive) motor, and the torque is applied by a 75 HP, 1750 rpm AC (absorption) motor. The MDTB can test single and double reduction industrial gearboxes with ratios from about 1.2:1 to 6:1. The gearboxes are nominally in the 5–20 HP range. The motors provide about two to five times the rated torque of the selected gearboxes; thus, the system can provide good overload capability for accelerated failure testing.

The gearbox is instrumented with accelerometers, acoustic emission sensors, thermocouples, and oil quality sensors. Torque and speed (load inputs) are measured within 1% on the rig. Borescope images are taken during the failure process to correlate degree of damage with measured sensor data. Given a low contamination level in the oil, drive speed and load torque are the two major factors in gear failure. Different values of torque and speed will cause different types of wear and faults. Figure 28.7 illustrates potential regions of failures.[35]

FIGURE 28.6
The Penn State ARL mechanical diagnostics test bed is used to collect transitional failure data, study sensor optimization for fault detection, and evaluate failure models.

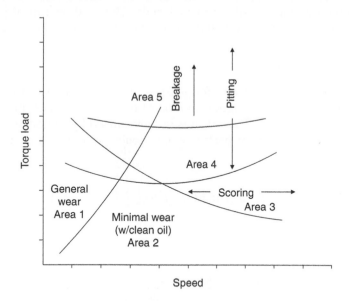

FIGURE 28.7
Regions of gear failures. (From Townsend, D.P., *Dudley's Gear Handbook: The Design Manufacture and Application of Gears*, McGraw Hill, New York, NY, 1992.)

28.5.1.1.2 *Gearbox Failure Conditions*

Referring to Figure 28.7, in Area 1, the gear is operating too slowly to develop an oil film, so adhesive wear occurs. In Area 2, the speed is sufficiently fast to develop an oil film. The gears should be able to run with minimal wear. In Area 3, scoring is likely because the load and speed are high enough to break down the existing oil film. Area 4 illustrates the dominance of pitting caused by high surface stresses that result from higher torque loads. As the torque is increased further, tooth breakage will result from overload and stress fatigue, as shown in Area 5.

On the basis of earlier discussion, the MDTB test plan includes test runs that set the operating drive speed and load torque deep into each area to generate transitional data for each of the faults. These limits, of course, are not known exactly *a priori*. Areas 4 and 5 are the primary focal points because they contain critical and difficult-to-predict faults. Being able to control (to a degree) the conditions that affect the type of failure that occurs allows some control over the amount of data for each fault, while still allowing the fault to develop naturally (i.e., the fault is not seeded). If a particular type of fault requires more transitional data for analysis, adjustment of the operating conditions can increase the likelihood of producing the desired fault.

28.5.1.1.3 Oil Debris Analysis

Roylance and Raadnui[36] examined the morphology of wear particles in circulating oil and correlated their occurrences with wear characteristics and failure modes of gears and other components of rotating machinery. Wear particles build up over time even under normal operating conditions. However, the particles generated by *benign* wear differ markedly from those generated by the *active* wear associated with pitting, abrasion, scuffing, fracturing, and other abnormal conditions that lead to failure. Roylance and Raadnui[36] correlated particle features (quantity, size, composition, and morphology) with wear characteristics (severity, rate, type, and source).

Particle composition can be an important clue to the source of abnormal wear particles when components are made of different materials. The relationship of particle type to size and morphology has been well characterized by Roylance,[36] and is summarized in Table 28.1.

28.5.1.1.4 Vibration Analysis

Vibration analysis is extremely useful for gearbox analysis and gear failures because the unsteady component of relative angular motion of the meshing gears provides the major source of vibratory excitation.[37] This effect is largely caused by a change in compliance of the gear teeth and deviation from perfect shape. Such a modulated gear-meshing vibration, $y(t)$, is given by

$$y(t) = \sum_{n=0}^{N} X_n[1 + a_n(t)]\cos[2\pi f_m t + \phi_n + b_n(t)] \tag{28.9}$$

TABLE 28.1

Wear Particle Morphology—Ferrography Descriptors

Particle	Description
Rubbing	Particles, 20 μm chord dimension and ~1 μm thick. Results from flaking of pieces from mixed shear layer mainly benign
Cutting	Swarf-like chops of fine wire coils, caused by abrasive cutting action
Laminar	Thin, bright, free-metal particles, typically 1 μm thick, 20–50 μm chord width. Holes in surface and uneven edge profile. Gear-rolling element bearing wear
Fatigue	Chunky, several micrometers thick from, for example, gear wear, 20–50 μm chord width
Spheres	Typically ferrous, 1–2 μm diameter, generated when micro-cracks occur under rolling contact fatigue condition
Severe sliding	Large/50 μm chord width, several micrometers thick. Surfaces heavily striated with long straight edges. Typically found in gear wear

where $a_n(t)$ and $b_n(t)$ are the amplitude and phase modulation functions, respectively. Amplitude modulation produces sidebands around the carrier (gear-meshing and harmonics) frequencies and is often associated with eccentricity, uneven wear, or profile errors. As can be seen from the Equation 28.9, frequency modulation will produce a family of sidebands. These will typically occur in gear systems, and the sidebands, which may either combine or subtract to produce an asymmetrical family of sidebands.

Much of the analysis has focused on the use of the appropriate statistical processing and transforms to capture these effects. A number of figures of merit or features have been used to correlate mechanical faults. Moreover, short-time Fourier, Hilbert, and wavelet transforms have also been used to develop vibration features.[38,39]

28.5.1.1.5 *Description of Features*

Various signal and spectral modeling techniques have been used to characterize machinery data and develop features indicative of various faults in the machinery. Such techniques include statistical modeling (e.g., mean, RMS, kurtosis), spectral modeling (e.g., Fourier transform, cepstral transform, autoregressive modeling), and time frequency modeling (e.g., short-time Fourier transform, wavelet transform, wide-band ambiguity functions). Several oil and vibration features are now well described. These can be fused and integrated with knowledge of the system and history to provide indication of gearbox condition. In addition to the obvious corroboration and increased confidence that can be gained, this approach to using multiple sensors also aids in establishing the existence of sensor faults.

Features tend to organize into subspaces in feature space, as shown in Figure 28.8. Such subspaces can be used to classify the failure mode. Multiple estimates of a specific failure mode can be produced through the classification of each feature subspace. Other failure

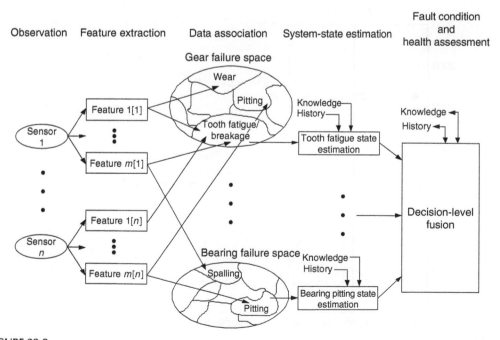

FIGURE 28.8
Oil/vibration data fusion process. (From Byington, C.S. et al., *Proceedings of the 53rd Meeting of the Society for MFPT*, 1999.)

mode estimates can be processed at the same time as well. Note that a gearbox may deteriorate into more than one failure mode with several critical faults competing.

During 20+ run-to-failure transitional tests conducted on the MDTB, data were collected from accelerometer, temperature, torque, speed, and oil quality/debris measurements. This discussion pertains only to Test 14. Borescope imaging was performed at periodic intervals to provide damage estimates as ground truth for the collected data. Small oil samples of a ~25 mL were taken from the gearbox during the inspection periods. Postrun oil samples were also sent to the DoD Joint Oil Analysis Program (JOAP) to determine particle composition, oil viscosity, and particle-wear type.

The break-in and design load time for Test 14 ran for 96 h. The 3.33 ratio gearbox was then loaded at three times the design load to accelerate its failure, which occurred ~20 h later (including inspection times). This load transition time was at ~1200 (noon), and no visible signs of deterioration were noted. The run was stopped every 2 h for internal inspection and oil sampling. At 0200 (2:00 a.m.), the inspection indicated no visible signs of wear or cracks. After 0300, accelerometer data and a noticeable change in the sound of the gearbox were noted. On inspection, one of the teeth on the follower gear had separated from the gear. The tooth had failed at the root on the input side of the gear with the crack rising to the top of the gear on the load side (refer to Figure 28.9). The gearbox was stopped again at 0330, and an inspection showed no observable increase in damage. At 0500, the tooth from the downstream broken tooth had suffered surface pitting, and there were small cracks a millimeter in from the front and rear face of the tooth, parallel to the faces. The 0700 inspection showed that two teeth had broken and the pitting had increased, but not excessively, even at three times design load. Neighboring teeth now had small pits at the top-motor side corners.

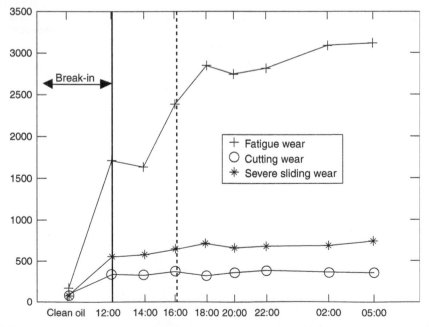

FIGURE 28.9
Number of particles/milliliter for fatigue, cutting, and sliding wear modes (>15 μm) collected at various times.

On shutdown at 0815, with a significant increase in vibration (over 150% RMS), the test was concluded, eight teeth had suffered damage. The damaged teeth were dispersed in clusters around the gear. There appeared to be independent clusters of failure processes. Within each cluster a tooth that had failed as a result of root cracking was surrounded by teeth that had failed due to pitting. On both clusters, the upstream tooth had failed by cracking at the root, and the follower tooth had experienced pitting.

Figure 28.9 shows the time sequence of three types of particle concentrations observed during this test run: fatigue wear, cutting wear, and severe sliding wear. Initial increases in particle counts observed at 1200 reflect debris accumulations during break-in. Fatigue particles manifested the most dramatic change in concentration of the three detectable wear particle types, nearly doubling between 1400 and 1800, suggesting the onset of some fault that would give rise to this type of debris. This data is consistent with the inspections that indicated pitting was occurring throughout this time period. No significant sliding or cutting wear was found after break-in.

Figure 28.10 illustrates the breakdown of these fatigue particle concentrations by three different micrometer size ranges. Between 1400 and 1800, particle concentrations increased for all three ranges with onset occurring later for each larger size category. The smallest size range rose to over 1400 particles/mL by 1800, whereas the particles in the midrange began to increase consistently after 1400 until the end of the run. The largest particle size shows a gradual upward trend starting at 1600, though the concentration variation is affected by sampling/measurement error. The observed trends could be explained by hypothesizing the onset of a surface fatigue fault condition sometime before 1600, followed by steadily generated fatigue wear debris.

Figures 28.11 and 28.12 show different features of the accelerometer data developed at Penn State ARL. The results with a Penn State ARL enveloping technique (Figure 28.11) clearly show evidence of some activity around 0200. The dashed line represents the

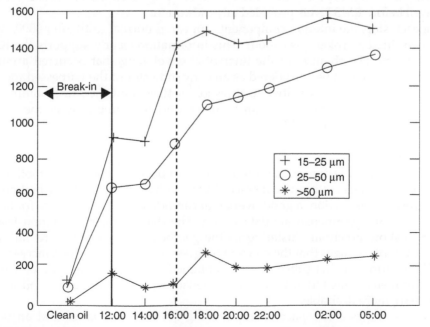

FIGURE 28.10
Number of fatigue-wear particles/milliliter by bin size collected at various times.

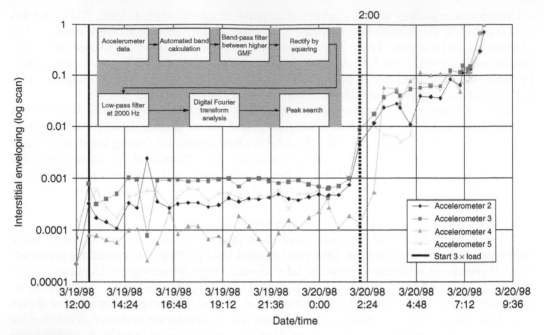

FIGURE 28.11
Interstitial enveloping of accelerometer.

approximate time the feature showed a notable change. This corresponds to the time when tooth cracking is believed to have initiated/propagated. The wavelet transform[40] is shown in Figure 28.12. It is believed to be sensitive to the impact events during breakage, and shows evidence of this type of failure after 0300. The processed indicators seem to indicate activity well before RMS levels provided any indication.

During each stop, the internal components appeared normal until after 0300, when the borescope verified a broken gear tooth. This information clearly supports the RMS and wavelet changes. The changes in the interstitial enveloping that occurred around 0200 (almost 1 h earlier) could be considered as an early indicator of the witnessed tooth crack. Note that the indication is sensitive to threshold setting, and the MDTB online wavelet detection threshold triggered about an hour (around the same time as the interstitial) before that shown in Figure 28.12.

28.5.1.1.6   Feature Fusion

Although the primary failure modes on MDTB gearboxes have been gear tooth and shaft breakage, pitting has also been witnessed.[41] On the basis of previous vibration and oil debris figures in this section, a good overlap of candidate features appears for both commensurate and noncommensurate data fusion. The data from the vibration features in Figure 28.13 show potential clustering as the gearbox progresses toward failure. Note from the borescope images that the damage progresses in clusters, which increase on both scales. The features in this subspace were obtained from the same type of sensor (i.e., they are commensurate). Often two noncommensurate features—such as oil debris and vibration—are more desirable.

Figure 28.14 shows a subspace example using a vibration feature and an oil debris (fatigue particle count) feature. There are fewer data points than in the previous example because the MDTB had to be shut down to extract an oil sample as opposed to using

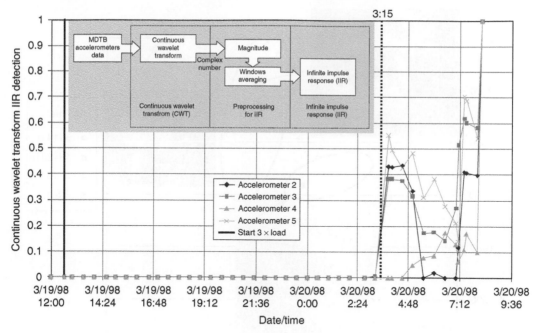

FIGURE 28.12
Continuous wavelet transform (IIR count).

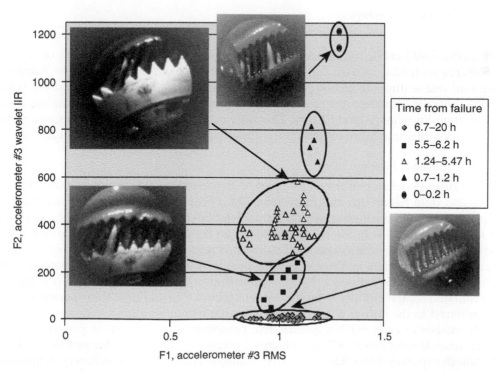

FIGURE 28.13
Feature subspace classification example.

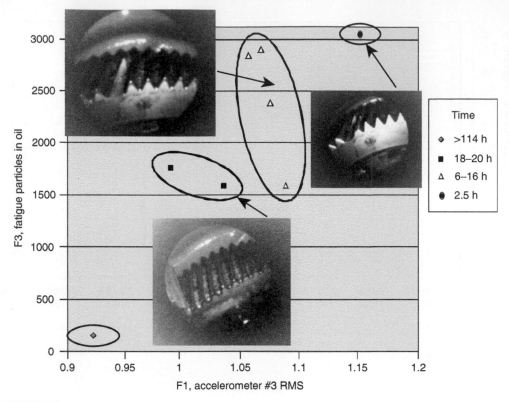

FIGURE 28.14
Noncommensurate feature subspace.

on-demand, DAQ collection of accelerometer data. During the progression of the run, the features seemed to cluster into regions that are discernible and trackable. Subspaces using multiple features from both commensurate and noncommensurate sources would provide better information for classification, as would the inclusion of running conditions, system knowledge, and history. This type of association of data is a necessary step toward accomplishing more robust state estimation and higher levels of data fusion.

28.5.1.1.7 Decision Fusion Analysis

Decision fusion is often performed as a part of *reasoning* in CBM systems. Automated reasoning and data fusion are important for CBM. Because the monitored systems exhibit complex behavior, there is generally no simple relationship between observable conditions and system health. Furthermore, sensor data can be very unreliable, producing a high false alarm rate. Hence, data fusion and automated reasoning must be used to contextually interpret the sensor data and model predictions. In this section, three automated reasoning techniques, neural networks (NNs), FL, and expert/rule-based systems, are compared and evaluated for their ability to predict system failure.[42] In addition, these system outputs are compared to the output of a hybrid system that combines all three systems to realize the advantages of each. Such a quantitative comparison is essential in producing high-quality, reliable solutions for CBM problems; however, it is rarely performed in practice.

Although expert systems (ESs), FL systems, and NNs are used in machinery diagnostics, they are rarely used simultaneously or in combination. A comparison of these techniques and decision fusion of their outputs was performed using the MDTB data. In particular,

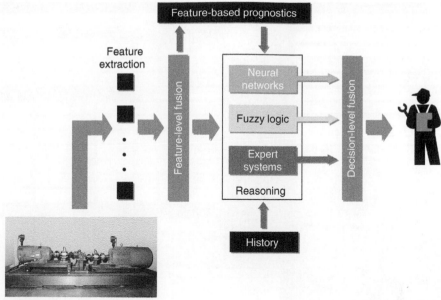

Mechanical diagnostic test bed (MDTB)

FIGURE 28.15
Flow diagram for comparison of three reasoning methods with mechanical diagnostic test bed data.

three systems were developed (ES, FL, and NN) to estimate the RUL of the gearbox during accelerated failure runs (Figure 28.15). The inputs to the systems consisted of speed, torque, temperature, and vibration RMS in several frequency bands.

A graphical tool was developed to provide a quick visual comparison of the outputs of the different types of systems (Figure 28.16). In this tool, colors are used to represent the relative levels of the inputs and outputs and a confidence value is provided with each output. The time-to-failure curves for the three systems and the hybrid system are shown in Figure 28.17. In this example, the FL system provided the earliest warning, but the hybrid system gave the best combination of early warning and robustness.

28.5.2 Fluid Systems

Fluid systems comprise lubrication,[20] fuel,[43] and hydraulic power application examples. Some efforts at evaluating a model-based, data fusion approach for fluid systems are discussed in the following sections. Such fluid systems are critical to many navy engine and power systems and, clearly, must be a part of the CBM solution.

28.5.2.1 Lubrication System Function

A pressure-fed lubrication system is designed to deliver a specific amount of lubricant to critical, oil-wetted components in engines, transmissions, and like equipment. The primary function of a lubricant is to reduce friction through the formation of film coatings on loaded surfaces. It also transports heat from the load site and prevents corrosion. The lubricating oil in mechanical systems, however, can be contaminated by wear particles, internal and external debris, foreign fluids, and even internal component (additive) breakdown. All of these contaminants affect the ability of the fluid to accomplish its mission of producing a lubricious (hydrodynamic, elastohydrodynamic, boundary, or mixed) layer between mechanical parts with relative motion.[44,45]

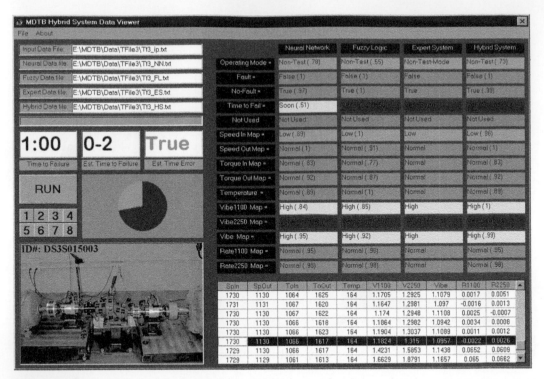

FIGURE 28.16
Graphical viewer for comparing the outputs of the reasoning systems.

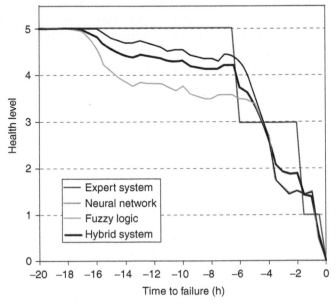

FIGURE 28.17
Time-to-failure curves for the reasoning systems with mechanical diagnostic test bed data.

TABLE 28.2

Lubricant and Wetted Component Faults

Lubricant Faults	Gear Faults	Bearing Faults
Viscosity breakdown	Plastic deformation	Surface wiping
Oxidation	Pitting	Fatigue
Emulsification	Heavy scuffing	Fretting
Additive depletion	Chipping and tooth crack	Foreign debris
Sludge formation	Tooth breakage	Spalling
Fluid contamination	Case cracking	Inadequate oil film
External debris contamination	Surface fatigue	Overheating
Internal debris contamination	Abrasive wear	Corrosion
System leakage	Chemical wear	Cavitation erosion

Lubricant contamination can be caused by many mechanisms. Water ingestion through seals (common in marine environments) or condensation will cause significant viscosity effects and corrosion. Fuel leakage through the (turbine fuel–lube oil) heat exchanger will also adversely affect lubricity. Moreover, fuel soot, dirt, and dust can increase viscosity and decrease the oil penetration into the loaded surface of the gears or bearings.[46] An often overlooked, but sometimes significant, source of contamination is the addition of incorrect or old oil to the system. Table 28.2 provides a list of relevant faults that can occur in oil lubrication systems and some wetted components' faults.

Many off-line, spectroscopic and ferrographic techniques exist to analyze lubricant condition and wear-metal debris.[47–51] These methods, although time-proven for their effectiveness at detecting many types of evolving failures, are performed at specified time intervals through off-line sampling.[52] The sampling interval is driven by the cost to perform the preventive maintenance versus the perceived degradation window over an operational timescale. The use of intermittent condition assessment will miss some lubricant failures. Moreover, the use of such off-line methods is inconvenient and increases the preventive maintenance cost and workload associated with operating the platform.

28.5.2.2 Lubrication System Test Bench

A lubrication system test bench (LSTB) was designed to emulate the lubrication system of a typical gas turbine engine.[53,54] The flow rate relates to typical turbine speeds, and flow resistance can be changed in each of the three legs to simulate bearing heating and differences between various turbine systems. To simplify operation, the LSTB uses facility water rather than jet fuel in the oil heat exchanger. The LSTB is also capable of adding a measured amount of foreign matter through a fixed-volume, dispensing pump, which is used to inject known amounts of metallic and nonmetallic debris, dirty oil, fuel, and water into the system. Contaminants are injected into a mixing block and pass through the debris sensors, upstream of the filter. The LSTB provides a way to correlate known amounts of contaminants with the system parameters and, thus, establishes a relationship between machinery wear levels, percentage of filter clogged, and viscosity of the lubricant.

The failure effects are in the areas of lubricant degradation, contamination, debris generated internally or externally, flow blockage, and leakage. These effects can be simulated or directly produced on the LSTB. Both degradation and contamination will result in changes in the oil transport properties. Water, incorrect oil, and sludge can be introduced in known amounts. Debris can be focused on metallic particles of 100 μm mean diameter, as would be produced by bearing or gear wear. Flow blockage can be emulated by restricting the flow through the control valves. Similarly, leakage effects can be produced by actual leaks or by opening the leg valves. Alternatively, seal leakage effects can cause air to flow into the lube system. This dramatically affects the performance and is measurable. In addition, the LSTB can be used to seed mechanical faults in the pump, relief valve, and instrumentation. In the case of mechanical component failure, vibration sensors could be added.

28.5.2.3 Turbine Engine Lubrication System Simulation Model and Metasensors

Note that the association of failure modes to sensor and fused data signatures remains a hurdle in such CBM work. Evaluation of operational data gathered on the gas turbine engine provided some association to believed faults, but insufficient data on key parameters prevented the implementation of a fault tree or even an implicit association. Given the lack of failure test data and the limited data available on the actual engine, a simulation model was developed. The turbine engine lubrication system simulation (TELSS) output was used to generate virtual or metasensor outputs. This data was evaluated in the data fusion and automated reasoning modules.

  The TELSS consists of a procedural program and a display interface. The procedural program is written in C code and uses the analog of electrical impedances to model the oil flow circuit. The model contains analytical expressions of mass, momentum, and energy equations, as well as empirical relationships. The interface displays state parameters using an object-oriented development environment. Both scripted and real system data can be run through the simulation. A great deal of effort was expended to properly characterize the Reynolds number and temperature-dependent properties and characteristics in the model. TELSS requires the geometry of the network, the gas generator speed, and a bulk oil temperature to estimate the pressures and flows throughout.[55]

28.5.2.4 Data Fusion Construct

The initial approach for lubrication system data fusion is summarized in Figure 28.18. This example follows the previous methodology of reviewing the data fusion steps within the context of the application. There are five levels in the data fusion process:

1. *Observation.* This level involves the collection of measured signals from the lubrication system being monitored (e.g., pressures, flow rates, pump speed, temperatures, debris sensors, and oil quality measurements).

2. *Feature extraction.* At this level, modeling and signal processing begins to play a role. From the models and signal processing, features (e.g., virtual sensor signals) are extracted; features are more informative than the raw sensor data. The modeling provides additional physical and historical information.

3. *Data association.* In this level, the extracted features are mapped into commensurate and noncommensurate failure mode spaces. In other words, the feature data is associated with other feature data based on how they reveal the development of different faults.

4. *System state estimation.* In this level, classification of feature subspaces is performed to estimate specific failure modes of the lubricant or oil-wetted components in the form of a state estimate vector. The vector represents a confidence level that the

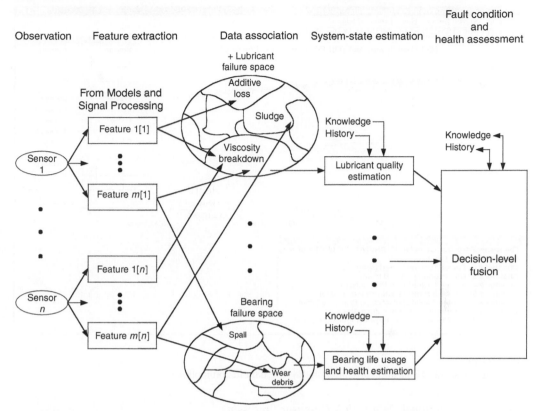

FIGURE 28.18
Lubrication system diagnostics/prognostics data fusion structure.

system is in a particular failure mode; the classification also includes information about the system and history of the lubrication system.

5. *Fault condition and health assessment.* For this level, system health decisions are made based on the agreement, disagreement, or lack of information that the failure mode estimates indicate. The decision processing should consider which estimates come from commensurate feature spaces and which features map to other failure mode feature subspaces, as well as the historical trend of the failure mode state estimate.

28.5.2.5   Data Analysis Results

28.5.2.5.1   Engine Test Cell Correlation

Engine test cell data was collected to verify the performance of the system. The lubrication system measurements were processed using the data fusion toolkit to produce continuous data through interpolation. Typical data is seen in Figure 28.19. This data provided the opportunity to trend variables against the fuel flow rate to the engine, gas generator speed and torque, and the power turbine speed and torque. Ultimately, through various correlation analysis methods, the gas generator was deemed the most suitable regression (independent) variable for the other parameters. It was used to develop three-dimensional maps and regressions with a measured temperature to provide guidelines for normal operation.

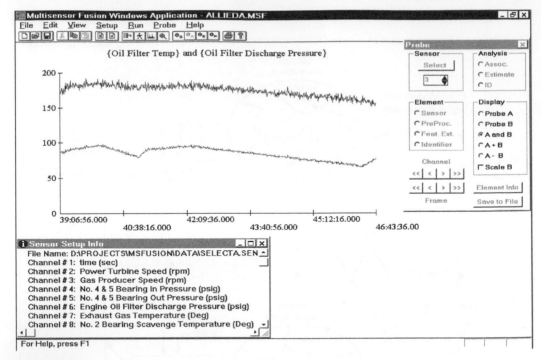

FIGURE 28.19
Processing of test stand data and correlation in the ARL data fusion toolkit.

28.5.2.5.2 Operational Data with Metasensor Processing

Operational data was made available by a navy unit that uses the gas turbine engines. The operational data was limited to the production engine variables, which consisted of one pressure and temperature. The TELSS model was embedded within the multisensor fusion toolkit. The TELSS interface for an LPAS run is shown in Figure 28.20. Because the condition of the oil and filter was unknown for these runs, the type of oil and a specified amount of clogging was assumed. The variation of oil and types of filters can vary the results significantly. Different MIL-L-23699E oils, for many of which the model possesses regressions, can vary the flow rate predictions by up to 5%. Similar variation is seen when trying to apply the filter clogging to different vendors' filter products.

The TELSS simulation model can be used to simulate different fault conditions to allow data association and system state estimation. Figure 28.21 illustrates the output of the metasensor under simulated fault conditions of filter clogging with debris. Filter clogging is typically monitored through a differential pressure measurement or switch. This method does not account for other variables that affect the differential pressure. The other variables are the viscosity, or the fluid resistance to flow, which is dependent on temperature and the flow rate through the filter.

With this additional knowledge, the T-P–mdot relationship can be exploited in a predictive fashion. Toward the mid to latter portion of the curves, the pressure increases slightly, but steadily, as the flow rate remains constant or decreases. Meanwhile, the temperature increases around 1000 s and then decreases steadily from 1500 until the engine is shut off. Let us investigate these effects more closely. The increasing pressure drop from about 1200 to 1500 s occurs whereas the temperature and flow are approximately constant. This is one indication of a clogging effect. From 1500 through 2100, the flow rate is at a lower level, but the pressure drop rises above its previous level at the higher oil flow rate. Looking only at

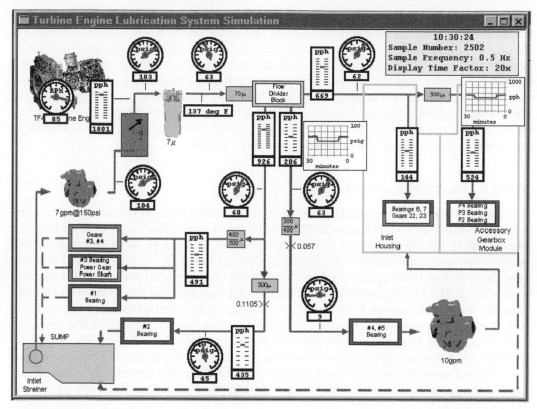

FIGURE 28.20
TELSS processing of operational data to produce metasensors. Massflow in pounds per hour is shown in bar scaled objects. Pressure throughout the circuit in pounds per square inch is illustrated by pressure gauge objects.

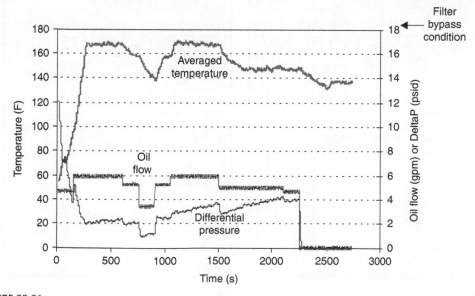

FIGURE 28.21
TELSS run illustrating the relationship between the system variables that can be fused to produce filter clogging association and estimates.

these two variables could suffice; however, deeper analysis reveals that during the same timeframe, the temperature decreases, which means the viscosity (or resistance to flow) of the oil increases. This lower temperature would indicate that higher pressure drops could be expected for the same flow rate. This effect (increased viscosity due to lower tempera- ture) is the reason why the pressure drop is so high at the beginning of the run. Conse- quently, this consideration actually adds some ambiguity to an otherwise crisp indication. The model analysis indicates, though, that the additional pressure drop caused by higher viscosity does not comprise the entire difference. Thus, the diagnosis of filter clogging is confirmed in light of all of the knowledge about the effects.

28.5.2.6 Health Assessment Example

The output from the TELSS model and multisensor toolkit was processed using an auto- mated reasoning shell tool (Figure 28.22). The output of a shell that could be used to detect filter-clogging fractions is shown in the figures below. An ES, a FL association, and a NN perform the evaluations of filter clogging. The flow, temperature, and differential pressure were divided into three operational ranges. The ES was provided set values for the frac- tion clogged. The FL was modeled with trapezoidal membership functions. The NN was trained using the FL outputs.[55,56] For the first case shown, the combination of 4.6 gpm, 175°F, and 12 psid, the reasoning techniques all predict relatively low clogging. In the next case, the flow is slightly less, whereas the pressure is slightly higher at 12.5 psid. The NN evalu- ation quickly leans toward a clogged filter, but the other techniques lag in fraction clogged. The ES is not sensitive enough to the relationships between the variables and the signifi- cance of the pressure differential increasing while the flow decreases markedly. This study

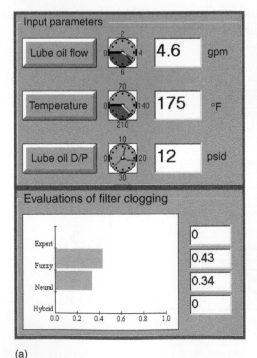

(a)

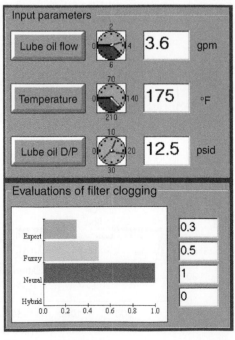

(b)

FIGURE 28.22
Hybrid reasoning shell evaluation (cases a and b).

and others conducted at ARL indicated that a hybrid approach based on decision fusion methods would allow the greatest flexibility in such assessments.

28.5.2.7   Summary

The objective of this fluid systems research was to demonstrate an improved method of diagnosing anomalies and maintaining oil lubrication systems for gas turbine engines. Virtual metasensors from the TELSS program and operational engine data sets were used in a hybrid reasoning shell. A simple module for the current-limited sensor suite on the test engine was proposed, and recommendations for enhanced sensor suites and modules were provided. The results and tools, although developed for the test engine, are applicable to all gas turbine engines and mechanical transmissions with similar pressure-fed lubrication systems.

As mentioned in Section 28.5.2.5.2, the ability to associate faulted conditions with measurable parameters is tantamount for developing predictive diagnostics. In the current example, metasensors were generated using model knowledge and measured inputs that could be associated to estimate condition. Development of diagnostic models results from the fusion of the system measurements as they are correlated to an assessed damage state.

28.5.3   Electrochemical Systems

Batteries are an integral part of many operational environments and are critical backup systems for many power and computer networks. Failure of the battery can lead to loss of operation, reduced capability, and downtime. A method to accurately assess the condition (state-of-charge, SOC), capacity (amp-h), and remaining charge cycles (RUL) of primary and secondary batteries could provide significant benefit. Accurate modeling characterization requires electrochemical and thermal elements. Data from virtual (parametric system information) and available sensors can be combined using data fusion. In particular, information from the data fusion feature vectors can be processed to achieve inferences about the state of the system.

This section describes the process of computing battery SOC—a process that involves model identification, feature extraction, and data fusion of the measured and virtual sensor data. In addition to modeling the primary electrochemical and thermal processes, it incorporates the identification of competing failure mechanisms. These mechanisms dictate the RUL of the battery, and their proper identification is a critical step for predictive diagnostics.

Figure 28.23 illustrates the model-based prognostics and control approach that the battery predictive diagnostics project addresses. The modeling approach to prognostics requires the development of electrical, chemical, and thermal model modules that are linked with coupled parameters. The output of the models is then combined in a data fusion architecture that derives observational synergy, while reducing the false alarm rate. The reasoning system provides the outputs shown at the bottom right of the figure. Developments will be applicable to the eventual migration of the diagnosis and health monitoring to an electronic chip embedded into the battery (i.e., intelligent battery health monitor).

28.5.3.1   The Battery as a System

A battery is an arrangement of electrochemical cells configured to produce a certain terminal voltage and discharge capacity. Each cell in the battery is comprised of two electrodes where charge transfer reactions occur. The anode is the electrode at which an oxidation (O)

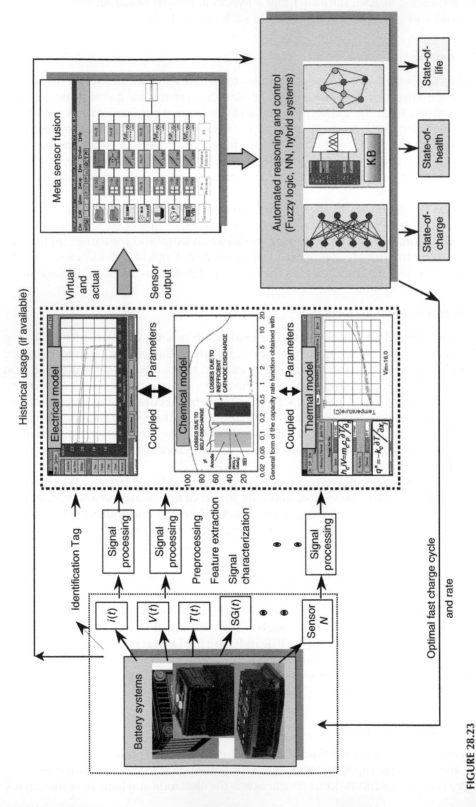

FIGURE 28.23
Model-based prognostics and control approach.

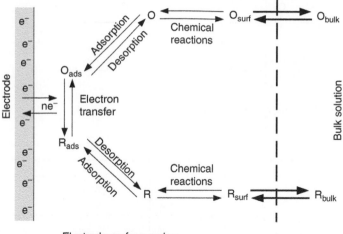

FIGURE 28.24
Electrode reaction process. (From Berndt, D., *Maintenance-Free Batteries*, Wiley, New York, NY, 1997; Bard, A.J. and Faulkner, L.R., *Electrochemical Methods: Fundamentals and Applications*, Wiley, New York, NY, 1980.)

reaction occurs. The cathode is the electrode at which a reduction (R) reaction occurs. The electrolyte provides a supply of chemical species required to complete the charge transfer reactions and a medium through which the species (ions) can move between the electrodes. Figure 28.24 illustrates the pathway ion transfer that takes place during the reaction of the cell.[57,58] A separator is generally placed between the electrodes to maintain proper electrode separation despite deposition of corrosion products.[57] Electrochemical reactions that occur at the electrodes can generally be reversed by applying a higher potential that reverses the current through the cell. In situations where the reverse reaction occurs at a lower potential than any collateral reaction, a rechargeable or secondary cell can potentially be produced. A cell that cannot be recharged because of an undesired reaction or an undesirable physical effect of cycling on the electrodes is called a primary cell.[57]

Changes in the electrode surface, diffusion layer, and solution are not directly observable without disassembling the battery cell. Other variables such as potential, current, and temperature are observable and can be used to indirectly determine the performance of physical processes. For overall performance, the capacity and voltage of a cell are the primary specifications required for an application. The capacity is defined as the time integral of current delivered to a specified load before the terminal voltage drops below a predetermined cut-off voltage. The present condition of a cell is described nominally with the SOC, which is defined as the ratio of the remaining capacity and the total capacity. Secondary cells are observed to have a capacity that deteriorates over the service life of the cell. The term state-of-health (SOH) is used to describe the physical condition of the battery, which can range from external behavior, such as loss of rate capacity, to internal behavior, such as severe corrosion. The remaining life of the battery (i.e., how many cycles remain or the usable charge) is termed the state-of-life (SOL).

28.5.3.2 Mathematical Model

An impedance model called the Randles circuit, shown in Figure 28.25, is useful in assessing battery condition. Impedance data can be collected online during discharge and charge to capture the full change of battery impedance and identify the model

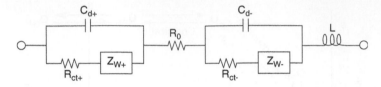

FIGURE 28.25
Two-electrode randles circuit model with wiring inductance.

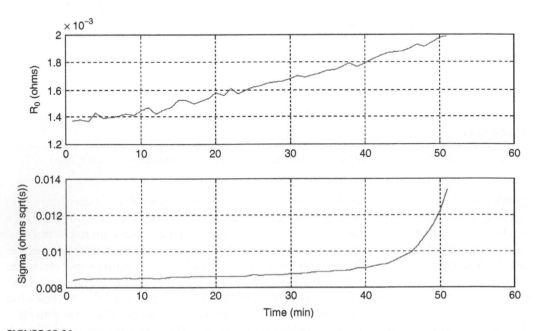

FIGURE 28.26
Nickel–cadmium model parameters over discharge cycle.

parameters at various stages of SOC, as well as over multiple cycles of the battery for SOH identification. Some of the identified model parameters of the nickel–cadmium battery are shown in Figure 28.26 as the batteries proceed from a fully charged to a discharged state. Identification of these model-based parameters provides insight and observation into the physical processes occurring in the electrochemical cell.[59,60]

28.5.3.3 *Data Fusion of Sensor and Virtual Sensor Data*

The approach for battery feature data fusion is summarized in Figure 28.27. There are five levels in the data fusion processes: *observation* (data collection), *feature extraction* (computation of model parameters and virtual sensor features), *data association* (mapping features into commensurate and noncommensurate feature spaces), *system-state estimation* (estimation of failure modes and confidence level), and *fault condition and health assessment* (making system health decisions).

Figure 28.28 illustrates the sensor and virtual sensor input to the data fusion processing. The outputs of the processing are the SOC, SOH, and SOL estimates that are fed into the automated reasoning processing. After the data association processing, an estimate of the failure mechanism is determined.

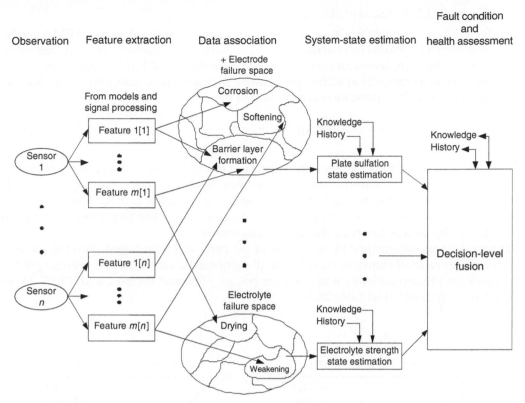

FIGURE 28.27
Battery diagnostics/prognostics data fusion structure.

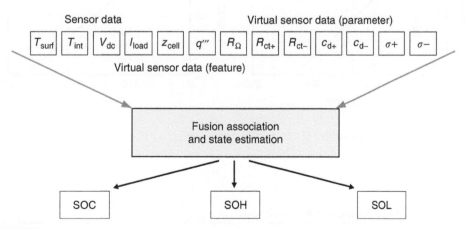

FIGURE 28.28
Generalized feature vector for battery predictive diagnostics.

Two approaches for SOC prediction are described in Sections 28.5.3.3.1 and 28.5.3.3.2. Each performs a kind of data fusion to utilize physically meaningful parameters to predict SOC. The definition of SOC is the amount of useful capacity remaining in a battery during a discharge. Thus, 100% SOC indicates full capacity and 0% SOC indicates that no useful capacity remains.

28.5.3.3.1   ARMA Prediction of State-of-Charge

An effective way to predict SOC of a battery has been developed using ARMA model methodology.[61] This model has performed well on batteries of various size and chemistry, as well as at different temperatures and loading conditions. ARMA models are a very common system identification technique because they are linear and easy to implement. The model used in this application is represented by

$$y(t) = aX(t) + bX(t-1) + c_0 y(t-1) \tag{28.10}$$

where y represents SOC; X represents a matrix of model inputs; and a, b, and c_0 represent the ARMA coefficients. Careful consideration was exercised in determining the inputs. These were determined to be V_D, I_D, R_Ω, θ, C_{DL}, and T_s, and output is SOC. The inputs were smoothed and normalized to reduce the dependence on noise and to prevent domination of the model by parameters with the largest magnitudes.

The model was determined by using one of the runs and then tested with the remaining runs. Figure 28.29 shows the results for all eight size-C batteries. The average prediction error for lithium batteries was <3%, for NiCad <5%, and for lead acid <10%. These results are summarized in Table 28.3.

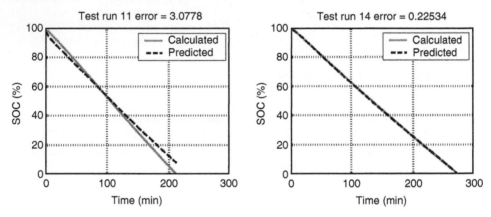

FIGURE 28.29
ARMA SOC prediction results for size-C lithium $(CF)_x$ batteries.

TABLE 28.3

Results of ARMA Model SOC Predictions

Chemistry	Size	Number of Cells	Type	Prediction Error (%)
Lithium[a]	C	1	Primary	2.18
Lithium[a]	2/3 A	1	Primary	2.87
NiCad[b]	D	1	Secondary	3.17
NiCad[b]	C	1	Secondary	4.50
Lead acid	12 V	6	Secondary	9.13

[a] Poly carbonmonofluoride lithium (spiral-type).
[b] Nickel–cadmium.

28.5.3.3.2 Neural Networks Prediction of State-of-Charge

Artificial NNs have been used successfully in both classification and function approxima-tion tasks.[61] One type of function approximation task is *system identification*. Although NNs very effectively model linear systems, their main strength is the ability to model nonlinear systems using examples of input–output pairs. This was the basis for choosing NNs for SOC estimation. NN SOC estimators were trained for lithium batteries of sizes C and 2/3 A under different loading conditions. For each type of battery, a subset (typically 3–6) of the available parameter vectors was chosen as the model input. Networks were trained to produce either a direct prediction of battery SOC or, alternatively, an estimation of initial battery capacity during the first few minutes of the run.

All networks used for battery SOC estimation had one hidden layer. The back propaga-tion, gradient-descent learning algorithm is used, which utilizes the error signal to opti-mize the weights and biases of both network layers. The inputs to the network were a subset of I_D, V_D, R_Ω, θ, and C_{DL}. As for the ARMA case, the inputs were smoothed and normalized. This led to smaller networks, which tend to be better at generalization. For time-delay NNs, the selection of the number of delays and the length of the delays is crucial to the performance of the networks. Both short and long delays were tried during different training runs. The short delays gave better performance, which indicates that the battery SOC does not involve long time constants. This is also evident from the ARMA examples. Several types of NNs were trained with battery data and extracted impedance parameters to directly predict the battery SOC. Among the several training methods that were used, the Levenberg–Marquardt (L-M) provided the best results. The size of the net-work was also important in training. An excessively small network size results in inade-quate training, and a larger than necessary network size leads to over-training and poorer generalization.

NNs trained to directly provide the battery SOC provided consistently good perfor-mance. However, the performance was much better when the battery's initial capacity was first estimated. These networks were also trained to estimate the initial capacity of the bat-tery during the first few minutes of the test. Then, the measured load current could easily be used to predict SOC because the current is the rate of change of charge. This is even more useful for SOH and SOL prediction because as the secondary batteries are reused they start at different initial capacity each time.

This method can also be used as a powerful tool in *mission planning*. Hypothetical load profiles can be used to predict whether a battery would survive or fail a given mission, thus preventing the high cost and risk of batteries failing in the field. The networks tend to slightly underestimate the battery SOC. This is a very important practical feature, since it results in a conservative estimate and avoids unscheduled downtime.

The results using radial basis function (RBF) NNs were the best and are summarized in Table 28.4. The SOC plots are shown for size-C lithium batteries in Figure 28.30. The results are quite remarkable, considering that very little training data are required to produce the

TABLE 28.4

Error Rates for SOC Prediction Based on Initial Capacity Estimation with RBF Neural Networks

Battery Size (Number of Hidden Neurons)	Average Training Error (Training Set)	Average Testing Error	Maximum Testing Error (Run Number)
Size C[6]	0.6%[13,14,16]	2.9%	6.8%[15]
Size 2/3 A[12]	0.8%[17,18,20]	2.9%	8.2%[23]

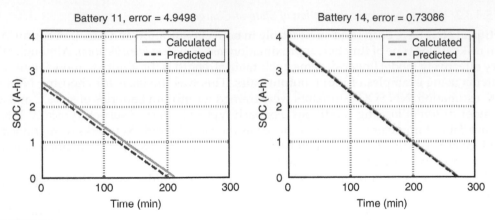

FIGURE 28.30
SOC prediction for size-C lithium batteries using initial SOC estimation with RBF neural networks.

predictors. As more data are collected and several runs of each of level of initial battery SOC become available, the robustness of the predictors will likely improve. In addition, the NN predictors have smaller error on outliers and provide a conservative prediction (i.e., they do not over-predict the SOC). Both of these advantages are very important in practical systems where certification and low false alarms are not just requirements, but can make the difference between a system that is actually used or shelved.

28.6   Concluding Remarks

The application of data fusion in the field of CBM and predictive diagnostics for engineered systems is rich with opportunity. The authors fully acknowledge that only a small amount of what is possible to accomplish with data fusion has been presented in this chapter. The predictive diagnostics application domain is relatively new for data fusion, but the future is bright with the many analogies that can be drawn between more mature data fusion applications and the current one. The authors anticipate that continuing developments in actual and virtual feature fusion, as well as hybrid decision fusion of condition assessments, will tend to dominate the research field for some time.

Acknowledgments

The authors gratefully acknowledge support from Dr. Thomas McKenna and Dr. Phillip Abraham from the Office of Naval Research on grant numbers N00014-95-1-0461, N00014-96-1-0271, N00014-97-1-1008, and N00014-98-1-0795, under which most of this work was originally performed. The authors would also like to acknowledge the contributions of several past and present colleagues at the Applied Research Laboratory: Mr. Bill Nickerson for his vision, help in understanding a framework for the CBM problem, and contributions to the definitions and systems thinking sections; Mr. James Kozlowski for his work on gearbox and battery diagnostics data fusion; Mr. Derek Lang for his work on the multisensor toolkit and insight into decision fusion; Mr. Dan McGonigal for his help in programming the lubrication system example and automated reasoning work on the gearbox failure tests; Mr. Jerry Kasmala for his development and assistance in the use of the data fusion toolkit; Ms. Terri Merdes for her contributions to the oil-vibration fusion analysis and discussion; Mr. Thomas Cawley for his development of a battery mathematical model; Mr. Matthew Watson for his work on ARMA modeling for SOC prediction; Mr. Todd Hay for his contributions to the neural network prediction section; Ms. Natalie Nodianos for her insightful editing; and Dr. David Hall for his mentoring and pointing the compass for the CBM data fusion effort at Penn State ARL.

References

1. Hansen, R. J., Hall, D. L., and Kurtz, S. K., A new approach to machinery prognostics, *Journal of Engineering for Gas Turbines and Power*, 117(2), 320–325, DN: Presented at 39th International Gas Turbine and Aeroengine Congress and Exposition, June 13–16, 1994, The Hague (NL), PBD, April 1995.
2. Kotanchek, M. E. and Hall, D. L., *CBM/IPD Process Model*, ARL TM 95-113, November 1995.
3. Byington, C. and Nickerson, G. W., Technology issues for CBM, *P/PM Technology Magazine*, 9(3), June 1996.
4. Byington, C., Monitoring the health of mechanical transmissions, *NCADT Quarterly*, No. 2, 1996.
5. Chamberlain, M., *U.S. Navy Pursues Air Vehicle Diagnostics Research*, Vertiflite, Vol. 40, pp. 16–21, March/April 1994.
6. Douglas, J., The maintenance revolution, *EPRI Journal*, 207(9), 6–15, May/June 1995.
7. Condition-directed maintenance, Electric Power Research Institute, *Compressed Air Magazine*, August 1990.
8. Hall, D. L., Hansen, R. J., and Lang, D. C., The negative information problems in mechanical diagnostics, in *Proceedings of the International Gas Turbine Institute Conference*, ASME, Birmingham, England, June 1997.
9 Elverson, B., Machinery fault diagnosis and prognosis, M.S. Thesis, The Pennsylvania State University, 1997.
10 Erdley, J., Data fusion for improved machinery fault classification, M.S. Thesis, The Pennsylvania State University, May 1997.
11 Dempster, A. P., Generalization of Bayesian inference, *Journal of Royal Statistical Society*, 30(2), 205–247, 1968.
12 Shafer, G., *A Mathematical Theory of Evidence*, Princeton University Press, Princeton, NJ, 1976.
13 Thomopoulos, S. C. A., Theories in distributed decision fusion: Comparison and generalization, in *Proceedings of the SPIE 1990 on Sensor Fusion III: 3-D Perception and Recognition*, Boston, MA, pp. 623–634, November 5–9, 1991.
14. Hall, D. and McMullen, S. A. H., *Mathematical Techniques in Multisensor Data Fusion*, Artech House, Inc., Norwood, MA, 1992.
15. Waltz, E. and Llinas, J., *Multisensor Data Fusion*, Artech House, Inc., Boston, MA, 1990.
16. Bayes, T., An essay towards solving a problem in the doctrine of chances, *Philosophical Transactions of the Royal Society of London*, 53, 370–418, 1763.
17. Freund, J. E. and Walpole, R. E., *Mathematical Statistics*, 3rd. Ed., Prentice-Hall, Englewood Cliffs, NJ, 1980.
18. Pearl, J., *Probabilistic Reasoning in Intelligent Systems*, Morgan Kaufmann Series in Representation and Reasoning, San Francisco, CA, 1988.
19. Natke, H. G. and Cempel, C., Model-based diagnosis: Methods and experience, in *Proceedings of a Joint Conference, the 51st Meeting of the Society for Machinery Failure Prevention Technology (MFPT)* and the *12th Biennial Conference on Reliability, Stress Analysis and Failure Prevention* (RSAFP Committee of ASME) (Henry C. Pusey and Sallie C. Pusey, Eds.), Virginia Beach, VA, pp. 705–719, April 14–18, 1997.

20. Byington, C. S., *Model-Based Diagnostics for Turbine Engine Lubrication Systems*, ARL TM98-016, February 1998.
21. Hall, D. L. and Kasmala, G., Visual programming environment (Toolkit) for multisensor data fusion, in *Proceedings of the SPIE AeroSense 1996 Symposium*, Orlando, FL, Vol. 2764, pp. 181–187, April 1996.
22. Hall, D. L. and Garga, A. K., Pitfalls in data fusion (and how to avoid them), in *Proceedings of 2nd International Conference on Information Fusion (FUSION 99)*, Vol. I, pp. 429–436, 1999.
23. Begg, C., Merdes, T., Byington, C. S., and Maynard, K. P., Dynamics modeling for mechanical fault diagnostics and prognostics, in *Maintenance and Reliability Conference*, 1999.
24. Mitchell, L. D. and Daws, J. W., *A Basic Approach to Gearbox Noise Prediction*, SAE Paper 821065, Society of Automotive Engineers, Warrendale, PA, 1982.
25. Ozguven, H. N. and Houser, D. R., Dynamic analysis of high speed gears by using loaded static transmission error, *Journal of Sound and Vibration*, 125(1), 71–83, 1988.
26. Randall, R. B., A new method of modeling gear faults, *Transactions of the ASME, Journal of Mechanical Design*, 104, 259–267, 1982.
27. Nicks, J. E. and Krishnappa, G., Evaluation of vibration analysis for fault detection using a gear fatigue test rig, C404/022, in *IMEC E Proceedings*, 1990.
28. Zakrajsek, J. J., An investigation of gear mesh failure prediction techniques, NASA Technical Memorandum 102340 (89-C-005), November 1989.
29. Tandon, N. and Nakra, B. C., Vibration and acoustic monitoring techniques for the detection of defects in rolling element bearings: A review, *The Shock and Vibration Digest*, 24(3), 311, March 1992.
30. Li, C. J. and Wu, S. M., On-line detection of localized defects in bearings by pattern recognition analysis, *Journal of Engineering for Industry*, 111, 331–336, 1989.
31. Rose, J. H., *Vibration Signature and Fatigue Crack Growth Analysis of a Gear Tooth Bending Fatigue Failure*, Presented at 44th Meeting of the Mechanical Failure Prevention Group (MFPG), April 2–5, 1990.
32. Mathew, J. and Alfredson, R. J., The condition monitoring of rolling element bearings using vibration analysis, *Journal of Vibration, Acoustics, Stress, and Reliability in Design*, 106, 447–453, 1984.
33. Lebold, M., McClintic, K., Campbell, R., Byington, C., and Maynard, K., Review of vibration analysis methods for gearbox diagnostics and prognostics, in *Proceedings of the 53rd Meeting of the Society for Machinery Failure Prevention Technology*, Virginia Beach, VA, pp. 623–634, May 2000.
34. Byington, C. S. and Kozlowski, J. D., Transitional data for estimation of gearbox remaining useful life, in *Proceedings of the 51st Meeting of the Society for Machinery Failure Prevention Technology (MFPT)*, Virginia Beach, VA, Vol. 51, pp. 649–657, April 1997.
35. Townsend, D. P., *Dudley's Gear Handbook: The Design Manufacture and Application of Gears*, 2nd Ed., McGraw Hill, New York, NY, 1992.
36. Roylance, B. J. and Raadnui, S., The morphological attributes of wear particles: Their role in identifying wear mechanisms, *Wear*, 175, 115–121, 1995.
37. Brennan, M. J., Chen, M. H., and Reynolds, A. G., Use of vibration measurements to detect local tooth defects in gears, *Sound and Vibration*, November 1997.
38. Wang, W. J. and McFadden, P. D., Application of the wavelet transform to gearbox vibration analysis, in *Structural Dynamics and Vibrations*, American Society of Mechanical Engineers (ASME), in *Proceedings of the 16th Annual Energy Sources Technology Conference and Exposition*, Vol. 52, pp. 13–20, January 31–February 4, 1993.
39. Ferlez, R. J. and Lang, D. C., Gear tooth fault detection and tracking using the wavelet transform, in *Proceedings of the 52nd Meeting of the MFPT*, 1998.
40. Byington, C. S. et al., Vibration and oil debris feature fusion in gearbox failures, in *Proceedings of the 53rd Meeting of the Society for MFPT*, 1999.
41. Byington, C. S. et al., Fusion techniques for vibration and oil debris/quality in gearbox failure testing, *International Conference of Condition Monitoring*, Swansea, pp. 113–128, 1999.

42. McGonigal, D. L., Comparison of automated reasoning techniques for CBM, M.S. Thesis, The Pennsylvania State University, August 1997.
43. Campbell, R. L. and Byington, C. S., Gas turbine fuel system modeling and diagnostics, in *COMADEM Proceedings, 13th International Congress on Condition Monitoring and Diagnostic Engineering Management*, Houston, TX, pp. 257–266, December 2000.
44. Toms, L. A., *Machinery Oil Analysis: Methods, Automation and Benefits*, 2nd Ed., Coastal Skills training, Virginia Beach, VA, 1998.
45. Lubrication Fundamentals: Fundamental Concepts and Practices, *Lubrication Engineering Magazine*, 26–34, September 1997.
46. Eleftherakis, J. G. and Fogel, G., Contamination control through oil analysis, *P/PM Technology Magazine*, 62–65, October 1994.
47. Toms, A. M. and Powell, J. R., Molecular analysis of lubricants by FT-IR spectroscopy, *P/PM Technology Magazine*, 58–64, August 1997.
48. Richards, C., Oil analysis techniques advance with new information technology, *P/PM Technology Magazine*, 60–62, June 1995.
49. Neale, M. J., *Component Failures, Maintenance, and Repair*, Society of Automotive Engineers, Inc., Warrendale, PA, 1995.
50. Anderson, D. P., Rotrode filter spectroscopy: A method for the multi-elemental determination of the concentration of large particles in used lubricating oil, *P/PM Technology Magazine*, 88–89, September/October 1992.
51. Stecki, J. S. and Anderson, M. L. S., Machine condition monitoring using filtergram and ferrographic techniques, *Research Bulletin: Machine Condition Monitoring*, 3(1), September 1991.
52. Troyer, D. and Fitch, J. C., An introduction to fluid contamination analysis, *P/PM Technology Magazine*, 54–59, June 1995.
53. Byington, C. S., Model-based diagnostics of gas turbine engine lubrication systems, *Joint Oil Analysis Program International Condition Monitoring Program*, April 20–24, 1998.
54. Byington, C. S., Intelligent monitoring of gas turbine engine lubrication systems, ARL TN 98-016, February 1998.
55. Garga, A. K., A hybrid implicit/explicit automated reasoning approach for condition-based maintenance, in *Proceedings of ASNE Intelligent Ships Symposium II*, 1996.
56. Stover, J. A., Hall, D. L., and Gibson, R. E., A fuzzy-logic architecture for autonomous multi-sensor data fusion, *IEEE Transactions of Industrial Electronics*, 43(3), 403–410, June 1996.
57. Berndt, D., *Maintenance-Free Batteries*, 2nd Ed., Wiley, New York, NY, 1997.
58. Bard, A. J. and Faulkner, L. R., *Electrochemical Methods: Fundamentals and Applications*, Wiley, New York, NY, 1980.
59. Boukamp, A., A nonlinear least squares fit procedure for analysis of immittance data of electrochemical systems, *Solid State Ionics*, 20(31), 31–44, 1986.
60. Kozlowski, J. D. and Byington, C. S., Model-based predictive diagnostics for primary and secondary batteries, in *Proceedings of the 54th Meeting of the MFPT*, Virginia Beach, VA, May 1–4, 2000.
61. Kozlowski, J. D. et al., Model-based predictive diagnostics for electrochemical energy sources, in *Proceedings of IEEE Aerospace Conference*, Vol. 6, pp. 63149–63164, Big Sky, MT, March 2001.

29

Adapting Data Fusion to Chemical and Biological Sensors

David C. Swanson

CONTENTS

29.1 Introduction

This chapter provides background information on the challenges and issues related to detection and identification of biological pathogens and chemical agents in the environment. Although laboratory-based methods can be used for effective identification, these techniques are generally unusable for first responders in a field environment, because they require significant time and expertise of a laboratory technician. The chapter introduces the issues and challenges in detection and identification, and also the concept of using strategies based on biological entities. In particular, an approach using multiple sensors and a fuzzy logic inference procedure is used to mimic how insects detect and characterize biochemical agents. The resulting data fusion technique provides effective detection and identification of complex chemical and biological agents in real-world environments.

29.2   Characterizing the Complexity of Detecting Chemical Agents and Biological Pathogens

Detection of harmful chemical vapors or biological pathogens has obvious value to society but also presents a number of problems. First, automated sensors are needed to provide a *detect-to-warn* capability. There exists a broad range of laboratory techniques that can separate and uniquely identify harmful agents, but these techniques take too much time and require rare human skills for analysis to be useful in an emergency. To be of use to first responders or those on the battlefield, chemical and biological sensors need to be both real time and completely automated. Second, although it is desirable to have *stand-off* sensors that can detect an agent without physical contact from a distant position (say hundreds of meters or even a kilometer), these often-active sensors are expensive, may require the use of non-eye-safe lasers, and also require rare human analytical skills to interpret the measured data. Third, we live in an environment with literally millions of chemical vapor mixtures making the false alarm problem, not the detection problem, which is the major concern. Fourth, for biological sensors, the false alarm problem is extremely difficult to detect since all bacteria and viruses are composed of the same amino acids and nucleic acids, but are arranged in different order. However, for both chemical and biological agents, we are most interested in a relatively short list of chemicals and pathogens (in the hundreds) compared to the millions of interferants naturally found in our environment.

To understand a useful strategy for data fusion to improve the performance of automatic detection with a low false alarm rate, one should take advantage of a diverse array of sensor technologies such that each sensor uniquely responds to the presence of, and concentration of, the agents of interest. This does not suggest *orthogonal* sensors, that is, sensors that only respond to one thing. Even if orthogonality could be guaranteed, it would lead to the need for too many sensors for a practical system. Therefore, our data fusion approach is based on having a *diverse* array of sensors where for the agents of interest we have a subset of sensors responding with unique sensitivities. Sensor responses will be categorized as *qualitative* for true-false types of detection and *quantitative* where the amount of response is functionally related to chemical concentration or the number of colony-forming units (NCU) for biological spores. We can use both kinds of sensor information, although the quantitative ones are more useful because they can be used for corroboration of vapor concentration (or NCU) and rejection of false alarms. This is because the diverse sensors each respond uniquely to a given chemical or biological agent.

Consider how nature has solved the chemical detection and identification problem with olfaction, and the biological detection and identification problem with the immune response system. Olfaction in humans and mammals is quite complex to study, and there are ethical barriers to neurological response experiments on live animals. However, insects have the most sensitive chemical detection capability of all animals and can demonstrate robustness to experimentation. The brain and antenna neurons will function for hours after the head has been removed from the body, and there has been more than 50 years of detailed scientific study to support the development of safer and more effective pesticides.

An excellent reference detailing the biology of insect olfaction can be found in Kaissling.[1] In short, insects have around 60 classes of *sensilla*, or olfactory sensors, totaling approximately 100,000 sensilla per antenna. Two antennae are used to better associate the chemical plume detected with wind direction to navigate to the source. Each sensilla is visually different but all have one or more pores with chemical coatings to control the types of chemical vapors that can pass on to the neural receptor beneath the pore. The biochemistry involved is truly remarkable, but for our purposes, it is sufficient to note that each sensilla responds a little bit to any large molecule that can pass through the coating. Air comprises mostly small, diatomic (N_2, O_2), and triatomic (H_2O, CO_2, O_3, and so on) molecules due to solar radiation; larger molecules precipitate out from reactions with water and sunlight. The large molecules make their way into the sensilla pore, contact the nerve dendrite, and are then broken down by enzymes and removed. To protect against too much stimulation, there are structures on the antenna and in the pore itself to reduce uptake in much the same way a human iris closes in response to bright light. The sensilla self-cleaning process takes a few seconds whereas the detection process takes only a few milliseconds. In this way, all olfaction sensing has a similar process: fast change detection, slow adaptation to changing backgrounds, and self-cleaning capability. Furthermore, the chemical functionality of each sensilla is not orthogonal, but has an overlapped response based on parameters like solubility, ion mobility, diffusivity, and even mechanical filtering by lipids and proteins. Each sensilla class is neurologically wired to a specific area of the insect's brain, called the mushroom body, which receives all of the corresponding sensilla neuron pulses in an avalanche when a vapor passes by the antenna. The mushroom body structures on the brain for all the sensilla classes look like a bunch of grapes. Several individual mushroom bodies respond to an odor stimulus with electrical activity and, when the insect recognizes the odor, a projected neural response is seen throughout the rest of the brain. This is often further observed as an associated behavioral response by the insect. This general response is seen for pure chemicals as well as chemical mixtures.

The biological process of detecting chemicals is quite different from the way we approach chemical identification. Generally, one first separates chemical mixtures into the constitutive molecules, and then one systematically identifies each molecule based on properties such as molecular weight, number of carbon atoms, mass spectroscopy, and so on. This is very reasonable for skilled analysts in a laboratory setting, but not easily automated in a real-time detect-to-warn sensor. So what is nature teaching us by the way olfaction sensing happens in insects? It is basically the same process in all life. To mimic the biological system for chemical detection and identification, we can summarize a few requirements:

- Use a diverse set of sensors with overlapping, but unique responses
- Treat pure (neat) chemicals and chemical mixtures similarly in terms of sensor response patterns
- Allow the responding classes of sensors to vary with concentration
- Ensure the sensors are self-cleaning
- Associate the presence and concentration of each specific chemical or chemical mixture with the pattern of responses from the diverse sensor array

Consider the task of pathogen detection.[2] There are literally millions of bacteria and viruses all around us. Most are harmless to humans and animals, and we have acquired immunity to many so long as the pathogen is of low concentration (low NCU). Our immune response system can give us guidance in terms of developing a sensor and data fusion strategy analogous to the biomimicry of the olfaction system for chemical detection.

When a harmful virus, bacteria, or foreign protein pass through the cell walls of our skin into the bloodstream, it is called a *pathogen*. All vertebrates have a defense system called the immune response system that produces substances called *antigens* that support the killing and removal of pathogens. To simplify, there are generally two types of cells involved: B-cells that originate in the bone marrow and T-cells that originate in the thymus gland. The B-cells mark the pathogen, whereas the T-cells destroy it. The B-cells have materials called *immunoglobulin* molecules (antibodies) that bind to a specific pathogen. When this

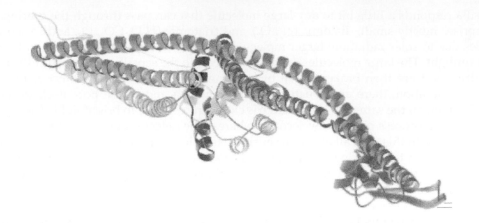

FIGURE 29.1
Colicin Ia is a transmembrane protein from *Escherichia coli* that creates a pore in a neighboring cell in which to secrete toxins and to extract food until the cell dies.

binding occurs, helper T-cells stimulate the particular B-cell variant to reproduce. Some of these cloned B-cells produce soluble antibodies that freely move about marking the corresponding pathogens they encounter. Other cloned B-cells provide *memory* for the immune response to facilitate a rapid counterattack should the same pathogen appear in the future. The key fact here is that the natural antigens in our bodies can form almost infinite combinations of amino acids that bind to unique sites on invading pathogens. Much of this natural immunity is part of our DNA but still these antigens can mutate to acquire immunity to a new pathogen. The memory B-cells retain the biological code to fight off frequent reinfections of the same pathogen at a later time.

If one approaches biological agent detection the same way as chemical detection, there will be great difficultly identifying a specific pathogen by means of molecular mobility, solubility, Raman spectrum, or even fluorescence spectrum.[3,*] This is because the characteristic protein materials associated with a given pathogen are very large (often thousands of amino acids long), the molecule has many folds and a generally complex structure, and there are large numbers of similar proteins associated with harmless bacteria, viruses, and foreign proteins. These complexities lead to a far more difficult challenge of *specificity* in biological detection. Given a sample of a suspected pathogen, one can visually identify it either from microscope images or by separating and identifying pieces of the DNA or unique proteins.

For example, the well-known *flesh-eating* bacteria, *Escherichia coli,* has a unique protein that protrudes out into the surrounding area of the cell and attacks other cells. Figure 29.1 shows a rendering of the *Colicin Ia* transmembrane protein that does this dastardly work.

The various ribbon-like sections (peptides) of the protein are drawn with colors and shapes to specifically call attention to biochemical properties, such as binding (or sticking) to other well-known reagents. The protein in Figure 29.1 is a long chain of amino acids, of which there are 20 types. Each type is given a letter symbol (A–V) and the entire sequence

---

* These are discussed in detail in Section 29.3.

TABLE 29.1

Amino Acid Sequence for *Colicin Ia* in *Escherichia coli*

MSDPVRITNPGAESLGYDSDGHEIMAVDIYVNPPRVDVFHGTPPAWSSFGNKTIWGGNEWVDDSP
TRSDIEKRDKEITAYKNTLSAQQKENENKRTEAGKRLSAAIAAREKDENTLKTLRAGNADAADIT
RQEFRLLQAELREYGFRTEIAGYDALRLHTESRMLFADADSLRISPREARSLIEQAEKRQKDAQN
ADKKAADMLAEYERRKGILDTRLSELEKNGGAALAVLDAQQARLLGQQTRNDRAISEARNKLSSV
TESLNTARNALTRAEQQLTQQKNTPDGKTIVSPEKFPGRSSTNDSIVVSGDPRFAGTIKITTSAV
IDNRANLNYLLSHSGLDYKRNILNDRNPVVTEDVEGDKKIYNAEVAEWDKLRQRLLDARNKITSA
ESAVNSARNNLSARTNEQKHANDALNALLKEKENIRNQLSGINQKIAEEKRKQDELKATKDAINF
TTEFLKSVSEKYGAKAEQLAREMAGQAKGKKIRNVEEALKTYEKYRADINKKINAKDRAAIAAAL
ESVKLSDISSNLNRFSRGLGYAGKFTSLADWITEFGKAVRTENWRPLFVKTETIIAGNAATALVA
LVFSILTGSALGIIGYGLLMAVTGALIDESLVEKANKFWGI

for a known protein can be found in many Web sites referred within protein data banks. Table 29.1 provides the sequence for *Colicin Ia*.

To gain specificity, biologists generally use an array of antibodies where each antibody is attached to a quenched fluorescing molecule, which shows fluorescence only when the antibody binds to the specific pathogen. Dyes are also used where the stain *sticks* if the pathogen is present and washes away if not. This *sticking* is actually a weak atomic force where a section of the protein is compatible with the sticking section of the antibody. An array of wells containing the antibody markers is called an immunoassay *sensor* but is really more like a microscope slide than an electronic chemical detector. One separates the proteins of the pathogen and applies a series of markers to identify the protein. Usually, a bioassay will have a large number of markers such that specific combinations would indicate one pathogen or another. This works well in the laboratory. But in the field, there is a broad spectrum of pathogens along with harmless bacteria in the air. So the real-world test is: can one identify a particular pathogen in a mixture of other biological material? This is why data fusion techniques are of high importance to biological detection.

Another biological identification approach is to isolate the pathogen's DNA and identify the DNA sequence. This is very often used in criminology or forensic science and is a useful technique to identify a captured pathogen. Unlike proteins, DNA is composed of sequences of only four different nucleic acids and serves the purpose of information storage only. A triplet of any three nucleic acids makes an amino acid. There are two basic steps to identifying a pathogen by DNA. First, one needs to replicate (amplify) the DNA once separated from the cell proteins, using polymerase chain reactions (PCR), to clone a unique section of the DNA for the pathogen of interest. The next step is to identify the sections of interest by measuring the mobility (using electrophoresis) of the DNA sections relative to known DNA. Only a few sections need to match to have high confidence and specificity. However, one drawback is that the PCR process can also amplify DNA from contaminants, leading to false positives.

In short, we have a fairly broad selection of chemical detectors which, if used in combination with data fusion, can provide both low false negatives (false rest, or no response to the actual agent) and low false positives (different sensing modalities will not corroborate false alarms among them). This is consistent with biological olfaction systems. For biological pathogen sensing, there are many more complex challenges even beyond the mechanics of wet biochemistry and pathogen isolation. This is because all biological materials are made of the same chemicals in long chains. The difference between a killer pathogen and bread yeast is only in the order of the amino and nucleic acids. Data fusion will be a very important tool in the automated analysis of biological sensing in real-world environments.

29.3　Chemical Sensors

Our purpose here is to explain briefly the typical types of real-time chemical detectors available to illustrate the challenges for false alarm reduction and automation. It should be clear that a skilled technician could use a host of laboratory equipment to quickly isolate and identify chemicals and mixtures. However, that is not the point here. We need to *automate* the detection and identification decision using specific concepts in data fusion based on science. Barring any new fantastic sensor technologies such as Mr. Spock's tri-corder,* we will briefly discuss a collection of typical chemical sensors based on first principles of physics, each of which provides an aspect of the full atomic signature of the chemical. Each sensor alone has vulnerabilities to false alarms, but since they respond differently to the same molecule, each offers important information for fusion into a robust automated detection and classification system. Most of these technologies have been around for more than 50 years and are not likely to ever be replaced. We make no distinction between point sensors, which must physically capture the chemical, or stand-off sensors, which optically interrogate a parcel of air from a distance. The important strategy of our discussion here is to fuse information from a variety of physically different sensors, expecting that a good data fusion algorithm will preserve the best detection performance while suppressing false alarms from each sensor individually.

29.3.1　Ion Mobility Spectrometer

The ion mobility spectrometer (IMS) sensor is most commonly seen in home smoke detectors. It uses either a radioactive isotope, or a strong electrical corona discharge to generate ions of the molecules in the air. Ions are molecules energized to have either additional or fewer electrons. Molecular ions therefore have a net electric charge and can be accelerated by an external electric field. The ionized molecule also has a physical size that will cause drag forces from collisions with the other molecules in the air. The IMS sensor generally has a *drift tube* where the ions are forced by electric field against a clean air flow toward a collection plate called the Faraday plate, which functions as a capacitor that provides a current signal proportional to the ions making the trip down the drift tube. The IMS cycles through hundreds of *mobility races* triggered by an electric grid *starting gate* at the ion-generation end with drift times measured at Faraday plate end.[4] The detected current (typically in pico-amperes) as a function of time is called a *plasmagram* where a peak in the curve at a particular time corresponds to a particular molecule. Highly charged small ions move fast whereas less-charged large ions move more slowly through the drift tube. Drift times are typically 10–20 ms or less in drift tubes around 10 cm in size; so, hundreds of plasmagrams can be averaged per second to enhance chemical signals and suppress background noise. There can be a positive field and a negative plasmagram if the IMS employs two drift tubes. There also can be calibration vapors added to confirm drift times as well as *doping vapors* added to amplify some molecular ions of interest. IMS technology is only about 25 years old and clearly offers a broad array of uses for this fast-responding sensor. *Detection* of a target chemical is essentially the appearance of a peak at the right drift time in the plasmagram. Although IMS is extremely sensitive, it does suffer from false alarms from other chemicals that share the same drift time.

---

* The science officer from Star Trek could detect anything with this device under any false alarms. With great prejudice we interpret the *tri* in *tri-corder* to imply automated data fusion with at least three sensor channels.

29.3.2 Surface Acoustic Wave and Electrochemical Cells

These devices are typically based on polymer materials doped with chemicals designed to repel all but one specific chemical of interest. The dopant is generally highly soluble. This *chemical functionalization* of the polymer substrate would ideally allow absorbing of exclusively the chemical of interest. However, these functionalized polymers generally allow similar chemicals to absorb, but at differing rates depending on solubility and diffusivity. Behind the polymer is a detector,[5] which in the case of electrochemical (EC), is usually a field-effect transistor (FET) that amplifies the voltages generated by oxidation–reduction reactions. The gate of the FET acts as a switch to open or close a much larger signal current through the source and drain leads of the FET. So, a very small voltage from the chemical ions in the polymer, or a small change in conductivity of the polymer due to absorbed chemical, can switch the EC from open (no chemical detected) to closed (target chemical detected) states. For the surface acoustic wave (SAW) case, there is an acoustic feedback path from a piezoelectric actuator, across the polymer surface, to a piezoelectric sensor, and back through an amplifier to complete the circuit. This circuit naturally oscillates at a frequency proportional to the wave speed in the polymer. As the polymer absorbs more chemical, the wave speed decreases causing the oscillation frequency to decrease. Since this frequency can be measured accurately, the SAW detector is also useful for quantitative measurements. Heating is used to clean out the polymer for reuse and a typical SAW or EC might absorb for say 10 s, report the result, then desorb for 10 or more seconds with heating to drive off the absorbed chemicals. The SAW is effective for absorption (chemical infusion into the polymer), adsorption (surface-only chemical collection), and sorption (both adsorption and absorption). Although the SAW and EC are relatively selective and low cost, the polymers have a limited lifetime and the calibration of the device is dependent on temperature and humidity.

29.3.3 Flame Photometric Detection

Flame photometric detection (FPD) is one of the most reliable sensors for atomic elements. Generally, FPD uses hydrogen as a flame source due to its well-defined optical spectrum in the red and its clean by-product (water). The temperature of the flame causes all of the atoms in the molecule to emit their characteristic photon frequencies. For chemical warfare agents, one looks for sulfur (for mustard gas) or phosphorus (for nerve agent) by looking for a unique spectral frequency associated with their respective target atom. The FPD is very quantitative, repeatable, self-cleaning, and simply does not false rest, which implies that if the target atoms are present, they are detected. The only problem, besides the logistics of handling compressed hydrogen, is that there are many chemicals and mixtures that have atoms of the same type; therefore, there is an inherent false alarm problem with FPD.

29.3.4 Photoionization Detection

Photoionization detection (PID) is widely used as a natural gas detector because there is no open flame. A high-intensity light source is used to ionize the molecules in the air and the resulting current to ground is measured. One chooses an ionization bulb energy (measured in electron volts, eV) strong enough to ionize the target molecules of interest. This is about the extent of selectivity of the PID. In general, the PID is very good at detecting hydrocarbons. Combine it with an EC cell for oxygen, and one has a very effective sensor array for detecting explosive atmospheres, such as those found in empty fuel tanks.

One of the drawbacks of a PID is keeping the bulb light source in a constant state of cleanliness. Dirt and carbon deposits that buildup on the bulb surface will decrease the ionization current and can lead to a loss of more than 30% sensitivity per day in constant use. One method to help alleviate this problem is to cycle the PID through a frequent clean-out mode. Air is trapped around the bulb allowing ozone to be generated, which assists in keeping the carbon and dirt buildup on the bulb to a minimum. The PID simply measures an ionization current; therefore, one needs to know the specific chemical (methane, propane, and so on) required to convert the current into a vapor concentration.

29.3.5 Spectrographic Methods

Spectrographic detection techniques identify the unique signature of the molecular bonds of a chemical by either the emission spectra, absorption spectra, or scattered spectrum. Emission spectra are the photons emitted by the molecule as electrons in the atom's transition from an excited state to a ground state. This can be the result of high temperature (such as in FPD) or fluorescence. Because the energy necessary to cause the atoms to emit characteristic photons can be quite high, the molecular bonds are often broken. Therefore, high-energy stimulated emission generally produces spectral lines associated with the atoms in the molecule, not the unique bond within the molecule. This makes emission spectra difficult for specific molecule identification, especially if mixtures of molecules are involved. Fluorescence is somewhat different in that a high-energy photon absorbed into the molecule causes lower-energy photons to be emitted. Proteins excited by ultraviolet (UV) light and emitting visible greens and yellows are good examples of fluorescing molecules. Fluorescence is quite complicated, to the extent that it is very difficult to distinguish one fluorescing material apart from another.

Absorption spectroscopy is often used in the infrared (IR) band because it is much easier to measure molecular resonances from IR photon absorption. What makes a given molecule unique are the specific bonds and atoms that create a molecular structure with unique normal modes of vibration and rotation.[6] Using a broadband IR source, one can compare the incident and transmitted spectrum to identify frequency bands of absorption unique to the molecule. The amount of absorption will increase for high concentrations of the target chemical. Spectral absorptions of several compounds are mixed independently, but in proportion to allow simultaneous mixture identification. One could use principal component analysis to estimate which combinations of molecules are present and in what concentrations. However, mixtures are best separated into components to simplify spectral identification.

One form of IR absorption spectroscopy that is quite popular is Fourier transform IR (FTIR). An FTIR splits the incoming light into two paths where the path length of one is modulated with a moving mirror. This modulation effectively increases the resolution of the spectrometer by physically widening the aperture. By mixing the light from the moving mirror path with the original, modulations of intensity are produced. By Fourier transforming these modulations and knowing the modulation frequency and amplitude of the movable mirror, one can coherently combine a broader range of light wavelengths into the absorption spectrum, thereby increasing the signal-to-noise ratio (SNR). An FTIR produces a higher resolution and higher SNR than a straightforward IR spectrometer, but it does require a fast computer to process the optical data.

Light scattered from a molecule can undergo frequency shifts due to interactions of the incident photons with the electron energies throughout the molecule. Some photons will pick up energy and scatter at a higher energy (shorter wavelength) whereas most other photons will lose some energy to the molecule and scatter at a lower energy (longer wavelength). To see this clearly, one uses a laser as a constant wavelength photon source.

The scattered spectrum is called the *Raman* spectrum.[7] The Raman spectrum is quite weak in amplitude where the lower-frequency Stokes scattering is stronger than the higher-frequency anti-Stokes scattered light. Roughened surfaces on metal substrates can be used to significantly enhance the Raman spectrum amplitude. The Raman spectrum for a molecule is a unique *fingerprint* for identification. One can use principal component analysis to estimate components of mixtures, but as with the IR absorption, it is much easier to identify pure chemicals (sometimes called *neat* chemicals) rather than mixtures.

The kinds of features that come from spectrographic methods of chemical identification are optical energy (or lack thereof) at specific wavenumbers of the optical spectrum. For a given neat chemical, one could create a statistical feature vector where each element represents a spectral range of wavelengths and the amplitude corresponds to the measured spectrum for the chemical. The number of elements in this spectral feature vector could be as high as the resolution of the spectrometer, or a smaller set of large optical bandwidths in key areas of the spectrum. Spectrographic sensors clearly offer a precise feature set for high specificity on pure chemicals. But on mixtures of unknown chemicals, this complexity becomes a combinatorial problem. Spectrographic sensing is most effective on neat chemicals in clean air.

29.3.6 Colorimetric Sensing

This kind of detection is very low cost and requires physical contact of a chemically activated dye or fluorescing agent that binds or reacts with the target chemical of interest, giving a visible color change to indicate concentration. Many biological-based sensors use this technique, and there is a wide range of detecting methodologies to choose from. Most require some degree of manual manipulation of the target chemicals, such as swiping a collection pad across a surface suspected of having the chemical and then rinsing the residue onto a paper card with the color-activated stripes. Popular examples of this are blood sugar tests, pregnancy tests, pH strips, and so on. New technologies are emerging for dyes in clothing, windows, wall paints, and other materials that would change visible color in the presence of hazardous vapor.

29.4 Biological Sensors

Since this book is addressed to those not focused on the biological sciences, we need to discuss several important terms from biology to understand better biological sensors and the real information they produce. We will then proceed to discuss the major categories of biological sensors. This is essential to the systems engineering behind a sensor data fusion algorithm for automatically interpreting biological sensor patterns. Biological organisms can divide every 20 min in a fermentation reactor so that a single colony-forming unit (CFU) can become more than 1 billion cells in about 10 h! Sizes of biological materials[8] are also important to keep in mind. A good approximation to the size of an atom is around 0.1 nm. The air is mostly small diatomic (O_2, N_2) and triatomic (H_2O, CO_2) molecules, whereas chemical agents with 10–30 atoms are about 1–3 nm in size. Proteins are large collections (dozens to thousands) of amino acids and can be more than 1 μm in length, but weak atomic forces will tend to fold and bunch them up like a loose pile of string of approximately 10 nm in size. Viruses are collections of protein, which can pass into a cell and replicate. Most viruses are harmless whereas some interfere with cellular functions, leading to disease and are between 40 and 100 nm in size. Single-cell bacteria tend to be between 1 and 10 μm in size and exist in the air as aerosols of dry spores that can be in clumps of several CFU per particle. Aerosol particles in the 1–2 μm size range are called respirable because the particles are small enough to make it deep into a human lung but large enough to get trapped in the alveoli sacks of the lung, leading to infection.[9]

The goal of biological detection systems is to detect viruses, bacteria, and toxins before they infect or poison humans. This is generally done by detecting the shape of proteins on the surfaces of the virus or bacteria spore, although other methods can also be used. However, biological detection becomes difficult as earth is full of harmless biological materials made of the same amino acids and very similar proteins.

Biological materials of interest are pathogens, which can be viral or bacterial, and biotoxins, which are chemicals produced by pathogens that are toxic. Biotoxins are a crossover threat. They are a toxic chemical that can poison the environment the same way a toxic industrial chemical or chemical warfare agent can, but the difference is that a life form produces the chemical rather than a reaction plant. Examples of biotoxins are *Ricin* and *Botulinum Toxin* (Botox).[10] Ricin is produced as a by-product of processing castor beans for oil, whereas Botox is produced by the rod-like bacteria *Clostridium botulinum*, which itself is harmless. Ricin is one of the most toxic chemicals known and simply kills any cell it comes in contact with. Botox causes paralysis in muscles and small quantities injected under the skin are used medically to remove facial wrinkles (essentially by paralyzing facial muscles). *C. botulinum* occurs naturally in uncooked foods (such as shellfish) and if ingested it can live in the intestine and is extremely difficult to get rid of, thus leading to severe paralysis and death. Rapid detection of biotoxins is also extremely difficult, but there are laboratory procedures to isolate and identify these agents. These *procedures* could be automated using data fusion as a decision aid in the future.

Aerosol collection and concentration is used to capture and concentrate particles in a specific size range that exhibit general properties of interest, such as electrostatic charge, fluorescence, or mobility.[11] The standard performance measure for these devices is the concentration of agent-containing particles per liter of air (ACPLA). Although for some pathogens, only a single CFU is needed to eventually infect and kill a human, for most pathogens, a sizable number of CFU are needed to be lethal. For anthrax, approximately 10,000 spores are needed to infect a person's lungs. Anthrax is quite common on livestock farms but farmers are only exposed to small concentrations and their immune response prevents a serious infection. In general, if someone manufactured a biological agent, they would work very hard to have a particle size in the 1–2 μm range by mixing the spores with materials that would neutralize electrostatic charges that naturally would cause the particles to agglomerate into bigger clumps of particles. This would help the spores easily disperse and eventually be trapped in the lungs and quickly infect a person. Biological aerosols can be distinguished from fine dust or metal aerosols (found in subway systems) in that the biological aerosol will almost certainly fluoresce into the visible light wavelength range when illuminated with UV light. The nature of the fluorescence spectrum will vary somewhat by pathogen species,[3] but there are always a large number of things that in various combinations can produce a similar fluorescence spectrum. The challenge is to collect and concentrate biological aerosols with properties consistent with a man-made biological weapon and then package the material for further identification processing.

The *bioassay* is any device used to measure a defining aspect of a biological material. The term has its roots in minting coins, where an *assayer* certifies the purity of the metal from a series of chemical processes.[12] There are many types of bioassays, the discussion of which is well beyond the scope of this discussion. In general, the bioassay deposits the biological material in whole spores, vegetative bacteria, hosted viruses, enzyme-stripped

proteins, or nucleic acids in small wells, each with a binding antigen, stain, or fluorophore*
marker that biochemically binds to the pathogen or its components. The concentrated bio-
material must, in general, be preprocessed and distributed onto the assay, which requires
subsequent washing to remove unbound markers, and then the assay is evaluated to see
which combination of markers were bound to the suspected pathogen. In the laboratory, a
homogeneous sample of pathogen distributed in the assay produces a repeatable color pattern from the various markers so that a broad set of markers in a single assay can be used
to identify a number of pathogens. But this does not mean that a mix of pathogens can be
assayed simultaneously. The assay pattern is based on a single pathogen being present,
typically from a cultured sample. However, literally thousands of wells can be placed on a
glass slide, each with a specific antigen or marker, and automatically scanned to produce
an image for pattern analysis. The intensity of each marker can then be used as qualitative
(yes or no) data fusion information or quantitative (gray scale) data fusion information if
appropriate.

Techniques such as PCR are used to *amplify* small amounts of DNA for subsequent
analysis. PCR is commonly used to identify individual people through trace amounts of
their DNA. Electrophoresis is a DNA detection technique where the mobility of various
molecules forced by an electric field allows repeatable separation based on the molecular
weight, electric charge of the molecule, and the size of the molecule. This sieving effect
allows DNA from one sample to be compared to another, thus allowing a positive identification. Therefore, it is very essential to have a library of DNA patterns for each pathogen
of interest and simply compare the unknown sample to see whether there is a match to
a pathogen of concern. In a controlled laboratory, this technique for DNA identification
works very reliably. However, the challenge is to start with a pure DNA source, because
the PDR will amplify everything in the sample. An air sample could contain hundreds or
even thousands of DNA sources. One must sort the samples as much as possible to be sure
of the results.

The *bio-sensor* output for data fusion is really the color response from binding markers
in a large number of assay wells. These are typically analyzed using neural networks, or
simple pattern match scores for the wells that correspond to a particular pathogen class.
From a data fusion perspective, these sensor *channels* can be interpreted qualitatively (true
or false) or quantitatively (on a scale). Since there is a broad range of potentially available
information from various sensors, the use of fuzzy logic can be employed to combine data
such as aerosol properties (particle size, shape, and charge) as well as protein marker tag
data. This logically leads one to inferenced-based logic of the form "if this is pathogen X,
the set Y of sensor channels should have a matching qualitative response, and the set Z of
sensor channels should have corroborating quantitative responses for the given concentration of pathogen X."

29.5 Developing Quantitative and Qualitative Information

We have discussed very briefly a range of chemical and biological sensor technologies
that are necessary to understand to create a scientific basis for sensor fusion. Why do we
assume some sort of fusion processing is required? First, we know that no single sensor
can respond to everything we are interested in detecting. Not even our own olfactory and

---

* A fluorophore is a material that fluoresces only when it is successfully bound to a target protein.

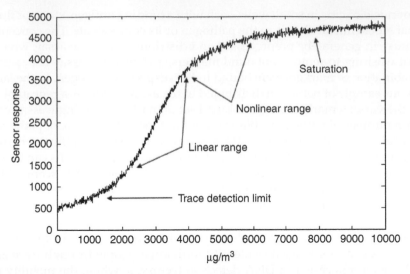

FIGURE 29.2
General sensor responses show detection limit, linear range, nonlinear range, and saturation.

immune systems have such a sensor. But rather, biomimicry suggests that a combination of diverse sensors is necessary where a subset of the sensors is all that is needed for any particular chemical or biological agent detection. Figure 29.2 shows a completely general sensor response as a function of concentration.

To understand the difference between quantitative and qualitative information, Figure 29.2 shows four general response ranges. If a sensor has a very narrow concentration range of linear response, we call it a *qualitative sensor* because its information is really just a true-false indicator. If a sensor has a fairly broad range of linear response, we call it a *quantitative sensor* because its output over the linear range can be used to accurately estimate concentration or the quantity of the chemical or biological agent present. When fusing multiple sensors, both types of information are very useful, but for different reasons that are explained in the next section.

The lower *trace detection limit* in Figure 29.2 is the region where the sensor started to respond to the agent. In this region, one generally sets a detection threshold and corresponding false alarm rate for deciding whether the sensor output is signal (agent present) or noise. For a given signal level, one can evaluate a number of detection thresholds (and the corresponding false alarm rates) and detection probabilities to trace out a curve called the receiver operating characteristic, or ROC curve. An ROC curve can be generated for a wide range of signal levels, allowing one to better evaluate the detection threshold for overall system performance. The ROC curve comes from automated detection systems in radar and sonar, which are traditionally low clutter environments (meaning that there is no need for specificity, just detection). There is a lot of interest in ROC curves for chemical and biological sensors because of false alarms, but the real cause for false alarms in chemical and biological sensors is not noise, but a lack of specificity in the detection output of the sensors. However, the lower trace detection limit is an important boundary for the data fusion algorithm to consider using the sensor output, or lack of output, as information.

The upper *nonlinear* and *saturation* ranges are less useful for quantitative information because the conversion into concentration is more complicated and much more prone to error. This is particularly true for the saturation region because the flatter slope means a small difference in sensor output that corresponds to a large difference in estimated concentration.

Furthermore, in the saturation region, the sensor could become coated with agent and require cleaning before readings would return to normal after the agent has gone. Generally, we use known limits for detection and the linear range to define where the sensor information is reliable. This can be done generally, or for each agent of interest, to give the user an idea of the data quality range. For some chemicals, a given sensor would not respond whereas others may cause saturation and even contamination (permanent saturation). The latter can be used to turn off sensors that might be damaged by high concentrations of certain agents. A good example is smoke detectors that are ruined by volatile chemicals in paint. A data fusion algorithm could only expose such sensors for short periods, if at all, to dangerous chemicals to prevent contamination. Since diverse (heterogeneous) sensors will be affected by contamination differently, data fusion algorithms could be used to identify and mitigate potential contamination and are very important to maintaining sensor health.

29.6   Inferencing Networks for Heterogeneous Sensor Fusion

We have presented a very brief discussion of biomimetics, olfaction, chemical and biological sensing, and qualitative and quantitative sensor information for data fusion. In this section, we focus on the data fusion algorithms directly, based on the science of the sensing. Sensor data fusion can use any operationally appropriate algorithm that combines different information from various sources to enhance confidence in the estimated pattern. For example, a pattern recognition algorithm such as a neural network could use inputs from multiple sensors, and one could train the network to associate various sensor patterns to corresponding situation outputs. This can work quite well when the sensor inputs are very well defined. The problem with chemical and biological sensor fusion is that we are looking for very specific sensor patterns among an approximately infinite number of chemical or biological backgrounds. So in our case, the inputs are not well defined in general, but the inputs for the situations of interest are so. One might consider a statistical classifier, or Bayesian algorithm, where the sensor outputs are part of a feature vector and one attempts to detect the situation pattern by a statistical measure of the feature vector relative to the mean and covariance of a corresponding training set. This, too, is problematic because the sensor amplitude responses can be nonlinear with concentration, as well as vary by agent or agent mixture. The third approach to pattern recognition is called a syntactic algorithm,[13] or rule-based approach. Although neural networks and Bayesian classifiers are rich in mathematics and very satisfying to derive and evaluate, the rule-based approach is often overlooked, primarily because one generally does not know the rules or syntax needed. In our case, for chemical and biological data fusion, we are seeking a consistent sensor response for physically diverse sensors on a given agent. Since the sensors are each measuring an aspect of the agent that is unique, we fully expect each sensor output to be uniquely representative of the agent observed. We will give ourselves a little *wiggle room* in the rules for the syntax by using fuzzy, rather than hard, logic for the situation pattern recognition. This will allow variations in sensor quantitative calibration and low levels of interference from other chemical or biological materials that get into the sensor signals.

The *Inference* or question for our syntax rules to answer is simply:

> If we are observing agent X at concentration Y, are the diverse sensor signals consistent and at what concentration?

This straightforward question makes it relatively easy to design fuzzy logic to fuse information from a variety of sensor sources to reject false alarms of agents that are not the chemicals or pathogens of interest. This rule applies for situations where at least some of the sensors respond to the agent in question. It is not an algorithm that enhances SNR by integrating responses across sensors. We will use both quantitative and qualitative sensor information, and combine it in a way that will only permit false alarms to slip through at very low concentrations, where only the qualitative information is present. As the concentration is increased, more sensor channels report, and the quantitative information then dominates the false alarm rejection because the sensors must report a consistent concentration to pass the syntax. This is an effective means to reject false alarms without trading off detection sensitivity because the criterion narrows as the concentration increases. A reported false alarm at a very low concentration is much less of a nuisance than a false alarm at a higher concentration, which requires immediate action.

The inference approach to fuzzy logic is not an obvious biomimicry of our own thought process. We are taught by the scientific method to collect information, apply known reduction methods, and evaluate results for a conclusion. This is actually a very complicated task to make software perform. However, if we turn the process around and say "if conclusion X is true, what must the sensor inputs be?" the algorithm amazingly forms a well-defined logic network that is generally quite simple to implement in software.[14] We are focused on this approach because both chemical and biological sensors suffer from very high false alarm rates and are exposed to an approximately infinite range of background clutter, but can detect a limited number of agents of interest at concentrations low enough to be useful as a warning system. We will discuss chemical and biological sensors in the same general terms of either qualitative or quantitative information and present a general fuzzy continuous inference network logic algorithm (CINet), that actually can extend generally to a broader range of sensors.

To describe our CINet in the most general terms, we will use hypothetical data from three different sensors for a range of calibrated exposures to two different chemicals. Figure 29.3 shows the measured responses versus concentration, normalized to a common scale of 0–1.

Figure 29.3 shows good qualitative information from sensor 1 and good quantitative information from sensors 2 and 3. The fact that each chemical will produce a unique response for physically different sensing methods (IMS, PID, SAW, spectroscopy, etc.) is what we are exploiting to recognize a target chemical and reject false alarms from *clutter* chemicals that may fool some or all of the sensors.

Figure 29.4 arranges the sensor data by chemical. Now a number of simple possibilities can be easily defined. For example, if only sensor 1 is responding, it could be a low

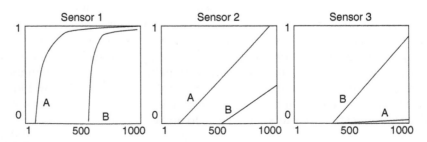

FIGURE 29.3
Sensors 1, 2, and 3 responses to chemicals A and B show diversity, nonlinearity, and indicate good prospects for false alarm rejection by data fusion with inference networks.

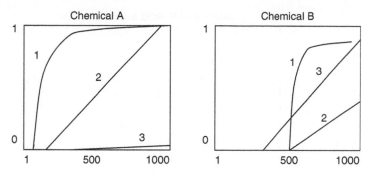

FIGURE 29.4
Sensor responses versus concentration from Figure 29.3 arranged by chemical.

concentration of chemical A, but not B. If only sensors 1 and 2 are responding, it could be a medium concentration of chemical A, but not B at any concentration. If only sensor 3 is responding, it could be chemical B at a low concentration, but not A, and so on. At higher concentrations where all three sensors are responding, the estimated concentration from the two quantitative sensors (sensors 2 and 3) should be used to compare consistency. Since their concentration slopes are quite different, they will only agree over a range of concentrations when the target chemical is present. It should be noted that for a small set of sensors and chemicals, one might find a few specific concentrations where an ambiguity exists. However, these ambiguities would be known in advance from analysis and the actual concentration for a real measurement would naturally fluctuate significantly over time, making this false alarm a rare and quantified event.

29.6.1 The Blend Function

The usefulness of fuzzy logic in capturing the desired behavior of the classification algorithm for diverse and nonlinear sensor responses is that we can *blend* information according to our human expert knowledge. We know that sensors can drift from perfect calibration. We also know that a defined state or situation could involve combinations of information with varying confidence, randomness, and accuracy. The inputs to our fuzzy logic could be outputs from other algorithms, such as neural networks, Bayesian classifiers, or other inference logic networks. So we need to blend this information together using either objective metrics of confidence or our own performance-based metrics of confidence. Algorithmically, we will use a blend function of the form given in the following equation:

$$B(a,b,c,d,x) = \begin{cases} b, x \leq a \\ = \frac{1}{2}\left[d + b + (d - b)\sin\left(\frac{\pi(x - a)}{c - a} - \frac{\pi}{2}\right)\right], a < x < c \\ d, x \succ c \end{cases}$$

Figure 29.5 shows the blend response function. This particular blend function would be for asserting *true* if x is lower than ~3 and false if higher than ~3. The blend range is from 1.5 to 4.5 and the maximum confidence for true is 0.9 and minimum confidence for false is 0.1. Basically, we are just spline fitting a sinusoid to make the blend. There are of course many

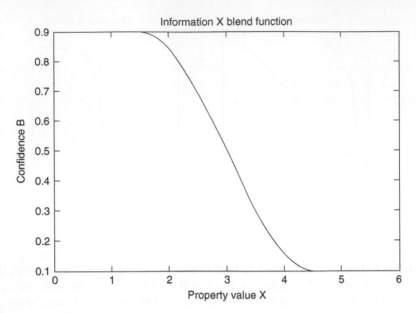

FIGURE 29.5
A blend function where $\langle a, b \rangle = \langle 1.5, 0.9 \rangle$ and $\langle c, d \rangle = \langle 4.5, 0.1 \rangle$.

options, such as arctangents, polynomials, and so on. For cases where *true* corresponds to a narrow range of the input "*x*," one can simply put two blends of the mentioned type together, or use functions such as Gaussians or polynomials to make the blend. In the literature for fuzzy logic, the blend function is usually called the *fuzzy set membership function*. At the output of the fuzzy logic, one makes a decision based on the confidence being greater or lower than some threshold, such as 0.5. This is called the *crisp* in the fuzzy logic literature. The general idea behind fuzzy logic is to tolerate a *gray scale* in logical operations until the final decision to achieve a desired *behavior* of the software algorithm is reached. We need this degree of nonlinearity and complexity because there are different scientific metrics that apply for different agents at different concentrations.

29.6.2 Qualitative Information Transformation

We can use the blend function to take a sensor output that may be either a raw voltage output or the logical output of a built-in classification algorithm, and turn it into a fuzzy *true or false* determination. Consider a sensor (designated as sensor 1) with a raw 8 bit digital output giving numbers from 0 to 255. Let us say the inherent background noise has an RMS value of 10 counts. If we choose a detection threshold of 10, we might have a false alarm rate of nearly 50%. Using techniques such as ROC curves we might pick a detection threshold of, say, 30 and have an acceptable false alarm rate of, say, 3%. This false alarm rate is the likelihood that noise has exceeded the detection threshold and not our target chemical. We can use this *hard* or *crisp* threshold to switch between a state of *not responding*, when the sensor output is less than 30 to responding for the sensor output more than 30, but there are advantages to using a blended transition.

For example, a completely objective statistical metric would be to integrate the probability density function for the sensor signal from the detection threshold to infinity. The mean of this density function is the signal level and the standard deviation is the RMS noise level if no signal were present. The result of this integral is called the probability of detection, P_d, and it approaches unity as the signal level is increased above the detection threshold. One could build a simple blend function that approximates P_d or one could add a little more flexibility by transitioning the blend over a wider range of sensor output values. For example, one could set the blend from $\langle a, b \rangle = \langle 10, 0.1 \rangle$ to $\langle c, d \rangle = \langle 50, 0.95 \rangle$ so that the maximum confidence stays at 95% above signal levels of 50, and the minimum confidence is limited to 10% for sensor outputs below 10. The confidence corresponds to the inference *Is the sensor responding?* The blend output is typically scaled to lie within a $\langle 0, 1 \rangle$ range where the minimum and maximum are not necessarily 0.0 and 1.0, respectively.

29.6.3 Qualitative Information Transform

As we depicted in Figure 29.2, quantitative sensor information is more restrictive than qualitative because we like to see an approximately linear sensor response to the agent concentration. It does not have to be perfectly linear. The best way to quantify this response is to build a simple calibration table. We will call this the sensor response model (SRM). The SRM is expected to be unique for each target agent. A simple table of concentrations and corresponding responses can then be used to estimate the concentration from the sensor's response in real time. For more than one quantitative sensor response, the concentrations can then be compared for consistency. If the sensors are diverse (based on physically different molecular measurements), their responses should be unique for each target agent of interest. The *Is the sensor quantitative?* blend should ramp up at around the same threshold as the Is the sensor responding? threshold, but ramp back down where the response becomes nonlinear and flattens out. For example, for sensor 1 and chemical A the quantitative response might extend from approximately 0.1 to 0.6 (concentrations of 100–150), but for sensor 2 and chemical A, the range might be from 0.01 to 1.0 (concentrations of 150–950), as seen in Figure 29.4. This simply requires a *double blend* or two blend functions of the type seen in Figure 29.5 to provide a true range for quantitative use of the sensor information.

29.6.4 Concentration Consistency

The sensor output has an inherent noise uncertainty, described typically as a Gaussian density function, where the mean is the sensor output and the standard deviation is the RMS noise level. When we use this sensor output to estimate concentration, it is reasonable to map the signal uncertainty to concentration uncertainty. Given the local slope of the concentration response in the SRM, the concentration error $\Delta conc$ due to noise scales can be represented by

$$\Delta conc = \frac{N_{rms}}{slope}$$

where N_{rms} is the noise standard deviation and *slope* is the local slope of the SRM curve. This is why we prefer not to use the SRM when the slope flattens out due to nonlinearity or saturation. Besides risking contamination of the sensor and erroneous data, the estimated concentration error can become too large outside of the *Is quantitative* range.

We can make a *floating* double blend centered at the SRM-estimate concentration with the blend width scaled according to the SRM slope and sensor noise. Since we can make the blend objectively scale with the sensor noise, or by some other metric, we have a way to compare two or more sensor concentrations for consistency. Consider the likelihood that two or more physically different sensors would provide an exact match in concentration to be exceedingly small. By having the floating double blend enveloping each concentration estimate, we have a method to measure the amount of overlap taking into account each sensor's noise and concentration uncertainty. The output of the floating concentration blend represents the confidence we have as a function of concentration for the current sensor output. This is a clear example of the power of fuzzy logic when the syntax has a scientific basis. Figure 29.6 depicts these floating blends for concentrations where the sensor array is consistent, and for the inconsistent case, the resulting net confidence is found from simply multiplying (AND-ing) the sensor confidences together.

Fuzzy AND's, OR's, and NOT's conversion of logical AND and OR into fuzzy operators generally involves following the operator with a blend function. There are many variants, including applying individual weights to operator inputs,[15] which has mathematically the same effect as applying a blend to each of the inputs to the operator. We will keep our discussion here limited to unity weights on the inputs. AND operators (symbolized by "$\wedge$") are essentially multipliers, OR operators (symbolized by "$\vee$") essentially sum the inputs, and NOT operators (symbolized by an over-bar) subtract the input from unity. For chemical A, one can define the following rules:

- If sensor 1 responds TRUE and sensor 2 responds FALSE and sensor 3 responds FALSE, then chemical A could be present, but at a low unknown concentration.

- If sensor 1 responds TRUE and sensor 2 quantitative TRUE and sensor 3 responds FALSE, then use sensor 2 SRM to report concentration of chemical A.

- If sensor 1 responds TRUE and sensors 2 and 3 are quantitative TRUE, and sensors 2 and 3 are consistent in concentration, report chemical A with high confidence.

Note that the qualitative *Is responding* blend outputs have a powerful veto in the detection logic. For the chemical A inference, sensor 1 must be responding, as the confidence collapses to 0 regardless of what the other sensors are reporting.

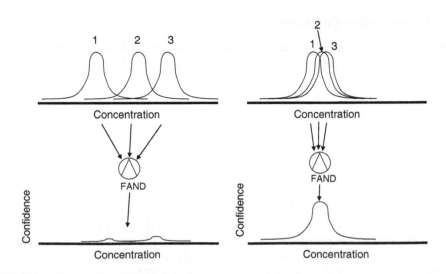

FIGURE 29.6
Applying a fuzzy AND function for three sensor concentration estimates showing inconsistency (left) and consistency (right).

Consider chemical B. There is hidden power in the logic for the quantitative responses as well. For chemical B, the following rules apply:

- If sensor 3 is quantitative and the concentration is less than 500, chemical B could be present.
- If sensor 3 concentration is more than 500, and sensor 1 responding TRUE and sensor 2 quantitative TRUE, and sensors 2 and 3 are consistent in concentration, report chemical B with high confidence.

Implied for chemical B is that sensor 3 must be responding for detection.

Once these rules are mapped out for each target agent of interest, it is significant to implement the fuzzy logic using any of a wide variety of available techniques or toolsets. The difficulty with syntactic fuzzy logic is that one needs to do some basic science, like careful calibration of the sensors for each target agent of interest, including determining the fuzzy blends for Is responding and Is quantitative and if quantitative, the floating concentration blends width from the sensor noise and the slope of the SRM. One should remember the fact that for diverse sensor physics, each target chemical will have a unique calibration response, or SRM, for each sensor. This creates the opportunity for sensor fusion to uniquely detect the target agent with extremely high selectivity, especially as the concentration increases and more sensors respond.

29.7 The Path Forward

This brief look at a methodology to perform heterogeneous sensor fusion for chemical and biological agents is based on using a diversity of sensors that are well calibrated for the target agents of interest. It can be seen that a sensor *manifest* or agent library is then needed containing the calibration SRM tables and response thresholds for each agent in the library. This would allow a remote process to fuse the information from an array of sensors exposed to the agent in question. The inference-based logic can then employ only those sensors and sensor channels with useful responses for a given agent inference, and each sensor's contribution to the overall fusion confidence would be independent of the other sensors. If the sensor has a response to the agent, it contributes to the inference logic by providing context for the sensor signal output. As signals from multiple sensors are fused, this context looks for consistency using fuzzy logic metrics. For small concentrations of a given agent, there is a higher possibility of a false alarm, but since the concentration is small, this is much less of a concern compared to higher concentrations that carry greater danger to people exposed. Higher concentrations would naturally bring in more sensors and sensor channels with higher SNR signals. The CINet logic captures this by making it more and more difficult for a false alarm to persist because of the ability to look for consistency across the sensor information.

This approach to false alarm rejection may be most interesting in the medical applications area, where qualitative signals from bioassay wells can be combined with more quantitative metrics such as volatile chemicals in the blood and breath. The CINet logic can automate many metabolic states based on highly objective metrics and calibrated sensor response libraries. The final diagnosis/prognosis would still require a medical doctor or pathologist, but the degree of sensor automation will likely produce very good decision aids. For example, current technology in 2007 can measure more than 1000 blood chemicals. Bioassays can have more than thousands of wells with binding proteins that fluoresce to varying degrees. It would not be difficult to also execute thousands of well-thought-out CINets to quickly produce a short list of likely situations. This methodology is only an automation of information processing that parallels human knowledge from the scientific method. But, as with human knowledge, we need diversity of information, repeatable calibration, and a reasoned syntax to fuse the information into a situation recognition.

References

1. K.E. Kaissling, Physiology of pheromone reception in insects, *ANIR-AVNP*, 6(2), 2004, pp. 73–91.
2. C.K. Mathews, K.E. van Holde, and K.G. Ahern, *Biochemistry* (San Francisco: Addison Wesley Longman, 1999) 3rd ed., pp. 241–244.
3. P.J. Hargis et al., UV detection of biological species, Edgewood Chemical Biological Center, ECBC Report Number CR-028, August 2000.
4. G.A. Eiceman and Z. Karpas, *Ion Mobility Spectroscopy* (Boca Raton: Taylor and Francis, 2005).
5. Y. Sun and K.Y Ong, *Detection Technologies for Chemical Warfare Agents and Toxic Vapors* (New York: CRC Press, 2005), pp. 184–194.
6. Q. Cui and I. Bahar, ed., *Normal Mode Analysis: Theory and Applications to Biological and Chemical Systems* (Boca Raton: Chapman and Hall/CRC, 2006).
7. E.B. Hanlon et al., Prospects for in vivo Raman spectroscopy, *Phys. Med. Biol.*, 45, 2000, R1–R59.
8. C.K. Mathews, K.E. van Holde, and K.G. Ahern, *Biochemistry* (San Francisco: Addison Wesley Longman, 1999) 3rd ed., p. 14.
9. W.C. Hinds, *Aerosol Technology* (New York: Wiley, 1999).
10. J. Ali, L. Rodrigues, and M. Moodie, *Jane's US Chemical-Biological Guidebook* (Coulsdon, Surrey, UK: Jane's Information Group, 1998), Chapter 3.
11. C.S. Cox and C.M. Wathes, *Bioaerosols Handbook* (Boca Raton: CRC Press, 1995).
12. http://en.wikipedia.org/wiki/Assay "Assay" definition.
13. R. Schalkoff, *Pattern Recognition: Statistical, Structural, and Neural Approaches* (New York: Wiley, 1992).
14. J.A. Stover and R.E. Gibson, Modeling confusion in automated systems, SPIE, vol. 1710, Science of Artificial Neural Networks, 1992, pp. 547–555.
15. D. Swanson, *Signal Processing for Intelligent Sensor Systems* (New York: Marcel-Dekker, 2000), pp. 478–489.

30

Fusion of Ground and Satellite Data via Army Battle Command System

Stan Aungst, Mark Campbell, Jeff Kuhns,
David Beyerle, Todd Bacastow, and Jason Knox

CONTENTS

30.1 Introduction

Purpose, method, and *endstate* are components of the concept of a commander's intent purported in the 1996 version of the U.S. Army's document titled, Field Manual 100-5 (FM 100-5).[1] The basic premise is that once military commanders state their intent, all other matters are subordinate and become supplementary information. Intent is a format or framework for command and information operations in the field.

In Force XXI, the U.S. Army is concentrating on developing a tactical command and control (C2) structure from the ground up taking into consideration all previous C2 nodes and echelons.[2] Previously, the U.S. Army utilized a theater C2 structure. This type of rigid structure is no longer feasible because recent events suggest that the mission be global and flexible.

Vital to this flexible C2 structure is a reliable, secure communications infrastructure that integrates forces and assets utilizing a variety of sensors connected by mission-specific support systems. It is the authors' intent to introduce this architecture, the application areas, and its evolution. A flexible C2 organizational structure is recommended, which can be expanded or contracted depending on the event. We will discuss a joint command and control (JC2) situational awareness (SA) in which real-time information and decision making use data fusion to form the basis for an effective communication infrastructure. The effective implementation of fusion technology across all agencies, military and civilian, will also decrease response and recovery time, be that in the battlefield or a natural disaster.

Our method is to describe the current state of technology for SA, the common operating picture (COP), and its evolution. We begin with the introduction of the Army's Battle Command System (ABCS) in Section 30.2, which contains the subsystems such as SA, COP, information fusion and decision making, and JC2.

In Section 30.3, we discuss the evolution of the ABCS to include remote sensing, unarmed and armed aerial sensors (unmanned aerial vehicles, UAVs), and ground sensors. Next, in Section 30.4, the authors suggest that a flexible command and control (JC2) and ABCS system and some software engineering recommendations could provide a framework for mitigating, responding, and minimizing recovery from a disaster.

30.2 Description of the Army Battle Command System

In the 1980s a decision was made within the military to aid the warfighter through the use of automation. The idea was to incorporate a multitude of sensors, networks, and software to increase the responsiveness and collaboration between all echelons of the military. The system has to be an integral component of all phases of an operation. From movement, logistics, and maneuver, there has to be some level of automation and information sharing.[3] The ABCS[4] was introduced to address this need. The system is composed of several sensor systems and networks (stationary and mobile) with software covering the battlefield operating systems (BOS). The advantage of these systems is the automated fusion of sensors, planning, positioning, logistics, and other important information systems that facilitate decision making. Information is distributed in all directions throughout the systems, both vertical and horizontal, to develop a consistent SA picture.

30.2.1 Situational Awareness

SA is an extremely important capability, necessary to effectively plan and execute operating orders, under rapidly changing variables beyond the control of the command organization. Possessing the precise knowledge to respond to evolving situations is only part of the process needed to control and successfully complete a mission. The ability to accurately and quickly combine (fuse) all the data from heterogeneous sources is equally important. Fusing this data allows an organization to be more proactive, while limiting the number of dynamic variables outside their control. Additionally, the versatility of heterogeneous data allows one to apply many different analysis techniques (signal and image processing, simulation, and mathematical modeling, etc.) to provide more accurate predictions of future events. The primary purpose of this fused data is to give a more accurate view of the battlefield by providing organizations with a COP and datasets that could be archived or used for training purposes.

30.2.2 Common Operating Picture

The COP represents a subset of the heterogeneous data collected and subsequently fused to give the individuals, across all organizations involved, the same near-real-time view of the event. The COP provides all individuals with the ability to proactively respond and reactively adjust to situations to comply with the intent of both the higher organization and their own organization. The COP includes key aspects of the situation such as allowing individuals to see key events that have occurred in near real time. Input data may come from ground, aerial, satellite, remote sensing, or human interaction. These heterogeneous sources must be fused together to form reliable information for real-time decision making in the field.

30.2.3 Information Fusion and Decision Making

Multisensor information fusion seeks to combine information from multiple sensors and sources that are not feasible from a single sensor or source. The proliferation of micro- and nanoscale sensors, wired and wireless communication, and ubiquitous computing enables the assembly of information from sensors models and human input for a variety of applications such as Department of Defense (DoD) applications, environmental monitoring, crisis management, disaster management, medical diagnosis, monitoring and control of manufacturing processes, and intelligent buildings. A key problem is how to integrate or fuse information from heterogeneous sources. Techniques for such information are drawn from a broad set of disciplines including statistical estimation, signal and image processing, artificial intelligence, cryptography, software engineering, computer engineering, and information sciences. Major issues involve communicating across an architecture of distributed sensing and processing sites, selection and integration of disparate types of algorithms, adding in the role of human-in-the-loop for analysis and decision making, and defining the degree of automation and computer-aided cognition. Information fusion and JC2 provide an effective combination for flexible battlefield management.

30.2.4 Joint Command and Control

The concept of *joint* in the military involves situations where organizations develop mechanisms for bridging organizational differences and extracting strategic value from inter-organization cooperation. This could involve considerations such as the operation, communication infrastructures, standard protocols and tactical structures, processes and human expertise. JC2 is defined as an interrelated system of command links or nodes integrating maneuver forces and strike assets, informed by a variety of sensors (ground, satellite, or other communications and data links). An important step is to create JC2 structures that function on the operational level to assist warfighters to respond quickly to events in regional commands.[3]

A C2 structure provides a framework for organizations (a system of systems) that operate under different levels of responsibilities. Each level contributes to the overall framework, providing both higher and lower organizations with key support assets. These interrelated systems allow a particular situation (e.g., within the battlefield) to be controlled, synchronized, and supported logistically from the top down. Each organization's leader has his/her own requirements to achieve objectives, which must fit within the constraints of the higher organizations leader. In addition, the higher-level organization must give the subordinate elements the support they need to achieve their aspect of the mission. The higher-level organization must task each subordinate organization and then synchronize the completion of these tasks. This synchronization acts as a force multiplier and provides a more effective synergistic effort. SA allows the leaders at a higher level to see accurately and in near real time how the situation is unfolding and how their intent is being met. It allows them to adjust quickly and stay proactive in the decision-making cycle. The COP allows each individual at any level in the organization to visualize their environment. It provides a validation that they are meeting the higher-level organizations intent while ensuring that they are staying within the constraints placed on them by the higher organization.

The ability to maintain C2 at every level requires that a communications infrastructure fuse data from heterogeneous sources, distribute fused information, and provide SA and COP to the precise levels of the organization in a timely, accurate, secure, and reliable manner. This command structure is usually enforced from the top down with the highest level of the organization providing the key components pertinent, accurate information. This C2 flexible structure enables the lower elements to successfully communicate to higher and lower subsystems of the organization, enabling the whole organization to effectively maintain C2 at each level. Simultaneously, appropriate area experts (e.g., G2, S2, etc.) analyze, fuse, and send the data to the organization leaders in a form that provides an accurate, near real-time SA to facilitate critical decision making. Under these conditions, information fusion must function within a secure, reliable, scalable communications infrastructure to facilitate decision making at each level and enable a more effective JC2 infrastructure.

The potential for fusing a data-intensive system supported by an information infrastructure is very promising. If performed effectively, it would enable organizations to expand and contract their assets quickly and efficiently. However, the designers (e.g., software engineers) must also be cognizant of the requirements to operate within a secure, reliable communication system, and a common interface for the purpose of greater autonomy and distribution of information to the battlefield.

30.2.5 The Global Command and Control System-Army

The Global Command and Control System-Army (GCCS-A) (Figure 30.1) is an organizational component of the Joint Global Command and Control System (GCCS-J).

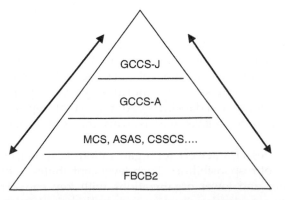

FIGURE 30.1
Diagram of information inclusiveness.

The Army Tactical Command and Control System (ATCCS) is a subsystem of the GCCS-A, which provides an information network that supplies seamless information flow from brigade to corps level of operations. ATCCS receives information from Force XXI, Battle Command Brigade-and-Below (FBCB2) and facilitates the flow of information from brigade through corps units to support critical decisions. This same information that is aggregated and analyzed by ATCCS is then transmitted to the GCCS-A providing information flow horizontally and vertically.

30.2.6 Force Battle Command Brigade-and-Below

FBCB2 provides SA and C2 to the lowest tactical echelons. This permits a bidirectional flow of data providing the COP discussed earlier.

The system is composed of the components summarized in Table 30.1.

The FBCB2 system provides the user with the general capabilities given in Table 30.2.

It should be noted that FBCB2 utilizes a common messaging protocol to communicate between systems for interoperability. To send a message to a destination, FBCB2 uses a

TABLE 30.1

FBCB2 Components

Components	Purpose
Processing unit	Provides processing power
Software	FBCB2
Display unit	Allows user to see the COP
GPS source	Provide position and timing information
Interface	Connection with data medium

TABLE 30.2

FBCB2 System Capabilities

Functional Areas	FBCB2 Capabilities
Communications	• Connect networks in the field • Troubleshooting and maintenance • Filtering and alert response • Message filing and organization
Visualization	• Situation awareness (SA) • Relays threats graphically to mobile units and main unit • Identification of friend and foe (IDFF) • Map overlays • Identifies obstacles and terrain • Graphically depicts SA • Common operating picture (COP)
Planning and preparation	• SA • Provides predefined military standard graphics for use in the planning phase • Terrain analysis • Perimeter defense planning • Logistical planning and support • Update overlays
Information exchange	• Standardize messaging • Confirms receipt of critical messages (user or hardware device) • Prepares and manages messages including graphics
Mobility	• Planning and executing movement • Facilitates decision making concerning maneuverability • Call for fire support

tactical internet that permits a redundant and routable data link, critical for battlefield operations. This standardized protocol increases flexibility and interoperability because it is based on an accepted standard.

The Defense Information Infrastructure Common Operating Environment (DII COE)[5] is a program that aids in defining standards for preexisting technologies. The use of standards ensures that FBCB2 and other ABCS systems can interoperate.

Common operating standards improve the quality of information processed, refined, and analyzed, aiding the decision-making process for battlefield management. Some of this information can be supplied by sensors or humans.

30.3 Evolution of the Army Battle Command System

30.3.1 Remote Sensing, Ground-Based Systems (Image and Nonimage)

Remote sensing technologies, together with other geospatial technologies such as geographic information systems (GIS), geographic positioning systems (GPS), and position navigation and timing (PNT) systems, play a significant role in the improvement of the nation's homeland security mission, disaster management, and critical infrastructure protection. Remote sensing from space combines a broad synoptic view with the ability to detect changes in surface features quickly and routinely. Remote sensing from aircraft and UAVs have the ability to examine areas in great detail from below the clouds, whereas ground-based systems make possible the close-in observation of events in real time. Each of these areas of remote sensing technology can contribute significantly in homeland security and disaster management.[6]

Examples of the use of remote-sensed data to support nuclear power plant facilities and transportation security proactively and reactively include the following capabilities:

* Detect, classify, and analyze temporal and spatial changes in surface features
* Develop accurate digital terrain models and 3D surface features as a means for modeling landforms along rights of way
* Visualize terrain from different perspectives, with the potential for developing threat cones and view sheds
* Classify vegetation types along transportation lifelines as a possible deterrent to concealment
* Identify facilities where topography or identifiable hazards (e.g., nuclear, chemical, fuel facilities) place communities at risk
* Analyze environmental factors quickly and effectively
* Merge real-time sensor output (video, biochemical sensors) with archived geospatial data
* Identify, characterize, and analyze a wide variety of risks to transportation networks through a gradual program of gathering image intelligence along rights of way
* Create detailed maps of an area that has suffered attack to assist in response[7,8]

30.3.2 Tactical Unmanned Aerial Vehicles and Aerostats (Sensor Networks)

The U.S. military's inventory includes a number of UAV platforms with a suite of sensors to aid in military sensing.

Larger UAVs have a greater payload capacity, including weapon systems. Typically, sensors enable surveillance (full-motion video cameras, thermal and infrared [IR]) communications intercept, retrans and jamming (EW) (Figure 30.2).

Nuclear biological chemical (NBC) agent detectors are mass spectrometers, gas analyzers, radiation sensors, etc.

An aerostat is a static, tethered airship similarly with fitted sensor payload. The joint land attack cruise missile defense elevated network sensor (JLENS) is an example. The aerostat can deliver robust performance from its array of sensors (Figure 30.3).

FIGURE 30.2
Unmanned aerial vehicle.

FIGURE 30.3
Joint land attack cruise missile defense elevated network sensor.

30.3.3   Ground Sensors

Numerous ground-based sensors have been deployed including sensor networks with acoustic, seismic, IR, and video capability. The rapid evolution of low-cost sensor networks, with each node containing GPS self-location capability, local wireless communications, and local processing, provides the ability to deploy sensor networks for detecting humans, animals, vehicles, chemical and biological entities, and local environmental conditions. These ground sensors may involve unattended sensor networks, mobile robotic sensors, and sensors worn by individual soldiers. Indeed, the soldiers themselves can become a multisensor system involving their own observations and the observations of sensors that they carry or wear.[9] Issues of the use of ground-based sensors include challenging observing environments (e.g., the effects of weather, terrain, and line-of-sight on observations), power requirements for observation and communication with other sensors, and issues regarding how to task and manage the sensor resources.[10]

30.4   Discussion and Implications for Disaster Management

In the twenty-first century, emergency management coordinators and managers face unprecedented challenges and threats. Increasing demands are being placed on agencies to maintain the existing communications infrastructure and new responsibilities to improve systems safety, communications, performance, and security policies. Presently, a variety of advanced technologies are involved to enhance planning, designing, managing, operating, communicating, and maintaining all facets of the nation's public and private utilities (e.g., nuclear power plants) and critical transportation systems. Aerial, ground, and satellite remote sensing including UAVs and information fusion represents areas of rapid development that can be leveraged to address these challenges. These technologies have significant and unique potential for application to a number of natural disasters and emergencies including homeland security applications.

A fundamental problem with a disaster is that it does not respect boundaries that including political, regional, organizational, geographical, professional, sociological, and consideration.[11] In addition, a disaster does not have a definite beginning and end point. It is a dynamic process that has one definite characteristic—chaos. Situational reports can be seen to characterize several concerns that inhibit optimal decision making during a disaster situation:

- Disasters and critical emergency events can overwhelm communication
- Coordination among agencies may be difficult
- Information may be fragmented
- The reliability of data could be questionable due to data overload such that decision makers may receive misinformation to base their decisions
- Data or information fusion may not be fully available

Considerable efforts are ongoing throughout the disaster management community to articulate concerns and characterize the dynamics between organizations. An important guide for this effort is found in the recommendations made by the Board on Natural Disasters (BOND) in their report to the National Research Council.[12]

The BOND's primary goals are to

1. Improve decision making before, during, and after emergencies to ensure better access and quality of data and information
2. Identify users and their needs
3. Provide information products specifically designed to meet users' need
4. Promote efficiency and cost-effectiveness
5. Stimulate and facilitate mitigation

These are important components in any fully functional, communications system for comprehensive disaster management. It is also important that any software engineering process focus on commercially available software in developing solutions that reflect the needs of the disaster management community. Software engineering should focus at the lower, more informative operational levels in the systems development life cycle (SDLC); it in fact forms the basis for the development of the software and hardware stipulated by the BOND.[12] As stated by the BOND, the assessment process will also be required to identify germane components associated with specific disaster events and scenarios. These components will then form the basis for future technical requirements and deliverables.

Information management and policy are important issues related to disaster and emergency management. Disaster management is an exercise in information fusion and information processing, logistics, simulation, intelligence analysis, and decision making. To effectively undertake these tasks requires a thorough understanding of disaster information requirements and the characteristics associated with the unique disaster event. Disasters come in different sizes, have different behavior, and can be categorized on the basis of their impact on natural resources, transportation systems, communities, and so on. They can also be discriminated and categorized along a number of dimensions such as impact, severity, duration, geographic setting, and advance warning.[12] To develop the information technology and data fusion architecture and infrastructure for case scenario application(s) and response, it is essential to understand the disaster event from the perspective of those responsible for assimilating the data and making critical decisions. The effective use of this information for producing operational plans and policies that respond to the disaster event and its recovery is critical to minimizing loss of life, property, and environmental impact.

As previously stated, it is important that the disaster management community identify commercial off-the-shelf products (COTS) and implement solutions that reflect the complex needs of disaster management. It is also equally important that the hardware and software engineers design the hardware and software to reflect the needs of the disaster management life cycle (DMLC) of mitigation, response, and recovery. Thus,

$$\text{software engineering} = \text{disaster life cycle management}$$

In addition, the procedure to fuse data (information fusion) requires accurate, reliable data (data accuracy) to facilitate decisions. Fusing unreliable data is not the answer. The old acronym GIGO (garbage in, garbage out) is still applicable for data fusion. Information that arrives late (timeliness) from disparate sources and is transmitted over nonstandard formats (data consistency) requires a significant effort to recompile into a coherent picture (COP and data understandability) of the disaster area (SA). Timely information fusion can be a key concept that ensures immediate responses are taken to minimize response time and recovery operations.

We need to consider software principles that aid in the improvement of the DMLC. Some recommendations for disaster management software engineers are as follows:

- Understand the information requirements before fusing the data
- Be cognizant of the unique disaster event and its impacts
- Utilize live simulations, virtual simulation, and constructive simulation for preparation and training for disasters[13]
- Effectively use information fusion to aid in the production of operational plans and policies for mitigation, response, and recovery
- Utilize event warning systems (e.g., ground sensors, remote sensing, and aerial sensing UAVs, etc.) that aid in the preparedness, response, and recovery phases of the DMLC
- Ensure that the communication infrastructure is scalable and within and between agencies
- Select a common interface for each agency with common message formats

The open geospatial consortium (OGC) is developing a standard sensor description model (metamodel) as part of its sensor web enablement (SWE) effort. The OGC is an international industry consortium organized to develop publicly available geoprocessing specifications. OGC members are developing a standard XML encoding scheme for describing sensors, sensor platforms, sensor-tasking interfaces, and sensor-derived data. The goal is to make web-enabled devices discoverable and accessible using standard services and schemas. The sensor model language (SensorML) is a component that provides sensor information necessary for discovery, processing, and georegistration of sensor observations.

The geolocation of sensor data has required software specifically designed for that sensor system. The availability of a standard model language for describing platform position, as well as instrument geometry and dynamics, allows for the development of generic multipurpose software that can provide geolocation for potentially all remotely sensed data. The availability of such software in turn provides a simple, single application programming interface (API) for software developers to incorporate sensor geolocation and processing into their application software. This allows the development of software libraries that can parse these files and calculate required look angles and timing for each sensor pixel.

The SensorML provides an XML schema for defining the geometric, dynamic, and observational characteristics of sensors and provides a standard XML encoding scheme for observations and measurements of all kinds. This research also has led to specifications for open interfaces for

- *Sensor collection services.* A software service that provides observed values by seeing what sensors of a specified type are available in a specified region.
- *Sensor planning services.* A software service that enables acquisition requests and notification of relevant events.
- *Sensor registries.* A catalog that enables discovery of sensors and observed values.

There are other issues to consider for data fusion to occur: first, it must have a physical path for data to flow. This data path must be designed to handle a variety of situations. Different organizations respond to different situations and each can be loosely categorized—the data path must be designed around the characteristics of these categories. One organization may have many subunits and the paths may be decomposed and designed around these subordinate units. The categories are levels of coverage area, mobility, flexibility, security, and bandwidth. Some organizations have only one function and cannot deviate. In this case flexibility is categorized as low. However, this same organization may need to be on the move constantly to react to a developing situation. The unit may also have a large coverage area. This can present a problem. How to develop a system that can communicate on the move, over a large geographic area? This situation may call for a wireless communication medium not limited by line-of-sight but most likely using a satellite link. This same link may require different levels of security? Is there a threat of jamming or a need to overcome it? Does the data being transmitted need to be encrypted or does it require a nonsecure link?

In the military the data link can be categorized as highly secure, mobile, and flexible. There is also a large coverage area and applications that are bandwidth intensive. There are many echelons to the military that require each of these categories to be decomposed to the lowest level. Security is a constant requirement within the military. The need for a high level of flexibility is also a constant. There is a wide range of military missions that force the military to adapt to a situation to complete the mission. The remaining categories vary at different levels of the military. In general, the higher the level the greater the need for bandwidth, and the larger the coverage area. Mobility is actually the reverse and the lower the level the larger the need for mobility.

Disaster management organizations have a wide range of communication assets. Given the multitude of possible scenarios, disaster management organizations may require a highly flexible data system. This system must be flexible enough to handle any type of reorganization that might occur, which may be achieved through interoperability between different communication assets.

Software can be designed to perform special functions or it can aid in developing the COP and thereby providing standard SA. These specialized functions must take raw data and fuse into information that is provided from the specific functional areas. This allows decisions in these areas to be executed with more confidence while maintaining the ability of the organizations or individuals responsible for these areas to stay as proactive as possible, minimizing the need to be reactive. This software must share some type of common protocol for passing and interpreting the data being passed. The common protocol aids in the efficient use of physical resources as well as interoperability between systems.

The military has several different software packages they use to ensure near real-time SA and all provide at a minimum the COP. There are systems dedicated to the logistical aspects of their missions, as well as fire support, air defense, and maneuver control aspects. The primary communication infrastructure to provide a COP and SA to the lowest possible level is FBCB2.[14]

Preparation for disaster management can be accomplished through simulation, intelligence analysis, data mining, and visualization, which may preclude a disaster or at least minimize their impact.

30.4.1 Simulation

There are basically three types of simulation methods, which could be easily integrated into training to prepare for disasters.

1. *Live simulation* with communication equipment in a real-life environment out in the elements (rain, snow, sleet heat, etc.), with assigned equipment and simulated

disasters. An example would be a digital field exercise (DFX) using state-of-the-art communications equipment, sensors, and UAV technology.

2. *Virtual simulation* places individual first responders, EMS personnel, and disaster management teams (e.g., Hazmat, CBR, etc.) in simulators (VR) that replicate actual disasters and homeland security scenarios as if the first responders were actually in the field.

3. *Constructive simulation* entails large-scale computer simulation that represents battalion size units and above.[3]

30.4.2 Intelligence Analysis, Data Mining, and Visualization

In addition to simulation exercises, disaster managers will be required to answer questions not yet posed. Intelligence analysis, data mining, and visualization will assist disaster managers in answering these questions. The ability to predict future hurricane events is enhanced by daily, seasonal, and annual data mining queries via data mining algorithms. Mathematical techniques such as fuzzy logic and genetic algorithms assist the data mining queries to interpret the intelligence contained in the images. Airport runways and possible path obstructions, underpasses, over-pass bridges, pipelines, borders, and port facilities are all candidates for virtualization.

It is our belief that information fusion is applicable to disaster management and homeland security, both proactively (mitigation and preparedness phase) and reactively (response and recovery phases). Some of the goals of information fusion are congruent with the BOND in their report to the National Research Council. These points are covered in Section 30.4. These are important components in an operational infrastructure for comprehensive disaster management.

Finally, it is important to prepare for disasters via simulations, intelligence analysis, and visualization methods and to be proactive to potential disasters, preempting them before they occur.

Disaster management is an exercise in logistics and communications. Disaster agencies and personnel (especially managers) rely on their hardware and telecommunications; both agency personnel and software engineers must have an understanding of the information needs (requirements) and the characteristics of the disaster event and its uniqueness. This knowledge will facilitate effective architectures and technologies that meet the needs of the disaster management community. There must also be a precise understanding of the DMLC of mitigation, preparedness, response, and recovery. Traditional software engineering methodologies must match this life cycle and also be cognizant that each disaster is unique. In addition, a common interface should be designed for each agency.

30.5 Summary and Final Recommendations

Information fusion for situation assessment and the development of a COP requires a system of systems that integrate the capabilities of unmanned aerial and ground sensors, satellite communications, remote sensing capabilities, and a joint C4ISR (command-control-communication-coordination-intelligence-surveillance-reconnaisssance) capability along with a common interface to a reliable communication infrastructure. With such a capability, regional commanders and disaster management personnel can operate independently and be confident in his/her ability to respond to any situation.

Glossary of Key Terms

COP: The common operational picture provides a fused synopsis of the heterogeneous data collected to give the individuals across all organizations involved, the same near-real-time view of the event.

DII COE: The Defense Information Infrastructure Common Operating Environment[5] is a program for the purpose of reducing cost and ensuring capability. It defines a standard for using preexisting technologies that have already been proven.

FBCB2: Force Battle Command Brigade and Below provides SA and C2 to the lowest tactical echelons.

SA: Situational awareness is an extremely important component to effectively plan and execute operating orders during a crisis involving rapidly changing variables that are beyond the control of the command organization.

Acknowledgments

We gratefully acknowledge the assistance of LTC Tim Purcell, Battalion Commander, for permitting Dr. Stan Aungst to attend classes, observe and ask questions at 3 DFX at Fort Indiantown Gap. Dr. David Hall, Associate Dean of Research, College of IST, Penn State University.

References

1. US Army Field Manual 100-5.
2. Force XXI: Division Redesign, Army Times, 22 June 1998.
3. MacGregor, D. A., *Command and Control for Joint Strategic Action, Digital War, The 21st Century Battlefield*, edited by Robert L. Bateman, i-books, 1999.
4. *Army Battle Command System (ABCS)*, 21 February 1999. Online. Internet. http://www.fas.org/man/dod-101/sys/land/abcs.htm.
5. *Defense Information Infrastructure Common Operating Environment (DII COE)*, 11 January 2007. Online. Internet. http://www.sei.cmu.edu/str/descriptions/diicoe.html.
6. Tobin, G. A. and Burrel, E., *Natural Hazards: Explanation and Integration*, London, NY, Guilford Press, 1997.
7. National Simulation Center, *Training and Simulation*, Fort Leavenworth, KS, Combined Arms Center, 1996.
8. NRC Report, *Information Infrastructure for Managing Natural Disasters*, Board on Natural Disasters, Washington DC, National Academic Press, 1998.
9. Magnuson, S., Army wants to make "every soldier a sensor", *National Defense Magazine*, May 2007.
10. Avasala, V., Mullen, T., and Hall, D., A comprehensive sensor management approach based on market-oriented programming, *Proceedings of the IEEE/WIC. ACM International Conference on Intelligent Agent Technology*, Hong Kong, 18–22 December 2006.
11. NRC Report, *Information Infrastructure for Managing Natural Disasters*, Washington DC, National Academic Press, 2000.

12. Statement of Bruce Baughman Office of National Preparedness, Federal Emergency Management Agency (FEMA), Before the Committee on Transportation and Infrastructure, Subcommittee on Economic Development, Public Building and Emergency Management, US House of Representatives, 11 April 2002.
13. Roper, W., *Geospatial Technology Support to the Nation's Navigation System*, Transportation Research Board, Washington DC, National Research Council, January 1999.
14. *Force XXI Battle Command, Brigade-and-Below (FBCB2)*, 12 September 1998. Online. Internet. http://www.fas.org/man/dod-101/sys/land/fbcb2.htm.

31

Developing Information Fusion Methods for Combat Identification*

Tod M. Schuck, J. Bockett Hunter, and Daniel D. Wilson[†]

CONTENTS

---

* Portions reprinted, with permission, from *Proceedings of the IEEE National Aerospace and Electronics Conference (NAECON) 2000*, © October 2000; *Proceedings of the 6th International Conference on Information Fusion*, © July 2003; and *Proceedings of the IEEE Aerospace Conference*, © March, 2004.

[†] The author's affiliation with the MITRE Corporation is provided only for identification purpose and is not intended to convey or imply MITRE's concurrence with, or support for, the positions, opinions, or viewpoints expressed by the author.

31.1 Introduction

The goal of *Combat Identification* (*Combat ID* or simply *CID*) fusion is to combine information at the appropriate information levels to derive a positive identification (ID) according to a classification structure. This in turn can help determine the *allegiance* of an object such as Friend, Neutral, or Hostile. Specifically, the overarching goal of CID is to attain an accurate characterization of detected objects in the joint battlespace so that high-confidence, timely application of military options and weapon resources can occur. A visualization of this concept is shown in Figure 31.1.[1]

Figure 31.1 illustrates the essential problems with understanding, processing, and fusing information in this domain. No other fusion domain exists where not only are there multiple complexities such as friendly, hostile, and neutral relationships* to the identifying agent[†] in addition to environmental confusor effects, but there may be objects that are actively attempting to deceive and evade detection and classification. This makes the design of a fusion process more difficult for CID than possibly for any other domain.

The decision of whether to engage or not engage an object may only occur very infrequently over a long period of time, especially if the identifying agent is not participating in an active or simulated battle-like situation. However, the results of poor CID are painfully well known. The *USS Vincennes* Airbus incident, the U.S. Army Black Hawk helicopter fratricide, and the Operation Iraqi Freedom (OIF) U.S. Army Patriot missile fratricides are some of the best-known examples. Many more occur in air-to-ground and ground-to-ground

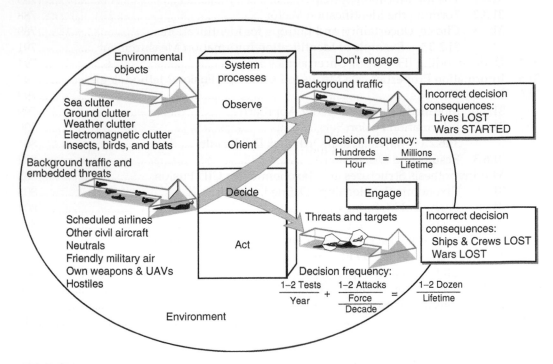

FIGURE 31.1
Warfighting domain complexities.

---

* Hereto described as *battlespace objects* or simply *objects*.
† The platform or node attempting to identify a detected object.

combat operations that are not as well documented, but make up the majority of the historical 7–25% rate of fratricides in major wars fought in the twentieth century.

Referencing Kebodeaux,[2] one of the many goals on the battlefield throughout history has been to reduce friendly fire (fratricide) incidents. These misidentifications are, not surprisingly, a result of what Clausewitz termed the *fog of war*, which is another way of describing *uncertainty*. The problem is exacerbated by the raw speed of modern warfare, where the time between a weapon engagement decision and target impact is exceedingly short. Improving awareness is the same as improving information processing; thus, this reduces fogginess that provides improvement in decision quality. To adequately represent the warfighting domain as an abstraction that can be captured computationally and reproduced, a systems engineering approach to the design of a fusion/track management/CID/sensor management command and decision (C&D) system must be taken. Internal to this C&D system is the CID function that has historically not been considered properly from a systems and information engineering perspective. CID has traditionally been treated with disastrous oversimplification by emphasizing what was practical, rather than correct. Figure 31.2 illustrates the attribute information domain that must be exploited properly to achieve CID success.

Figure 31.2 illustrates the application of *information engineering* to generate an *information model*–based design which is a radical change for the concept of CID which has historically been rule based and developed from a cooperative-space information domain (due to the lack of sensors and information from other domain spaces). In this figure, it can be seen that kinematics and cooperative information form only a subset of the total information space that should be used to derive a CID. If disparate sets of sensors are used to provide high fidelity attribute data to a fusion process, then the more encompassing information domain space can be mastered and high-confidence positive Hostile, Suspect, and Neutral

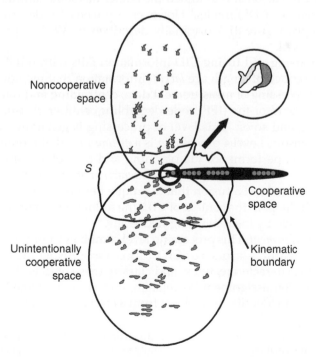

Noncooperative space

S

Unintentionally cooperative space

Cooperative space

Kinematic boundary

FIGURE 31.2
Attribute information space.

CID declarations (not just Friend) can be developed. This domain space consists of *cooperative*, *noncooperative*, and *unintentionally cooperative* information sets and spaces, which are expounded upon in this chapter.

The purpose of performing CID fusion is to obtain correct and timely object identification. To achieve this, it is necessary to optimize the baseline of physical observations (attribute and kinematic measurements) to capture the entire object information domain space to the greatest extent possible, and then reconstruct a representation of the object. The human senses work in a similar fashion. The human brain fuses sight, hearing, smell, touch, and taste to form a representation of an object (such as spoiled fish—certainly unfriendly), which is then used as input to determine how the body should react. The behavior response to a car moving toward you with an engine that is *roaring* is very different than the same car without the same aural characteristic.

This approach also avoids the generation of CID conflicts that may arise when multiple platforms work from sparse information to derive identifications that will probabilistically differ due to violations of the principle of information entropy.* Further, by dominating the information sample space *S* in Figure 31.2 as completely as possible, it is possible to extract CID information even when sensor responses are conspicuously absent or contradictory, such as the case when an air object is in EMCON (Emission Control) or when spoofing is attempted. This is the concept of negative information and can be exploited when the state of the CID domain space is well known.

31.2 Mapping CID to JDL Levels

For the purpose of CID fusion, it is assumed the reader has some familiarity with the Joint Directors of Laboratories (JDL) model.† However, a purposeful description is needed to clarify some concepts. Figure 31.3, originally described by White,[3] presents a graphical description of the model.

The problem of parsing and fusing CID information falls within JDL levels 1, 2, and 3. Other levels not described at length are *level 0* information that includes such operations as coherent signal processing of measurement data, centroiding and filtering of kinematic data, identification friend-or-foe (IFF) code degarbling, emitter classification, local association and tracking, and so on. This level of processing is generally performed entirely within individual sensors. Level 4 processing is a *metaprocess* that monitors and optimizes the overall data fusion performance via planning and control, not estimation as in JDL levels 1 through 3. *The level 5 human–computer interaction* at the right of Figure 31.3 is also sometimes referred to as level 5 fusion that is unique because it adds a human operator/ decision maker in the loop. In the context of CID, the following discussion defines the type of information processed by the remaining JDL levels.

JDL level 1—object refinement. This processing level combines information from the results of level 0 processing within sensors. For CID, this is about fusing the attributes of objects so as to detect, locate, characterize, track, and classify these objects. This level of processing involves information assignment/correlation, and what is defined as *taxonomic CID*. This means assigning an identity to an object from a *taxonomy*—a set of mutually exclusive

---

* A measurement of information entropy, in a CID context, can be described as a means to capturing and quantifying the amount of information, choice, and uncertainty associated with the output of a sensor source and a CID fusion process.

† It is described in other chapters.

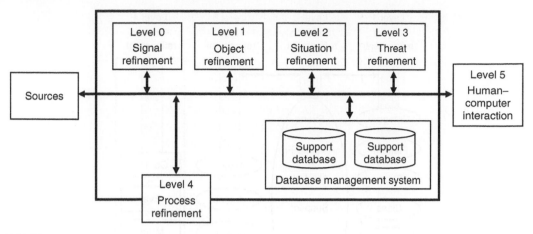

FIGURE 31.3
Joint Directors of Laboratories fusion model representation.

alternatives. Examples are the hierarchical taxonomies: category, platform, type, and class. Other examples are unit and nationality—although nationality is often considered to be a level 3 category.

Examples of level 1 declarations include tank, M-1 Abrams, fighter, Destroyer, 737-300, and so on. The goal, therefore, of JDL level 1 processing is to determine exactly what an object is, not its relationship to the identifying node or platform. Taxonomic CID processing includes (1) information alignment such as time synchronization, and gridlock and bias removal, (2) information correlation (platform–platform, type–platform, class–type, etc.), and (3) probabilistic/evidential attribute estimation (Bayesian, Dempster–Shafer, etc.). Generally, to fuse information at this level, multiple hypotheses must be tracked. This is because taxonomic CID information from sources is available at different levels and often not sufficient to discriminate with certainty. Automating the fusion and declaration decision process flexibly across these levels, as information becomes available, without throwing away valuable information and context is challenging.

JDL level 2—situation refinement. This processing level includes the ability to establish contextual relationships of objects declared in level 1 processing with their environment. This includes situation refinement using some sort of assessment to declare an object as (1) *Friend*, (2) *Assumed Friend*, (3) *Neutral*, (4) *Suspect*, (5) *Hostile*, and (6) *Unknown*. This is also referred to as defining an object's *allegiance*. Level 2 processing uses some sort of decision methodology such as if-then-else logic, voting fusion, or even Bayesian decision logic to *derive* the object state and establish the CID. For example, if an object is determined to have a high-confidence taxonomic CID of an F-16, this in itself offers no information on whether it is friendly toward the platform or node performing the identification. However, inserting a level 2 assessment decision will enable the proper CID declaration to occur based on such rules as country-of-origin, flight profile, intelligence information, and so on. In this example, doctrine (which historically defines some sort of rule set) could derive a CID of *Assumed Friend* if no hostile entities in the theater are known to have any F-16s in their inventories. Some sensor/source information may also be *directly fused* at this level, for example, secure cooperative information such as some IFF modes and data link self-identifications, if available. The presence of this type of secure information can be directly associated with the existence of a *Friend*, although the lack of it cannot normally imply a *Hostile* or *Unknown* designation. It is important to note that level 2 categories are drawn from a single taxonomy, so the structures for fusion are different from the ones best

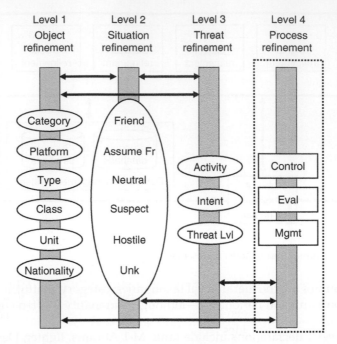

FIGURE 31.4
Multihypothesis structures with Joint Directors of Laboratories levels.

suited for the multiple taxonomies of level 1. This will be explained in the discussion of Figure 31.4 in Section 31.2.1.

JDL level 3—threat refinement. This processing level attempts to interpret a situation from a dynamic behavior point-of-view and involves evaluating hypotheses concerning the future actions of an object and the potential consequences of those actions. This includes threat analysis and the assessment of intent if it is possible to get into the adversary's decision cycle (as in the Boyd OODA loop*). For CID, this processing level can take the results of levels 1 and 2, in addition to independently processing information, and will determine if the identified object is a candidate for engagement. For example, using a self-defense scenario, if an object is positively identified as a *Hostile*, but is flying away from any defended assets and poses no threat, then the object's intent may not be threatening and the identifying platform or node may choose not to engage it. Depending on the refinement in the level 1 processing, if platform type, class, unit, nationality, and activity are known, then this may be used along with the CID output at level 2 for level 3 processing. As an example, if an object is identified as *Hostile*, with a taxonomic identification of a military reconnaissance aircraft on a surveillance mission, then that object may not qualify for engagement because it has no offensive weapons capability (again depending on doctrine and situation). The methods of fusing and processing information at this level include Bayesian networks, neural networks, and Markov gaming.

31.2.1 Multihypothesis Structures

An information model for multihypothesis structures is shown in Figure 31.4. This figure stipulates that there are six hypothesis taxonomies for level 1, one for level 2 (that has six

---

* The OODA (Observe-Orient-Decide-Act) was first described by Col. J.R. Boyd in the 1950's and is described by T.A. Hightower in "Boyd's O.O.D.A. Loop and How We Use It." (http://www.tacticalresonse.com/d/node/226).

possible states), and three for level 3. This can be expanded as needed for any domain. The level 4 structure is included for completeness, but will not be referenced in this discussion. For each taxonomy, the set of hypotheses for an object is populated from various sensor inputs, and not all hypotheses are included all the time. This is determined by the amount and type of information present for every object associated with a track store (or file). A *declaration* is made for a taxonomy when one alternative can be presumed to be the correct choice. Figure 31.4 represents the following:

- Each hypothesis category has an associated probability distribution that is derived as a function of the observed sensor/source parameters. Information primarily moves between hypotheses within an individual level (vertical bars) with some information moving between levels 1, 2, and 3. Level 4 primarily accepts requests for optimization and provides feedback to levels 1, 2, and 3.

- Each taxonomy hypothesis has a unique threshold value for declaration that is dictated by mission, doctrine, and available information. Some means to automate a decision threshold is required, which is the subject of future work. A good level 2 example is the decision between a declaration of *Hostile* or *Neutral*. Obviously the *Neutral* CID hypothesis requires less information to declare than the life-critical *Hostile* CID; so, a higher decision threshold for a *Hostile* declaration probably is reasonable to consider.

- Each hypothesis structure requires decision logic to determine how to arrive at a decision, given a set of observed sensor/source parameters.

- Each hypothesis declaration includes taking into account the probability of a wrong decision (or nondecision) and its consequences.

31.2.1.1 JDL Level 1 Structures

The level 1 structure is quite different from the level 2 and 3 variants. The level 1 hypothesis structures of *category*, *platform*, *type*, and *class* form a taxonomic refinement series, whereas *unit* and *nationality* are related to all four of the taxonomies in the series. Figure 31.5 illustrates this example.

Every sensor type that produces information will provide those that span across the three JDL levels as well as the hypothesis taxonomies within these levels. For example, suppose the following sets of level 1 information are received from two very good sensors:

- Sensor 1—F-14, F-15, F/A-18
- Sensor 2—F-14A, F-15E, F/A-18E, F/A-18F

Sensor 1 is providing *type* information whereas sensor 2 *class* information. However, both sensors are providing information that will improve all the hypothesis structures, especially sensor 2, which is providing specific subsets of objects declared by sensor 1. Sensor 2

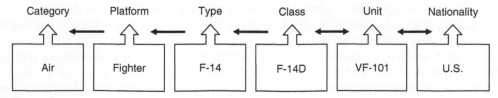

FIGURE 31.5
Combat identification level 1 information flow.

TABLE 31.1

Sensor Type versus Class Reporting

Type	Possible Class	Class Elements
F-14	F-14A	3
—	F-14B	—
—	F-14D	—
F-15	F-15A	5
—	F-15B	—
—	F-15C	—
—	F-15D	—
—	F-15E	—
F/A-18	F/A-18A	6
—	F/A-18B	—
—	F/A-18C	—
—	F/A-18D	—
—	F/A-18E	—
—	F/A-18F	—

has defined one instance of object F-14, one instance of object F-15, and two instances of object F/A-18 related to sensor 1. To support both hypothesis structures, object mappings can be performed between them.

In the case of sensor 1 *type* information, Table 31.1 represents the possible maps to the *class* information provided by sensor 2.

So for F-14, sensor 1 provides three objects to the hypothesis structure of sensor 2; for F-15, five objects; and for F/A-18, six objects. The probability of each new possible *class* element is the probability of the class as reported (or derived) from the sensor divided by the number of possible objects (i.e., 3, 5, and 6, respectively in this case) available in the *a priori* database. Since sensor 1 can only report to the *type* level in this case, in the absence of additional information, entropy requires that all classes within that type be equiprobable. For the opposite case where sensor 2 can contribute to a *class* hypothesis, a mapping can occur between the declared aircraft in Table 31.1, and their respective *type*. So F-14A with its associated probability confidence (which is equiprobable to F-15E, F/A-18E, and F/A-18F) can be mapped to F-14 and processed within the type hypothesis.

A more comprehensive relationship between level 1 object structures can be seen in the following Bayesian network example in Figure 31.6 from Paul[4] built using the Netica® software package from Norsys. In Figure 31.6, the relationships between the various level 1 structures are immediately clear. The relationships between objects in this network were constructed from various open sources and entered into the model, which is how the discrete probabilities were obtained. Figure 31.6 therefore represents the *a priori* state of the universe for F-14, F-16, F/A-18, and Boeing 737 aircraft. The assumption in this example is that a series of aircraft object classes are returned by a set of unbiased attribute sensors. For this example, there is a high-quality, complex sensor that returns information that an object is a fighter (0.959) or a commercial aircraft (0.041), and that it is from either Israel (0.309), the United States (0.676), Indonesia (0.0025), or Spain (0.0126). The resulting Bayesian network is shown in Figure 31.7.

The grayed areas of Nationality and Platform in Figure 31.6 represent where the new sensor information was read. The *ID* node in Figures 31.6 and 31.7 is a mirror of the *class* node in this example; however, it can reflect any node of interest. There are many complexities with building a tactical Bayesian network for air object taxonomic CID that are beyond the scope of this discussion. This includes assigning probabilities to large amounts

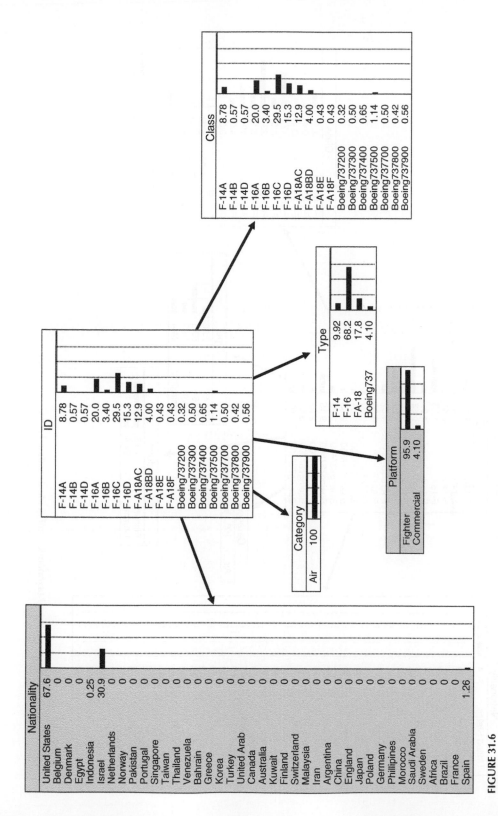

FIGURE 31.6

Taxonomic combat identification Bayesian network after information processing.

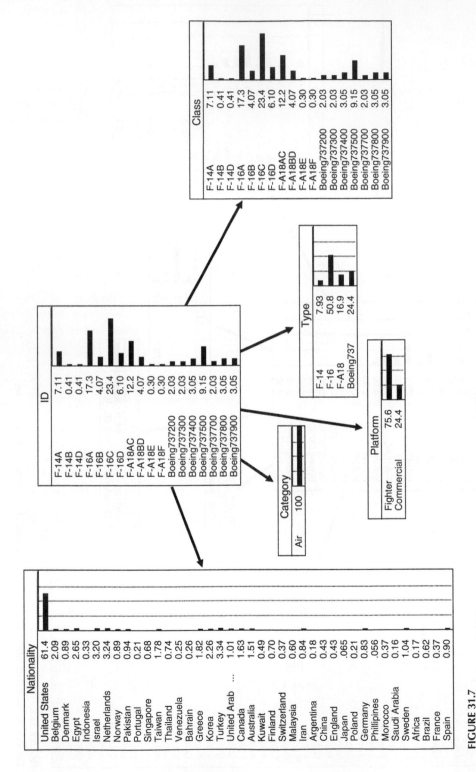

FIGURE 31.7

Joint Directors of Laboratories level 1 taxonomic combat identification Bayesian network.

of information and handling ambiguous or corrupted information. However, this example demonstrates the necessity of developing relationships between JDL level 1 information structures to minimize information loss.

31.2.1.2 JDL Level 2 Structures

Unlike level 1 constructs, each level 2 entity has little (or no) relationship to other level 2 entities because level 2 information is really contained within a single hypothesis category. If an object is declared *Hostile* via level 2 hypothesis, this hypothesis will have no contribution to a level 2 *Friend* hypothesis, other than to test for conflicts from other declaration sources. Referring to Figure 31.4, level 2 information is both measurement based and derived from level 1 information. Measurement-based level 2 information is provided only from secure sources of information that usually involve cryptologic methods to convey trusted information that is resilient against spoofing and compromise. However, only direct evidence of a *Friend* is possible with these systems. No other CID declaration is "directly" measurable. So the CID declarations of *Assumed Friend, Neutral, Suspect, Hostile, Unknown, and Pending* are normally derived states, because no positive measurement information exists beyond a perfectly cooperating friendly object via secure information transfer.

Derived level 2 information providers include all level 1 information sources. The level 1 information discussed previously can be used to declare a level 2 CID of *Friend, Hostile*, and so on, after the application of doctrine or equivalent processing techniques. For example, a detected emitter that is correlated to an adversary's platform would be a candidate for refinement to *Suspect* or *Hostile* after application of CID doctrine.

31.2.1.3 JDL Level 3 Structures

Referring to Figure 31.4, level 3 information is both measurement based and derived from level 1 and 2 information. Level 3 information consists of an object's *activity* (antisubmarine warfare [ASW], intelligence, combat air patrol [CAP], etc.), *intent* (threat, nonthreat, etc.), and *threat level* (lethality). The ability to estimate with accuracy the future actions of an opponent and the possible consequences of those actions is the primary goal of an application of the JDL level 3 structures. This has been described in the literature as similar to consulting a clairvoyant; only recently has there been more dependable methods reported in the literature such as the work by Chen et al.,[5] using Markov Game theory to predict adversary intent. The JDL level 3 *threat refinement* structures provide more than a means to look at hostile forces; they also provide for a clear assessment of all theater forces including those that are friendly and neutral. Therefore, it is the goal of this level to provide the following capabilities, derived from Ref. 6:

- Estimation of aggregate force capabilities—includes *Hostile, Friend*, and *Neutral* forces as determined by level 2 processes
- Identification of threat opportunities—includes the mission planning process, force vulnerabilities, and probable hostile force actions and scenarios
- Prediction of hostile intent—includes analysis of information, actions, events, and communications
- Estimation of implications—For every hypothesized force (friendly and hostile) action, estimates of timing, prioritization, and opportunities can be made

These tasks represent a classical fusion of inferences drawn from dissimilar sources based on direct observation. As the taxonomy addressed for CID is fused based on both the physical characteristics of the object itself as well as its behavior (including actions, missions, and apparent intent), there exist essentially two basic forms of reasoning and information available from situational awareness (SA) for JDL level 3 threat refinement. These two forms are as follows:[7]

1. *Observation-driven SA reasoning*—provides an evaluation of the situation based on direct observation, that is SA based on what a potential threat is doing.

2. *Mission-driven SA reasoning*—provides an evaluation of the situation based on how well it matches to a specific anticipated threat mission.

Level 3 fusion requires both activities to be performed in concert. A more detailed discussion of level 3 information is found in Section 31.8.

31.3   CID Information and Information Theory

Concentrating on the core JDL level 1 CID fusion problem,[8] it is necessary to look at the basic behavior of classification information and how it closely parallels communications theory. More than 50 years ago, the seminal paper *A Mathematical Theory of Communication* laid the foundation of communications theory.[9] Claude Shannon, while at Bell Labs in 1948, developed his theories of communication based on the work of Nyquist and Hartley who preceded him by including the effects of noise in a channel and the statistical nature of transmitted signals. The Shannon analysis of communications system properties provides a way to describe CID methodologies and resultant information.

Shannon defines the fundamental problem of communications as "that of reproducing at one point either exactly or approximately a message selected at another point." Further, he states that the messages "refer to or are correlated according to some system with certain physical or conceptual entities." For a subsurface, surface, airborne, or space-based object, the following correspondence definition can be made:

Correspondence 1

> Identification of an object using some form of sensor information is the process of reproducing either exactly or approximately that object at another point.

Shannon's message in a CID context is that the information received from a sensor (or sensors) describes an object with certain physical traits. Examples include whether the object has the intrinsic characteristics of rotors or fixed wings, a classifiable type of radar or communications system, a categorical thermal image, and so on. For identification purposes, the information in a message contains features that allow attributes to be assigned to an unknown object that can be used to form an abstraction of the object at some level of approximation. Thus, the use of the term *identification* refers to a taxonomic identification that describes what an object is (JDL level 1—F/A-18, etc.) as opposed to its relationship (or allegiance) to the identifying node (JDL level 2—Friend, Hostile, etc.). For most types of objects, the complete set of possible attributes that can be derived is dependent on the number, quality, and type of sensor information providers assigned to the identification task. In essence, whether a detected object can be classified as an aircraft or ship, bomber or airliner, B-1 or 747, and so on is dependent on these sensor characteristics

and their ability to form the abstraction and any subsequent fusion processes. This relates identification to a communications link that will vary in effectiveness depending on its fidelity and number of paths. This leads to a second correspondence definition:
Correspondence 2

> Each identification message that is received from a sensor is one that is selected from a set of possible identification messages, which can describe one or more possible objects or set of objects depending on the information content of the message.

The number of possible messages is finite because the number of possible objects that can be reproduced by a sensor is also finite. So the selection of one message can be regarded as a *measure of the amount of information produced about an object when all choices are otherwise equiprobable.* This is significant because it allows an assessment of whether enough information exists to adequately describe an object, based on the number and types of identification messages (sensor outputs). The measure of *information content* is what enables an automated process or human operator to determine if enough information exists to make a decision. A derivative of the Shannon information entropy measurement, which is described later in this chapter, is used to measure the information content of a message or series of messages.

31.3.1 The Identification System

The one-way Shannon communication system is schematically represented in Figure 31.8, with modifications to incorporate elements of the sensor domain for identification.[10]
Referencing Figure 31.8, the five parts can be described as follows:

1. An information source corresponds to something that either produces or reflects energy that is captured by a receiver and consists of a series of deterministic entities such as reflected spectra or electromagnetic emissions. For identification, the information source can provide multiple channels of information (often

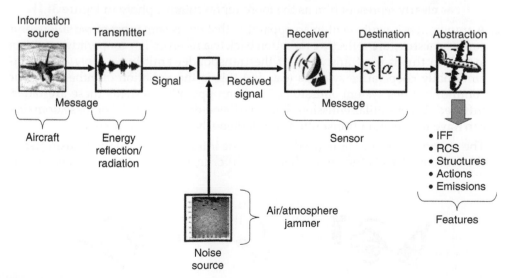

FIGURE 31.8
One-way diagram of a general identification system process.

orthogonal) that can be correlated. The three information domains (as first listed in the introduction) consist of *cooperative information* (e.g., IFF), *unintentionally cooperative information* (e.g., electronic support [ES]), and *noncooperative information* (e.g., high range resolution [HRR] radar). IFF systems, such as used for air traffic control (ATC) all over the world are considered cooperative information because the information source willingly discloses information about itself to a requestor. ES is considered to be unintentionally cooperative information because the information source, in the course of its normal operations, unknowingly discloses information about its identity based on the characteristics of its emissions. Noncooperative information (often termed noncooperative target recognition [NCTR]) requires no cooperation from an information source other than its physical existence to derive features associated with its identity. Each of these sources can be considered as a unique discrete function $f_n[t]$, $g_n[t]$, and $h_n[t]$ where the subscript, n, indicates multiple sensor types from each information domain of cooperative, unintentionally cooperative, and noncooperative, respectively. Together these functions form an *identification vector* that contributes to the generation of the *abstraction* of the original information source. This is illustrated with the series of features (similar to information domains) of a famous celebrity shown in Figure 31.10 (derived from Refs 6 and 11).

> Each set of features in Figure 31.9 represents different information types at similar levels of abstraction (in this case). Each of these features can be considered as part of an identification vector for the image (caricature) shown in Figure 31.10.

> Figure 31.10, in turn, is almost universally recognized as an abstraction of the photo of Bob Hope (the object) in Figure 31.11.

> In this example, the human brain fuses these feature vectors to determine the identity of the object. Notice that not all of the information about the object, Bob Hope, is present in Figure 31.9 (there is no information abstraction of the ear). The assembled feature identification vectors in Figure 31.10, even though they are an exaggerated abstraction of the object Bob Hope, can just as clearly represent him as the more representative photo in Figure 31.11.

2. A transmitter is equivalent to an apparatus that emits some sort of radiation, or it could be a structure reflecting radiation back to a receiver. For cooperative systems (e.g., IFF) this is the transmitter of the transponder emitting a reply. For unintentionally cooperative systems (e.g., ES) this is the radiation of either radar or communications system emissions. For noncooperative systems these are radar, infrared (IR) or similar emissions being radiated or reflected. All of these domains can have information transmitted simultaneously or asynchronously.

3. The channel is the medium used to carry the information from the transmitter to the receiver. This is the atmosphere for airborne, space, and surface objects and

FIGURE 31.9
Abstracted information features.

FIGURE 31.10
Correlated information features (identification vector).

FIGURE 31.11
Bob Hope (object).

water for undersea (and surface) objects. Noise sources also exist that change the channel characteristics and include target noise, atmospheric noise, space noise, random charge noise, and so on. In a tactical environment there also exists the possibility of intentional channel modification or destruction in the form of spoofing or jamming.

4. The receiver is the device used for converting the transmitted or reflected energy from the information source and passing it on to a destination. Each set of CID sensor information domains has a unique receiver type optimized to extract signal energy in its respective domain.

5. The destination is the process that gathers the information from the information source via the receiver and processes it to extract the feature vector. This is generally the processing performed within the sensor that results in a *message* about the information source. In the example using the caricature of Bob Hope, one sensor type might extract the *hairline*, whereas another type might extract the *chin*, and still another his distinctive *nose*.

31.3.2 Forming the Identification Vector

For each sensor information domain, the communications system is slightly different. In the case of Mk XII IFF cooperative information, there are essentially two different waveforms, one each for the uplink and downlink at 1030 and 1090 MHz, respectively. For unintentionally cooperative ES, there is a single channel where an object emits a signal from radar, sonar, or a communications system, which is the information source that provides the sensor with its input. For NCTR, the paths are generally the same (with the exception of IR which is a single path like ES) with the return path being the most interesting because it contains the feature information of interest to the destination processing. Each of these supports the formation of the object abstraction through some sort of fusion process.

For each source of information, just as in the principles from Shannon's discrete noiseless channel system, there exists a sequence of choices from a finite set of possibilities for a given object. For Mk XII IFF it is the set of possible reply codes (such as 4096 octal Mode 3/A codes). For ES it is the set of all possible emitters that can be correlated to a physical object. For NCTR a similar set of features can be correlated to physical objects. Each of these choices is defined by a series of unique parameters (Shannon's *symbols*) S_i that are defined by their domain. As an example for ES, S_i could describe one of a couple of dozen possible parameters such as frequency, pulse width, pulse repetition frequency (PRF), and so on. If the set of all possible sequences of parameters S_i, $\{S_1, \ldots, S_n\}$ is known and its elements have duration $t_1, \ldots, t_n$ then the total number of sequences $N(t)$ is

$$N(t) = N(t - t_1) + N(t - t_2) + \cdots + N(t - t_n) \tag{31.1}$$

which defines the channel capacity, C,

$$C = \lim_{T \to \infty} \frac{\log N(T)}{T} \tag{31.2}$$

where T is the duration of the signals.

Following Shannon's pattern, it is correct to consider how an information source can be described mathematically and how much information is produced. In effect, statistical knowledge about an information source is required to determine its capacity to produce information. A modern identification sensor will produce a series of declarations based on a set of probabilities that describe the performance of that sensor. This is considered to be a stochastic process, which is critical in the construction of the identification vector.

Cooperative, unintentionally cooperative, and noncooperative information sources all independently contribute to the identification vector. The mathematical form of each type is defined by a modulation equation that is bounded by Shannon information limits. Therefore, a finite amount of information content is available from each sensor type. For an Mk XII IFF interrogation, this information is related to the pulse position modulation (PPM) equation:[12]

$$x_{IFF}(t) = \sum_{n=0}^{N} |A\cos(t\omega + nt_n\omega)| \qquad (31.3)$$

where N is the number of cosines necessary for pulse shaping, A the pulse amplitude (constant), and t_n the pulse pair spacing depending on mode (1, 2, 3/A, C).

From this type of modulation, it is possible to get various octal codes that correlate to specific aircraft object types.

For ES, a typical signal can be of the type (among others):

$$x_{ESM}(t) = A_c[1 + k_a m(t)\cos(2\pi f_c t)] \qquad (31.4)$$

where A_c is the carrier amplitude, k_a the modulation index, $m(t)$ the message signal, and f_c the carrier frequency.

These signal characteristics can describe an emitter frequency, mode, PRF, polarization, pulse width, coding, and so on. From this information, emitter equipment associations can be made that can then be associated with object types.

For NCTR, one possible method in the case of helicopter identification exploits the radar return modulation caused by the periodic motion of the rotor blades. The equation for radar cross section (RCS) as a function of radar transmit frequency and rotor blade angle (θ) is shown in the following equation:[13,14]

$$RCS(\theta) = \exp(i\omega t)\frac{c}{2i\omega\tan(\theta)}\left(1 - \exp\left(\frac{2i\omega l}{c}\sin(\theta)\right)\right) \qquad (31.5)$$

From the spectra described by this equation, it is possible to determine the main rotor configuration (single, twin, etc.), blade count, rotor parity, tail rotor blade count and configuration (cross, star, etc.), and hub configurations.

The purpose of these examples using Equations 31.3 through 31.5 is to show that all sensors function like a communications system and it is necessary to understand the amount of information that can be produced by these processes.

31.3.3 Choice, Uncertainty, and Entropy for Identification

For each CID domain, there is a need to measure (1) the amount of information present in an identification vector and (2) the amount of dissonance between its components before and after applying it to a fusion process.

The issue is resolved by Shannon when he states that if the number of messages (or *features*) in the set is finite then this number or any monotonic function of this number can be regarded as a measure of the information produced when one message is chosen from the set, all choices being equally likely. So, still following Shannon, let $H(p_1, p_2, ..., p_n)$ be a measure of

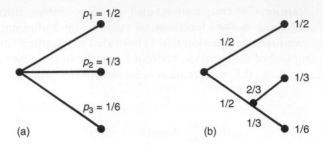

FIGURE 31.12
Decomposition of choice.

how much *choice* there is in a selection of an event or *feature*. This should have the following properties:

- H is continuous in the probabilities (p_i).
- If $p_i = 1/n$, then H is a monotonic increasing function of n. Thus with equally likely events there is more choice (uncertainty) when there are more possible events.
- If a choice is broken down into two successive choices, the original H should be the weighted sum of the individual values of H. This is illustrated in Figure 31.12.[9]

Referring to Figure 31.12b, if one choice is F/A-18, successive choices of F/A-18A, F/A-18D, and F/A-18E can be made. The three probabilities in (a) are (1/2, 1/3, 1/6). The same probabilities exist in (b) except that first a choice is made between two probabilities (1/2, 1/2), and the second choice between (2/3, 1/3). Since both of these figures are equivalent, the equality relationship is shown as

$$H\left(\frac{1}{2},\frac{1}{3},\frac{1}{6}\right) = H\left(\frac{1}{2},\frac{1}{2}\right) + \frac{1}{2}H\left(\frac{2}{3},\frac{1}{3}\right) \tag{31.6}$$

Shannon concludes with H the measure of information entropy of the form:

$$H = -K\sum_{i=1}^{n} p_i \ln p_i \tag{31.7}$$

where K is the positive constant.

The Shannon limit (average) is the ratio C/H, which is the entropy of the channel input (per unit time) to that of the source. However, a problem still exists in determining how to apply entropy to disparate information sets. Sudano[15] derived a solution described as the probability information content (PIC) metric that provides a mechanism to measure the amount of total information or knowledge available to make a decision. A PIC value of 0 indicates that all choices have an equal probability of occurring and only a chance decision can be made with the available information set(s) (maximum entropy). Conversely, a PIC value of 1 indicates complete information and no ambiguity is present in the decision-making process (minimum entropy). If there are N possible hypotheses (choices) $\{h_1, h_2, ..., h_N\}$ with respective probabilities $\{p_1, p_2, ..., p_N\}$, then the PIC is defined as

$$\text{PIC} \equiv 1 + \frac{\sum_{i=1}^{N} p_i \ln p_i}{\ln N} \tag{31.8}$$

This is essentially a form of Shannon entropy in Equation 31.7 normalized to fall between 0 and 1. The following example demonstrates the utility of the PIC for identification and incorporates the supporting use of a conflict measure for quantifying information dissonance.

31.3.3.1 Example of Identification Information Measurement

This example uses the modified Dempster–Shafer (D-S) methodology first described by Fixsen and Mahler[16] and then implemented by Fister and Mitchell.[17] A set of attribute sensor data is given in Table 31.2.

The following formulas are used to derive the combined distributions and agreements. First, the combined mass function m_{12} is defined as

$$m_{12} = m_1(a_1)m_2(a_1) \tag{31.9}$$

where $m_1(a_1)$ and $m_2(a_1)$ are the singleton mass functions from two separate sensors describing object a_1.

The combined agreement function $\alpha(P_1, P_2)$ is

$$\alpha(P_1, P_2) = m_{12} \frac{N(P_1 \wedge P_2)}{N(P_1)N(P_2)} \tag{31.10}$$

The following is an explanation of the terms in this equation:

- P_1 is proposition 1 from sensor 1. Its truth set with masses is $P_1(a_i)$ = {(F/A-18, 0.3), (F/A-18C, 0.4), (F/A-18D, 0.2), (unknown, 0.1)}
- P_2 is proposition 2 from sensor 2. Its truth set with masses is $P_2(a_j)$ = {(F/A-18, 0.2), (F/A-18C, 0.4), (F-16, 0.2), (unknown, 0.2)}
- $N(P_1)$ and $N(P_2)$ are the number of elements in the truth set of P_1 and P_2, respectively
- $N(P_1 \wedge P_2)$ is the number of elements in the truth set that satisfies the combination (denoted by $\wedge$) of P_1 and P_2

The *normalized combined agreement function* r_{ij} is

$$r_{ij} = \frac{\alpha(P_1(a_i), P_2(a_j))}{\alpha(B, C)} \tag{31.11}$$

TABLE 31.2

Attribute Sensor Data from Two Sources with Computed Belief/Plausibility Intervals

	Sensor 1	Sensor 2
Reported	F/A-18 (0.3)	F/A-18 (0.2)
Mass	F/A-18C (0.4)	F/A-18C (0.4)
Distribution	F/A-18D (0.2)	F-16 (0.2)
—	Unknown (0.1)	Unknown (0.2)
Belief	F/A-18 [0.9, 0.9]	F/A-18 [0.6, 0.6]
Plausibility	F/A-18C [0.4, 0.7]	F/A-18C [0.4, 0.6]
Evidential	F/A-18D [0.2, 0.5]	F-16 [0.2, 0.2]
Intervals	F/A-18C or F/A-18D [0.6, 0.9]	—
—	Unknown [0.1, 0.1]	Unknown [0.2, 0.2]

TABLE 31.3

Dempster–Shafer Combined Distributions

Sensor 1		Sensor 2			
		(1, 1, 1, 0, 0) F/A-18 $m_i = 0.2$ $N = 3$	(0, 1, 0, 0, 0) F/A-18C $m_i = 0.4$ $N = 1$	(0, 0, 0, 1, 0) F-16 $m_i = 0.2$ $N = 1$	(0, 0, 0, 0, 1) Unknown $m_i = 0.2$ $N = 1$
(1, 1, 1, 0, 0) F/A-18 $m_j = 0.3$ $N = 3$	Comb Object $m_{ij} =$ $\alpha(B_i, C_j) =$ $r_{ij} =$	(1, 1, 1, 0, 0) 0.2×0.3 $3/3 \times 3$ $.02/.28 = 0.071$	(0, 1, 0, 0, 0) 0.4×0.3 $1/1 \times 3$ $.04/.28 = 0.14$ 2	(0, 0, 0, 0, 0) 0.2×0.3 $0/1 \times 3$ 0	(0, 0, 0, 0, 0) 0.2×0.3 $0/1 \times 3$ 0
(0, 1, 0, 0, 0) F/A-18C $m_j = 0.4$ $N = 1$	Comb Object $m_{ij} =$ $\alpha(B_i, C_j) =$ $r_{ij} =$	(0, 1, 0, 0, 0) 0.2×0.4 $1/3 \times 1$ $.027/.28 = 0.09$ 6	(0, 1, 0, 0, 0) 0.4×0.4 $1/1 \times 1$ $.16/.28 = 0.57$ 1	(0, 0, 0, 0, 0) 0.2×0.4 $0/1 \times 1$ 0	(0, 0, 0, 0, 0) 0.2×0.4 $0/1 \times 1$ 0
(0, 0, 1, 0, 0) F/A-18D $m_j = 0.2$ $N = 1$	Comb Object $m_{ij} =$ $\alpha(B_i, C_j) =$ $r_{ij} =$	(0, 0, 1, 0, 0) 0.2×0.2 $1/3 \times 1$ $.013/.28 = 0.046$	(0, 0, 0, 0, 0) 0	(0, 0, 0, 0, 0) 0	(0, 0, 0, 0, 0) 0
(0, 0, 0, 0, 1) Unknown $m_j = 0.1$ $N = 1$	Comb Object $m_{ij} =$ $\alpha(B_i, C_j) =$ $r_{ij} =$	(0, 0, 0, 0, 0) 0	(0, 0, 0, 0, 0) 0	(0, 0, 0, 0, 0) 0	(0, 0, 0, 0, 1) 0.2×0.1 $1/1 \times 1$ $.02/.28 = 0.071$

TABLE 31.4

Total Object Mass and Belief/Plausibility Intervals

Object	Total Mass	Evidential/Credibility Interval
F/A-18	0.071	[0.93, 0.93]
F/A-18C	$0.142 + 0.096 + 0.571 = .809$	[0.81, 0.88]
F/A-18D	0.046	[0.05, 0.12]
Unknown	0.071	[0.07, 0.07]

where the *normalizing factor* $\alpha(B, C)$ (the summation of all of the combined mass functions) is

$$\alpha(B, C) = \sum_{1}^{n} \alpha(P_1, P_2) = \sum_{i,j=1}^{n} \alpha(P_1(a_i), P_2(a_j)) \tag{31.12}$$

The combined distributions are contained in Table 31.3.

The ordered elements for each entry (F/A-18, F/A-18C, F/A-18D, F-16, Unknown) show the membership each element has with the other elements, as shown in Figure 31.5. For example, the F/A-18 is also composed of F/A-18C and the F/A-18D, so its truth set is (1, 1, 1, 0, 0). The total mass and belief/plausibility for each platform type/class is calculated from Table 31.3 and shown in Table 31.4.

If the mass assignments in Table 31.3 are converted using a Smets pignistic probability[18] transform (assuming that multiple independent sensor reports of information identical to Table 31.4 are available), then the following taxonomic identifications, PICs, and conflict measures are produced for the F/A-18C with truth set (0, 1, 0, 0, 0) in Table 31.5.

TABLE 31.5

Probabilities, PICs, and Conflict Measures for Object F/A-18C

Iteration	Probability of (0, 1, 0, 0, 0)	PIC	Fister Self-Conflict PD	Fister Inconsistency-B
0	.5000	.6161	0.3400	0.4757
1	.8333	.6161	0.2098	0.5767
2	.9496	.8512	0.0820	0.4497
3	.9822	.9396	0.0327	0.3062
4	.9930	.9730	0.0136	0.2111

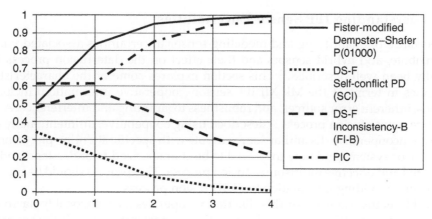

FIGURE 31.13

Chart of probabilities, PICs, and conflict measures for object F/A-18C.

This information is represented in Figure 31.13.

The solid line in the graph represents the probability that the object being reported by sensors 1 and 2 is an F/A-18C. After iteration 4, the cumulative probabilities level out at a high probability of occurrence. At the same time, the PIC also approaches unity as more evidence is accumulated.

The self-conflict index (SCI) and Fister Inconsistency-B (FI-B) in the above-mentioned example are measures of conflict in the evidence. A conflict metric is valuable because it is possible to have a low entropy (uncertainty) and also an unsatisfactory state of confusion. For example, if the set of possible objects is reduced from 100 to 2 in successive fusion iterations, then the PIC will approach unity, which represents low entropy. However, the entropy of a set of objects does not consider the relationship between the objects. If one of the two objects is an F/A-18 and the other an high mobility multi-purpose wheeled vehicle (HMMVEE), there is a serious conflict between the two conclusions that still needs to be resolved; this conflict propagates to all levels of fusion in the JDL model. The SCI measures the amount of conflict in the information sets that support F/A-18C without regard to evidence for other objects; it is a self-similarity measurement. The FI-B measures the amount of information conflict across the set of taxonomic identification probabilities of the F/A-18C to the F/A-18D, F/A-18, F-16, and Unknown.*

---

* The SCI and FI-B curves shown here are based on the results from algorithms that are still under development. These will be presented in later publications.

A priori or dynamic thresholds can be applied to these information sets to determine when enough information is held and when conflict has been reduced sufficiently to declare the taxonomic identification of an object. Methods are now emerging beyond the simple if-then rule sets traditionally employed. This is important because disciplined decision theory has not been well implemented in practice. Powell[19] suggests the use of Bayesian methods for complex decisions of this type. A method by Haspert[20] proposes to tie in the relative *cost* of the possible selections as a means to threshold a decision using Bayesian methods.

31.4   Understanding IFF Sensor Uncertainties

The importance of quantifying and modeling sensor uncertainties associated with kinematic, attribute, and hybrid sensors and their effect on the data fusion process has not historically been well described.[12] This section explores some of the characteristics and uncertainties in terms of the Mk XII IFF sensor (cooperative information) including its limitations, inherent error sources, and robustness to jamming and interference. A general multisource sensor fusion process is described using cooperative, unintentionally cooperative, and noncooperative dissimilar source inputs with specific attention placed on realizable IFF sensor systems and how they need to be characterized to understand and design an optimized and effective multisource fusion process. This process should be undertaken for all sensors providing information to a CID fusion process.

Mk XII IFF, as the example for this illustration, operates as a reasonably narrow-band, directional communications system deteriorating in performance with the square of the range (R^2) to the cooperating object of interest using two distinct uplink/downlink frequencies (1030/1090 MHz) and can provide both medium- and high-quality kinematic and CID information. Uncertainties are similar to radar in some respects, whereas the selective identity feature (SIF) octal codes generated have unique error functions that are described in the following sections.

The mechanics of a multisource integration (MSI) fusion process (that is supported by similar source integration [SSI] and dissimilar source integration [DSI] processes) are illustrated in Figure 31.14.[21] $AHS(t_v)$ is the updated active hypotheses support at a time of validity (t_v) that consists of a set of CID recommendations $\{H_i\}$ and their confidence values $B(\{H_i\})$.

Specifically, the MSI input values of

$$\sum_K = \text{Sum of Kinematic Uncertainty}$$

(31.13)

and

$$\sum_\alpha = \text{Sum of Attribute Uncertainty}$$

(31.14)

where

$$\sum_{K,\alpha} = \sigma^2_{K,\alpha} = \sigma^2_{K1,\alpha1} + \sigma^2_{K2,\alpha2} + \cdots + \sigma^2_{Kn,\alpha n}$$

(31.15)

is the sum-square error that is of concern when characterizing IFF or any other similar type of sensor uncertainties.

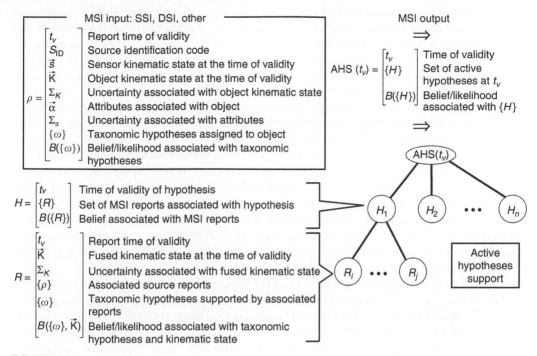

FIGURE 31.14
MSI input/output associations with uncertainty.

A general description of IFF principles is beyond the scope of this section; however, for more information the reader should refer to the standard text by Stevens.[22] However, common IFF error and uncertainty sources include the following:

- Mechanical (interrogator antenna rotating equipment including wind loading and pitch and roll)
- Antenna pattern degradation (mechanical and phased array)
- Timing/stability/radar pretrigger jitter
- m of n reply failures and split targets
- Round reliability/internal–external suppression of transponder (including intentional and unintentional jamming)
- Friendly replies uncorrelated-in-time (FRUIT)
- Garble code, interleaved pulses, and echoes
- False (phantom) targets, FRUIT targets, ring-around targets, and inline targets
- Antenna blockage (structures and dynamic maneuvering if on aircraft)
- Transponder reply generation variability (Δ time)
- Colocated interference systems (primarily affects transponder)
- Multipath and altitude lobing
- Mode C (barometric) altimeter errors (nonlinear diaphragms, shaft eccentricity, etc.)
- Equipment alignment
- Tracking filter characteristics and cost functions
- Deceptive IFF code stealers/repeaters

Specific values for these uncertainty sources will vary with the specific IFF system used and the environment where it resides (i.e., land-based, shipboard, or aircraft-based). For this illustration, a land/sea-based interrogator with airborne transponder(s) is assumed. The measurement of these error sources individually may not always be feasible, but combining them into total uplink/downlink characterizations will make them more manageable. A detailed analysis of this is found in Ref. 12.

Understanding how sensor systems are installed and their subsequent environmental limitations is vital for building a comprehensive CID fusion process. For example, for shipboard IFF interrogator installations, mast blockage can reduce IFF range across several degrees of bearing by many miles depending on the IFF antenna and the extent that it is blocked. There is no set amount of degradation, it depends on the superstructure configuration, the placement of the IFF antennas, and their tilt in altitude. The degradation of IFF signal replies due to obstructed areas must be accounted for in the fusion process.

Blockages with aircraft antennas are more problematic to predict, due to obscurant problems that are dependent on flight dynamics and aircraft stores on some military aircraft. Ground vehicles may exhibit similar problems depending on their movement, environmental obscurants (like trees), and sensor look angle.

Vertical L-band multipath nulls occur at all IFF ranges that results in a significant loss of detection. Multipath in IFF, such as in radar, is the result of the interrogation energy from horizontally polarized fan beam antennas following two separate paths to the target—one direct and the other reflected off a surface. The results of the different path lengths create a lobing effect where areas of constructive and destructive interference exist. This effect is readily seen in Figure 31.15, which shows the measured IFF data from the AN/SPS-40 aboard the USNS Capable.[23]

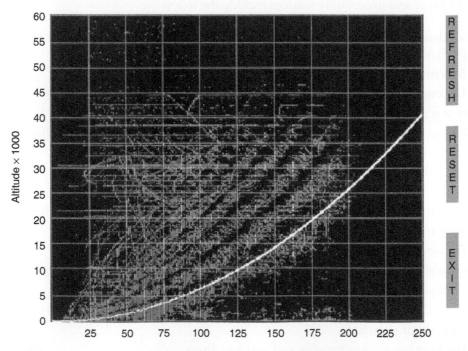

FIGURE 31.15
AN/SPS-40 IFF multipath on stalwart class USNS Capable (T-AGOS 16).

In Figure 31.15, the raw IFF target replies are only present in the regions of multipath maxima. The in-phase range doubling maxima and out-of-phase vertical multipath nulls are clearly seen around the areas of IFF target replies. The long white curve is a representation of the standard atmospheric model. An interesting note concerns the IFF replies approximately 200 nautical miles (nm) [abscissa] at less than 2500 feet [ordinate]. These are actual replies attributed to extreme L-band ducting during testing. The in-phase maxima double the effective IFF range, whereas the minima severely reduce it. These nulls can present a significant problem for IFF track continuity, start or stop track functions, and can contribute to multiple track declarations.

The robustness of IFF to jamming either intentional or accidental is questionable. Non-encrypted modes or *SIF* modes (1, 2, 3/A, and C) were never designed to be resistant to any non-IFF form of pulse modulated RF because of their rather isolated 1030/1090 MHz frequencies at the time of design. Assuming jamming is directed toward a transponder (also true for an interrogator), since IFF is a communications system, it follows that the jamming-to-signal power ratio (*J/S*) follows the form[24]

$$\frac{J}{S} = \left(\frac{\text{ERP}_J}{\text{ERP}_R} \right) \left(\frac{R_B^2}{R_{BJ}^2} \right) \left(\frac{G_{BJ}}{G_{BR}} \right) \tag{31.16}$$

where ERP_J, ERP_R are the effective radiated power of the jammer and IFF interrogator, respectively, R_B, R_{BJ} the propagation path lengths of interrogator-to-transponder and jammer-to-transponder, respectively, and G_{BJ}, G_{BR} the transponder antenna gain in the direction of the jammer and interrogator, respectively.

This discussion on IFF is not meant to be exhaustive or complete. The important issue is that the knowledge of sensor operation and domain space is critical to understanding how a sensor can affect the fusion process. Although difficult to necessarily characterize and use, this understanding must be sought for each sensor across the cooperative, unintentionally cooperative, and noncooperative domain spaces.

31.5 Information Properties as a Means to Define CID Fusion Methodologies

Sensor information resides in what Alberts et al.[25] term the *information domain*. All sensors observe some physical parameter(s) of an object or entity and then generate an abstraction of the observed object as discussed previously. This abstraction will always contain less information about an object than if the object were to be physically realized. This results from the finite ability of any single sensor or sensor group to measure all possible physical parameters, which as discussed is analogous to Shannon information entropy. Special difficulties arise when sensor information is potentially corrupted by an intelligent object that purposely deceives the sensor, when multiple types of objects cannot be discriminated (forming a *confusion class* of objects), or when a sensor is predisposed to make type I or II errors when observing certain classes of objects.

As an example, a simple sensor might be able to roughly categorize the RCS of aircraft according to their *bigness*, such as small, medium, and large. The *large* characteristic would not be able to distinguish between an airliner and a military refueling aircraft, but it may work well in distinguishing between a fighter and larger aircraft. However, if a small

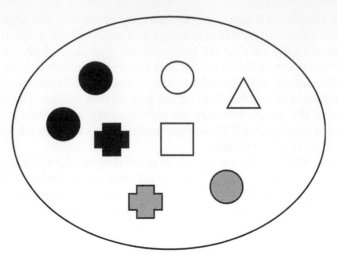

FIGURE 31.16
The benign urn model.

general aviation aircraft (or the fighter) were to use a series of corner reflectors to increase its RCS, then this sensor may not be able to catch the deception. Stealthiness, of course, works in the other direction, to reduce the target's RCS.

Following this example, the classic Polya urn model using a variety of shaded shapes can illustrate the problem of information fusion for a *hostile* environment. Figure 31.16 shows an urn with four sets of shapes (circles, squares, crosses, and triangles) filled with the colors black, gray, and white, in a set space S (as discussed in the introduction of this chapter). This figure is termed the *benign urn model*.

This urn has objects that work in traditional probability space; therefore, the *a priori* probabilities are well defined (for instance, they may all be equiprobable). Given that a selector randomly takes an object, the likelihood that a black circle is taken is 1/4. The likelihood that a triangle is selected is 1/8. For the conditional probability problem, given that it is known that a gray object will be picked, then the likelihood that it is a circle is 1/2. Here the selector has a perfect understanding of the environment.

Another urn that appears to be identical, the *nonbenign urn model*, is shown in Figure 31.17. The objects in the nonbenign urn differ from the benign urn in the following ways:

1. The objects have intelligence, they are aware of their color and shape.
2. The objects are capable of deceiving the selector (sensor) as to their color and shape, or either of these.
3. The selector may not be able to distinguish between some colors or shapes.
4. The total number of objects may not be known.
5. The selector may not be able to correctly perceive which objects it is selecting.
6. The selector's ability to distinguish some objects may be predicated on which type of object is being selected because of a relationship between the object and selector.

Military classification of aircraft, ground vehicles, ships, submarines, land mines, and biohazard agents resemble the nonbenign model, not the benign one. A fusion process must be able to work within such environments, else classification will be misleading or incorrect.

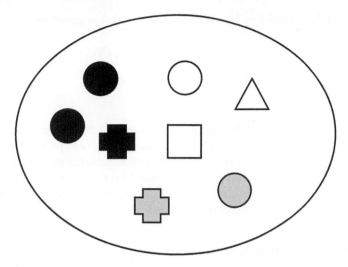

FIGURE 31.17
The nonbenign urn model.

31.6 Fusion Methods

In this section, fusion techniques are discussed using a set of simple data sets that represent a multisensor classification problem. These data sets represent the expected outputs of classification types of sensors, but are not meant to be complete or exhaustive.

The fusion methods themselves as represented by Bayes and Dempster–Shafer approaches are not meant to be exclusive to this discussion. Other fusion approaches certainly are pertinent. In fact, a work by Friesel[26] on Bernoulli methods to fuse information for CID classification provides another way to look at handling uncertain information. Any fusion process for CID must be able to handle uncertain information.

31.6.1 Modified Dempster–Shafer Approach

Classic D-S tends to measure the *absence of conflict* (e.g., $B \cap C = \varnothing$). Modified D-S focuses on the degree of *agreement*, much like Bayesian inference. The following is an expansion of the modified D-S technique from Section 31.3.3.1.

Modified D-S emphasizes terms in the classic D-S numerator sum whose corresponding pairs of hypothesis are in relatively good agreement with each other. It weights each term with a new *agreement function* defined as

$$a(B,C) = \frac{v(B \cap C)}{v(B)\,v(C)} \tag{31.17}$$

where $v(B)$ is the number of elements in a set B, and so on. If B and C are much the same set, the agreement will be larger than if they have relatively few elements in common. The denominator is chosen so that the resulting masses will sum to 1. If the combined agreement function is defined to be

$$\alpha(B, C) = m_1(B)m_2(C)a(B, C)$$

$$= m_1(B)m_2(C)\frac{v(B \cap C)}{v(B)\,v(C)} \tag{31.18}$$

then the Fister–Mitchell modified D-S fusion of m_1 and m_2 is given by

$$m_1 *_{\text{FMDS}} m_2(A) = \frac{\displaystyle\sum_{B \cap C = A} \alpha(B, C)}{\displaystyle\sum_{B \cap C \neq \varnothing} \alpha(B, C)} \tag{31.19}$$

Fister–Mitchell D-S fusion can be further modified by mapping the noncommon elements of each pair of subsets to an *unknown* class, which is typical of real-life analyses of alternatives. This stratagem prevents alternatives not mentioned in one response, but present in another, from being completely discarded by the fusion process. This prevents occurrences in a nonbenign environment, intelligent deception for example, from zeroing out the mass corresponding to the correct conclusion.

31.6.2 Information Fusion Sets

The simulated data sets represent outputs from sensors, such as an ES system with a classifier and a synthetic aperture radar (SAR) system with automatic target recognition (ATR). The measurement and processing of various parameters, such as pulse repetition frequency/interval (PRF/PRI), frequency, modulation, and Doppler spectrum, seldom leads to a single atomic member of a class of objects. Rather, a set of possible objects is declared and refined as additional information is processed, such as that caused by a detected sensor mode change.

The following confusion class (or set) of objects (A, B, C, D) in Table 31.6 were used for input into the previously discussed fusion processes in the order presented by alternating inputs between the two ES and SAR sensors.

The first row represents the *a priori* information on each class of objects. Each row provides an update to the class probabilities. The priors are assumed to be equiprobable due to a lack of *a priori* information. An unknown class (Z) is added that contains the complement of the set of objects referenced in the sensor response, and whose probability is $1 - P_{cc}$, where P_{cc} is the probability of correct classification. These two hypothetical sensors are quite good (as $P_{cc} = .99$). Equivalently, each object (A, B, C, D) in Table 31.6 also has a probability assigned to it that assumes entropy across the data sets $(1/n)$. In other words,

TABLE 31.6

Fusion Input Data Sets with Probabilities

Update	A	B	C	D	Z
Prior	.2475	.2475	.2475	.2475	.01
1	.33	.33	.33	0	.01
2	.33	.33	0	.33	.01
3	.495	.495	0	0	.01
4	.495	.495	0	0	.01
5	.495	.495	0	0	.01
6	.495	.495	0	0	.01
7	.495	.495	0	0	.01
8	.495	.495	0	0	.01
9	.495	.495	0	0	.01
10	.495	.495	0	0	.01
11	.495	.495	0	0	.01

the distribution of possible objects is not known. With the exception of the priors, these probabilities are usually only used for the Bayesian approach, because D-S works on the class of objects as a set (BBA). Here the D-S process is modified to work on each element of the class to compare outputs.

31.6.2.1 Bayesian and Orthodox D-S Results

The results of the Bayesian and D-S fusion processes are shown in Table 31.7. The results are identical for both methods because the individual elements of the sets were operated on. Accordingly, this reflects the findings by Fixsen and Mahler[17] and Sudano[27] concerning the mathematical equivalency of Bayesian and D-S methods.

The noncommon objects C and D are quickly removed from the fusion process as expected. However, as discussed, this may not be desired in nonbenign environments. The results of the modified D-S process, which maintains evidence, applied to a Smets pignistic probability transform are shown in Table 31.8.

The results of mapping the noncommon objects into the unknown (Z) class are apparent in this example. Initially, after update 1, some of the mass from the first noncommon object, D, is mapped into the unknown class (cf. Table 31.6). This is assigned via Equation 31.19 and preserves its mass for future iterations. After update 2, the absence of C

TABLE 31.7

Bayesian and Dempster–Shafer Fusion Results
(Probabilities)

Update	A	B	C	D	Z
1	.3332	.3332	.3332	0	.00040
2	.4999	.4999	0	0	≈0
3	.5	.5	0	0	≈0
4	.5	.5	0	0	≈0
5	.5	.5	0	0	≈0
6	.5	.5	0	0	≈0
7	.5	.5	0	0	≈0
8	.5	.5	0	0	≈0
9	.5	.5	0	0	≈0
10	.5	.5	0	0	≈0
11	.5	.5	0	0	≈0

TABLE 31.8

Modified Dempster–Shafer Fusion Results (Probabilities)

Update	A	B	C	D	Z
1	.2495	.2495	.2495	.1488	.1025
2	.2815	.2815	.1776	.1776	.0818
3	.3469	.3469	.1231	.1231	.0600
4	.3985	.3985	.0810	.0810	.0410
5	.4358	.4358	.0510	.0510	.0265
6	.4608	.4608	.0310	.0310	.0164
7	.4766	.4766	.0184	.0184	.0099
8	.4863	.4863	.0108	.0108	.0059
9	.4920	.4920	.0062	.0062	.0035
10	.4954	.4954	.0036	.0036	.0020
11	.4974	.4974	.0020	.0020	.0012

causes it to lose mass to both D and Z, while classes A and B begin to grow in likelihood. Finally, after the 11th update, after successive fusion iterations with only objects A and B, the results approach those in Table 31.7.

31.6.3 Results and Discussion

The results shown in Table 31.6 are not robust to a nonbenign information environment. A possible scenario is that the real answer is object C (or D), which was spoofing the two sensors during updates 2 through 11. The mass movement associated with the modified D-S process allowed for the preservation of the noncommon objects. These could have allowed for a confused condition (or correct response) if C had been detected a few more times in subsequent updates, thus preventing the wrong conclusion from the fusion process. If the real answer is object A or B, then a potential drawback of this approach is the number of updates or iterations required to arrive at this decision compared to the other methods. However, there are other modifications of D-S that can allow for faster convergence than given in this example. Again, regardless of particular issues, the handling of uncertainty is a critical element in designing a robust fusion process.

31.7 Multihypothesis Structures and Taxonomies for CID Fusion

As we have discussed in this chapter, one of the greatest difficulties in developing a fusion process for CID is determining the type, quantity, and quality of the information provided.[28] Even when this is accomplished, the utility (relationship) or implications of the information is often difficult to establish. Often numerous sources provide information, but the relationship between the implications of different data is not well described, or is ambiguous or inconsistent. This deficiency leads to poorly constructed fusion architectures and methodologies because information is either ignored or improperly combined in the fusion process. Using the JDL information fusion model as a guide, this section addresses the movement of attribute information across multiple hypothesis classes as it relates to developing the identification of different objects, and how it can be combined both within and between JDL fusion levels to construct the identification vector. This analysis leads to an information architecture that is naturally adaptive to information regardless of quality, level, or specificity.

This section looks at the implications of sensor information in the context of the level 1 taxonomies described in Section 31.2. The methods are, however, more generally applicable to any situation which can be described in terms of many taxonomies. It develops a method for finding the implications of information in one taxonomy for another. A simple CID example is the use of aircraft type, a level 1 attribute, to arrive at conclusions on a threat level taxonomy, a level 3 attribute. The basic structure is a set of relationships among taxonomies that can be used to map the implications of sensor information from one taxonomy to another, and from one JDL level to another. This gives a structure for an integrated multitaxonomy multihypothesis analysis that can span JDL levels, thus providing a context for situation awareness (SA) analysis. A basic concept is response mapping, described further in this section, that makes it possible to develop the complete implications of a sensor response in all related taxonomies. For simplicity, the exposition in this discussion is in the context of air CID.

31.7.1 Taxonomic Relationships Defined

A *taxonomy* is a classification scheme for objects of interest, which parallels the study of *ontologies*. It is a set of mutually exclusive labels. An example is the JDL level 2 CID taxonomy {Friend, Assumed Friend, Neutral, Pending, Unknown, Suspect, and Hostile}. The *Nationality* taxonomy is {United States, Russia, United Kingdom, France, Iraq, Iran, Zimbabwe, ...}. Other examples are the *Category* taxonomy {Exoatmospheric, Air, Surface, Subsurface, Land}, the *Platform* taxonomy {Fighter, Bomber, Transport, ...}, the *Type* taxonomy {F-14, F/A-18, F-22, Typhoon, Viggen, E-3, ...}, and the *Class* taxonomy {F-14A, F-14B, F-14D, F/A-18A, F/A-18B, F/A-18C, F/A-18D, ...}. The Category, Platform, Type, and Class taxonomies are successive refinements of predecessor taxonomies. Given a Class label, a Type label can be inferred; given a Type label, a Platform label can be inferred; and given a Platform label, a Category label can be inferred. (There are some exceptions, for example, a C-130 might be an attack aircraft or a transport.) More precisely, taxonomy A is an *f-refinement* of taxonomy B if f is a function $f: A \to B$ such that if $b_1 \neq b_2$ then $f^{-1}(b_1) \cap f^{-1}(b_2) = \varphi$, where φ is the empty set. An example is given in Figure 31.18, in which it can be seen that $f^{-1}(F - 16) \cap f^{-1}(F/A - 18) = \varphi$. If f is an obvious function, as it is for example for the taxonomies Type and Class, then it can be said simply that taxonomy A is a *refinement* of taxonomy B. Other collections of taxonomy do not show such a relationship—an F/A-18 might have any of a dozen or more Nationalities, and each Nationality can have many different aircraft Types.

If a taxonomy A is an *f-refinement* of taxonomy B and $a \in A$, $b \in B$, and $f(a) = b$, it can be said that a is an *f-refinement* of b. If f is an obvious function, then a is a refinement of b. For example, F/A-18A in the Class taxonomy is a refinement of F/A-18 in the Type taxonomy.

Given a set S of objects, a taxonomy imposes a partition on the set. Each element of the partition is the set of all elements of S for which a single element of the taxonomy is the appropriate name. An example of an element of the partition imposed on aircraft by the Type taxonomy is the set of all F-15s. Another is the set of all 747s. A taxonomy T_1 is a refinement of another taxonomy T_2 if the partition imposed by T_1 is a refinement of the partition imposed by T_2. Figure 31.19 shows the *f*-refinement of Figure 31.18 as a partition refinement.

A *taxonomic refinement series* is a set of taxonomies, $\{T_i\}_{i=1}^{n}$, such that T_{i+1} is a refinement of T_i. An example is the series Category, Platform, Type, and Class. The problem of interest is how to use information about an object from different taxonomies to categorize the object in one of those taxonomies, or in another, completely different, taxonomy. It is a practical problem since some sensors categorize an object in the context of one or more taxonomies. For example, an electronic support (ES) sensor might be able to discern that the object might be an F-14 or an F/A-18D on the basis of emission characteristics. The information is from both the Type and Class taxonomies. It can be used to infer that, in the Type taxonomy, the object is either an F-14 or F/A-18. It can be used to infer that the object is an

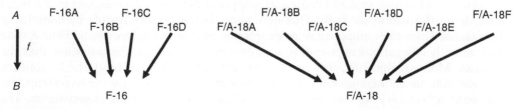

FIGURE 31.18
An example of *f*-refinement taxonomy.

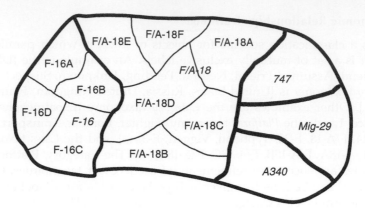

FIGURE 31.19
An *f*-refinement as a partition refinement.

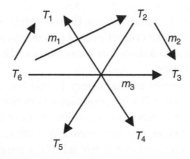

FIGURE 31.20
An example of non-Canonical mapping.

Air object in the Category taxonomy, and that it is not a Chinese aircraft in the Nationality taxonomy. Canonical mappings provide a way to exploit these taxonomic relationships.

31.7.2 Canonical Mappings

As shown in Figure 31.20, two taxonomies might be related through mappings in more than one way. This shows how sets T_6 and T_3 are related through both the mapping m_3 and the composite mapping $m_2 * m_1$. Since these mappings in general will not be equal for any particular application, a set of canonical mappings must be defined between any two related taxonomies (the canonical mapping from a taxonomy to itself is, of course, the identity mapping). In the case of a collection of taxonomies that are successive refinements, the canonical mappings reflect the hierarchical nature of the taxonomies themselves.

It is conceivable to have multiple canonical mappings between two sets, and to maintain parallel state information for both. This may allow better quality results by combining the two states, since each mapping is an expression of domain information. Thus a state arrived at with one canonical mapping encapsulates background information that the other lacks. More significantly, it is conceivable that the two states might be in conflict. This might indicate an inconsistency in the sensor inputs used to infer the two states, but it also could reflect varying uses or ambiguous interpretations of the observations. The gain from using more than one canonical mapping might or might not be worth the extra complexity.

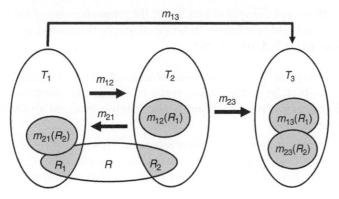

FIGURE 31.21
Response mapping in a refinement series.

31.7.3 Response Mapping

Response mapping is a way to interpret a response with elements from one taxonomy in terms of another. It also provides a means of interpreting a response with elements from more than one taxonomy in the various referenced taxonomies. Level expansion, which is a case of response mapping among the taxonomies, is of particular interest because of existing sensors that yield a response with elements from both the Type and Class taxonomies (shown by R in Figure 31.21). The objective is to make it possible to maintain parallel states in multiple taxonomies. As each response is received, it is interpreted as described in the following in the taxonomies of interest. The state for each taxonomy is then updated.

This concept is best explained by example. Referring to Figure 31.21, let R be a response from a source of information. It is composed of a set of attributes, potentially from different taxonomies. Let the canonical mapping from taxonomy T_i to taxonomy T_j be m_{ij}. Each taxonomy potentially has elements that are part of the response (R_1 and R_2 in the figure), as well as elements that are the images, under a canonical mapping, of elements in other taxonomies ($m_{12}(R_1)$, $m_{21}(R_2)$, $m_{13}(R_1)$, $m_{23}(R_2)$). When the implications of the canonical mappings and the response R have been taken into account, each taxonomy T will experience a response R_T which is the union of the $R \cap T$ and $m_i(R) \cap T$, where m_i is a mapping to T from another taxonomy T_i with elements referenced in R.

31.8 Multihypothesis Structures, Taxonomies, and Recognition of Tactical Elements for CID Fusion

This section extends the previous discussions on the determination of activity, intent, and threat level and develops an extended approach to the JDL model for level 3 SA and CID.

31.8.1 CID in SA and Expansion on the JDL Model

CID in the context of SA is invariably referenced, but far less often is implemented into CID algorithms as an explicit taxonomic process. Hanson and Harper[29] demonstrate that situation assessment (for threat refinement) is strongly related to data fusion. A general definition of SA is given by Endsley:[30] *Situation awareness is the perception of the elements in the environment within a volume of time and space, the comprehension of their meaning, and the projection of their status in the near future.* In terms of levels, this definition can be structured in a way analogous to the JDL hierarchy:

- Level 1 SA: *Perception of the environmental elements.* The identification of key elements of *events* that, in combination, serve to define the situation. This serves to semantically tag key situational elements for higher levels of abstraction in subsequent processing.
- Level 2 SA: *Comprehension of the current situation.* This combines level 1 events into a comprehensive holistic pattern (or tactical situation). This serves to define the current status in operationally relevant terms to support rapid decision making and action.
- Level 3 SA: *Projection of future status.* The projection of the current situation into the future, so as to predict the course of an evolving tactical situation. Time permitting, this supports short-term planning and option evaluation.

A direct comparison of these three levels of SA and JDL data fusion show that the functions are clearly distinct at level 1, since JDL data fusion focuses on the *numeric* processing of tactical elements to provide identification and tracking, whereas SA focuses on the *symbolic* processing of these entities, to identify key *events* in the current situation. At level 2, the definitions are virtually identical to yield the conventional definition of SA (that of generating a holistic pattern of the *current* situation). At level 3, the SA definition is more general than the pure data fusion definition, since the former also includes projection of ownship*/aircraft/battalion/etc., and friendly intent, and capability in addition to threat intent and impact assessment.

Although mission-driven awareness focuses on level 3 SA, due to the desirability of a concurrent multilayered data fusion/SA approach, mission-driven awareness can be used to accelerate level 2 SA and hence focus the fusion process more quickly. Such a benefit may be crucial in the common case of time critical targets of interest.

31.8.2 Recognition of Tactical Elements

Mission-driven awareness allows us to infer from the results of lower level knowledge fusion the mission (or activity) component of CID. Such identification of the mission or activity implies a SA of the decisions made by both sides of an adversarial encounter. This is analogous to the Boyd OODA loop discussed in Section 31.1. This CID capability is not only essential to generate a needed predictive situational awareness (PSA) to minimize fratricide, but is also broadly characteristic of the class of fusion algorithms which can recognize behaviors and project from those behaviors the intent and probable courses of action (CoAs) available to hostile commanders. In short, these fusion algorithms may be represented in the familiar JDL Fusion Model as level 3 fusion to address threat refinement (impact assessment) both predicatively and deductively:

- PSA projects CoAs to determine potential impacts (evaluate utility of CoAs to a hostile commander in terms of the impact achieved)
- Determination of intent deduces the hostile commander's intent from the evaluation of the CoAs that best corroborate observed behaviors

---

* Ownship can refer to one's self, platform, organization, etc.

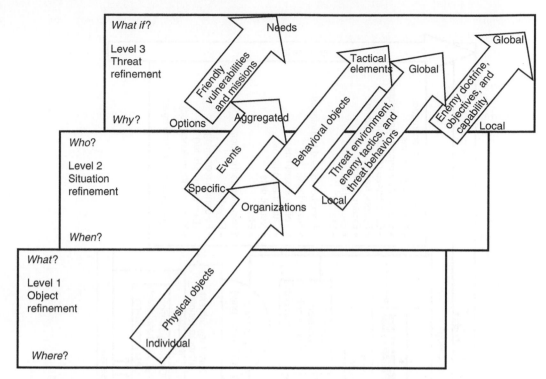

FIGURE 31.22
Tactical elements employed by fusion level.

However, these decisions are not generally made at the level of the unit doing the observation (typically a platform such as an *aircraft, vehicle,* etc.), but rather at a higher *decision element* that is tasked to perform a given tactical mission. Therefore, to enable *decision fusion* at an actionable level, recognition of the tactical elements* that represent the decision-level units becomes essential.

This suggestion is based on the observation that although newly developed CID algorithms show great promise for target identification and classification, their utility for determination of intent is limited by their focus on track fusion (JDL levels 1 and 2). Use of a CID architecture structured hierarchically helps implementation of inference processing for level 3 fusion (Figure 31.22 as modified from the 2000 Data and Information Fusion Group).[31]

As may be suggested from Figure 31.22, to reproduce the decision-making process, it is desirable for the decision fusion to be able to recognize and characterize not only individual physical objects (bottom of figure), but also the organizations, events, tactics, objectives, missions, and capabilities (middle to top of figure). Most of these recognition schemes are domain specific and typically either require or at least benefit from use of *a priori* information (e.g., *intelligence*).

Accordingly, the JDL model is reorganized in Figure 31.23 to show information flow with the inference level increasing from bottom up. This corresponds loosely to the techniques typically used as discussed in this chapter and shown on the right side of the figure.

---

* Often an organizational entity such as an *armored column, flight of aircraft, convoy of vehicles, terrorist cell,* and so on.

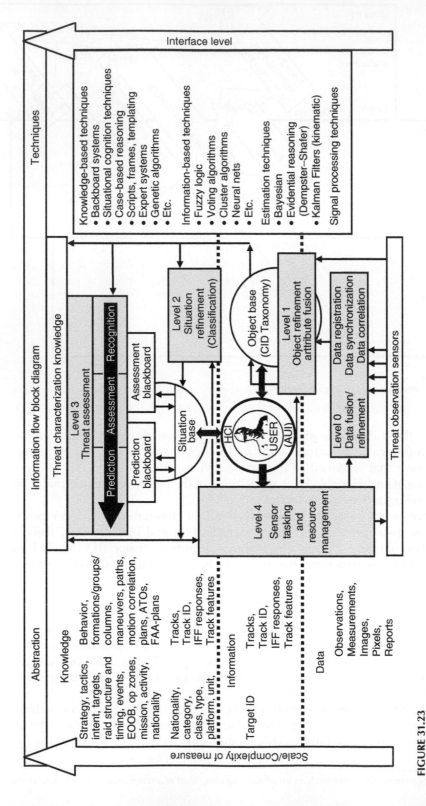

FIGURE 31.23
Aggregate of proposed level 0–4 fusion methodologies.

Data structures are displayed on the left side of Figure 31.23. Data structures representative of knowledge fusion as identified in the upper left of the figure include the tactical elements with raid structure and timing, missions, activities, and operational zones. The critical process of generating a *situation base*, as shown in the center of the figure, represents *recognition of tactical elements*. The purpose of this section is not to characterize this process so much as to identify the architectural characteristics of a CID structure that would support such a capability. Rather, it is to identify the information and knowledge needed to build such a structure.

Implementation of such an aggregative recognition of tactical elements is facilitated by use of the corresponding architectures for multihypothesis structures and taxonomies for CID. Use of such architectures may in turn facilitate a *power to the edge* approach to decision making which enables edge units by providing these units with CID information that is structured, traceable, and displayed to the level of recognizable tactical elements for which decisions are made.

The *power* that would be provided to *edge units* would be in the form of enhanced CID, earlier CID when contextual cues can discriminate between alternative CID hypotheses, and integration of CoA assessments to evaluate the CID of various threatening elements with the generation of CoAs for mission planning. This information can enable a *self-synchronization* as described by Alberts and Hayes[32] as well as the more obvious support of effects based operations as described by Smith.[33]

Determination of a hostile commander's intent typically requires the representation of his decision-making process. By returning to the dual-view characterized previously that addresses impact assessments in both a predictive and deductive form, one can propose a functional flow as shown in Figure 31.24. In this example, a recognition-primed decision-making (RPD) process is employed as described by Klein.[34] As a result, for both forms one can observe the familiar recognition/assessment/evaluation model. For the predictive form at the top of the figure, the system recognizes threatening tactical elements and generates threat assessments and impact predictions. For the deductive form at the bottom of the figure, the system recognizes potential threat intents and generates assessments of intent and their associated CoAs. Acting in tandem, the dual activity model thus proposed provides a useful tool for the determination of intent coupled with the recognition of the critical tactical elements that can be integrated with existing CID assessment tools.

31.9 Conclusions and Future Work

Regardless of the method used to fuse information across the JDL levels, the numerical result of the fusion process (or the confidence in the declaration of *Neutral*, as an example) does not necessarily provide all of the information necessary to make a decision. This is especially critical for CID where a wrong answer can have disastrous effects. The *USS Vincennes* incident in July 1988, the Black Hawk shootdown in 1994, and the Patriot Missile battery misidentification in 2003 are all reminders that the context, type, timeliness, quantity, and quality of information must be understood before making a decision. For expansion of work in this area, measures of information value, completeness, and decision cost that complement this multitaxonomic approach for information fusion must be developed. Other future work of interest includes contextual reasoning approaches with extensibility beyond a given domain, realization of the recognition of tactical elements, multiobjective

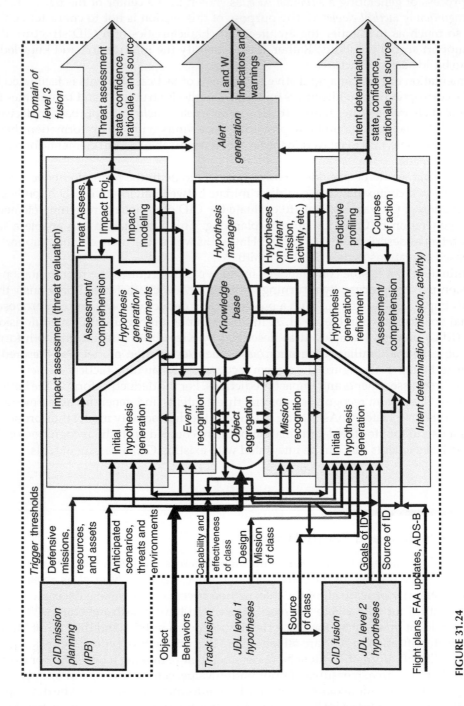

FIGURE 31.24
Potential functional flow for a proposed threat evaluation tool for combat identification.

collaborative mission planning and decision making under uncertainty, and PSA integration approaches. Finally, suitable strategies for JDL level 4 and 5 constructs must also be considered, with many approaches available in the literature.

References

1. McCants, T., CC&D ORD KPP analysis, internal presentation, 22 May 2001.
2. Kebodeaux, C.E., *Reducing the Fog of War in the Single Integrated Air Picture through Improved Data Fusion*, Thesis to the Air Force Institute of Technology (AFIT), Wright-Patterson AFB, March 2005.
3. White, F.E., Jr., *Data Fusion Lexicon*, Joint Directors of Laboratories, Technical Panel for C3, Data Fusion Sub-Panel, Naval Ocean Systems Center, San Diego, CA, 1987.
4. Paul, M., *Bayesian Belief Networks*, Lockheed Martin MS2 internal white paper, August 2002.
5. Chen, G. et al., Game theoretic approach to situational awareness and threat prediction, *Proceedings of the 9th International Conference on Information Fusion*, Florence, Italy, July, 2006.
6. Hall, D., *Lectures in Multisensor Data Fusion and Target Tracking*, CD lecture notes section 7.3-272, Artech House, MA, 2001.
7. Schuck, T., Hunter, J. B., and Wilson, D., Multi-hypothesis structures and taxonomies for combat identification fusion, *9th International Command and Control Conference Research and Technology Symposium (ICCRTS)*, 2004.
8. Schuck, T., A mathematical theory of identification for information fusion, *7th Annual International Command and Control Research and Technology Symposium (ICCRTS)*, Quebec City, Canada, September 2002.
9. Shannon, C., A mathematical theory of communication, *The Bell System Technical Journal*, 27, July and October 1948, pp. 379–423 and pp. 623–656.
10. All information associated with Shannon is reproduced or derived from previous Shannon reference.
11. Haak, J., personal communication, March 2002.
12. Schuck, T., Shoemaker, B., and Willey, J., Identification friend-or-foe (IFF) sensor uncertainties, ambiguities, deception, and their application to the multi-source fusion process, *IEEE National Aerospace and Electronics Conference (NAECON) 2000*, October 2000.
13. Bullard, B. and Dowdy, P., *Pulse Doppler Signature of a Rotary-Wing Aircraft*, Georgia Tech Research Institute, 1991.
14. Misiurewicz, J., Kulpa, K., and Czekala, Z., Analysis of recorded helicopter echo, radar 97 (Conf. Pub. No. 449), *Proceedings of the IEEE 1997*, Edinburgh, UK, October 1997, pp. 449–453.
15. Sudano, J., The system probability information content (PIC) relationship to contributing components, combining independent multi-source beliefs, hybrid and pedigree pignistic probabilities, *Proceedings of the 5th International Conference on Information Fusion*, 2002.
16. Fixsen, D. and Mahler, R., The modified Dempster-Shafer approach to classification, *IEEE Transactions on Systems, Man, and Cybernetics-Part A*, 27(1), January 1997, pp. 96–104.
17. Fister, T. and Mitchell, R., Modified Dempster-Shafer with entropy based belief body compression, *Proceedings of the 1994 Joint Service Combat Identification Systems Conference (CISC)*, Naval Postgraduate School, CA, August 1994.
18. Sudano, J., Pignistic probability transforms for mixes of low and high probability events, *4th International Conference on Information Fusion*, Montreal, Canada, August 2001.
19. Powell, M., *Decision and Risk Analysis for Complex Systems*, Course at the Stevens Institute of Technology, Hoboken, NJ, 2006.
20. Haspert, K., *Optimum ID Sensor Fusion for Multiple Target Types*, Institute for Defense Analysis Document D-2451, March 2000.

21. Athena multi-source integration—systems engineering team (MSI-SET) final report, 1998.
22. Stevens, M., *Secondary Surveillance Radar*, Artech House, MA, 1988.
23. Shoemaker, W.B., IFF on the AN/SPS-40 Multipath Diagram, USNS Capable testing, 1998.
24. Lothes, R.N., Szymanski, M.B., and Wiley, R.G., *Radar Vulnerability to Jamming*, Artech House, MA, 1990.
25. Alberts, D. et al., *Understanding Information Age Warfare*, DOD Command and Control Research Program (CCRP), CCRP Press, Washington DC, July 2002.
26. Friesel, M., *A Bernoulli Analysis for Multiple Observations of a Single Object*, Lockheed Martin internal document, 2004.
27. Sudano, J., Equivalence between belief theories and naïve bayesian fusion for systems with independent evidential data: Part 1 the theory, *Proceedings of the Sixth International Conference on Information Fusion*, July 2003.
28. Schuck, T., Hunter, J., and Wilson, D., Multi-hypothesis structures and taxonomies for combat identification fusion, *Proceedings of the IEEE Aerospace Conference*, Big Sky, MT, March 2004.
29. Hanson, M.L. and Harper, K.A., An intelligent agent for supervisory control of teams of uninhabited combat air vehicles (UCAVs), *Unmanned Systems 2000 Conference*, July 11–13, 2000.
30. Endsley, M.R., Toward a theory of situation awareness in dynamic systems, *Human Factors*, 37(1), 1995, pp. 32–64.
31. Data & Information Group, Data & information fusion: Status of U.S. data & information fusion, *Fusion 2000 Conference*, Paris, France, 10 July 2000.
32. Alberts, D.S. and Hayes, R.E., *Power to the Edge*, DOD Command and Control Research Program (CCRP), CCRP Press, Washington DC, June 2003.
33. Smith E.A., *Effects Based Operations: Applying Network Centric Warfare in Peace, Crisis and War*, DOD Command and Control Research Program (CCRP), CCRP Press, Washington DC, 2002.
34. Klein, G.A., Naturalistic decision making: Implications for design, *Gateway*, 5(1), 1994, pp. 6–7, also Recognition-primed decisions, *Advances in Man-Machine Systems Research*, 5, 1989, pp. 47–92.

Index

A

Abnormal cognition and perception, human-computer interaction, 544

Absence of change, top-down data fusion automation, 152–153

Abstracted information features, combat identification (CID) mapping, Shannon communication analysis, 786–788

Access issues, geospatial data fusion, 103

Action information fusion (ACT-IF), information fusion design, 516–517

Adaptive architectures, situation and threat assessment, 473–474

Adaptive filtering, elastic transformations, data registration, 129–130

Adaptive maximum likelihood-probabilistic data association estimator, basic principles, 236–240

Adaptive model matching, automatic target recognition (ATR), feature-level fusion, 97–98

Ad hoc models
  hypothesis enumeration, 603
  hypothesis evaluation problem space to solution space mapping, 606–607

Advanced Geospatial Intelligence (AGI) techniques, geospatial intelligence spatial data fusion, 110

Agreement function, combat identification (CID) mapping, modified Dempster–Shafer method, 799–800

Airborne imagery, image and spatial data fusion, 90–92

Air target tracking, top-down data fusion automation, 161–162

Algorithms. *See also* Data association algorithmics
  covariance intersection, 324–327
  database management system design
    intuitive algorithms, 626
    performance evaluation, 626
  data registration, 120–122
  multiple-frame assignments, 309–314
    complexity, 313
    fine gating, 309–310
    future research issues, 315
    improvement, 313

Lagrangian relaxation algorithm, 311–313
    preprocessing, 309–311
    problem decomposition, 310–311
  multisensor data fusion, 12
  multisensor data fusion survey, 692–699
  net-centricity and, 34–37
  situation and threat assessment, 493

Alignment, image and spatial data fusion, 94

All-source, top-down data fusion automation, 154

Ambiguously generated ambiguous (AGA) measurements
  Bayesian filtering, 380–381
  generalized likelihood functions, 388–389

Ambiguously generated unambiguous (AGU) measurements, Bayesian filtering, 380–381

Amino acid sequencing, chemical and biological sensors, complexity analysis, 742–743

Amplitude information, target motion analysis, maximum likelihood-probabilistic data association, 210–212

AND rule
  chemical and biological sensors, concentration consistency, 756–757
  distributed decision fusion, global decision rules, 173–175

A posterior belief
  condition-based maintenance, 707–708
  situation and threat assessment, 486–488

A priori information
  combat identification (CID) mapping, 793–794
  condition-based maintenance, 707–708
  contextual knowledge, 157–158
  data registration, 115–116
  information fusion design, 513–514
  information-processing cycle, 30–32
  single-target tracking, Bayesian formulation, 269
  top-down data fusion automation, puzzle-solving metaphor, 140–142

Architectures
  adaptive architectures, 473–474
  centralized-fusion, 301–302, 417–418

Printed and bound by CPI Group (UK) Ltd, Croydon, CR0 4YY

22/10/2024

01777631-0002